Sustainability Management in Plastics Processing

Strategies, targets, techniques and tools

Robin Kent

Tangram Technology Ltd

First edition 2022

British Plastics Federation (BPF)

6 Bath Place, Rivington Street, London, EC2A 3JE.

British Library Cataloguing-in-Publication Data

A catalogue record for this book is available from the British Library

ISBN: 978-1-3999-1160-3

Contents

Introduction by the BPF

British Plastics Federation

Sustainability is extremely important to the world and to the plastics industry. Plastics have undoubted benefits in terms of sustainability but there is always more that we can, and should, do. The BPF alongside other trade associations around the world is working hard to provide resources and information for the industry, legislators and the general public .

We are working on a variety of initiatives to reduce energy use, increase recycling and prevent pellet escape and litter. Some of these projects are:

- BPF Energy – a voluntary agreement with targets to increase energy efficiency and reduce CO_2 emissions.

- Operation Clean Sweep – a programme to help industry to contain plastic pellets.

- For Fishes Sake – a campaign to reduce litter entering The Thames by using fun interactive methods to help people understand how litter reaches the Thames.

- Bincentives – a school's campaign to encourage young people not to drop litter and to recycle.

- Vinyl Plus - a PVC industry scheme to reduce the environmental impact of PVC.

- Recovinyl – a PVC recycling scheme.

- Recomed – a PVC recycling scheme in hospitals.

- Recocard – a PVC recycling scheme for gift cards.

- Marine Litter Solutions - a series of global projects as part of the Declaration of the Global Plastics Associations for solutions on Marine Litter.

- Ecodesign - the BPF have carried out a great deal of work in this area including the creation of guides and courses. More information can be seen at www.bpf.co.uk/ecodesign

- Net Zero - in 2021 the BPF established a Net Zero Working Group and its activities can be seen at www.bpf.co.uk/netzero

All of these projects involve working with the industry and stakeholders to improve the sustainability of the industry.

Many of you will already know Robin Kent from his work in the plastics industry not only in the UK but around the world. His interest and experience in led him to produce this book to help plastics processors in their efforts to improve sustainability. He has produced this book specifically for the plastics processing sector of the industry and it is aimed directly at processors who want to know what to do next and how to go about it. He offered this work to the BPF and we gratefully accepted the opportunity to publish the book as a guide to the industry.

We are delighted to be able to provide this resource to all members of the BPF to signpost the many things that the industry can do to improve sustainability in areas.

Philip Law
Director-General
British Plastics Federation
London, 2021.

Preface

This volume is a companion to three previous books, published by Elsevier, titled respectively 'Cost Management in Plastics Processing', 'Energy Management in Plastics Processing' and 'Quality Management in Plastics Processing'. These books all dealt with aspects of sustainability in some shape or form but did not cover the complete subject. Sustainability management has, however, become a vital topic, and deserves more attention than was possible within the constraints of the previous books.

There are many books, articles, reports and publications on sustainability and its importance in the modern world. Some of these are general, some are specific to the use (or abuse) of plastics, some are supportive of the plastics processing industry and some demonise plastics, some are well-written and unbiased whereas others have a very definite bias for or against plastics. However, there is still a lack of practical information for the plastics processor who wants to improve the sustainability of their business. This book aims to provide plastics processors with the information necessary to achieve improved internal and external sustainability and to demonstrate this to the outside world.

As with the previous books, this book provides a structured approach to the techniques of sustainability management and covers the main topics of relevance to plastics processors. It is designed as a workbook for practical use and not as an academic textbook. It does not cover all aspects of the selected topics but focuses on the key management and technical issues for each topic. Each topic is dealt with in a single 2-page spread, (to misquote the New York Times – 'all the news that fits the space') and most can be read independently of each other – this is not a 'cover-to-cover' book. It should be easy to understand and the actions recommended should be easy to undertake for most people in the plastics processing industry. Sustainability management is not 'rocket science'; simply good management and engineering with an eye to the long-term effects of our actions.

My own introduction to sustainability management began in 1972 when I was at university and working with environmental groups looking at the difficulties of municipal solid waste (MSW) disposal in Melbourne, Australia. At this time recycling did not really exist and all MSW was land-filled. Melbourne was running out of 'holes to put it in' and there were increasingly long road journeys, costs and environmental impacts simply to put waste in a hole. Things have changed dramatically since then and the challenges have grown. A lack of holes is no longer the most important issue.

I would like to dedicate this book to my first grand-daughter, Chloe Rose Bradbury, who brightens up the life of the solitary author and every so often helps him out in the office.

Robin Kent
Hitchin, 2022
rkent@tangram.co.uk
www.tangram.co.uk

"It would be disingenuous of us to disguise the fact that the principal motive which prompted this work was the sheer fun of the thing".

Alan Turing

Chapter 1

Introduction to sustainability management

The topic of sustainability cannot be ignored by any plastics processor or the wider plastics industry. Our products are extraordinarily useful to mankind and have enabled the massive increases in living standards that we have achieved over the past 60 years. Without plastics products, modern life would simply not be possible. It would be, in the words of Thomas Hobbes, 'solitary, poor, nasty, brutish, and short'.

Despite this, there is increasing criticism of both the industry and our products and increasing legislative activity on a range of environmental issues. Some of these criticisms cannot be denied or countered and it is undeniable that the plastics processing industry, along with many other industries, needs to improve its sustainability performance dramatically to become more socially acceptable.

When I sat down to write this book, I imagined that it would be similar to the experience of writing my previous books. After all, I had been involved in various aspects of plastics processing and sustainability for many years. However, in the previous books the engineering and social aspects were relatively straightforward and uncontentious. This meant that the writing, although sometimes painful, was also relatively straightforward. In contrast the topic of sustainability is not at all straightforward and many of the issues are very contentious. There are many models, theories and concepts that are often mutually exclusive and contradictory. Some of the theories are even internally contradictory.

As an engineer, it is difficult to deal with some of the available material especially when one book boldly and proudly proclaims "feelings are facts". This makes it difficult for engineers, who are used to dealing in real facts, to do the right thing for a range of stakeholders who have different 'feelings' and therefore different 'facts'.

The bulk of this book is made up of the concrete things that engineers can do to improve sustainability in plastics processing. This chapter tries to put sustainability into context by looking at the main available models, their relevance to the industry and the many ways in which the industry helps the world to achieve the UN Sustainable Development Goals.

"Writing is easy: All you do is sit staring at a blank sheet of paper until drops of blood form on your forehead."

Gene Fowler

"Writing is easy: All you have to do is cross out the wrong words."

Mark Twain (Samuel Clemens) in "The Adventures of Tom Sawyer."

1.1 Where we are going

The process

Nobody in the plastics industry wants it to be unsustainable, therein lies the path to the demise of our industry. If the route forward was clear and unambiguous then there would be no discussion and we could all just get on with delivering a sustainable industry.

However, even within the industry there are diverse opinions on what sustainability is, how to best achieve it and what it all means to us. Admittedly, this is sometimes clouded by the inevitable commercial interests but I have no doubt that everybody believes that what they are doing is right and is the best route forward for a sustainable industry.

Some of the things in this book will not please everyone and there will be some who read this book and think that I have got the emphasis wrong (hopefully I got all the facts right). This is because 'sustainability' is a developing concept and no sooner is one issue agreed and solved than a new issue presents itself for discussion and solution.

In the words of Ricky Nelson "... you can't please everyone, so you've got to please yourself".

Sustainability is not a destination and there is no journey, only an iterative process of getting better and solving the issues as they present themselves. We will get things wrong through the 'law of unintended consequences' but that is no excuse for inaction and the industry must start to act at the most basic level. It is our world too and this is about everybody, not just the 'Sustainability Manager'.

The road-map

The road-map shown on the right covers the contents of this book and identifies the range of actions and tools necessary to improve sustainability for plastics processors and even this is limited by space. The road-map shows the type of things that you will have to do to achieve and prove improvements in sustainability. It is not simply about management systems, recycling or any other single thing. Specifically, it is not about simply proclaiming that you are 'green'; the evidence must be there. This makes it about the whole company and it is about a mind-set that says 'sustainability is everybody's job' and that we are all responsible for improving sustainability.

The road-map covers the main areas of a company's operations from the management focus through to reporting progress.

Improving sustainability will impact on all areas of the company. This is not for the faint-hearted but the rewards are more than worth the effort.

- **Tip** – None of the actions in the road-map are ever completed. Get used to continual improvement in every part of the business as we try to make a sustainable future for the plastics processing industry.
- **Tip** – This is a marathon and not a sprint. Take your time, move wisely and, above all, plan for the future.

Where are you starting from?

Many companies will already have completed some of the basics for sustainability but it will help to think about the following questions and get the answers ready:

- What sustainability objectives are required by law, e.g., emission controls, and which have already been achieved?
- What would excellence in sustainability look like in your company?
- What would excellence in sustainability look like in your sector of the industry?

It is perhaps unfair to ask these questions at this stage of the book but they are worthwhile thinking about at this stage and coming back to after you have finished the book, the answers will be a lot easier then.

To check if you are ahead of the rest of the industry, answer the following four questions:

- Is sustainability expressly stated as part of the company's mission or strategic vision?
- Is sustainability expressly stated as part of the company's goals?
- Is sustainability part of the company's strategy and processes?
- Is sustainability reflected in the company and personal KPIs?

If you answer 'Yes' to all these questions then you are probably ahead of 90% of the industry.

Major areas	Processes/Tools		Benefits	Results
Management focus	Value chain mapping	Lifecycle analysis	Understanding of impacts	Responsible business management
	Product lifecycle	UN SDGs	Quantification of impacts	
Management systems	Environmental management	ISO 14001	Effective systems	Structured approach
	Energy management	ISO 50001	Compliance	
	Health &Safety management	ISO 45001		
	Risk assessment		Reduced management risk	
Design	Product Design Specification	Resource efficiency	Reduced lifecycle impacts	Sustainable product portfolio
	Product planning	Product innovation	Reduced cost to user	
Raw materials	Recycled materials		Reduced cost	Sustainable raw materials
	Bio-based materials	Biodegradable materials	Reduced impact	
Manufacturing	Processing	Internal re-use	Reduced waste and material use	Minimised and efficient resource use
	Product quality	Pellet control	Reduced social impact	
	Sustainable procurement	Sustainable suppliers	Reduced reputational risk	
	Sustainable distribution		Reduced transport impact	
	Energy	Improved energy management	Reduced energy use	
	Water	Improved water management	Reduced water use	
	Waste minimisation	Improved waste management	Reduced materials use	
Use and end-of-life	Very short-life products		Elimination	Environmental impact eliminated
	Short-life products	Improved recyclability	Material recovery	Environmental impact reduced
	Medium-life products	Simplification and identification		
	Long-life products	Reusability	Product re-use	
Social responsibility	Labour relations	Professional careers	Improved staff satisfaction	Social reputation improved
	Workplace health and safety	Staff equality	Improved social reputation	
	Staff awareness			
Reporting	Agreed standards	Clarity of goals	Focus on improvement	Clear metrics for progress assessment
	Reliable data & reporting		Linkage to finance	

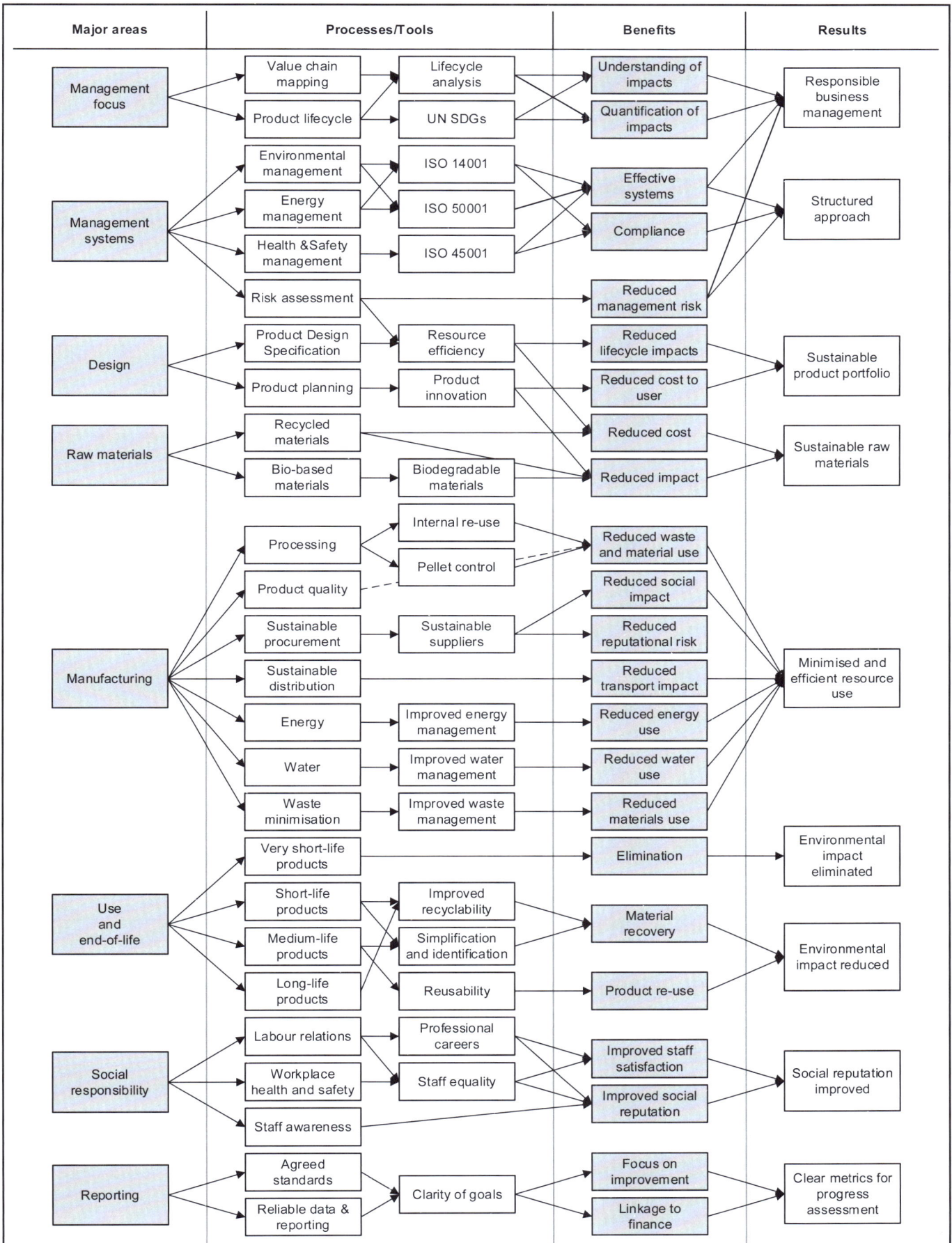

The sustainability road-map

Improving sustainability is complex and it needs a road-map of the available processes, tools and actions. That is what this book seeks to provide.

1.2 What is sustainability?

What is sustainability?

The large-scale environmental movement effectively started with the publication of Rachel Carson's 'Silent Spring'[1] in 1962 but sustainability moved to the fore in 1987 with the publication of 'Our Common Future'[2], also known as 'The Brundtland Report'.

Previously, most commentators had been largely negative about economic development in terms of sustainability but Brundtland recognised that economic progress was fundamental to sustainability and that poverty reduces sustainability. This new recognition of the balance between economic growth and sustainability led to the famous Brundtland definition of sustainability:

"Sustainable development seeks to meet the needs and aspirations of the present without compromising the ability to meet those of the future. Far from requiring the cessation of economic growth, it recognises that the issues of poverty and underdevelopment cannot be solved unless we have a new era of growth in which developing countries play a large role and reap large benefits."

This is a broad definition that is widely recognised and powerfully links the environmental, social and economic aspects of sustainability. Unfortunately, in many cases only the first sentence is quoted and these links are missing in some of the 'hair shirt' approaches to sustainability promoted by some environmentalists.

These concepts are also linked in a definition produced by the US Department of Commerce[3]. This defines sustainable manufacturing as "The creation of manufactured products that use processes that minimize negative environmental impacts, conserve energy and natural resources, are safe for employees, communities, and consumers and are economically sound."

Whichever definition is used, the vital question still remains for plastics processing companies seeking to become 'sustainable': 'What is it that we can actually do to become sustainable in environmental, social and economic terms?' This question is not answered by broad statements of intent or by 'greenwashing' (see Section 1.20) but by concrete actions to improve sustainability, however measured.

The elements of sustainability

Sustainability has three essential elements, environmental, social and economic and these are shown in the diagram on the right. Each element has specific issues but there are also issues that are common between the elements, e.g., climate change is an issue that links the environmental and social elements. A selection of these linking issues are also shown in the diagram.

- **Tip** – I have included some of the linking issues but there are more to be considered. Why is nothing ever simple?

- **Tip** – Just as sustainability is more than the environment, it is also about more than the product. Products are important for plastics processors but they are not the only thing, do not think that just because you have dealt with the product issues that it is all over. The social issues are much broader than the simple product issues.

Feel free to make up your own definition of 'sustainability', you will be in good company, but be prepared to justify it to the world.

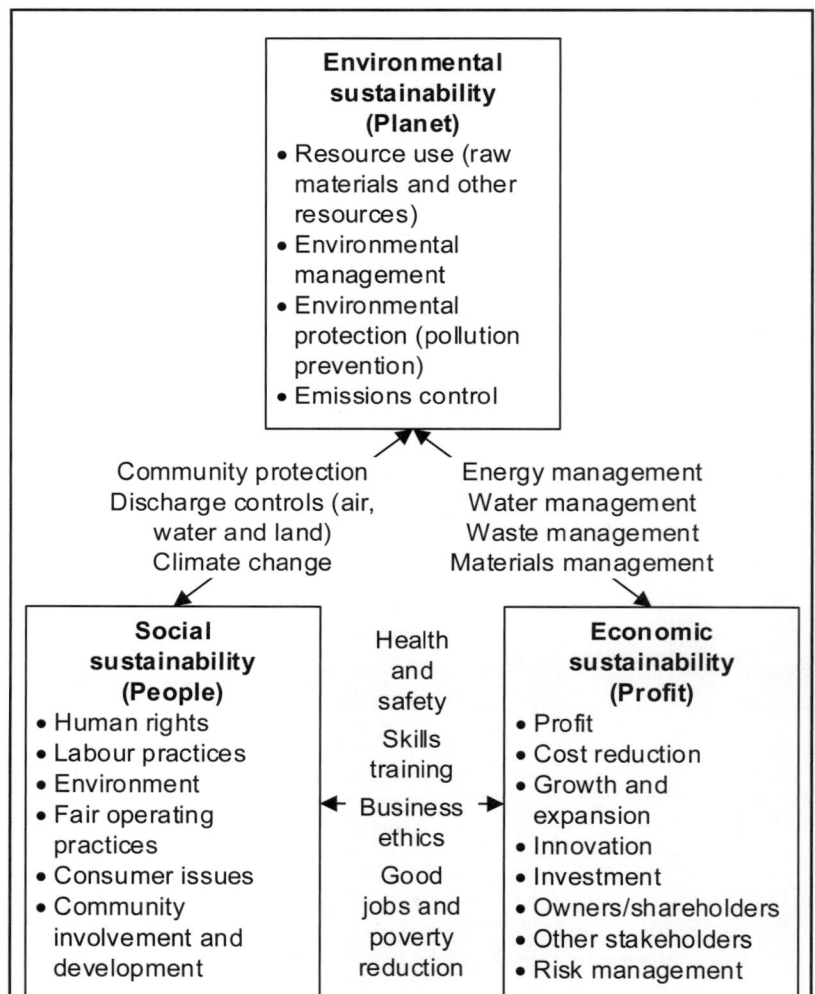

Environmental sustainability (Planet)
- Resource use (raw materials and other resources)
- Environmental management
- Environmental protection (pollution prevention)
- Emissions control

Community protection
Discharge controls (air, water and land)
Climate change

Energy management
Water management
Waste management
Materials management

Social sustainability (People)
- Human rights
- Labour practices
- Environment
- Fair operating practices
- Consumer issues
- Community involvement and development

Health and safety

Skills training

Business ethics

Good jobs and poverty reduction

Economic sustainability (Profit)
- Profit
- Cost reduction
- Growth and expansion
- Innovation
- Investment
- Owners/shareholders
- Other stakeholders
- Risk management

The three elements of sustainability

Environmental sustainability is only one facet of overall sustainability. It may be the first one that people think of but social and economic sustainability are equally important. All three are needed for a company to be sustainable.

Chapter 1 – Introduction to sustainability management

Environmental

The environmental elements of sustainability are perhaps the most familiar to readers. This element covers the familiar topics of reducing resource use, protecting the environment and controlling emissions. In many cases, there are legal requirements for compliance but these should always be regarded as the minimum to be achieved.

- **Tip** – Environmental sustainability cannot be divorced from the other elements of sustainability.

Social

Many of the social sustainability issues fall within what was once termed Corporate Social Responsibility (CSR) and some companies will be more familiar with this idea than with the idea of social sustainability. The main issues are shown in the diagram on the left and are covered in more detail in Chapter 11.

- **Tip** – Many companies already do great work in social sustainability but do not record it as such.

Social to environmental links

The environmental impacts of a company are generally externalised even if they have financial implications. This means that there are inevitable links between the social and environmental elements, e.g., emissions from a site can impact on the local community and emissions as a result of excess energy use can impact on the whole world through climate change.

Economic

The issue of economic sustainability is not often discussed but the quickest way to reduce poverty is to have strong economic growth and this needs financially viable companies. The main issues and the route to economic sustainability are covered in more detail in another book[4].

Economic to social links

Economically sustainable companies do not exist in a vacuum, they provide good and safe jobs for people, provide skills training, work ethically and enhance the community by being good neighbours.

Economic to environmental links

Achieving economic sustainability does not mean sacrificing environmental sustainability. Efforts to reduce energy use, water use, waste and materials use not only benefit the environment but also reduce costs and improve economic sustainability.

- **Tip** – The three elements of sustainability are all important. It is OK to focus on one of these at the start but this should never be exclusive.

It is broader than you thought

Many readers can be forgiven for being surprised at the breadth of the concept of sustainability. It is truly broader than you thought. The good news is that most companies are already doing a great deal towards being sustainable. They may not record it as such but in most cases sustainable business is simply good business.

The simple fact that you are reading this book indicates that you are interested in sustainability, concerned about the implications for your business or want to know how to go forward. The motives don't matter and the definition doesn't really matter either, except as a useful tool to try to understand what it is that you can do to get better.

- **Tip** – Motives don't matter. You can be motivated by the desire to do something for the world, by fear or by greed. The important thing is that you start soon to make the industry more sustainable.

- 1. Carson, R.L. 1962. 'Silent Spring', Houghton Mifflin.
- 2. Our Common Future: Available at sustainabledevelopment.un.org/content/documents/5987our-common-future.pdf.
- 3. OECD Sustainable Manufacturing Toolkit: Available at www.oecd.org/innovation/green/toolkit/aboutsustainablemanufacturingandthetoolkit.htm.
- 4. Kent, R.J. 2018. 'Cost management in plastics processing', Elsevier.

"Leaders make choices that keep them up at night. If you are sleeping well then you are not doing your job."

Dean Fulton 'New Amsterdam'

OK, it is a TV programme and the choice he made in this instance was wrong but the sentiment is correct.

"It's not easy being green."

Kermit the Frog

Sustainable businesses balance the environmental, social and economic pressures for responsible, ongoing success.

It is just as important for small companies

It may be thought that small companies are able to ignore the sustainability agenda, keep below the parapet and still survive. This is no longer true:

- ■ Everybody must comply with the same regulations no matter what your size. It may be more difficult for small companies but nobody cares about your size anymore.

- ■ Small companies can be quicker to respond and gain a competitive advantage. 'Corporate inertia' is reduced in small companies and they can respond quicker to the pressures.

- ■ With the rise of the Internet, small companies are just as visible as large companies and sometimes make easier targets for activists and legislators.

- ■ Small companies may feel the pressure of large customers more quickly and be forced to respond more quickly.

1.3   The pressures for sustainability

It is only going to get worse

It is almost impossible for any plastics processor to be unaware of the increasing pressure in the past years to become sustainable. This started with consumers and governments but now appears to be coming from every stakeholder in the business. Indeed, this concept of stakeholders has broadened considerably as the pressure increases to internalise all impacts that the industry traditionally, and perhaps wrongly, externalised. As noted in the introduction to this chapter, some of these stakeholders also believe that 'feelings are facts' and it has traditionally been difficult for a technical industry to communicate with these stakeholders. The industry responds with facts which do nothing to change their feelings. It is almost as if there are two separate conversations being held and that they have no common ground. Science and evidence are no longer treated as facts and a set of testable hypotheses but as a series of opinions, where yours is as valid as mine.

The reality is that even otherwise good companies can be surprised and caught unawares by sustainability issues generated by somebody's feelings or by issues that they had never considered.

- **Tip** – You can try to appeal to the 'feelings' but be very sure that you have your facts right too.
- **Tip** – The development of the scientific method drove the greatest increase in living standards ever. It was accompanied by liberalisation and huge advances in society. It was driven by evidence and not by feelings, but that was then.

Sustainability is no longer a 'nice to have', it is vital and getting sustainability issues wrong can destroy markets, companies and careers overnight. Failing can cost you your reputation and real money very quickly.

The best companies see sustainability issues not simply as 'avoiding disaster' but as a way to establish competitive advantage by strategically managing and minimising their inevitable issues. As an example, note the number of companies who now promote their products as 'energy saving'. What was relatively unimportant in 1990 is now seen by all as a competitive advantage. Issues such as energy use, recycling and plastics in the ocean are now shaping markets and directly affecting companies and their profitability.

Where is the pressure coming from?

The pressure is coming from the complete range of stakeholders (however defined) and some of these are:

- Customers – these are potentially the most important group, without customers there is no business and no economic, or any other type, of sustainability. Today's customers are much more aware of sustainability issues and are driving change in the way products are designed, marketed and dealt with. A company that has a poor reputation for sustainability will be shunned by customers with an inevitable result.
- Community – the wider community is an extremely powerful agent in driving change. They may not be direct customers but they can be stakeholders and be affected by, and affect, operations, e.g., they may not buy your products but if they see them littered then they will be affected by them and can bring pressure to bear on the company.

Community also includes the physical neighbours who can have a powerful

> "Over time, companies and countries that do not respond to stakeholders and address sustainability risks will encounter growing scepticism from the markets, and in turn, a higher cost of capital. Companies and countries that champion transparency and demonstrate their responsiveness to stakeholders, by contrast, will attract investment more effectively, including higher-quality, more patient capital."
>
> **Larry Fink**
> **CEO of Black Rock**
> **2020 client letter**

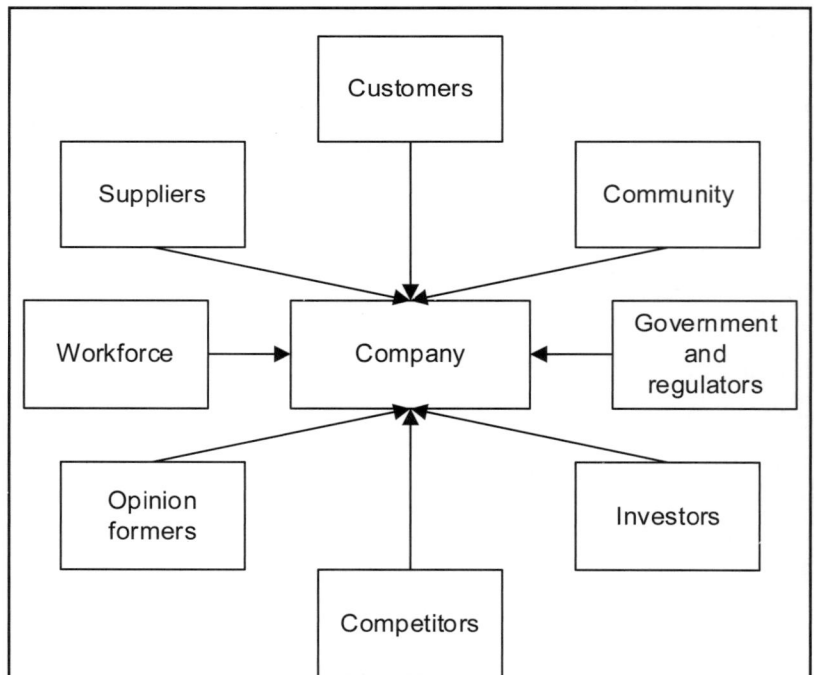

The pressures for sustainability

Many groups are applying pressure for sustainability and, unfortunately, they don't all want the same things, have the same knowledge or 'feelings'. This often makes it difficult to know what to do but having a plan means that you can approach the idea in a logical manner.

impact on operations and costs, e.g., one company had to spend over £500,000 on noise baffles to reduce noise from the site even though the site was there for many years before the houses were built.

- **Tip** – Similarly, it always intrigues me when city people move to the country and promptly complain about the smell of animals or the church bells. As a country boy, these are evocative smells and noises.

- **Tip** – Trade associations are also part of the community. They can be a great help in dealing with sustainability issues by providing leadership and advice. Being with the wrong one can also affect you too.

- Government and regulators – this is not so much a pressure as a requirement that no individual company can resist. If the regulations change then you must change with them or have your licence to operate removed.

- **Tip** – Failing to track and understand changes (and proposed changes) in regulations and laws can put a company at a competitive disadvantage. It is an essential for sustainable operations.

- Investors – for many companies, investors provide a powerful driver for improved sustainability (see sidebar on the left). Investors have a global view and there is a trend for ethical investing and rating of companies for their sustainability performance, e.g., Sustainalytics (www.sustainalytics.com) provides risk ratings of companies in terms of environmental, social and governance issues, carbon risk ratings and a range of other sustainability topics. Investors see the transition to a low-carbon and sustainable economy as both a risk and a potential reward. They will move their money to the rewards.

- Competitors – competitors are a pressure and will go where the customers are going, even if only by 'greenwashing' (see Section 1.20). Sustainability has been 'weaponised' by some companies to put pressure on competitors and drive change (and profits).

- Opinion formers – this includes a range of NGOs, some are single-issue with strongly held feelings and some are more general, but they are all active in applying pressure for sustainability improvements.

- **Tip** – Opinion formers include the media. Managing relations with the media is vital in communicating progress. Sometimes they will even print good news.

- Workforce – the workforce is often neglected as a stakeholder and pressure group in sustainability. They can be a powerful ally in driving change. Good worker relations and social sustainability can have an enormously beneficial effect with the wider community (or vice versa).

- Suppliers – suppliers are not often seen as a pressure for sustainability but they are also driven by the market and are working to improve their products or services. They will work with you or with your competitors.

"If you think about Generation X and all those other generations coming next, they will care about environmental issues and they will spend money on being sustainable. And they will opt out of companies that are not doing it."

Fridrik Larsen

Don't just think about sustainability as a risk. For some companies, this could be the greatest opportunity ever.

What can processors do to help? (and thus both survive and prosper)

The sub-title here is just as important because we need an industry that is sustainable by any metric and that will continue to deliver benefits to the world.

For a start, every processor needs to:

- Understand the impact of their operations and products in all areas of sustainability.
- Identify the actions necessary to reduce the impact of their operations and products.
- Work to reduce the impact of their operations and products.
- Identify the actions that will affect other areas in the wider world.
- Work both up and down the supply chain to reduce the impact of their operations and products.
- Be prepared to pivot away from unsustainable products and towards sustainable products.

This is not going to be easy for many companies but there is no alternative. If your products stop being relevant and are judged (on whatever criteria) to be unsustainable then you will stop producing them (either willingly or unwillingly).

Whatever your own views on sustainability, this is going to be the defining issue for the next 20 years.

1.4   Approaches to sustainability

A multitude of approaches

There are many frameworks and definitions for sustainability, some are very general and are primarily definitions, e.g., the Brundtland definition (see Section 1.2), some focus purely on the environmental issues, e.g., life cycle assessment (LCA, see Section 1.5) but there are three approaches that have had a real impact in terms of practical application. In alphabetical order, these are:

Cradle to cradle™ (regenerative design)

Cradle-to-cradle™ (C2C) is based on the work of William McDonough (an American architect) and Michael Braungart (a German chemist)[1] and aims to design products and systems to mimic natural metabolic processes (www.cradletocradle.com). All materials are divided into either technical or biological 'nutrients':

- Technical nutrients are inorganic or synthetic materials that have no negative effects on the environment and can be used continuously without losing effectiveness, e.g., metals and plastics, provided they are not 'downcycled' or lost as waste.

- Biological nutrients are organic materials that will decompose in the natural environment and provide food for bacteria and other microorganisms.

There are five criteria for C2C:

- Material health – the materials that make up the product are identified as either:
  - High risk and harmful, e.g., those covered by REACH and RoHS (see Section 3.6).
  - Moderate risk but acceptable.
  - Low risk and safe for use.
  - Incomplete data.
- Material re-utilization – material recovery and recycling at the end of product life.
- Carbon management and renewable energy – the amount of energy needed for production and the source of the energy, e.g., renewable energy.
- Water stewardship – use and discharge.
- Social fairness – labour practices.

These criteria can be used to assess and certify the product. Until 2012, certification was only available from McDonough and Braungart's companies. This restricted the growth of C2C and the concept underperformed. Since 2012, certification is available from the non-profit 'Cradle to

Cradle Products Innovation Institute' although McDonough and Braungart still control the term 'Cradle to Cradle' and offer consultancy on implementing cradle to cradle.

The framework has been implemented in many areas around the world, particularly in buildings, with some success. The process can be related to the environmental aspects of LCA but C2C does not consider the impacts during the use phase of the product life cycle and is not an 'open' standard.

The Natural Step

The Natural Step (TNS) framework was devised by Karl-Henrik Robèrt (a Swedish doctor)[2] in 1989 and sets out the system conditions for a sustainable society (www.thenaturalstep.org). TNS is essentially 'systems thinking' applied to sustainability.

There are five levels to the framework (5LF) that can be used to analyse complex systems. If the 5LF is applied to sustainability then it is called The Natural Step Framework or the Framework for Strategic Sustainable Development. The five levels are:

- System – defining how a society influences the socio-ecological system. TNS considers the influences in terms of scientific principles, e.g., thermodynamics, ecological systems, social systems and the human influences on these.

- Success – defining the system conditions needed so as not to affect the system negatively. TNS defines three ecological principles and five social principles of what is not acceptable and this allows the definition of what must be done. This is:
  - Eliminate the progressive build-up of substances extracted from the Earth's crust.
  - Eliminate the progressive build-up of chemicals and compounds produced by society.
  - Eliminate the progressive physical degradation and destruction of nature and natural processes.
  - Eliminate structural obstacles to health, influence, competence and impartiality.
- Strategic – strategic planning to achieve the required system conditions. TNS uses a concept called 'backcasting from first principles' which starts with a vision of success in the future and moves backwards to what needs to be done today to get to the vision.

Changes in legislation can lead to essential raw materials becoming unavailable or unusable, e.g., Pb-based stabilisers in PVC-U. Knowing what is happening can allow companies to plan for the necessary resource changes to stay ahead of the game instead of playing catch-up.

"There are no inherently sustainable or unsustainable materials. Though certain materials are easier than others to manage (financially and technically) within sustainability constraints, the important distinction to make is the one between sustainable and unsustainable management of materials."

Lindahl et al
Journal of Cleaner Production 64 (2014) 98-103

Can we please not talk about sustainability as a journey?

"Unless there is a ticket and a ticket inspector involved then it is not a journey – just a load of PR guff".

Celia Walden
Daily Telegraph 11/08/2020

- Actions – defining the priority actions to achieve the required system conditions. The strategic planning defines the actions that are needed to deliver the strategy within the system.
- Tools – defining or developing the appropriate tools to achieve the required system conditions.

TNS has been a not-for-profit organization since the start and is not only a planning model but also a set of freely available tools to allow companies to learn about and contribute to sustainable development.

TNS has been applied around the world by large numbers of cities and companies and forms the basis of the European PVC industry response to sustainability issues.

Triple bottom line

The Triple Bottom Line (TBL) is an extension of the Brundtland definition and closely resembles the elements of sustainability outlined in Section 1.2. The TBL concept was more fully developed by John Elkington (a British business author)[3] in 1997. This sets out a sustainability-based accounting framework based on three bottom lines (environmental, social and financial) that allows companies to measure performance on more than simple financial numbers (johnelkington.com).

The three bottom lines are:

- Planet – a measure of environmental or natural capital performance. Companies should try to benefit the natural order or at a minimum do no harm and minimise the environmental impact and natural capital use. TBL supports using LCA or 'cradle to grave' assessment to assess the true environmental costs and supports internalising environmental impacts rather than externalising them.

- People – a measure of human capital performance. Companies should consider not only their conventional shareholders but also have a responsibility to a wider community of stakeholders. TBL strongly supports the concept of stakeholders and companies should try to provide benefits to all stakeholders and not to exploit them. This is effectively social sustainability (see Chapter 11) and reporting can be by the GRI standards (see Chapter 12).

- Profit – a measure of economic capital performance. TBL does not consider this to be the same as the traditional definition of profit, i.e., in TBL the profit is the economic benefit of the host society and not simply the internal profit made by a company.

TBL has had limited affect in companies due to the lack of concrete implementation and

evaluation advice and guidance. Despite this, the concept has gained in strength and still provides a robust framework for considering sustainability issues.

Which one to use?

The three approaches differ quite significantly but all have good aspects. Our advice for any company aspiring to become sustainable is to examine all the approaches and to then to play 'mix and match' with the concepts and ideas that are most suitable for them. A model that is suitable for one processor may be totally unsuitable for another. It all depends on where you are starting from.

The important thing is to avoid knee-jerk reactions, e.g., changing materials for those that are touted as being 'eco-friendly' (see Section 1.20). Sustainability needs planning, consideration and is too important and too multi-layered for 'quick fixes' (and they will probably fail anyway).

Renewable can also be considered sustainable but sustainable is not necessarily renewable.

Natural capital is an evolving concept and in 2012 the UK government established a Natural Capital Committee to 'advise the Government on the state of natural capital in England.'

The Committee recommended a long-term, 25 Year Environment Plan (25 YEP) to protect and improve the environment.

See: UK Government Natural Capital Committee, UK, 2020 (www.gov.uk/government/publications/natural-capital-committees-end-of-term-report).

- 1. McDonough, W.A. and Braungart, M. 2002. 'Cradle to Cradle', North Point Press.
- 2. Robèrt, K-H. 2008. 'The Natural Step Story: Seeding a Quiet Revolution', New Catalyst Books.
- 3 Elkington, J. 1997, 'Cannibals with Forks – The Triple Bottom Line of 21st Century Business', Capstone.
- 4. American Chemistry Council. 2016 'Plastics and Sustainability', (plastics.americanchemistry.com/Plastics-and-Sustainability.pdf).

Natural capital valuation

An approach that considers only the environmental impact on the 'natural capital'. This can use 'natural capital' valuation techniques to measure and communicate the environmental impacts in monetary terms.

This has been done for plastics products versus alternative materials[4] and the results show that, for the vast majority of products, the natural capital cost of plastics products is less than the cost for products manufactured from alternative materials.

1.5   Life cycle assessment (LCA)

An environmental assessment

One increasingly important technique for assessing environmental sustainability is life cycle assessment (LCA) and at the product level this is a valuable approach. This was not considered in Section 1.4 because an LCA only looks at the environmental impacts and there is no consideration of the social or economic impacts.

Not materials but products

LCA identifies the energy, materials and waste flows of a product over its life cycle to allow evaluation of the environmental impacts, e.g., the emissions and wastes released to the environment.

- **Tip** – An LCA is only valid for a product, it is not valid for a material alone.

An LCA measures all the relevant factors over the life cycle of the product and covers:

- Raw materials.
- Manufacturing.
- Use.
- End-of-life.
- Transport.

The phases of an LCA

An LCA to ISO 14040[1] has five phases and the relationship between these is shown in the diagram on the right. It is important to consider producing an LCA as an iterative process rather than as a linear process. This means there will inevitably be changes and revisions to each of the phases during the process, e.g., nothing is agreed until it is all agreed. The five phases of an LCA are:

Goal and scope definition

This is where the purpose of the LCA and what it is going to cover are determined. The goal and scope can be defined by the answers to:

- Which product is going to be covered?
- Is the LCA going to be used for comparison, e.g., similar products from different materials, processes or manufacturers, or is it going to be used to assess a single product?
- What is the application and can another product meet the application requirements?
- What are the requirements and are there standards that must be met?
- Are there specific environmental impacts of concern to the company or stakeholders?

The answers to these questions will help to formulate an explicit statement of the goal of the LCA.

The scope of the LCA requires definition of the system boundaries. A full LCA would include all upstream and downstream processes associated with a product. In reality, the scope can, and should be, adjusted to exclude processes that will not change the overall conclusions of the LCA, i.e., an appropriate cut-off point or boundary must be defined.

The scope should also define the 'functional unit' being studied by the LCA to provide a reference for comparison with other products, e.g., 1 kg of product or 1 metre of product.

Scoping also involves screening of the initial data to ensure that the goal is achievable.

- **Tip** – The goal and scope will almost certainly be changed as the LCA progresses and more data is available.

Life cycle inventory analysis (LCI)

This is the data gathering phase for all data on the energy, material and waste flows associated with the selected product. This is simply a list of the inputs (energy and materials) and outputs (emissions and waste) at each stage of the life cycle from

> An LCA is only about environmental sustainability. It does not consider social or economic sustainability.

> LCAs must be read very carefully to check that you are comparing 'like with like'.

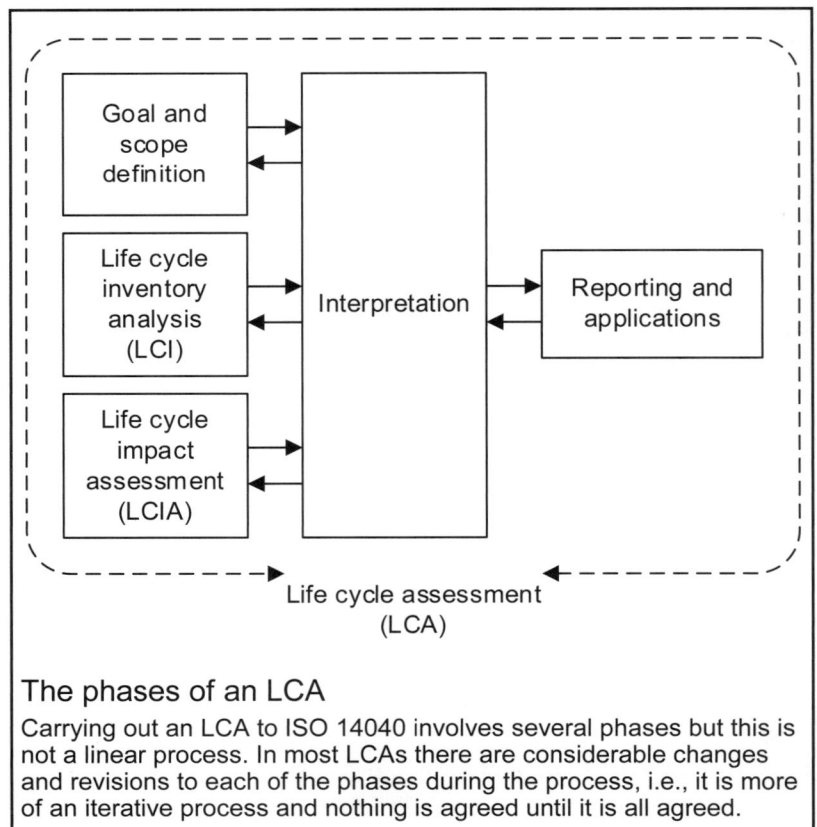

The phases of an LCA

Carrying out an LCA to ISO 14040 involves several phases but this is not a linear process. In most LCAs there are considerable changes and revisions to each of the phases during the process, i.e., it is more of an iterative process and nothing is agreed until it is all agreed.

raw materials to end-of-life. This is similar to the process flow chart described in Sections 2.5 and 8.3.

Typical data in the LCI will be:

- Energy used (by the process and services).
- Other resource consumption.
- Gases emitted to air, e.g., SO_2 and CO_2.
- Water input to the process and wastewater disposed of (see Chapter 8).
- Solid wastes sent for recycling or to landfill.

Deciding on the LCI data to be collected and then collecting, collating and validating the LCI data is the most a time-consuming task but companies who have collected data for an EMS (see Section 2.5) or waste minimisation (see Section 9.3) will have already collected most of the data.

- **Tip** – LCIs can be generated using established databases and special LCA software, e.g., SimaPro™ (simapro.com).
- **Tip** – Depending on the goal and scope, it is possible to stop after the LCI phase. The result will not be compliant with ISO 14040 but the LCI may reveal how to improve the product or process without needing an LCIA.

Life cycle impact assessment (LCIA)

Where the LCI identifies a large number of potential impacts then interpretation and comparison can be difficult and impact assessment is necessary. LCIA is where the flows associated with the product are related to the actual environmental burdens.

In most LCAs, the environmental burdens with the same environmental impact are grouped together in a 'classification' process and an estimate is made of their individual contribution to that particular environmental impact. This is 'characterisation' and allows the most significant burdens to be identified. LCIA is typically assessed on factors such as:

- Acidification of water.
- Climate change.
- Ecotoxicity to land.
- Ecotoxicity to water.
- Eutrophication (where water is enriched with minerals and nutrients).
- Fossil fuel depletion.
- Greenhouse gases/global warming potential.
- Human toxicity.
- Photochemical ozone creation (summer smog)
- Stratospheric ozone depletion.
- Waste disposal.

- Water extraction.

Not all factors need to be reported and the reported factors will vary with the product and the LCA results.

Interpretation

An LCA is an iterative process and as it is carried out it is necessary to review and, if appropriate, revise the scope of the LCA by considering the results of the LCI and LCIA.

This is interpretation and it aims to identify issues such as:

- Data gaps.
- Data quality issues.
- Validation of the scope and goals of the LCA.

Where an LCA is being used to compare products, the relative performance in each impact category can be assessed. Rarely does one product perform better in all areas than another and prioritisation is needed. This means decisions about which impact category is most important to the environment, company, stakeholders, customers or regulators. This activity normally would form part of the goal and scope definition phase but this will inevitably change during the process.

Reporting and application

Having read many product LCAs, the issue of reporting is one that needs most work. Reporting is an essential part of an LCA and a good report will address all the phases of the study but, most importantly, it will be fully transparent in terms of the value judgements and choices made.

An LCA is designed to find the environmental impacts of a product, so decisions can be made in the product design to evaluate alternative products and to minimise the environmental impact.

- **Tip** – An LCA is a rigorous scientific analysis. If value judgements are made then they must be recorded and justified.
- **Tip** – A partisan, biased or incomplete LCA report undoes all the hard work in an LCA. If you want to talk about 'feelings' then do it somewhere else. This is the place for verifiable facts.
- **Tip** – LCA is a tool to aid decision-making and not to make them. There is a world of difference.

An LCA will never give a 'single number' result, it is a collection of numbers and must be assessed carefully.

An LCA for Christmas trees

As a festive note, but maybe not as you read this:

The LCA for Christmas trees shows that the environmental burden shifts after 4.7 years, i.e., an artificial Christmas tree has a lower impact than a real tree if it is used for > 4.7 years.

I am not sure how they got that accurate as 'Christmas comes but once a year'.

Materials World December 2019 (55-57)

- 1. ISO 14040:2006: Environmental Management – Life Cycle Assessment – Principles and Framework.
- 2. ISO 14044:2006: Environmental management – Life cycle assessment – Requirements and Guidelines.

LCAs allow quantification of environmental impacts and allow us to make better environmental decisions.

1.6   Mapping actions on time

Time is of the essence

Sustainability is not only about today; it is also about protecting the future. If a company does not make profits today then it is not sustainable, no matter how 'green' it is, i.e., survival tomorrow needs survival today.

Sustainability is not a single point objective; it is a process of getting to the future and we need to consider both the short and the long-term impacts and the associated risks. Risks and impacts can be short-term or long-term but most approaches to sustainability do not deal with the concept of time. TNS (see Section 1.4) deals with this implicitly by discussing issues such as extraction rates for renewables and non-renewables and the build-up of materials in the environment at a faster rate than they can be degraded or broken down.

Considering the time element of sustainability allows examination of the impacts in terms of the short and long-term.

- **Tip** – A risk assessment is always more accurate for short-term impacts.
- **Tip** – The precautionary principle says that the best decision is the one with the least anticipated damage for a wrong decision. We need to include time in our considerations if we are in this for the long-term future.

Environmental impacts

The environmental impacts (in alphabetical order) are:

Short-term (< 5 years)

- Hazardous emissions.
- Photochemical ozone creation (summer smog).
- Pollutant emissions (winter smog).
- Toxic air, land and water emissions.
- Waste disposal.

These pose an imminent risk to health or the environment; they may, or may not, have a long-term risk but the primary risk is short-term.

Long-term (> 5 years)

- Eutrophication.
- Greenhouse gases/climate change.
- Stratospheric ozone depletion
- Resource depletion.
- Water extraction.

These do not pose an imminent risk to health or the environment but the long-term risk is significant; they may have a short-term risk but the primary risk is long-term.

Social impacts

The social impacts (in alphabetical order) are:

Short-term (< 5 years)

- Accidents and injuries.
- Job losses.
- Ethical and governance failures.
- Short-term health risks to stakeholders (toxicity).
- Skills shortages.

These pose an imminent risk to social order or to the stakeholders; they may have a long-term risk but the primary risk is short-term.

Long-term (> 5 years)

- Industrial disease, e.g., asbestosis.
- Long-term skill shortages.
- Loss of social cohesion.

These do not pose an imminent risk to the social order or the stakeholders but the long-term risk is significant; they may have a short-term risk but the primary risk is long-term.

Economic impacts

The economic impacts (in alphabetical order) are:

Short-term (< 5 years)

- Cost reduction.
- Profitability.

These pose an imminent risk to a company's economic stability and continued operations; they may also have a long-term component but the primary risk is short-term.

Long-term (> 5 years)

- Innovation failure.
- Investment failure.
- Reputational brand damage

These do not pose an imminent risk to a company's economic stability and continued operations but the long-term risk may be very significant; they may also have a short-term risk but this is minor compared to the long-term risk.

Not materials but products

The assessment of impacts and risks should always consider products and not materials. Materials are only part of any product and

If 'perfect is the enemy of good' then sometimes short-term solutions are the enemy of long-term sustainability.

Assessing an impact as short or long-term is always contentious and there is room for discussion, i.e., feel free to move them around. As an example, I think climate change is a serious long-term impact but some commentators assign climate change as a reason for every short-term weather event.

There is a difference between climate and weather (see Sense about Science – archive.senseabouts cience.org/ resources.php/10/ making-sense-of-weather-and-climate.html).

most products are only part of a system.

- **Tip** – Cars contain plastics but they are only part of the 'transportation system' and the plastics in food packaging are only part of the 'food preservation and delivery system'.

There are few absolutes in sustainability and products must be considered in terms of the sustainability of the complete system in both the short and the long-term.

It is possible for a product's environmental impact to be high in the short-term but to be low (or even beneficial) in the long-term. An example of this might be food packaging which reduces food waste where the carbon impact of food waste is much higher than that of the packaging (see Section 10.3). The short-term environmental impacts from poor end-of-life handling and resource consumption may be rated as high but this is considerably less than the risk associated with the long-term impact of climate change. It is equally possible for a product to have low impacts in the short-term but to have high impacts in the long-term.

The diagram on the right shows a simple grid based on the short-term impacts (< 5 years) and the long-term impacts (> 5years). The grid allows us to divide products into four categories and to look at the short and long-term sustainability impacts of products.

Unsustainable products

These products have high impacts and risks in both the short and the long-term. They are typically very short-life products with a functional use timescale of < 1 day (see Section 10.2). These products are under threat throughout the world[1] and it is difficult to justify their continued production and use.

- **Tip** – Processors of these products need to seriously consider exiting the market.

Sustainable products

These products have low impacts and risks in both the short and the long-term. They are typically long-life products with a functional use timescale of > 15 years (see Section 10.8). These products are widely accepted as being sustainable throughout the world and it is easy to justify continued production and use.

- **Tip** – Processors of these products should invest and grow markets to improve the sustainability of the industry.

'Investigation needed' products

These products have high impacts and risks in the short-term but low impacts and risks in the long-term. They are typically short-life products with a functional use timescale of 2-15 years (see Section 10.3). These

products are not yet under threat but improvement options need investigation[1].

- **Tip** – Processors of these products need to understand the value-added by the products.

Value products

These are products that have low impacts and risks in the short-term but high impacts and risks in the long-term. They are typically medium-life products with a functional use timescale of 2-15 years (see Section 10.6). It is unlikely that these products will find themselves under threat in the next 40 years.

- **Tip** – Processors of these products should act to reduce long-term impacts and risks.

- 1. 'Eliminating Problem Plastics', Version 3 December 2019, WRAP, www.wrap.org.uk.

Good companies do not seek simply to comply with legislation, they want to achieve an advantage.

The protection of the environment is an essential precondition for being in business but we must learn to think in both the short and the long-term.

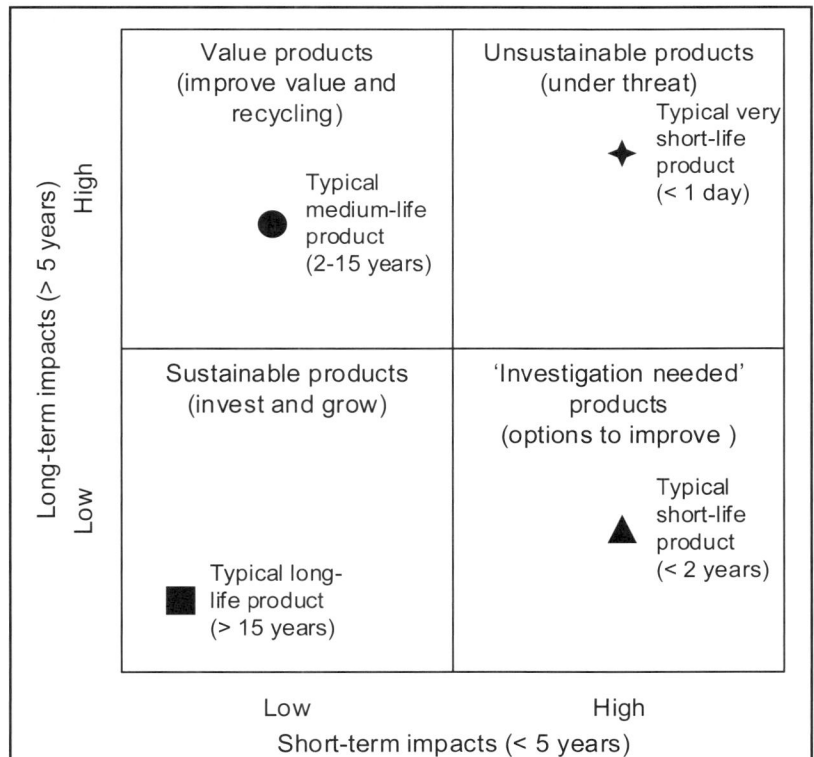

Sustainability must consider short and long-term impacts

Sustainability not just about short-term impacts, it is also about the long-term impacts. These two types of impacts vary with the product use and end-of-life timescale (see Section 10.1). Products need to be examined in terms of both types of sustainability impact.

1.7   Mapping actions on the product life cycle

Impacts change through the life cycle

Looking at the short and long-term impacts of products (see Section 1.6) helps to understand products and time but companies also need to decide what to do to improve company and product sustainability. This is best done by considering potential sustainability actions through the product life cycle because these will change through the product life cycle. Mapping the potential actions in each of the three main sustainability strands to the product life cycle clarifies thinking about which issues are important and where action can be taken.

This has been done in the diagram on the right where a selection of typical actions has been mapped to the relevant sustainability strand and the product life cycle phase.

- **Tip** – Always consider the complete life cycle; issues, impacts and risks will change.

Companies and products together

Mapping actions through the product life cycle should focus not simply on products but also on the company. Some of the actions listed in the diagram are related to specific products and some are related to the company, e.g., energy in the use phase is product related but fair business practices is company related. Depending on the product range, companies are advised to complete a grid like that shown at several levels, e.g.,

- Company (general).
- Product A.
- Product B.
- etc.

The grids will help to identify the general and specific actions.

- **Tip** – Some of actions for similar products will be the same. It may help to think of 'product groups' to reduce the workload.
- **Tip** – If a processor only makes a component for a customer then the grid can be simplified to exclude phases over which they have no control. It is still worthwhile discussing the complete grid with the customer to see if actions can help them. Customers will be ready for the discussion as they are facing the same type of pressures (if they haven't asked for it already).

It is OK to choose

The mapping exercise will inevitably identify a wide range of actions, i.e., unlimited work, but every company has limited resources and must choose the actions to carry out. This is about identifying those projects that can be reasonably expected to deliver benefits and those that may never be completed.

Do the legal stuff first!

Legal requirements are essential and compliance is non-negotiable. Any action related to legal compliance must be the highest priority. Great companies stay ahead of the legal requirements, they anticipate them and address them early and effectively to gain an advantage.

- **Tip** – Legal requirements should always be the minimum requirement. Try to get ahead of the curve and do better, it is almost certain that the requirements will be higher in the future.

Specify the goals and set the limits

As with any process, sustainability needs set goals (what you want to achieve) and this should be part of the implementation process (see Section 1.18). Setting the limits is about defining what you can realistically achieve with the available resources.

- **Tip** – Be realistic in selecting which actions to start.

Select the actions

Not all actions are equal in costs and benefits but companies must still select which actions to carry out. Selection can be made easier by answering the questions:

Is it relevant and significant in impact?

Any action undertaken should be relevant to the company's stakeholders and significant in its impact on the company's operations. The concept of 'materiality' (see Section 1.8) can be used as part of the selection process by prioritising the projects that are most relevant to the stakeholders and have the highest impact.

- **Tip** – Not all actions are equally relevant or significant.

Is it a risk or an opportunity?

Sustainability actions should not always be seen as dealing with risk in the negative sense, action can also present an opportunity, e.g., reducing energy use (see Chapter 6) or water use (see Chapter 8) in

Companies will inevitably have to choose which actions to focus on and where to expend their efforts. This is not a bad thing, a small programme that is completed and then extended is better than a large programme which is never completed.

Starting something means nothing, you only get points for what you finish.

"A good hockey player plays where the puck is. A great hockey player plays where the puck is going to be."
Wayne Gretzky

Group projects to get the best return in sustainability terms, i.e., retrofit all the lights not just those that have a positive ROI.

Chapter 1 – Introduction to sustainability management

the manufacturing phase will not only benefit the environment but also reduce costs. Identifying an opportunity rather than focusing on risks and costs can make many sustainability actions profitable. Look for this type of project and try for "Yes and ..." (see sidebar).

- **Tip** – Improved waste management at a site will reduce disposal and regulatory costs and improve environmental sustainability (see Chapter 9).

What do we already do?

Many companies already carry out significant sustainability actions but do not recognise or publicise them as such. This is particularly true in social sustainability (see Chapter 11) but it is also true in other areas. This is not an excuse for 'greenwashing' (see Section 1.20) but simply a recognition of the good things that are already being done.

- **Tip** – Check for actions that you already carry out but which can be improved or should be 'reclassified' as part of the sustainability actions.

- **Tip** – After you have completed the selected actions then it is time to start again. Sorry, but that is the way it is.

Area \ Phase	Raw materials	Manufacturing	Use	End-of-life	Logistics
Environmental sustainability	• Sustainable materials. • Carbon footprint of raw materials. • 'Conflict free' materials. • Recycled materials use.	• Energy use. • Water use. • Air and water emissions. • Waste. • Carbon footprint. • Design for sustainability.	• Energy used in use phase. • Water used in use phase. • Recyclable, reusable or compostable packaging.	• Disposal options. • Take back programme. • Designed for disassembly and recycling. • Labelling. • Biodegradable materials.	• Sustainable distribution methods. • Sustainable distribution systems (including transit packaging).
Social sustainability	• Fair business practices. • Sustainable suppliers (CSR through the supply chain). • Support for local suppliers.	• Working conditions. • Health and safety. • Diversity and inclusion. • Employee development (training). • Ethics and governance.	• Consumer/user health and safety. • Product safety. • Packaging to reduce waste.	• Working conditions. • Health and safety.	• Working conditions. • Health and safety.
Economic sustainability	• Local suppliers. • Raw materials selection. • Recycled and/or renewable materials use. • Brand reputation protection.	• Profit. • Cost reduction. • Investment. • Continual improvement. • Staff cohesion and satisfaction. • Sustainable new products.	• Product and service quality. • Brand reputation protection.	• Product recycling potential.	• Sustainable distribution.

Mapping actions through the product life cycle

Sustainability actions vary with the phases of the product life cycle and it is possible to map the major potential actions through the product life cycle. Not all actions will be relevant for every product or company and companies should also use materiality (see Section 1.8) to determine which actions are important to them.

1.8   Mapping actions on materiality

Materiality = relevant and important

Materiality is an accounting concept used in reporting where information that is material (or relevant) to the company must be reported. This reporting element of materiality is covered in Section 12.4 where the focus is on what a sustainability report should include so that it is 'material' in terms of external company reporting.

The concept and tools used to assess materiality are also useful in deciding the actions to take in a sustainability programme. The focus in this section is on using materiality to prioritise which internal actions should be taken to improve sustainability.

- **Tip** – Reporting is essential but getting the right actions to report upon is even more important.

Choosing sustainability actions needs to consider not only the economic aspects but also the needs of other internal and external stakeholders, i.e., environmental and social needs, and materiality offers a method of explicitly doing this.

- **Tip** – Material matters are the most significant for performance assessment and therefore the most important for action.

- **Tip** – Sustainability is not about one thing. There is no 'single number' for sustainability. Get used to complexity.

The materiality grid

A materiality grid for the selection of actions is a simple plotting of the relevance of the action to stakeholders versus the relevance of the action to the company and an example grid is show on the right. Relevance to stakeholders and to the company for a proposed action is simply rated as 'low', 'medium' or 'high' and plotted on the grid to give a rapid assessment of the importance of various actions.

In the example grid, we have differentiated and marked the actions according to the sustainability area, i.e., governance, environmental, social and economic. This is to highlight the need for a balanced programme of sustainability actions.

- **Tip** – The grid used for reporting (see Section 12.4) uses the axes: 'relevance to the stakeholders' and 'significance of impact'. Here we use 'relevance to stakeholders' versus 'relevance to the company' as this is more appropriate for selecting actions.

- **Tip** – This is an example only; every company needs to do their own analysis.

Assessing materiality

Assessing materiality can be a large exercise but it does not have to be repeated every year. It can simply be re-validated on a yearly basis by a short review process. Major changes are only required if the internal or external conditions change significantly, e.g., new legislation or product changes. This does not mean that the actions can stand still even if the materiality assessment does not change, there is always more to do than there are available resources, e.g., in the example grid energy management is highly relevant for both stakeholders and the company – there is always more to do to improve energy management and achieving good results does not mean that efforts should stop.

Materiality needs to be assessed from the perspective of the stakeholders (external) as well as from the perspective of the company

> Assessment of relevance is always an estimate, this is not an exact science.

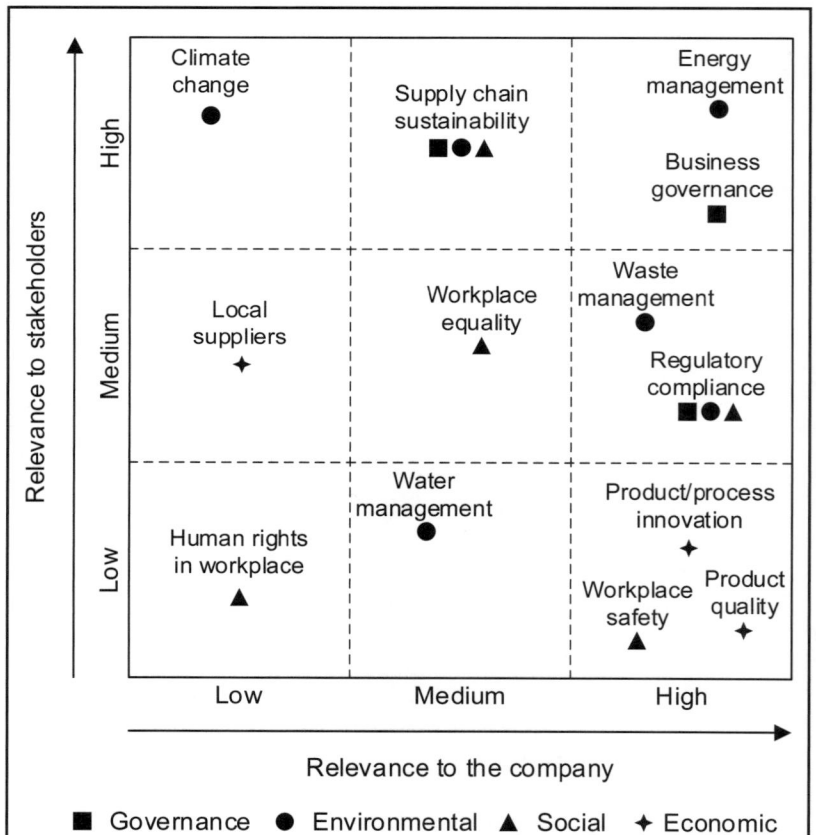

A simplified materiality grid

The materiality grid allows assessment of the significance and relevance of a range of sustainability actions. The initial scan should consider all the possible material actions but only those considered material to the company's operations should be included in the grid.

Chapter 1 – Introduction to sustainability management

(internal). Every stakeholder will have a different view on what is relevant and the degree of the relevance.

Relevance to the company

It is always easier to assess relevance to the company because the company has a higher degree of control over the process. The easiest way to do this is:

- Review:
  - Existing company strategies and targets for ISO 9001, ISO 14001, etc.
  - What the company is already doing in sustainability.
  - Current and potential future legislative requirements.
  - Current and potential future customer requirements.
  - Current and potential future standards requirements.
  - Current MSDS data provided by suppliers.
  - The wider industry and society, e.g., trade associations and competitors, for general sustainability trends.
- Prepare a list of potential actions that the company can take to improve in sustainability. Use the actions from Section 1.7 as a guide for potential actions.
- **Tip** – Leave space for people to add actions that you may not have thought of.
- Arrange for management staff to rate each action in terms of relevance to the company Use 'low', 'medium' or 'high' to make it easy.
- Collate the results to give an ordered list.
- **Tip** – Legal compliance must always be of high relevance to the company.

Relevance to the stakeholders

This is always more difficult to determine because of the difficulty in contacting and working with a disparate group of external stakeholders. The easiest way to do this is:

- Distribute the internally generated ordered list of actions (see above) as widely as possible to identified stakeholders such as:
  - Internal staff.
  - Neighbours.
  - Suppliers.
  - Customers.
  - Local communities.
- **Tip** – The list may need to be modified to be suitable for external distribution.
- **Tip** – Leave space for people to add actions that you may not have thought of.
- **Tip** – It is possible to use online surveys to get feedback but control the process and be aware of data protection issues.

- Ask the stakeholders to rate each action in terms of relevance to the them. Use 'low', 'medium' or 'high' to make it easy.
- Collate the results to give an ordered list.

The stakeholder relevance ratings can then be combined with the company relevance ratings to generate a complete materiality grid.

- **Tip** – It is OK to include actions that are high priority to the company but are low priority to the stakeholders. The important thing is to get the external viewpoint.

Using materiality to select actions

Using materiality as a filter is about choosing the best actions with limited resources and provides an easy way to assess projects.

Actions rated as 'high' for both the stakeholders and the company are the highest priority but an action can also be material if it ranks highly in only one of the factors, i.e., it does not have to rank highly in both.

- **Tip** – Do not forget to use risk analysis (see Section 2.17) to assess the risks and opportunities of actions.

Link the actions to the SDGs

In Section 1.16 and Appendix 1, we discuss the UN SDGs and their value in demonstrating sustainability and in communicating sustainability progress and actions to a wider audience, e.g., external stakeholders. Using materiality as a filter for actions also allows linkage to the UN SDGs for a broader perspective on the actions. If help is needed in assessing the linkages between actions and the SDGs then the 'Inventory of Business Indicators' (sdgcompass.org/business-indicators) is an excellent tool.

- **Tip** – Companies should select between 4 and 6 of the UN SDGs and relate their sustainability actions to these SDGs.

Would a project to remove single-use sugar sachets in a company processing 100,000 tonnes of plastic per year pass the materiality test?

If you are already working on sustainability projects then presumably the company considers these relevant and material.
But still check!

When compiling the list of potential actions, try to think as broadly as possible.
In the words of United States Secretary of Defence Donald Rumsfeld, "..there are also unknown unknowns – the ones we don't know we don't know. And if one looks throughout the history of our country and other free countries, it is the latter category that tends to be the difficult ones."

1.9   Are we running out of materials?

It depends on your viewpoint

The pessimist will answer "Yes, we are running out of materials" and the optimist will answer "No, we are not running out of materials".

The concept of sustainable development was first raised in 1798 when Thomas Malthus published 'An Essay on the Principle of Population'. He stated that a population increases in a geometrical ratio and that food production increases in an arithmetical ratio. This predicts that as a population grows it will outstrip food capacity until there is no longer enough food for the population. Living conditions will then decrease and the population will oscillate as more people are taken by famine and disease. Unfortunately, this prediction was made just before the industrial and farming revolutions dramatically changed our ability to produce food and support populations.

In 1968, Paul Ehrlich's 'The Population Bomb'[1] predicted world-wide famine and societal upheaval due to unsustainable population growth. Ehrlich's ability to predict the future is indicated by a speech in 1971 when he said, "By the year 2000 the United Kingdom will be simply a small group of impoverished islands, inhabited by some 70 million hungry people". Again, this prediction was made just before another revolution in food productivity and the predicted collapse failed to occur.

This was followed by the 1972 Club of Rome report 'The Limits to Growth'[2] and, when their predicted collapse did not occur, they released 'Beyond the Limits: Confronting Global Collapse'[3] in 1992.

Similar neo-Malthusian arguments continue to be advanced to predict the collapse of society due to over-population and resource scarcity. Although these arguments are superficially attractive, there is statistical evidence (www.gapminder.org/) that life continues to get better for most people in the world in almost every facet of life. This is driving world-wide decreases in fertility, e.g., between 2020 and 2100, 90 countries are expected to lose population. Rosling[4] discusses why we do not see or appreciate these huge increases in living standards even when the data are clear and we should be optimistic (or, as Hans Rosling puts it, 'possibilistic').

What about resources?

Despite things getting better, there is still the issue of resource use and resource scarcity. It is variously calculated that for everybody to have the same living standards as the typical Western person then we would need 1.5-2.5 Earths to satisfy the demand.

When Julius Caesar was assassinated in 44 BC, the molecules expelled in his last breath dispersed around the world and some calculations show that you will breathe one of these molecules in your next breath. These calculations may be disputed, but almost all the atoms around at the start of the Earth are still here. We don't lose many, we don't gain many and we will never run out of atoms.

- **Tip** – We transform them through resource use and, whilst we have enough, some of them end up in the wrong place, i.e., the destination is wrong (see Section 1.10).

Simon[5] and Lomborg[6] show that natural resources have historically become more abundant due to improved technology, e.g., 'The Limits to Growth'[2] predicted that we would run out of zinc in 1990 and yet in 2020 the existing reserves were estimated to be sufficient for 55 years. The major issue in these calculations is the difference between proven reserves and the economics of recovery. Uneconomic reserves can become economic through advances in technology or increasing prices, e.g., it is possible to debate the wisdom of 'fracking' but the USA went from an energy importer to an energy exporter in the space of a few years. In general, the most important reason for increasing reserves is technological improvement, which is why Simon's book is titled 'The Ultimate Resource', i.e., it is people, and why the price of most resources has also historically decreased (see sidebar).

Despite all of this, we need to reduce resource use (efficiency) and improve resource use (effectiveness) to make the best of what we have. This is only made possible by technology. The potential for improvement here is illustrated by von Weizsäcker et al[7] in 'Factor Four' which is sub-titled 'Doubling wealth, halving resource use' and the proposal to raise the ambition to 'Factor Ten'.

- **Tip** – We have the atoms. They may be unevenly distributed and may cost more but we are not running out of resources.

Petrochemicals and plastics

The main raw materials for plastics production are petrochemicals (oil and gas) and plastics production represents ≈ 4-6% of the world's petrochemical use with the majority ($\approx 87\%$) being used for transport, energy and heating.

The traditional model for use and depletion of oil and gas reserves (the Hubbert Peak) predicted that oil production in the USA would peak in 1970 and decline thereafter. This model showed when the known reserves would peak (and decline) and is remarkably accurate until ≈ 2000. It then fails completely as it takes no account of unconventional reserves, i.e., shale gas, or improved technology for extraction. In a sense, it is a Malthusian approach to petrochemical resources and fails for the same reasons, i.e., models cannot predict the future, only extrapolate the past.

The rapid growth of renewables as sources for transport (electric vehicles), energy (solar and wind power) and heating means that there are now two different stories:

- The world is going to run out of petrochemicals.

- Petrochemical companies are going to be left with 'stranded assets', i.e., existing oil and gas assets, that may not be recoverable due to a shift to renewable energy based on lower costs or legislative action. The scale of the change is indicated by the fact that $\approx$ 50% of the identified petrochemical reserves need to stay in the ground to meet existing climate change agreements and in the UK is it already cheaper to generate electricity using renewables than to use petrochemicals.

You can make your own choice but I prefer the second story, i.e., we will not run out of petrochemical feedstocks to produce plastics. The feedstocks may become more costly if the plastics industry becomes the main user of petrochemicals due to the economics of the industry but this, as always, will be a case of supply and demand.

Bio-based materials

One constraint on the price of petrochemical feedstock would be the rise in the use of bio-based polymers (see Section 4.14). The raw materials for most plastics can also be sourced from biomass and, whilst these may be currently more expensive, the technology is readily available and developing rapidly. The chemical engineering is different but the resulting materials can be the same (or even better) than those derived from petrochemical feedstocks.

- **Tip** – It is also possible to take over-abundant molecules such as CO_2 from the atmosphere and to make materials such as ethylene (C_2H_4) and then polyethylene ($C_2H_4)_n$. This has been demonstrated in the laboratory but commercialisation is not likely soon due to the process economics.

Recovery

Even if feedstock materials supply is not an issue, the plastics processing industry needs to rapidly reduce and improve resource use. We need to increase the utility of the molecules that we already have and we need to recover and re-use the materials we already have by:

- Increasing mechanical recycling to recover materials.

- Increasing chemical recycling to recover polymer chains and feedstocks.

- Recognise that landfill can be considered 'storage for the future'.

This is at the heart of sustainability and these technologies are covered in Chapter 4.

- **Tip** – Running out of materials is not an issue for plastics processors (although the cost implications may be).

- **Tip** – There are probably already enough polymer chains out there to supply the plastics industry indefinitely. All we must do is collect and process it (and maybe get used to paying a lot more for it).

- **Tip** – One thing you can guarantee: If we ever get to a flying car then it won't be made from metal, it will be mainly plastic. Plastics are too useful and valuable for trivial uses.

- 1. Ehrlich, P.R. 1968, 'The Population Bomb', Sierra Club.

- 2. Meadows et al. 1972, 'The Limits to Growth', Potomac Associates.

- 3. Meadows et al. 1992, 'Beyond the Limits: Confronting Global Collapse', Chelsea Green Publishing.

- 4. Rosling et al, 2019, 'Factfulness', Sceptre.

- 5. Simon, J.L. 1996, 'The Ultimate Resource 2', Princeton University Press.

- 6. Lomborg, B. 2001, 'The Skeptical Environmentalist', Cambridge University Press.

- 7. von Weizsäcker et al, 1998, 'Factor Four', Routledge.

You may not believe any of this.

Man is not a rational creature; he is a rationalising creature.

We make the decision (almost instantly) and then seek the evidence. If you are a pessimist then that is the evidence that you will find.

"The Stone Age did not end because the world ran out of stones, and the oil age will not end because we run out of oil".

Variously attributed to Don Huberts, Sheikh Ahmed Zaki Yamani and about 6 other people.

1.10 Are we running out of destinations?

Yes!

Most of the current issues with sustainability and plastics are about the outputs overwhelming the capacity of the destinations.

The earth can break down everything given enough time, although dinosaur fossils from ≈ 240 million years ago are still being found. The issue is that if we put them into the biosphere quicker than they can be broken down then we will overwhelm the destination. This is covered by Robèrt in The Natural Step (see Section 1.4) where one of the rules is that the biosphere is not subjected to increasing concentrations of substances from the Earth's crust, e.g., CO_2 in the atmosphere or metals such as Pb or Cd in the biosphere.

Air

The main destination issue with the air (atmosphere) is the increasing amount of CO_2. This is a factor in global warming and can be explained by the carbon cycle (shown in the diagram on the right).

The carbon cycle

There are two main elements to the carbon cycle and these are:

- Short (biogenic) carbon cycle – this is the natural carbon cycle from organic matter and trees. This is the carbon recycled through the biosphere by plants and animals. It may not be in complete equilibrium (few systems are) but it is broadly stable.

- Long (fossil) carbon cycle – this is the carbon that has been taken from the biological cycle over millions of years by fossilisation of biological matter to form fossil fuels and $CaCO_3$. The increasing use of fossil fuels is adding this long carbon to the short carbon cycle and risks overwhelming the atmosphere with resulting long-term impacts such as climate change.

- Tip – The ocean carbon cycle is not shown in the diagram. If this fills up then it may become a big issue but ocean carbon is a very complex and poorly understood issue.

The plastics industry contributes to the transfer of carbon from the long cycle to the short cycle as a result of:

- The basic raw materials (which unless bio-based – see Section 4.14) are long carbon.

- The high carbon intensity of the power needed for processing.

The world-wide CO_2 output of the plastics processing industry (processing only and not embodied carbon) is estimated at ≈ 400 Mtonnes/year (see Section 7.8).

- Tip – This is slightly more than the 2018 emissions of the UK and is $\approx 1.1\%$ of the world-wide emissions.

Carbon intensity (kg CO_2/kWh) is decreasing as energy systems are being decarbonised, but it is still an issue for all industries.

The production of plastics from petrochemicals uses long cycle carbon, although this carbon is still locked up in the plastic and it is only if it is incinerated that the carbon is released (as CO_2) into the short carbon cycle.

The destination of plastics in Melbourne, Australia in 1972 was my first introduction to sustainability.

It seems like yesterday.

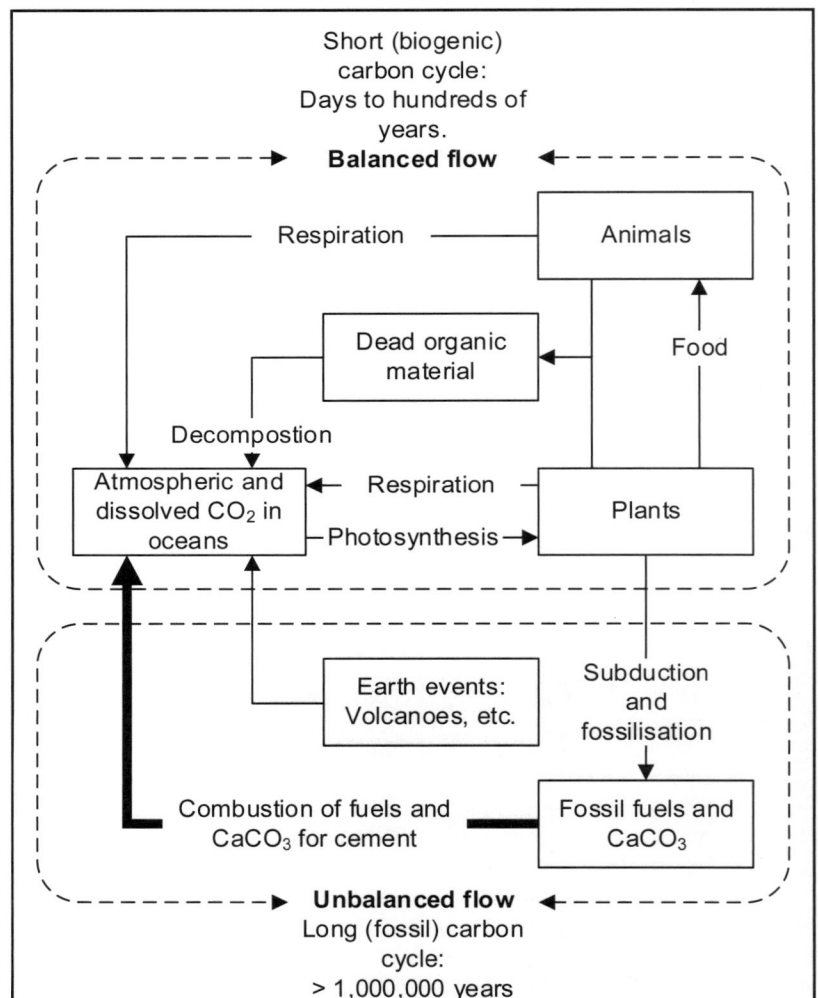

Short (biogenic) carbon cycle: Days to hundreds of years.
Balanced flow

Respiration — Animals

Dead organic material — Food

Decompostion

Atmospheric and dissolved CO_2 in oceans — Respiration — Plants — Photosynthesis

Earth events: Volcanoes, etc. — Subduction and fossilisation

Combustion of fuels and $CaCO_3$ for cement — Fossil fuels and $CaCO_3$

Unbalanced flow
Long (fossil) carbon cycle: > 1,000,000 years

Note: This diagram does not include the ocean carbon cycle which is $\approx 1,000$'s of years. This is one of the largest carbon sinks.

The two main carbon cycles

The short (biological) and long (geological) carbon cycles determine the amount of CO_2 in the atmosphere and oceans. The short cycle has been 'balanced' for many years but industrial activity means that long cycle carbon is being released at an unsustainable rate.

Chapter 1 – Introduction to sustainability management

- **Tip** – Decarbonization of the energy system will reduce indirect CO_2 emissions for the plastics industry but this may also increase pressures on the incineration of plastics waste (see Section 4.11) as this will remain a net CO_2 addition to the atmosphere. This does not affect bio-based materials (which take and return carbon to the short carbon cycle).

Apart from CO_2 as a result of energy use, the major atmosphere related issues with plastics at the processing stage are:

- GHGs released from refrigerant leakage in chillers and driers (see Section 7.2).
- VOCs from paint processes unless these are treated properly (see Section 9.11).

Land

Litter is an unacceptable destination for any plastic product but this is a sensibility that is not always shared in countries which do not have the same ethos or a well-established waste management infrastructure, e.g., if you litter in a city centre in Colombia then be prepared for a robust discussion but similar behaviour in other countries (unnamed) would simply show that you are a 'local'.

Landfill (see Section 4.12) is a destination in many countries but there are increasing pressures on landfill. To give an illustration of the issue: the USA generates $\approx$ 300 million tonnes of waste/year, MSW has a density of between 600-1200 kg/m^3, and using 1,000 kg/m^3 this represents a cube of waste of $\approx$ 700 metres on each side. This may not seem much in the USA (apart from the physical distances) or in countries with no pressure on land use, but in most countries, landfill is not a viable long-term destination.

Watercourses and oceans

Watercourses and oceans (hydrosphere) are also an unacceptable destination for plastics. Plastics are too valuable to be in watercourses or the oceans. The major causes are littering (in the developed world) and poor waste management (in the developing world). This is not only a bad destination for plastics it is also a bad destination for the biosphere.

Much of the public discussion centres on plastic gyres or 'garbage patches' in the ocean. These gyres exist in various areas of the oceans and some areas are denser than other but much of the debris is made of microplastics. These are small and are rarely visible to the naked eye. Research[1] shows a maximum of 580,000 particles/km^2; this is about 1 particle for every 2 square metres. The US National Oceanic and Atmospheric Administration (NOAA) states "It's more like pepper flakes swirling in a soup than something you can skim off the surface. You may come across larger items, like plastic bottles and nets, but it's possible to sail through some areas of a garbage patch and not see any debris at all."[2].

This is not exactly a 'plastics soup' but it is still unacceptable as a destination for plastics waste.

- **Tip** – The 'plastics soup' is more like water than soup despite the emotive photographs.

None of this means that we should accept watercourses and oceans as acceptable destinations for plastics but we also have to recognise that the majority of these plastics enter the ocean from a minority of sources (see Section A1.14) where waste management has failed or is virtually non-existent.

It doesn't matter where you look

We are not at risk of running out of materials (see Section 1.9) but the destinations question is far more harmful, immediate and public.

Whether you consider air, land or watercourses and oceans, there is increasing pressure on destinations.

- 1. Law et al. 2010, 'Plastic Accumulation in the North Atlantic Subtropical Gyre', Science, 329. 1185-8. 10.1126/science.1192321.
- 2. NOAA: marinedebris.noaa.gov/info/patch.html (accessed April 2021).

Counter-intuitive?

Increasing the use of plastics would reduce the consumption of non-renewable fossil fuels and reduce overall GHG emissions.

Reducing the use of plastics would increase the consumption of non-renewable fossil fuels and reduce overall GHG emissions.

(See Section 10.3)

Nature does not seem to have the same abhorrence about concentrating materials in the lithosphere and has a history of doing this, e.g., most of the oil is in the Middle East and most of the cobalt is in the Democratic Republic of the Congo.

Can we be sensible here?

Sustainability is often clouded by a generalised fear of the chemical industry and a multitude of chemical scares and misinformation on these topics. The following shows how this works:

Roger Bate, Director of the European Science and Environment Forum stood outside a London Underground station and asked 123 people:

"The chemical industry routinely uses a chemical di-hydrogen monoxide in its processes. It is used in significant ways and often leads to spillages and other leaks and it regularly finds its way into rivers and into our food supply. It is a major component of acid rain. It contributes to erosion. It decreases the effectiveness of automobile brakes. In its vapour state it is a major greenhouse gas. It can cause sweating and vomiting. Accidental inhalation can kill you. It has been found in the tumours of terminal cancer patients.

Should this chemical be strictly regulated or, even banned, by the British Government or the EU?"

- 76% said 'Yes'.
- 19% said 'Don't Know'.
- 5% said 'No' (probably the scientists).

Di-hydrogen monoxide is water (H_2O).

Everything Roger Bate said was true, but it was not the whole truth. In most 'scares' involving the plastics industry, there is some truth in the story but it is not the whole truth.

"Di-hydrogen monoxide, now there is a real killer."

Matt Ridley, Daily Telegraph 15 Sept. 1997.

1.11 Energy and climate – where are you now?

Where are we starting from?

Understanding the current situation provides the basis for an improvement strategy and many of the basic actions necessary for successful sustainability implementation.

The next sections in this chapter provide a series of self-assessment charts designed to assess your current position.

The charts are easy to complete but we suggest copying the relevant pages before completing the forms.

Completing the chart

Each chart has several columns which cover various aspects of the main topic.

To complete a column read the descriptions in the column cells and select the cell that is closest to the current situation at your site.

It is unlikely that every part of the description in the cell will fully describe your specific situation but choose the cell that has the most appropriate description. This will give a score ranging from 0 to 4, mark this at the base of the column.

After all the columns have been scored, transfer the scores to the radar chart for the relevant columns/axes. This gives a rapid visual assessment of the current situation for the specific topic.

If you don't know where you are starting out from then it is unlikely that you will end up where you want to get to!

This is a team effort!

Completing the chart on your own is not recommended. It is much better to either complete the chart as a group – you will be amazed at the divergence of opinions – or to get several people in the company to complete the chart separately and then to compare the results.

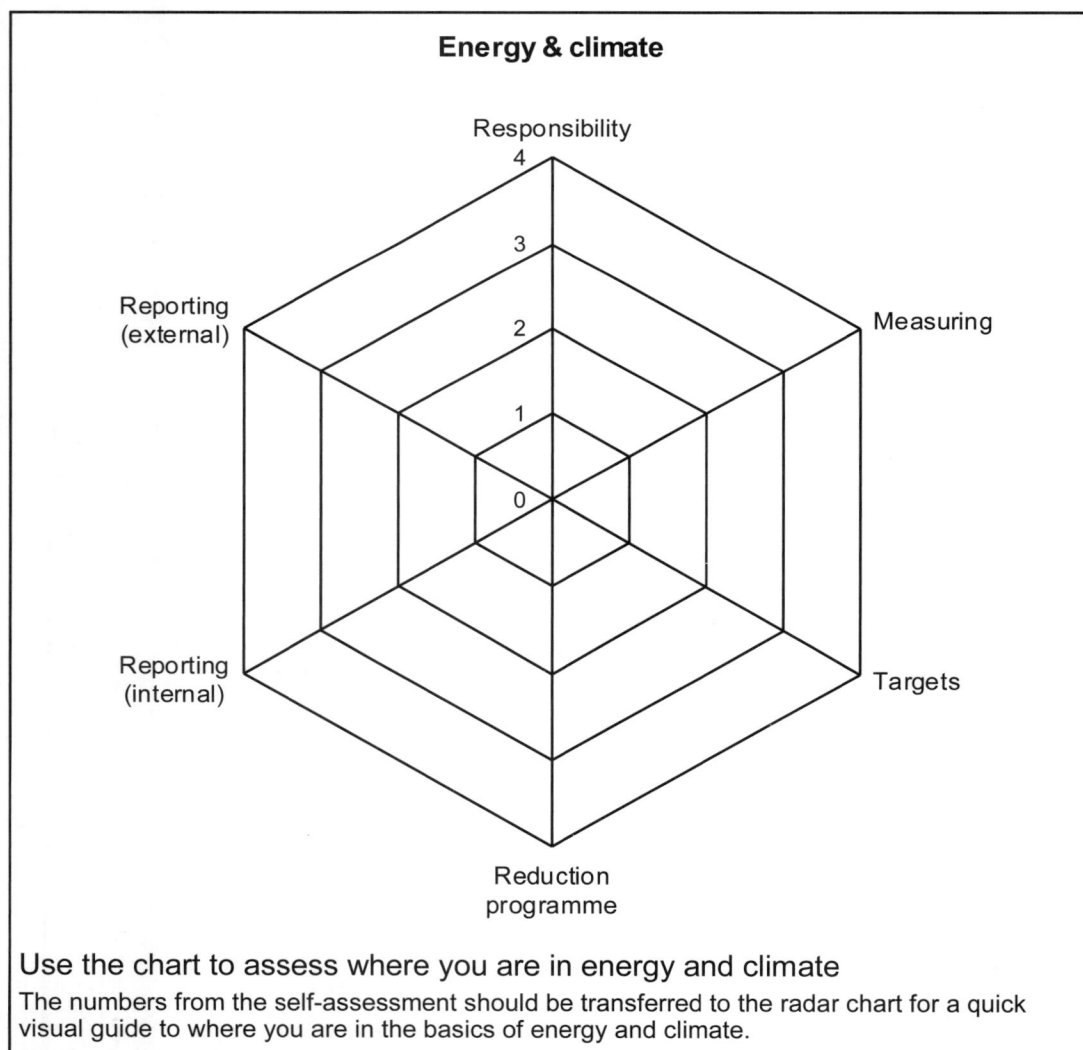

Energy & climate

Responsibility
Measuring
Targets
Reduction programme
Reporting (internal)
Reporting (external)

Use the chart to assess where you are in energy and climate

The numbers from the self-assessment should be transferred to the radar chart for a quick visual guide to where you are in the basics of energy and climate.

Assessing the results

Ideally, a site would have balanced score with all columns/axes in the same broad area. This is rare and in most cases, sites will show strengths in certain areas and weaknesses in others. The axes with low scores are the areas that the site needs to work on to improve the overall score.

				Energy & climate		
Level	Responsibility	Measuring	Targets	Reduction programme	Reporting (internal)	Reporting (external)
4	Main board director responsible for energy & climate issues. Regular reporting to Main Board.	Greenhouse gas emissions calculated for all scopes. Updated yearly. Excellent measurements & methods used.	Greenhouse gas reduction targets set & agreed with Main Board. Progress towards targets regularly monitored.	Formal greenhouse gas reduction programme produced & agreed by Main Board.	Monthly reporting of key indicators, e.g., energy use. Comparison with targets based on activity or condition drivers.	Regular & validated external (publicly available) reporting.
3	Main board director responsible for energy & climate issues. No regular reporting to Main Board.	Greenhouse gas emissions calculated for all scopes. Updated yearly. Good measurements & methods used.	Greenhouse gas reduction targets set & agreed with Main Board. Progress towards targets monitored irregularly.	Formal greenhouse gas reduction programme produced but not agreed by Main Board.	Quarterly reporting of key indicators, e.g., energy use. Comparison with poorly defined targets.	Regular external (publicly available) reporting. Not fully validated.
2	Mid-level manager responsible for energy & climate issues. No regular reporting to Main Board.	Greenhouse gas emissions calculated for scopes 1 & 2 only. Updated yearly. Good measurements & methods used.	Greenhouse gas reduction targets set but not fully agreed or supported by Main Board. No progress monitoring.	Formal greenhouse gas reduction programme produced at low level with no support or agreement from Main Board.	Annual reporting of some key indicators but mainly for accounting purposes. Some comparison with budget.	Regular external reporting only via Annual Report, i.e. not fully public. Not fully validated.
1	Low-level manager responsible for energy & climate issues. No regular reporting to Main Board.	Greenhouse gas emissions calculated more than 1 year ago for scopes 1 & 2 only. No updating carried out.	Some informal greenhouse gas reduction targets set by lower management. Not agreed or supported by Main Board. No progress monitoring.	Informal greenhouse gas reduction programme available but it has no support or agreement from Main Board.	Annual reporting of some key indicators but only for accounting purposes. No comparison with targets.	External report only available on request, i.e., not public.
0	No designated person responsible for energy & climate issues.	Greenhouse gas emissions not calculated.	No greenhouse gas reduction targets set.	No greenhouse gas reduction programme.	No internal reporting of any key indicator of greenhouse gas emissions.	No external reporting.
Score						

1.12 Material effectiveness – where are you now?

Using it wisely

Using all materials effectively is a key element of sustainability. This means reducing waste outputs (of any type) to ensure that all materials, not simply plastics, entering the site are used effectively and that all outputs from the site (apart from saleable product) are minimised and, where possible, recycled. This not only reduces the site's environmental impact but also improves the financial performance of the site. Less waste and the correct treatment of any waste generated means reduced costs for the materials and reduced disposal costs.

Sustainability is not simply about removing or minimising any environmental impacts, it is also about creating and growing a business that can grow and prosper to provide employment and clean outputs in the future. Material effectiveness is a fundamental in achieving this.

Completing the chart

This chart is completed and assessed as for the previous charts.

Effectiveness versus efficiency

Concentrate on effectiveness rather than efficiency. It is useless being efficient at something that you shouldn't be doing in the first place.

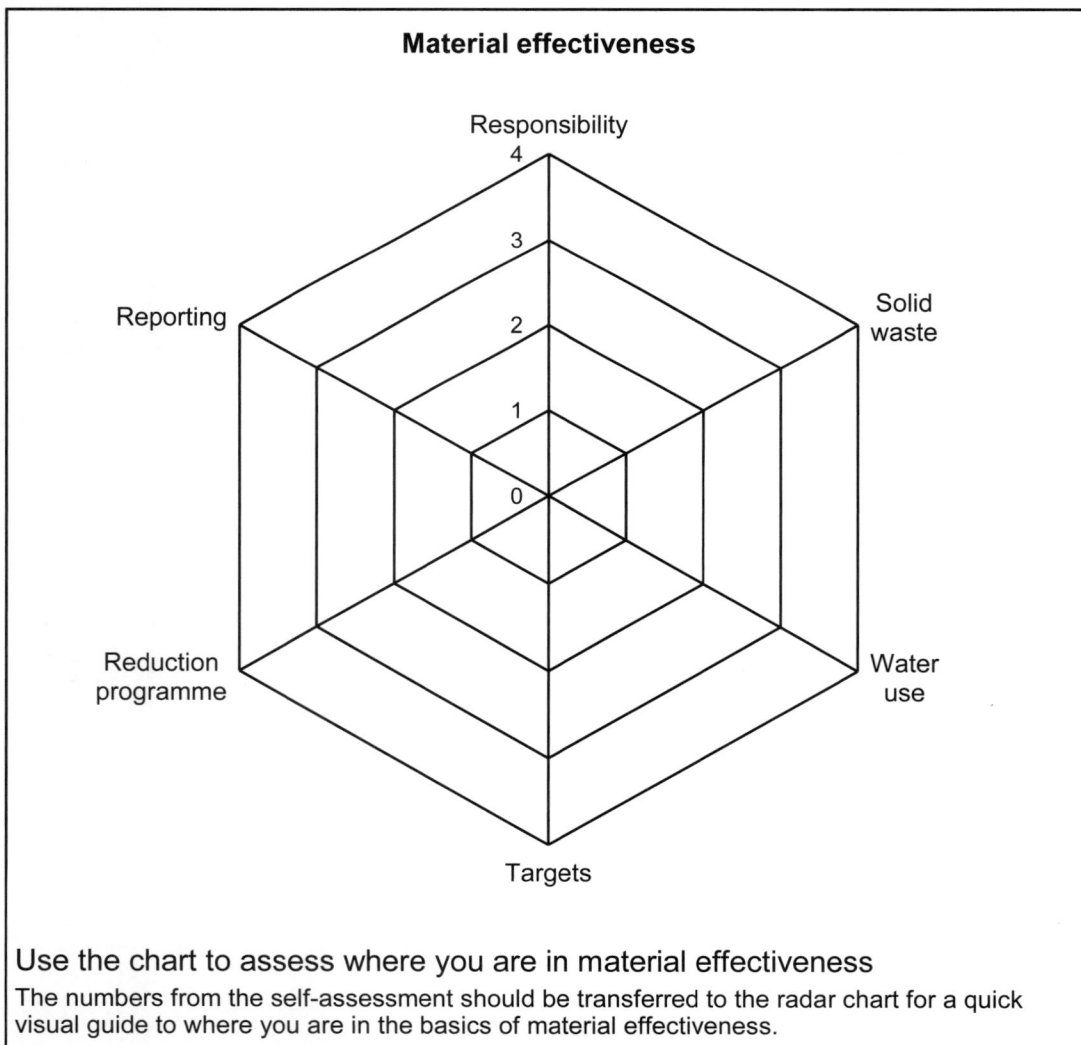

Material effectiveness

Use the chart to assess where you are in material effectiveness

The numbers from the self-assessment should be transferred to the radar chart for a quick visual guide to where you are in the basics of material effectiveness.

Creating a sustainable business is not simply about the environment. Sustainability is the ability to exist into the future and for a business this implies making sufficient profits to continue to employ staff and invest in the business.

A company that does not make profits is unsustainable and will perish. Its place may well be taken by a company which is more financially sustainable but less environmentally sustainable.

		Material effectiveness				
Level	Responsibility	Solid waste	Water use	Targets	Reduction programme	Reporting
4	Main board director responsible for solid waste & water use. Regular reporting to Main Board.	Solid waste reliably measured for all areas & materials. Excellent understanding of the sources & destinations of all solid waste.	Water use reliably measured for all areas. Excellent understanding of the source & destination of water (including recycling).	Solid waste & water use reduction targets set & agreed with Main Board. Progress towards targets regularly monitored.	Formal solid waste & water use reduction programmes produced & agreed by the Main Board.	Monthly reporting of key indicators. Comparison with targets from activity or condition drivers. External reporting.
3	Main board director responsible for solid waste & water use. No regular reporting to Main Board.	Solid waste measured for most areas & materials. Good understanding of the sources & destinations of all solid waste.	Water use measured for most areas. Good understanding of the source & destination of water (including recycling).	Solid waste & water use reduction targets set & agreed with Main Board. Progress towards targets monitored irregularly.	Formal solid waste & water use reduction programmes produced but not agreed by the Main Board.	Quarterly reporting of key indicators. Comparison with poorly defined targets. External reporting.
2	Mid-level manager responsible for solid waste & water use. No regular reporting to Main Board.	Solid waste measured for some areas & materials. Average understanding of the sources & destinations of all solid waste.	Water use measured for some areas. Good understanding of the source & destination of water (including recycling).	Solid waste & water use reduction targets set but not fully agreed or supported by Main Board. No monitoring.	Formal solid waste & water use reduction programmes produced at low level with no support or agreement from the Main Board.	Annual reporting of key indicators, but mainly for accounting purposes. Some comparison with budget targets.
1	Low-level manager responsible for solid waste & water use. No regular reporting to Main Board.	Solid waste measured for few areas & materials. Poor understanding of the sources & destinations of solid waste generated.	Water use measured for very few areas. Poor understanding of the source & destination of water (including recycling).	Some informal solid waste & water use reduction targets set at low level. Not agreed or supported by Main Board. No monitoring.	Informal solid waste & water use reduction programmes available but they have no support or agreement from the Main Board.	Annual reporting of key indicators but mainly for accounting purposes. No comparison with targets.
0	No designated person responsible for solid waste & water use.	Solid waste not measured for any areas or materials. No understanding of the sources & destinations of any solid waste.	Water use not measured for any area, i.e. global use only. No understanding of the source & destination of water (including recycling).	No solid waste & water use reduction targets set.	No solid waste & water use reduction programmes.	No internal reporting of any key indicator of solid waste & water use.
Score						

1.13   Natural resources – where are you now?

Where does it come from and go to?

Every processing operation has material inputs that are transformed in the process. Responsible and validated sourcing of these inputs is an important factor in sustainable processing. Sites need to be aware of where materials are coming from, that the supplier complies with good practice in all relevant areas and that third-party certification is available where it is relevant. Sites also need to be aware of any restrictions on materials use or legislatory requirements before materials can be used.

The plastics processing industry is fortunate in using a material that, in many cases, can be effectively and economically re-used internally. It is therefore important, for both sustainability and for financial performance that as much of the input material is converted into good product as possible. The re-use and retention, i.e., preventing material escape, of valuable raw materials is key to sustainable processing.

Completing the chart

This chart is completed and assessed as for the previous charts.

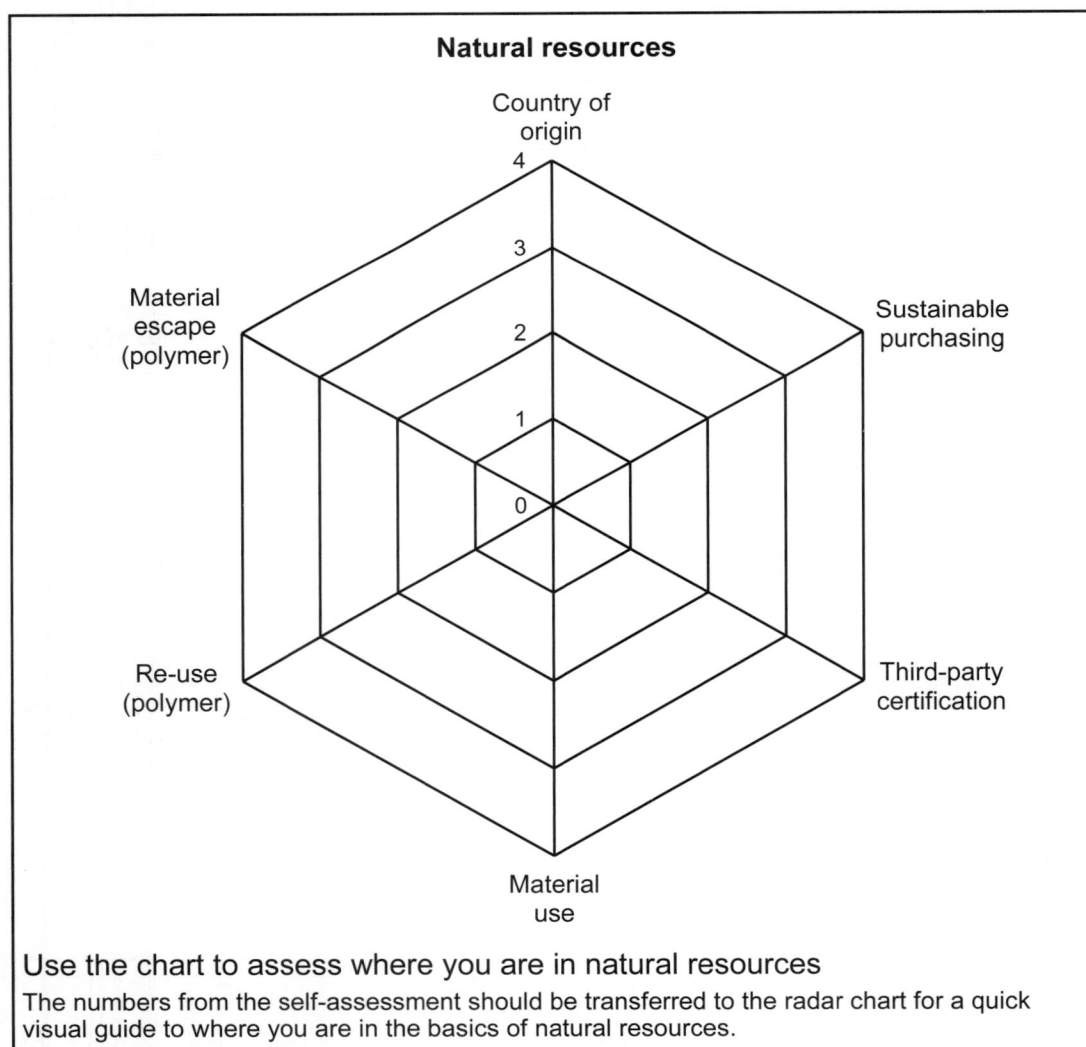

Natural resources

Use the chart to assess where you are in natural resources

The numbers from the self-assessment should be transferred to the radar chart for a quick visual guide to where you are in the basics of natural resources.

			Natural resources			
Level	Country of origin	Sustainable procurement	Third-party certification	Material use	Re-use (plastics)	Material escape (plastics)
4	Country of origin known for 100% of the materials & products used in production.	Publicly available sustainable procurement guidelines for all suppliers covering environmental, employment & product safety.	Third party environmental certifications available (or declared as not needed) for all products.	All materials used comply with the RoHS & REACH requirements with full & easily accessible documentation available to prove this.	All plastics scrap re-used internally. Good handling processes to preserve value & cleanliness of scrap.	Excellent precautions to prevent escape from all processes. Containment is excellent & very low chance of material escape.
3	Country of origin known for most (>50%) of the materials & products used in production.	Internal sustainable procurement guidelines available to most suppliers covering environmental, employment & product safety.	Third party environmental certifications available (or declared as not needed) for most products.	Good internal knowledge of RoHS & REACH requirements but limited documentation available to prove compliance.	Most (>50%) plastics scrap re-used internally. Reasonable handling processes to preserve value & cleanliness of scrap.	Good precautions to prevent escape from most processes. Containment not complete & some areas show escape potential.
2	Country of origin known for some (<50%) of the materials & products used in production.	Internal sustainable procurement guidelines available for some suppliers covering environmental, employment & product safety.	Third party environmental certifications available (or declared as not needed) for some products.	Poor internal knowledge of RoHS & REACH requirements & poor documentation available to prove compliance.	Little (<50%) plastics scrap re-used internally. Poor handling processes to preserve value & cleanliness of scrap.	Average precautions to prevent escape from a few processes. Containment average & some areas show escape potential.
1	Country of origin known for very few (<10%) of the materials & products used in production.	Informal sustainable procurement guidelines available but these do not cover all issues.	Third party environmental certifications available (or declared as not needed) for very few products.	Little internal knowledge of RoHS & REACH requirements & very little documentation available to prove compliance.	No internal treatment of plastics scrap. All plastics waste sold or sent for recycling.	Poor precautions to prevent escape from any process. Containment poor & many areas show escape potential.
0	Country of origin not known for any of the materials & products used in production.	No sustainable procurement guidelines available.	No third party environmental certifications available for any product produced.	No internal knowledge of RoHS & REACH requirements & no documentation available to prove compliance.	No internal treatment of plastics scrap. Plastics waste treated as solid waste & disposed of via solid waste channels.	No precautions taken to prevent escape. All areas show escape potential.
Score						

1.14   People and community – where are you now?

We have to contribute too

Sustainability is not simply about materials and products. It is also about investing in and building a community. Our workers need good jobs that are safe and conform to, or exceed, all the relevant social requirements. However, it is not enough to concentrate solely on our own staff, the industry needs to ensure that all our suppliers also meet the relevant social requirements and have plans to continuously improve compliance.

Our community is not simply our own staff. Every site is part of many diverse communities, these can be local and based on the site, countrywide and based on the industry or world-wide and based on the speciality. Contributing to these communities increases and reinforces the sustainability of the business, it provides a driver for improvement and increases the reputational capital of the business.

Community development can be multi-faceted but it is never wasted.

Completing the chart

This chart is completed and assessed as for the previous charts.

A business that is isolated and divorced from the community can easily suffer reputational damage because of a lack of goodwill in the community.

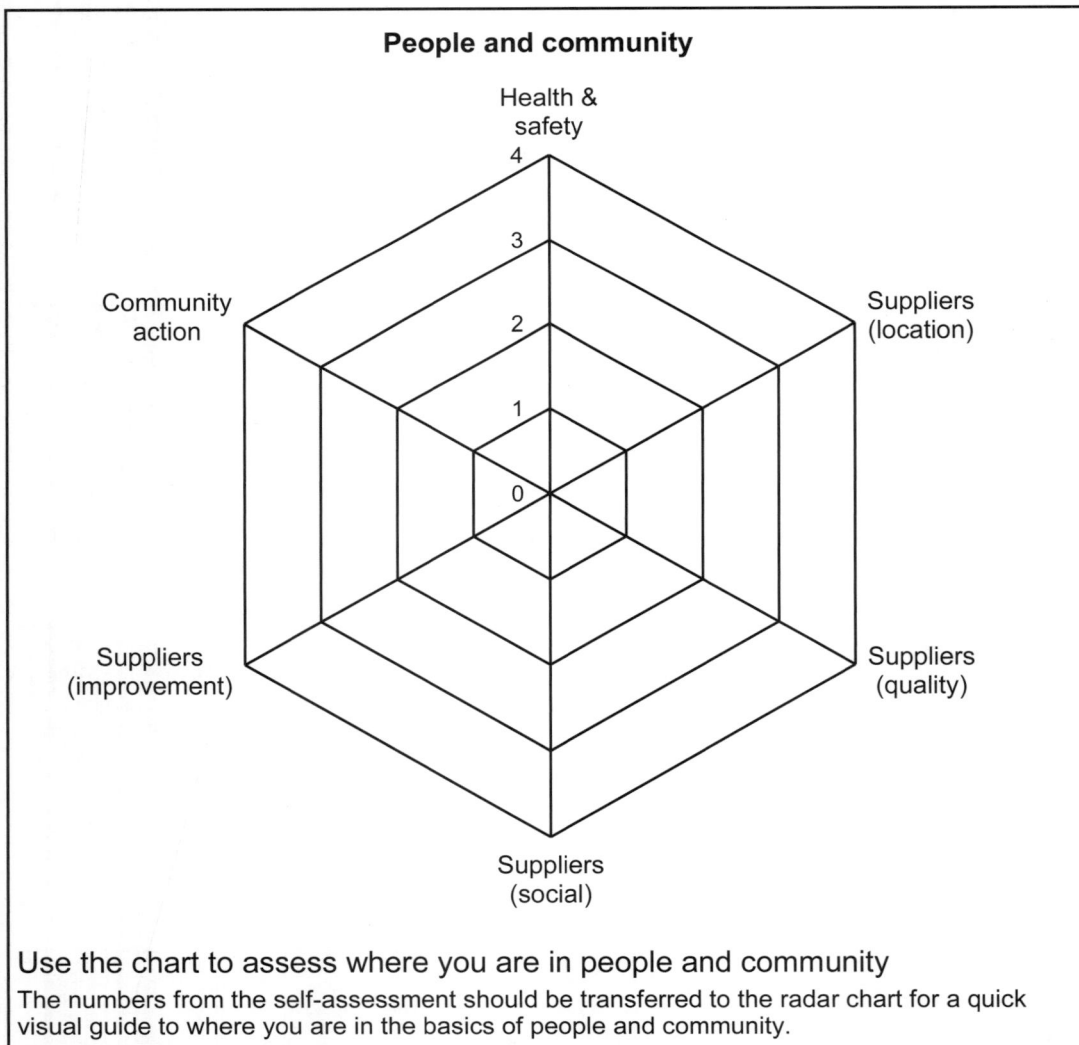

People and community

Use the chart to assess where you are in people and community

The numbers from the self-assessment should be transferred to the radar chart for a quick visual guide to where you are in the basics of people and community.

Business is always a chain of suppliers, that is why it is called a 'supply chain'.

Looking after the supply chain is essential in creating a sustainable business.

		People and community				
Level	Health & safety	Suppliers (location)	Suppliers (quality)	Suppliers (social)	Suppliers (improvement)	Community action
4	Risk assessments carried out. Assessments follow well defined process & are well documented.	Location of production sites known for 100% of the products used in production.	Quality, capability & capacity formally assessed for all suppliers before supplier selection & trading relationship established.	Social compliance formally assessed for all suppliers before supplier selection & trading relationship established.	Supplier social compliance correction & improvement programme produced & agreed.	Well defined & funded investment in community development activity in relevant locations.
3	Risk assessments carried out. Assessments follow poorly defined or inappropriate process but are well documented.	Location of production sites known for most of the products used in production.	Quality, capability & capacity formally assessed for most suppliers before supplier selection & trading relationship established.	Social compliance formally assessed for most suppliers before supplier selection & trading relationship established.	Supplier social compliance correction & improvement programme produced but not agreed.	Good investment in community development activity.
2	Risk assessments carried out. Assessments follow poorly defined or inappropriate process & are poorly documented.	Location of production sites known for some of the products used in production.	Quality, capability & capacity assessed for some suppliers before supplier selection & trading relationship established.	Social compliance assessed for some suppliers before supplier selection & trading relationship established.	Low-level supplier social compliance correction & improvement programme with no support or agreement.	Poor investment in community development activity.
1	Risk assessment carried out for some areas but informal & poorly documented.	Location of production sites known for very few of the products used in production.	Quality, capability & capacity assessed for few suppliers before supplier selection & trading relationship established.	Social compliance assessed for few suppliers before supplier selection & trading relationship established.	Informal social compliance correction & improvement programme that has no support or agreement.	Little investment in community development activity.
0	No health & safety risk assessment carried out at any stage. NOTE: This could contravene local legislation.	Location of production sites not known for any of the products used in production.	Quality, capability & capacity not assessed for any supplier before trading relationship established.	No assessment of social compliance before trading relationship established.	No social compliance correction & improvement programme.	No investment in community development activity.
Score						

1.15 Product life cycle – where are you now?

Understanding the cycle

The new product life cycle (see Section 1.7) needs to be understood to minimise environmental impacts at all stages of a product's life and to improve the sustainability of the industry. It is no longer enough to focus simply on the manufacturing step and to assume that everything that happens afterwards is external.

The product lifecycle is an outstanding opportunity for plastics processors not only to get ahead of the regulatory demands and reduce costs but also to establish an ethical lead in the market.

Changes in legislation and markets will force many of this on processors whether they like it or not, but by becoming pro-active, processors also improve sustainability and achieve cost reductions.

This is the start of things to come.

Completing the chart

This chart is completed and assessed as for the previous charts.

> "Some people change their ways when they see the light, others when they feel the heat."
>
> **Caroline Schoeder**

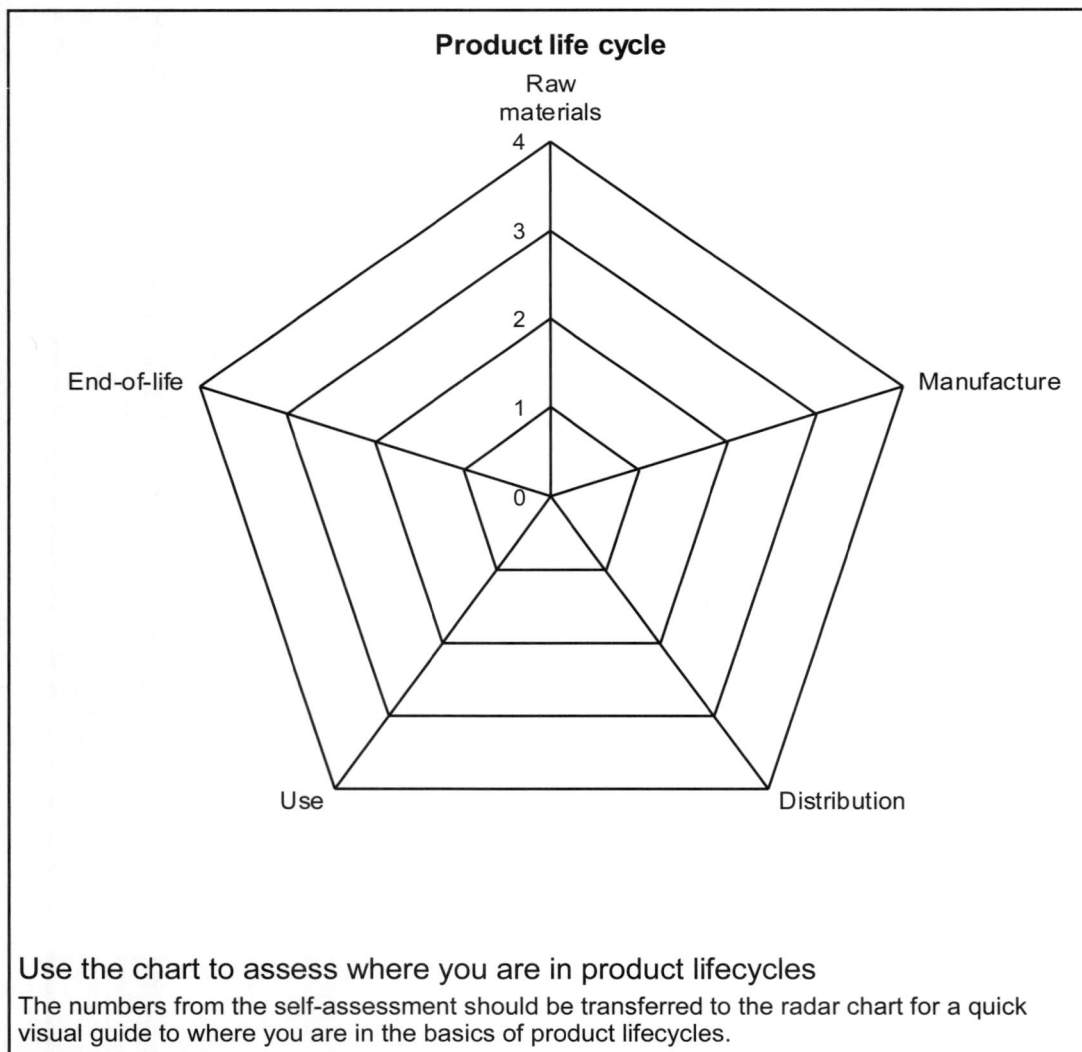

Product life cycle

Use the chart to assess where you are in product lifecycles

The numbers from the self-assessment should be transferred to the radar chart for a quick visual guide to where you are in the basics of product lifecycles.

Product life cycle					
Level	Raw materials	Manufacture	Distribution	Use	End-of-life
4	Use & cost of raw & recycled materials is an integral part of process & product design. Targets set & achieved.	Resource use & environmental impacts are an integral part of process & product design. All benchmark resource use targets known & achieved.	Distribution considered as an integral part of process & product design. Distribution cost targets are known & targets achieved.	Resource use & environmental impacts in use stage are an integral part of process & product design. All benchmark resource use targets known & achieved.	Disposal options & routes are an integral part of process & product design. Cost of disposal targets are known & achieved with well-defined disposal routes.
3	Use & cost of raw & recycled materials are known & targets achieved.	Resource use & environmental impacts considered in process & product design. Most benchmark resource use targets available & achieved.	Distribution considered in process & product design. Distribution costs available but not always achieved.	Resource use & environmental impacts in use stage considered in process & product design. Most benchmark resource use targets available & achieved.	Disposal options & routes considered in process & product design. Cost of disposal targets & disposal routes considered but not well defined.
2	Use & cost of raw & recycled materials considered in process & product design.	Resource use considered in process & product design. Limited benchmark resource use targets available & achievement is variable.	Distribution costs poorly considered in process & product design. Limited distribution cost targets available & achievement is variable.	Resource use in use stage considered in process & product design. Limited benchmark resource use targets available & achievement is variable.	Disposal options & routes considered in process & product design. Cost of disposal targets & disposal routes not considered.
1	Use & cost of raw materials targets available but not always achieved.	Resource use considered only for cost reduction element of process & product design. No benchmarks for resource use available or considered.	Distribution costs considered only for publicity purposes. No serious benchmarks for distribution costs available or considered.	Resource use in use stage considered only for publicity purposes. No serious benchmarks for resource use available or considered.	Disposal options & routes poorly considered in process & product design. No cost of disposal targets set and disposal routes.
0	Use & cost of raw & recycled materials poorly considered in process & product design.	Resource use in manufacturing is not considered in the process & product design.	Resource use in distribution is not considered as part of the process & product design.	Resource use in use stage is not considered in process & product design.	Disposal options & routes not considered in process & product design. No cost of disposal targets set and disposal routes not considered.
Score					

1.16 The United Nations Sustainable Development Goals (SDGs)

Setting targets for sustainability

The Millennium Summit of the United Nations in 2000 led to the adoption of the UN Millennium Declaration where every UN Member State committed to achieve the 8 Millennium Development Goals by 2015. Unlike previous aspirational declarations, each goal had specific measurement methods, targets and dates for achieving the targets.

The Millennium Development Goals (MDGs) – 2015

The 8 top-level MDGs and their targets were:

Goal 1: Eradicate extreme poverty and hunger

- Halve the proportion of people living on less than $1.25 a day.
- Halve the proportion of people who suffer from hunger.

Goal 2: Achieve universal primary education

- Ensure a full course of primary schooling for all.

Goal 3: Promote gender equality and empower women

- Eliminate gender disparity in primary and secondary education by 2005, and at all levels by 2015.

Goal 4: Reduce child mortality rates

- Reduce the under-five mortality rate by two-thirds.

Goal 5: Improve maternal health

- Reduce the maternal mortality rate by three quarters.
- Achieve universal access to reproductive health.

Goal 6: Combat HIV/AIDS, malaria, and other diseases

- Halt and begin to reverse the spread of HIV/AIDS.
- Halt and begin to reverse the incidence of malaria and other major diseases.

Goal 7: Ensure environmental sustainability

- Integrate the principles of sustainable development into country policies and programs; reverse loss of environmental resources.
- Reduce biodiversity loss and achieve a significant reduction in the rate of loss.
- Halve the proportion of the population without sustainable access to safe drinking water and basic sanitation.
- Achieve a significant improvement in the lives of at least 100 million slum-dwellers by 2020.

Goal 8: Develop a global partnership for development

- Develop an open, rule-based, predictable, non-discriminatory trading and financial system.
- Address the special needs of the Least Developed Countries (LDCs).
- Address the special needs of landlocked developing countries and small island developing states.
- Deal with the debt issues of developing countries through national and international measures to make debt sustainable in the long-term.
- Provide access to affordable, essential drugs in developing countries.
- Make available the benefits of new technologies, especially information and communications.

Progress towards achieving all of these goals was mixed but the remarkable thing was that by 2018 the majority of the world's population lived in what were 'middle-income' countries with dramatically improved quality of life and life expectancy. Despite the large number of 'doom and gloom' stories, the data shows that life is getting better for most people.

Success?

- The proportion of the world population living in extreme poverty has halved in the last 20 years.
- The average life expectancy in the world today is approximately 70 years.
- 80% of the world's population have some access to electricity.
- 80% of the world's 1-year old children have been vaccinated against some disease.
- In low-income countries 60% of girls finish primary school.

These are remarkable numbers that remain unrecognised by most people.

The world is getting better for most people.

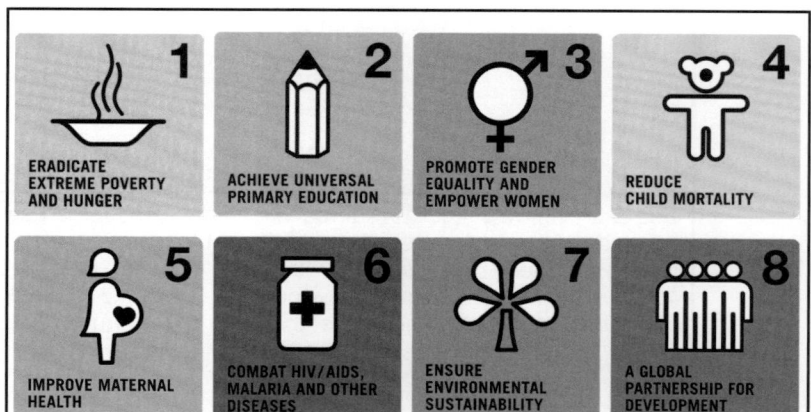

The 8 Millennium Development Goals – 2015

The 8 Millennium Goals were designed to be achieved by 2015. Progress was reported on the basis of defined targets and specific metrics for each MDG. The goals were largely achieved by the target date of 2015.

The Sustainable Development Goals (SDGs) – 2030

At the end of the Millennium Development Goal period in 2015 the UN moved to a new set of goals that focus on sustainable development in a broader context.

The 17 SDGs provide a template for sustainable development until 2030. The goals cover a broad range of social and economic development aims and, unlike the previous MDGs, are designed to be applicable to all parts of the world.

The issue of sustainability is a world-wide issue and the SDGs and the related targets provide a consistent framework for sustainable development throughout the world. As with the MDGs, each goal has separate metrics and targets to allow evaluation of progress.

Plastics products have contributed significantly to the development of our society and have brought huge benefits to mankind. Despite the obvious benefits, the public perception of plastics has never been high and has greatly decreased in the last few years due to increased volumes of waste plastics reaching the wider environment. The longevity of plastics products, previously one of the strengths, is now seen as a weakness if the material leaches into society.

This is largely been a result of:

- A failure to deal with product disposal at the 'end-of-life' stage. Plastics are easily recycled but industry developments and growth have not kept pace with societal developments.

- A failure or absence of waste management infrastructure and systems in parts of the world which allows products to reach the environment rather than being captured as potentially valuable materials.

- A failure or absence of recycling infrastructure and systems in parts of the world which means that, even if the materials are captured, there is insufficient recycling.

- A failure of society to deal with littering.

- Poor waste management practices.

Plastics products have much to contribute to improving the quality of life and the sustainability of the planet. They are not inherently 'bad' and neither are they inherently 'good'.

As with anything, their correct use can improve lives throughout the world but their incorrect use can also be detrimental. The industry is working hard to decrease the impact of plastics on the world but equally there is a need to recognise the significant beneficial effects plastics can have in achieving the UN SDGs.

A detailed examination of the 17 UN sustainability goals from the point of view of the plastics industry is given in Appendix 1. This shows how the plastics industry already contributes to achieving the SDG goals and that the industry is a vital partner in achieving the UN SDGs.

- **Tip** – Select 5 of the UN SDGs (see Appendix 1) as a focus for sustainability efforts and link your actions to these.

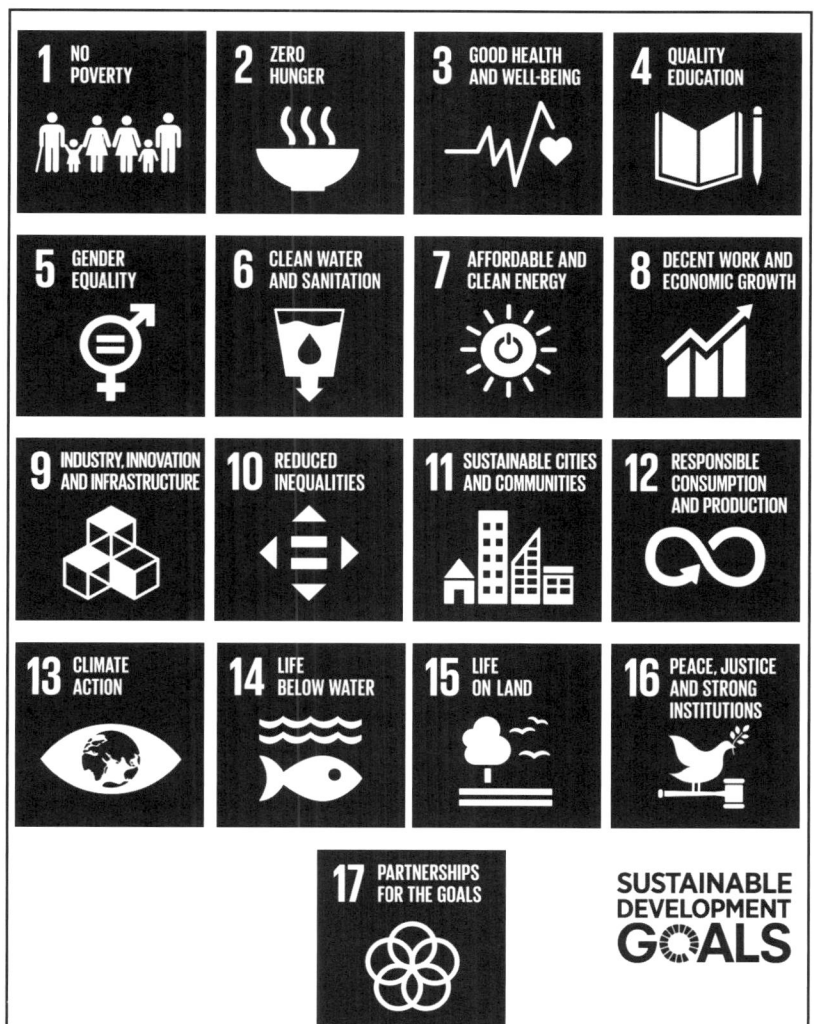

The 17 Sustainable Development Goals – 2030

The 17 Sustainable Development Goals extend the Millennium Goals to reflect changes in the world since 2000. There is a greater focus on world-wide sustainability due to the rise in importance of sustainability and sustainability thinking since 2000.

1.17   UN SDGs – where are you now?

A comprehensive framework

At the start of efforts to improve sustainability, the concept often appears too broad and nebulous to grasp apart from the fact that we know it is 'a good thing'. This means defining concrete actions that will have a positive result is also difficult.

Setting targets, measurements and reporting at a company level is good practice (see Chapter 12) but the targets should be aligned with a global concept of sustainability and to fit in with a simple framework. The UN SDGs provide this essential world-wide framework and cover all facets of sustainability. They allow actions and progress to be measured without falling into the 'it isn't obvious' trap and avoid the 'law of unintended consequences'. They define the concrete actions that will have a positive result in a complex world.

The UN SDGs provide a global framework for action at a local level.

Completing the chart

This chart is completed and assessed as for the previous charts.

The UN SDGs provide a framework and measurements for action to improve sustainability. They allow companies to prioritise actions within a consistent framework that is recognised around the world.

UN SDGs

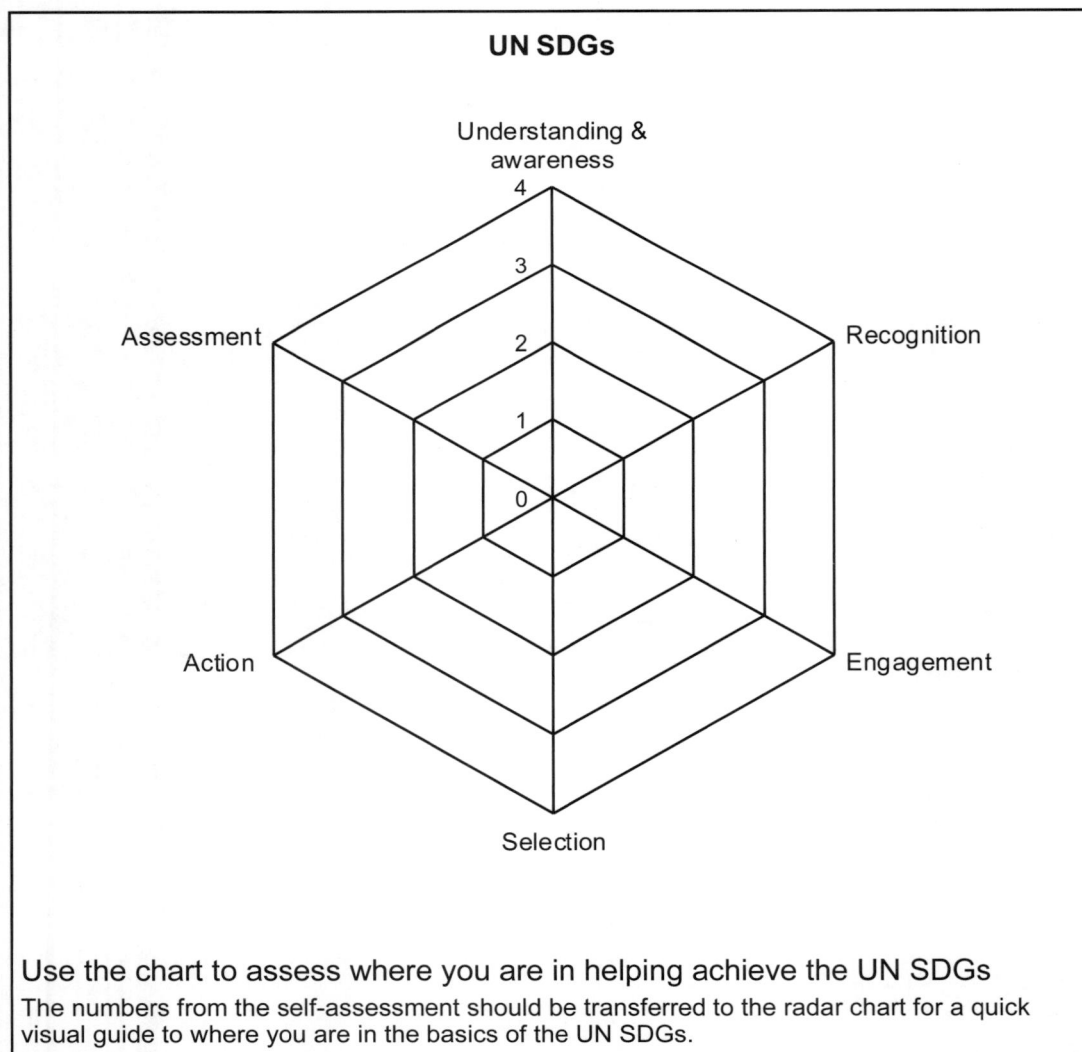

Understanding & awareness

Recognition

Assessment

Engagement

Action

Selection

Use the chart to assess where you are in helping achieve the UN SDGs

The numbers from the self-assessment should be transferred to the radar chart for a quick visual guide to where you are in the basics of the UN SDGs.

Oceanographers will generally agree that the crucial issue for the seas is climate change but this is difficult to grasp, quantify or see.

The result is that the press and public focus on the more visible issue of 'plastics in the oceans' which has good graphics and is easier to understand (even if most people get the numbers wrong).

This is not to say that plastics in the oceans is not important, simply that climate change is a much more important issue.

The UN SDGs provide the essential framework for assessing the issues.

UN SDGs						
Level	Understanding & awareness	Recognition	Engagement	Selection	Action	Assessment
4	Full understanding & awareness of the UN SDGs at all of levels of the company.	Top management sees the UN SDGs as an opportunity with benefits to society as well as reputational & financial benefits.	Top management & all staff have full engagement with achieving the UN SDGs.	Top management has chosen 5 UN SDGs to focus attention & activity. UN SDGs have been discussed & agreed with staff.	Action taken for all the selected UN SDGs.	Excellent assessment of all actions for effectiveness. Assessment checks benefits to society & reputational & financial benefits.
3	Good awareness of the UN SDGs at most of levels of the company.	Top management sees the UN SDGs as an opportunity but purely in reputational & financial benefits.	Top management engaged with achieving the UN SDGs but little engagement of other staff attempted or achieved.	Top management has chosen 3 UN SDGs to focus attention & activity. UN SDGs have been discussed & agreed with staff.	Action taken for the majority of the selected UN SDGs.	Good assessment of some actions taken to check effectiveness. Assessment is only in terms of reputational & financial benefits.
2	Average understanding & awareness of the UN SDGs at some levels of the company.	Top management sees the UN SDGs as an opportunity but purely in reputational benefits.	Staff have good engagement with achieving the UN SDGs but this is not reflected by any top management engagement.	Top management has chosen 2 UN SDGs to focus attention & activity. UN SDGs have not been discussed or agreed with staff.	Action taken for the minority of the selected UN SDGs.	Some assessment of actions taken to check effectiveness.
1	Poor understanding & awareness of the UN SDGs at any level of the company.	Poor recognition that the UN SDGs are an opportunity or that they have any benefits to the company.	Engagement with the UN SDGs is limited to the marketing who see this as a 'good news' story. Engagement is 'greenwashing' exercise.	Top management has chosen a single UN SDG for action. UN SDG has been not been discussed or agreed with staff.	Action taken for only 1 of the selected UN SDGs.	Assessment is only in terms of reputational & financial benefits.
0	No understanding or awareness of the UN SDGs. Company unaware of the existence of the SDGs &/or their relevance.	No recognition that the UN SDGs are an opportunity or that they have any benefits to the company.	No staff at any level has any engagement with the UN SDGs.	No selection of any relevant UN SDG for action.	No action on any of the UN SDGs.	Little assessment of actions taken to validate effectiveness.
Score						

1.18   Project selection and management

Project management

Sustainability improvements will only come about through the successful identification and implementation of projects. Implementing sustainability is a change programme and the identification, assessment and successful completion of projects is at the heart of any change programme. This needs good project management to be effective.

Project selection

A precursor to a successful project is good selection of the project to be undertaken. If this book succeeds in its aims then there will be a wide range of potential projects and it is best to choose a limited number of projects and succeed at these rather than to start many projects and never to complete any.

Sections 1.6 to 1.8 dealt with selecting projects based on the time, life cycle and materiality but it is still likely that there will be competing projects. In this case, project selection can be based on a simple 2×2 'ease-relevance' matrix as shown in the diagram on the right. Projects can be ranked quickly based on:

- The relevance to the company – this is decided by the materiality assessment (see Section 1.8).
- The ease of implementation.

Projects with a high relevance that are easy to implement are preferred and these will be in the A-segment of the matrix. These should be the first projects attempted. Projects in the D-segment of the matrix are, realistically, never going to start.

- **Tip** – When faced with a competing group of projects with comparable 'ease-relevance' indices then the project with the shortest time to completion should be chosen.

Resource bottlenecks

Every company has resource bottlenecks, these can be about staff – there are always more things to do than there are time and people or about finance – there are always more demands on capital than there is money in the bank. Whatever the resource bottleneck, follow these simple rules to get the best results:

- Set a limit on the number of sustainability management projects that can be active at any time – do not start new projects unless a current project is completed or suspended. Make no exceptions, otherwise staff will not know where to focus efforts or spending.
- Never start projects that you cannot finish no matter how attractive they may appear in terms of time or return. Unfinished projects are a waste of time, effort and money.
- If the resource bottleneck is finance, then insist that the sustainability programme is self-funding. This should not be a problem if 'A' projects are started first.

Project planning

There are three approaches to project planning:

- No planning – We'll do it!
- Simple planning methods.
- Complex computer-based planning methods.

Project planning is essential for successful sustainability management but simple planning methods are far preferred. Sustainability management projects can be broken down into small elements to reduce investment and time.

> "Our experience has taught us that it is harder to teach people to think independently than to obey."
>
> **Thomas Bata**

> If you don't do this right then you will not be here in 10 years' time. This applies to the complete industry as well as to individual companies.

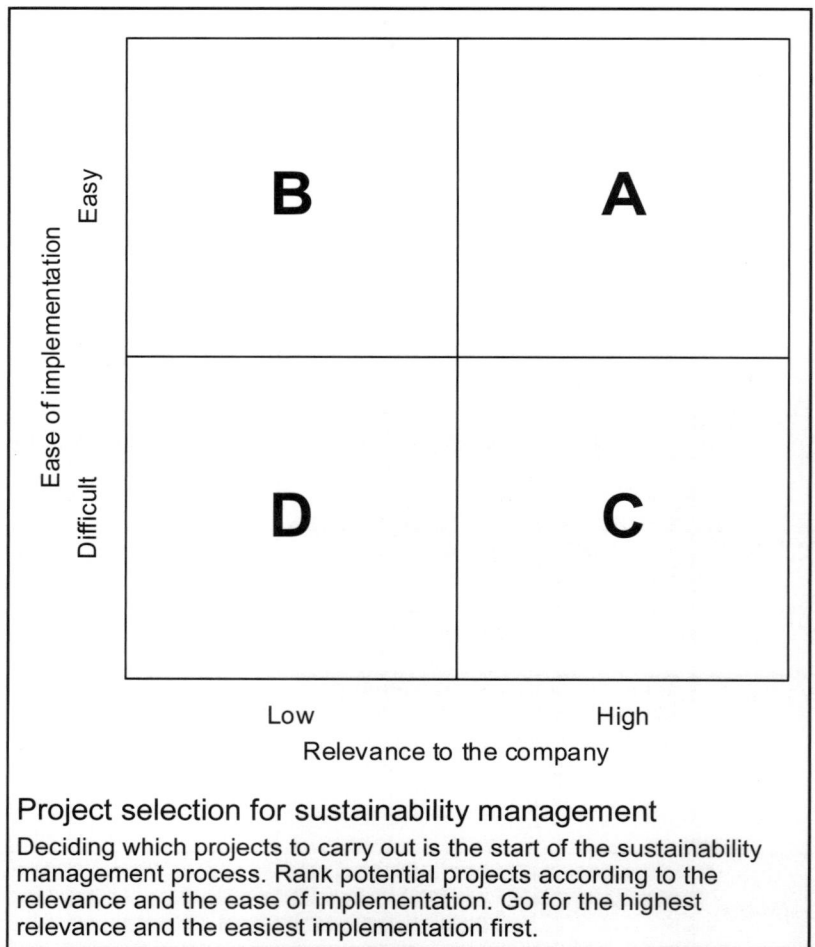

Project selection for sustainability management
Deciding which projects to carry out is the start of the sustainability management process. Rank potential projects according to the relevance and the ease of implementation. Go for the highest relevance and the easiest implementation first.

One method is to use top-down planning and Post-it notes. This method is described in the box below and, despite the apparent simplicity, is very powerful and flexible for small project planning. The method encourages an open approach to planning where the whole process is visible, in contrast to computer-based methods where the project plan is controlled by the software operator.

Whichever method of project planning is chosen, every project plan must have the following elements:

Aims and objectives

These are the clearly stated and agreed aims and objectives of the project. These must be measurable and achievable to allow performance assessment. Starting a project without clear aims and objectives is the surest way to fail.

- **Tip** – One aim may include several objectives.

Milestones

These are dates (from the actual start of the project, not from the date that approval was sought) that show when particular tasks are to be completed. Milestones allow assessment of project plan/time results.

Budget

All projects should have an initial allocated budget.

Assessment

Projects must be assessed after the aims and objectives have been completed or when the project manager decides that no more progress can be made. Assessment is a review of the achievements against the agreed aims.

- **Tip** – Assessment should not only consider the financial aspects such as return on investment but also other non-financial benefits of the project.

Closure

All projects should be formally closed after the assessment phase.

Project management

Project teams

- Project teams need a leader or 'project champion'.
- Delegate control and accountability to the project manager.
- Project teams can make decisions without fear of being over-ruled later (over-rule the project team and you become the team leader).
- Project teams are free to innovate.

- Project teams are assessed on their results.

Meetings

- Project teams meet regularly.

Schedules

- Check progress (and slippage) against agreed time and financial targets.

Communication

- Project teams report on progress via a pro-active reporting and communication plan.
- **Tip** – The One-Page Project Manager[1] method is an excellent tool for both planning and reporting.

- 1. Campbell, C.A. and Campbell, M. 2013. '*The One-Page Project Manager*'. Wiley.

Home truths of project management

- What you don't know hurts you.
- Any project can be accurately estimated for cost – after it is finished.
- Nothing is impossible to the person who doesn't have to do it.
- What is not on paper has not been said.
- If you can keep your head while all around you are losing theirs then you haven't understood the plan.

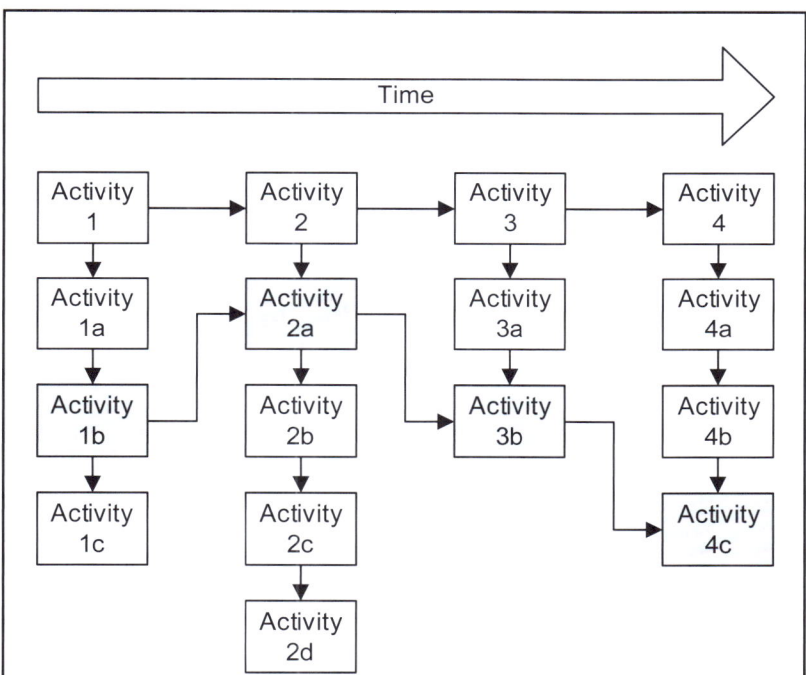

Project planning using Post-it notes

Top-down project planning for cost management can be carried out using Post-it notes and a flip-chart.

Write each task and the approximate time needed on a Post-it note. Move the notes around and group them according to the major tasks. Combine or divide tasks as the process continues.

Leave the chart in view and add, subtract or move the notes around as the plan develops. Finally, move the notes to overlap activities that can be done at the same time and reduce the total time taken for the project (simultaneous engineering). The critical path is easily seen from the sum of the individual tasks.

1.19 Sustainability management projects – where are you now?

The sustainability management process

Choosing between sustainability management projects will always be difficult. There will always be too many projects competing for too few resources.

Companies need to rapidly assess the potential gains and difficulty of implementing any potential project before rushing into a complex project that has a relatively low potential to improve sustainability.

Project selection is a key to successful sustainability management.

After projects have been selected then an effective project management system is an essential to delivering projects and achieving the potential gains.

Cross-functional teams are an invaluable tool for sustainability management due to the organisation of most companies.

Completing the chart

This chart is completed and assessed as for the previous charts.

A Technical Manager reporting to me always had too many projects running at any one time and was judged to be 'failing'. This was because the Managing Director constantly introduced new projects and changed the priorities.

We set a limit of 8 projects that could be active at any one time.

Any potential new project had to pass a monthly review and to displace an existing project before it could be considered for action.

Any displaced existing project was labelled as being 'on hold' and not considered for assessment.

The Managing Director was therefore forced to prioritise the active projects and could not randomly introduce new projects.

The results:
- A more stable project list.
- More completed projects.
- A successful Technical Manager.

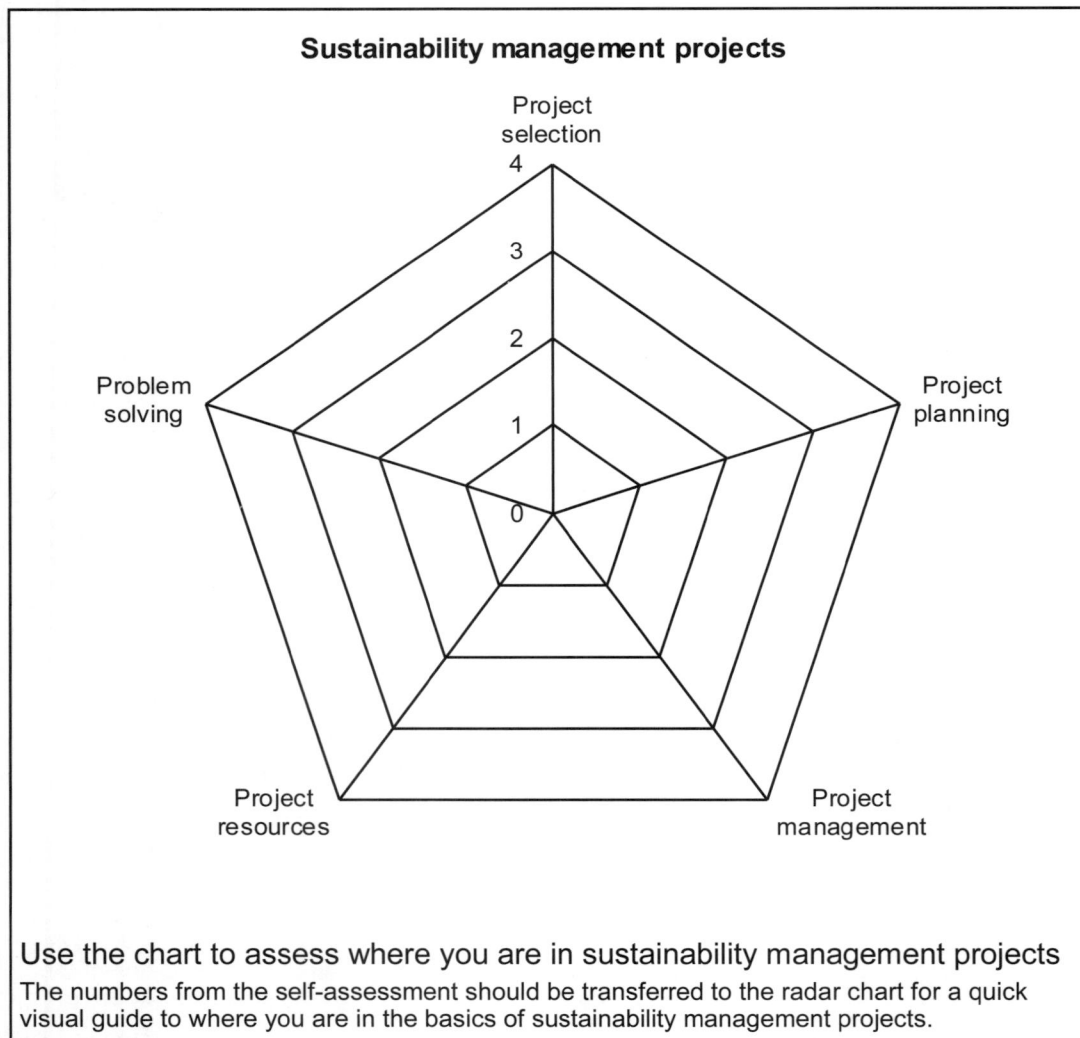

Sustainability management projects

Project selection

Project planning

Project management

Project resources

Problem solving

Use the chart to assess where you are in sustainability management projects

The numbers from the self-assessment should be transferred to the radar chart for a quick visual guide to where you are in the basics of sustainability management projects.

Sustainability management projects

Level	Project selection	Project planning	Project organisation	Project resources	Problem solving
4	All relevant sustainability opportunities identified & prioritised for action.	Formal project definition & project plan necessary for any project. Progress is regularly reported & post-project assessment is carried out.	Excellent sustainability project management system used in all cases. Projects have clearly defined management & sustainability benefits.	Project resources defined & allocated before project start. Projects are rarely delayed due to resource constraints.	Firmly embedded culture of improvement & problem solving through planning, action & review. Root causes identified & resolved.
3	Most available sustainability reduction opportunities identified but not prioritised for action.	Formal project planning carried out for all projects but control, reporting & assessment are variable. Failed projects are sometimes hidden & no lessons learnt.	Good sustainability project management system but use is variable. Good integration across departments but many projects have poor sustainability benefit definition.	Project resources defined but not allocated at project start.	Problem solving is largely reactive with focus on solving root causes. Solutions developed but not always fully implemented.
2	Some sustainability reduction opportunities identified but no real planning process.	Project planning carried out for most projects but control, reporting & assessment are poor or rarely carried out. Failed projects are often hidden & no lessons learnt.	Sustainability project management system available but not used. Some integration of projects across departments & poor sustainability benefit definition.	Project resources poorly defined at project start.	Problem solving is largely reactive; solutions are developed but rarely fully implemented. Focus on dealing with urgent effects & not on solving root causes.
1	Few sustainability reduction opportunities identified via unplanned process.	Cursory & undocumented project planning but no formal project planning or monitoring. Projects can become dormant & remain unfinished.	No sustainability project management system. Some integration of departments for projects that clearly cross departmental boundaries.	Project resources rarely considered at project start.	Problem solving is purely reactive & focused on dealing with urgent effects & not on solving the root cause.
0	Significant sustainability reduction opportunities ignored due to 'urgent' daily pressures.	No effective project planning. Actions are ad hoc & driven by events. Action is seen as more important than planning.	No sustainability project management system. Every project is 'different'. Projects are run by departments with little input from other departments.	Projects often started without adequate resources (due to poor planning) or starved of resources during project. Urgency is rated more highly than strategic importance.	Problems are ignored until they go away.
Score					

1.20   The perils of greenwashing

We see it every day

The Oxford English Dictionary defines 'greenwash' as 'disinformation disseminated by an organization so as to present an environmentally responsible public image'. 'Greenwashing' is actively misleading consumers or purchasers about the environmental features or benefits of a product or the environmental practices of a company. In some cases, the claims are simply misleading, in other cases they are simply untrue.

- **Tip** – When a company spends more time and money redoing their logo to include green tints than they do on actually improving their sustainability then you know they are indulging in greenwashing. Just because it is coloured green doesn't make it 'green'.

One of first major uses of greenwashing was in the hotel industry, we all remember those little signs saying 'Preserve the environment, reduce water use and detergent use – if you don't need a clean towel then put it on the towel rail'. In most cases, the objective was not to save the environment but to reduce costs for the hotel. For the cost of some small cardboard signs the hotel industry saved washing millions of towels and lots of money. Still, the guests felt 'socially responsible' so it wasn't all bad.

Green is currently the trend and it is inevitable that companies will try to present themselves as 'green' to sell more products. However, unsubstantiated and false claims must be resisted because greenwashing will eventually be discovered and result in:

- Damage to the company image and reputation.
- Reduced stakeholder confidence in the company.
- Increased complaints about misleading advertising.

Greenwashing is a contagious disease, if the competition is greenwashing there is always a temptation for any company to also use it. This temptation must be strongly resisted.

- **Tip** – There are many guides to making sustainability statements but one of the best is ISO 14021: 2016 'Environmental labels and declarations – self-declared environmental claims'. This is specifically for environmental claims but it is equally relevant for all types of claims.

Tip – Get a copy of ISO 14021 and read it carefully, it is full of great advice.

The sins of greenwashing

The list of the 'sins' of greenwashing was first developed by TerraChoice in 2007 (now part of UL – www.ul.com). The seven sins are:

- Hidden trade-off.
- No proof.
- Vagueness.
- Irrelevance.
- Lesser of two evils.
- Fibbing.
- Worshipping false labels.

Do not be guilty of any of these sins.

The rules of making claims

All self-declared claims should be:

- Accurate – unless claims are accurate (and not misleading) then they are worthless and are effectively greenwashing.
- Verifiable – claims should be supported by verifiable documented proof. The proof should either be public information or disclosed on request. Claims should not be

My new aluminium water bottle (a gift from my nephew) proudly proclaims that it is 'BPA free'.

This is correct as it is entirely aluminium. This is also an irrelevance and is gratuitous greenwashing aimed at people who have heard of BPA but have no idea of what it is or where it is found (or not found).

It also makes me think less of the manufacturer (a well-known sporting goods manufacturer based very near where I was born and surfed for many years).

Any idea what these logos mean?

These are, naturally, all green and were picked off the internet in seconds (all are copyright free). They all mean nothing apart from 'communicating' green to consumers and are ideally suited for use in greenwashing. Look out for similar logos.

made using confidential information and documentation should include:

- The standard or test method used.
- The test results to verify the claim.
- Other evidence if the claim cannot be made by testing the complete product.
- Third-party verification of the results.

- Relevant to the product.
- Not misleading – claims such as 'BPA free' (see sidebar) are difficult to verify and are potentially misleading (the material may be present due to contaminants). The preferred wording is 'No added BPA' which can be verified.
- Specific – claims should not include vague and non-specific terms such as 'green', 'eco-friendly' or 'environmentally friendly'. Such words should only be used if they are qualified with specific verifiable claims.
- Unlikely to result in misinterpretation or misunderstanding – if a claim can be misinterpreted or misunderstood then it should include an explanation with the claim unless the claim is always valid.
- Consider the final product – if claims are being made about a product then they should be true for the final product and look at the complete product life cycle, e.g., if you save energy in producing a product then this does not mean that the product can be marked 'energy saving'.
- **Tip** – Some specific environmental terms, e.g., degradable, compostable, recycled content etc., are defined in ISO 14021 and the rules for using them are given in the standard.

Control the marketing

Scientists and engineers are not immune to the temptations of greenwashing but marketing departments and PR consultancies are perhaps more susceptible. They are desperate to tap into the green wave and look for the slightest positive message to amplify. Whilst this is understandable, it is still not acceptable if the rules are broken.

Instead of using greenwashing, all areas of the company should treat being green as being 'normal' and simply 'the way it is around here'. This normalises sustainability in the business and it is much easier to report when everybody regards it as normal.

- **Tip** – Get a second copy of ISO 14021 and give it to the Marketing Department. Test them to make sure they have read it and adhere to it. This is not an option; it is only common sense.
- **Tip** – Bio-based (see Section 4.14) and biodegradable plastics (see Section 4.15) are particularly susceptible to

greenwashing and European Bioplastics has produced an excellent guide 'Claims on biodegradability and compostability on products and packaging'. This is available free from www.european-bioplastics.org/news/publications/. It is recommended reading to show how claims can be made without indulging in greenwashing.

What about the reverse?

The perils of greenwashing can sometimes lead companies into the reverse predicament of 'greenblushing', which is where a company is doing good work in sustainability but fails to effectively communicate their progress or the benefits of their products. This is particularly evident in the area of social sustainability (see Chapter 11) where many companies do not know how to report or communicate their performance (even if they are doing quite well).

This is a fine line to tread but applying the rules against greenwashing can help to clarify what is acceptable and what is not.

- **Tip** – If you do good work then report it and don't be ashamed of what you are doing. Start small, communicate what you actually do and get the dialogue going.

The final words

Do not greenwash products, processes, materials or companies, now or ever. The temptation to inflate green credentials will be strong but, in the end, it is not only fruitless but the potential for reputational damage is huge.

Do not buy products from people who do it (and tell them why).

- **Tip** – Now that you have read this, have a look in your kitchen and other cupboards and see how many fake and misleading logos you see. Feel free to send them to me, I am always on the lookout for the more egregious examples of greenwashing. Perhaps we can, together, shame the perpetrators into stopping greenwashing. We have got to start somewhere.

I love the term 'astroturfing' which is used to describe a fake 'grassroots' campaign.

Not to give anybody any ideas but this involves using fake organisations or people to hide who is really trying to get the message across.

Who knew people could be so devious?

I am just a simple engineer looking for solutions.

My wife pointed out to me the existence of 'gluten-free' shampoos. I couldn't believe it, I thought gluten was something you ate (or didn't) but a quick search showed that these are out there.

One site says: "..if you ever get shampoo in your mouth or touch your hair and put your fingers in your mouth, you risk getting glutened unless all your hair care products are gluten-free". (sic)

OK, I give up, it is everywhere.

Pointing the finger at other companies and processes is not the way to engage with the public or the competition. It simply tarnishes both your reputations.

As Mercutio said in Shakespeare's 'Romeo and Juliet'; "A plague on both your houses".

Key tips

- Sustainability is far broader than most people think, it is not simply about environmental sustainability.
- Sustainability includes environmental, social and economic sustainability.
- Many activities are on the borders of the three groups, e.g., energy management has benefits in both environmental and economic areas. These can be the most profitable part of a well-managed sustainability management programme.
- Your motives for starting work on sustainability do not matter but you do need to start work soon.
- The pressures for sustainability come from many directions and are not going to go away. This is going to be around for the next 20 years or more.
- Don't think of sustainability only in terms of costs and risks. It could be the greatest opportunity ever (if you do it right).
- There are many approaches to sustainability but companies should feel free to adopt a 'mix and match' approach and use the best ideas from all of them. The important thing is to do what works for you.
- Life cycle assessment provides an assessment of the environmental impacts of a product – it is not suitable for comparing materials alone and should only be used for well-defined products.
- Products and processes have both short and long-term impacts. The best product in the short-term may also be the worst product in the long-term. We need to look at time in sustainability as well as the impacts. The short and long-term impacts can be used to assess products and product groups.
- Product and process impacts need to be mapped through the product life cycle to establish the most effective actions. These are not always in manufacturing but can be in other parts of the product life cycle.
- Product and process impacts need to be examined for materiality to include the external viewpoint on the impacts. Sustainability needs to consider the external stakeholders as well as the company view.
- We are not running out of materials for plastics in the short or the long-term.
- We are running out of destinations for the materials and resources that we use and do not return to the system through recycling or other materials recovery methods. This includes the air, land and sea.
- The 17 UN SDGs provide a template for sustainable development across the world and also provide a robust but achievable set of goals for the world.
- Plastics have contributed to the achievement of the UN MDGs and will continue to contribute to many of the SDGs. As an industry we still need to improve in certain areas.
- Sustainability will be achieved by successful projects and we need to select and manage these well to get the best out of the available resources.
- 'Greenwashing' is never acceptable. It is not simply 'spin', it is damaging to everybody. At the worst, it is simply lying.

Chapter 2

Management systems

Business as a whole and the broader sustainability area has recently seen an explosion in the number of management systems and related standards for auditing and certification. This began with the introduction of BS 5750 in the UK in 1979 and on the international stage with the introduction of the ISO 9000 series of standards in 1987 when the ISO series largely followed the BS 5750 series in requirements and layout.

The ISO 9001 series of standards are used in virtually every country in the world and this worldwide movement towards formal quality management systems moved rapidly from the established economies of the West to the developing economies of the East.

When the original quality system standards were developed, they were designed for manufacturing industry but the use of the standards for non-manufacturing industries led them to evolve into what are now termed 'Management Systems Standards' (MSS). This was particularly true of the substantial revisions that took place in 2000 when the ISO 9001 standard was revised to take a more generic approach based on business processes. ISO 9001:2015 continued this process and introduced some new and updated concepts, e.g., the concept of risk-based thinking to support the process approach, changes to terminology and a reduction in emphasis on 'documents' and a change to 'documented information'.

The biggest change for this revision was the adoption of a consistent high-level structure (Annex L) for all ISO MSS. This means that all ISO MSS, e.g., quality, environment, energy, health and safety etc. share a common structure, identical core text and common terms and definitions. When companies meeting one MSS come to install another MSS meeting Annex L then the key components will already be in place.

This common structure was implemented for ISO 9001 and ISO 14001 in 2015 and for ISO 45001 and ISO 50001 in 2018. The role-out of the Annex L high-level-structure is continuing for a wide range of ISO standards

Nobody can deny that the wide range of MSS has been successful in terms of market penetration, revenues generated for standards-setting bodies, revenues generated for certification bodies and revenues generated for consultants advising companies seeking registration. What is less certain is the benefit to the companies who have spent large amounts of money on consultants and on achieving certification.

This chapter looks at the various MSS, their impact on sustainability, their effectiveness and how they can best be implemented.

"Every system is perfectly designed to get the results it gets."

Arthur Jones

2.1 Management systems

The rise of the external systems

Companies can change or remove internal systems depending on their needs but, with external systems, the general requirements, but not necessarily the detail, of the system are specified by an external standard or body. External systems can also have defects that actually prevent people from delivering the desired aim of the system. These are:

- Inflexibility ('The standard requires this') – this is where a specific interpretation of the general requirements of a standard is translated into an edict that must be obeyed even when everybody can see that the result is a system that either doesn't deliver any benefits or, worse still, is detrimental to the overall stated aim of the system.

- Atrophy ('We can't change this') – this is where the interpreted requirements of the standard are regarded with an almost religious passion and become a dogma that cannot be changed. The result is a system that becomes engraved in stone, that nobody pays any attention to and eventually leads to disillusionment with the whole system.

- Tick the boxes ('We have to show that we have done this') – this is where the system is acknowledged as not delivering and management no longer sees any benefits but the threat of losing certification is worse than the alternative of attending meetings, producing reports and ticking all the required boxes.

External MSS can deliver excellent benefits provided companies have the courage to use them to the best effect, to modify or discard them if they are not working and to refuse to accept the boilerplate solutions that are commonly delivered. Companies need to:

- Cut through the jargon and use the MSS to deliver real and profitable improvements.

- Use the systems that they already have. In many cases, particularly for quality systems, if a company has been operating for 5 years or more then the internal systems probably meet most of the requirements of the MSS. If they didn't then the company probably wouldn't still be trading, i.e., natural selection would have taken place.

- Carry out a business appraisal of the case for installing the MSS. Systems are a business decision. The decision may be influenced by a customer saying 'If you don't have ISO X, then we won't buy from

you' and if this is the case then the company needs to expend the appropriate amount of effort.

External systems are often used to justify an extraordinary amount of work for a very small reward. Do not let this happen to you.

Most MSS follow the 'Plan-Do-Check-Act' (PDCA) model:

- **Plan** – Identify the requirements and set the policy.

- **Do** – Decide on the procedures needed and implement them.

- **Check** – Set targets and objectives, and assess achievement.

- **Act** – Continually improve the system.

- **Tip** – The PDCA model is sometimes referred to as the Deming or Shewhart cycle.

Quality management systems (ISO 9001)

Quality systems were the first of the MSS to be developed and this was the first attempt to use the standards process, traditionally used for products, to define a management system. A quality MSS does not attempt to

It almost appears that many of the systems were designed to allow non-technical people to control technical processes.
With predictable results.

It will come as no surprise that there is a 'standard' for MSS. Annex L is the standard high-level structure for all ISO management systems standards. It does not enforce any new requirements but simply makes the standards consistent in numbering and format.

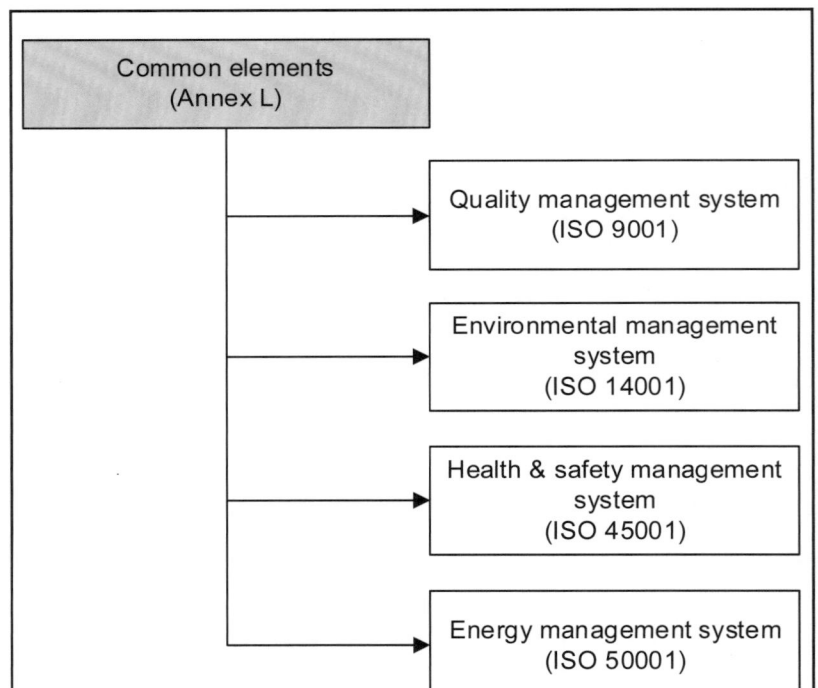

Many of the external systems have common elements

Where external systems have common elements, e.g., training, management review, auditing and non-conformance reporting, then these should be combined into a single standard company-wide process.

define the requirements of the product except by reference to the customer's 'requirements' and 'satisfaction'. This concept of an MSS is still widely misunderstood by many people who think that an MSS is some type of product standard and assume that it involves some type of testing or certification of the product. An MSS is a 'Management Systems Standard' and not a product standard.

The MSS approach can be valuable and can form a basis for quality management. However, the cost savings due to the implementation of a 'quality MSS', as opposed to the introduction of 'quality management', have been widely overstated. They are not the same thing and are often joined together for marketing reasons.

Most plastics processors will already be certified to ISO 9001 and will be familiar with this MSS. Quality as part of sustainability is covered in Section 5.5.

Environmental management systems (ISO 14001)

BSI released BS 7750 in 1992 and similar to the development of ISO 9001 from BS 5750, BS 7750 was used as the basis for ISO 14001 (see Sections 2.3-2.7 for more details of ISO 14001).

Given the similarity of many of the issues, it proved relatively easy to generate the same type of MSS for environmental management as for quality management. Where an element was common to the standards, e.g., training, calibration, management review, etc. then the requirements were designed to be compatible to avoid conflict.

Energy management systems (ISO 50001)

Energy management was the next management system to become standardised with the publication of ISO 50001:2011. This again used the PDCA model and had many requirements that were common to the other MSS (see Sections 2.8-2.11 for more details of ISO 50001).

Energy management (the process but not the standard) is covered in greater detail in Chapter 6).

Health and safety management systems (ISO 45001)

Health and safety systems grew up over time through the introduction of legislation to protect both workers and the general public. Whilst the detailed requirements and effectiveness of enforcement varied from country to country there was still felt to be a need for an MSS for health and safety. As a result, OHSAS 18001 was developed by an international working group and released in 1999 as BS OHSAS 18001. This was the internationally recognised specification for occupational health and safety management systems until the release of ISO 45001 in 2018 and companies certified to OHSAS have until 2021 to migrate from OHSAS 18001 to ISO 45001.

ISO 45001 was written to conform to the Annex L structure from the start and is compatible with ISO 9001, ISO 14001 and ISO 50001 (see Sections 2.12-2.15 for more details of ISO 45001).

More systems standards?

The MSS approach does not end with the systems listed. There are many more MSS available. The rise of MSS continues.

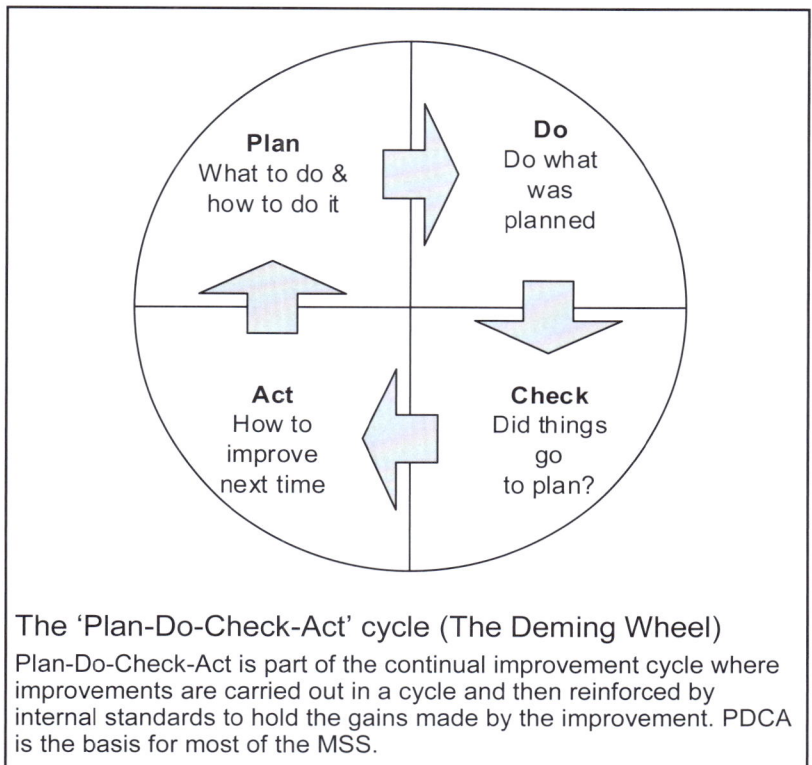

The 'Plan-Do-Check-Act' cycle (The Deming Wheel)

Plan-Do-Check-Act is part of the continual improvement cycle where improvements are carried out in a cycle and then reinforced by internal standards to hold the gains made by the improvement. PDCA is the basis for most of the MSS.

Do they deliver?

One issue that is largely absent from many of the discussions of this wide variety of 'systems standards' is the answer to the unasked question 'Do they deliver'? Many companies are driven into the management systems standards approach by their own insecurity and need to present some external verification of their systems.

In some companies, the introduction of the standards is primarily an exercise in 'window-dressing', where current and failing systems are simply documented and engraved in stone. This response will inevitably fail and the systems will not deliver.

In other companies, the introduction of this type of standard forces a re-evaluation of the systems that are currently in place. This can be a powerful catalyst for change and improvement. The standard can act as a reference for 'best practice' and provide guidelines for the development of internal systems.

Do they deliver?

The answer depends on you.

2.2   Management systems – where are you now?

Getting ready for systems

Developing and installing an effective MSS is not an easy task and the biggest issue is that this inevitably involves changes in the way people work. These changes can be transformational or disastrous depending on how they are managed.

Most managers believe that they have excellent systems, after all they usually designed them, However, a few minutes of investigation will often show that the systems are old, do not work properly and get in the way of the staff doing the things that we actually want them to do. Before installing any MSS, the company needs to examine if it is ready for the changes.

Getting the systems right can quickly improve performance, improve staff satisfaction and improve sustainability. To do this, companies must be ready for change; they must have systems in place to manage the changes and must provide appropriate support structures for staff during the changes.

How much of this are you doing?

Completing the chart

This chart is completed and assessed as for the previous charts.

> "If you have always done it that way, it is probably wrong."
> **Charles Kettering**

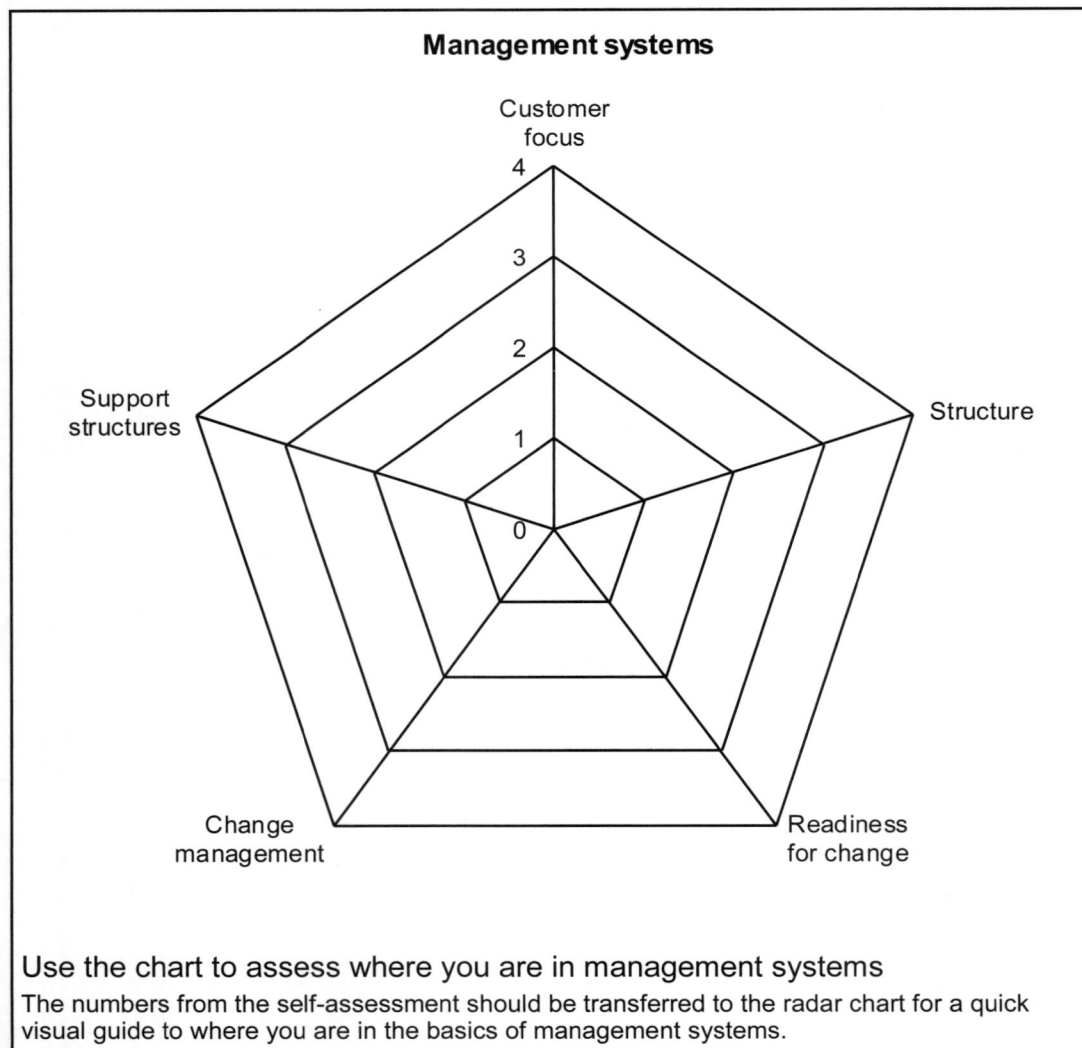

> "If actions speak louder than words, there are an awful lot of quiet people in business."
> **Octavius Black**

Management systems

Radar chart with axes: Customer focus, Structure, Readiness for change, Change management, Support structures. Scale 0 to 4.

Use the chart to assess where you are in management systems

The numbers from the self-assessment should be transferred to the radar chart for a quick visual guide to where you are in the basics of management systems.

> "We don't see things as they are, we see things as we are."
> **Anaïs Nin**

Management systems

Level	Customer focus	Structure	Readiness for change	Change management	Support structures
4	Internal & external customers are the highest priority. They are seen as the only reason for the existence of the operations. Staff are happy with their ability to serve the customer.	Structure encourages all staff to identify & solve problems. It encourages collaborative work across departments to solve problems & capitalise on opportunities.	High readiness for change at all levels. Company in constant state of change to adapt to changing markets. All staff see change as normal & examine systems for improvements.	Change management has a history of success even for significant changes. Change management is proactive, communicated & managed well.	Staff well supported by management in executing changes to systems. Management actively supports & encourages suggestions for changes to systems & operations.
3	External customers are seen as important but internal customers are not. Staff feel moderately able to deal with external customers due to internal systems & constraints.	Structure encourages information sharing but provides limited opportunity for collaborative work across departments to solve problems & capitalise on opportunities.	Moderate readiness for change at most levels. Key employees are negative in response to change & prefer security of business as usual.	Change has been well managed in the past but primarily for small changes. Experience of large changes is not universally positive but small changes communicated & executed well.	Staff well supported by management in executing changes to systems. Management support & encouragement for changes to systems is passive.
2	Internal issues & systems take precedence over external customers. Staff feel unable to deal with external customers due to internal systems & constraints.	Structure encourages information sharing but does not encourage collaborative work across departments. Moderate departmental 'Tribe' culture.	Low readiness for change at many levels. Middle management has poor expectations of success in change implementation & does not see this as their role.	Change has been moderately managed in the past but only for small changes. Little experience of major change but small changes communicated & executed well.	Staff moderately supported by management only in small changes to systems. High-level management approval needed for even minor changes.
1	Internal & external customers are tolerated. Internal systems & constraints positively hinder efficient dealings with customers.	Structure discourages information sharing & collaborative work across departments. Strong departmental 'Tribe' culture.	Poor readiness for change at all levels. Low expectations of success for any change in systems.	Change has been poorly managed in the past. Change is primarily a reaction to noise with little communication. Change management has been minimal or ineffective.	Staff poorly supported by management & systems. Systems do not help staff carry out tasks. Only changes suggested by management are authorised.
0	Internal & external customers are regarded as an imposition on normal working. Employees treat internal & external customers as 'the opposition'.	Structure & dynamics of business encourages a 'not my job' attitude. Staff are disinterested in their job & office politics is a costly & consistent problem.	Organisation is stagnant. All efforts to change meet with resistance & 'we tried that before' attitude. Previous change efforts have always failed.	Change management non-existent in the past. Management makes significant changes based on perception not facts, without communication & without attempting to manage the process.	Staff have no support from managers & systems. Systems stop them getting the job done. Management appears to have no interest in helping them to succeed.
Score					

2.3 Environmental management systems – the basics

Clean business = good business

Environmental Management Systems (EMS) are rapidly becoming an important issue in sustainability and particularly in the plastics industry. Some companies are under pressure from their customers and society whilst others recognise the environmental impacts of their business and want to minimise these. Perhaps the most important reason is that companies who have implemented an effective EMS have often not only improved their environmental and sustainability performance but have also achieved substantial cost reductions. An EMS with strong emphasis on minimising waste and continual improvement will help a company to reduce costs.

A good EMS is a practical management tool to:

- Identify, assess and manage the environmental consequences of operations.
- Reduce waste and operating costs.
- Gain a competitive advantage.
- Establish and show a system for continual environmental improvement.
- Demonstrate legal compliance.
- Improve the public image.

Waste minimisation and EMS

An EMS focused on reducing physical waste and emissions (see Chapter 9) will not only improve sustainability but will also give cost reductions from reduced scrap, reduced rework and reduced energy use. The average UK plastics processing first-time yield rate is just under 95%, this represents an average first-pass rejection rate of over 5% (see Section 5.2). This rate increases operating costs and reduces capacity from the lost opportunity to produce saleable product.

Eliminating or reducing waste gives environmental benefits by reducing the use and waste of resources and also reduces costs.

EMS basics

An effective EMS will include:

- An assessment of the environmental aspects and impacts of the company's activities, products, processes and services (see Section 2.5).
- An environmental policy.
- An environmental improvement programme with objectives and targets (see Section 2.4).
- Identified roles and responsibilities for all employees.
- A training and awareness programme.
- Written procedures to control activities with a significant environmental impact.
- A control system for 'documented information'.
- A programme of regular auditing.
- A formal review process for the EMS.
- **Tip** – Many of these will already be in place as a result of ISO 9001.

Approaches to EMS

An EMS can be developed to comply with the ISO 14001 model but it is also possible to follow the EU Eco-Management and Audit Scheme (EMAS) or even to develop an in-house EMS. Companies using the first two approaches can obtain formal certification to ISO 14001 or verification to EMAS. EMAS is slightly more demanding than ISO 14001 but does offer more benefits.

- **Tip** – Look at EMAS as an alternative to ISO 14001. The publication of the 2017 and 2019 EMAS Annexes means that EMAS now includes all the requirements of ISO 14001 and transition has been made much easier.
- **Tip** – EMAS uses the term 'verification' but ISO 14001 uses the term 'certification'. They are effectively the same thing.

It is not necessary to get external recognition to obtain many of the benefits but the formal approach increases the commitment to continual improvement and to identifying opportunities for improvement and cost savings. External recognition increases the credibility of an EMS and provided the EMS has been systematically and properly implemented then certification does not require much more effort, although it does increase the cost.

Key factors for success

Gain senior management commitment

Strong senior management commitment is essential for the successful implementation and operation of an EMS. The benefits and aims of the EMS should be explained to senior managers before starting the implementation process. Convincing senior managers will require a project plan and a detailed estimate of the potential costs and the potential cost savings from adopting an EMS.

For details of EMAS see: ec.europa.eu/environment/emas. EMAS registration means that the company complies with all the requirements of ISO 14001 and the environmental verifier can issue an ISO 14001 certificate.

Certification and accreditation are not the same thing.

A company is certified, registered or approved by a 'Certified Body'. These will generally be 'for profit' companies and there will generally be many in each country.

A 'Certified Body' will be accredited to issue certification by the national 'Accreditation Body'. These will generally be national bodies and will be backed by the government. There will normally only be one 'Accreditation Body' in each country and they agree to accept each other's judgements and accreditations.

Build on existing systems

As shown in Section 2.1, there are many links between existing quality, health and safety and other MSS developed to meet the Annex L structure. Using these links and processes from other MSS can reduce the effort needed to implement an EMS.

- **Tip** – It is environmentally good to re-use, so do it with procedures and processes that are common across the standards.
- **Tip** – There is no requirement to have separate systems or documents for common areas. If your systems meet one MSS then they will meet another.

Getting certified

To be ready for certification to ISO 14001, an EMS should have been fully operational for at least three months and at least one management review should have been conducted. For initial registration, participants need to have a fully operational EMS with an audit programme already in place and to produce an initial and validated Environmental Statement.

Many companies use the same certification body for their EMS as for their QMS. However, it is important to check that the auditor is also accredited for ISO 14001 certification.

- **Tip** – Check that the proposed auditor has relevant experience in the plastics industry.

Auditors use a range of methods for certification. Be sure to understand the different stages of the proposed process and what the auditor will be looking for at each stage. Ask the chosen auditor to run through the process of certification.

Before the auditor visits for the first time, hold a meeting to ensure everyone knows about the certification and what it will entail.

An 'Initial Review' (see Section 2.4) will help to gather the data that will give a 'snapshot' of the environmental status. Regular reviews will help to quantify the savings made and maintain the momentum.

Formal certification of an EMS is a significant milestone but not the end of the journey. Every EMS needs continued attention to deliver further improvement and savings. This must be appreciated by senior managers – otherwise the initial enthusiasm for the EMS may decline after certification is achieved.

What to do next

Implementing an EMS with a focus on waste minimisation and continual improvement will reduce costs and improve environmental performance. The basic practical steps in implementing an EMS are:

- Understand the main elements of an EMS and become familiar with the standard's requirements.
- Appoint someone to manage implementation and operation.
- Develop an environmental policy.
- Identify the company's environmental aspects, evaluate their significance and draw up a register of significant aspects (see Section 2.4).
- Identify legislative requirements and draw up a Register of Legislation.
- Set objectives and targets.
- Assign responsibility.
- Develop employee awareness and conduct training.
- Prepare procedures to deliver operational and documented information control.
- Regularly monitor and measure significant aspects, e.g., waste, water and energy.
- Develop an internal audit mechanism and timetable.
- Review progress and, if necessary, revise the policy, objectives and targets.

In the UK, Envirowise Programme produced excellent information on environmental management but sadly these are no longer available.

The best publication (GG251 – 'EMS for the plastics industry') is available as an archived copy from: www.tangram.co.uk/TI-EMS_for_the_Plastics_Industry.html.

This is dated (produced in 2000 and pre-Annex L) but is still a good free resource for the plastics industry.

Before starting ISO 14001, ask yourself the critical question 'Did ISO 9001 actually deliver quality to our company?'

If the answer is 'No' then ISO 14001 will probably not deliver environmental management to your company.

The benefits of an EMS

Management and financial

- Structured approach to environmental issues.
- Keeping ahead of environmental legislation.
- Identification and reduction of waste.
- Increased profits.
- Reduced risk of fines.
- Reduced insurance premiums.

Productivity

- Improved process control.
- Reduced use of raw materials and consumables.
- Reduced waste and rejects.

Sales and marketing

- Improved products.

Public relations

- Improved community relations and public image.

Personnel and training

- Improved working environment.
- Reduced potential for environmental incidents.
- Improved employee motivation.

2.4 Environmental management systems – starting out

Implementation planning

Planning and organisation are vital to successfully implementing an EMS. It is important to involve a range of people in implementation, particularly when the EMS overlaps with their normal roles or functions, as it will normally do.

A formal implementation project team (see Section 1.18) will help to keep the EMS on track and identify and remove obstacles to progress. The team should include representatives from:

- Top management.
- Production.
- Quality.
- Environmental/health and safety.

Representatives from the procurement, finance and personnel departments may also need to be involved from time to time.

- **Tip** – An EMS 'champion' should be made responsible for implementing the EMS and coordinating the efforts of the implementation team.

The team should agree a common and collaborative approach and share out the work. To ensure progress is made, it is essential that team members be allocated enough time and resources.

The team should meet regularly – perhaps fortnightly – with adequate administrative support to ensure minutes are taken and, most importantly, action plans are updated.

To keep the whole project on track, the EMS champion should review any action plans weekly. The EMS champion may also find it useful to set up separate teams to tackle specific issues such as energy efficiency (see Chapter 6), water use (see Chapter 8), waste minimisation and packaging use (see Chapter 9). These teams should always involve employees from all levels of the business.

- **Tip** – Use a top manager to steer the team. This will facilitate progress and give good communication with top management.

Timescale for implementation

Implementing an EMS normally takes around 12-18 months but can be shorter if the company already has a similar MSS, e.g., ISO 9001, because of the ability to re-use procedures and documents. Where customers are demanding an EMS, they will often accept a reasonable timescale provided it is accompanied by a good, realistic implementation plan.

Initial review of operations

An Initial Review will help to assess how the company operations affect the environment and will provide benchmark data to help achieve continual improvement.

- **Tip** – ISO 14001 does not require a formal Initial Review, but it does require an assessment of environmental issues and impacts whereas EMAS requires a formal Initial Review.

Carrying out an Initial Review will help to:

- Gain an overview of the company attitude to waste and the environment.
- Prepare/revise the environmental policy (see below).
- Identify the environmental aspects of activities and their impacts.
- Assess relevant legislation.
- Identify opportunities for improvement.
- Set objectives and targets.

The main tasks in an Initial Review are data gathering and analysis. Checklists and worksheets provided in GG251 (see sidebar in Section 2.3) can be used to identify and locate the documents needed. Information will need to be collected about:

- The site and its environmental history.
- Raw material consumption and storage.
- Utility consumption and costs.
- Solid waste amounts and management.
- Emissions to atmosphere.

The environmental policy

After completing an Initial Review, it is possible to write an effective environmental policy. The policy must be reasonable, practical and match the business needs. The policy may commit the company to different management approaches and both customers and members of the public may want to see it. It should be reviewed regularly and, if necessary, revised to take account of developments in the EMS.

The policy should refer to the aims for significant environmental aspects, refer to continual improvement (through objectives and targets) and compliance with legislation. It could also refer to:

- Training and awareness for employees.
- Working with the supply chain.
- Planning for emergencies.
- Relations with neighbours and regulators.
- Sustainability in the broader context.

Project manager

Implementing an EMS crosses company boundaries. You will need a Project Manager or EMS Champion to successfully complete the project.

The documents gathered for the Initial Review should be filed for future reference.

They can also be useful for assessing materiality in reporting terms (see Section 12.4).

The environmental policy should be signed and dated by a senior manager and made available to all employees, customers and other stakeholders.

Objectives and targets

Setting objectives and realistic targets is the best way to achieve continual improvement and maximum savings from an EMS.

Objectives

These should aim to give improvements in:

- The significant environmental aspects.
- The environmental policy.
- Technical options.
- Financial, operational and other business requirements.

Targets

Targets should always be SMART (Specific, Measurable, Achievable, Relevant and Time-limited) and can be one of three types:

- **Measurement** – improvement targets cannot be set without base-line measurements.
- **Improvement** – measuring an aspect and then identifying the scope for improvement allows improvement targets to be set (quantify the cost/benefits for senior managers).
- **Control** – after improvements have been made, control targets can be used to 'hold the gains'.

Owners should always be identified for targets to ensure that the workload is shared out, that individuals are clearly responsible for different issues and that they know where to focus their efforts.

Note: Objectives and targets must be set for continual improvement.

- **Tip** – Set a target for carbon footprint reduction as an overall objective but make it SMART (see Section 7.1).

Legal requirements and EMS

Compliance with the law is a key part of any EMS and appropriate controls are needed to be sure of full compliance. It is necessary to:

- Identify a source of guidance to all environmental legislation.
- Identify the legislation relevant to the site and operations.
- Get copies of the Acts, Regulations or Codes of Practice as necessary.
- List the appropriate legislation and how it applies to the site (the compliance obligations).

The method of identifying the legal requirements should be part of a procedure within the EMS. This procedure should require at least an annual review/update of compliance obligations and the review should be linked to an annual assessment of compliance. When the compliance obligations are updated, key changes should be summarised and relevant employees should be notified. If you are not sure which legislation, regulations and codes of practice apply to your site, you should seek specialist advice.

Checklist:

- An environmental policy.
- A process for setting objectives and targets.
- Written objectives and targets.
- Records of previous objectives and targets and a summary report of performance.
- A process to identify current and future compliance requirements for environment-related activities.
- A list of compliance obligations and proof of updating.
- All permits, authorisations, etc., required by legislation.

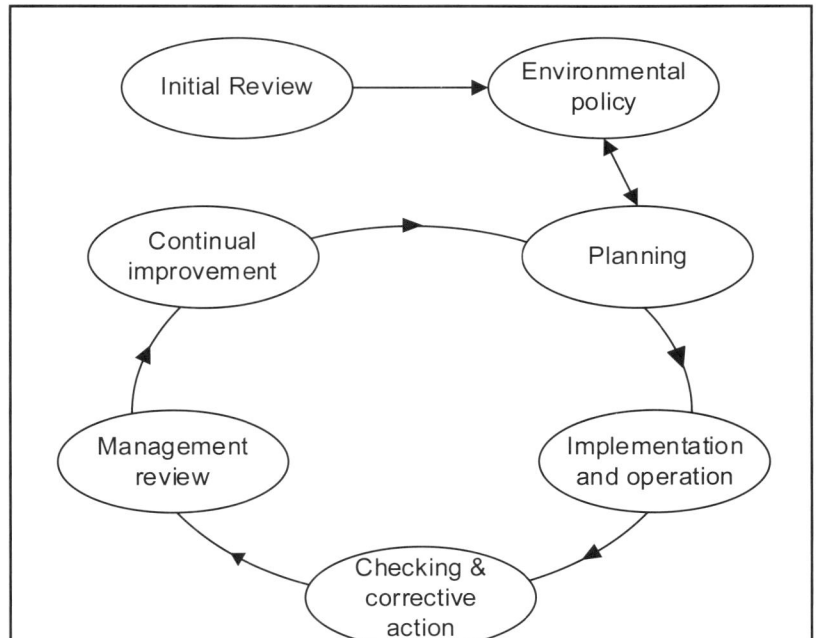

Starting out with an EMS

Implementing an EMS follows a simple process that is repeated to continuously improve performance, reduce waste and improve profits. This is the PDCA cycle applied to environmental issues and PDCA is explicitly referred to in ISO 14001.

2.5 Environmental management systems – interactions and risk

What matters?

One of the most difficult things in implementing an EMS to ISO 14001 or EMAS is understanding the concepts of 'aspects' and 'impacts' and how these work.

Identifying and understanding interactions with the environment is generally considered in terms of 'aspects' and 'impacts'. Aspects are the cause of an environmental 'impact' or effect. Environmental aspects may also include measures you have already taken to prevent or reduce pollution.

Note: Planning for aspects and impacts and compliance obligations is required and all documented information should be retained.

The process matters

The easiest way to start to assess aspects and impacts is to produce a process flow chart for each of the main processes. This is very similar to the process flow chart used in Section 8.3 for waste minimisation. Process flow charts should be created for each activity on the site, e.g., manufacturing, utilities, stores, maintenance and office processes. It is important to consider all emissions to air, water, and land (as waste or through spills) in the initial process flow charts – however small they may be.

- **Tip** – If you have created process maps for ISO 9001 (or any of the many purposes for which they are vital) these can be re-used as the basis for the ISO 14001 process flow charts.

- **Tip** – An aspect may later be discounted as very low risk but it should be in the plans to show that it has been considered.

- **Tip** – Do not focus only on normal operations, i.e., also consider what happens under abnormal situations such as start-ups, shutdowns and cleaning, as well as the potential for incidents and accidents.

Remember to include:

- Non-core processes.
- Refrigerants in cooling and air conditioning.
- PCBs in electrical transformers.

Normally these will not escape into the environment, but the EMS should have procedures for dealing with them during maintenance and final disposal.

Identifying aspects

From the process flow charts, decide which inputs and outputs may interact with the environment. These are the environmental aspects.

- **Tip** – Do not focus only on those aspects which are covered by legislation. It may not be covered by legislation but it may be significant in impact terms.

Identifying impacts

Impacts cannot be directly controlled – they are generated by the previously identified aspects. An aspect can generate more than one impact and many aspects have indirect impacts. Electricity use (an aspect) has three indirect impacts, i.e., potential climate change from CO_2 emissions, air pollution from acid gas emissions and resource depletion through fossil fuel use. Think beyond the obvious at the initial stages.

Compiling a list of environmental aspects and impacts and assessing their significance is often the most difficult stage of implementing an EMS.

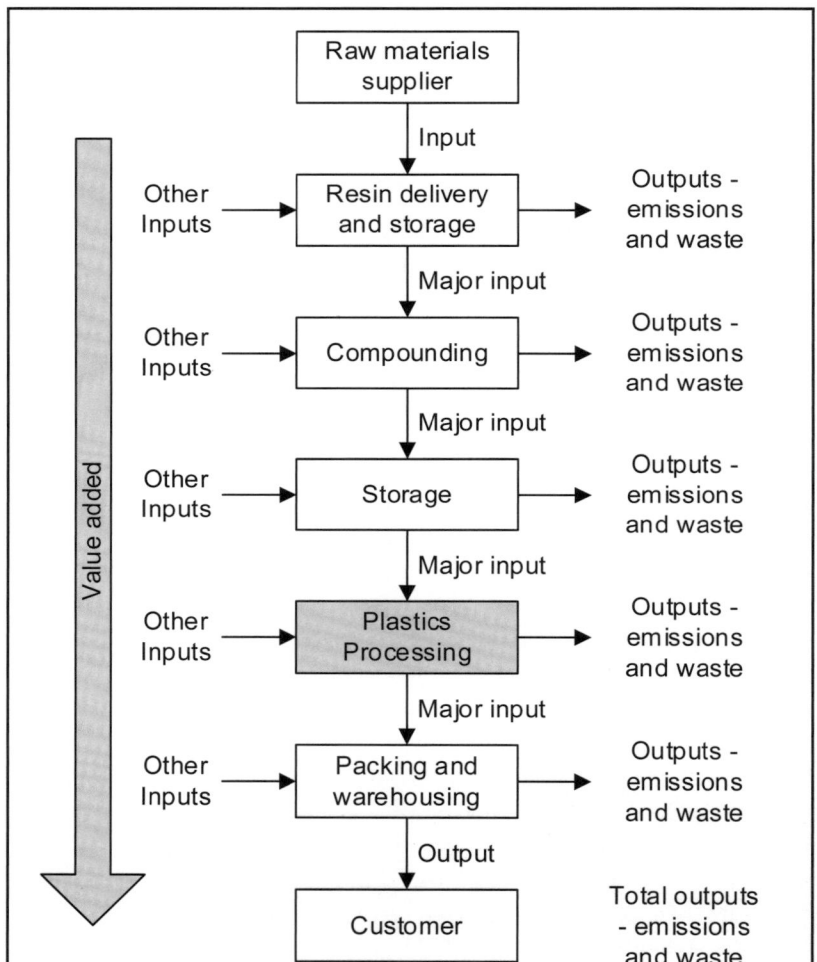

A general process flow chart for plastics processing

The process flow chart gives an overview of the aspects and impacts from a process. This is very similar to the process flow chart shown in Section 8.3 and the general guidelines for process flow charts, e.g., ISO 9001, should be followed.

Assessing significance

Having identified the aspects and impacts, it is necessary to assess which of the aspects are 'significant' to the organisation so that these can be managed by the EMS. The Initial Review should have revealed which activities are covered by legislation and/or have a high potential cost. Improvements in these activities will have a high beneficial environmental effect and can significantly reduce costs.

Assessing significance through a formal procedure makes effective use of limited resources and avoids having to try to deal with all the potential impacts (including the insignificant ones).

ISO 14001 requires planning to address risks and opportunities of significant aspects and impacts. However, there is no requirement for a formal risk management process and the standard does not specify a method for assessing or quantifying these risks and opportunities, i.e., the company can select the method. Whichever method is chosen to assess significance, it should be an approach that is appropriate to the company. The keys to success are:

- A consistent approach that allows each issue to be clearly treated in the same way.
- An ability to demonstrate and justify the methodology used.
- A full systematic record for future reference of decisions taken.
- The use of criteria that provide a rational basis for the rest of the EMS.

Risk assessment method

One of the easiest methods is risk assessment through a formal FMEA method (see Sections 2.17 and 2.18). This approach uses proven risk assessment methods to predict the likelihood and severity of outcomes or events. This is similar to other risk assessment methods used for quality and health and safety management. In all these methods, ratings of severity, likelihood and detection are individually assessed and then combined to produce an overall assessment of the risk.

A risk factor rating is assigned to each potential impact after considering the following:

- Hazardous properties.
- Size.
- Frequency or likelihood of occurrence.
- Presence of sensitive environmental receptors, e.g., people, a watercourse and/or site of special scientific interest.
- Presence or absence of environmental controls, e.g., techniques designed to control or prevent the impact.

For each impact, decide the degree of severity (minor, moderate, major) and how likely it is to occur (unlikely, likely, very likely). A total risk assessment, the Risk Priority Number (RPN), is obtained by combining the severity of the consequences with the likelihood of occurrence for each impact. A numerical rating is given to each, with a higher RPN indicating a higher risk of adverse impact.

What is significant?

After assessing significance, an impact is considered significant if the score is above an internally set threshold value. It is up to the company to set the RPN threshold value over which impacts are considered significant but the reasons for the decision should be recorded.

- **Tip** – Full details of this process are given in Section 2.17.

Recording decisions

The reasons for all decisions should be recorded in a systematic manner for future reference and examination by auditors.

Procedure for evaluating significance

The procedure and output used to identify aspects and assess them for significance must be recorded and produce consistent results for each site.

Risks and opportunities

The collection of lists of environmental aspects, their impacts and an evaluation of their significance makes up the 'documented information' on risks and opportunities. This should give details of the company's environmental aspects, together with an analysis of their impacts. It should indicate whether an aspect is considered significant and how significant environmental aspects are linked to the EMS.

The documented information should be regularly reviewed and updated to take account of any changes in legislation or operations.

Checklist:
- Process flow charts and evaluation tables (the proof of the process).
- A method of identifying environmental aspects and evaluating those with significant environmental impacts.
- A record of the environmental risks and opportunities.

Process flow charts for the EMS can also be used for waste minimisation (see Section 9.3) to reduce the cost of waste.

They can also be used for quality and energy management to highlight losses to the system.

Understanding the flows in a process is the key to improving performance.

Remember to assess new projects according to the chosen method and to link the evaluation procedure to the capital expenditure application and authorisation process.

2.6 Environmental management systems – ISO 14001

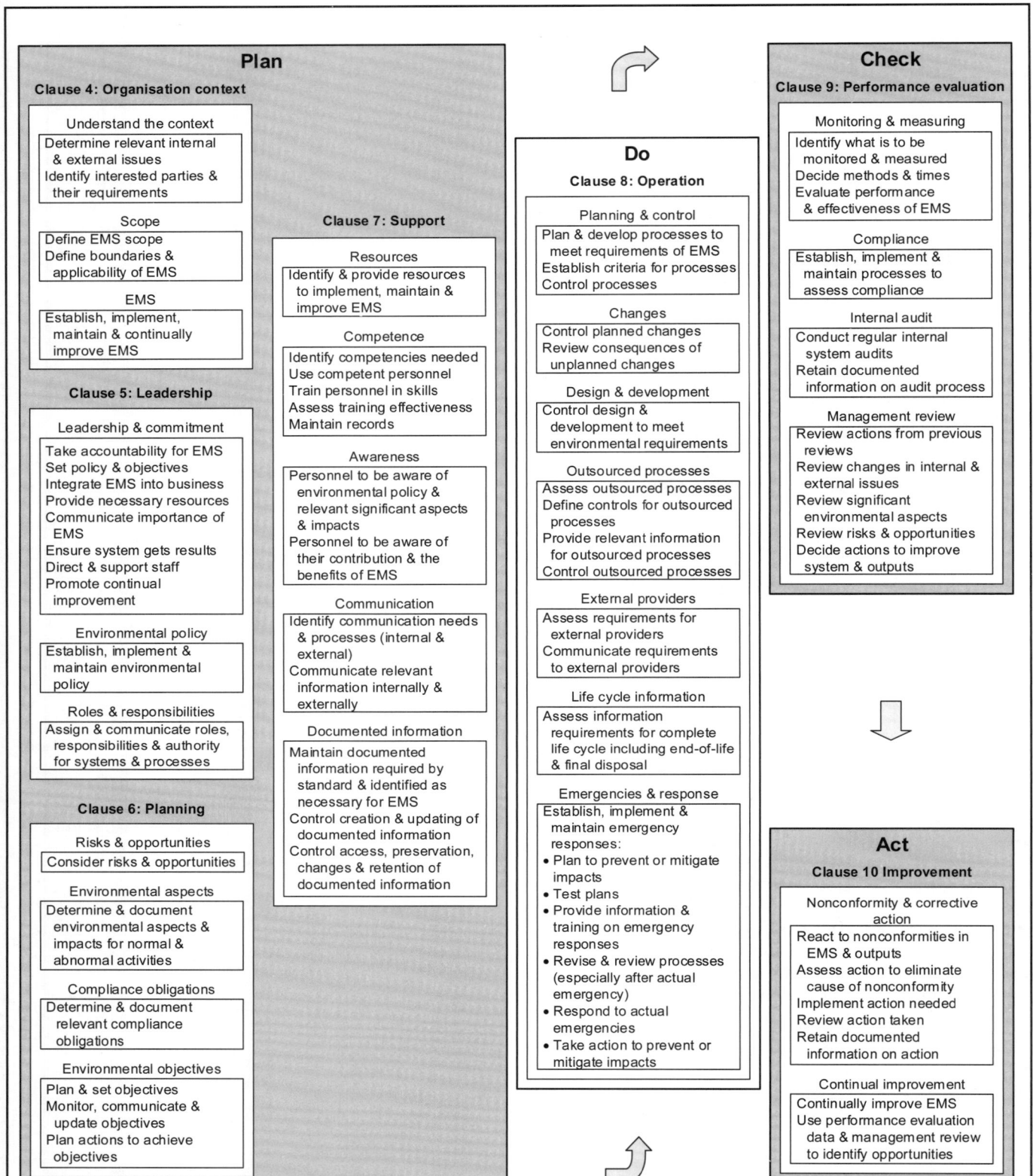

Plan

Clause 4: Organisation context

Understand the context
- Determine relevant internal & external issues
- Identify interested parties & their requirements

Scope
- Define EMS scope
- Define boundaries & applicability of EMS

EMS
- Establish, implement, maintain & continually improve EMS

Clause 5: Leadership

Leadership & commitment
- Take accountability for EMS
- Set policy & objectives
- Integrate EMS into business
- Provide necessary resources
- Communicate importance of EMS
- Ensure system gets results
- Direct & support staff
- Promote continual improvement

Environmental policy
- Establish, implement & maintain environmental policy

Roles & responsibilities
- Assign & communicate roles, responsibilities & authority for systems & processes

Clause 6: Planning

Risks & opportunities
- Consider risks & opportunities

Environmental aspects
- Determine & document environmental aspects & impacts for normal & abnormal activities

Compliance obligations
- Determine & document relevant compliance obligations

Environmental objectives
- Plan & set objectives
- Monitor, communicate & update objectives
- Plan actions to achieve objectives

Clause 7: Support

Resources
- Identify & provide resources to implement, maintain & improve EMS

Competence
- Identify competencies needed
- Use competent personnel
- Train personnel in skills
- Assess training effectiveness
- Maintain records

Awareness
- Personnel to be aware of environmental policy & relevant significant aspects & impacts
- Personnel to be aware of their contribution & the benefits of EMS

Communication
- Identify communication needs & processes (internal & external)
- Communicate relevant information internally & externally

Documented information
- Maintain documented information required by standard & identified as necessary for EMS
- Control creation & updating of documented information
- Control access, preservation, changes & retention of documented information

Do

Clause 8: Operation

Planning & control
- Plan & develop processes to meet requirements of EMS
- Establish criteria for processes
- Control processes

Changes
- Control planned changes
- Review consequences of unplanned changes

Design & development
- Control design & development to meet environmental requirements

Outsourced processes
- Assess outsourced processes
- Define controls for outsourced processes
- Provide relevant information for outsourced processes
- Control outsourced processes

External providers
- Assess requirements for external providers
- Communicate requirements to external providers

Life cycle information
- Assess information requirements for complete life cycle including end-of-life & final disposal

Emergencies & response
- Establish, implement & maintain emergency responses:
  - Plan to prevent or mitigate impacts
  - Test plans
  - Provide information & training on emergency responses
  - Revise & review processes (especially after actual emergency)
  - Respond to actual emergencies
  - Take action to prevent or mitigate impacts

Check

Clause 9: Performance evaluation

Monitoring & measuring
- Identify what is to be monitored & measured
- Decide methods & times
- Evaluate performance & effectiveness of EMS

Compliance
- Establish, implement & maintain processes to assess compliance

Internal audit
- Conduct regular internal system audits
- Retain documented information on audit process

Management review
- Review actions from previous reviews
- Review changes in internal & external issues
- Review significant environmental aspects
- Review risks & opportunities
- Decide actions to improve system & outputs

Act

Clause 10 Improvement

Nonconformity & corrective action
- React to nonconformities in EMS & outputs
- Assess action to eliminate cause of nonconformity
- Implement action needed
- Review action taken
- Retain documented information on action

Continual improvement
- Continually improve EMS
- Use performance evaluation data & management review to identify opportunities

An overview of the ISO 14001 requirements

The initial issues of ISO 14001 were daunting to many in the plastics industry but, after the introduction of Annex L, the close correspondence of all the ISO MSS means that much of the language and many of the requirements are now common between the various MSS. This diagram shows the main activities that must be carried out to conform to ISO 14001.

It is not that difficult

An overview of ISO 14001 is shown on the left and the requirements are relatively straightforward for anybody who has already dealt with ISO 9001. There is, however, certain information that is required to prove conformance with the standard and this is referred to in the standard as a requirement for 'documented information'. Some of these specific requirements and a list of 'good practice' information are given in the box on the right. If you have all of this information then proving compliance is not difficult, getting the information may well be.

Monitoring and measuring

ISO 14001 is not simply intended to monitor aspects and impacts or to provide a framework for emergency responses. It is intended to provide a framework for improving environmental performance and this needs monitoring and measuring.

The EMS should include processes for monitoring and measuring the critical factors for the significant aspects identified by the site. Obtaining reliable and effective data is the key to generating information that can be used for management action. Data collection and analysis is a vital tool in reducing resource use, minimising waste and improving sustainability.

Although ISO 14001 does not specify the frequency of monitoring and measuring it is unlikely that anything longer than annual measurement would be acceptable for significant aspects or impacts. More frequent measurement is often necessary to identify variations and opportunities to reduce environmental impacts and costs. The sooner corrective action is taken, the less impacts are created and the more cost savings will be achieved. The measurements can be used in the management review, displayed internally to report success and used in the full Sustainability Report (see Chapter 12 for full details of reporting sustainability).

• **Tip** – The sustainability report can also link to social responsibility issues (see Section 11.6).

Typical parameters that could be measured include:

• Production levels.
• Waste generated.
• Water use.
• Energy use.
• Emissions to air.

Waste and utility data should always be related to a measure of production, e.g., tonnes of waste per tonne of product or tonnes of waste per number of units. Some data will be affected by a 'base load' and may need more sophisticated data analysis. This is the case for utilities data (see Section 6.4).

• **Tip** – Calibrate all measuring equipment.

What you will need for ISO 14001:

4 – Context of the organisation
■ An assessment of the internal and external issues and records of the scope and boundaries of the EMS.

5 – Leadership
■ Demonstration of leadership and commitment to the EMS.
■ An environmental policy that is communicated to the staff and available to interested parties.

6 – Planning
■ An assessment of the risks and opportunities to the company.
■ An assessment of the environmental aspects and impacts and their significance.
■ A compliance assessment process for environmental aspects.
■ Plans to reduce aspects, improve compliance and reduce risks.

7 – Support
■ A process to assess competencies, deliver training to meet competence needs and assess effectiveness.
■ Processes for internal and external communications.
■ Records of internal and external communications.
■ A process for control of documented information.

8 – Operation
■ Planning and processes to control the environmental impact at all stages of the product life cycle.
■ Planning and processes for emergency responses to reduce environmental impacts of all situations where their absence could lead to adverse impacts.
■ Proof of tests of emergency response processes.

9 – Performance evaluation
■ Processes to measure significant environmental aspects such as raw material use, solid waste, water use, releases to water/sewer, emissions to air, energy use, etc.
■ Assessment of the company's environmental performance.
■ An audit process, programme and audit reports.
■ A management review process including management review agenda and minutes of management review meetings.

10 – Improvement
■ Processes for non-conformance and corrective/preventative action.
■ Recording of non-conformances and corrective/preventative action, i.e., reports on follow-up actions.

The broad ISO 14001 requirements

ISO 14001 requires 'documented information' in several areas to prove compliance with the standard. Search the standard for the phrase 'documented information', anything listed is an absolute requirement for conformance.

2.7 Environmental management systems where are you now?

Clean, green and low cost

Environmental management is becoming more and more important as customers and legislators demand improvements in environmental performance. Major customers are already signalling that they see environmental performance and sustainability as key factors in where they place their business.

Some companies see this as a negative and fail to see that good environmental management can not only achieve and reduce the costs of meeting these demands but also reduce overall costs by reducing waste and improving operations. Forward-looking companies also see the considerable PR and other benefits of improving their environmental performance and being able to promote this (without resorting to greenwashing). Simple environmental measurements such as the carbon footprint (see Section 7.1) are already being used to report performance on a wider scale and companies need to be ready for these changes.

Completing the chart

This chart is completed and assessed as for the previous charts.

Sustainability

A common and growing theme of successful businesses is having good environmental and sustainability credentials.

An EMS can:

- Improve sustainability.
- Improve environmental performance.
- Be installed alongside a conventional QMS for relatively low cost.
- Deliver a wide range of benefits.
- BUT ....

It must be administered properly and effectively!

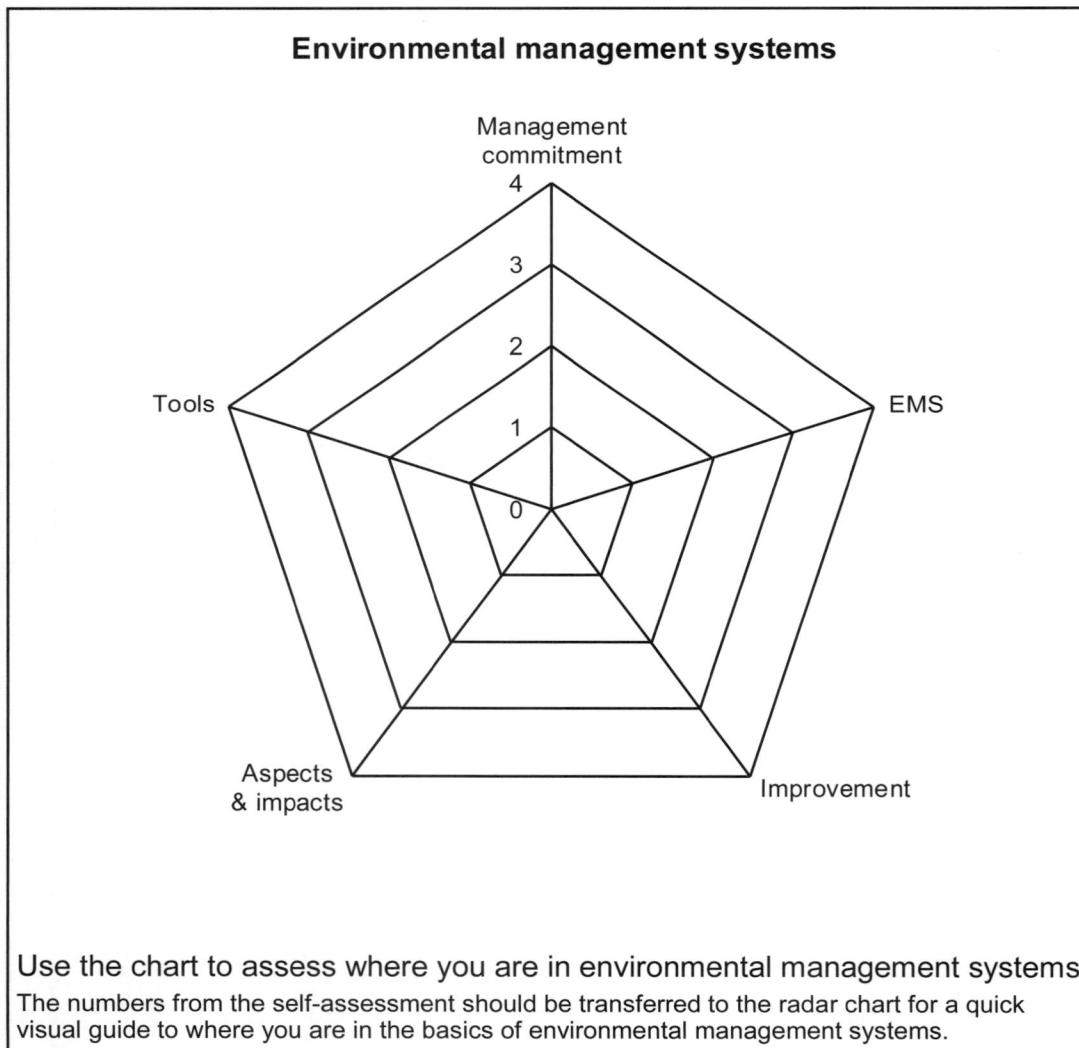

Environmental management systems

Use the chart to assess where you are in environmental management systems

The numbers from the self-assessment should be transferred to the radar chart for a quick visual guide to where you are in the basics of environmental management systems.

If you think that this is an idle academic exercise then reflect on what happened to the market share, share price and senior executives at BP as a result of the Gulf of Mexico oil spill.

Failing to get environmental management right can have severe consequences.

		Environmental management systems			
Level	Management commitment	EMS	Improvement	Aspects & impacts	Tools
4	Management is totally committed. Environmental policy is integral part of business, all resources provided, staff trained & have delegated authority.	Formal EMS in place with full external verification of system. No major non-conformances found in last 2 years.	Environmental improvement is a fundamental business goal. Improvement techniques used whether concerns present or not.	Full aspects & impacts assessment carried out. Active efforts to reduce major aspects & impacts.	Full range of improvement tools used to identify concerns, to determine root causes & to assess rectification actions.
3	Management has moderate commitment. Majority of requirements are in place but enforcement is sporadic.	Formal EMS in place with full external verification of system. No major non-conformances found in last year.	Environmental improvement is an important business goal. Improvement techniques only used when concerns are present & visible.	Partial aspects & impacts assessment carried out. Some efforts to reduce major aspects & impacts.	Good knowledge & use of improvement tools in environmental analysis & problem solving.
2	Management has low commitment & only really involved when problems occur. Basic requirements are in place but not enforced.	Formal EMS in place with full external verification of system. Significant major non-conformances found in last year.	Environmental improvement is a minor goal. Improvement techniques sometimes used when concerns are present & visible.	No formal aspects & impacts assessment carried out. Some efforts to reduce main perceived aspects & impacts.	Some knowledge of improvement tools & often used for analysis. Problems often solved but key concerns remain unsolved & reappear.
1	Management not committed. Some aspects of environmental management are in place due to middle management dedication but few resources available.	Formal EMS in place but no external verification of system.	Environmental improvement is not seen as a goal. Improvement techniques not used even when concerns are present & visible.	No formal aspects & impacts assessment carried out. Some efforts to reduce visible aspects & impacts (but possibly misdirected).	Little knowledge of improvement tools & rarely used. When used they are not fully followed through to completion. Same concerns return time & again.
0	Management not committed. No environmental policy, no resources, no training & no delegated authority.	No formal EMS in place.	Environmental improvement is not seen as a goal. Getting the product out the door is the only goal.	No concept of aspects & impacts of operations.	No knowledge or use of improvement tools.
Score					

2.8   Energy management systems – the basics

Setting the structure

One of the major benefits of an energy management system (EnMS) is that it encourages setting up a permanent structure to manage energy use.

In many companies, energy becomes a 'flavour of the month' when prices rise and then fades in importance as the site and management become used to the increased costs. This leads to random initiatives to reduce energy use but no long-lasting energy use reduction programme or focus. An EnMS provides the structure and processes for long-term work and long-term energy use and cost reductions.

The requirements for an EnMS are quite straightforward:

Policy

Every site needs an energy policy that gives:

- A detailed statement of commitment from top management.
- Performance targets from internal or external benchmarking.
- Short-term site goals (1 year).
- Medium-term objectives (3 years).
- Long-term corporate goals (5 years).

This policy should be provided to all staff to raise awareness of energy use and the environmental and financial benefits of improved energy management. The policy is the framework for developing and establishing good energy management practice and should establish the importance of energy management to all staff levels.

Personnel

Operating the energy management system should be the clear responsibility of a nominated person, the 'energy manager', who will act as the facilitator, expert adviser and project manager for energy management projects.

- **Tip** – ISO 50001 (see Section 2.9) refers to 'energy management team' in terms of responsibilities and actions. We prefer to have a nominated 'energy manager' so that we know who is in charge.
- **Tip** – This may not be a full-time role but the responsibility and authority should be clear.
- **Tip** – For the same reasons that the quality manager is never the production manager, it is best if the energy manager is not the person responsible for using the energy, i.e., the production manager.

Information

Any management system needs information to operate and energy management is no different. Much of the required data are probably already being collected in various places and by various people at most sites and this simply needs to be collected and formatted so that it is easily used for energy management.

Information is needed to allow for monitoring, targeting and reporting. This is generally easily available but systems are needed that will regularly and automatically generate the information. The system should be target based and allow cost allocation and performance assessment. This will drive responsibility down to those who actually use the energy.

- **Tip** – Targeting is the key action – what gets measured gets done.

Information is also needed for project identification. At most sites, the opportunities to reduce energy use are greater than the resources available. Actions must therefore be prioritised for the quickest and easiest return. Potential projects should be assessed in terms of the ease and cost of implementation and the size of the potential energy use reduction. Projects should be prioritised for the maximum effect – there is no shame in starting with the 'low hanging fruit', in fact, it is the best way to generate enthusiasm for the whole process of energy management.

Planning

Project planning is needed even for simple projects. This should identify timings, resources and benefits to be achieved. The system must include a planning process, even at a basic level, so that projects are delivered as required.

Resources

An energy management project is the same as any other project and planning will identify the resources needed (time and/or money) for the project to be completed. Sites need to allocate the necessary resources to achieve the desired results.

- **Tip** – Attempting to manage energy without allocating any resources is doomed to failure.

Energy management will almost certainly need more resources than currently provided at most sites. However, given the magnitude of the rewards this can easily be financially justified.

Set the energy management structure …

Then use the structure to set targets and implement projects to achieve the targets.

Before starting ISO 50001, ask yourself the critical question 'Did ISO 9001 actually deliver quality to our company?'

If the answer is 'No' then ISO 50001 will probably not deliver energy management to your company.

Training

Staff training is essential in implementing an energy management system. The policy and other actions will have little effect unless the staff know the targets, the action being taken and how they can contribute.

Staff training may be limited due to the automated nature of most plastics processing and the lack of ability of the staff to affect the outcomes. Training should be no longer than 40 minutes for existing staff and should be integrated with other induction procedures for new staff.

- **Tip** – Training sessions should be followed by a 'go-see' exercise where the staff go to their area and identify energy improvements.
- **Tip** – This training is for the general staff, there will be additional training needed for staff who are needed to identify opportunities and implement them.

Auditing

Any system needs auditing to ensure that progress continues to be made and that the system continues to operate. This is best achieved by regular site audits by site staff. This will ensure that the system is operating and will discover new areas for improvement. At least one person per site should be trained to a basic level in energy auditing to allow regular audits to be carried out. Internal auditors are carrying out two functions:

- They are checking that the system is functioning as it was designed to function.
- They are looking for new profitable projects to implement.

Energy management is not a single task but a series of continuing actions; only continuous identification of new actions and resolving these issues will allow energy use and costs to be reduced.

- **Tip** – Site audits should be carried out using standard auditing methods. The audits should identify areas where actions do not meet the energy policy or where items from this workbook are noted. Site audits should also create non-conformance reports to initiate improvement actions – these must be closed out when completed. Non-conformance reports must lead to action.
- **Tip** – Unsurprisingly, there are also two ISO standards (ISO 50002 and ISO 50003) dealing with energy audits.

Reporting

An energy management system must also incorporate a reporting function. Reporting should be to both the management (to show the value of the work being done) and to the staff (to show the progress being made).

Reporting can also be to external stakeholders to show the corporate social responsibility of the site in reducing greenhouse gas emissions.

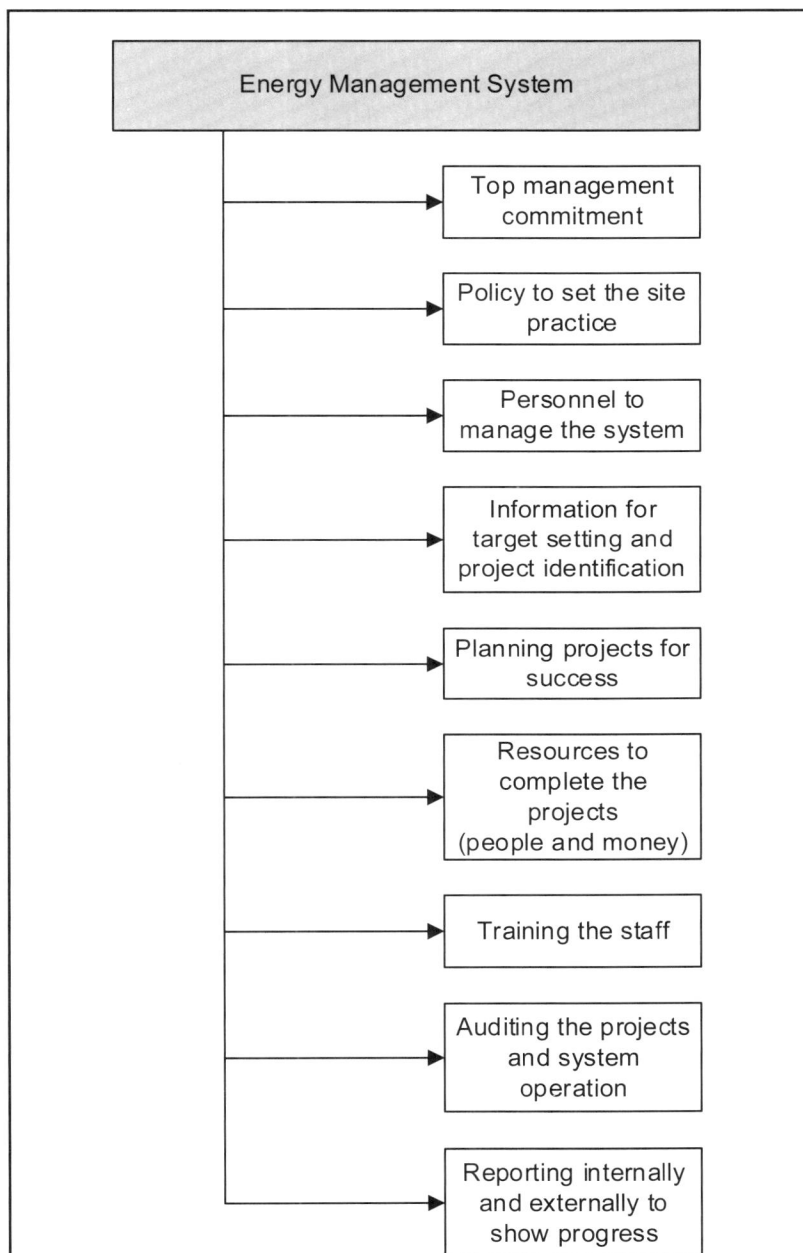

```
                 ┌─────────────────────────────────┐
                 │   Energy Management System      │
                 └─────────────────────────────────┘
                          ├──→  Top management
                          │     commitment
                          ├──→  Policy to set the site
                          │     practice
                          ├──→  Personnel to
                          │     manage the system
                          ├──→  Information for
                          │     target setting and
                          │     project identification
                          ├──→  Planning projects for
                          │     success
                          ├──→  Resources to
                          │     complete the
                          │     projects
                          │     (people and money)
                          ├──→  Training the staff
                          ├──→  Auditing the projects
                          │     and system
                          │     operation
                          └──→  Reporting internally
                                and externally to
                                show progress
```

The components of an EnMS

An EnMS is needed to identify, plan and complete projects to deliver energy savings and to report on these to top management. Without a system, the energy management activities will be poorly coordinated and will eventually fall into disrepute.

2.9　Energy management systems – the standard

ISO 50001 – the MSS for energy management systems

ISO 50001 was first released as an MSS in 2011 to provide organisations with guidance on implementing energy management. The standard was originally written in the 'old' management systems standard format. The introduction of Annex SL in 2015 (renamed to Annex L in the 10th edition in 2019) meant that the standard had to be revised to bring it into line with other existing MSS. This was completed in 2018 as part of the standard ISO review process and ISI 50001:2018 now fully conforms to Annex L.

As with the other MSS based on Annex L, ISO 50001: 2018 is based on the 'Plan-Do-Check-Act' model and the common format increased the range of activities that were in common with existing quality and environmental management systems. This allows companies to make use of many existing structures, documents and processes and allows energy management to be seen as part of the normal business processes.

Despite the common format, ISO 50001 is very different to the other MSS in that it is much more 'data driven' and quantitative. MSS, in general, are designed for use by a wide range of organisations and the metrics used are also quite wide ranging, i.e., they are specific to the organisation. In contrast, ISO 50001, whilst still trying to be general, is much more specific about the data that it requires and the requisite data analysis. It demands analysis of the variables and the current energy performance for each significant energy use (SEU). This requires the use of techniques such as energy mapping (see Section 6.2) to identify the significant energy users and statistical techniques to determine the current energy performance.

- **Tip** – Energy management is always data-driven and good data handling skills are needed to get the best out of any EnMS.

- **Tip** – The requirement to identify SEU will automatically improve understanding of a site's energy use. Knowing what uses energy is one of the first steps in reducing energy use.

What does ISO 50001 do?

ISO 50001 provides a very useful broad framework of requirements and good practices that will enable plastics processors to:

- Develop an energy policy for more efficient use of energy use.

- Use risk-based thinking to assess risks and opportunities.
- Plan the implementation of a programme to meet the policy.
- Set targets and objectives to meet the policy.
- Review current energy use in the company.
- Set energy performance indicators and baselines relevant to the company.
- **Tip** – This is very useful in measuring performance improvements because a site knows where it is starting from and what is being used to measure performance but be sure to get the metrics right.
- Identify and prioritise energy saving opportunities.
- Plan data collection that will validate the targets and objectives and check the effectiveness of implemented opportunities.
- Use the collected data to give a better understanding of energy use and to make better decisions about energy use.
- Measure the results of implemented opportunities.
- Review the effectiveness of the energy policy.
- Continually improve energy management.

These are all good things and ISO 50001 can be extremely useful in the process.

- **Tip** – If a site has ISO 50001 certification in Europe then it automatically complies with the EU Article 8 Energy Efficiency Directive and does not have to have a separate audit. This directive is known by various names in the EU, e.g., in the UK it is ESOS (Energy Savings Opportunities Scheme).
- **Tip** – ISO 50001 is not designed to manage a site's carbon footprint (see Section 7.1). However, the information required for ISO 50001 will make the calculation easy.

What does ISO 50001 not do?

For all the good things that ISO 50001 does, it is designed as a broad framework for energy management that is equally applicable to all organisations. This means that ISO 50001:

- Does not provide guidance for any specific industry or sector, i.e., it is very general and provides no guidance for plastics processors or any other industry sector.
- Does not provide specific targets or benchmarks for any industry or sector. These are determined by the company and,

Choosing to obtain ISO 50001 may or may not reduce energy use but that is no reason not to start on the essential journey of reducing energy.

ISO 50001 requires actions to address risks and opportunities but the risk management requirements are far less stringent than in other management systems standards.

Risk-based thinking in relation to an EnMS is more related to how the system functions and how to avoid system failures rather than operational risks such as environmental incidents or workplace accidents.

A simple risk analysis similar to that for health and safety systems (see Section 2.19) for the system should be enough.

in our experience, many companies choose the wrong metrics (see Section 6.5).

- Does not provide techniques or tools to reduce energy use apart from the PDCA cycle. The decisions on these are left to the company to decide and implement.

- Does not provide any guidance on projects to actually reduce energy use.

- **Tip** – Do not think that implementing ISO 50001 will automatically reduce energy use. This needs projects to be identified, quantified and implemented. For further guidance on this area see Chapter 6 and the author's book on this subject.[1]

ISO 50001 and ISO 14001

The data driven aspect of ISO 50001 means that it is much more quantitative than ISO 14001. ISO 14001 can certainly consider energy use and performance as environmental aspects and impacts but this consideration tends to be in a more qualitative manner.

At most plastics processing sites, the major environmental aspect and impact is associated with energy use and ISO 50001 will provide a much more focused approach to environmental improvement. The fact that the cost of energy is also likely to be high for most plastics processors also means that any financial benefits of ISO 50001 are likely to be clearer.

There is nothing to stop sites implementing both ISO 14001 and ISO 50001. The additional implementation cost is not likely to be large but the additional certification costs may be a barrier.

- **Tip** – Some sites use ISO 14001 to manage the overall environmental issues and ISO 50001 for specific energy issues. If this is done then the integration of the MSS can be used to minimise costs.

Since the introduction of ISO 50001 in 2011, there has also been a range of new standards produced. These include:

- ISO 50002:2014 – Energy audits. Requirements with guidance for use.

- ISO 50003:2014 – Energy management systems. Requirements for bodies providing audit and certification of energy management systems.

- ISO 50004:2014 – Energy management systems. Guidance for the implementation, maintenance and improvement of an energy management system.

- ISO 50006:2014 – Energy management systems. Measuring energy performance using energy baselines (EnB) and energy performance indicators (EnPI). General principles and guidance.

- ISO 50007:2017 – Energy services.

Guidelines for the assessment and improvement of the energy service to users.

- ISO 50015:2014 – Energy management systems. Measurement and verification of energy performance of organizations. General principles and guidance

- ISO 50021:2019 – Energy management and energy savings. General guidelines for selecting energy savings evaluators

- ISO/TS 50044:2019 – Energy saving projects (EnSPs). Guidelines for economic and financial evaluation.

- ISO 50046:2019 – General methods for predicting energy savings.

- ISO 50047:2016 – Energy savings. Determination of energy savings in organizations.

These standards expand on the usefulness of ISO 50001 but again, all are written to be non-specific in terms of industry or sector.

What to do?

For plastics processors starting the journey towards energy management, ISO 50001 provides a good MSS but the lack of project guidance is one of the things that decreases the usefulness of ISO 50001 to plastics processors. Finding and choosing the right projects is an essential part of energy management – if the projects were that obvious then most processors would already have implemented them.

Actions and information for ISO 50001

- How much energy are we using?
- When are we using energy?
- Where are we using it?
- Who influences energy use?
- What are the energy use drivers?
- What is our energy baseline?
- What are our energy indicators?
- Are there any legal or other requirements?
- What are our objectives and targets?
- What is our action plan?
- Do we have the necessary resources and knowledge?

Free tools

The US DoE has some excellent tools and resources for implementing ISO 50001. These are available at navigator.lbl.gov.

- 1. Kent, R.J. 2018. 'Energy management in plastics processing', Elsevier.

2.10 Energy management systems – ISO 50001

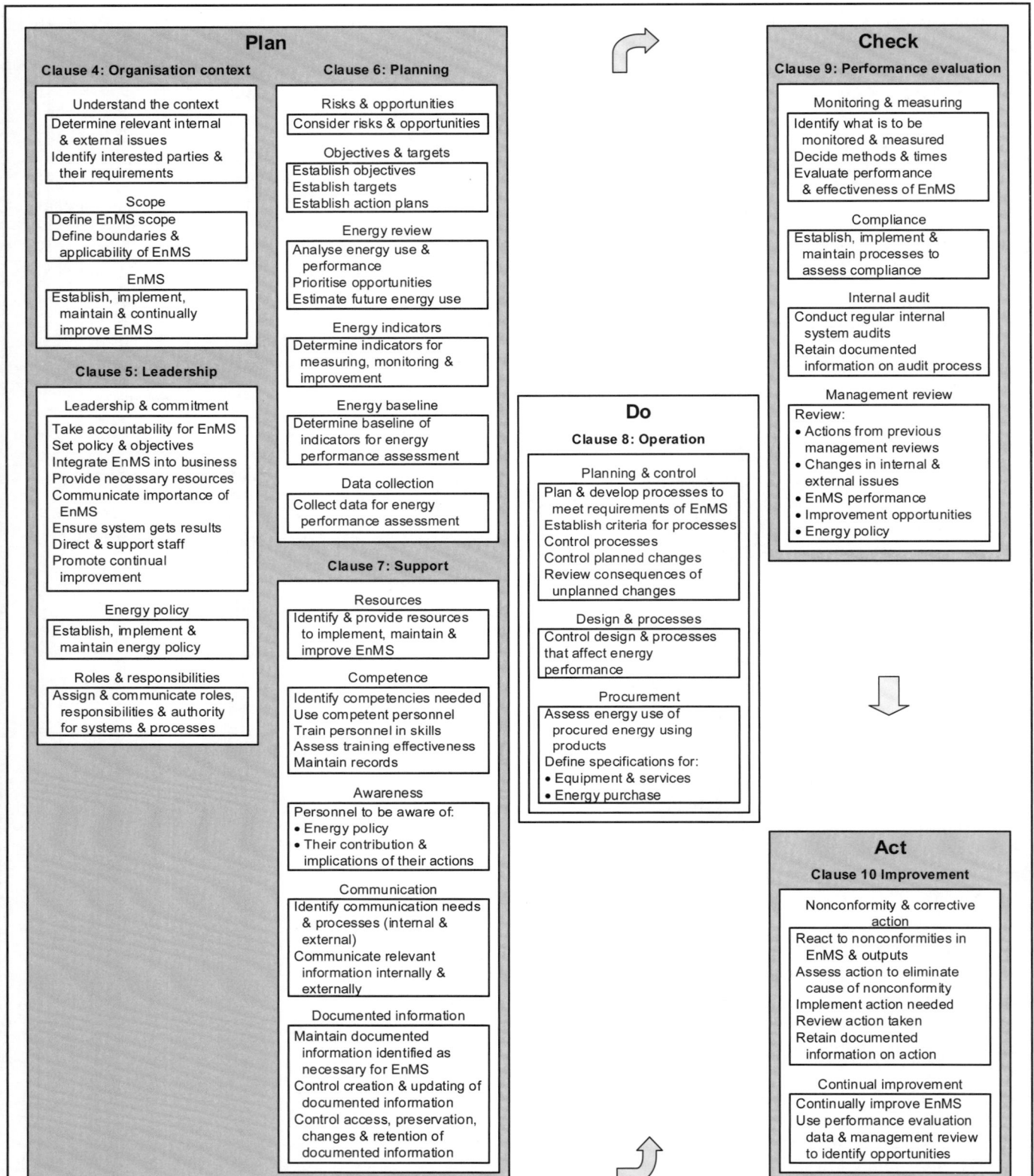

Plan

Clause 4: Organisation context

Understand the context
- Determine relevant internal & external issues
- Identify interested parties & their requirements

Scope
- Define EnMS scope
- Define boundaries & applicability of EnMS

EnMS
- Establish, implement, maintain & continually improve EnMS

Clause 5: Leadership

Leadership & commitment
- Take accountability for EnMS
- Set policy & objectives
- Integrate EnMS into business
- Provide necessary resources
- Communicate importance of EnMS
- Ensure system gets results
- Direct & support staff
- Promote continual improvement

Energy policy
- Establish, implement & maintain energy policy

Roles & responsibilities
- Assign & communicate roles, responsibilities & authority for systems & processes

Clause 6: Planning

Risks & opportunities
- Consider risks & opportunities

Objectives & targets
- Establish objectives
- Establish targets
- Establish action plans

Energy review
- Analyse energy use & performance
- Prioritise opportunities
- Estimate future energy use

Energy indicators
- Determine indicators for measuring, monitoring & improvement

Energy baseline
- Determine baseline of indicators for energy performance assessment

Data collection
- Collect data for energy performance assessment

Clause 7: Support

Resources
- Identify & provide resources to implement, maintain & improve EnMS

Competence
- Identify competencies needed
- Use competent personnel
- Train personnel in skills
- Assess training effectiveness
- Maintain records

Awareness
- Personnel to be aware of:
  - Energy policy
  - Their contribution & implications of their actions

Communication
- Identify communication needs & processes (internal & external)
- Communicate relevant information internally & externally

Documented information
- Maintain documented information identified as necessary for EnMS
- Control creation & updating of documented information
- Control access, preservation, changes & retention of documented information

Do

Clause 8: Operation

Planning & control
- Plan & develop processes to meet requirements of EnMS
- Establish criteria for processes
- Control processes
- Control planned changes
- Review consequences of unplanned changes

Design & processes
- Control design & processes that affect energy performance

Procurement
- Assess energy use of procured energy using products
- Define specifications for:
  - Equipment & services
  - Energy purchase

Check

Clause 9: Performance evaluation

Monitoring & measuring
- Identify what is to be monitored & measured
- Decide methods & times
- Evaluate performance & effectiveness of EnMS

Compliance
- Establish, implement & maintain processes to assess compliance

Internal audit
- Conduct regular internal system audits
- Retain documented information on audit process

Management review
- Review:
  - Actions from previous management reviews
  - Changes in internal & external issues
  - EnMS performance
  - Improvement opportunities
  - Energy policy

Act

Clause 10 Improvement

Nonconformity & corrective action
- React to nonconformities in EnMS & outputs
- Assess action to eliminate cause of nonconformity
- Implement action needed
- Review action taken
- Retain documented information on action

Continual improvement
- Continually improve EnMS
- Use performance evaluation data & management review to identify opportunities

An overview of the ISO 50001 requirements

The issue of ISO 50001:2018 means that it conforms to the Annex L structure for MSS and is now similar in structure to the other MSS such as quality (ISO 9001:2015), environment (ISO 14001:2015) and health and safety (ISO 45001:2018). This means that much of the language and many of the requirements are now common with the other MSS. This diagram shows the main activities that must be carried out to conform to the standard.

It is also not that difficult

An overview of ISO 50001 is shown on the left and the requirements are relatively straightforward for most companies. There is, however, certain information that is required to prove conformance with the standard and this is referred to in the standard as a requirement for 'documented information'. Some of these specific requirements and a list of 'good practice' information are given in the box on the right. If you have all this information then proving compliance is not difficult.

Monitoring and measuring

ISO 50001 is not simply intended to monitor energy use. It is intended to provide a framework for improving energy performance and this needs monitoring and measuring.

The EnMS should include processes for monitoring and measuring the energy use and energy performance and this is one of the areas where companies without an energy management system perform poorly. Good data collection and analysis is a vital tool in reducing energy use and improving sustainability.

Although ISO 50001 does not specify the frequency of monitoring and measuring it is unlikely that anything greater than monthly measurement would be acceptable or provide enough information for good energy management (see Section 6.4). Our experience is that weekly energy use data provides the ideal amount of information. It is enough to see where performance can be improved but is not so short that random fluctuations are seen.

- **Tip** – A weekly energy report, similar to those for raw materials or direct labour use is strongly recommended.
- **Tip** – The data from a weekly energy use report can easily be totalled and linked to the annual Sustainability Report (see Chapter 12) and to social responsibility issues (see Section 11.6).
- **Tip** – Be very careful when selecting the reporting metrics for energy use. Single numbers metrics such as 'kWh/kg' might have some validity over a long time frame but as weekly or monthly metrics they are fundamentally flawed for plastics processors (see Section 6.5).

What you will need for ISO 50001:

4 – Context of the organisation
- An assessment of the internal and external issues and records of the scope and boundaries of the EnMS.

5 – Leadership
- Demonstration of leadership and commitment to the EnMS.
- An energy policy that is communicated to the staff and available to interested parties.
- Establishment of the EnMS and conformity to ISO 50001.

6 – Planning
- An assessment of the risks and opportunities to the company.
- Objectives, targets and action plans consistent with the policy.
- An energy review to assess energy use and current energy performance.
- A prioritised set of opportunities for improvement.
- Estimated future energy use.
- A set of energy performance indicators and relevant baseline values to allow performance assessment.
- A data collection/measurement plan for relevant variables.

7 – Support
- A process to assess competencies, deliver training to meet competence needs and assess effectiveness.
- A training plan to raise awareness of the policy and it's implications.
- Processes for internal and external communications.
- Records of internal and external communications.
- A process for control of documented information.

8 – Operation
- Planning and processes to control energy use and to implement the identified opportunities for improvement.
- Consideration of energy use in the design of processes that affect energy use.
- A specification of energy use in procured equipment and services.
- A specification for the purchase of energy.

9 – Performance evaluation
- A measuring/monitoring plan to measure the effectiveness of the action plan.
- An evaluation of the energy performance of the company.
- An evaluation of legal and compliance performance.
- An audit process, programme and audit reports.
- A management review process including management review agenda and minutes of management review meetings.

10 – Improvement
- Processes for non-conformance and corrective/preventative action.
- Recording of non-conformances and corrective/preventative action, i.e., reports on follow-up actions.
- Demonstrated continual improvement.

The broad ISO 50001 requirements

ISO 50001 requires 'documented information' in several areas to prove compliance with the standard. Search the standard for the phrase 'documented information', anything listed is an absolute requirement for conformance.

2.11 Energy management systems – where are you now?

Getting the basics right

Energy management is a new skill for many companies. The cost of energy has not previously been an issue and it is only in the last 10-15 years that energy has become a major financial cost. Energy cost rises are a feature all over the world and this is being driven not only by supply issues but also by taxation issues.

Cost is not the only driver for reducing the amount of energy used. The rise of new concepts such as 'carbon footprint' and 'sustainability' have also been drivers for energy use reduction.

These new issues are an opportunity for companies to not only become 'greener' but to also reduce costs. There is no conflict, you can be green and reduce costs!

Energy management systems can provide the basic structure for sustainability improvements by reducing energy use.

Completing the chart

This chart is completed and assessed as for the previous charts.

Note: See Sections 6.20 to 6.22 for more comprehensive charts in the area of energy management.

Sustainability

A common and growing theme of successful businesses is having good environmental and sustainability credentials.

An EnMS can:

- Improve sustainability.
- Improve environmental performance.
- Be installed alongside a conventional QMS for relatively low cost.
- Deliver a wide range of benefits.
- BUT ....

It must be administered properly and effectively!

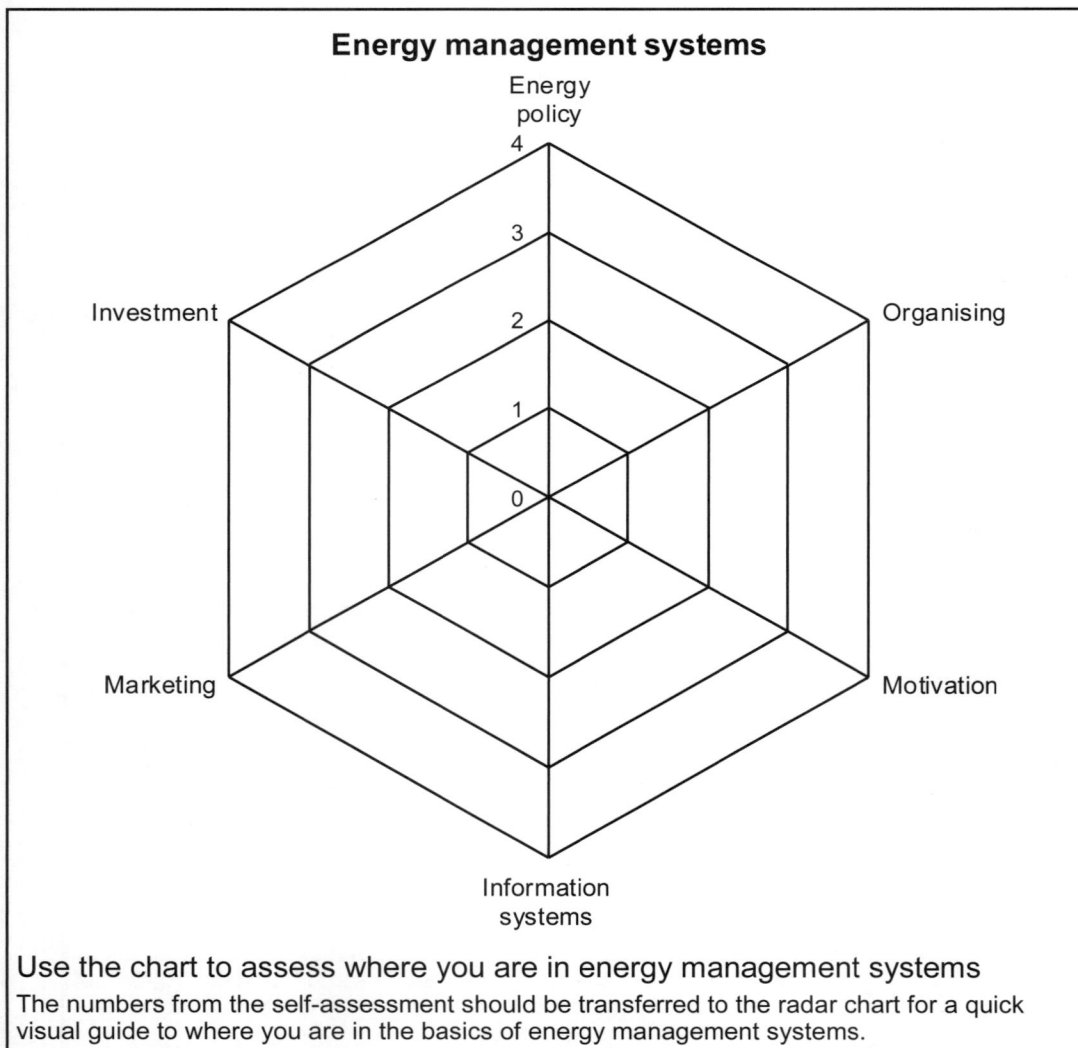

Energy management systems

If you don't know where you are starting out from then it is unlikely that you will end up where you want to get to!

Energy management is one of the last great unexplored frontiers of cost and sustainability management.

The rewards are great for the pioneers.

Use the chart to assess where you are in energy management systems

The numbers from the self-assessment should be transferred to the radar chart for a quick visual guide to where you are in the basics of energy management systems.

	Energy management systems					
Level	Energy policy	Organising	Motivation	Information systems	Marketing	Investment
4	Energy policy, Action Plan & regular review have commitment of top management as part of an environmental strategy.	Energy management fully integrated into management structure. Clear delegation of responsibility for energy consumption.	Formal & informal channels of communication regularly exploited by energy manager & energy staff at all levels.	Comprehensive systems used to set targets, monitor consumption, identify faults, quantify savings & provide budget tracking.	Marketing of energy efficiency & energy management performance both internally & externally.	Positive discrimination in favour of 'green' schemes with detailed investment appraisal of all opportunities.
3	Formal energy policy, but no active commitment from top management.	Energy manager accountable to energy committee representing all users, chaired by a member of the managing board.	Energy committee used as main channel together with direct contact with major users.	Monitoring & targeting reports for individual premises are based on sub-metering. Savings not reported effectively to users.	Program of staff awareness & regular publicity campaigns.	Same payback criteria employed as for all other investment.
2	Unadopted energy policy set by energy manager or senior departmental manager.	Energy manager in post, reporting to ad hoc committee, but line management & authority are unclear.	Contact with major users through ad hoc committee chaired by senior departmental manager.	Monitoring & targeting reports based only on supply meter data. Energy unit has ad hoc involvement in budget setting.	Some ad hoc staff awareness training.	Investment using short-term payback criteria only.
1	An unwritten set of guidelines.	Energy management is the part-time responsibility of someone with limited authority or influence.	Informal contacts between engineering staff & a few users.	Cost reporting based on invoice cost details only. Engineer compiles reports for internal use within technical department.	Informal contacts used to promote energy efficiency.	Only low-cost measures taken.
0	No explicit policy.	No energy management or any formal delegation of responsibility for energy consumption.	No contact with users.	No information system. No accounting for energy consumption.	No promotion of energy efficiency.	No investment in increasing energy efficiency.
Score						

2.12 Health and safety management systems – the basics

Required by legislation and good sense

Health and safety have gained a poor reputation in some areas due to over-zealous application of the principles. Despite this, health and safety are a key sustainability and business issue. A good health and safety system will control risks, cut costs and provide a business advantage. A poor health and safety system will do little to control risks, will raise costs and make a company inflexible in operations. The health and safety of staff and other stakeholders is a key part of social responsibility and sustainability (see Section 11.5).

Every company has a general legal responsibility for the health and safety of everyone affected by the business. This will include not only direct employees but also subcontractors, visitors, customers and members of the public affected by company operations.

An effective health and safety system will ensure compliance with legal requirements and ensure that everybody in the company knows the right way to do things. An effective system provides a consistent and structured approach to managing health and safety.

- **Tip** – The current MSS for Occupational Health and Safety (OH&S) is ISO 45001:2018 (see Section 2.13) but even if this is not used then every company must have some type of health and safety management system. This is not negotiable.

Set and promote the policy

Every company needs a health and safety policy covering four key areas:

- It should make a commitment to health and safety management, to the reduction of accidents and to the welfare of all employees and stakeholders.

- It should set out the goals of the system with specific targets and objectives for accident reduction, incident reduction and improvements in performance.

- It should describe how the system is organised and clearly describe who is responsible for each activity.

- It should cover the practical arrangements, such as safety procedures and staff training.

As with any other policy, the health and safety policy should describe what the business actually does and not what it would like to do. This is a working policy and not simply a piece of paper. The policy also needs to be understood by employees and seen as a commitment to their welfare and not simply as part of 'meeting the legal requirements'. They will need to follow the correct procedures and understand that the company is serious about their health and safety.

Quantify the risks

Regular risk assessments are necessary to identify and control health and safety risks. The risk assessments may result in no action being required but the formal assessments still need to be carried out and the results recorded (see Section 2.19).

- **Tip** – Risk assessment can help to assess the insurance cover needed for workers and other stakeholders.

Reduce the risks

After carrying out risk assessment, the risks involved should be minimised. Many people will immediately go for the PPE (Personal Protective Equipment) route and issue ear protection, safety glasses or other equipment. This is not the recommended method and the control hierarchy is shown in the box on the right. The most effective action is to eliminate the source of the risk and the hierarchy should be followed to the least effective action – PPE. PPE should always be the last resort and only used when

> The plastics industry accident record is worse than the UK manufacturing industry average.
>
> **UK Health and Safety Executive**

> **Where are the accidents?**
> - During manual handling – 34% of all injuries.
> - Struck by something (including knife-cuts) – 14% of all injuries.
> - Slipping or tripping – 25% of major injuries.
> - Machinery related – 17% of major injuries.
>
> **UK Health and Safety Executive**

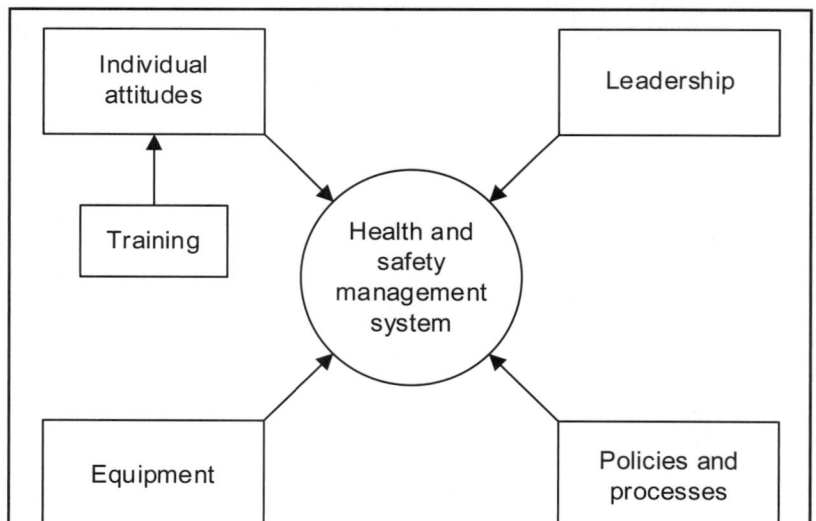

The components of an OH&SMS

An OH&SMS is needed to identify, reduce, control and manage risks to the staff and other stakeholders. This has several components and all of these must be present for an effective H&SMS. If you are missing one component then the system will not work.

the source of the risk cannot be eliminated, substituted, isolated or controlled.

Establish the procedures

Reducing risks also means establishing and monitoring procedures to ensure safe systems of work. It is not enough to simply establish the systems; it is also necessary to ensure that they are complied with on a daily basis. A system that is ignored is worthless and will provide a false sense of security.

- **Tip** – Procedures must be developed, implemented and policed.

Manage incidents

Even if the risks are minimised and all the procedures are followed, then it is still possible for incidents to occur. These incidents need to be managed to get the best result and a good health and safety system will include procedures for managing minor and major incidents and also accidents such as fires and major spillages, e.g., if your factory is on fire then what is your plan for the future?

Incident management should also include communications planning for internal company communications, press communications and general public relations. In the event of a major incident, effective communications can make all the difference to a company's reputation.

No amount of planning will eliminate all risks but good planning will minimise the impact of an incident on the stakeholders and on the business.

- **Tip** – Incident management should also include disaster recovery planning for data and business systems.

Keep records

Keeping records of health and safety incidents is generally a legal obligation. In most countries it is a requirement to report major injuries, diseases and dangerous occurrences. Systems for reporting these are needed not only for the relevant authorities but also to assess if the health and safety system is working effectively.

Good records will enable assessment of the achievement of the policy targets and objectives and if changes need to be made, e.g., if manual handling incidents are 34% of the incidents then there is a need for further training or procedures.

Stay up to date

Legislation is constantly changing in all areas of the world and the system should include a process or procedure to ensure that relevant new or changed legislation is identified promptly so that it can be complied with. For some plastics processing operations there will be a need to keep up to date with legislation on potentially dangerous substances (see Section 3.6).

Review the system

As with any system, a regular system review is needed to check that the system is working, that targets are being achieved, that risks are being controlled, that procedures are being followed and that costs are being controlled.

Use the employees to help draw up the health and safety procedures. They know what works, what is unsafe and can often suggest improvements to reduce risks.

This increases their involvement and the systems will be more 'user-friendly'.

Clearly document the safe operating procedures for all machines and operations.

Make sure that safe operating procedures are clearly displayed and followed by all staff.

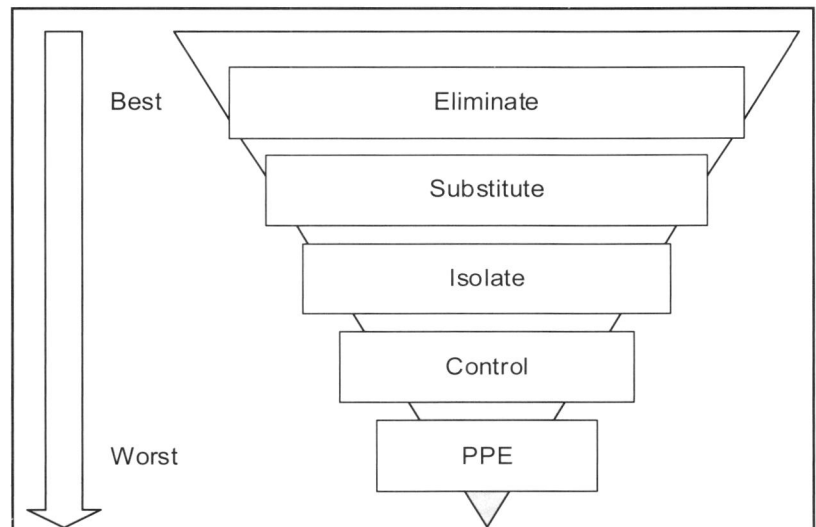

The control hierarchy

The control hierarchy starts with elimination of the risk and continues through substitution of the process, isolation of people from the risk, e.g., barriers, and control by changing the way people work. PPE is the last and least effective action in the hierarchy.

The benefits of a health and safety management system

- It will reduce the costs of accidents and incidents.
- It will reduce the possibility of committing an offence against legislation.
- It will improve employee morale as they see that the company is actively looking after their welfare.
- It will improve the public perception of the company.
- It can potentially reduce insurance premiums by demonstrating effective risk management and control.

2.13 Health and safety management systems – the standard

ISO 45001 – the MSS for occupational health and safety management systems

ISO 45001:2018. replaced the previous standard (OHSAS 18001) in 2018 and ISO 45001:2018 conforms to the Annex L structure.

This has many common features with the other MSS and this reduces the amount of new documentation required by increasing the range of activities that are common to existing quality, environmental and energy management systems. This makes use of existing structures and allows health and safety management to be seen as part of the normal business processes.

As with any MSS, this will provide a framework for the health and safety system but it will not specify which standards need to be complied with or how to comply with these, e.g., in Europe, injection moulding machines manufactured after 2009 are covered by EN 201:2009 and CE marking.

- **Tip** – OH&S is not simply a legal requirement; it is part of sustainability and social responsibility through the ISO 26000 (see Section 11.5).
- **Tip** – Reporting OH&S can easily be done via GRI 403 (see Section 12.10). This is a consistent and well-recognised standard for reporting OH&S information.
- **Tip** – Meeting the legal minimum requirements is just that, the minimum requirement. It is not what leaders aspire to.

ISO 45001 and OHSAS 18001

ISO 45001 differs from the now superseded OHSAS 18001 in several areas and these are:

- A requirement to understand the context of the organisation and internal/external risks and opportunities.
- A requirement to understand the needs and expectations of interested parties (internal and external).
- An increased requirement for leadership in the company and for the top management to demonstrate this leadership.
- An increased requirement for worker participation in health and safety management, e.g., competence, training and roles and responsibilities.
- An increased focus on planning, e.g., resources, measurements and evaluation.
- A focus on hazard identification and a risk-based approach to health and safety management (in common with the other management systems standards) and the need for companies to identify and manage risk. This concept of risk management replaces the previous concept of 'preventive action'.
- The replacement of the concept of 'documents' with 'documented information' and record retention.
- A focus on continual improvement to improve health and safety management and worker safety.
- An increase in the integration of health and safety management with other MSS, e.g., quality, environment and energy management systems.

For companies with existing an OHSAS certification there was a 3-year transition period from the 'old' to the 'new' formats and this transition should have been completed by March 2021.

What does ISO 45001 do?

ISO 45001 provides a very useful broad framework of requirements and good practices that should enable plastics processors to:

- Develop an effective OH&S policy to identify hazards, control risk and reduce incidents.
- Set targets and objectives to meet the policy.
- Develop processes that consider risks and opportunities as well as legal requirements.
- Use risk-based thinking to assess hazards and risks in the context of the broad organisation.
- Ensure that staff are part of the OH&S process.
- Plan the implementation of a programme to meet the policy by eliminating hazards or by establishing controls to minimise any potential effects.
- Review the effectiveness of the health and safety policy and system.
- Continually improve health and safety management.

The ISO 45001 MSS and framework allows the OH&S management system to be integrated with a company's other MSS and improve the system effectiveness and the company's reputation.

- **Tip** – Being certified to ISO 45001 can sometimes reduce insurance premiums.

ISO 45001 should not be regarded as a cost-saving opportunity. It is more of a 'risk reduction' opportunity to reduce the cost of poor health and safety management.

This is not simply about staff health and safety.

The OH&S risks considered must include all stakeholders potentially affected by the company.

What does ISO 45001 not do?

ISO 45001 is equally applicable to all sizes and type of organisations. This means that ISO 45001:

- Does not give any specific criteria for the performance of the OH&S system. This is decided by the company after considering the local legal requirements and the company's needs.

- Does not specify how the MSS is designed, only what it is required to achieve. The system design is again decided by the company and should be appropriate for the company's needs.

- Does not provide guidance for any specific industry or sector.

- Does not provide guidance on local regulations or requirements (although knowing these is part of the requirements of ISO 45001).

- Does not provide projects to improve health and safety. These will need to be generated as part of the continual improvement process.

- Does not consider product safety or environmental impacts unless these are also a risk to the stakeholders.

- **Tip** – Do not think that implementing ISO 45001 will automatically improve health and safety in a company. This is a management framework for the really hard work of improvement.

- **Tip** – ISO 45001 is not, in any way, legally binding. It is an MSS.

- **Tip** – Since the introduction of ISO 45001 in 2018, there have not been as many supplementary standards produced as for the other MSS. There are two standards in development and these are:

  - ISO/AWI 45002 – Occupational health and safety management. General guidelines for the implementation of ISO 45001:2018. It is forecast that this will follow the outline of BS 45002:2018 which also covers general guidelines for the application of ISO 45001.

  - ISO/WD 45003 – Occupational health and safety management. Psychological Health and Safety in the Workplace. Guidelines

These are both relatively new projects and, whilst they will expand on the usefulness of ISO 45001, they will again be written to be non-specific in terms of industry or sector.

What to do?

For plastics processors working to improve OH&S management, ISO 45001 provides a good methodology but the vagueness of the requirements does tend to reduce the utility of the MSS.

You can have a perfectly good and functioning occupational health and safety system without implementing ISO 45001 but having ISO 45001 means that you are certain to have covered all the important areas of health and safety management.

For this alone it is worthwhile.

The systems are called occupational health and safety for a reason. Sustainability means also being concerned about the health of the stakeholders.

It is not simply about safety and it pays to take a wide view of what 'health' means, i.e., worker wellness and wellbeing should also be considered.

The incident management section of a health and safety management system should share a common approach with incident management for environmental issues.

2.14 Health and safety management systems – ISO 45001

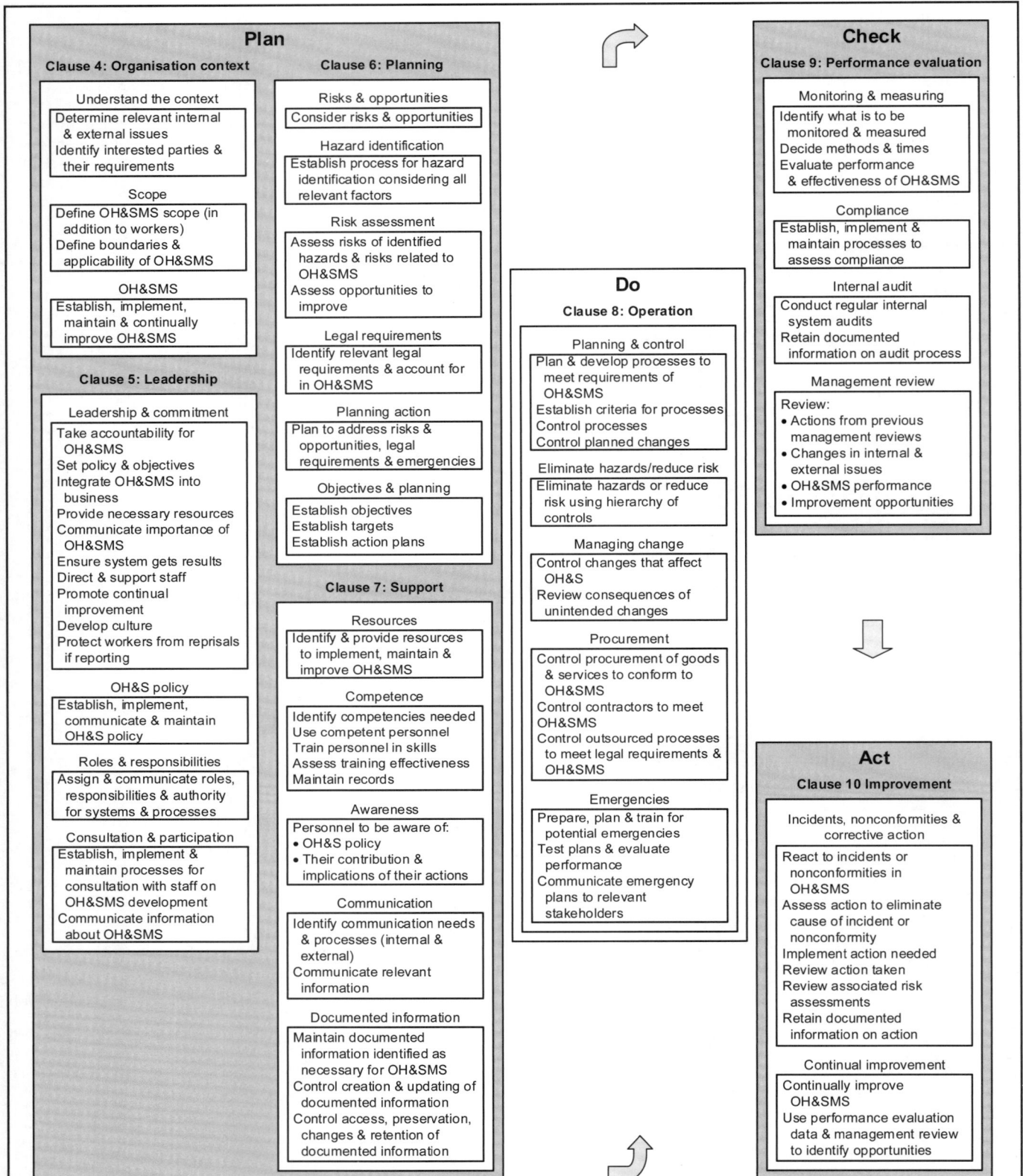

Plan

Clause 4: Organisation context

Understand the context
Determine relevant internal & external issues
Identify interested parties & their requirements

Scope
Define OH&SMS scope (in addition to workers)
Define boundaries & applicability of OH&SMS

OH&SMS
Establish, implement, maintain & continually improve OH&SMS

Clause 5: Leadership

Leadership & commitment
Take accountability for OH&SMS
Set policy & objectives
Integrate OH&SMS into business
Provide necessary resources
Communicate importance of OH&SMS
Ensure system gets results
Direct & support staff
Promote continual improvement
Develop culture
Protect workers from reprisals if reporting

OH&S policy
Establish, implement, communicate & maintain OH&S policy

Roles & responsibilities
Assign & communicate roles, responsibilities & authority for systems & processes

Consultation & participation
Establish, implement & maintain processes for consultation with staff on OH&SMS development
Communicate information about OH&SMS

Clause 6: Planning

Risks & opportunities
Consider risks & opportunities

Hazard identification
Establish process for hazard identification considering all relevant factors

Risk assessment
Assess risks of identified hazards & risks related to OH&SMS
Assess opportunities to improve

Legal requirements
Identify relevant legal requirements & account for in OH&SMS

Planning action
Plan to address risks & opportunities, legal requirements & emergencies

Objectives & planning
Establish objectives
Establish targets
Establish action plans

Clause 7: Support

Resources
Identify & provide resources to implement, maintain & improve OH&SMS

Competence
Identify competencies needed
Use competent personnel
Train personnel in skills
Assess training effectiveness
Maintain records

Awareness
Personnel to be aware of:
• OH&S policy
• Their contribution & implications of their actions

Communication
Identify communication needs & processes (internal & external)
Communicate relevant information

Documented information
Maintain documented information identified as necessary for OH&SMS
Control creation & updating of documented information
Control access, preservation, changes & retention of documented information

Do

Clause 8: Operation

Planning & control
Plan & develop processes to meet requirements of OH&SMS
Establish criteria for processes
Control processes
Control planned changes

Eliminate hazards/reduce risk
Eliminate hazards or reduce risk using hierarchy of controls

Managing change
Control changes that affect OH&S
Review consequences of unintended changes

Procurement
Control procurement of goods & services to conform to OH&SMS
Control contractors to meet OH&SMS
Control outsourced processes to meet legal requirements & OH&SMS

Emergencies
Prepare, plan & train for potential emergencies
Test plans & evaluate performance
Communicate emergency plans to relevant stakeholders

Check

Clause 9: Performance evaluation

Monitoring & measuring
Identify what is to be monitored & measured
Decide methods & times
Evaluate performance & effectiveness of OH&SMS

Compliance
Establish, implement & maintain processes to assess compliance

Internal audit
Conduct regular internal system audits
Retain documented information on audit process

Management review
Review:
• Actions from previous management reviews
• Changes in internal & external issues
• OH&SMS performance
• Improvement opportunities

Act

Clause 10 Improvement

Incidents, nonconformities & corrective action
React to incidents or nonconformities in OH&SMS
Assess action to eliminate cause of incident or nonconformity
Implement action needed
Review action taken
Review associated risk assessments
Retain documented information on action

Continual improvement
Continually improve OH&SMS
Use performance evaluation data & management review to identify opportunities

An overview of the ISO 45001 requirements

The issue of ISO 45001:2018 means that it conforms to the Annex L structure for MSS and is now similar in structure to the other MSS such as quality (ISO 9001:2015), environment (ISO 14001:2015) and energy (ISO 50001:2018). This means that much of the language and many of the requirements are now common with the other MSS. This diagram shows the main activities that must be carried out to conform to the standard.

If you've got this far

If a company has installed several of the other MSS then ISO 45001 should be familiar in terms of what is required. An overview of ISO 45001 is shown on the left and the requirements are very similar to the those for the other standards. As for the other standards, certain information is required to prove conformance with the standard and this is referred to in the standard as a requirement for 'documented information'. Some of these specific requirements and a list of 'good practice' information are given in the box on the right. If you have all this information then proving compliance is not difficult.

Hazard identification and risk assessment

Unlike the other MSS which relate to things that are happening and use varying amounts of data, ISO 45001 deals largely with identifying events that may happen and then finding methods of totally preventing them, reducing the risk or of mitigating the results should an event occur. This means that hazard identification and risk assessment are key to an effective OH&SMS, i.e., prevention is far better than cure.

The standard gives good advice in the body of the standard but Annex A (see A.6.1.2 in the standard) gives further guidance on hazard identification and risk assessment. However, the standard does not recommend any specific method for hazard identification or risk assessment. In the unfortunate event of an incident or non-conformance, the hazard identification and risk assessment processes (as well as any actions taken) may well be the subject of legal action, either personal or governmental. These are areas where documenting the process and output is critical.

- **Tip** – The Bowtie process (see Section 2.16) is an excellent tool to show a clear process of hazard identification, risk assessment and risk mitigation. It can easily be integrated with a 5x5 matrix of risk probability and risk severity to provide risk rankings.

- **Tip** – The FMEA process (see Section 2.17) is a proven method for risk assessment in both product and process design. It has been used throughout the world in automotive design for many years and can be also used with Bowtie to provide risk rankings.

What you will need for ISO 45001:

4 – Context of the organisation
- An assessment of the internal and external issues.
- Record of the scope and boundaries of the OH&SMS.

5 – Leadership
- Demonstration of leadership and commitment to the OH&SMS.
- An OH&S policy that is communicated to the staff and available to interested parties.
- Defined roles and responsibilities at all levels of the company.
- Evidence of consultation and participation of workers.

6 – Planning
- Assessment of the risks and opportunities to the company.
- A hazard identification process.
- A risk assessment process for identified hazards and risks.
- Identification and assessment of legal requirements.
- A plan to address risks and opportunities, legal requirements and emergencies.
- A set of OH&S objectives and targets and a plan to achieve these.

7 – Support
- A process to assess competencies, deliver training to meet competence needs and assess effectiveness.
- A training plan to raise awareness of the policy and its implications.
- Processes for internal and external communications.
- Records of internal and external communications.
- A process for control of documented information.

8 – Operation
- Planning and processes to meet the OH&SMS.
- Records of process criteria and control.
- A process to eliminate hazards or reduce risks.
- A process to manage temporary or permanent changes that may impact on OH&S performance.
- Control of procurement and contractors to meet the OH&SMS.
- Plans and training for emergencies.
- Testing and evaluation of emergency plans.

9 – Performance evaluation
- A measuring/monitoring plan to measure OH&SMS effectiveness.
- Evaluation of OH&S performance.
- Evaluation of legal and compliance performance.
- An audit process, programme and audit reports.
- A management review process including management review agenda and minutes of management review meetings.

10 – Improvement
- Processes for non-conformance and corrective/preventative action.
- Recording of non-conformances and corrective/preventative action, i.e., reports on follow-up actions.
- Demonstrated continual improvement.

The broad ISO 45001 requirements

ISO 45001 requires 'documented information' in several areas to prove compliance with the standard. Search the standard for the phrase 'documented information', anything listed is an absolute requirement for conformance.

2.15　Health and safety management systems – where are you now?

Safe and sound

Health and safety systems are covered by legislation in most areas of the world and certain aspects will be mandatory. However, the benefits of a good health and safety management system are much more than simple compliance with legislation.

A good health and safety management system will protect a company's investment in their staff and also protect the general public.

If the health and safety system fails (for whatever reason) then the result can be either a minor or a major incident and a good health and safety management system will not only seek to prevent incidents but also include procedures for dealing with them if they occur.

Prompt and effective incident management can not only reduce the seriousness of an incident but also control and reduce the impact on the business.

Health and safety are also part of sustainability in terms of social responsibility (see Section 11.5) and reporting sustainability (see Section 12.10).

Completing the chart

This chart is completed and assessed as for the previous charts.

This is serious

A major Health and Safety incident can destroy a company faster than almost any other event.

The things we remember about companies are not their profits or their expansion but the disasters that befall them. These inform our impression of the sustainability of the company.

To check this, think of the following (in no particular order):

- Flixborough.
- Three Mile Island.
- Chernobyl.
- Texas City Oil Refinery.
- Deepwater Horizon.
- Piper Alpha.
- Bhopal.
- Seveso.
- Herald of Free Enterprise.

All of these were major events and most resulted in significant loss of life (internal and external to the company).

All of these also cost a lot of people their jobs and in some cases destroyed the company.

Health and safety are not simply about legislative compliance!

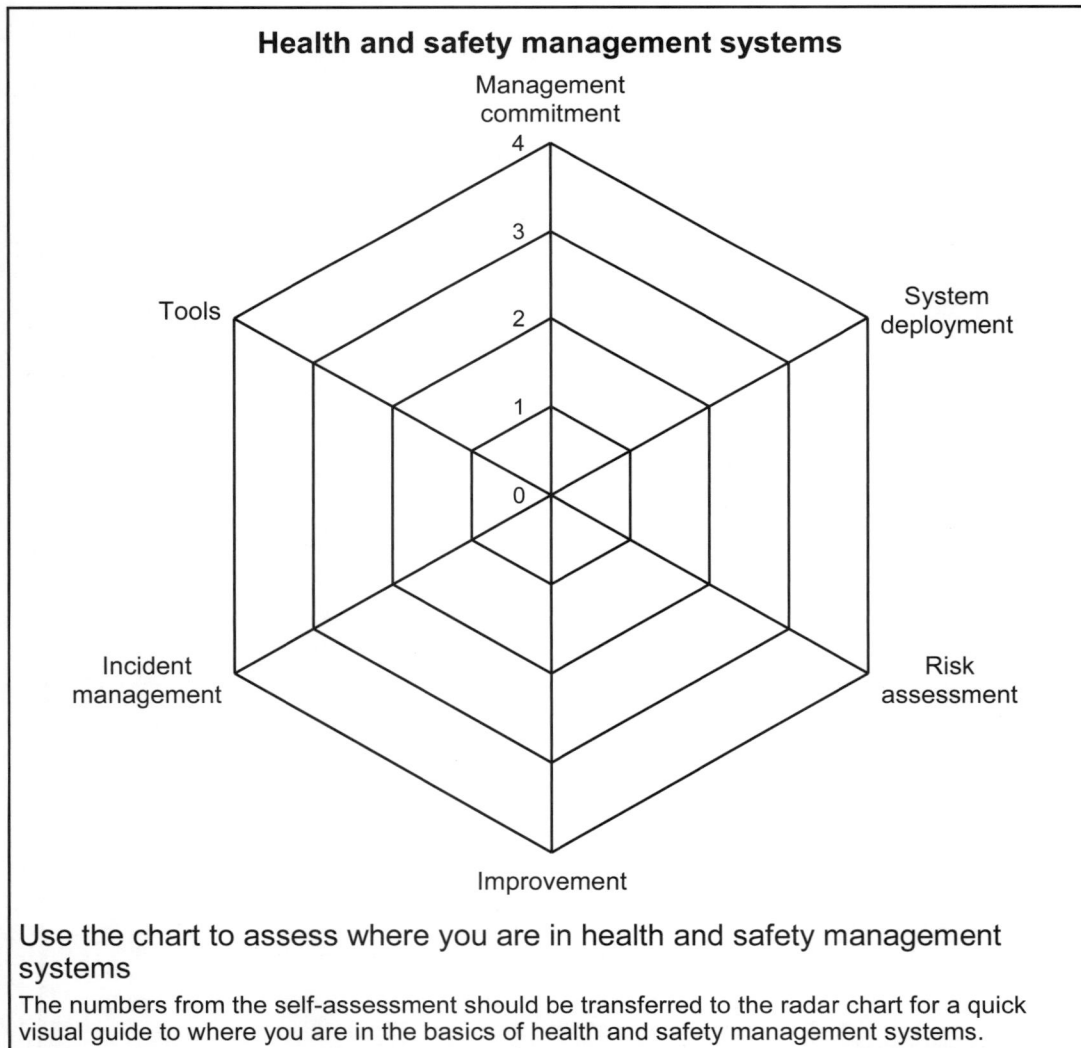

Health and safety management systems

Use the chart to assess where you are in health and safety management systems

The numbers from the self-assessment should be transferred to the radar chart for a quick visual guide to where you are in the basics of health and safety management systems.

						Health and safety management systems					

Level	Management commitment	System deployment	Risk assessment	Improvement	Incident management	Tools
4	Management is totally committed. H&S is integral part of business, all resources provided, staff trained & have delegated authority.	Formal H&S system in place with full external verification of system. No major non-conformances found in last 2 years.	All processes (major & minor) covered by full risk assessments. Action taken to minimise all risks identified.	H&S improvement is a fundamental business goal. Improvement techniques used whether concerns present or not.	Comprehensive incident management program in place. All potential incidents are covered by incident management plan.	Full range of improvement tools used to identify concerns, to determine root causes & to assess rectification actions.
3	Management has moderate commitment. Majority of requirements are in place but enforcement is sporadic.	Formal H&S system in place with full external verification of system. No major non-conformances found in last year.	All major processes covered by full risk assessments. Action taken to minimise most risks identified.	H&S improvement is an important business goal. Improvement techniques only used when concerns are present & visible.	Good incident management program in place. Most potential incidents are covered by incident management plan.	Good knowledge & use of improvement tools in identifying & reducing risks.
2	Management has low commitment & only really involved when problems occur. Basic requirements are in place but not enforced.	Formal H&S system in place with full external verification of system. Significant major non-conformances found in last year.	Some major processes not covered by risk assessments. Some identified actions not taken to minimise risks.	H&S improvement is a minor goal. Improvement techniques sometimes used when concerns are present & visible.	Poor incident management program. Few potential incidents are covered by incident management plan.	Some knowledge of improvement tools & often used for analysis. Problems often solved but key concerns remain unsolved & reappear.
1	Management not committed. Some aspects of H&S management are in place due to middle management but few resources available.	Formal H&S system in place but no external verification of system.	Most major processes not covered by risk assessments. Few actions taken to minimise risks.	H&S improvement is not seen as a goal. Improvement techniques not used even when concerns are present & visible.	No incident management program in place. Some informal procedures exist but not agreed or widely available.	Poor knowledge of improvement tools, rarely used & when used are not fully followed through to completion. Same concerns return time & again.
0	Management not committed. No H&S policy, no resources, no training & no delegated authority.	No formal H&S system in place.	No risk assessments carried out. Actions taken to minimise risks are minimal.	H&S improvement is not seen as a goal. Getting the product out the door is the only goal.	No incident management program in place. Any incident comes as a surprise. Reactions are unplanned & uncoordinated.	No knowledge or use of improvement tools.
Score						

2.16   Risk assessment – bow tie analysis

A visual assistant

Bow tie analysis[1] is a relatively recent development in risk assessment and provides an excellent visual representation of the risk assessment and control process. It is one of a range of techniques for risk assessment, e.g., HAZID, HAZOP, Tripod Beta and FMEA (see Section 2.17), but the highly visual nature of the tool makes it easy to use and, more importantly, to easily communicate the results to a wide range of levels in the company, especially to top management.

Bow tie analysis is an emerging technique that deserves wider attention and recognition. The tool was originally designed for use in the oil and gas industry but is now widely used in the aviation industry (www.caa.co.uk/Safety-Initiatives-and-Resources/Working-with-industry/Bowtie/), the rail industry and has wider applications in safety, environmental and, in fact, any area where risk assessment and controls are needed.

What is in it?

A model bow tie diagram is shown below and all bow ties have common elements:

- Hazard – the hazard is the thing that has the potential to cause harm. This is the first thing to be identified and is central to the bow tie.
- Incident – this is what happens when control of the hazard is lost or if the hazard is released.
- Threats – these are typically on the left-hand side of the diagram (the prevention side). These are the potential causes of the loss of control or release.
- Consequences – these are typically on the right-hand side of the diagram (the mitigation side). These are the potential results of the loss of control or release.
- Prevention barriers – these are the actions or systems that are designed to stop the threat from developing into the incident, i.e., what is being done to prevent the loss of control or release.
- Mitigation barriers – these are the actions or systems that are designed to reduce the severity, minimise the effect or stop the consequence after the incident has occurred.
- Tip – Although the word 'barrier' is used, prevention and mitigation do not have to

Although bow tie analysis is not designed for hazard identification, the visual nature and clarity of the process makes identification easier.

Bow tie analysis not only shows today's controls but shows how to manage them to ensure that they stay effective in the future.

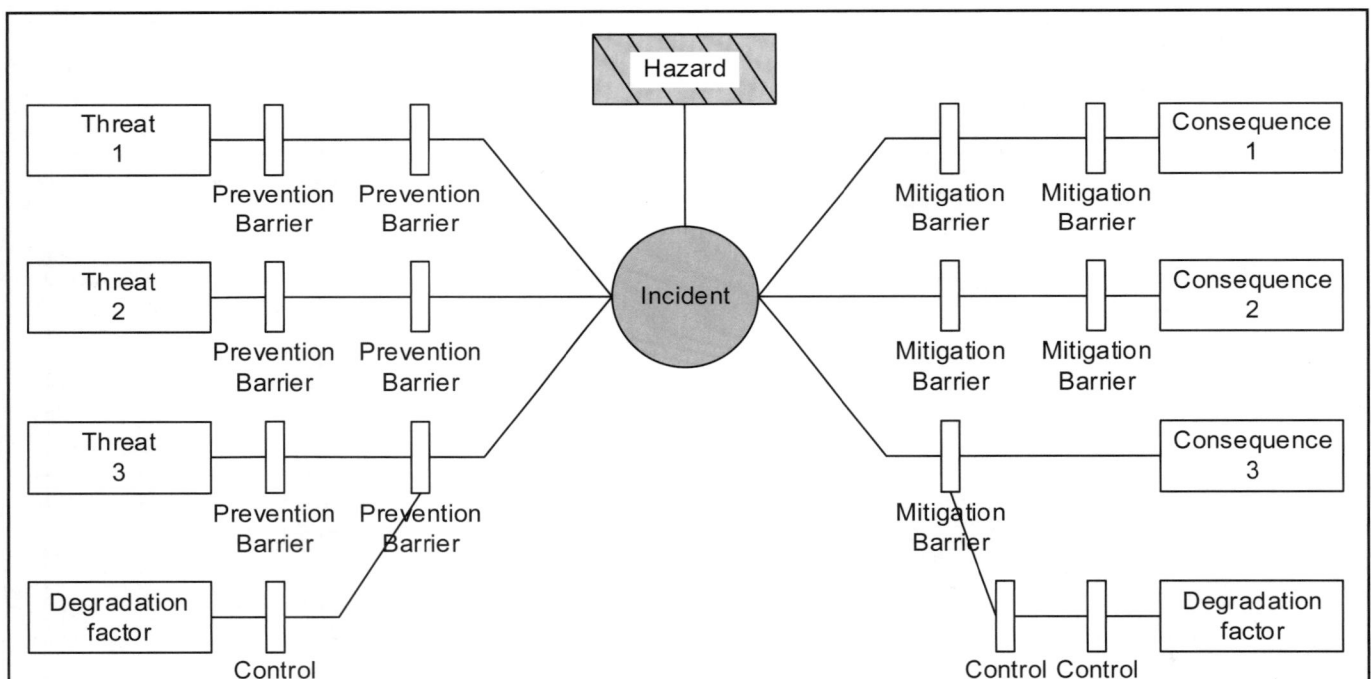

A model bow tie analysis showing the threats, the consequences, the barriers and the controls

Bow tie analysis visually represents the threats, the prevention barriers that can stop the threat becoming an incident and the degradation factors that degrade the effectiveness of the prevention barriers. If a threat makes it through the prevention barriers and becomes an incident then the mitigation barriers are designed to reduce the scale of or the effects leading to the undesirable consequences.

Chapter 2 – Management systems

be physical barriers, they can be work practices, training, awareness campaigns or any action that is taken to prevent the threat from becoming an incident or any action that is taken to mitigate the incident after it has occurred.

- Degradation factors – these are the factors that can lead to reduced effectiveness or failure of a barrier (prevention or mitigation), i.e., how a barrier can fail or be ineffective.
- Degradation controls – these are the actions or systems that are designed to control degradation and retain the effectiveness of the barrier (prevention or mitigation), i.e., how a barrier is protected from degradation.

The visual nature of the bow tie diagram makes it an excellent tool to communicate the existing controls (or lack thereof), to identify any gaps in the current controls and to provide a systematic overview of the risk.

- **Tip** – Bow tie diagrams are particularly useful in risk assessment and control of processes and operations although they can also be used for design to identify design improvements.

Taking the bow tie further

Quantifying the risk

Bow tie diagrams are not quantitative and are best used for communication of the risk rather than quantitative risk assessment. This does not mean that they cannot be used to provide a quick assessment of risk. The table on the right shows a simple risk matrix to assess a risk (threat) in terms of severity and probability of occurrence. It is relatively quick and easy to assign a severity and a probability to a risk (threat) from the table and use this assessment to prioritise control actions.

- **Tip** – A quick sketch bow tie diagram can be used to assess the overall risk and if it is low (bottom right-hand corner of the matrix) then it may not be necessary to complete a full bow tie. In these cases, the assessment should still be recorded to show that it has been carried out.
- **Tip** – Bow tie diagrams can be prepared rapidly by hand and are particularly useful in risk assessment and control of processes and operations although they can also be used for design to identify design improvements.

Actions

Preparing the bow tie itself is not enough, there will inevitably be actions that are needed to improve barriers or to control degradation of the system. The output of process should therefore include:

- A list of identified improvements/changes to barriers or controls.
- Who is responsible for each identified improvement and a timescale for completion.
- A list of maintenance actions for barriers or controls to ensure that they remain effective.
- Who is responsible for each maintenance action and plan for regular maintenance action.
- A date for system review.

There is specialised software available for bow tie analysis. This can help preserve the output in a logical manner but getting the relevant people together to generate a visual map of the risks and controls is the important thing.

- 1. Center for Chemical Process Safety (CCPS) and the Energy Institute. 2018. 'Bow Ties in Risk Management: A Concept Book for Process Safety', Wiley.

Analysis is not the important thing. It is the actions taken to manage the risks that are revealed as a result of the analysis.

Risk/Threat probability	Risk/Threat severity				
	Catastrophic	Hazardous	Major	Minor	Negligible
Frequent	Not acceptable	Not acceptable	Not acceptable	Review	Review
Occasional	Not acceptable	Not acceptable	Review	Review	Review
Remote	Not acceptable	Review	Review	Review	Acceptable
Improbable	Review	Review	Review	Acceptable	Acceptable
Extremely improbable	Review	Acceptable	Acceptable	Acceptable	Acceptable

A matrix approach to risk (threat) assessment

Assessing the risk severity and probability can be quite rapid and risk increases as you move from extremely improbable events with negligible severity to frequent events with catastrophic severity. The three major regimes are highlighted. Deal with the upper left area first!

2.17   Risk assessment – FMEA

Quantifying risk

The bow tie diagram (see Section 2.16) provides a visual communication tool for risk assessment but the current development is not strictly designed for quantitative assessment. Each of the MSS discussed in this section involves assessing 'risk', i.e., a process of logically assessing the risks involved in a process to determine the actions necessary to reduce or control the risks.

Risk assessment can be carried out in a variety of ways but one of the easiest methods is Failure Modes and Effects Analysis (FMEA). This is a technique developed in the automotive industry to assess both design and development risks and process risks. In the automotive industry the type of FMEA used has been strictly differentiated between Design and Process FMEA. However, the basic techniques are very similar and the methodology is proven to quantify and reduce risk.

Risk assessment through a formal FMEA-style procedure enables:

- Concentrating on action to reduce major risks.
- Effective use of resources.
- Avoiding having to try to deal with all risks (including insignificant ones).

The process

The process involves assessing three or more factors separately and then combining the individual assessment of each factor to provide an overall assessment of the risk via a Risk Priority Number (RPN) to allow targeted actions and risk reduction.

Assessment of individual areas

For a typical assessment the three factors to be assessed are:

Severity: What is the severity of the effect? A failure inevitably creates an effect and the severity of the effect is judged on a scale of 1 to 10. A rating of 1 indicates a low effect severity should a failure occur and a rating of 10 indicates a very high effect severity should a failure occur.

Probability: What is the probability of the failure occurring? The probability of failure occurring is judged on a scale of 1 to 10. A rating of 1 indicates a low probability of failure and a rating of 10 indicates a very high probability of failure.

Current controls: What is the likelihood

that a failure will be detected before it becomes critical? A potential failure may be easily detected and avoided or be very difficult to detect and avoid. A rating of 1 indicates a high probability of detection before failure and a rating of 10 indicates a very low probability of detection before failure.

Each factor is assessed individually to allow a considered judgement on the basis of the individual factors. The separation of the severity, probability and current controls factors makes assessment of the overall risk easier and less judgemental than attempting to assign a single risk number. The separation also provides an insight into areas that require improvement and/or preventative measures.

When assessing significance:

- Be consistent – develop a consistent approach that allows each issue to be clearly treated in the same way.
- Be able to demonstrate and justify the methodology used – use criteria that provide a rational basis for the rest of the assessment.
- Record the method and decisions in a systematic manner.

Creating the RPN

The overall Risk Priority Number (RPN) is simply the product of the ratings for the individual factors:

RPN = Severity factor × Probability factor × Current controls factor.

This gives a simple single number to assess the overall risk associated with the design feature, process event, activity or environmental aspect.

Example:

A designer is producing a design FMEA for a braking system in a car. The process (for this single factor) would be:

Severity factor: Assessed as a value of 10, i.e., brake failure is a severe effect and the severity would be high.

Probability factor: Assessed as a value of 3, i.e., brake failure is not likely to occur given the robust design of the braking system.

Current controls factor: Assessed as a value of 3, i.e., the braking system has detectors to indicate possible failure and wear.

The overall RPN for this factor would then be: **RPN = 10 × 3 × 3 = 90.**

Note: The absolute RPN is not in itself a

Risk assessment is carried out in many areas of industry such as:

- Design and development: To assess methods of reducing failure risks and the hazards associated with design.
- Production processes: To assess and reduce the risks associated with process failure or product failure.
- Environmental: To assess and minimise the risks associated with processes and possible outputs.
- Health and safety: To assess the risks to affected parties (staff, contractors and the general public) of various processes and activities.

Variations on a theme

There are variations on the theme of risk assessment. Some methods use a 1-5 rating instead of 1-10. Some use poor (3), fair (2) and good (1). Some methods use different terms, e.g., 'likelihood' for probability or 'consequence' for severity.

This is not important; the important thing is that the process of risk assessment is carried out in a logical and justifiable manner.

meaningful number. It is only meaningful when compared to other RPNs to allow relative assessment of the risk and actions to be taken to reduce the highest risks.

The individual areas of concern are assessed (preferably by a team or group of people to allow discussion and to prevent personal bias) and RPNs assigned to the various factors.

The RPNs can then be prioritised (deal with the highest first) to define corrective action to reduce the RPN. The individual RPN factors provide strong guidance on the actions to take to reduce the RPN.

For the example above, an obvious area for action would be the severity factor. Reducing the severity factor would greatly reduce the RPN. Actions to take might include:

- Fitting seat belts to the car.
- Fitting warnings to prevent motion without the seat belts being fitted and operational.
- Fitting air bags to the car.

Taking these actions could reduce the severity factor to 7 and reduce the overall RPN to 63. This is a considerable reduction in the RPN and would increase the chances of survival in the event of a failure of the brake system.

Setting the RPN limits

The setting of the RPN limits for action is an entirely judgemental decision. Each assessor will evaluate each factor differently but the result of the judgement will be a set of RPNs that prioritise the criticality of each risk.

The reality is that the assessor or management sets the RPN threshold for action. The only critical factor is that the assessor or management must be prepared to justify their actions in the setting of the RPN threshold. This is made easier using an RPN than with many other broader judgemental methods. The logic of the RPN makes decisions on thresholds easier to make and justify.

Experience and RPN

Where there is experience of failures, risks or emissions then historical records can be used to set the RPN factors and to justify these. Where data are available, standard methods, e.g., Pareto analysis, can be used to rate the critical factors and generate the RPN.

Historical data can be based on:

- Previous designs, products or events.
- Similar designs, products or events.
- Analogous designs, products or events.

The amount of historical data available is almost always underestimated and the first action should be to survey existing warranty claims or experience data to allow a judgement for the individual factors. The use of historical data improves decision-making and makes subsequent justification of allocated values easier should this be necessary.

Using the factors for improvement

The rigour of the decision-making in assessing RPN factors also allows clarity in assessment of the benefits of any improvements (theoretical and practical). Any proposed improvement can be assessed in terms of the RPN factors and an assessment of the likelihood and magnitude of the improvement generated.

The living document

A risk assessment is not a static document that is produced and then forgotten. The document is a 'living document', the initial focus will be on reducing the highest RPN factors but this will simply mean that other (and currently lower) RPN factors will become top of the list. These 'top of the list' factors are then subject to analysis and improvement to lower the overall risk profile. This is part of the reason for avoiding an absolute RPN threshold. The tasks to work on are always those at the top of the list and reducing these will reduce the risk profile of the process or task.

A risk assessment should be under continuous review but at the very minimum should have a specified review date. Risk assessments should also be reviewed after any significant change in circumstances.

2.18   Risk assessment – environmental

Introduction

Assessing the significance of environmental aspects is a key requirement of an Environmental MSS and ISO 14001 requires identification of significant aspects (those that have a significant impact on the environment) using a formal procedure. ISO 14001 does not specify a set method for assessing the significance of environmental aspects. However, the procedure used to assess significance should be recorded in a systematic manner for future reference.

This is again a good application for a formal risk assessment using a slightly modified version of the technique used for other issues (see Section 2.17).

The environmental aspects that are judged to be significant are the ones that will be managed by the Environmental Management System. These activities will generally be revealed by the 'Initial Review' (see Section 2.4) and will also be the activities which are covered by legislation and/or have a high cost. These will also be areas where improvement activities will have a high beneficial environmental impact and even reduce costs.

Each potential impact is assigned a rating after considering the following:

- Hazardous properties (**H**) – this relates to the potential environmental release or event, 1 is non-hazardous and 10 is very hazardous.
- Size (**S**) – this defines the potential size of the aspect, 1 is very small and 10 is very large.
- Frequency or likelihood of occurrence (**F**) – this is how often the aspect may occur, 1 is low frequency or likelihood of occurrence

and 10 is high frequency or likelihood of occurrence.

- Presence of sensitive environmental receptors, e.g., people, a watercourse and/or site of special scientific interest (**R**) – 1 is little presence of sensitive environmental receptors and 10 is high presence of sensitive environmental receptors.
- Presence or absence of current controls, e.g., techniques designed to control or prevent the environmental impact (**C**) – 1 is high presence of current controls and 10 is low presence of current controls.

For each aspect, the individual factors are rated to produce a composite RPN (**H** × **S** × **F** × **R** × **C**) and a total risk assessment for the impact.

The impacts with the highest RPN values are targeted for initial action and the RPN provides a structured method of prioritising action.

An example of a completed environmental risk assessment is shown on the opposite page for information. This example includes a recalculation of the RPN after improvement action has been taken.

FMEA allows risk assessment of aspects and impacts to meet the requirements of ISO 14001.

The assessments should be retained as documented information.

The actions to reduce the RPN can be seen as part of the drive to continual improvement.

Hazardous properties (H)	Ranking
Major hazard	10
Extremely high hazard	9
Very high hazard	8
High hazard	7
Moderate hazard	6
Low hazard	5
Very low hazard	4
Minor hazard	3
Very minor hazard	2
Non-hazardous	1

Size of event (S)	Ranking
Very large	10
	9
Large	8
	7
Moderate	6
	5
Minor	4
	3
Small	2
Very small	1

Frequency of event (F)	Ranking
Repetitive	10
	9
High	8
	7
Moderate	6
	5
Low	4
	3
Rare	2
None	1

Environmental receptors (R)	Ranking
Major environmental receptors	10
Extremely high environmental receptors	9
Very high environmental receptors	8
High environmental receptors	7
Moderate environmental receptors	6
Low environmental receptors	5
Very low environmental receptors	4
Minor environmental receptors	3
Very minor environmental receptors	2
No environmental receptors	1

Current controls (C)	Ranking
Absolute uncertainty	10
	9
Remote	8
	7
Low	6
Moderate	5
	4
High	3
	2
Almost certain	1

Wentworth Site	Date	05/10/2020
Prepared By	Robin Kent	
Notes	No consideration of system interactions made at this stage	

Risk aspects	Current impacts													Recommended actions				Action results					
	Hazardous properties	(H)	Size of event	(S)	Frequency of event	(F)	Environmental receptors	(R)	Environmental controls	(C)	Prevention	Detection	R.P.N.	Recommended actions	Responsible	Target completion date	Actions Taken	(H)	(S)	(F)	(R)	(C)	R.P.N.
Leak from sludge tank	High	7	Small until sensors sound	4	Low	5	Moderate in adjoining river	6	Sensors	5	Regular checking of sludge tank.	Checking of sensors.	4200	Improve sensors. Provide bund for overflow.	Works Manager	Immediate	Done	7	4	2	6	5	1680
Leak from refrigeration system	High	7	Small until system empty	3	Very minor	2	Few in vicinity	3	Chiller alarm	8	Regular checking of chillers	Chiller alarms	1008	Wire chiller alarm to general alarm	Works Manager	Immediate	Done	7	3	2	3	5	630
General waste disposal	Low	2	Moderate	3	Repetitive	10	Few in landfill area	3	None	10	Improve recycling rate for paper and cardboard	None	1800	Place recycling bins in all areas	Sustainability Manager	Immediate	Done	2	3	10	3	10	1800
Petrol use in vehicles	Emissions	4	Minor	4	Repetitive	10	Dependent on area	5	Catalytic convertors fitted to all vehicles and lead free, ULS petrol used in all vehicles	3	Improve servicing checks	Check catalytic convertor operation during servicing	2400	Evaluate EV options. Evaluate LPG options.	Transport Manager	6 months from date of issue. (05/04/2020)		2	4	10	5	3	1200

Example of environment-related risk analysis using SnapSheets XL ™ to generate Risk Priority Numbers

This is not a complete example and the actual risk analysis is much longer and detailed. It is shown only to illustrate the process and the output. The risk events are the main environmental aspects and the risk analysis looks at the impacts of the aspects.

2.19 Risk assessment – health and safety

Introduction

Risk assessment is a vital tool in health and safety and risk assessments must be carried out to comply with legal requirements, e.g., UK Health and Safety at Work Act.

Each potential risk is assigned a rating after considering the following:

- Severity (S) – 1 is low severity and 10 is high severity.
- Probability (P) – 1 is low probability and 10 is high probability.
- Current controls (C) – 1 is high current controls and 10 is low current controls.

For each risk, the individual factors are rated to produce a composite RPN (S × P × C) and a total risk assessment for the risk.

Risk assessment for health and safety is simply an assessment of what could cause harm to people and what precautions are being taken or could be taken to minimise the risk. The important decisions are what hazards are significant and what precautions have been taken to minimise the risk. Risk assessment documents all the decisions taken, evaluates the risks and assesses the current precautions. In assessing the risks, the assessments and controls must be 'suitable and sufficient', which is not necessarily the same thing as perfect. It is suitable and sufficient if reasonable precautions are taken so that the remaining risk is low. The use of RPN can document these actions and the residual risk.

- **Tip** – There is no such thing as absolute removal of risk. Risks can only be managed to be 'As Low As Reasonably Practical' (ALARP).

When evaluating health and safety risks do not forget to include risks to the all the other stakeholders, e.g., general public, site visitors and contractors, as well as the risks to employees on the site. All are covered under most legislation.

As with all risk assessments it is necessary to review and revise the risk assessments with the passage of time ('living document') to reflect new equipment, processes, substances and procedures.

If you think this is difficult then consider the emergency services.

The fire and rescue services carry out 'dynamic risk assessment' when fighting fires. They need to assess if it is safe for fire-fighters to enter buildings or carry our similar tasks.

The Officer-in-Charge must carry out his 'dynamic risk assessment' at the scene and then take action. No time for spreadsheets but the actions may need to be justified later so the thought process is carried out.

It is called 'dynamic' because it is constantly changing as the fire progresses.

Severity of event (S)	Ranking	Probability of event (P)	Ranking	Current controls (C)	Ranking
Hazardous: without warning	10	Very high: event is inevitable	10	Absolute uncertainty of detection	10
Hazardous with warning	9		9	Very remote chance of detection	9
Very high	8	High: Repeated events	8	Remote	8
High	7		7	Very low	7
Moderate	6	Moderate: Occasional events	6	Low	6
Low	5		5	Moderate	5
Very low	4		4	Moderately high	4
Minor	3	Low: Relatively few events	3	High	3
Very minor	2		2	Very high	2
None	1	Remote: Event is unlikely	1	Almost certain to detect	1

Fabrication	Date	05/10/2020
Prepared By	Robin Kent	
Notes	No consideration of system interactions made at this stage	

Risk event	Potential failure mode	Potential effect of failure	Sev (S)	Class	Current controls						Recommended actions				Action results				
					Potential cause	Prob (P)	Prevention	Detection	Controls (C)	R.P.N.	Recommended actions	Responsible	Target completion date	Actions Taken	Sev (S)	Prob (P)	Controls (C)	R.P.N.	Class
Accessible powered clamps (all types of machinery)	Trapped fingers and hands	Potential lost limb or extremities.	8		Risk of trapping fingers and hands through operator error.	2	Restricted stroke (6 mm or less), two-hand control, guarding of the clamps and low pressure approach to within 6 mm of the workpiece. Clamp retracts if an obstruction is detected during descent. Approach pressure on the clamp is not adjustable by the user.	QC checks on routine inspection.	3	48	None				8	2	3	48	
Workpiece movement	Cuts and lost limbs	Potential lost limb or extremities	8		Movement of workpiece during cutting.	3		QC checks on routine inspection.	2	48	None				8	3	2	48	
Contact with saw blades	Large cuts and possible loss of limb	Severe	7		Over-ride of safety mechanisms.	3	Moveable guarding encloses the saw blade only in cutting position.		6	126	Fixed guarding provided to enclose the saw blade in both its cutting and retracted position. Improve and maintain training.	Works Manager	Immediate	Complete	7	1	3	21	
Contact with saw blades where blade movement is lateral	Contact with blades and loss of limbs	Potential loss of limbs	9		Operator over-ride of safety guards.	1	Interlocked guards fitted to prevent access to the blades in cutting and retracted positions.		3	27	None				9	1	3	27	
Large automated equipment (e.g., cutting and welding centres)	Operator inside machine.	Potential loss of life from contact with machine	10		Interlocks not used correctly. Guarding fails or is over-ridden.	6	Lack of compliant guarding gives high probability		9	540	Fixed guarding. Person-sensing devices, limited movement devices deployed within the perimeter fencing for access with the machine under power. Improve and maintain training	Works Manager	Immediate	Complete	10	1	2	20	

Example of health and safety-related risk analysis using SnapSheets XL to generate Risk Priority Numbers

This is not a complete example and the actual risk analysis is much longer and detailed. It is shown only to illustrate the process and the output. The potential risk events and the consequences are evaluated to determine actions to reduce the risk.

2.20 Risk assessment – where are you now?

Failing to plan is planning to fail

An assessment of the risks and opportunities is a standard part of the Annex L structure and is included in every MSS. Risks and opportunities need to be assessed for two functions:

- The first function is assessing the operation of the system itself, i.e., what are the risks and opportunities of the system delivering the intended results. This is the main concern for an EnMS but is still important for an EMS and an OH&SMS.

- The second function is assessing the potentially harmful effects that the system is attempting to control, i.e., what are the risks and opportunities that the system in trying to manage. This is the main concern for an EMS and an OH&SMS but is less of a concern for an EnMS.

Risk assessment and management of the identified risks are essential skills in improving sustainability.

Completing the chart

This chart is completed and assessed as for the previous charts.

'Risk' is defined by Annex L as 'the effect of uncertainty'. This can be good or bad but risk implies that we do not know.

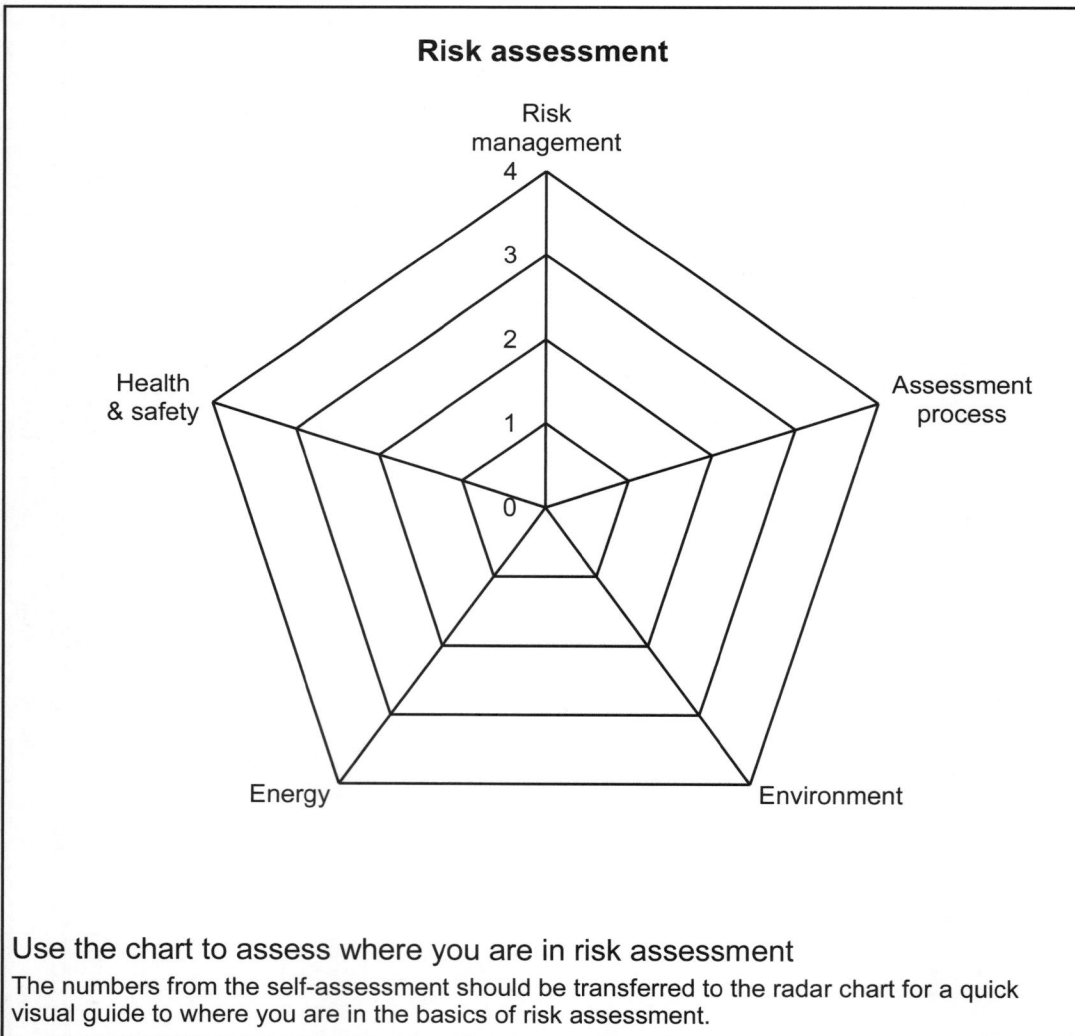

Risk assessment

Risk management — Assessment process — Environment — Energy — Health & safety (scale 0–4)

Use the chart to assess where you are in risk assessment

The numbers from the self-assessment should be transferred to the radar chart for a quick visual guide to where you are in the basics of risk assessment.

Level	Risk management	Assessment process	Environment	Energy	Health & safety
			Risk assessment		
4	Comprehensive risk reduction & management program in place. Plans are up to date & appropriate.	Formal & well documented risk assessment process available & used extensively.	Full environmental risk assessments carried out. Assessments follow well defined process & are well documented.	Planning process identifies risks & opportunities in energy management system. Assessments follow well defined process & are well documented.	Full health & safety risk assessments carried out. Assessments follow well defined process & are well documented.
3	Limited risk reduction & management program in place for very specific events, e.g. environmental issues. Plans are up to date & appropriate.	Formal risk assessment process available but rarely used.	Full environmental risk assessments carried out. Assessments follow poorly defined or inappropriate process but are well documented.	Planning process identifies risks & opportunities in energy management system. Assessments follow poorly defined or inappropriate process but are well documented.	Full health & safety risk assessments carried out. Assessments follow poorly defined or inappropriate process but are well documented.
2	Limited risk reduction & management program in place for very specific events, e.g. environmental issues. Plans are out-of-date or inappropriate.	Informal risk assessment process in place but used extensively or for a majority of areas.	Full environmental risk assessments carried out. Assessments follow poorly defined or inappropriate process & are poorly documented.	Planning process identifies risks & opportunities in energy management system. Assessments follow poorly defined or inappropriate process & are poorly documented.	Full health & safety risk assessments carried out. Assessments follow poorly defined or inappropriate process & are poorly documented.
1	No risk reduction & management program currently in place but plans in place for implementation.	Informal risk assessment process in place but rarely used or used for a minority of areas.	Informal environmental risk assessment carried out for some areas but poorly documented.	Informal risk assessment of energy management carried out for some areas but poorly documented.	Informal health & safety risk assessment carried out for some areas but poorly documented.
0	No risk reduction & management program in place & not planned.	No risk assessment process in place for any area.	No environmental risk assessments carried out.	No risk assessment carried out.	No health & safety risk assessment carried out at any stage. **NOTE:** This could contravene local legislation.
Score					

Key tips

- Management systems standards are becoming increasingly important, if only for the reason that many customers are saying "If you don't have ISO XXXX, then we won't buy from you".

- Management systems standards (MSS) for the sustainability area, e.g., environment, energy and health and safety are now harmonised into a common framework. This avoids duplication of effort and information.

- The common MSS is based on the Plan-Do-Check-Act (PDCA) cycle.

- Before starting to implement an MSS, check that the company will actually benefit and can use the MSS effectively.

- An Environmental Management System (EMS) focused on reducing physical waste and emissions not only improves sustainability but also reduces cost.

- An Initial Review is an essential part of an EMS to determine the important factors that should be managed.

- An EMS is based on 'aspects' and 'impacts' and process mapping is used to identify the aspects and impacts.

- Process mapping is an essential skill – learn how to do it now.

- An EMS can include energy use as an 'aspect' but it is better to use an Energy Management System (EnMS) to provide the specific focus on energy.

- ISO 50001 (the MSS for energy management) is not sector or process specific and does not provide sector specific guidance, benchmarks or targets.

- Energy management needs projects to be generated and delivered to be successful.

- An EnMS system is highly data based and needs large amounts of data to deliver. Good data handling skills are essential.

- Managing health and safety is a legal requirement almost everywhere in the world.

- An Occupational Health and Safety Management System (OH&SMS) provides a framework for health and safety management.

- ISO 45001 (the MSS for occupational health and safety management) is not sector or process specific and does not provide sector specific guidance, benchmarks or targets.

- Implementing a health and safety management system requires hazard identification, risk assessment, risk control and emergency procedures.

- Risk assessments are easy to carry out and document for environmental and health and safety purposes.

- Risk assessments provide a formal method for recording the decisions we make every day.

- The bow tie method is an easy-to-use visual tool for communicating and assessing controls and their effectiveness at all levels of a company.

- The bow tie method can be used for basic quantification of risk.

- Failure Modes and Effect Analysis (FMEA) can provide a rigorous risk assessment and improvement tool.

- FMEA can be used to formally record all the judgements made in the risk assessment process.

- There is no such thing as absolute removal of risk. Risks can only be managed to be 'As Low As Reasonably Practical' (ALARP).

> "Everything must be made as simple as possible. But not simpler."
>
> **Albert Einstein**

> "Dangers lurk in all systems. Systems incorporate the unexamined beliefs of their creators. Adopt a system, accept its beliefs, and you help strengthen the resistance to change."
>
> **Frank Herbert**
> **God Emperor of Dune**

Chapter 2 – Management systems

Chapter 3
Design

This chapter is not titled 'Design for sustainability' or 'Eco-design' for a very specific reason. There should no longer be any need to refer to sustainability in the design process, it should be an integral part of the process, it should not be an 'add on' to the process or a separate later consideration, it should simply be 'part of the process' and considered as much as the type of material, production process or any of the other design decisions.

Therefore, in this chapter I will try not to refer to 'design for sustainability' but simply 'design'. I cannot guarantee that I will be successful because many things need to change in many companies and in the design process.

I am a judge for the 'Design Innovation In Plastics' competition, the longest running student plastics design award in Europe and one of the submission requirements is:

"6. Sustainability: a clear explanation as to how the design addresses environmental, economic and social issues".

This is acceptable guidance for design students where we are telling them what we, as judges, expect to see in a submission. For experienced designers in industry there should never be a need to specify that designs should be sustainable. Sustainability is no longer about the environment; it is the environment.

It is estimated that 80% of the environmental impact of a product is defined by the very first decisions that are made at the design stage (see Section 3.2). This is similar to the amount of cost tied up in the product by the same first design decisions and designers should be conscious that early thought and planning will not only reduce the environmental impact of a product but also reduce the cost of the product.

The design and development process locks environmental impact and cost into the product at a very early stage. The visual methods that most engineers and designers use to think and evaluate ideas often lead them to start to design a product before they have a full idea of what the product is going to do ('ready, fire, aim'). The design then becomes fixed at a very early stage and this fixes the basic impacts and costs early in the process.

Reducing environmental impacts and cost at the design phase is the easiest type of sustainability management and has a rapid payback for the effort involved.

'Design for sustainability' is no longer applicable – it is just 'design' now.

3.1   The design and development process

Managing design

The design and development process, whilst primarily an exercise in imagination and problem solving, is a manageable process. Failure to manage the process will lead to designs and products which have high environmental impacts and costs as well as inevitable project over-runs that will add additional costs and potential lost market opportunities. The financial impact of managing the design process is covered in another book[1] and in this chapter we will try to focus on the sustainability aspects.

Effective management of the design process is one of the most effective of all sustainability management measures. It can attack many of the concerns at source and preventing these is much easier than trying to clean up afterwards, prevention is always the cheapest option.

Design is at the heart of sustainable products. They do not happen by accident and by starting at the design phase we can ensure that the next generation of products are more sustainable than the previous generation. Design is part of the long-game for sustainability.

Over the wall products

The traditional method of product design was the 'over the wall' model (see diagram on the right):

- Design took some ideas informally fed back from Sales on the market demands and designed a new or improved product. The ideas were often vague, poorly defined and even more poorly interpreted. Having completed the design with little concern for manufacturability, tolerances or anything else, they threw their completed design over the wall to Production for manufacture.

- Production attempted to make something like the original design (having changed some things so that they could actually make it) and threw the finished product over the wall to Marketing.

- Marketing took the new or improved product, prepared the sales forecasts, brochures, publicity and marketing information and threw it over the wall to Sales.

- Sales didn't recognise the product from their original feedback on the market demands, had no idea what the product was supposed to do and vainly tried to sell it.

The result was that nobody had any real control, the product had no real ownership, and every time it was thrown over the wall it became 'their problem'. Using this method, it was little wonder that the management of costs, sustainability and most other things was lost every time the product was thrown over the wall.

Team products

In most companies, the over the wall model has now been replaced by team-based product development where a product team is made up of representatives of all the participating departments (see diagram on the opposite page). Team-based design or 'matrix management' uses an ad hoc team to develop the product and all the departments can feed in their knowledge and skills to deliver the product to market. The project team is responsible for the management and completion of the project, under the leadership of the Project Manager. The team has real control and ownership (see sidebar) to effectively develop and deliver products that meet real needs. Sustainability can be built into the product via the Product Design Specification (see Section 3.16) to ensure that consistency of the sustainability message is retained as the product is developed.

Project managers

The project manager's responsibilities should be matched by an equal amount of authority to execute those responsibilities. This authority must be expressed as clearly and as formally as the responsibilities.

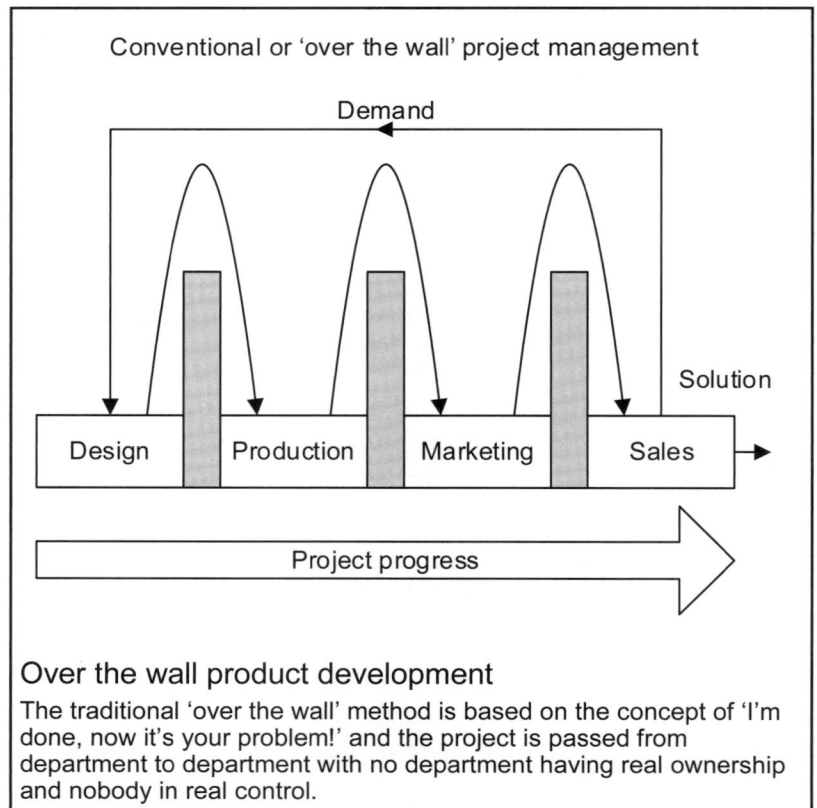

Conventional or 'over the wall' project management

Demand

Design   Production   Marketing   Sales

Solution

Project progress

Over the wall product development

The traditional 'over the wall' method is based on the concept of 'I'm done, now it's your problem!' and the project is passed from department to department with no department having real ownership and nobody in real control.

The project management system

Effective management of the design process should:

- Provide regular and planned new products to the market.
- Assess the merits of projects and decide which projects are started.
- Provide details of the requirements of each project/product early in the project.
- Manage the sustainability and environmental impacts of the product.
- Track progress of projects by proactive reporting.
- Delegate control and accountability to the project manager.
- Monitor progress (and slippage) against agreed time and financial targets.
- Review achievements against the aims.

Projects that do not have effective project management are extremely unlikely to be sustainable in the long-term.

Sustainability built in

Sustainability can be built into the design process by:

- Ensuring that sustainability is significant part of the Product Design Specification (see Section 3.16).
- Constantly focusing on the life cycle of the product, as well as carrying out the standard design activities.
- Ensuring that every final product design is assessed over the complete life cycle of the product and not simply against the internal requirements or sales forecasts. This means identifying, at the design stage, issues such as:
  - The product carbon footprint.
  - The product water footprint.
  - Other sustainability and environmental impacts such as product safety concerns or solvent use.

Building in sustainability is not an easy task for most product designers who have traditionally been concerned with issues such as 'fit and function' but it cannot be ignored simply because it is difficult or new. The rules of design have changed fundamentally for all products.

This means that every new design or product in the future will be assessed as much for sustainability as it is for the traditional functional and visual attributes.

The new rules for design

In addition to the sustainability hierarchy of eliminate, reduce, re-use, recycle, etc., designers in the future will need to become equally familiar with the concepts of re-design and re-imagine.

The sustainability improvement opportunities resulting from good design are exceptional but, in the future, these should be simply regarded as product improvement opportunities with no mention of sustainability.

Project management is not compulsory but then again neither is survival.

The benefits are:

- Projects delivered on time and to budget.
- Projects begin return on investment earlier and bring greater revenues earlier.
- Project costs are brought under control.

- 1. Kent, R.J. 2018. 'Cost management in plastics processing', Elsevier.

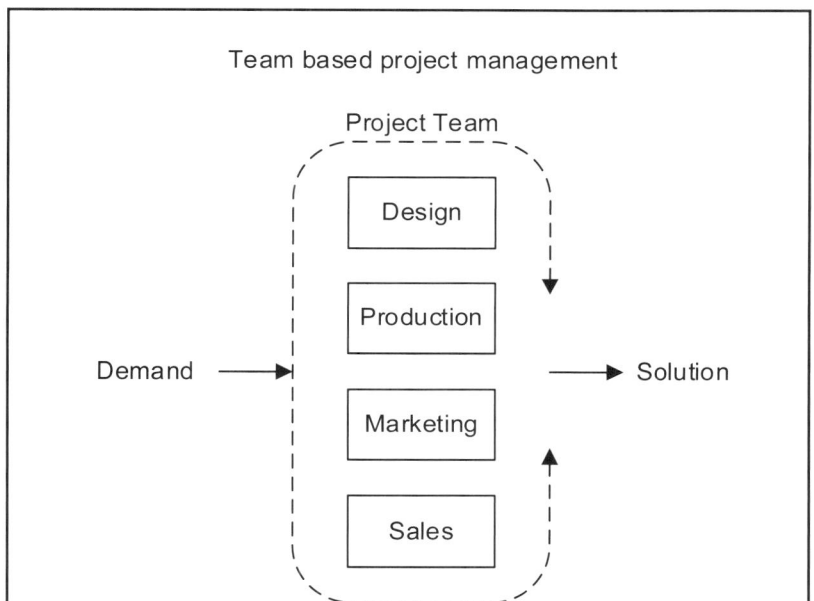

Team based project management

Project Team

Design

Production

Marketing

Sales

Demand → → Solution

Team-based product development

Team-based design or matrix management gives more staff satisfaction and gets results quicker. The project team is responsible for the management and completion of the project, under the leadership of a Project Manager.

Product development teams

Teams must learn to act as Bomber Crews. The project manager is the pilot but he does not work alone. The navigator gets them to the target, the bombardier gets the bombs onto the specified target, the gunners provide protection, the flight engineers keep the machinery running and the radio operator gets some instructions and tells other people where they are. During the flight to the target the Bomber Crew is independent but is working to complete a larger plan decided by Bomber Command, they cannot get much help from home, they are under constant attack and each member depends on the others to survive and to complete the mission.

Only teams that work together survive to fly again.

3.2 Sustainable from the start

The first lines on the paper

Product designers and their assumptions influence, indeed define, the environmental impacts, sustainability and product cost from the very start of the product development process.

The initial stages are where the very basic decisions on the production method, shape and design of the product are taken. The 'simple and obvious' decisions such as the type of material, the production method, the wall thickness and the rough outline dimensions effectively define the most of the overall environmental impact and cost of the product (see diagram on the upper right).

Define the length, width, height, wall thickness and material type and at least 80% of the product cost is already defined. This can only be marginally reduced whatever the detailed design.

At the same time, at least 80% of the environmental impact is also defined. The environmental impact can be reduced by considering the complete value chain, e.g., the use of recycled materials, how the product will be recycled and other aspects, but these actions rarely reduce the cost of the product and may increase it in the first instance.

- **Tip** – Sometimes you don't even have to make any decisions, just thinking of a product defines most of the environmental impact.

- **Tip** – Think broadly about what the product is designed to do and include intangible gains as well as adding value across the full chain.

With any design project the first 15 to 20% of the project involves very little actual spend but it defines and commits between approximately 80% of the environmental impact and between 80 and 90% of the final product cost.

This makes care and innovation at the start of the design process vital for both sustainability and profitability. However, there is almost always an unseemly haste by designers to get past the critical first stage and on with the more exciting stuff of actual detailed design.

There are two simple reasons for this:

- Product designers rarely understand the sustainability and cost impacts of their decisions.

- Sustainability professionals and accountants rarely understand the

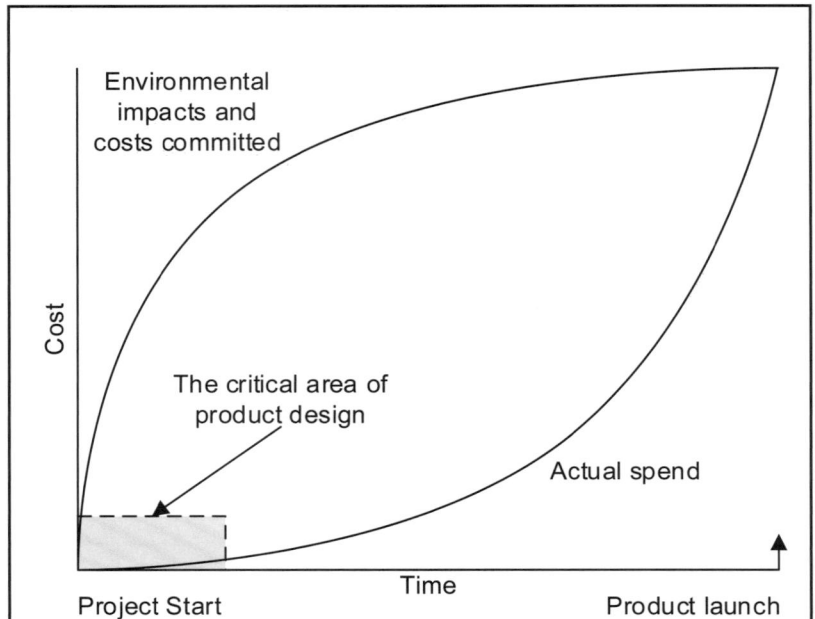

Design fixes the impacts and the costs

The basic decisions made at the early stages of the design process fix the environmental impacts and cost of the product for the rest of its life. Post-design impact and cost reductions are much more difficult to achieve than reductions made during the early stages of the design.

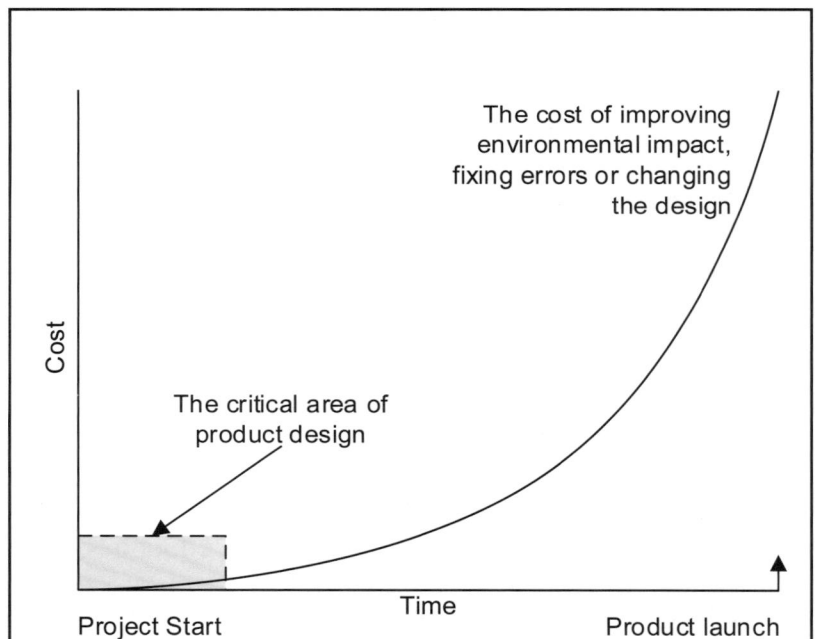

The cost of improving impacts, error fixes or changes

Improving impacts, error fixes or design changes during design and development cost real money. Early error detection reduces costs – extra time spent on validating the design to reduce the impacts is never wasted. Late design changes raise costs exponentially.

technical aspects of product design and how they can influence the design at an early stage rather than just calculating the impacts and the cost after it is all finished.

It may seem ironic, but design freedom is highest in the early stages of the design process and this is the very time at which the least is known about the detail of the design. Getting a good, and sustainable, start to the design is critical to sustainability management and the success of the project – there are rarely second chances in the product design process.

Sustainability and cost management both start with the design and designers must understand the consequences of their actions. Environmental and cost reduction efforts must be concentrated in the early part of the project to get the basics right. Get them wrong and the impacts and the costs are built into the product for life.

- **Tip** – Slow down!
- **Tip** – Sustainability professionals must learn to work with the product designers and product designers must learn to work with sustainability professionals to get designs right at the start of the process.

The costs of getting it wrong

The cost of improving environmental impacts, fixing errors or mistakes or making other changes to a design escalates dramatically as a project proceeds (see diagram on the lower left).

Changes in the initial stages of design are relatively easy to carry out but the cost of any change rises rapidly as actual expenditure is committed or takes place. Care and innovation at the start of a project will drastically reduce any changes necessary and through this reduce the cost of development.

- **Tip** – Sustainability and product development efforts must be concentrated more in the early part of the project to get the basics right.

Do the right things because they are part of the core values even if it is not profitable in the short-term.

Take the long-term test: What will my grand-children think of this decision?

Phase / Area	Raw materials	Manufacturing	Use	End-of-life	Logistics
Product performance	Product Design Specification.		Performance requirements.	Take back programme. Recycling potential.	
Health and safety	Raw materials selection	Working conditions (health and safety).	Health and safety.	Working conditions (health and safety).	
Processing	Raw materials selection	Processing efficiency.		Recycling potential	
Energy	Embodied energy.	Energy efficiency.	Energy efficiency.		Improve logistics efficiency.
Water	Embodied water (water footprint)	Water efficiency.	Water efficiency.		
Waste (materials and consumables)	Sustainable consumption.	Resource efficiency. Lightweighting.	Consumption in use.		
Social	Local suppliers.	Local suppliers.		Littering and dumping. Waste export.	

High impact	Medium impact	Low impact

Mapping sustainability and design across the value chain

Product design affects all the major phases of the product life cycle. It is possible to map the major effects by area. The size of the effect will vary with the product (see Section 7.7). This approach to design matches the approach used in other areas to improve sustainability (see Section 1.7).

3.3   Resource efficient design

The product life cycle

The previous sections covered the basics of resource efficient design from a purely design viewpoint, this section looks at the complete product life-cycle and how design has impacts over the complete life cycle.

The old product life cycle was a strictly linear process: Raw materials were manufactured and purchased, products were made, the consumer/industry used them and then the old products were thrown away (see diagram on the upper right). The bad things that happened along the way and at the end, e.g., pollution, waste, etc., happened to society as a whole and the manufacturers and users 'externalised' them, i.e., it was not the company's or individual's problem and they could be safely ignored.

The new product life cycle (the circular economy) is about the complete life cycle for products (see diagram on the lower right). It is about 'internalising' the bad things that happen, it is the 'polluter pays' principle. This circular cycle is being driven by resource constraints (both energy and materials), resource security (both energy and materials), environmental concerns, pressures on destinations (see Section 1.10), pressure groups and by Governments (at the supra-national and the national levels). In most cases, the drivers for this new product life cycle are not negotiable for industry and the plastics processing industry can only respond to ever-increasing directives and legislation.

The new product life cycle involves satisfying the customers' requirements, whilst using the minimum resources and creating the minimum environmental impact over the product's entire lifecycle. What used to be considered as 'external' is rapidly become 'internal'.

Design is a key element of this evolving transformation of the product life cycle and designers need to be aware of the new requirements and change their designs and thinking.

The industrial revolution used machinery to increase labour efficiency and reduce costs, the sustainability revolution is starting to use design and the new product life cycle to improve resource efficiency, reduce environmental impacts and reduce costs. Rapidly improving resource efficiency requires management and control of the complete product life cycle and design is the first and fundamental step towards meeting the new demands of the product life cycle.

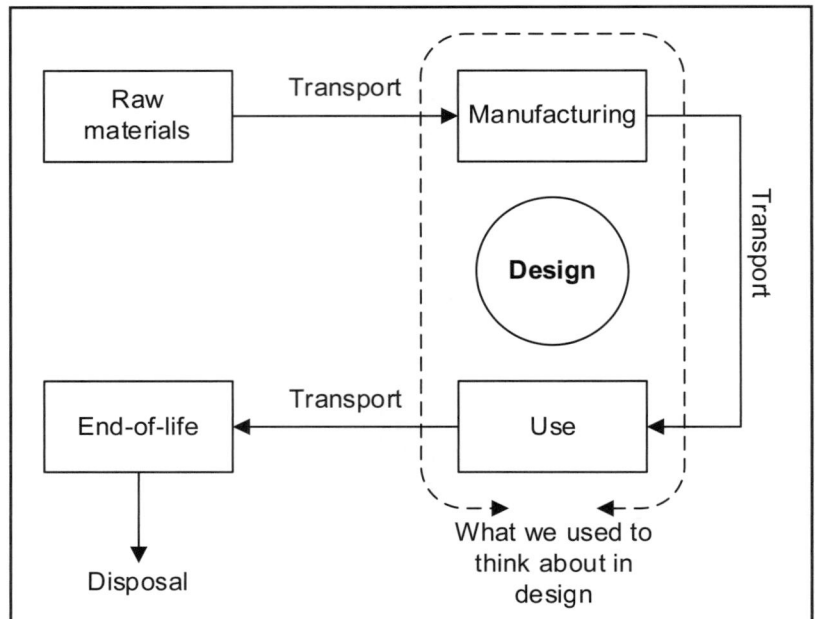

The old product life cycle – an unsustainable future
The old product life cycle was a linear process. Collect the raw materials, process, use and dispose of the remains. There was limited in-house recycling but post-use recycling was rare and materials exited the system totally at the end-of-life phase.

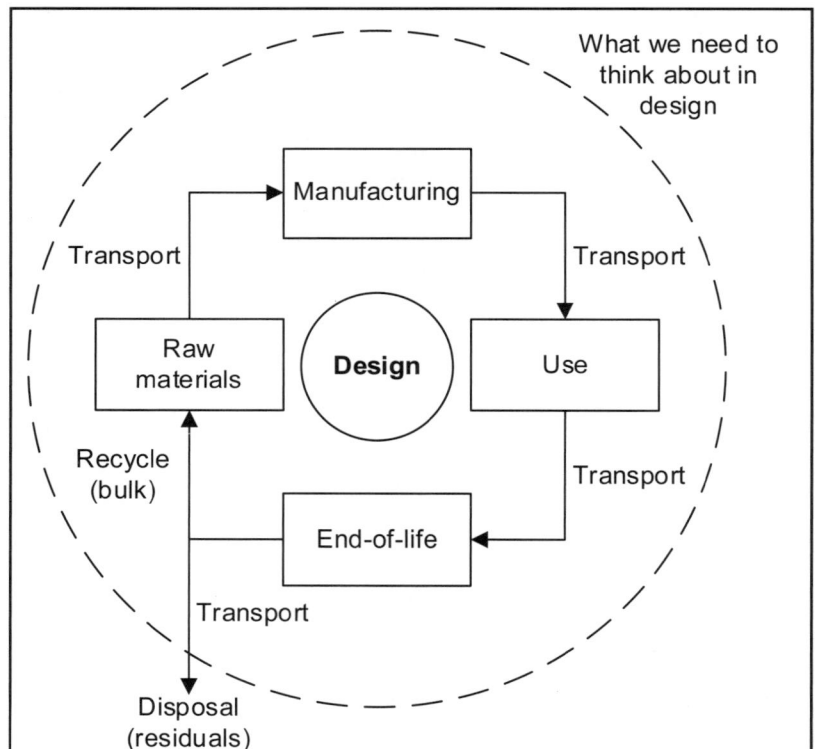

The new product life cycle – the key to the future
Resource efficient design is not a single task but a system for the whole life cycle. Materials only exit the system when recycling is not possible. This is sometimes referred to as the 'circular economy'. The design process is central to the circular economy and sustainability.

The life cycle issues of the future will be in the areas of:

- Manufacturing.
- Use.
- End-of-life.
- Raw materials.

Transport between all of the areas can also be related to resource efficient design and this is covered as part of sustainable distribution (see Section 5.17).

The key design technique will be the use of 'resource efficient design' and product designers need to understand that the rules have changed. Resource efficient design means:

- Reducing resource use and converting the resources that are used into higher added-value products in the most effective way.
- Reducing all environmental impacts.
- Achieving competitive advantage.

Why focus on products?

One study[1] has shown that:

- 93% of production materials do not end up in saleable products.
- 80% of products are discarded after a single use.
- 99% of materials used in the production of, or contained within goods, are discarded in the first six weeks.

Note: This is not simply for plastics but for all products and production materials.

Although this study is now over 20 years old, the problem has not gone away and in fact it appears to have become worse or, at the very least, more visible. Products consume and waste vast amounts of resources and at least 50% of a product's overall environmental impact is a result of the design decisions taken at the start of the design process (see section 3.2). Focusing on improving product design and using resource efficient design concepts at the start can produce remarkable improvements in resource efficiency and reduce a range of environmental impacts as well as improving profitability.

What is resource efficient design?

Resource efficient design is design to minimise the environmental impacts over the entire product life cycle and to meet customer requirements. It is a proactive tool to reduce resource depletion, waste, pollution and the environmental impacts of a product throughout the life cycle. In many cases, it will also reduce the product cost.

Resource efficient design involves meeting the customers' requirements whilst using the minimum amount of resources and creating the minimum environmental impact.

Resource efficient design aims to move from the traditional linear approach to product life (manufacture, use, dispose) to a more cyclical approach to allow material to be re-used or recycled at the end-of-life phase.

Design through the whole cycle

Sections 3.4 to 3.14 will consider each of the phases of the product life cycle in terms of:

- Drivers – what will drive the process of resource efficient design. These can be legislative or cost-based.
- Strategies – the overall strategic actions needed to implement resource efficient design. These will be the general responses to the drivers.
- Tactics – the detailed tactical actions to achieving resource efficient design. These will be the specific responses needed to achieve the strategies.
- Results – the financial and other benefits from implementing resource efficient design.

- 1. von Weizsäcker et al, 1998, 'Factor Four', Routledge.

Start at the top to get a commitment to resource efficient design and environmental thinking across the company.

Phase	Key Task
Raw materials	Minimising inputs
Manufacture	Targeting efforts
Use	Optimising use
End-of-life	Minimising outputs

The key phases and tasks of resource efficient design

Each of the phases of resource efficient design has a key task within the new product life cycle. All these tasks must be completed to fully achieve the circular economy and to continue to be profitable into the future. Good design helps to achieve all the tasks.

3.4 Raw materials – minimising inputs

Minimising inputs

The key task in the raw materials phase of resource efficient design is minimising the inputs to the product life cycle. Finite resources, finite production capacity and increasing demands will automatically translate into rising raw material costs. The demands will increase even more rapidly as other nations aspire to and attain the living standards of the West. These are fundamental changes that will require huge amounts of raw materials and increase both the resource depletion rate and prices. Polymer supply is a world market and a buoyant market in the Far East always translates into rising prices in the West.

The challenge in the long term is to control costs by planning for circularity. This means designing to reduce the use of virgin materials and increase the use of recycled materials.

Current materials

Some materials currently used in products cannot be recycled and create significant environmental impacts during their production. A good starting point for assessing material suitability is to prepare a list of the materials used in the product. These can then be investigated to find alternatives with lower environmental impacts and costs. Suitable materials might be:

- Recycled or contain recycled materials.

- Obtained from suppliers that are environmentally conscious.

- Capable of recycling at the end-of-life stage.

- **Tip** – Suggestions for suitable alternative materials or potential opportunities for recovery, re-use and recycling of materials should be sought from suppliers and customers.

- **Tip** – As well as looking at the types of materials used in the product, it is important to look at the quantities and diversity of materials used. There may be opportunities to redesign the product to reduce the weight and thickness of components or to use one recyclable material for the entire product.

- **Tip** – Reducing the number of materials used will also improve the recyclability of a product. Unless absolutely necessary, design products to be made from PE, PP or PET. These are all widely and easily recycled. PA may be ideal for the

application but mechanical recycling for PA (see Section 4.6) is virtually non-existent and the only realistic options are landfill, incineration or chemical recycling.

Restricting materials choices at the design stage is easier than trying to substitute materials when the design is complete and the product is being produced.

Examination of current materials must be ruthless in the search for reduced environmental impact. In some countries, products made from recycled materials already attract a price premium; in the future this may become the norm rather than the exception. The markets are changing and plastics processing must change to meet the consumer demands.

Less material

The first target for reducing environmental impacts and costs should always be to use less material. This reduces materials costs, resource use, transportation and the amount of waste for treatment when the product reaches end-of-life.

Reduce the materials used in the product by:

- Analysing how the main product function is delivered and whether it can be delivered with less material or even without the material at all. This can often be achieved, without compromising quality, through a detailed understanding of the product function and improvements in manufacturing technology.

- Retaining the current form and reducing

The use of recycled material can save material and environmental costs, but there is a need to invest in post-consumer structures for material collection and processing.

Companies such as Lego™ are already embarking on the search for new lower environmental impact materials.

They are getting ahead of the curve.

Drivers

- Consumer and legislative pressure.
- Compliance with environmental design standards (e.g., Integrated Product Policy and WEEE) will become a legislative requirement.
- EMS will effectively become mandatory for manufacturers.
- Competition and price for recycled raw materials will increase as demand and use increases.
- Raw material constraints, caused by both resource depletion and growing demand, will increase prices of both products and utilities.

Strategies

- Development and implementation of company strategy for purchase of recycled, renewable materials.
- Development and implementation of company strategy for use of renewable energy.
- Long-term and sustainable corporate environmental plans.
- Full implementation of resource efficient design principles.

Drivers and strategies for resource efficient design in the raw materials phase

material use by thinner sections or reduced numbers of fixings.

- Reducing the part count by combining parts but still retaining functionality.
- Using the product design team to identify areas where material can be used more efficiently.

The process of 'lightweighting' or 'dematerialisation' not only brings environmental benefits, but also reduces manufacturing and transport costs and increases profits (see Section 4.2).

Less environmental impact

The second target is to reduce the environmental impact of the materials used in both the product and the production process. This will reduce the costs and environmental impacts associated with the product life cycle.

Reduce the environmental impact of the materials used in the product by:

- Using renewable materials and recyclates instead of virgin materials to reduce resource depletion and create opportunities and markets for using waste, thus diverting it from disposal.
- Using materials that have less environmental impact during their production, i.e., use less energy or cause less pollution during production, will reduce a product's environmental impact and can also reduce the need for expensive controls during production.
- Eliminating or replacing hazardous substances from both the product and the production process. This will reduce the costs and environmental impacts associated with the product life cycle.

Many plastics processors and their customers are already carrying out this work by:

- Increasing and investigating the use of recycled materials.
- Developing 'black' lists (banned substances) and 'grey' lists (substances whose use should be limited) for use by component suppliers.

  Some customers are also developing RAG (Red, Amber, Green) lists of materials and product formats, e.g., Tesco.
- Developing materials declaration tools to help suppliers document the material content of their products.

The future

In the future, the raw materials used will define the environmental impact of the product even more than today. Incorrect materials choices will increase the impact at all stages of the life cycle.

The correct materials choice will only be possible by knowing the impact and costs of the materials used over the complete product life cycle. This can be achieved by:

- Collecting information on possible material substitutes that are:
  - Less hazardous.
  - From renewable or recycled sources.
  - Produced with less environmental impact.
- Identifying materials databases that contain information on environmental impacts.
- Requiring suppliers of materials and components to provide detailed materials declarations as part of their supply contract.
- **Tip** – Use a formal materials declaration list to collect the information.
- **Tip** – Initially it will be difficult to obtain information about every part of every component but as the requirements become more common it will become easier.
- **Tip** – Ask suppliers to provide proof of any assertions they make.

Design for constraints and reduced environmental impact involves a transition to resource efficient design that addresses the complete product life cycle.

A focus on 'lightweighting' and materials selection in packaging is essential to reduce environmental impact.

Lightweighting can reduce manufacturing costs by reducing the amount of energy needed to process the plastic.

Raw materials should also be considered in terms of REACH and RoHS (see Section 3.6) and WEEE (see Section 3.14).

Tactics

- Work with customers to define the real product needs.
- Work with customers to reduce the amount and number of materials used.
- Work with customers to remove hazardous materials from products.
- Work with customers to introduce recycled and renewable materials.
- Work with customers to gain acceptance of new life cycle of all products.
- Promote and sell provable environmental benefits to the marketplace – no 'greenwashing' allowed.
- Introduction of sustainable technology.

Results

- Winners and losers.
- Transformation of the marketplace.

Tactics and results for resource efficient design in the raw materials phase

---

3.5 Raw materials – design for recycling

Design for the future

The correct selection of materials at the design stage is key in terms of sustainability. Getting the materials selection right at the design stage makes the future recycling of the product much easier. The increasing societal and legislative pressure to recycle needs to be considered at the design stage. This is about considering the complete product life cycle and preparing for recycling at the design stage.

Design for recycling is important but designers should not prejudice the design intent just to allow recycling. In the past, recycling was rarely considered at all, now it is simply a normal part of the overall design brief.

- **Tip** – Design for recycling and sustainability, but not at the expense of function or service life.

Reduce the number of materials

Whatever material is chosen, sorting and recycling is easiest if only one basic polymer is used and it is even easier if the number of grades of material has been reduced. Designers should therefore reduce variety and optimise the number of different plastics within a single product.

- **Tip** – Reduce the types and grades of plastic used and use versatile materials with a wide range of applications.
- **Tip** – If it is not possible to reduce the number of materials used then the design should allow the different materials to be easily separated (see Section 3.13).
- **Tip** – The design stage is also the best time to design in the use of recycled materials.

Choose materials to maximise mechanical recycling

Having reduced the number of materials, designers should attempt to use plastics that are easily and commonly recycled. This not only increases the possibility of recycling at the end-of-life but also increases the potential for using recycled materials.

- The plastics that have the highest proportion of mechanical recycling are PE-HD, PP and PET.
- **Tip** – PVC is also widely recycled although this is primarily in closed-loop industrial schemes rather than as Post-Consumer Waste (PCW).
- The films that have the highest proportion of recycling are PE-LD, PE-LLD, PE-HD and PP.

These materials are all widely available on the market and are preferred to improve the potential for recycling at the end-of-life.

Sorting at most recycling centres uses near-infrared (NIR) sorting and this is extremely efficient (> 95%) for clear plastics or coloured plastics using NIR detectable colours. For very dark or black products or those using non-NIR detectable colours, NIR can fail to separate the plastics. In these cases, flotation techniques based on the density of the material can be used to separate the materials if these are available at the site.

- **Tip** – Use clear, uncoloured material whenever possible to maximise recycling potential (and potentially reduce costs).
- **Tip** – If black or dark colours must be used then ensure that there is a density gap of at least 0.15 g/cm^3 to allow separation by flotation techniques (if used).

Choose compatible materials

If it is not possible to reduce the number of polymers then the polymers used should, at the very least, be compatible and capable of forming a polymer blend when reground and recycled. Some polymers are very poor at blending with other materials, e.g., PE has very poor compatibility with most other polymers (except naturally PP) whereas ABS is compatible with some other polymers (see the compatibility chart on the lower left). Selecting compatible materials can allow bulk regrinding and compounding rather

> Design for recycling is not a linear process. You may have to go through this process several times to get the best results.

> There are other free resources on polymer properties, polymer processing, the 'Periodic Table of Thermoplastics' and more at:
>
> www.tangram.co.uk.

		Primary material							
		ABS	PA	PC	PE	PET	PP	PS	PVC
Secondary material	ABS	✓	?	✓	?	?	?	?	✓
	PA	?	✓	X	?	?	?	?	X
	PC	✓	X	✓	?	✓	?	?	X
	PE	x	?	?	✓	X	✓	X	?
	PET	✓	?	✓	?	✓	?	?	X
	PP	X	?	X	?	X	✓	X	?
	PS	?	?	?	?	?	?	✓	?
	PVC	✓	X	X	?	X	?	?	✓

Compatibility chart for common thermoplastics

Compatibility varies depending on whether the material is the base material or the minor material.

✓ = Compatible.

? = Depends on the volume ratio, i.e., OK at low volumes (<1%).

X = Incompatible.

After Bayer AG (95)

than needing disassembly and sorting.

If incompatible materials are recycled together then the mixed recyclate will have reduced physical properties and may be rejected from the recycling process.

- **Tip** – If possible, use a single material for the complete product.
- **Tip** – If using a single material is not possible then use compatible materials to reduce dismantling and sorting.
- **Tip** – If using compatible materials is not possible then reduce the percentage used for the minor material to < 1% (2% is the upper limit).
- **Tip** – If using compatible materials is not possible and the volume of the incompatible components must be > 2% then design the product to allow physical separation of the incompatible components (see Section 3.13).

Remove other materials

If an assembly is to be recycled then the presence of any non-plastics materials, e.g., metal inserts, screws and clips, labels, adhesives and paints makes recycling difficult and where possible these should be eliminated at the design stage (this will also reduce cost). If there are metals present then the product will have to be disassembled before the plastic can be recycled (see Section 3.13).

- **Tip** – Painting not only makes recycling difficult but can have other environmental impacts such as VOCs (see Section 9.13).

Recycling any plastic is more complicated and difficult if the material contains significant amounts of fillers or additives. Some are good for recycling, e.g., fillers such talc, $CaCO_3$ and TiO_2, but others are poor, e.g., glass or vegetable fibres (>10%), nano-particles, flame retardants and any additive containing heavy metals (see Section 3.6).

- **Tip** – Use the minimum amount of fillers and additives.
- **Tip** – Ensure that any fillers used are compatible.
- **Tip** – Minimise the use of flame retardants and any additive containing heavy metals.

Identify the materials

One of the difficulties in mechanical recycling is in identifying the basic raw material even with NIR sorting. This can be aided by identifying the base material with the ISO standard symbol that includes the internationally recognised abbreviation for the base polymer (see diagram below).

- **Tip** – The primary material should be marked on all plastics parts, using the standard symbols and abbreviations.

Standard mould inserts are readily available and should be used to identify the base material. Alternatively, mark the material type with stamping or laser printing.

If the material cannot be recycled then, in many cases, the energy contained in material can be recovered via incineration (see Section 4.11).

Another reason for avoiding the use of brominated flame retardants and any additive containing heavy metals.

RecyClass (recyclass.eu) has a product certification scheme to assess and certify the recyclability of products. This works through accredited Certification Bodies in various countries.

Additional resources

Some excellent, comprehensive and free guides to raw materials selection (primarily for packaging but still relevant) are:

- 1. The BPF has a directory to most of the tools, guides and LCAs at: ecodesign.bpf.co.uk/resources. This is an excellent resource.
- 2. 'Recyclability by Design', Recoup, 2020, www.recoup.org.
- 3. 'PackScore', BPF 2020, www.bpf.co.uk/design/packscore/packscore.aspx.
- 4. 'Design for the Environment Guidelines', Plastics New Zealand, 2006, www.plastics.org.nz.
- 5. 'APR Design Guide for Plastics Recyclability', Association of Plastics Recyclers, 2018, www.PlasticsRecycling.org.
- 6. 'Design for Recycling Guidelines', RecyClass, 2020, www.recyclass.eu.

If you want a good laugh then consider the comment that I ran across on the Internet that said: 'The identification numbers indicate how many times the plastic can be recycled'.

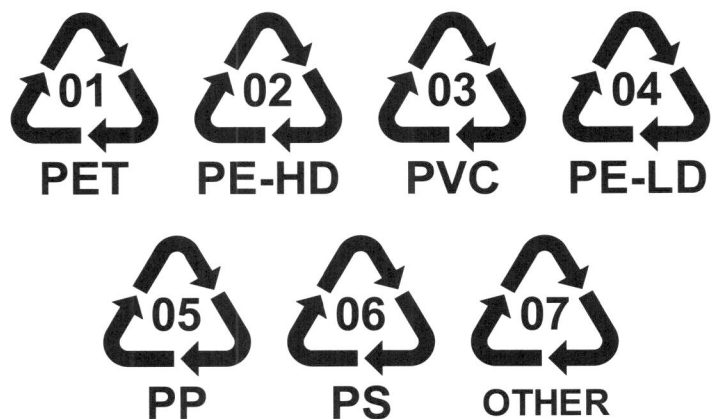

The standard plastics codes

The standard codes were developed by ASTM but adopted by the EU. They now cover materials and items ranging from steel (40 FE) to lithium batteries (12 Li) and even glass (70 GL if it is clear to 79 GL if it is gold backed glass). Plastics are not alone.

3.6 Raw materials – REACH and RoHS

The legislation

There are two main EU Directives that affect raw materials and hence many product designs. These are:

- REACH (Registration, Evaluation, Authorisation and restriction of Chemicals) requires that any company manufacturing or importing into the EU a chemical substance has to register it (depending on certain conditions) with the European Chemicals Agency (ECHA).

  With the departure of the UK from the EU at the end of 2020, the REACH legislation was replicated into UK law as UK REACH. This has very similar requirements but the regulations operate independently of one another. To minimise disruption, it is possible for GB-based holders of EU REACH registration to 'grandfather' these approvals into UK REACH approval.

  This is, however, a developing situation and UK-based companies should check the current requirements at www.hse.gov.uk/reach.

- RoHS (Restriction of Hazardous Substances) restricts the use of certain hazardous substances in electrical and electronic equipment at the design and manufacturing stage.

REACH

Reducing the environmental impact of a product starts at the design phase where eliminating or reducing the use of hazardous substances is most effective.

EU REACH and GB REACH are very similar to the American Toxic Substance Control Act (TSCA). All require that manufacturers and suppliers should ensure that their products do not contain any material that is a 'critical or hazardous' substance and that suitable controls are in place. REACH sought to 'tidy up' over 40 existing directives into one regulation. Any company manufacturing, importing or distributing in Europe and the UK must comply with EU REACH and UK REACH. REACH covers everything from cars to clothing.

The stages of REACH and the process

REACH was introduced in stages from June 2007 and since 01 June 2018 requires registration for substances supplied at ≥ 1 tonne/year.

REACH places the burden of proof on producers and the process for REACH approval for any substance is:

- Registration – this is done with the ECHA (in the EU) or the HSE (UK) and requires a technical dossier and chemical safety report.

- Evaluation – registered substances are evaluated for potential hazards or risks.

- Authorisation – this does not mean that substances are authorised for use, it means that they must be authorised (for a specific use) before they can be used. Substances 'authorised' under REACH are listed in Annex XIV of REACH and this currently contains 31 substances of very high concern (SVHC).

The existing Annex XIV list has been retained under UK REACH.

Note: There are separate lists of REACH restricted substances (which in July 2020 contained 54 substances) and a candidate list for authorisation (which in July 2020 contained 209 substances).

- Restriction – substances that have unacceptable or uncontrollable risks will be restricted from the EU.

- **Tip** – Authorisation under REACH for any application applies to all fields of application.

- **Tip** – When REACH was first introduced there was an exemption for recycling of

'Producer' means any person who:

- Manufactures and sells own brand of product.
- Resells equipment produced by other suppliers under his own brand.
- Imports or exports products into an EU Member State.

'Distributor' means any person who provides products on a commercial basis to a party who is going to use it.

Fines and penalties for using 'authorised' substances in the wrong way vary from country to country and with the level of the breach but be assured that it will be expensive.

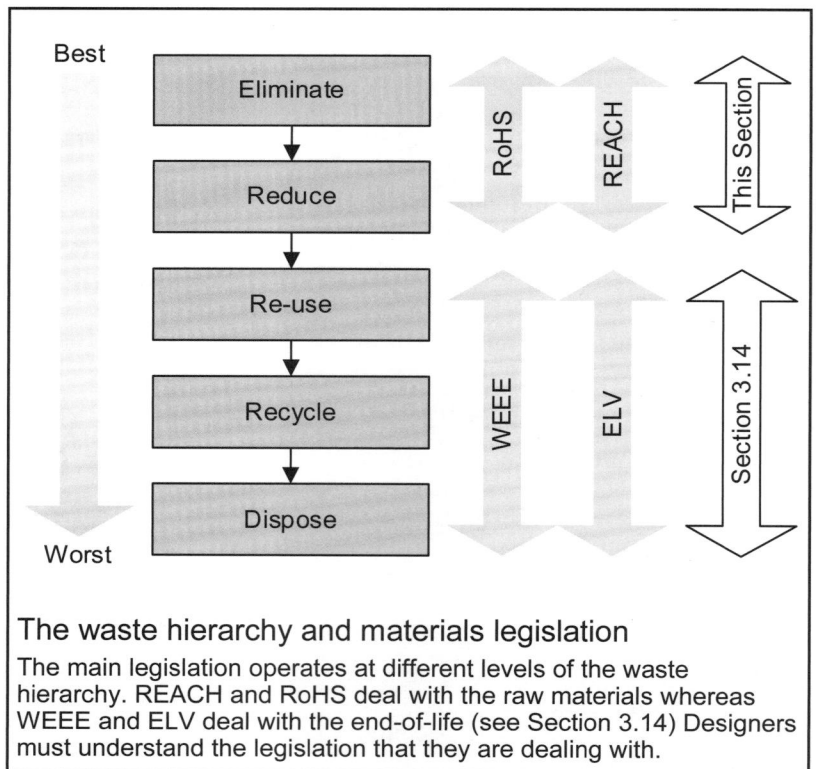

The waste hierarchy and materials legislation

The main legislation operates at different levels of the waste hierarchy. REACH and RoHS deal with the raw materials whereas WEEE and ELV deal with the end-of-life (see Section 3.14) Designers must understand the legislation that they are dealing with.

PVC products. In February 2020, the European Parliament voted against a proposed derogation to allow lead-based stabilisers in recycled PVC (despite ECHA advice recommending the derogation). This decision threatens the PVC recycling industry and the credibility of the European strategy for plastics in a Circular Economy. The best is truly the enemy of the good.

Materials Safety Data Sheets

Under REACH, it is the responsibility of the manufacturer or importer to know the chemical composition of their products and to be able to prove the safety of all the substances used in it. The relevant Materials Safety Data Sheet (MSDS) should provide all this information. Designers must ask suppliers for these for any material being considered as part of the design process and suppliers must be prepared to supply MSDS for any substance supplied.

- **Tip** – Don't just ask, make sure you get it.

RoHS

RoHS deals with the elimination and reduction aspects of the waste hierarchy but only for electrical and electronic equipment.

The affected substances

The current restricted materials under RoHS are:

- Lead (Pb).
- Mercury (Hg).
- Cadmium (Cd).
- Hexavalent chromium (Cr^{6+}).
- Polybrominated biphenyls (PBB).
- Polybrominated diphenyl ether (PBDE).
- Bis(2-ethylhexyl) phthalate (DEHP).
- Butyl benzyl phthalate (BBP).
- Dibutyl phthalate (DBP).
- Diisobutyl phthalate (DIBP).

Manufacturers need to demonstrate that their products do not contain more than the maximum permitted levels of any of the substances listed above. The maximum permitted levels are < 0.01% by weight for cadmium and < 0.1% by weight for the other substances in a homogeneous material. A 'homogeneous material' is a single substance such as a plastic, e.g., the plastic used in the insulation of a wire is considered to be a single substance. The assembly of 'wire + insulation' is not considered to be a single substance but a component.

Some of the substances included in the RoHS list have been used extensively in the plastics in the past, e.g., lead-based stabilisers in PVC-U, cadmium colorants in red pigments and PBB and PBDE as flame-retardants in plastics.

- **Tip** – In general RoHS takes precedence over REACH for the 10 listed substances but RoHS 2 stated that REACH took precedence for phthalates in toys.

Where are the affected substances likely to be used in plastics?

In most cases the affected materials have already been phased out from plastics but designers of components and products covered by RoHS are still advised to confirm with suppliers that the affected substances are either not used or are used in less than the maximum permitted levels.

- **Tip** – Despite the phase out of these materials, there are still significant issues in recycling materials containing these chemicals.

Proving compliance and policing

Proving compliance requires a producer to 'self-declare' compliance. The simplest approach for producers involves two steps:

- Obtain an assurance from all suppliers that no banned substances are present (except where they are exempt from the requirements) and keep a permanent record to show that 'reasonable' steps have been taken to comply with RoHS.
- Carry out an analysis of products to verify the supplier's declarations or to provide assurance where the supplier's declarations are either unavailable or unreliable. Where there may be a high probability of the product or component containing one of the banned substances, e.g., PVC products or red/orange coloured plastics, then analysis may need to be carried out more frequently.
- **Tip** – The act of placing a product on the market with the CE mark is a declaration that the product complies with RoHS.
- **Tip** – There are some limited exemptions to the RoHS regulations based on substances and applications, but the detailed lists given in the RoHS Annexes should be consulted.
- **Tip** – RoHS, or similar, schemes operate around the world, e.g., USA and China.

Summary

REACH and RoHS are part of the legislative response to controlling the raw materials stage in the product life cycle. They are only the first of these and designers need to adopt a design process to reduce future costs resulting from the 'producer pays' or 'polluter pays' principle.

RoHS has a distinct list of controlled chemicals (although this may be increased in future).

REACH is intended to apply to all chemicals throughout the EU.

Check the web sites for the EU-approved lists – there are some lists created by industry groups but the only ones that count are the official EU lists.

REACH fundamentally changes the legislation. Previously you were 'innocent until found guilty', i.e., the burden was on public authorities to prove that substances were harmful. Now, you are 'guilty until found innocent', i.e., the burden is on producers to prove that substances are safe for human health and the environment.

This is a much higher hurdle because 'absence of evidence is not the same thing as evidence of absence'.

Under Article 2 (2), REACH does not apply to waste. Waste for recycling is exempt if the status of the waste is retained in the process and if it is converted directly into a product.

If recyclate is placed on the market then REACH applies.

3.7　Manufacture – targeting efforts

Targeting efforts

The key task in the manufacturing phase of resource efficient design is targeting efforts that are currently misaligned with the realities of the real environmental impacts and costs.

Targeting the efforts means changing the emphasis from designing to reduce labour costs to increasingly more important areas. It means keeping the labour and sacking the kilograms (materials), kilowatt-hours (utilities) and environmental impacts.

Targeting the efforts means designing the product so that the manufacturing process uses less process materials, energy, water and other resources. Resource efficient design focuses on the product to reduce all waste and pollution during manufacture.

This increases resource efficiency and reduces the environmental impacts of production.

Sustainable manufacturing

The design defines the manufacturing process used and all modern production processes have significant environmental impacts. However, all processes are not equal in terms of resource use, e.g., injection moulding is not as energy efficient as extrusion, or in terms of the use of other consumables, e.g., water use.

Resource efficient design in the manufacturing phase is design for both manufacture and assembly to:

- Reduce utilities use – designers need to consider not only the raw materials physically incorporated into the product (see Section 3.4) but also the utilities consumed during manufacture, e.g., energy and water (see Chapters 6 and 8).

- Reduce hazardous materials use – designers need to consider hazardous materials not only in terms of those incorporated in the product (see section 3.6) but also to use design to reduce or eliminate the use of hazardous materials during production.

- Reduce consumables use – not all consumables are hazardous and designers need to use design to reduce the use of all consumable during production. These can be identified by a waste minimisation programme (see Chapter 9).

- Improve the reliability of manufacture – designers should always use the standard detailed design rules (see Section 3.8) will make products easier to manufacture and

to reduce issues with product quality both during manufacture and use.

- Reduce the number of components and materials in the product – this will reduce raw material and assembly costs and increase the recyclability of the final product.

- Reduce the time taken to assemble and disassemble the product – good design practice can (see Sections 3.9, 3.10 and 3.13).

- Reduce waste sent to landfill at end-of-life.

- Lower production costs.

Designing the product so that less pollution and waste occur during manufacture will also reduce local environmental impacts and may lead to safer working conditions for employees.

Designers can no longer live in isolation from the manufacturing of the product. They need to use resource efficient design in the manufacturing phase as a tool to contribute to the sustainability of the product.

What do designers need to know?

Designers should answer the following questions to identify potential areas for improvement:

- How much energy do the process use and what are the opportunities for reducing energy use?

- How much waste does the process produce?

Carbon footprint

The carbon footprint of a product is the total amount of CO_2 (or the equivalent amount of CO_2 for other emissions) emitted by a product during the complete product life cycle.

It is an approximate measure of the climate change impact of a product over the product life cycle.

See Section 7.6 for details of product carbon footprinting.

Drivers

- Increasing environmental legislation from the supra-national (e.g., EU), national and local levels.
- Increasing resource costs, e.g., Climate Change Levy, as an incentive for reducing resource use.
- Increasing disposal costs, e.g., Landfill Taxes, as an incentive for reducing resource disposal.
- Integrated Product Policies driving consistent approaches to resource use and products.
- Continued profitability and survival.

Strategies

- Improve relative resource efficiency for all resources.
- Reduce the amount of resources used in absolute terms, e.g., materials, energy and water.
- Introduce resource efficient design concepts at the design level to reduce future costs.
- Set demanding objectives for reduced impacts and costs.

Drivers and strategies for resource efficient design in the manufacturing phase

- How can the amount of waste be reduced?
- How many different types of waste does the process produce?
- How can the number of types of waste be reduced?
- Is any of the waste produced by the process classed as hazardous (special waste)?
- What natural resources (e.g., water and fossil fuels) are used?
- Can the use of resources be reduced?
- How can the use of consumables, e.g., mould cleaners, lubricants etc., be reduced.
- Can improved technology be used to reduce resource use?
- **Tip** – Ask the production director for information and talk to operators about the sources of unnecessary waste.
- **Tip** – Calculate the resource consumption and environmental impacts for different components or production processes. Review the product design to reduce these.

What are the results?

Manufacturing and the environmental impacts it generates are not separate processes in resource efficient design; they are integral to the overall design process.

- Products with reduced environmental impacts during their life cycle.
- Improved product function and quality.
- Longer product design life.
- Cost savings.

Targeting the efforts

Targeting the efforts means a renewed focus on the environmental impacts of manufacturing. The short-term actions should be:

- Set demanding but realistic objectives for reduced impacts and costs.
- Use the existing company records and product specifications to research products and resource use.
- Start internal work to reduce energy use (see Chapter 6).
- Start internal work to reduce water use (see Chapter 8).
- Start internal work to reduce waste (see Chapter 9).
- Start internal work on resource efficient design to reduce environmental impacts and costs.
- Dismantle current products (both internal and competitor's products) to see how easy they are to recycle.
- Start to use life-cycle assessment (LCA – see Section 1.5) and product specific checklists.

- Benchmark your product's environmental performance against previous products or competitors' products but be careful to compare like with like products.
- Keep abreast of forthcoming changes to legislation.

> Minimising the number of manufacturing steps allows a company to stay in business by taking out cost while keeping functionality.

> Resource efficient design is an integral part of good design to improve environmental performance, maintain competitive advantage and reduce costs. It is no longer an option – it is the environment.

Tactics
- Survey, measure and target the real resource use costs.
- Work with customers to reduce materials use and costs.
- Benchmark real resource costs against competitors to set future targets for reduced resource use.
- Invest in improved technology to reduce resource use.
- Reduce waste at all levels of the supply chain.

Results
- Energy bills reduced by 10-20% from no-cost and low-cost measures (see Chapter 6).
- Water use reduced by 10-20% from no-cost and low-cost measures (see Chapter 8).
- Waste reduced by 75% from no-cost and low-cost measures (see Chapter 9).
- Cost savings from improved resource efficiency and profits improved by 25-30%.

Tactics and results for resource efficient design in the manufacturing phase

3.8　Manufacture – detailed design

The basics

A product is not sustainable if it does not function correctly or fails in service. This makes detailed design critical for plastics products. Most of the fundamental design rules are the same whatever the process and these are relatively easy to set out and learn.

However, there are some essential differences between plastics and other materials that need to be clearly understood by anybody designing or using plastics products. These important differences are:

- Plastics are visco-elastic materials and are very sensitive to temperature and the rate of loading. In fact, there is a principle in polymer physics called the 'time-temperature superposition'. This was developed by Williams, Landel and Ferry (WLF) to explain the changes in plastics properties with short-term and long-term loads and with low and high temperatures. Simply stated, the time-temperature superposition means that a plastic product will broadly react the same under a long-term load at a low temperature as it will under a short-term load at higher temperature, i.e., the time and the temperature are roughly equivalent.

- Product designs for metal or other materials cannot be transposed to plastics. The behaviour of the plastics product will be very different at high temperatures or long times (whatever the data sheet says) and failure may result.

- The chemical resistance of plastics, particularly for environmental stress cracking is highly affected by increased temperature. Raising the temperature increases the potential for environmental stress cracking dramatically.

Understanding these differences is basic to the design of high-quality products that are capable of good sustainable long-term service.

Product Design Specification conformity

Whatever design rules applied, conformance to the Product Design Specification (PDS – see Section 3.16) is a basic prerequisite. If the design does not meet the PDS then it is not a good one even if it conforms to all the rules. The PDS is the guide for the overall design intent and every design must meet it.

The design rules

The design rules for the various processes are relatively material-independent, i.e.,

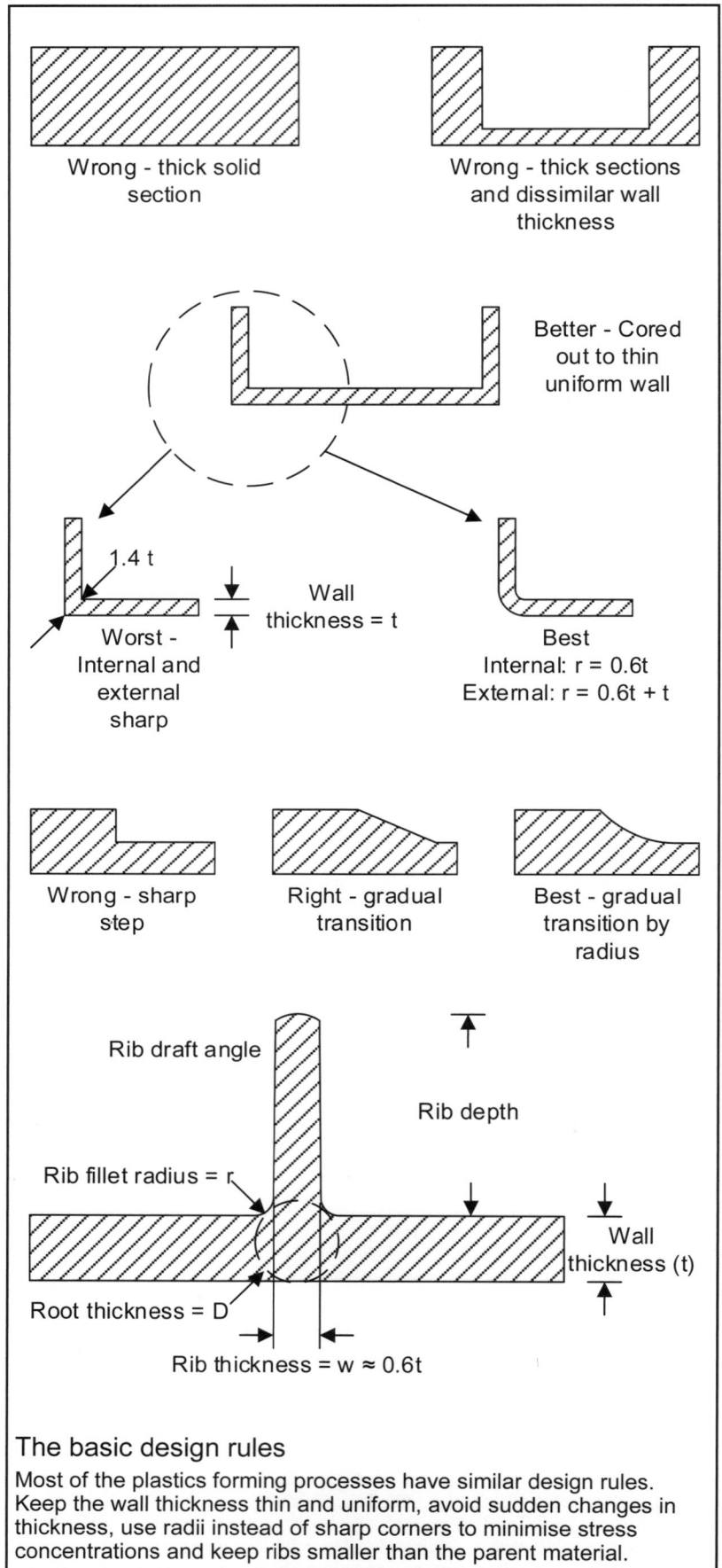

Wrong - thick solid section

Wrong - thick sections and dissimilar wall thickness

Better - Cored out to thin uniform wall

1.4 t

Wall thickness = t

Worst - Internal and external sharp

Best
Internal: r = 0.6t
External: r = 0.6t + t

Wrong - sharp step

Right - gradual transition

Best - gradual transition by radius

Rib draft angle

Rib depth

Rib fillet radius = r

Root thickness = D

Wall thickness (t)

Rib thickness = w ≈ 0.6t

The basic design rules

Most of the plastics forming processes have similar design rules. Keep the wall thickness thin and uniform, avoid sudden changes in thickness, use radii instead of sharp corners to minimise stress concentrations and keep ribs smaller than the parent material.

they apply to all the thermoplastics and to most thermosets. The thermoplastics will vary in mechanical and chemical properties, e.g., tensile strength and shrinkage rates, and some rules are more important for some materials than for others but they are all important and should only be ignored at your own peril.

- **Tip** - It is strongly recommended that designers use a checklist approach for detailed design to ensure conformance with the basic design rules.

Injection moulding design rules

The main design rules for injection moulding are:

- Keep wall thickness as uniform as possible throughout the part.
- Avoid sharp changes in the wall thickness, i.e., if wall thickness must change then use gradual transitions between thick and thin sections.
- Wall thickness must suit both function and process.
- Avoid sharp internal or external corners.
- Internal radii should be at least 0.5 and preferably 0.6 to 0.75 times wall thickness.
- Keep corner wall thickness as close as possible to nominal wall thickness. Ideally, external radii should be equal to internal radii plus wall thickness.
- Rib thickness should be 50-75% of wall thickness.
- Fillet radius should be 40-60% of rib thickness.
- Rib root thickness should not be more than 25% greater than wall thickness.
- Rib depth should not be more than 5 times rib thickness.
- Taper ribs for mould release.
- Always think about creep (deformation under low loads and long times).
- Always look at the environment.
- **Tip** - Further design tips for injection moulding are available in 'Design Guides for Plastics' at www.tangram.co.uk.
- **Tip** - Perhaps the best texts on injection mould design are those by Pye and by Menges and Mohren. They are classics. Anything by Glenn Beall is also invaluable.

Extrusion design rules

The main design rules for extrusion are very similar to those of injection moulding regarding wall thickness, gradual transitions and ribs. This should not be a surprise given that in many cases we are working with the same material. Extrusion has some unique rules regarding internal walls where the wall position cannot be

guaranteed due to the lack of calibration. These are:

- Make uncalibrated internal spaces larger than necessary and adjust after tool tuning with small location pips.
- When using external open spaces then 'bridge' these with reinforcing areas to support the location.
- Avoid small 'closed' internal or external chambers as these can be difficult to calibrate and control.
- Be wary of small protrusions on the surface of the extrusion, these can be difficult to calibrate accurately. Use clearance on the calibrators and allow the protrusion to float.
- Be wary about excessive haul-off forces as they can crush small extrusions.
- Always think about creep (deformation under low loads and long times).
- Always look at the environment.
- **Tip** - Further design tips for extrusion are available in 'Extrusion design tips' at www.tangram.co.uk.
- **Tip** - The classic text for extrusion design is 'Polymer Extrusion' by Rauwendaal.

Thermoforming

Thermoforming has some specific design rules depending on whether it is thin-wall (packaging applications) or thick-wall (more properly vacuum forming).

- **Tip** - The classic text for thermoforming is 'Thermoforming Technology' by Throne.

Rotational moulding

Rotational moulding also has some specific design rules but in many ways is more flexible than other processes.

- **Tip** - The classic texts for rotational moulding are 'Rotational Moulding Technology' by Crawford and Throne and 'Rotational Moulding: Design, Materials and Processing' by Glenn Beall.

Not following the basic design rules or getting them wrong will almost certainly give a poor-quality and unsustainable product.

They are called 'rules' because they are meant to be followed.

Reduce material use

Reducing material use (lightweighting) is an obvious tactic for reducing environmental impact and product cost:

- Use good structural design to make walls thinner.
- Use good flow design to make walls thinner.
- Use gentle processing conditions to minimise polymer damage.

We will not consider tolerances at this stage even though they are basic for the production of a robust and manufacturable design.

There are some very good free resources available on the Internet. Highly recommended are:

- Bayer 'Engineering Polymers, Part and Mold Design - Thermoplastics'.
- GE Engineering Thermoplastics 'Guidelines for Injection Molded Design'.
- LANXESS 'Part and Mold Design - A Design Guide'.
- BASF 'Design Solutions Guide'.

3.9   Manufacture – Design for Manufacture

Design for Manufacture

Design for manufacture (DfM) is a design concept to improve the manufacturability of products and is often applied with simultaneous engineering (when the product and manufacturing processes are developed at the same time). The combination of DfM and simultaneous engineering includes a range of proven methods and tools to:

- Produce designs that can be reliably manufactured and have a long service life.
- Reduce development costs.
- Reduce time-to-market.
- Reduce manufacturing wastes and costs.
- Improve product quality, reliability, safety, sustainability and customer satisfaction.

For this book, the focus is on sustainability improvements that can be delivered by DfM. DfM allows sustainability to be 'designed in' to the product by considering manufacturing, quality and service life at the beginning of the design rather than designing a product and then hoping that manufacturing can produce it to the required standards.

DfM can be divided into four basic elements:

People and teams

DfM includes the concepts of project and design teams similar to those discussed in Section 3.1. It uses team-based project management where team members have a variety of technical and managerial skills.

Simultaneous engineering

Simultaneous engineering uses the team concept to develop the whole product system in parallel to greatly reduce time-to-market. The parallel strands are:

- Product design.
- Manufacturing method design.
- Manufacturing systems and support items.

This concept has been termed 'concurrent engineering' by some software vendors but the major issues are organisation and planning – no computer is needed. Simultaneous engineering is possible with a pen and paper, a good company structure and good people.

Tools

DfM uses the 'alphabet soup' of design and manufacturing tools. The main tools are:

- Failure modes and effects analysis (FMEA) at the design, process and machine levels (see Section 2.17).

- Taguchi analysis or Design of Experiments (DOE).
- Value engineering (VE).
- Design for assembly (DfA – see Section 3.10).
- Design for disassembly (DfD – see Section 3.13).
- Quality function deployment (QFD).
- Statistical process control (SPC).

Most of these tools are covered in more detail in other Workbooks[1,2]. DfM simply uses these tools in an integrated manner during design, development and production commissioning. The tools are not used in isolation but form part of the overall project management.

Management

DfM uses a matrix management approach that is based more on processes and projects than on the traditional functional divisions (see Section 3.1).

As a global approach, DfM can be adapted to the needs of individual companies. A fully implemented DfM system will reduce design and development costs but more importantly it will deliver robust high-quality sustainable products to market quickly and reliably.

- **Tip** – Train designers in DfM and in the principles of economic product design.

The basic operation of DfM

DfM has some basic operational principles and these are:

Understand the process

Every process has design guidelines (see Section 3.8) and ignoring these is a recipe for disaster. Process specialists must be part of the DfM team to highlight any potential process difficulties. Avoiding and designing-out process concerns at the start is far better than trying to solve them after the product is in production. This is also relevant for the assembly and disassembly processes (see Section 3.13).

Understand process tolerances

Process specialists can advise on achievable tolerances and limits of the process. More importantly, they can stop designers specifying unachievable tolerances. DOE can be used to understand how tolerance variations will affect product quality and how to optimise these to produce a robust design for high-quality and low-cost production.

DfM and DfA (see Section 3.10) are sometimes treated as a single 'method' and referred to as DfMA.

We have treated them separately for clarity but many of the concepts are similar.

Sometimes the phrase DfX is used where 'X' is the desired attribute, e.g., DfD = Design for Disassembly (see Section 3.13).

One design tool that is not as well-known as it should be is 'TRIZ'.

This is an excellent and well-developed tool for finding innovative solutions using a defined invention process.

Fascinating to see in action with many successful case studies.

Aim for wide tolerances that are rigidly enforced rather than tight tolerances that are loosely enforced.

It makes life a whole lot easier.

- **Tip** – Always involve process specialists, they are essential members of the DfM team.
- **Tip** – Tolerances are one of the keys to good DfM and process specialists can help to get the tolerances right.

Design the tooling at the same time

Designing the tooling at the same time as the product will highlight complexity, time and quality issues with any proposed tooling (production, assembly or disassembly).

- **Tip** – Tooling and process engineers also need to be part of the DfM team.

Learn from the past

Most products are similar in some way to a previous product and it is important that hard-won experience with previous products is incorporated into new designs.

- **Tip** – If you had a quality/time issue with a fastener in the last design then why are you using the same fastener in the new design? More importantly, why are you using a fastener at all?

DfM guidelines

DfM also has some basic guidelines for good designs and these are:

Reduce the number of parts

This is a key issue for DfM and DfA (see Section 3.13). Reducing the number of parts generally reduces manufacturing costs and the potential for quality concerns as well as reducing inventory and development costs. It can also improve the recycling potential.

- **Tip** – If a part does not need to move relative to any other part and does not need to be made from a different material then consider it for elimination.

Use modular designs

If the product design can be broken down into discrete modules then this helps with production, inspection, servicing, remanufacturing, re-use and recycling.

- **Tip** – Modules with good 'connection' design can also be re-used in other products and updated or remanufactured easily.

Design multi-functional parts

If parts can be designed to be multi-functional this will reduce the number of parts as well as improving their quality. The concept of multi-functionality can be used to check functionality, guide assembly, align and fix subsequent parts, aid inspection, provide user instructions and aid remanufacturing.

- **Tip** – If you have to include a part then make it as 'useful' as possible.

Design multi-use parts

Reducing the number of parts should not only consider a single product. If similar parts are being used in multiple products then reducing part variation can dramatically reduce development costs, time and improve quality.

- **Tip** – Why try to get two parts right when you can concentrate on one?
- **Tip** – Identify potential multi-use parts and consider what modifications would need to be made to use them in multiple products.

Design to minimise finishing

Finishing operations such as painting and chrome plating are costly and have high environmental impacts (see Section 9.13) as well as being areas for quality concerns. Always attempt to minimise finishing operations at the design stage.

- **Tip** – A two-shot moulding will be more consistent and reliable than a painting operation but try to keep the materials the same to enable recycling (see Section 3.5).

Design for assembly/disassembly from the start

Design for assembly/disassembly are an essential for DfM. Thinking about it at the start pays dividends.

Want to know more?

One of the best books on the subject of DfM is by David Anderson[3] (see quote above). It is packed full of practical hints, tips and case studies. You won't regret buying it.

- 1. Kent, R. 2018. 'Cost management in plastics processing'. Elsevier.
- 2. Kent, R. 2016. 'Quality management in plastics processing'. Elsevier.
- 3. Anderson, D. 2020, 'Design for Manufacturability'. Productivity Press.

"Never design a part you can buy out of a catalogue."

David Anderson

and select quality suppliers!

'Coffee Jar Design'

Some years ago, whilst working with PP injection mouldings we experienced severe quality issues with screw threads and jamming threads. Examining the product design, it was obvious that the thread form was virtually that of a machine thread when functionally it only needed to pull down and lock against a shoulder.

We changed the design to an open thread – similar to those on coffee jar lids – with huge tolerances. The issue went away and moulding was much easier.

3.10   Manufacture – Design for Assembly

Design for Assembly (and disassembly)

Many plastics products are used as part of an assembly and getting assembly (and disassembly) correct is an essential technique in designing products for recycling and sustainability.

The original 'design for assembly' (DfA) methodology was pioneered by Boothroyd and Dewhurst at the University of Rhode Island. The method is a design evaluation tool that enables designers to:

- Reduce the part count of an assembly.

- Design products for easier manual or automatic handling in assembly.

- Reduce the labour and time involved in assembly.

DfA is a tool that is concerned with reducing the product assembly cost by minimising the part count, the number of assembly operations needed to produce the part and by making these assembly operations as easy and fail-proof as possible.

DfA reduces the number of 'opportunities to fail' in any assembly operation. This automatically leads to improved design and manufacturing quality as well as significant cost reductions.

- **Tip** – Designers need to be familiar with DfA concepts and methods to reduce costs and improve quality.

The basics of DfA

DfA has some basic principles and these are:

- Minimise the part count in the assembly by combining part functions.

- Parts should have self-aligning and self-locating features so that they cannot be installed incorrectly.

- Parts should have self-fastening features.

- Parts should not be 'left-' or 'right- handed', even if this means adding unused features to the parts.

- Parts should either be symmetrical (for easy orientation during assembly) OR very asymmetric for clear identification (for easy detection and orientation).

- Parts should automatically check that all the previous parts are present and located correctly.

- Parts should be designed to reduce the need for re-orientation during assembly.

- Parts should be designed for handling and insertion (both manual and automatic).

- Parts should be designed for easy gripping and transfer in fixtures using well-defined registration locations.

- An assembly should have a 'base' part into which all other parts are located.

- Parts should be designed to be assembled from the top down onto the 'base' part to use gravity rather than fight against it. This can dramatically affect the ease of assembly.

- Fasteners should be avoided wherever possible.

The DfA method

Functional analysis

DfA first assesses the part functionality to determine if the part is a primary part or a secondary part. The primary part is the

Plastics processing is ideally suited to DfA. Using twin shot moulding or multi-layer extrusion can radically reduce the number of parts needed – but make sure that compatible plastics are used to allow easier recycling (see Section 3.5).

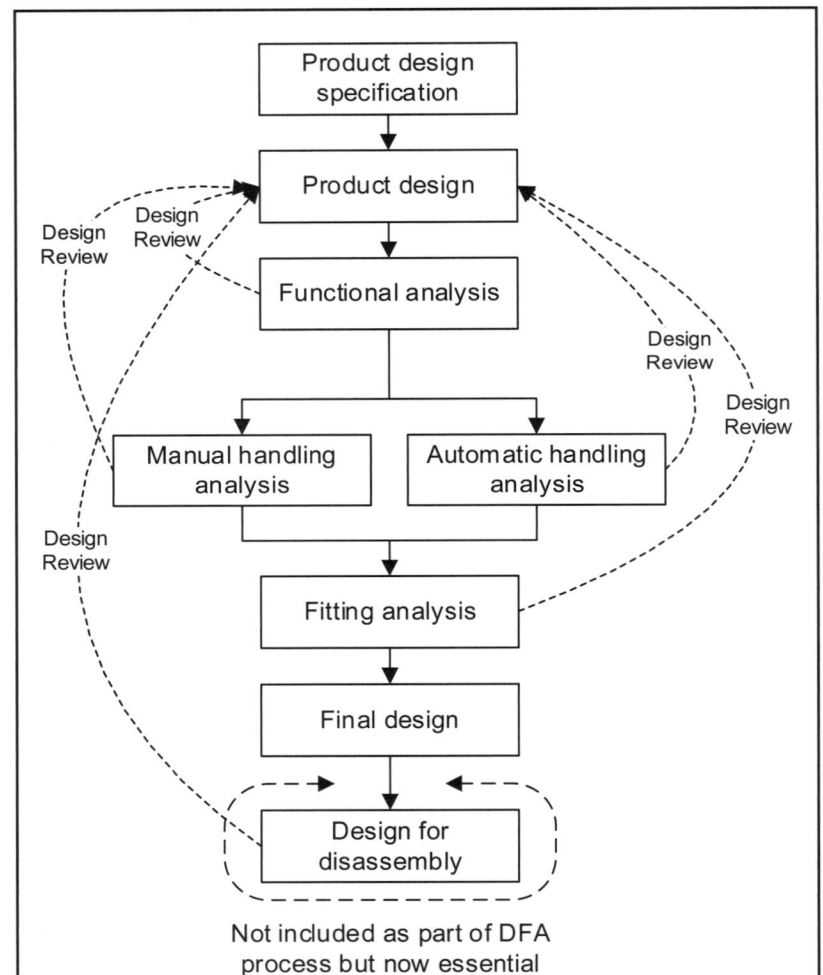

Not included as part of DFA process but now essential

DfA by the Boothroyd/Dewhurst method

The method is a process designed to assess and improve product design for assembly. It will reject less than optimum solutions and will improve designs for manual and automatic assembly. The method now needs to include consideration of design for disassembly.

'base' part of the assembly. A secondary part is potentially a non-essential part, i.e., a candidate for elimination.

Functional analysis should also:

- List the parts in the assembly and their order of assembly.
- List the number of interfaces between the individual parts.
- Calculate the theoretical minimum number of parts for the assembly by identifying any secondary parts that can be incorporated into the primary part.
- Identify any parts that can be standardised, e.g., if you must use fasteners then reduce the types and sizes needed to the absolute minimum.
- Look for part count efficiency and assess the relative part costs.
- **Tip** – A process flow diagram is a very useful tool to visualise assembly.
- **Tip** – Threaded fasteners are always a source of quality issues and should be avoided wherever possible.
- **Tip** – Quality is an essential part of the DfA process and designers should use the basic principles to 'mistake-proof' the assembly.

Functional analysis will almost always result in design modifications to reduce the part count and improve the basic product design.

Handling analysis

The handling analysis looks at the difficulty of controlling and orienting the part during both manual and automatic assembly.

Handling analysis looks at issues such as:

- Part size.
- Part weight.
- Part flexibility.
- Any other part details that could affect storage and handling, e.g., potential for nesting.
- Handling time and complexity.

All parts and the proposed assembly process should be examined in terms of the DfA basics (see above) to verify that handling is:

- Consistent.
- Easy for both people and machines.
- Capable of delivering high-quality products.

Handling analysis will almost always result in design modifications to improve the ease of orientation in the assembly and to reduce assembly times and costs.

Fitting analysis

Fitting analysis reviews the fitting methods and looks at issues such as:

- The attachment method used for each part.
- The assembly time needed for each part.
- The elimination of non-value-adding processes.

At each stage of DfA, scores are allocated to the design and the design must achieve a minimum score analysis to go forward to the next stage. DfA is supported by extensive selection tables and software elements to carry out the calculations and assessment.

DfA also helps to decide on the assembly method. Boothroyd and Dewhurst[1] give selection charts to determine the best assembly method but it should be noted that 90% of production is still best performed by manual or semi-manual methods with only 10% of products being recommended for fully automated assembly of any type.

- **Tip** – DfA needs more than designers, production engineers must be involved at the design level as well.

Design for Disassembly

The original DfA process did not include the additional process of design for disassembly, i.e., DfD (see Section 3.13). We have the choice of regarding DfD as an extension of DfA or of regarding this as a separate process. My preference is to regard these as linked processes (as shown in the diagram on the left). It is essentially an iterative process, i.e., find out the best way to put it together (DfA) and in many cases the best way to take it apart (DfD) will become remarkably obvious.

In earlier books, I stated that 'if it can be taken apart then it wasn't put together properly in the first place'.

I was wrong.

Times have changed and design now needs products that can be easily taken apart so that parts can be re-used and materials can be recycled.

Products still need to be assembled effectively but they also now need to be taken apart effectively.

I have changed my mind in the light of pressures and evidence.

- 1. Boothroyd, G. Dewhurst, P. and Knight, W. 2010. 'Product design for manufacture and assembly'. CRC Press.

3.11   Use – optimising use

Optimising use

The key task in the use phase of resource efficient design is optimising use and the challenge is to reduce the cost of ownership of the product. This requires an additional consideration in design – the new issue is 'design for use' where the focus is on the best use of plastics to meet the user requirements and prolong the life of the product (and the resources embedded in it). Planned obsolescence is no longer an acceptable strategy, if it ever was in the long term.

Consumers are increasingly aware of the resource efficiency of competing products and the 'energy rating' of products from light bulbs to refrigerators and windows is increasingly being used by consumers to inform and guide their choices. This consumer pressure is driving the development of better-quality or more efficient products with reduced running costs.

Some processors will think that this is not their problem; they will see themselves as 'converters', but this will not be sufficient for future survival. Successful processors are increasingly producing complete assemblies to add value and 'cost in use' is becoming a key indicator for success in any market. Any processor who ignores this is not planning for the future. The good news is that the unique properties of plastics give them exceptional advantages in 'cost in use' calculations compared to traditional materials.

Design can be used to optimise the resource efficiency of products throughout their life cycle and to reduce the environmental impacts of products and the manufacturing process.

Use patterns

It is important to look beyond simply what the product is designed to do and to look at how it is actually used. Information is needed from users about what they actually do with a product, i.e., their 'product habits'. One example of 'product habit' is to boil a and leave it, returning later to reboil the hot water. This is a habit that I am unfortunately guilty of (I get distracted easily) even though I know that this process uses a lot of energy. Design could help to reduce this energy waste in use by insulating the kettle to retain the heat, but most of them are simply a single moulding of PP that loses heat rapidly despite the excellent insulating properties of plastics.

Feedback from suppliers and/or customers can also reveal opportunities for reducing the functions or parts of the product or packaging that customers regard as unnecessary, e.g., surplus modules, features and attachments. Do we need all those features on a media recorder when all we want to do in most cases is 'time shift'?

The Brennan B2 (www.brennan.co.uk) is a brilliant music player, it allows you to simply record and play your old CD collection with no real fancy features, just simplicity, and I am listening to one right now.

Functional and use analysis of the product will become a key task in the design phase. Existing techniques such as customer surveys (now part of ISO 9001) can be used to produce a Product Design Specification (see Section 3.16) based on real 'critical' and 'desirable' functions rather than a design based simply on drawings and materials specifications.

Designers need to use design as an incentive and a tool to reduce the total resources used during the use stage of the product life, to tune the product design to the real needs and use patterns of the consumer and to increase the value added for the end-user.

Designing for use

Designers must focus on new issues such as:

- Using fewer resources in use – designing the product so that its use and maintenance requires less materials, consumables, energy and other resources, in order to reduce the adverse impacts of the product and the costs of using the product, e.g., electricity and water consumption.

Many consumer products (such as cars, white goods, brown goods and windows) use or lose more energy and resources during their use stage than during manufacture.

This is even true of many industrial products, e.g., running an electric motor for around 1,000 hours costs as much in electricity as the capital cost of the motor.

Drivers

- Increased market demands (from both customer and ultimate end-user) for improved and documented environmental performance.
- Continued growth of 'push-pull' taxes and legislative instruments such as climate change levies and landfill taxes.
- Introduction of 'Integrated Product Policy' requirements from EU.
- Potential for increased profitability.

Strategies

- Improve the design process and outputs to reduce the whole life cost of products, particularly the use costs.
- Improve the manufacturing process to reduce the whole life cost of products, particularly the use costs.

Drivers and strategies for resource efficient design in the use phase

- Causing less pollution and waste in use – designing the product so that it causes less pollution, produces fewer emissions, has less waste and has reduced environmental impact during use.

- Optimising functionality and service life – optimising functionality to reduce the need for additional products or resources to achieve the same task and making the product more efficient in use.

- Giving the product a longer service life – this will increase repeat sales, albeit at greater intervals, increase 'recommendation' sales and will require fewer products to be manufactured, thereby reducing the impacts associated with product production, delivery and disposal.

- Using customer surveys as an integral part of the design process – these can identify important areas of the product's efficiency and whether it has any redundant functions that can be removed to reduce production and use costs.

- Considering maintenance issues – reducing maintenance or making it easier to carry out will prolong the product life and improve the intrinsic value to the consumer.

The issue of resource efficiency during the use stage must be part of the Product Design Specification (see Section 3.16) and part of the basic design of any product.

Reducing environmental impact during use

Opportunities to reduce resource consumption and environmental impacts during the product's use can be identified during initial product research. Other general measures to improve resource efficiency and to reduce the environmental impact during use include:

- Providing instructions on how to use the product efficiently – never assume that users will automatically use the product in the most efficient way (or even in the way you thought they would).

- Fitting better controls (automatic or manual) to optimise energy and material use without user intervention.

- Improving the functionality and service life of the product through more durable design concepts such as making it easier to repair and service and making it adaptable to different tasks, related either by the technology or by the function.

- Asking customers to rank the product's efficiency and the various functions to inform design improvements in the future.

- Designing the product so that it lasts longer – a good place to start to identify weaknesses is to look at records of complaints and reasons for return.

- Improving the insulation of hot or cold elements.

Although such measures can sometimes (but not always) increase the product's cost, the user inevitably achieves long-term benefits from reduced energy and resource consumption. The rise of energy labelling schemes is making users more aware of the lifetime benefits of efficient products and driving consumer purchasing towards more efficient products.

The future

Resource efficient design and resource efficiency in manufacturing and use are largely internal issues – most of the improvements are driven by internal costs and the need to reduce these. This is changing and the major drivers will become largely external. They will be legislation and the cost results of legislation being used as a tool to internalise the social costs of products that were previously ignored by manufacturers.

It is not getting any easier as legislation forces producers to internalise the social costs of their products.

'Integrated product policy' is the overall EU programme to stimulate all the phases of a product's life to improve the environmental performance.

This does not cover a single, simple policy measure but a range of tools (voluntary and mandatory) to drive improvement.

For more details see ec.europa.eu/environment/ipp.

Tactics

- Implement resource efficient design as a formal part of the design process.
- Design products to be resource efficient during the use phase.
- Train product designers in:
  - ☐ Design for Manufacture.
  - ☐ Design for Assembly & Disassembly.
  - ☐ Life Cycle Assessment and similar techniques.
- Integrate potential use costs into the product costing calculations.
- Reduce manufacturing impacts and costs through resource efficient design and technology.

Results

- Improved focus on customer and consumer needs for reduced cost-in-use.
- Reduced product environmental impacts by improving resource efficiency at the design stage.
- Adoption of clean technology to give real cost savings from reduced design and manufacturing environmental impacts.

Tactics and results for resource efficient design in the use phase

3.12 End-of-life – minimising outputs

Minimising outputs

The key task in the end-of-life phase for resource efficient design is minimising the outputs or leakages from the product life cycle.

Disposal is becoming increasingly socially unacceptable and expensive for waste created at any stage of manufacture and this is particularly true for the end-of-life stage. The current trend in EU and other legislation, e.g., WEEE (for electronic equipment) and EOLV (for cars), is to both increase the cost of disposal and to allocate a large part of it to the original producer. The trend for the future is to make the producer responsible for end-of-life costs.

The issue is not simply one of resource depletion, it is also due to either shortages of destinations or unacceptable destinations for the outputs. The drive for reductions in CO_2 emissions is essentially a 'destination shortage' where the atmosphere cannot accept more CO_2. Packaging taxes and plastic bag bans are being used to 'nudge' consumer behaviour/littering where the destination is unacceptable, i.e., the general environment or the seas. This lack of external destinations is being internalised to producers through legislation and taxation, e.g., landfill taxes, CO_2 taxes and effluent taxes.

The key to improved environmental performance at end-of-life is to appreciate why the product is no longer used and what happens to the product at this stage. It is then possible to design products to minimise both the environmental impacts and the costs. Part of the design challenge is to improve the end-of-life options.

The choices

At the end of its 'first' life, the product (or parts of it) can be re-used, remanufactured, recycled, disposed of in an incinerator (to recover energy) or disposed of to landfill (see Chapter 4). This is the end-of-life hierarchy (see sidebar) and the further down the hierarchy the end-of-life option chosen is then the higher the environmental impact and cost.

In an environment where the 'producer pays' there is a need to improve control of the product during and after use to reduce costs. This can be achieved by:

- Labelling re-usable and recyclable parts.
- Using existing distribution channels to collect used products or components.

- Developing new distribution and more effective recovery channels to collect used products or components.
- Keeping up-to-date with developments in recovery and recycling to improve the options available.
- Discussing ways of recovering and recycling products with trade associations, waste management companies or companies offering similar products.

What is happening now?

Designers need to find out what is happening to their products now to provide the direction for future designs. The current methods of disposal can provide opportunities for increasing the product's recycling potential and decreasing the end-of-life costs. Typical questions are:

- Is the product typically disposed of to landfill? This can give rise to environmental impacts and will be subject to increasing costs in the future (see Section 4.12).
- Can the product be re-used or recycled instead of being sent to landfill?
- If products with only minor faults are typically discarded, is it possible to salvage some of the parts or components for re-use or remanufacture?
- Is there potential for re-using modules or parts of the product at the end-of-life?
- Does the product contain materials or components that can be easily recovered and recycled to reduce costs?

The end-of-life cost hierarchy

Re-use
Remanufacture
Recycle
Disposal with energy recovery
Disposal (landfill)

Increasing impact and cost

Drivers
- Increases in regulations for disposal of products and emissions, e.g., WEEE, ELV (see Section 3.14) and CFC regulations.
- An Environmental Management System (EMS) will become an essential qualification for business continuity.
- Market effects of product disposal costs impact on producers and increasingly on consumers, e.g., refrigerators and cars.
- Rising emissions charges, e.g., climate change levy and landfill taxes to reflect reductions in disposal sites.
- Increase in cost of disposal of products and emissions.

Strategies
- Plan to actively manage tradeable resource credits as they are introduced, e.g., carbon trading, packaging recovery notes (PRN).
- Formulate a 'take-back' strategy to deal with emerging product end-of-life requirements.
- Improve resource efficiency and reduce resource use to minimise effects of rising disposal costs.

Drivers and strategies for resource efficient design in the end-of-life phase

- Can parts be labelled to indicate their recycling potential? This is particularly relevant for plastics where standard material designations and labels are available for common materials (see Section 3.5).
- Can the product be designed for service or maintenance to increase the life span?
- Can a 'take-back' service be developed to reduce 'producer pays' costs?

Asking and answering these questions will help designers to identify the existing options for product re-use, remanufacture, recycling or disposal. These need to be improved to minimise the total cost of the product and to avoid end-of-life costs that were not accounted for in the 'first' cost of the product.

Making re-use, remanufacture and recycling easier

Designers need to consider the costs of the end-of-life stage at the design stage to avoid high disposal costs and they must:

- Make re-use, re-manufacturing and recycling easier for all products to reduce total costs by reducing raw material use and by diverting material away from landfill.
- Design the product for re-use in the current form (i.e., without reprocessing) to extend the useful life of the product. Product designs need to incorporate the requirements of subsequent uses, e.g., for packaging and containers, this may mean extra durability and the introduction of a re-use system suitable for the market.
- Design for product re-manufacture or recycling – this needs increased focus on the physical organisation of the product, i.e., the structure and the way in which components and materials are put together. Reducing the number of fastenings and making fastenings easier to undo will help to make the product easier to disassemble and recycle (see Section 3.13).
- Design to enable recycling by reducing the number of different materials used. Single material products are much easier to recycle.
- Design to eliminate materials that can be hazardous during re-manufacturing or make recycling difficult.

Resource efficient design in the future will not be easy but the alternative high end-of-life costs will be even less acceptable.

Reducing the impact of disposal

Disposal is destined to become the costliest option in the future but if there is no viable alternative then designers must reduce the costs and environmental impact of disposal.

To reduce the cost of disposal, designers should:

- Design the product to allow the volume to be reduced before disposal to reduce landfill charges.
- Choose materials, where possible and appropriate, to build in biodegradability or to make the product completely inert.
- Reduce or eliminate the use of hazardous materials in the product design to avoid additional 'special waste' charges.

Incineration with energy recovery provides an alternative 'disposal' option and plastics are excellent for producing energy during incineration. They do, however, require good control systems to reduce harmful emissions (see Section 4.11).

Whatever options are chosen for the product end-of-life stage, the costs will rise in the future and resource efficient design offers a unique opportunity to minimise these.

> Minimising the number of materials in a product will make it easier to recycle.

> End-of-life must be part of the standard Product Design Specification (PDS) for all products.

Tactics
- ■ Monitor resource intensity and follow legislation as a tool for success, not as a minimum compliance requirement.
- ■ Change or modify accounting systems and verify resource intensity to enable resource credit trading.
- ■ Form customer and end-user partnerships to enable 'take-back' strategies to be implemented when appropriate.

Results
- ■ Minimising the inevitable effects of increasing disposal and end-of-life requirements and costs.
- ■ Environmental design and control will become an essential cost control and marketing tools.

Tactics and results for resource efficient design in the end-of-life phase

3.13   End-of-life – Design for Disassembly

Make it easy to take apart

Design for Disassembly (DfD) is designing products so that they can be easily taken apart at end-of-life for either re-use or recycling. This section is not titled 'Design for recycling' because DfD is about extracting the maximum value from products at the end of their life. This can be via re-use of some (or all) of the components or via recycling of the materials.

Products designed with DfD in mind should be easily, cost-effectively and quickly taken apart so the product can be maintained or repaired and, at end-of-life, the components can be either re-used or recycled. Components that can wear out and need replacement should be designed for easy access, e.g., I have an iPad and the battery is not designed for replacement. When it, inevitably, fails it will be more waste to the WEEE pile (see Section 3.14) and such a shame for a beautiful device.

Disassembly is not simply the reverse of assembly; assembly (either manual or automated) will use special tools or fixtures to preserve the function and quality of the product whereas disassembly is likely to involve common tools or no tools at all. However, many of the design rules for DfA (see Section 3.10) apply for DfD and a design that is easy to put together is often also easy to take apart.

Materials

The guidelines given for raw materials selection (see Section 3.5) are essential in achieving DfD and these should always be followed: Minimise the number of materials, choose recyclable materials, choose compatible materials, remove other materials and identify the materials used. Simply following these rules will improve the capacity for DfD in any product.

- **Tip** – If non-recyclable materials must be used then try to locate these so that they can be easily removed and separated.

Components

The best solution for disassembly is to make the process unnecessary. DfM and DfA (see Sections 3.9 and 3.10) both recommend reducing the part count of an assembly and this can also be a benefit in DfM by reducing the time for disassembly. Integrating multiple functions into a single, well-designed multi-functional part made from one material reduces both assembly and disassembly costs, allows more efficient

recycling and has many other benefits, e.g., production capacity and stockholding.

- **Tip** – Achieving simplicity in design through multi-functional parts is elegant but difficult to achieve. Nobody ever said life was meant to be easy.

Joints

It is not always possible to design single component products and, in these cases, joints are inevitable. Unfortunately, in many cases, it is not clear where the joints are or what they do when it is time to disassemble the product. DfD should clearly signpost where the joints are and make them easily accessible for rapid disassembly with basic tools.

- **Tip** – Make joints easily accessible and easy to recognise.
- **Tip** – Colour coding of components can aid joint recognition and disassembly.

Disassembly instructions should be embossed or marked on the product so that it is clear what to do for disassembly at end-of-life.

- **Tip** – Provide disassembly guidance on the product if possible, the manual will not be with it at disassembly.

Plastics products can be joined by a variety of methods and the decision on the joint type can have a dramatic effect on the potential for re-use and recycling.

Fasteners – screws, inserts and clips

Whilst both DfA and DfD do not favour screw fasteners, it is sometimes not possible to use any other type of fastener. If screws, inserts or clips must be used then ensure that:

- Fasteners are standardised as much possible to reduce the number of types used.
- Screw heads are standardised to reduce the tools needed for disassembly.
- Fasteners are clearly marked to allow detection and removal.
- Fasteners are designed to be easily separated from the base plastic components using break points or similar.
- Fasteners are always carbon steel (or magnetic stainless steel) to allow for magnetic separation during recycling if they are not fully removed – fasteners based on aluminium or brass may not be detected and can damage recycling machinery.

Designers need to start thinking about product de-creation at the same time as they think about product creation.

Do it because it is right even if it is difficult.

Glues, solvents and welding

Plastics can be joined by a wide range of glues, solvents and welding or riveting methods. These are generally permanent joints that cannot be disassembled without major damage to the components that will prevent re-use of the components (although not recycling of the material). In disassembly terms, these should be avoided unless the components are manufactured from the same material.

- **Tip** – If the same materials are being joined then these methods can be acceptable for recycling.

Snap fits

Snap or press fits are the preferred jointing method in DfD terms as they can eliminate screws, glues and solvents and can be easily seen and broken by hand or by simple tools.

Depending on the product needs and detailed design, a snap fit can be two-way (reversible) or one-way (irreversible). Two-way snap fits are preferred as they allow re-use but should always be signposted and allow access for disassembly. If one-way (irreversible) snap fits are necessary for fit and function then, if possible, access points should be provided to allow the snap fit to be released or broken away using simple tools. This can be assisted by providing designed-in stress-concentrators (break points) to allow the product to be broken apart into separate materials or components.

- **Tip** – Use snap and press fits where possible.
- **Tip** – Use two-way snap fits if possible.
- **Tip** – If one-way snap fits must be used then design in 'break points' for easy disassembly.
- **Tip** – Use 'break points' to separate materials when disassembling.

Bi-injection moulding and co-extrusion

Bi-injection and co-extrusion can be used to produce products with unique properties, e.g., soft and hard together. However, disassembly is not possible because of the intimate bonding of the materials. These products must be used carefully and the materials chosen with care (see Section 3.5) so that the product can be recycled 'en masse' with no requirement for separation.

Modularisation

Complex products should be designed as modules that can be assembled to provide flexible functionality and easily disassembled for upgrading, repair, re-use or recycling. Modular assemblies can also improve production flexibility, reduce stockholding and improve product flexibility in the market, i.e., personalised products.

- **Tip** – Designers of complex products should always look at the product architecture and design these as modules to facilitate re-use and recycling. Modern construction methods are already looking at a modular approach to improve re-use and recycling at end-of-life.
- **Tip** – Modern computer code is also designed as modules (DLLs) to promote modular architecture and to allow code to be re-used efficiently.

Design for biodegradability?

DfD does not strictly consider design for biodegradability. Depending on the technology used (see Section 4.15), biodegradation can mean reduced opportunities to re-use product components or to recycle the materials.

Designers should try to arrange a visit to a recycling operation. Seeing how it is taken apart can be a revelation and an inspiration.

Active disassembly

An emerging trend in DfD is the use of 'active disassembly'. This is based on smart materials which use a 'stimulus' rather than a fastener, tool or machine to disassemble the product. This is achieved by using shape-memory polymers, e.g., PU, PET and some other block copolymers.

These materials can have a permanent shape and a temporary shape. A fastener made from a shape-memory polymer will be in the temporary shape and will hold the product together. Applying the required stimulus will allow the fastener to revert to the permanent shape and the product will automatically disassemble quickly, cleanly and non-destructively.

Thermal shape-memory plastics essentially have two or three phases with different thermal properties, heating the plastic to $> T_g$ (or T_m in some cases) allows one of the phases to soften and the product will revert to the permanent shape. Other shape-memory plastics use UV light as a stimulus to produce photo-activated cross-linking and cleaving to change between the permanent and temporary shapes and new plastics are being developed that use electricity as a stimulus for the transition.

These technologies can be used for snap and press fits, screws, bolts and rivets to design products for easy disassembly simply by raising the temperature, using UV light or potentially applying a current. Exciting stuff.

3.14   End-of-life – WEEE and ELV

Legislation

There are two main EU Directives that affect the end-of-life stage and hence many product designs. These are:

- WEEE (Waste Electrical and Electronic Equipment) covers the treatment and recycling of WEEE and is designed to encourage the re-use and recycling of WEEE and to reduce the amount of WEEE being disposed of. WEEE requires producers to pay for at least the collection of their products at end-of-life from central points and meet targets for re-use, recycling and recovery.

- ELV (End-of-Life Vehicles) covers the producer's responsibility at the end of a vehicle's life and is designed to encourage the re-use and recycling of waste from vehicles and to reduce the amount of waste being disposed of. ELV requires producers to have responsibility to take back vehicles they have introduced. This does not require that they physically do this and they can join networks or schemes to do this. As with WEEE, there are targets for re-use, recycling and recovery.

With the departure of the UK from the EU at the end of 2020, there are no immediate plans to change either WEEE or ELV requirements. These have, so far, been subject to a simple 'tidying up' exercise to refer to UK legislation rather than EU legislation. This is, however, a developing situation and UK-based companies should check the current requirements at www.hse.gov.uk.

If hazardous substances have been used in any product then it is best to re-use or recycle the waste to reduce the environmental impact. Only when these options have been exhausted is it acceptable to properly dispose of the waste. WEEE and ELV deal with the re-use, recycling and disposal aspects of the waste hierarchy.

The high volumes of plastics used in many WEEE and automotive products means that the plastics must also be recycled in order to comply with WEEE and ELV.

WEEE

The WEEE Directive covers the design and production of electrical and electronic equipment to aid the recycling of redundant electrical equipment and passes the responsibility for recycling back to the producer.

The EU WEEE requirements are:

- Member States must set up systems to encourage the separate collection of WEEE and systems that allow the free-of-charge return of WEEE.

- Member States must achieve an average WEEE collection rate of at least 45% of the weight of electrical and electronic equipment entering the market. This was changed from the original target of 4 kg per inhabitant per year in 2012 because of failures to meet the target.

- Member States must ensure that WEEE collected from private households is sent to authorised treatment facilities.

- Member States must ensure that producers set up systems to provide for recovery, re-use and recycling of WEEE to defined targets that are a proportion of the WEEE collected from private households.

- Member States must report WEEE targets on a regular basis.

There is no current requirement for individuals to separate WEEE at source and in most cases, this will be done as part of the standard waste collection or waste disposal.

Retailers must ensure that WEEE is taken back (on a one-for-one basis) when a new, equivalent product is sold but Member States can allow retailers to make other arrangements, provided that they are free of charge to the consumer.

The ELV Directive was issued in 1997 and was the first EU Directive to introduce the concept of 'extended producer responsibility' or the 'producer pays' concept. This is despite the fact that most of the environmental impacts from vehicles covered under ELV are actually the emissions from driving it (which is decided by the consumer).

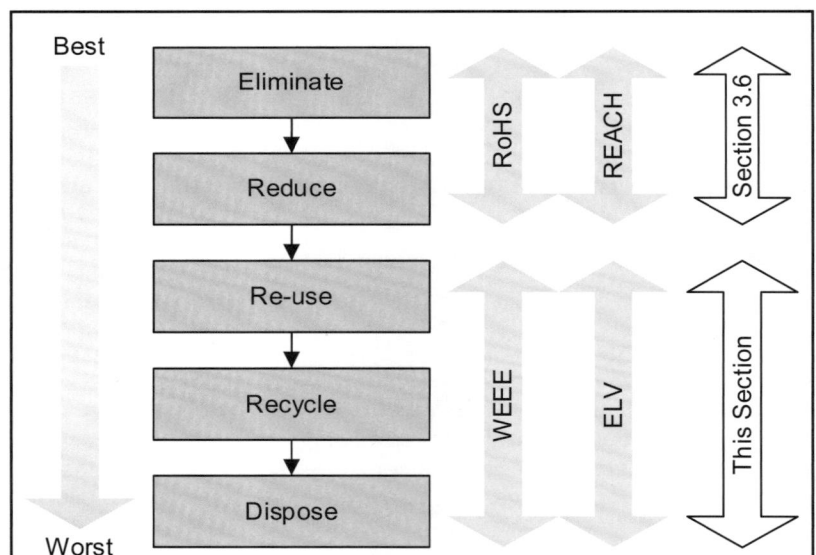

The waste hierarchy and materials legislation

The main legislation operates at different levels of the waste hierarchy. WEEE and ELV deal with the end-of-life whereas REACH and RoHS deal with the raw materials (see Section 3.6) Designers must understand the legislation that they are dealing with.

The producer requirements

The WEEE requirements at the producer level are:

- Producers who wish to market electrical and electronic equipment in any EU Member State must guarantee that the future costs for the collection of WEEE from central collection points and later treatment and recycling costs will be met, even if the company ceases to trade.

- Producers selling to commercial customers must provide systems for the collection, treatment and recycling of old products (on the sale of new products) and commercial sales must have contractual arrangements for WEEE.

- New products must be clearly marked with the producer's name and the WEEE symbol to indicate that it must not be disposed of in municipal waste collection.

- Producers must provide information on components and materials to allow treatment facilities to disassemble, re-use and recycle them.

- Producers must provide information to treatment facilities to identify components and materials that must be removed, e.g., capacitors containing PCB.

ELV

ELV puts responsibility on producers of passenger cars and light commercial vehicles to meet recycling and recovery targets for vehicles. As with WEEE, there are requirements at the Government level that are broadly similar to the requirements for WEEE.

The producer requirements

Under ELV, producers are responsible for achieving a re-use and recovery target of 95% and an 85% re-use and recycling target of the weight of the vehicles for which they have declared responsibility, together with any for which they have been given responsibility, and which are treated by the producer's collection network during 2015 and later years.

In addition, ELV:

- Places restrictions on materials used in vehicles. These are much the same as the RoHS materials (see Section 3.6), e.g., lead, hexavalent chromium, cadmium and mercury. However, some spare and replacement parts for older vehicles were initially excluded from this requirement.

- Requires marking of rubber and plastics parts over 200 grams with the relevant ISO identification codes to promote dismantling, re-use, recycling or recovery.

- Requires publication of design and dismantling information so that

components can be dismantled, re-used or recovered.

- Requires 'free take-back' at end-of-life of vehicles put on the market from 01 July 2002.

- Requires that producers declare the brands for which they are responsible.

- Requires licensing of treatment facilities used for ELV and their site and operating standards.

- **Tip** – There are lower targets for vehicles put on the market before 1980.

ELV does not require the producer to actually recover the vehicles themselves. Producers can be part of a network of authorised treatment facilities and collection points who will manage the process on their behalf. However, ELV does require that producers provide and publish all the information and keep this for a minimum of 4 years.

- **Tip** – Suppliers to the automotive industry may need to use the International Material Data System which is the automobile industry's material data system to ensure that materials are controlled.

Summary

WEEE and ELV are part of the legislative response to the end-of-life phase of the product life cycle. They are probably only the first of new regulations designed to cover 'extended producer responsibility'. This is where society demands that manufacturers are responsible for the complete product life cycle (including end-of-life).

Designers need to adopt resource efficient design to reduce future costs as similar legislation advances the 'producer pays' or 'polluter pays' principle.

- **Tip** – Although not yet part of any legislation, this 'extended producer responsibility' could be extended to make producers responsible for any of their products which are dropped as litter by the general public. Why worry about putting it in a bin when the producer is going to pick up the bill anyway?

The WEEE Symbol for Waste Electrical and Electronic Equipment:

The bar under the bin indicates that the product was placed on the market after 2005, i.e., it is 'non-historic' WEE.

Some WEEE is also classed as 'hazardous waste', e.g.,

- Lead-acid batteries.
- Uninterruptible power supplies.
- Cathode ray tubes.
- Fluorescent tubes.
- Laptop screen backlights.
- Equipment containing polychlorinated biphenyls (PCB).
- Fridges and freezers (due to the refrigerant).

The average UK individual generates ≈ 3.3 tonnes of waste electrical goods in a lifetime.

3.15 Resource efficient design – where are you now?

Resource efficient design

Resource efficient design represents an outstanding opportunity for plastics processors to not only get ahead of the regulatory demands and reduce costs but also to establish an ethical lead in the market.

Resource efficient design can provide an incentive for the design team to lead cost reduction throughout the complete product life cycle.

Changes in legislation and markets will force many of these changes on processors whether they like it or not, but by becoming pro-active processors can win through cost reductions in all areas.

Resource efficient design is a growing trend and sensitive customers at all points on the supply chain are starting to ask for the basics of resource efficient design, e.g., Walmart is already asking suppliers to complete their list of 15 sustainability questions which concentrate on issues such as energy use, material efficiency, natural resources and people and community. This is the start of things to come.

Completing the chart

This chart is completed and assessed as for the previous charts.

You can do it in your own time or you can do it when you are told to.

It is your choice.

Resource efficient design

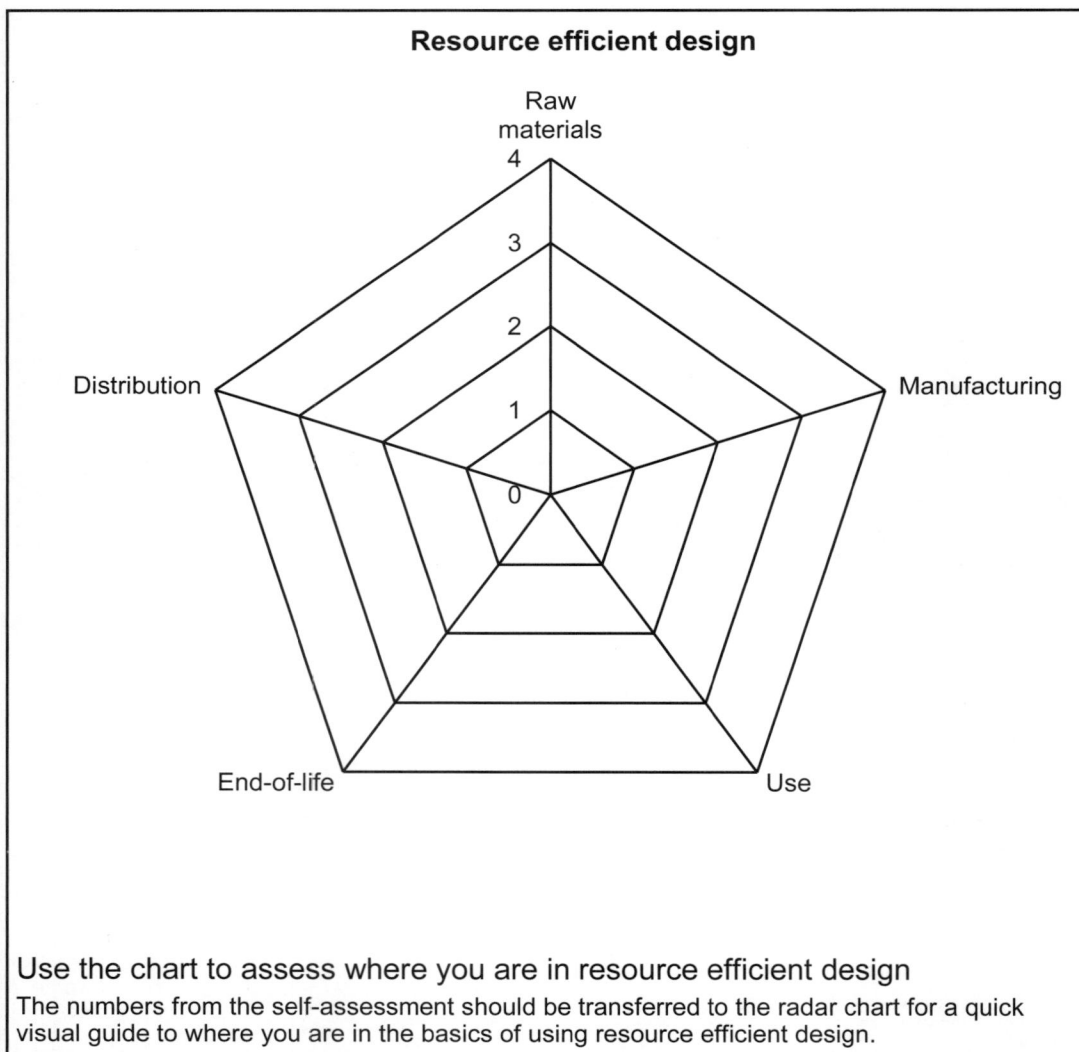

Use the chart to assess where you are in resource efficient design

The numbers from the self-assessment should be transferred to the radar chart for a quick visual guide to where you are in the basics of using resource efficient design.

A useful reference in all of this is:

Chapman, J. 2017. 'Routledge Handbook of Sustainable Product Design'. Routledge.

			Resource efficient design		
Level	**Raw materials**	**Manufacturing**	**Use**	**End-of-life**	**Distribution**
4	Use of raw & recycled materials is an integral part of design brief. Impact & cost of raw materials (all areas) are known & targets achieved.	Resource use & environmental impacts of manufacturing an integral part of design brief. All benchmark resource use targets known & achieved.	Resource use & environmental impacts in use stage an integral part of design brief. All benchmark resource use targets known & achieved.	Disposal options & routes are an integral part of design brief. Cost of disposal targets are known & achieved with disposal routes well defined.	Distribution considered as an integral part of design brief. Distribution cost targets are known & targets achieved.
3	Use of raw & recycled materials considered in design brief. Impact & cost of raw materials targets available but not always achieved.	Resource use & environmental impacts of manufacturing considered in design brief. Most benchmark resource use targets available & achieved.	Resource use & environmental impacts in use stage considered in design brief. Most benchmark resource use targets available & achieved.	Disposal options & routes considered in design brief. Cost of disposal targets available but not always achieved.	Distribution considered in design brief. Distribution cost targets available but not always achieved.
2	Use of raw & recycled materials poorly considered in design brief. Limited raw materials use targets available & achievement is variable.	Resource use in manufacturing considered in design brief. Limited benchmark resource use targets available & achievement is variable.	Resource use in use stage considered in design brief. Limited benchmark resource use targets available & achievement is variable.	Disposal options & routes poorly considered in design brief. Limited cost of disposal targets available & achievement is variable.	Distribution costs poorly considered in design brief. Limited distribution cost targets available & achievement is variable.
1	Use of raw & recycled materials considered only for publicity purposes. No benchmarks for impact & cost of raw materials available or considered.	Resource use in manufacturing considered only in cost reduction element of design brief. No benchmarks for resource use available or considered.	Resource use in use stage considered only for publicity purposes. No serious benchmarks for resource use available or considered.	Disposal options & routes considered only for publicity purposes. No serious benchmarks for cost of disposal available or considered.	Distribution costs considered only for publicity purposes. No serious benchmarks for distribution costs available or considered.
0	Resource use in raw materials is not considered in design brief.	Resource use in manufacturing is not considered in design brief.	Resource use in use stage is not considered in design brief.	Disposal options, routes & cost of disposal not considered in design brief.	Resource use in distribution is not considered as part of the design brief.
Score					

3.16   The product design specification

The essential definition

The product design specification (PDS) is the essential definition of what the product is required to provide.

The PDS is a statement of what the customer wants the product to achieve. In some cases, the customer is external and, in some cases, e.g., product range extensions for own-brand products, the customer is internal. In every case, a PDS must be prepared to act as a reference for the design objectives.

This is initial statement of the design intent which should be expanded by the design team to give all the technical detail and limits that are realistically achievable.

In the past, the PDS rarely included sustainability issues other than those that impacted on the product cost or function. This is no longer acceptable and consideration of sustainability at all stages of the product life cycle, particularly in the design phase is now essential.

It is not the design!

The PDS is *not* the design but the specification for the product and must be completed before any design work is started (see Section 3.2).

The PDS should be circulated to ensure that comment and criticism are given by all members of the project team and other interested parties.

If you don't do this then the risks of designing a 'product failure' dramatically increase.

- **Tip** – Product development teams are always in a hurry to develop products but getting the PDS right is a key activity in delivering a quality design for the market.

Preparing the PDS

The PDS is normally prepared by the design and development function for agreement/ sign-off by the external or internal customer and by the project team. The PDS sets out the technical detail of the requirements to be met to achieve a successful product.

- **Tip** – A PDS must not be confused with all the other specifications which refer to the product, i.e., the type of polymer to be used or the standards that the eventual product must conform to.

- **Tip** – A PDS is not sacred and may be changed but all changes must be noted and recorded. This is a 'living' document.

Creating the PDS – the checklist

The creation of the PDS involves asking the right questions – if the answers are easy it is probably because the difficult questions have not been asked. These questions should be answered by the internal or external customer in broad functional terms to allow the designer freedom to innovate. A broad checklist of general points is shown on the right. These are designed simply to stimulate thinking and not all the factors noted will apply to a given product but the subject headings should be scanned to give areas for profitable consideration. It is a series of questions to ask to enable the capture of the essential requirements for the product and not a rigorous design methodology.

Verifying the PDS

A PDS can be created by one person but should always be checked and agreed by the project team (especially by the customer). The full commitment of the project team is necessary before the PDS is issued.

Using the PDS

The PDS forms the basis for the design. The PDS is subject to change but should become firm at the pre-production/production stage. The PDS then forms the basic raw material for handbooks, manuals, sales literature and, in conjunction with the final drawings, can become the final specification for the product.

- **Tip** – Create other checklists and standard procedures to standardise later projects and processes. It is quicker and helps to remember to include all the factors.

The PDS for every project will be different. Creation of a standard format or generic PDS for specific company requirements will make completion of the PDS rapid and will ensure that most of the essential points are considered.

Slow down and get it right at the early stages rather than have to try to fix it up later.

Sustainability at all phases of the product life cycle must be considered at the PDS stage.

These are no longer 'afterthoughts' but integral parts of the design.

Product design specification checklist (example)

General

- ☐ Aesthetics
- ☐ Assembly
- ☐ Bought-in parts
- ☐ Complexity
- ☐ Competition
- ☐ Constraints
- ☐ Cost
- ☐ Customer
- ☐ Delivery requirements
- ☐ Design life
- ☐ Energy
- ☐ Ergonomics
- ☐ Existing designs
- ☐ Experts
- ☐ Export
- ☐ Fatigue
- ☐ Features
- ☐ Finish
- ☐ Friction
- ☐ Function
  - ☐ Main
  - ☐ Secondary
  - ☐ Tertiary
- ☐ Installation
- ☐ Lead time
- ☐ Legal
- ☐ Life in service
- ☐ Maintenance
- ☐ Manuals/information
- ☐ Marketing
- ☐ Modelling
- ☐ Noise
- ☐ Operator
- ☐ Outdoor exposure
- ☐ Packaging – delivery, transport
- ☐ Packaging – point of sale
- ☐ Patents
- ☐ Performance
- ☐ Portability

- ☐ Process
- ☐ Project schedule
- ☐ Project team
- ☐ Prototype
- ☐ Quality
- ☐ Quantity
- ☐ Reliability
- ☐ Safety
- ☐ Shelf life and storage conditions
- ☐ Size
- ☐ Standards
- ☐ Sub-contractors
- ☐ Targets
- ☐ Testing
- ☐ Timescales
- ☐ Tools
- ☐ Toxicity
- ☐ Training
- ☐ Use factors
- ☐ Vibration
- ☐ Waste
- ☐ Weight and wall thickness

Materials selection

- ☐ Production
  - ☐ Production volume
  - ☐ Weight
  - ☐ Shape and size
  - ☐ Tolerances
  - ☐ Surface finish
  - ☐ Inserts
  - ☐ Undercuts
  - ☐ Holes
- ☐ Acoustic property requirements
- ☐ Assembly requirements
- ☐ Carbon intensity of material
- ☐ Chemical resistance requirements
- ☐ Density
- ☐ Electrical property requirements
- ☐ Environmental factors: effect of the product on the environment

- ☐ Environmental factors: transport and storage environment
- ☐ Environmental factors: use environment
- ☐ Fire property requirements
- ☐ Insulation property requirements
  - ☐ Electrical
  - ☐ Thermal
- ☐ Magnetic property requirements
- ☐ Mechanical property requirements
- ☐ Optical property requirements
- ☐ Surface finish requirements
- ☐ Thermal requirements
  - ☐ Minimum operating temperature
  - ☐ Normal operating temperature
  - ☐ Maximum operating temperature
  - ☐ Thermal expansion

  These temperatures should include production, storage and transportation conditions.
- ☐ UV requirements
- ☐ Wear and friction requirements
- ☐ Miscellaneous requirements

Sustainability

- ☐ Raw materials
- ☐ Manufacture
- ☐ Use
- ☐ End-of-life
  - ☐ Remanufacture
  - ☐ Re-use
  - ☐ Recycling
  - ☐ Disposal
- ☐ Distribution
- ☐ Carbon footprint
- ☐ Life cycle assessment

Note: This list is not comprehensive. It should be expanded and modified to clarify the design intent.

3.17 Using the product design specification to improve sustainability

Getting started

Designing products for sustainability is not easy and initial progress will be slow. In most companies, the bulk of the environmental impacts will be from established products, new products will take time to reach production and even significant design improvements in new products will take time to feed through the system. This means that the most effective method is to start by looking at existing products to see how design changes can be made to reduce the environmental impact quickly and easily. Simple design changes can be a remarkably fast and effective way to reduce impacts and to test new ideas and concepts but you have to start now.

Select a product

The first step is to select a product for examination. In the workshops (see sidebar in lower right) we selected common generic products but for a company the product selection can be based on:

- The production volume – even nominally small improvements to a large volume product will be leveraged into a larger impact.
- The potential environmental impacts.
- The potential improvement opportunities.
- A product that is under threat from environmental activists, competitors or government legislation.

The reason for choosing the product doesn't really matter at this stage, the important thing is to choose a product and start the process.

- **Tip** – Only choose one product initially to get the process started.

Review the PDS

The PDS (see Section 3.16) contains the essential definition of the design requirements and reviewing this is a key step in assessing design changes. If no PDS was created for the original design then this should be created to define the requirements.

- **Tip** – It is impossible to assess designs unless the requirements are unambiguously specified.
- **Tip** – Reviewing the PDS clarifies the design freedom and the limits of change.

Look at the life cycle

It is possible to carry out a full quantitative assessment using an LCA but these are often complex and not relevant to the specific product, i.e., the LCA boundaries may be different. At this stage it is often easier to make a quick indicative assessment of the significant impacts to highlight the pressure points or areas where design changes can be most effective, e.g., see Section 7.7 for a quick assessment of the carbon footprint of various products.

A quick assessment can be carried out by looking at the life-cycle of the product using a process flow chart for the specific product (see Section 2.5 and Section 9.3) to identify the significant impacts at the various phases of the life-cycle and using a grid similar to that shown in Section 3.2 to map these onto the life-cycle of the product.

- **Tip** – This is about sustainable products and should include, where possible, not only the environmental aspects but also the social and economic aspects. A review such as this may throw up potential cost savings and there is nothing wrong with that.
- **Tip** – This process is akin to brainstorming. In the early stages, no idea should be evaluated, simply record the idea and carry out the evaluation when all the ideas are recorded.

Get the information

Having identified the source of the major impacts it is then necessary to make estimates of the potential benefits of any design change. These estimates can be based on internal or external sources but it is vital to put some numbers on the outputs, the

> The PDS clarifies the design intent. Changes to existing products must always retain the original design intent.

> We started running Cleaner Design workshops for Envirowise in 2002. These stripped-down existing products, e.g., telephones, kettles and torches, and examined them over the product life-cycle. The objective was to find quick and easy design changes to improve the sustainability of the products.
>
> The results were often remarkable as designers bought in to the process very quickly.

ISO/TR 14062: Integrating environmental aspects into product design and development

Naturally there is an ISO standard for designing sustainable products and ISO/TR 14062 presents a general method of integrating environmental aspects into product design and development. The standard covers all types of product design and is therefore necessarily broad, it is also not applicable as a specification for certification and registration.

The two main objectives of the standard are:

- To promote the adoption of a preventive approach to pollution rather than curative measures.
- To conserve resources (energy and materials).

This is to be achieved by integrating environmental issues covering the complete product life cycle as soon as possible into product design and tasking management with setting targets and objectives relating to the environmental aspects of products.

ISO 14062 is not a management systems standard (see Chapter 2) and ISO 14006 (Environmental Management Systems - Guidelines for incorporating ecodesign) gives guidance on how to integrate the concepts into an ISO 14001 EMS.

---

benefits, the costs of implementation and the ease of implementation.

- **Tip** – If detailed data is not available then a 'best informed guess' will have to do.

Review the options

The proposed design changes can then be reviewed for implementation feasibility and impact. When reviewing the options, score the potential projects on two criteria:

- The significance of the impact, i.e., will the change have a low, medium or high impact?
- The ease of implementation, i.e., is it difficult, medium or easy to implement?

These scores do not have to be exact assessment but both must be made for each proposal, they are simply a filtering method – there will be lot more discussion before committing to implement the proposed design changes.

Prioritise the options

It is inevitable that the review process will generate more opportunities than there is time or capacity to carry out. This means that the opportunities will need to be prioritised for implementation. This is most easily done using a 2 x 2 or a 3 x 3 grid similar to that used to assess materiality (see Section 1.8 and Section 11.4) and an example is shown on the lower right. This gives a quick filter for effective and worthwhile projects and those projects in the upper right (grey) area are the obvious candidates for the first stage of work.

The output of this process should be two or three potential design change projects that can be implemented with existing products to significantly reduce existing impacts.

- **Tip** – If there are several projects in the same general area for significance and ease of implementation then the choice can be made by assessing if the projects will remove or relieve an issue. When in doubt, go for the project that will remove the issue. As an example, if VOCs from paint (see Section 9.11) are an issue then it is possible to treat the VOC laden air or to design products that do not need painting. Obviously designing products that do not need painting is the preferred option.
- **Tip** – The selected projects will then need to be assessed by any existing project approval process before implementation.

Implement (and start again)

Project proposals do not improve sustainability, this is only achieved by completed projects. Therefore, the identified projects need to be delivered and monitored for effectiveness as with any project (see Section 1.18.

- **Tip** – It isn't what you start that counts, it is what you finish.

Reviewing existing products for design improvements and delivering these can quickly improve sustainability across a range of areas but the process does not stop there. The design team simply moves on to the next project on the list.

- **Tip** – The design improvements identified and delivered for existing products should also be implemented on all new projects to hold the gains and protect the future.

In the sustainability area, not all arguments are won by being right.

The facts may not be enough to win the argument against strong feelings but you still have to continue to develop products that are as good as they can be.

Don't wait, do it now!

A rough plan implemented quickly is better than a great plan that is never implemented.

Progress in design improvements may be slow initially but you have to start somewhere.

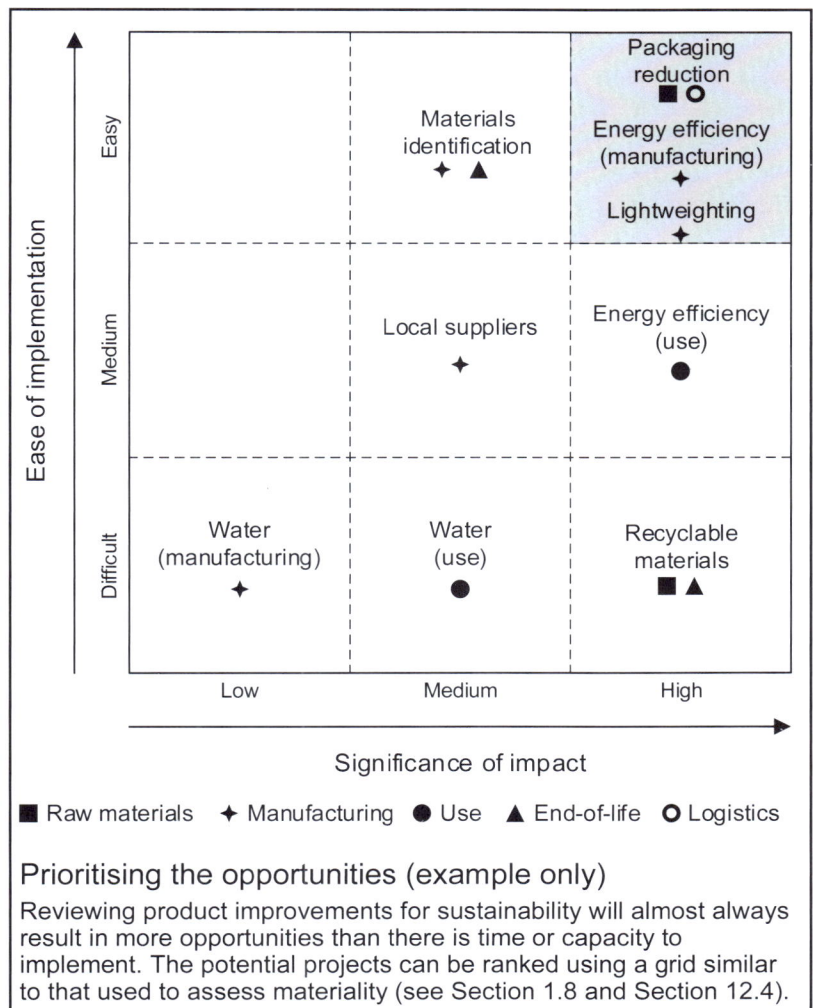

Prioritising the opportunities (example only)

Reviewing product improvements for sustainability will almost always result in more opportunities than there is time or capacity to implement. The potential projects can be ranked using a grid similar to that used to assess materiality (see Section 1.8 and Section 12.4).

Key tips

- Project management of the design and development process is the key to effective product development.
- Multi-disciplinary teams are fundamental to successful product development and project management.
- 'Over the wall' project management will always be ineffective.
- The design and development process sets the basic product environmental impacts.
- The basic impacts are difficult to reduce significantly once the design is complete.
- The cost of reducing impacts in the design process increases the later they are recognised.
- Correctly structured design and development processes can be used to manage environmental impacts.
- Design improvements need to be mapped across the complete life-cycle and value chain.
- Designers need to be aware of the environmental impact of their actions to avoid costs in the later phases of the life-cycle.
- Resource efficient design is a design tool to reduce product and social costs at all stages of the new product life-cycle.
- Resource efficient design is part of the circular economy.
- The actions to take are different for each phase of the life-cycle.
- At the raw materials stage the key task is to minimise the inputs to the process. This involves using less material and choosing materials with the lowest environmental impact.
- Design for recycling involves:
  - Reducing the number of materials.
  - Choosing recyclable materials.
  - Choosing compatible materials.
  - Removing other materials.
  - Identifying the materials used.
- REACH and RoHS control the materials that can be used.
- At the manufacturing stage the key task is targeting efforts to reduce resource use.
- 'Design for Manufacture' provides a set of tools for cost management during design and development.
- 'Design for Assembly' provides a method for reducing component numbers and easier assembly.

- At the use stage the key task is to design for efficient use of the product and to reduce the environmental impact of the use.
- At the end-of-life stage the key task is minimising the outputs. This can be achieved by designing for re-use, remanufacture and recycling.
- Re-use, remanufacture and recycling are all made easier by 'Design for Disassembly', this involves:
  - Reducing the number of materials.
  - Designing joints that can be taken apart easily.
- WEEE and ELV are legislative instruments to control the end-of-life of products. They are part of moves to 'extended producer responsibility'.
- The product design specification (PDS) is statement of the design intent and guides the decision-making process.
- Companies do not have to wait for new products to implement resource efficient design. They can start with existing products and review these (with the PDS) to identify opportunities for improvement.
- The legislative burden is going to be ever more difficult to meet and unless this is considered at the design stage then it will cost the company money throughout the product life cycle.

Chapter 4

Raw materials

Choosing and using the minimum amount of the right materials is fundamental to sustainability. As a materials conversion industry, the plastics processing industry converts raw materials into a finished product. In many companies there is little value added to the basic plastic material apart from the forming process. This means that raw materials are inevitably one of the largest factors in the sustainability and cost of a product and yet this is also one of the least considered or managed. There is much to do in the area of raw materials to improve the sustainability of the plastics processing industry.

The issue of the circular economy and recycling means that materials recovery can be considered either under 'raw materials' or under 'end-of-life'. Most plastics processors will internally re-use material but will not actually be recycling post-consumer waste materials although they may well be, and should be, using these materials in their processes. We have therefore chosen to consider materials recovery and the issues of bio-based plastics and biodegradability in this chapter because of the close association of these topics with raw materials.

The initial sections of this chapter look at methods to reduce the amount of materials used in the production of good products (lightweighting) and how to reduce use in the production area.

The latter sections of this chapter look at the issues of materials recovery including mechanical and chemical recycling (which recover the material), energy recovery (which recovers the embodied energy of the material) and landfill (which stores the material for the future).

The final sections of this chapter look at bio-based materials and biodegradability as these are important sources and sinks of raw material.

Note: The focus here is on methods to improve the sustainability of the plastic raw materials used and not on any other materials used. Materials such as packaging are dealt with in Chapter 9.

4.1 Raw materials – a key factor

The key factor

As a basic conversion process, raw materials use has one of the largest sustainability impacts of plastics processing. Historically there have been few formal attempts to manage these impacts through reducing the amount of material used or reducing the impact by using externally recycled materials. This is changing rapidly as legislators, raw materials suppliers and processors are all under pressure to develop and extend the concept of the circular economy (see Section 3.3). Despite these pressures, the actual progress on the ground appears to be slow, as opposed to the progress in the number of press releases (which is increasing exponentially).

In many companies there is a lack of consistent direction, focus and effort to proactively reduce the raw material impacts. So, who is really responsible for improving sustainability in the area of raw materials? In many ways this is similar to the design and development process (see Section 3.1) where materials are selected and controlled by ineffective 'over the wall' processes:

- The Technical Department designs the product to meet the specification set by the customer or sales. This rarely includes any sustainability metric (see Chapter 3).

- The Technical Department calculates how much material is required and specifies the material to meet the specification requirements.

- The Procurement Department negotiates a price for the specified materials within the specification and volume parameters set by the Sales Department.

- The Production Department makes the part from the specified material and attempts to keep waste to a minimum.

- The Finance Department records material use and material variances from production records.

The answer is that nobody is really in charge of this area of sustainability and everybody blames someone else.

It doesn't have to be this way.

This is not just about environmental sustainability, the cost of the raw materials in typical plastics products will vary between 45% of the total cost for a 'technical' product and 80% for a 'mass produced' product. Reducing materials content and use can therefore also improve the economic sustainability of a company.

- **Tip** – All types of bottles weigh a lot less today than they did 15 years ago. Lightweighting has delivered both environmental and cost benefits.

- **Tip** – If you don't do this to be 'green' then do it to become more profitable.

The Materials Team

In the same way that designing new products needs a team approach to improve sustainability, improving materials content and use also needs a team approach (see Section 3.1). Companies have whole departments, e.g., Human Resources, to look after the human impacts and costs but the materials impacts and costs are often treated as an afterthought. There is a real need for a cross-functional approach to managing impacts and costs in the area of materials and there needs to be a 'Materials Team' to look at the amount, type and impact of materials used in all current products and those proposed for new products.

The Materials Team should include staff from Design, Production, Quality, Finance, Procurement and major suppliers to ensure that all materials are used responsibly and cost-effectively. This is not an idle task, reducing materials use has a direct impact on environmental sustainability but an even greater impact on financial sustainability – a

> I once worked with a company which processed over 100,000 tonnes/year.
>
> One of their sustainability goals was to remove disposable cups and single-use stirrers from coffee rooms.
>
> Misdirected or what?

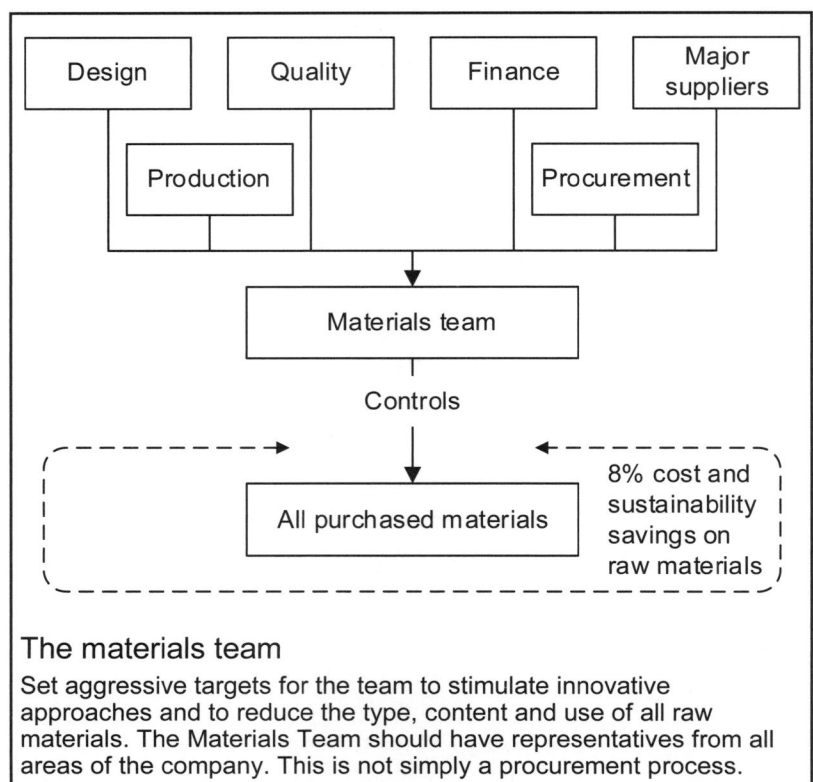

The materials team

Set aggressive targets for the team to stimulate innovative approaches and to reduce the type, content and use of all raw materials. The Materials Team should have representatives from all areas of the company. This is not simply a procurement process.

1% reduction in the cost of raw materials can have approximately the same effect on profits as a 10% increase in sales volume.

The Materials Team should be set aggressive materials cost reduction targets – an initial target of 8% reduction in the total materials bill is realistic and should force them to think hard.

Existing products

Section 3.17 showed how to use the PDS to review potential design changes to improve sustainability. This process should naturally consider material content and use.

Materials cost reduction needs open accounting information to be effective and to provide the focus for the impact and cost reduction efforts. The Materials Team will need accounting information to allow them to do their job effectively. They will need detailed breakdowns of all the materials' cost components. This will allow them to carry out value analysis on existing products and produce a 'materials use review' for the products. The materials use review forms the basis of the reduction strategy. The raw accounting information holds the key to materials cost reduction.

Note: When looking at the accounting information, ignore the allocated overheads based on machine hours etc., but consider some of the transaction costs, i.e., number and size of orders to each supplier. Try to concentrate on costs that are fully identifiable.

The Materials Team should:

- Identify the impact and cost of every materials selection decision, i.e., if a high-cost material is specified then this must be justified in terms of performance.
- Identify the real cost of every finish, operation and special feature.
- Justify every cost component or eliminate it.
- Remove features that are not required and reduce the price accordingly.
- Compare competitive products, strip them down and look for every cost saving – each one may be small but the total will be significant.
- Go for the big impacts and costs first and use a screening grid (see Section 3.17) to look for the easiest and most rewarding targets.

The Materials Team should have the freedom of action (both the power and the responsibility) to change anything and everything. This includes the product design, the raw materials and the manufacturing process. The potential gains are so great that nothing should be 'off-limits' in materials cost reduction.

Existing products are a particularly fertile area for materials cost reduction. The sales volumes and accounting data are well known and proven. This gives good data for impact and cost reduction and allows the Materials Team to prove that it is a worthwhile process.

- **Tip** – Use the output from a PDS review (see Section 3.17) and run a Materials Team exercise on it today. The results can be profit changing.
- **Tip** – If the current suppliers do not want to do this then find new suppliers who do. This is your money.
- **Tip** – Never forget that the customer has a stake in this too. Why not include them in the Materials Team? Making them part of the process (and sharing the rewards) can make it a much more effective process.

New products

Products in the early design stage are good candidates for consideration by the Materials Team because there are few 'committed' impacts, costs or ideas in the design (see Section 3.2).

It is easier to make changes to designs and specifications while in the drafting stage than at any other time. The Materials Team is in an ideal position to help reduce the materials decisions, i.e., type, content and use, and to work with suppliers to improve the materials choices.

Every new product design must be subjected to a full materials use review before the design is signed off and tooling manufacture begins.

- **Tip** – Good design in the early stage can make a real impact on the sustainability of a product.

This is not just about environmental sustainability.

Reducing material content and use has a direct impact on economic sustainability, i.e., profits.

Not all raw materials have the same environmental impact.

The embodied carbon and energy varies with the material but this is rarely considered at the design phase, .e.g., PE-HD has an embodied carbon of 1.93 kgCO$_2$/kg and PC has an embodied carbon of 7.62 kgCO$_2$/kg.

Source: circularecology.com/ embodied-carbon-footprint-database.html.

The total LCA results will also vary and the best source for these are the PlasticsEurope Eco-profiles (www.plasticseurope. org/en/resources/eco-profiles).

Make sure that all specifications are in terms of the function of the product and not in terms of the materials to be used.

Give suppliers a chance to demonstrate their technical capabilities and to take a share of the risk.

4.2 Reducing raw materials content (lightweighting)

The first opportunity

Reducing materials content at the design stage is vital to reducing the impact of materials (see Section 3.4). Designing materials out of the product is easiest at the design stage – it is the first and often the best opportunity to manage the product impacts.

Sales

Sales should always ensure that the primary and secondary product requirements are well defined and specified in terms of the function (it is preferable and helpful to use two word 'verb–noun' combinations rather than long sentences, e.g., 'support screen' or 'protect electronics') and not prescriptively in terms of materials or the quantity of materials.

Full information on the design intent is needed before starting design. This allows product designers and raw material suppliers the freedom to generate the most economic design and then to choose the most suitable material.

Sales is also responsible for ensuring that the completed PDS (see Section 3.16) is clarified and signed off by the customer.

- **Tip** – Sales has an important part to play in materials reduction; they have got to define the needs and sell the result.

- **Tip** – If a product feature adds cost or material but justifies no extra margin then it should be eliminated from the PDS unless it can be strategically justified.

Design

The design group should generate the full PDS and include all the essential and desirable features (based on functional analysis and not on materials). The PDS is provided to the sales team for customer approval and to the Materials Team for initial approval (see Section 4.1).

- **Tip** – Every PDS must have a 'materials design and use review' before it is signed off for detailed design.

The current PDS must be available to designers as the design is prepared to focus their minds on what the design is to achieve.

On completion, the detailed design must be checked against the PDS and a full design review carried out to ensure that all the requirements of the PDS and the materials design review have been met. This is the last chance to design out excess material.

Reducing the materials content

Basic items to check during this process are:

Project brief/PDS

- Set bold targets for materials content reduction. An initial target of 4% materials content reduction on current products should be achievable but similar work has achieved 10% materials content reduction.

- Ensure that the APQP[1] design methods are followed. These are a road map for successful product, process and tooling development. Reducing materials content is as much about management as it is about design – it is about giving designers the incentive, structure and opportunity to be innovative and to reduce materials content.

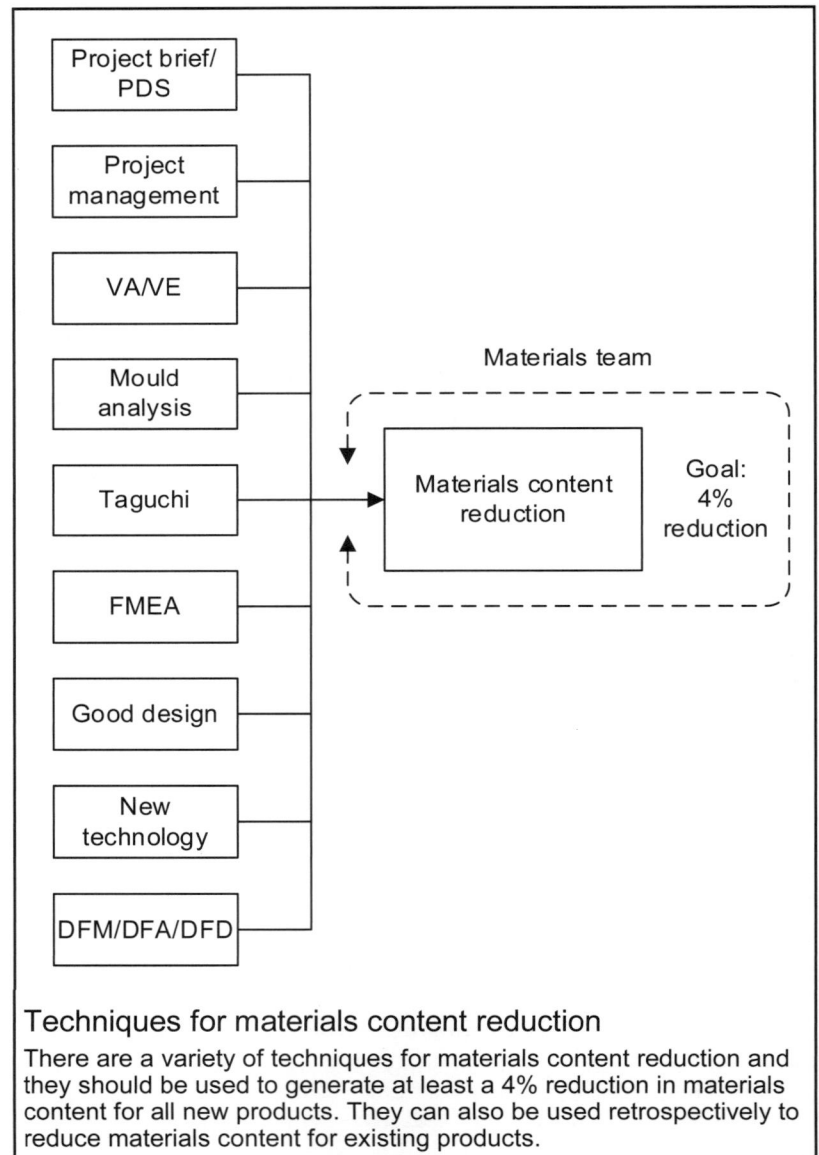

Techniques for materials content reduction

There are a variety of techniques for materials content reduction and they should be used to generate at least a 4% reduction in materials content for all new products. They can also be used retrospectively to reduce materials content for existing products.

Project management

- Use good project management practice to manage the product design process and to achieve the most economical design.
- Train designers in project management.

VA/VE

- If special parts are used then always question why the designers didn't use a cheaper part to meet the required function. The worst that will happen is that they will justify why they used the more expensive part. The best result is that the part cost can be reduced.
- The Materials Team should always question why the designers didn't use a 'stock' part. Stock parts include stock materials and finishes. The use of standard parts has a huge payback – no development costs and high volumes can reduce costs.
- Ruthlessly cut out 'over-designed' product features that add to cost but not value to the customer.

Mould analysis

- Always question the wall thickness for any plastic part. Use mould fill analysis to optimise and reduce wall thicknesses at all stages. Low-cost mould analysis will almost always reduce materials content and will also improve materials use during series production. For extruded products, wall thickness is equally important and should also be minimised to meet the functional requirements. Excessive wall thickness uses more material than necessary.
- Never accept the excuse of 'that is the way we did it last time'. Lightweighting is too important to be lazy at the design stage. Do the calculations, make models and carry out the trials – do not add material 'just in case'.

Taguchi

- Use experiment design techniques, such as Taguchi methods, to generate robust designs that have reduced failure rates.

FMEA

- Use design risk analysis methods such as FMEA to analyse potential failure modes and to build solutions into the design.
- Use the results of Taguchi and FMEA to refine designs and reduce materials content.

Design ideas

- Look for innovative ways to remove material by using good design principles – look at disposable razors to see how stiffness can be created by good design.

- Maintain a 'good practice' design library of parts that use good economic design techniques. These can be competitors' parts, parts from similar products or even radically different parts that have a good idea in them. I have a collection of over 60 plastic wind-up animals to remind me how innovation can work in strange places.

New technology

- Keep designers up-to-date with new technology that can be used to reduce materials content significantly.
- Current candidate new technologies are water-assisted moulding, foam moulding and highly filled wood plastic composites. All these techniques can be used to reduce the amount of plastic used and to replace it with either gases, voids or lower-cost fillers. Any of these options will reduce the amount of material used (but check the effect on the implications for recycling first).

DfM, DfA and DfD

- Train designers in DfM, DfA and DfD.
- Consider using 'dual-hardness' mouldings to meet multiple function requirements in a single component (but make sure that they are compatible materials – see Section 3.5).

The best opportunity

It is better to avoid putting material into the product than it is to try to take it out after the product has been designed and tooling has been manufactured. The design stage is the best opportunity to manage the product cost.

Note: It is essential that the primary function of the product is retained. Taking cost out of the product design should never result in taking customer value out of the product. Taking value out of the product will result in dissatisfied customers and product failure.

> **The moment of inertia can save material**
>
> The magic of 'bd³/12' (the second moment of area) means that you can build stiffness into a product but still reduce material use.

> **Extra cost**
>
> Extra effort at the start of the development process can greatly reduce the materials content in series production.

> The average weight of a yoghurt pot has halved in the last 30 years.

> New technology can reduce PET preforms by 1.0 g. This may not seem like a lot but over series production it is a huge amount of material.

- 1. AIAG. 2008. 'Advanced Product Quality Planning and Control Plan'. AIAG.

4.3   Reducing raw materials use in production

The last opportunity

The global material use of a product has largely been determined by the time the product reaches the production stage. Despite this, there are a range of techniques available to manage materials use during production and these will all significantly reduce materials use.

Production is where the material is actually used and is a key area for materials use reduction. The main opportunities for materials use reduction are:

Process settings

- Poorly set machines cost money in wasted material from increased scrap rates.

- Taguchi methods allow the optimum process parameters to be found to increase the reliability of production.

- Set-up sheets can get machines started quickly and with the right settings. Create set-up sheets for all products and keep them up-to-date.

- Institute SPC on every product to reduce materials content. Use capability studies and SPC (see Sections 5.7 and 5.8) to run products reliably at the lower end of the tolerance band to reduce materials use.

- Make changing process settings without written approval a dismissible offence. If you think this is harsh then how would you react if an operator let all the air out of your car tyres 'because he thought it would run better'? Operators changing process controls without authorisation costs real money in time, scrap and wasted materials.

- For injection moulding use scientific moulding techniques to find reliable and optimised settings.

Process controls

- Invest in process controls.

  - Gravimetric feed units will accurately control product weight and materials use for most processes. Accurate feeding will reduce materials use considerably.

  - Gear pump systems for extruders also give excellent materials use control and rapid payback.

  - Gauge controls for film extrusion can accurately control thickness and reduce the amount of material given away from inadequate control.

  - Parison thickness controls for extrusion blow moulding optimise the material

distribution in the product and reduce material use.

- Link control systems to inspection devices to reduce materials use. This can have very short payback times.

  - In extrusion, a saving of 3% of material on a pipe produced for a year at 800 kg/hr would save more than £25,000 per year per extrusion line. Typical systems control the wall thickness and material distribution and can be used to provide records for proof of conformance to specification. The systems allow control of tolerances to within 15% of the standards and can be adjusted to

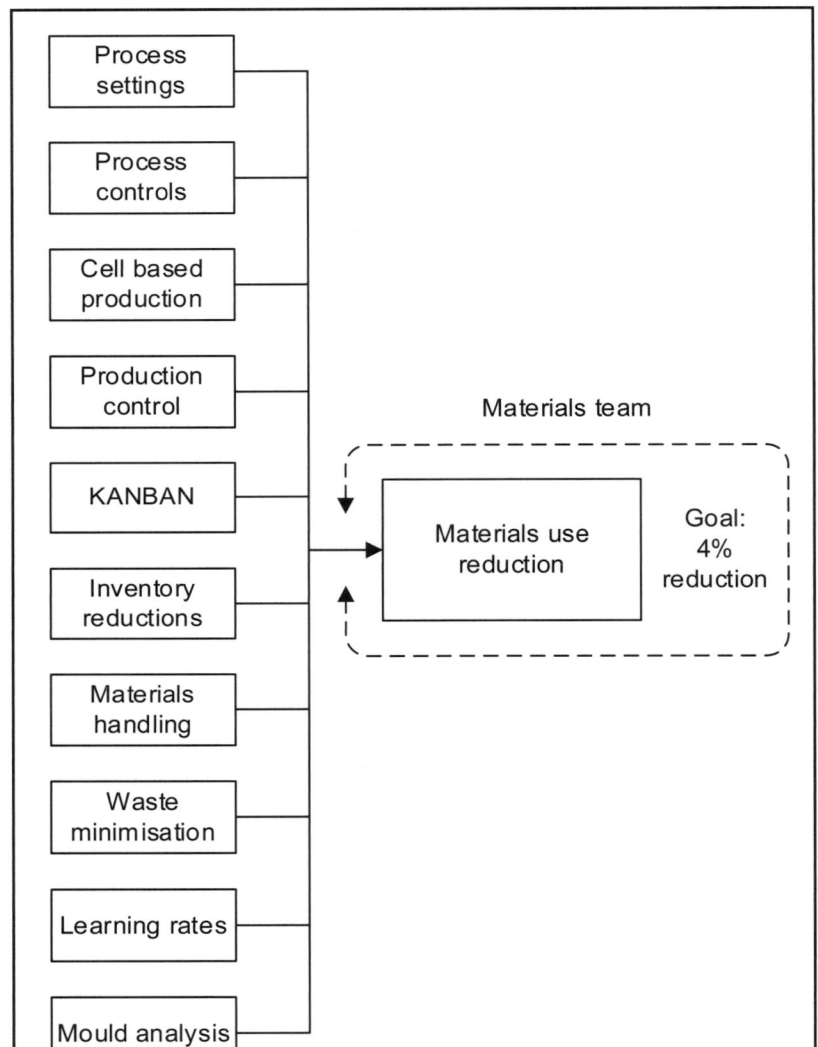

Techniques for materials use reduction

There are a variety of techniques for materials use reduction and they should be used to generate at least a 4% reduction in materials use for all products. They are best applied for new products but can also be used retrospectively for existing products.

minimise materials use and maximise output.

- For extrusion blow moulding, parison thickness controls can be linked to quality control testing to minimise material use by tuning the parison thickness to the minimum required.

Cell-based production

- Reduce set-up times to reduce scrap produced at machine start-up.

Production control

- Use simple control systems to produce to order and not to forecast. Products produced to forecast will sit in inventory and will 'shrink' due to inevitable damage and product obsolescence whilst in inventory.

KANBAN

- Use KANBAN containers to control production volumes and protect products from damage.

Inventory reduction

- Reduce raw materials inventory to decrease product 'shrinkage' and losses.

Materials handling

- Invest in 'closed system' materials handling equipment to ensure that raw materials are not spilt or lost onto the factory floor.
- **Tip** – This will also reduce pellets lost to the wild and subsequent pollution (see Section 5.4).
- Invest in automated scrap handling equipment that treats any scrap produced carefully and does not allow it to touch the factory floor. Producing scrap that can be reground is costly; producing contaminated scrap that is only fit for disposal is even more costly.
- **Tip** – Scrap that 'touches' the floor will be immediately reduced in value by up to 50% and will not be suitable for internal re-use.
- Invest in materials handling and transport systems that prevent finished products from touching the floor and losing value.

Waste

- Scrap is the result of inadequate production control and is an opportunity for materials cost reduction. Scrap, even when re-used, has consumed time, power and effort and has created unnecessary costs for the business.
- Reduce in-transit and final packaging needs in conjunction with the customer. They will help you with this because they pay for it and then they have to dispose of it (see Section 9.7).

- Invest in returnable packaging for both raw materials and finished product. Use returnable packaging and look after it. Negotiate a discount for returnable packaging.

Learning rates

- Operators learn how the process and the product works very quickly and often have useful ideas on how to improve the process. Use their experience to improve process yield, to reduce scrap and to increase good production (but don't let them adjust machines).

Mould analysis

- Mould analysis is often not used for existing tooling because it is assumed that the savings could never justify the costs of retooling. Despite this, the calculations sometimes show that even this 'unthinkable' option can be profitable over a short time scale – do the calculations if in doubt. When retooling is needed for other reasons it is logical to seek materials cost reductions rather than simply purchasing a replacement tool to the same design.
- Redesign (or preferably remove) sprues and runners and tops and tails to improve materials use.

Minimise materials use

Customers are not paying for plastic; they are paying for solutions. Minimise the amount of plastic used to achieve the solution and minimise the environmental impact (and maximise the profit).

Note: As with materials content cost management, it is essential that the primary function of the product is retained. Taking cost out of the product at production should never result in taking value (as seen by the customer) out of the product.

> Remember that customers are not paying for plastic, they are paying for solutions.

4.4 Reducing raw materials content and use – where are you now?

Managing materials use

Minimising materials use is a key to both environmental and economic sustainability, i.e., doing good can also be profitable. Many companies have recognised this and taken significant action but there is still more to do in reducing materials use. This is not the same as managing the costs by watching the polymer cost indices and adjusting your expectations or attempting to adjust your prices. Taking material out of the product at either the design or production stage permanently reduces the product cost whatever the raw material prices do.

This is a prize worth having and yet many companies fail to attack the issue with sufficient rigour or organisation. The materials content and use process crosses too many departmental boundaries for companies organised along functional lines. The Materials Team is one way to organise the company to manage materials use and content issues but companies must accept the need for a cross-functional approach to this concern.

The Materials Team must be target driven and an initial target of an 8% total reduction in materials content and use for the same output of saleable product is recommended.

Completing the chart

This chart is completed and assessed as for the previous charts.

> Reducing material use reduces the environmental impacts and improves profits.

> This is too important to be ignored and left to procurement, design and production to attack separately. There is a need for a team approach.

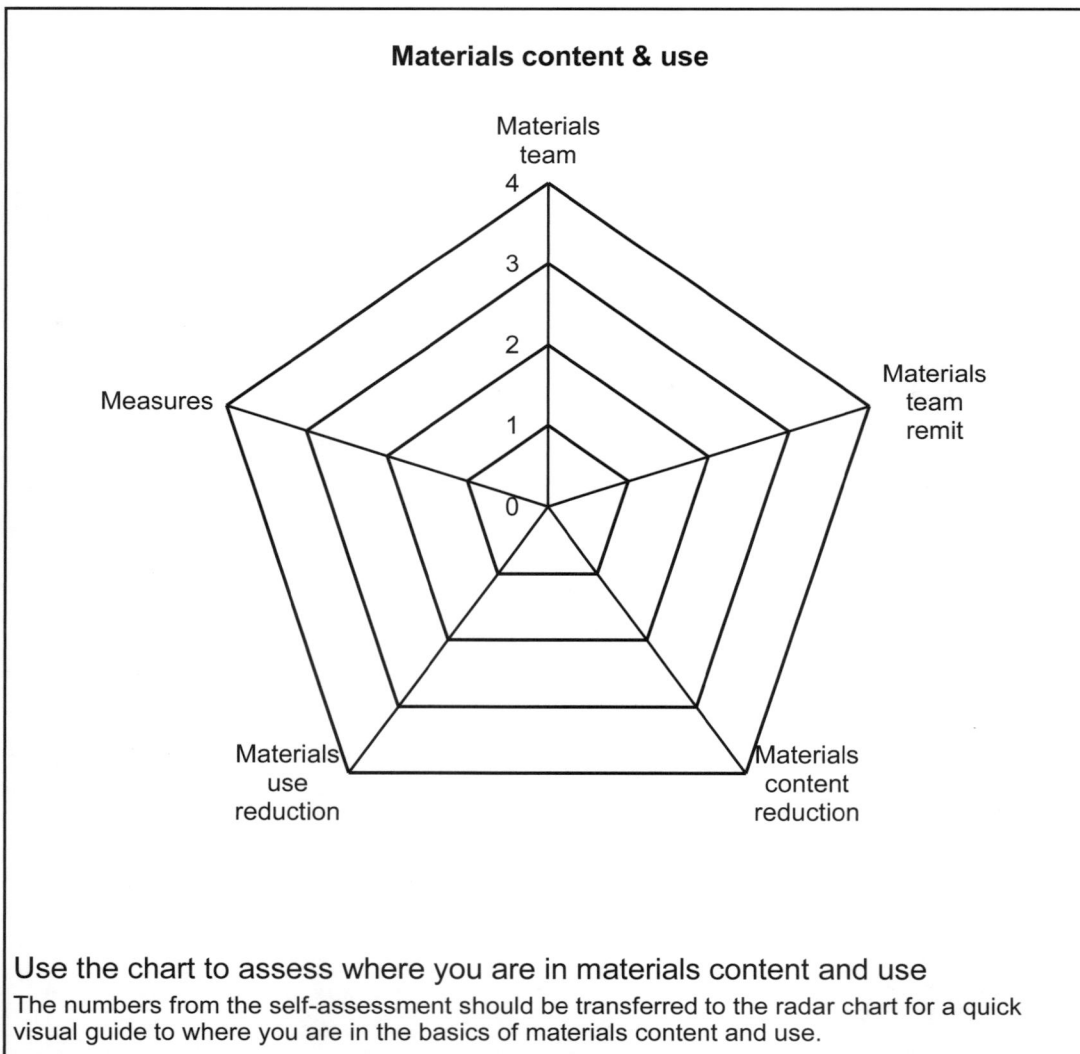

Materials content & use

Radar chart with axes: Materials team, Materials team remit, Materials content reduction, Materials use reduction, Measures. Scale 0 to 4.

> Many years ago, I was involved in the development of the European standard for PVC-U windows.
>
> Some countries wanted to specify a wall thickness of 3.0 mm (that is what their current designs and tooling were).
>
> We negotiated hard for a 'performance-based' standard, i.e., the wall thickness could vary as long as the final window met the required functional tests.
>
> We succeeded and enabled reductions in the wall thickness and significant materials savings for no loss in performance.

Use the chart to assess where you are in materials content and use

The numbers from the self-assessment should be transferred to the radar chart for a quick visual guide to where you are in the basics of materials content and use.

> At 10% margin a materials content or use reduction of £1 is the same as increasing profits by £1 and sales by £10.

Materials content and use

Level	Materials Team	Materials Team remit	Materials content reduction	Materials use reduction	Measures
4	Materials Team formed & active for both current & new products. Materials content & use extensively & rigorously controlled.	Materials Team has power & responsibility to make substantial changes to materials content & use.	Formal & aggressive materials content reduction target set (>8%). Target monitored & achieved.	Formal & aggressive materials use reduction target set (>8%). Target monitored & achieved.	Excellent measurement of materials cost reductions against aggressive targets. Excellent measurement of Materials Team performance against specific targets.
3	Materials Team for content reduction for new & existing products. Materials use for current products is production responsibility only.	Materials Team has power & responsibility to make only minor changes to materials content & use.	Formal but non-challenging materials content reduction target set (<4%). Target monitored but not achieved.	Formal but non-challenging materials use reduction target set (<4%). Target monitored but not achieved.	Good monitoring & targeting of materials cost reductions against moderate targets. Good monitoring of Materials Team performance against moderate targets.
2	Materials Team for new product content reduction by design team. Existing products not considered. Materials use for current products is production responsibility only.	Materials Team has responsibility for materials content & use but little power to actually implement decisions.	Informal & challenging materials content reduction target set. Target not monitored & rarely achieved.	Informal & challenging materials use reduction target set. Target not monitored & rarely achieved.	Some monitoring & targeting of materials cost reductions but against poorly defined targets. Few measurements of effectiveness of materials use & against poorly defined targets.
1	Materials content & use reduction is low priority & managed by single function.	Materials Team has advisory role only. Team makes recommendations only. Recommendations often overruled by other managers.	Informal but non-challenging materials content reduction target set. Failure to achieve target is regarded as normal & acceptable.	Informal but non-challenging materials use reduction target set. Failure to achieve target is regarded as normal & acceptable.	Poor monitoring & targeting for materials cost reductions. Only vague idea of effectiveness of materials use, i.e., some measurements available.
0	No central contact for materials content or use reduction.	No Materials Team in operation.	No targeting for materials content reduction at site.	No targeting for materials use reduction at site.	No monitoring & targeting for effective materials cost reductions. No cost monitoring or targeting for materials use, e.g., cost/purchase order.
Score					

4.5   Material destinations

It is circular

It was noted in the introduction to this chapter that materials can either be treated under raw materials or under end-of-life and that we will cover them under raw materials. For completeness, we will also cover the three processes where end-of-life materials exit the system, e.g., energy recovery, landfill and littering.

The framework

The methods used for materials recovery are shown in the flow chart on the right. This shows the flow of material from the petrochemical or biomass feedstock through the standard phases of the life cycle to the end-of-life. The most common materials recovery options and where these materials re-enter the system are also shown.

The flow chart also shows where material that is not recovered exits the system.

• **Tip** – Internal re-use is not regarded as recycling as the material has not left the site boundaries (see Section 5.3 and 8.2).

Product re-use

This is where the product is designed for more than a single life. This is rare for most long-life products, e.g., pipes or windows, but is sometimes used for short- or medium-life products such as bottles, e.g., refillable hand wash bottles where the consumer can buy bulk product to refill bottle at home. For single use drinks bottles, the economic and environmental impact of the sorting, washing and handling processes make it very rare even though a PET bottle can be re-used up to 25 times before it is no longer suitable for re-use.

Mechanical recycling

Mechanical recycling retains the basic polymer structure and is used to recover the complete material (including any additives or colours). Products at the end-of-life are collected, sorted, cleaned, and reprocessed into the raw material for new products.

The raw material can be recycled into similar products, e.g., PET bottle to PET bottle, or diverted to other products, e.g., PET bottle to PET fibre. Where the raw material is used for products which have a lower specification then this is sometimes referred to as 'downcycling', as if it is somehow inferior to recycling.

Mechanical recycling is generally limited in terms of the types of plastics recycled because of the need for large volume

collection to make sorting and processing economically viable, i.e., it is mainly used for PET, PE-HD and PP, that is collected from the Municipal Solid Waste (MSW) stream. Mechanical recycling is also extremely successful for PVC but this is generally through an industry run closed-loop system (see Section 10.9) rather than through MSW collection.

Mechanical recycling is covered more fully in Section 4.6.

Chemical recycling

Chemical recycling covers an extremely broad range of technologies that use different chemical processes to break the plastic down to varying degrees, e.g., the main polymer backbone may be retained, the polymer backbone may be broken down to

The recycling industry is complex, there are many processes and variations of these. These sections are designed to give plastics processors an overview of the many materials recovery processes and not as a textbook on recycling.

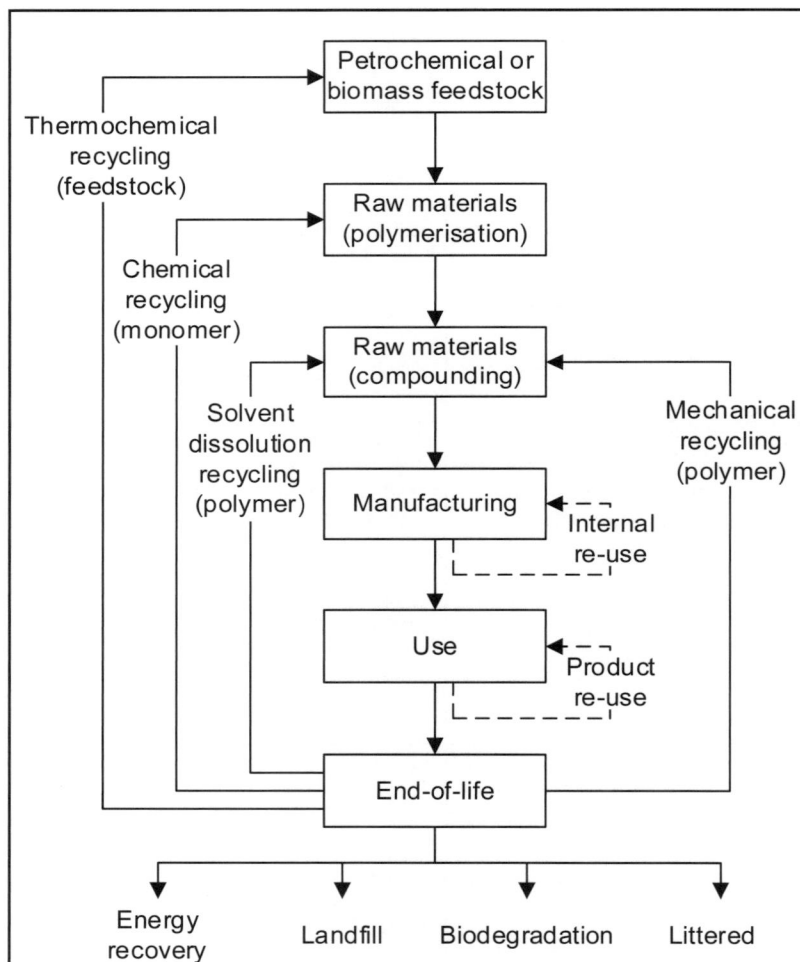

The routes for materials recovery

After end-of-life, there are several routes for recovery of the material, either as a polymer, as a monomer or as petrochemical feedstock. If the material is not recovered then it is either incinerated to recover the embodied energy, landfilled, composted or littered.

produce a monomer or the polymer backbone may be broken down completely to give a feedstock similar to that originally used to produce the monomer. The point that the recycled output re-enters the process depends on the degree that the polymer has been broken down. The three main types of process are:

- Solvent dissolution – the polymer is selectively dissolved in a solvent and the main polymer backbone is retained. The output is a 'clean' version of the original polymer that can be compounded as normal before processing.

  It is debatable whether solvent dissolution is a chemical recycling process or a mechanical recycling process as there is no change to the polymer backbone and the material can be re-compounded without further treatment. However, we will treat it as chemical recycling to avoid confusion with traditional mechanical recycling.

- Chemical recycling – the main polymer backbone is broken down (depolymerised). The output is the monomer or oligomer form of the input polymer which must undergo polymerisation again before it can be compounded for processing.

- Thermochemical (feedstock) recycling – not only is the main polymer backbone broken down but the material is split back to the basic components. The output is a mixed petrochemical feedstock, e.g., wax, oil and gas, that can be used to produce new polymers instead of petrochemical or biomass feedstock. The output can also be used for other chemical processes or as fuel.

The materials which can be recycled depend on the process used. Solvent dissolution and chemical recycling are more limited in the scope of applicable materials than feedstock recycling, e.g., pyrolysis or gasification. Whichever chemical recycling process is used, these technologies are generally more energy intensive and complex than mechanical recycling. As a result, chemical recycling in not used significantly but the sector is developing rapidly.

Chemical recycling is covered more fully in Section 4.10.

Energy recovery (incineration)

If raw materials cannot be recovered through mechanical or chemical recycling then it is possible to recover the embodied energy of the polymers through incineration with energy recovery.

- **Tip** – This is not the same thing as simple incineration. The important thing is that the embodied energy is recovered.

- **Tip** – Energy recovery is not recycling as the material is lost to the system forever and only the embodied energy is recovered.

The implementation of incineration with energy recovery varies across the world and is much more prevalent in some areas than in others.

- **Tip** – Mixed plastic waste generally has a higher calorific value than coal.

Incineration with energy recovery is covered more fully in Section 4.11.

Landfill

Where incineration is not widely used, the exit route for many plastics products is landfill. This is increasingly being phased out by rising public opinion, the moves towards a circular economy and rising costs. Depending on the country, waste products may be crushed and treated to reduce volume and pollutants before landfill disposal.

Landfill is covered more fully in Section 4.12.

Biodegradation

The rapidly developing area of biodegradable plastics offers an alternative method of exiting the system. This is a form of controlled degradation and offers a method of exiting the system where the materials are not recovered or recycled.

Bio-based plastics and biodegradability (and the confusion regarding these) are covered more fully in Sections 4.14 and 4.15.

Littering

In countries with a poor waste management system (or none at all) the exit route for most plastics is littering, although in many of these countries it is probably more accurate to call it 'open disposal' as the concept of littering is not well developed. This is the source of most of the plastics in the ocean that present a real challenge for the plastics industry.

Littering is a societal issue, even in countries with a well-developed waste management system and is not covered in this workbook.

Materials recovery in any form should be designed to prevent losses to the system (however defined).

Most plastics processors will not be involved in any of these processes but still need to know about them to prepare for the future.

4.6   Mechanical recycling

The heat is on

Mechanical recycling is sometimes termed 'secondary recycling' and is used to reclaim plastic waste as a source of material for new products. Mechanical recycling is the preferred option for the recycling of plastics in MSW and accounts for ≈ 33% of the plastic waste generated in Europe (> 42% for plastics packaging waste) but this is increasing rapidly.

Legislation is increasingly driving mechanical recycling and there are many schemes around the world designed to increase recycling rates, e.g., EU Directive 2018/852 requires a recycling rate of 50% of plastics packaging by 31 December 2025 and 55% by 31 December 2030 and EU Directive 2019/904 requires the use of recycled PET in bottles of 25% from 2025 and 30% from 2030 at the same time requiring separate collection of single use plastic products for recycling of 77% (weight) by 2025 and 90% (weight) by 2029. Some of these schemes are contentious as the infrastructure to achieve the requirements does not actually exist.

- **Tip** – Mechanical recycling retains the basic polymer structure and is used to recover the complete material (including any additives or colours). This can give contamination issues if the material is not properly sorted.

Mechanical recycling is primarily used for PET, PE-HD and PP.

- **Tip** – These plastics make up > 75% of plastics in MSW and are collected and sorted for recycling in many countries.

- **Tip** – Most other plastics can be mechanically recycled but this rarely done because of the volumes necessary for mechanical recycling.

Depending on the degree of contamination, the process used and the design of the product (see Section 3.5) the recycled material can be directly recycled into similar products, e.g., bottle to bottle, used for other processes, e.g., bottle to fibre, or used in conjunction with virgin material for new products, e.g., used as an internal layer for window profiles.

Note: PVC-U window profiles are extensively mechanically recycled (≈ 89%) via an industry run closed-loop system, i.e., installers of new windows collect and recycle the old windows (see Section 10.9). Similar systems operate to recover the lead and PP from car batteries and these have a > 98% recycling rate.

A multitude of processes

There are many competing technologies for mechanical recycling that vary with the material and the intended use of the recycled material. A general flow chart for mechanical recycling is shown below and the major steps (not necessarily in this order) are:

- Collection of MSW – depending on the location this is done by local authorities or by 'bottle banks'. Collection schemes vary widely and there is little consistency even within a country.
- Separation of the MSW – the waste is initially separated into waste streams such as paper, cardboard, glass and plastic for despatch to the plastic recycler.
- Initial sorting at plastics recycler – this can be automatic or manual and will involve separation of large contaminants, metals and films from the general plastics waste.
- NIR sorting of the plastic waste – this will sort the waste into the basic polymer types and colours but good product design (see Section 3.5) is needed to avoid incorrect identification, e.g., materials coloured with carbon black cannot be identified with NIR and large labels can make identification difficult.

The packaging sector contributes ≈ 60% of the plastic in MSW collection and almost all the plastic in the recycling stream.

In the packaging sector, using materials with a high recycled content is rapidly becoming necessary for the 'Right-to-Operate' and part of the proof of sustainability.

A significant issue for mechanical recycling is the number of agents in the value chain and their need to make profits when required to make large investments.

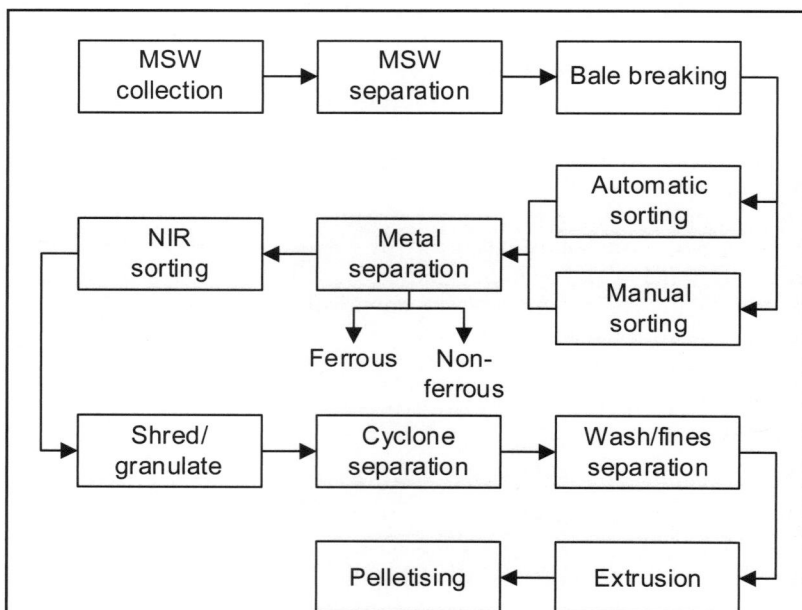

A general flow diagram for mechanical recycling
Most mechanical recycling will follow a process like this but the order may be different, i.e., some processes begin with washing and label removal and some begin with NIR sorting. PET recycling will also differ and can include special treatments for food grade products.

- Shredding/granulation of the waste – this reduces the waste to flakes for treatment.
- Cyclone separation – this separates the flakes based on weight and can be combined with sink/float separation to separate the plastics based on the density.
- Wash/fines separation – washing is designed to remove labels and any contaminants before extrusion.
- Extrusion – the clean flakes are extruded and pelletised to produce pellets for new product production. The extrusion process can also be used to:
  - Re-compound the base material for specific applications.
  - Provide a final filter of the melt to improve the purity of the material.
- **Tip** – These steps are not carried out by the same companies and the material may change hands several times in the process.

There are also many variations on this basic process, i.e., some processes begin with dry removal of labels and some begin with a caustic wash, but most processes will include these basic steps.

PET is a special case for mechanical recycling, it can be easily mechanically recycled but the output is not always suitable for food grade use and there are specific techniques for producing food grade (bottle to bottle) materials, e.g., Vacunite® by Erema which uses vacuum assisted solid state polycondensation (SSP) or P:REACT® by NGR which uses liquid state polycondensation (LSP).

These techniques will become increasingly important if the industry is to achieve the recycling targets already mandated by many governments.

- **Tip** – Mechanical recycling can recover only 60-75% of the input mass, even if it is well sorted. The remaining 25-40% will not be recycled and will either be sent for chemical recycling (see Section 4.10), energy recovery (see Section 4.11) or landfill (see Section 4.12).

Environmental benefits

The environmental benefits of mechanical recycling are significant. Some of these are:

- The use of mechanically recycled materials gives a considerable reduction in the carbon footprint of the final product – our calculations show that the total carbon impact from extraction to the processing site is 2.45 kg CO_2e/kg for virgin materials and from collection to the processing site is 0.46 kg CO_2e/kg for recycled materials. This means that the input carbon footprint of recycled materials is < 20% that of virgin materials.

- The use of mechanically recycled materials reduces the use of virgin materials – provided the recycled material is used to displace virgin material then using 1 kg of recycled material means that 1 kg of virgin material is not used. This is not the case where recycled material is used in products such as wood plastics composites where the recycled material is largely displacing the use of wood.

- Mechanical recycling, even with only 60% efficiency, significantly reduces the amount of material sent to landfill and the high costs associated with this.

The focus by governments on mechanical recycling and 'The Circular Economy' is increasing the pressure to collect more, to recycle more and to use more recycled material. This is not going to go away.

Design for recycling begins at the start. Leaving it until the end is too late.

Recycling and the use of recycled materials must be part of the standard PDS for all products.

Despite the desire for a circular economy, there is a limit to the number of times that a polymer can be recycled. There is some inevitable chain scission and a reduction in molecular weight with each processing step. This may prove a limiting factor with mechanical recycling and necessitate chemical recycling to 'refresh' the polymer chains or EfW to recover the embodied energy (see Section 4.11.

Recycled material exiting the extruder

Depending on the exact details of the recycling chain the cleaned and filtered material is extruded into strands, cooled and pelletised for despatch to and use by the processor as a substitute for virgin material. The material shown here is PE-HD from bottles.

4.7 Recycled materials in the supply chain

Legislation versus economics

For many years the competition between recycled and virgin materials was economics based but the rise in legislative requirements and environmental taxes on the industry is rapidly changing this balance. Decisions that were once made on the simple relative cost of the materials now must consider the legislative requirements – things just got more difficult.

Note: This section only covers mechanically recycled plastics.

- **Tip** – How waste is managed and materials are recycled, is becoming the single biggest issue for the plastics processing industry. This is not going to go away.

- **Tip** – Whilst the processors are the focus of this workbook, the threat to petrochemical companies from increased recycling rates are more direct and obvious, e.g., a kg of recycled material is a kg of virgin material not sold.

A complicated supply chain

The supply chain for recycled material is more complex and varied than the supply chain for virgin materials and an outline of the two supply chains is shown on the right.

Virgin material is produced in well controlled processes by a limited number of large chemical companies and follows a well-established and short supply chain that provided processors with very well-defined materials. In fact, this supply chain has lengthened in the last 20 years as many of the major chemical companies stopped direct sales to processors and started to use distributors for most of their sales. In contrast, the supply chain for recycled materials involves many thousands of companies, is far longer, involves mixed input materials that can be contaminated and poorly defined, and differs dramatically not only between countries but also within countries.

The outline of the mechanical recycling process given in Section 4.6 described the process, which is easier to describe than the fragmented supply chain where the many different agents are all trying to either make a profit or, at the very least, minimise costs. Mechanical recyclers exist at the intersection of two mature industries, waste management and plastics processing. The waste management industry wants to increase the prices of sorted material, the plastics processing industry wants to

decrease their materials costs and is constantly comparing the cost of recycled materials to that of virgin material in an era of low oil prices.

Material Recovery Facility (MRF)

Input material to the MRF must be sorted for sale to the plastics recycler. Excluding the cost of collection, the material entering the MRF is essentially free but sorting is not free and cost is added to the process. In some countries, consumers pre-sort the MSW into recyclables, e.g., paper, cardboard and plastics, but in other countries with less developed waste management systems the recyclables may be contaminated with organic material, e.g., food waste.

Mixed plastic waste has a low value (< £30/ tonne). Sorting adds value to the waste and this can be done to several different levels:

- A basic sort: PET bottles, PE-HD bottles and others (including rigids and films).

- A simple sort: PET bottles, natural PE-HD bottles, coloured PE-HD bottles, bulky rigids and others (including films).

Sorting is what adds value

High standards for sorting at an MRF increase the value of the plastics collected from MSW. Even small amounts of non-compatible materials (see Section 3.5) can decrease the utility and value of sorted plastics.

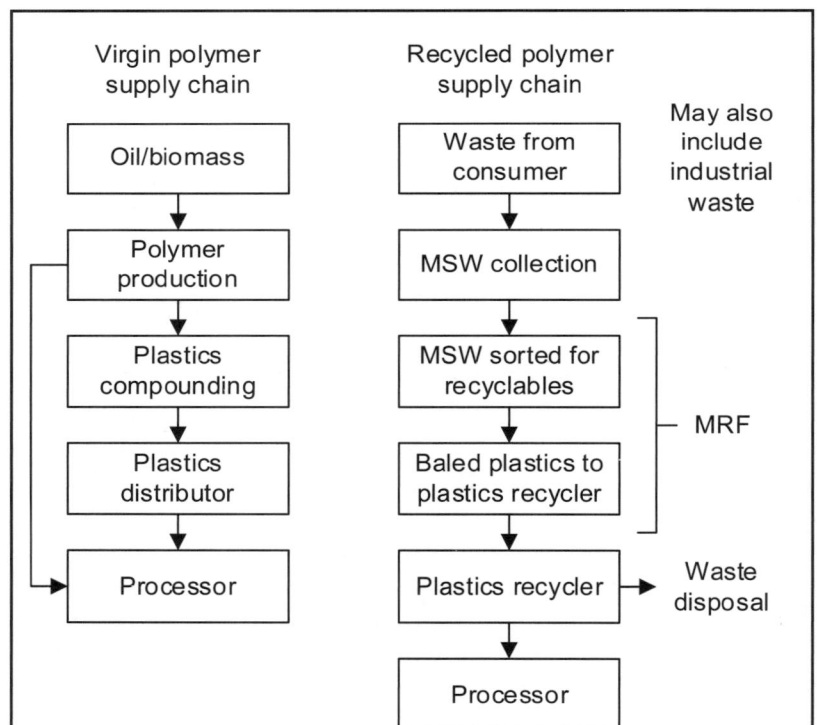

The supply chain for recycled materials

The supply chain for recycled materials is longer and more complex than the comparable supply chain for virgin materials. The multiple agents in the recycling chain compete against each other for limited profit margins and this constrains growth.

- A moderate sort: PET bottles, natural PE-HD bottles, coloured PE-HD bottles, PP, bulky rigids, others (including small rigids and films) and residuals for disposal.
- A high sort: PET bottles, natural PE-HD bottles, coloured PE-HD bottles, PP, PE-HD rigids, others (including small rigids and films) and residuals for disposal.

Each sort adds value to the input material and increases the bale price to the recycler.

- **Tip** – Industrial waste entering an MRF should already be well sorted. It already has more value.
- **Tip** – The Association of Plastics Recyclers (APR – www.plastics recycling.org) is a good source of information for the recycling industry in the USA. They have: a 'Guide for Plastics Sorting' that uses three different levels of sorting (1 to 3), a 'Sort for Value' calculator to show the financial benefits of good sorting and 'Model Bale Specifications' for various plastics.
- **Tip** – Plastics Recyclers Europe (PRE – www.plasticsrecyclers.eu) is a good source of information for the recycling industry in Europe.

Plastics recycler

The recycler buys bales of sorted material from the MRF and sorts (manual or automatic), separates any metals and further sorts to produce a clean input stream for extrusion and pelletising. The recycler depends on a clean and well sorted input stream to produce a good recyclate that is acceptable to processors.

However, the plastics recycler is constrained by two factors, the cost of the material from the MRF and the availability of suitable input material from the MRF. This has constrained the growth and success of recycling companies and the volume of recyclate in the market.

Closed and semi-closed loop systems

Closed loop systems, e.g., PVC-U windows and PP batteries, have proved successful in certain market sectors. These systems have better control on the input stream, have higher recycling rates and show higher yields.

Some companies have developed semi-closed loop systems, e.g., Axion in the UK segregated PS-HI from WEEE refrigerator liners to produce specific compounds, this approach produced a much more consistent output product which commanded higher prices from processors and improved traceability in the supply chain. Increasing the use of semi-closed loop systems offers recyclers the opportunity to improve sales prices to match those of virgin materials by

offering processors a greater degree of confidence in the material.

Processor

Processors need a constant supply of high-quality plastic with minimal contamination especially for demanding applications, e.g., food contact applications. In the past, processors who have purchased recycled materials generally did so because of the economic incentive, i.e., they pay less for recycled materials than for virgin materials.

Legislative requirements for the inclusion of recycled materials may well change this equation, i.e., the cost of not using recycled materials will increase and will potentially force some processors to use recycled materials (see Section 4.8).

- **Tip** – The future demand for recycled materials may depend not as much on the economics as on legislative compliance particularly for packaging companies.
- **Tip** – In the future, processors will need to examine their materials choices more carefully to ensure that they minimize the total cost of their raw materials, i.e., not just the simple purchase cost but also the costs of legislative compliance and Extended Producer Responsibility.

Ripe for consolidation?

If the plastics processing industry is going to meet the legislative targets for mechanical recycling then these conflicts need to be reconciled and investment made in increased recycling capacity. The mechanical recycling industry is ripe for consolidation from waste management companies moving downstream into plastics recycling where previously their focus has been on treating MSW and selling bales of poorly sorted materials.

The market for recycled materials is there and will increase in the future but the supply chain suffers from multiple issues that constrain growth.

- **Tip** – Countries without a national framework for recycling will under-perform because of the market conflicts. A national framework is not evident in most countries.

Residuals for disposal to landfill can cost the MRF money and increase bale prices. The residuals may also be plastics but the increased sorting costs for low volume materials generally means that these are sent to landfill.

Just as with machine purchases where the initial purchase cost is only part of the cost, the purchase cost of raw materials may well be only part of the cost.

Processors need to start looking at the total purchase cost of their materials.

This offers the potential not only to reduce costs but also to improve sustainability performance.

Extended Producer Responsibility (EPR) in Europe may change the face of recycling – there is neither enough capacity nor enough material in the supply chain to meet the projected demand.

There is no currently accepted scientific method of determining the amount of recycled material present in a plastic. They are all polymer chains.

4.8 The market for recycled materials

A growing market

Not a day goes by where we don't read of another company using recycled materials and developments in this field. This can be in automotive parts or consumer packaging but the direction of change is clear, the market for recycled materials is growing despite the pressures on recyclers. The good news for recyclers is that the legislative pressures and the recycling volume constraints mean that in some cases it is possible for the cost of recycled material to exceed the cost of virgin material.

This section looks at the market forces driving this and the potential markets.

Market readiness

Despite the issues regarding the supply chain (see Section 4.7), the market readiness for recycled materials is high and growing. This demand is created by environmentally conscious consumers and legislators who are reacting to their social demands and pressures. The resulting legislation and pressures are then driven up the supply chain from the brand owners to the processors, particularly packaging processors, and finally to the recyclers. The needs at each stage of the supply chain are shown in the diagram on the right.

Consumers

Consumer attitudes towards plastic have changed in the last 20 years. Plastics are no longer seen as good or neutral but as potentially bad. Consumers, particularly in Europe and the USA, are demanding a reduction in packaging, the ability to recycle packaging, clarity on what can be recycled and clear statements of the recycled content of the packaging that they purchase.

- **Tip** – There is a real knowledge gap at the consumer level about the benefits of plastics use in packaging and other areas.

Legislators

Legislators around the world are reacting to the social and consumer pressures and have been increasingly adopting the circular economy ethos. This has resulted in significant legislation to increase recycled content in Europe (see Section 4.6) and similar measures in the USA. Some of these schemes to increase recycled content are optimistic in the extreme and include proposals to implement fines for beverage bottles made from less than 100% recycled materials.

- **Tip** – The direction of travel is clear. Processors should start planning now.

Brand owners

Brand owners traditionally avoided disclosing the presence or amount of recycled material in their products to avoid any perception of lower quality but this has now changed and many brand owners are now declaring the amount of recycled material used and recycling information on their packaging. In some areas, there is almost an 'arms race' in terms of declaring more recycled material than the competition. The brand owners are now placing pressure on processors to increase the amount of recycled material in their products.

- **Tip** – Declarations should meet the guidance of ISO 14021:2016 (see Section 9.2).

Plastics processors

Processors have always internally re-used material, e.g., tops and tails for EBM and sprues and runners for IM, but have also traditionally been reluctant to use recycled materials because of the higher material variability and potential for contamination. Improved sorting and recycling techniques have reduced both concerns but processors still need confidence in the quality and

The mechanical recycling industry is going to have to embrace certification (see Section 4.9) to provide confidence at all levels of the supply chain.

This will add costs at a time when petrochemicals prices are at historically low levels.

In some areas, there is a price premium for recycled material due to legislation and the desire of brand owners to be 'green'. There are even cases of 'bootleg' virgin resins being sold as 'recycled' to get a higher price.

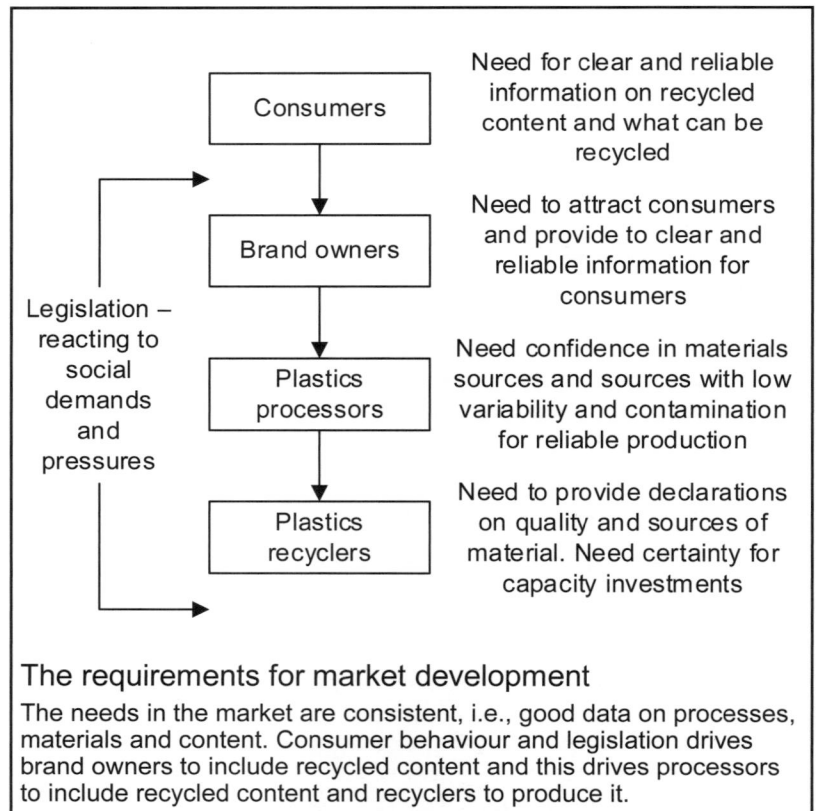

The requirements for market development

The needs in the market are consistent, i.e., good data on processes, materials and content. Consumer behaviour and legislation drives brand owners to include recycled content and this drives processors to include recycled content and recyclers to produce it.

consistency of recycled materials. This confidence can be gained by improved certification of both recyclers and recycled materials (see Section 4.9).

- **Tip** – Some processors are taking a lead here and are taking proposals to use recycled materials to the brand owners.

Plastics recyclers

Recyclers are under intense pressure because of the complexity of the supply chain (see Section 4.7) and the need to provide improved declarations on the source of the raw materials and the quality of the output recycled materials. This increased need for transparency will increase costs in a fragile infrastructure.

Markets

The market readiness described above is primarily in relation to packaging products but there are other markets that need to be considered to fully utilise recycled materials.

Food contact packaging (short life)

Food contact packaging has high visual and cosmetic requirements and there are also financial and supply constraints but the primary barrier to the use of recycled materials is the regulations governing food contact materials. These are heavily regulated in all parts of the world, e.g., the European Food Safety Authority (EFSA) in the EU, the Food and Drug Administration (FDA) in the USA and other local authorities. Food contact regulations will generally override any applicable environmental regulations.

The standards, in terms of contamination and additives, for recycled materials in food contact applications are the highest of any application but PET and PE-HD are extensively recycled for bottles and trays. This means that the residual materials tend to be higher, i.e., the processes have a lower yield.

Obtaining EFSA, FDA or local approval for processes and materials is very complicated and beyond the scope of this workbook.

Non-food packaging (short life)

General packaging of items such as domestic and industrial products is not subject to any regulation apart from standard health and safety regulations and the need to protect the product. Recycled plastics have a market opportunity here due to the less stringent requirements for non-food packaging.

Automotive (medium life)

Automotive parts are increasingly using recycled materials such as recycled PP for non-aesthetic parts, fibre from recycled PET for interior fabrics, recycled PE-HD for non-

structural parts and recycled foam for acoustic pads, e.g., Volvo have set a target to use recycled materials for 25% of plastics used in new cars after 2025.

Fibres (medium life)

The manufacturing of fibres from recycled PET is a well-developed market that can accept higher levels of contamination than food contact applications.

- **Tip** – PET fibres are referred to as 'polyester' (PET is a member of the polyester family) rather than as PET, e.g., Terylene (UK) and Dacron (USA).

Construction products (long life)

Construction products have a very long lifespan, e.g., PVC-U windows ≈ 40-50 years and pipes ≈ 100 years, and are easily recycled.

- Windows – recycled PVC-U can be coextruded for internal non-visible surfaces.
- Pipes – recycled PE-HD can be used for low pressure applications such as drainage and corrugated piping and performs as well as virgin material. This takes a 60-day product such as a shampoo bottle and converts it into a 100-year product. One company in Australia (RPM Pipes) uses recycled PE-HD materials for all their products.

Legacy additives

Legacy additives present a significant issue for long-life products and for food contact products. The main issues are with lead-based stabilizers for PVC-U windows, heavy metal colourants and phthalate plasticisers (see Section 10.9). For current products these have largely been eliminated or reduced to trace levels as a result of REACH and RoHS (see Section 3.6) but older products may contain significant levels of legacy additives. These can be treated in standard mechanical recycling and reduced to very low levels but some legislation is forcing these to be landfilled rather than recycled, even when this is against scientific advice.

Recycled materials must also conform to REACH and RoHS (see Section 3.6).

European Food Safety Authority approval requires that 95% of the recycled material for food contact applications has been sourced from food-contact applications, and there must be full and provable traceability through the supply chain.

The market for recycled materials is growing but will be constrained by both collection rates and capacity issues.

4.9   Qualifying recyclers and recycled materials

It is about proving it

The recycling industry deals with a highly variable raw material and controls are needed to ensure that recycled material is really recycled material and that the quality levels needed by processors are maintained. Controls are needed for the input material, the recycling process and for the output material. Good application of controls can give processors confidence in the quality and safety of recycled input material that they process. Good data from the recycler will also allow a processor to confidently calculate the recycled content of their products and declare this to brand owners and legislators.

The controls operate at several levels (see diagram on the lower right) but all aim to provide traceability, accountability and transparency in the recycling process and to increase and harmonise data flow.

There are differences between the process in Europe and the USA but there are many similarities and both areas are trying to improve the qualification process.

Qualifying input materials

The output material from the MRF is the input material to the plastics recycler and the MRF can add value through good sorting (see Section 4.7) and good bale preparation. Recyclers need to know the composition of the bale and what impurities are present and qualifying input materials is the start of good recycling and minimised waste.

The primary standard in Europe is EN 15347:2007 'Characterization of plastics waste'. This sets out the data that the waste supplier should make available to the recycler. The standard is very broad and the data is either 'required' or 'optional'. The 'required' data is minimal and this allows the supplier and purchaser to set their own standards for contaminants and other requirements but does decrease the usefulness of the standard. Some European countries, e.g., France, Germany and others, introduced their own classification or characterisation systems to make up for this deficiency.

Plastics Recyclers Europe (PRE – www.plasticsrecyclers.eu) has produced 'Recycling Input Characterisation Guidelines' to allow suppliers and purchasers to specify the main properties of input materials but, as for EN 15347, these consist of a set of blank input fields that allow the supplier and purchaser to

mutually agree the specification. This standard format can then be used to sort and check the quality of the input material.

The Institute of Scrap Recycling Industries (ISRI – www.isri.org) and the Association of Plastic Recyclers (APR – plasticsrecycling.org) represent the suppliers and the purchasers of plastics waste in the USA and have produced joint 'model bale specifications' for plastics.

These are available from either ISRI (as part of their 'Scrap Specifications Circular') or individually from APR. These specifications cover a wide variety of materials including film and PVC-U from pipes and windows. A real show of cooperation across the industry.

- **Tip** – Traceability adds value to recyclates and can allow processors and brand owners to avoid taxes based on the amount of recycled material in their products.

- **Tip** – There is nothing to stop a supplier and purchaser agreeing their own specifications but these are great for guidance.

- **Tip** – For computers and recycling, GIGO still applies: Garbage In = Garbage Out.

Qualifying recyclers

Qualification of recyclers covers a variety of topics designed to provide process approval,

There is an ISO standard (ISO 15270:2008 - 'Guidelines for the recovery and recycling of plastics waste') but it is very general.

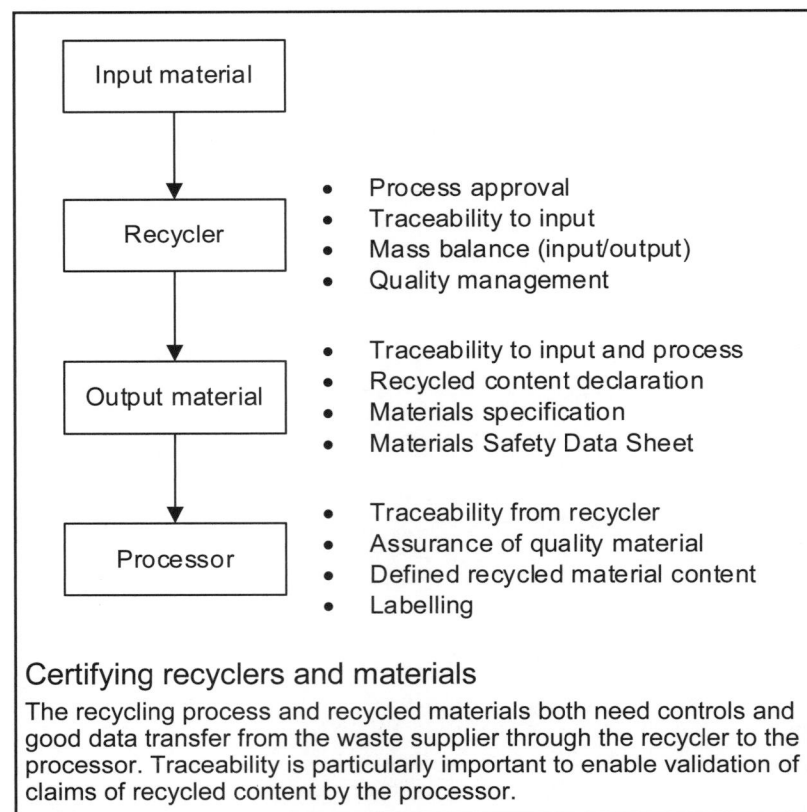

- Input material
  - Process approval
  - Traceability to input
  - Mass balance (input/output)
  - Quality management

- Recycler
- Output material
  - Traceability to input and process
  - Recycled content declaration
  - Materials specification
  - Materials Safety Data Sheet

- Processor
  - Traceability from recycler
  - Assurance of quality material
  - Defined recycled material content
  - Labelling

Certifying recyclers and materials

The recycling process and recycled materials both need controls and good data transfer from the waste supplier through the recycler to the processor. Traceability is particularly important to enable validation of claims of recycled content by the processor.

traceability, a mass balance for recycled content calculation and general management.

The primary standard in Europe is EN 15343:2007 'Plastics recycling traceability and assessment of conformity and recycled content'. This sets out the requirements for the control of input material, recycling processes and the characterisation of the final recyclate.

EuCertPlast (www.eucertplast.eu) is the result of a European project to establish a certification system for recyclers and uses EN 15343:2007 as the basis for the certification of ≈ 100 recyclers across Europe. This certification covers traceability in the supply chain to allow assessment of conformity to the standard and the amount of recycled content. The requirements of the EuCertPlast audit scheme and the associated quality management scheme are available from EuCertPlast.

APR operates a recycled material Certification Programme to ensure that recyclers are consistent in terms of the definitions used in ISO 14021:2016 (see Section 9.2). The APR requirements are not publicly available.

Both schemes use accredited auditors from third-party companies to audit the recyclers and either recommend or issue the certification.

- **Tip** – Both the EuCertPlast and APR schemes perform a mass balance calculation for the recycling process to validate the materials flows and allow processors to declare the recycled content.

In addition to the voluntary schemes, many countries have local rules although often these are more to comply with regulatory requirements for waste, e.g., the UK Environment Agency assesses and accredits recyclers and has quality protocols (based on the EN standards).

- **Tip** – Process approval is vital for recycling of food-grade materials. ACR provides a link to the US Food and Drug Administration 'No Objection Letters' where the FDA has issued a favourable opinion on a process for food contact materials.

Management systems

A common feature of any certification scheme is validating that the recycler has:

- The necessary permits for operation.
- Incoming material controls.
- Process controls.
- Product quality controls.
- An approved recycling process and associated mass balance calculation.

A recycler already certified to ISO 9001 would be expected to have most of these controls in place.

Qualifying output materials

The final step in qualification is to qualify the actual output recyclate to provide a processor with confidence in the material.

Materials specifications

In Europe, there are several established standards for various materials and these are

- EN 15342:2007 Characterization of PS recyclates.
- EN 15344:2007 Characterization of PE recyclates.
- EN 15345:2007 Characterization of PP recyclates.
- EN 15346:2007 Characterization of PVC recyclates.
- EN 15348:2007 Characterization of PET recyclates.

These provide the basis for testing and approving recyclates using standard test methods and can be used by any recycler. In most cases, however, the output materials specification will be determined by the process and the recycler and agreed with the processor.

- **Tip** – Placing recycled material on the market in the EU requires compliance with REACH and RoHS (see Section 3.6) and a Materials Safety Data Sheet (MSDS) should be available for all recycled materials.

Traceability

Traceability through the recycling process is vital in proving source and quality of the recycled material and validating claims of recycled content. In the EU, EN 15343:2007 provides the traceability framework and, in the USA, APR Certification provides the traceability framework.

Certification offers the opportunity for recyclers to increase cross-sectoral use of recycled material and increase volumes and margins.

Both APR in the USA and the BPF in the UK offer 'matching' services to put buyers and sellers of recyclates together.

Digital blockchain may be a technology to increase the traceability of materials in the future and to control the industry.

Brand owners need confidence in the declared amount of recycled content in products.

There is no current test for the amount of recycled material in a plastic. This makes mass-balance calculations (and traceability) vital to avoid 'greenwashing', i.e., using virgin material instead of recycled material.

4.10 Chemical recycling

A range of technologies

Chemical recycling is sometimes termed 'tertiary recycling' and there are a wide range of candidate processes. Not all of these will successfully make it to market and many are still at the pilot stage. As noted in Section 4.5, these processes break the plastic down to varying degrees, e.g., the main polymer backbone may be retained, the polymer backbone may be broken down to produce a monomer or the polymer backbone may be broken down completely to give a feedstock similar to that originally used to produce the monomer. Chemical recycling can be divided into three main types of process and these are shown in the diagram on the lower right.

Solvent dissolution

Solvent dissolution uses solvents to dissolve the input plastic back to the polymer chains without changing the polymer backbone in any way. The dissolution process is a physical change, not a chemical change, and the separated and precipitated polymer can be returned to the compounding process. This process removes all the additives, colours and impurities and the output is simply a clean, highly pure version of the original polymer that can be compounded as normal before manufacturing, i.e., there is no change in the base polymer itself. This effectively decontaminates the plastic and the process can produce food contact grades of polymer.

Solvent dissolution has been developed for PET, PE (all types), PP and PS (and EPS). The original solvent dissolution processes were only applicable if the input plastics were all the same, i.e., the materials needed to be separated before treatment because each polymer type needed a specific solvent. New developments make it possible to separate multi-layer materials back into the separate polymers and to treat mixed plastic waste.

The key to efficient operation of solvent dissolution is the recovery and cleaning of the solvent and even small solvent losses can render the process uneconomic.

The process has the lowest energy use, of any chemical recycling technology due to the low operating temperatures, although the energy needed to evaporate the solvent is high, and this is approaching the energy use of conventional mechanical recycling.

- **Tip** – Solvent dissolution is a technology to watch due to the low energy use, the fact that the output is already a polymer and purity of the output.

Chemical recycling

Chemical recycling is a depolymerisation process. It is not a single process but a range of processes which break the polymer backbone and turn the polymer back into monomers, oligomers or other intermediates. The output of the process must be polymerised again before it can be processed.

The driving force for chemical recycling is either chemical, or thermal or a combination of the two, although in some cases this can also be biological (see Section 4.15). Whichever driving force is used, these processes all need a high energy input for depolymerisation and the polymerisation energy is lost to the system. In the most efficient processes, the CO_2 benefits from chemical recycling are approaching those of mechanical recycling.

Chemical recycling offers the potential to turn waste plastic into new plastics that will compete with those made from petrochemical or biomass feedstocks.

The rate of technological change in chemical recycling sector is very high.

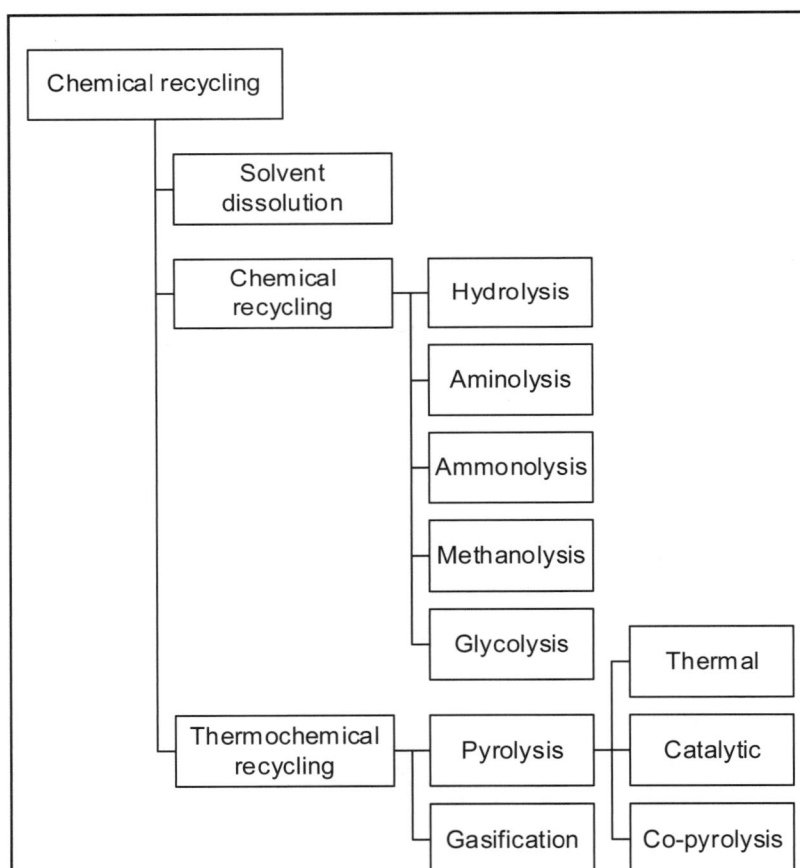

The options for chemical recycling

Chemical recycling is not a single process but a range of processes (over 60 at the time of writing) and there is considerable interest in the commercial development of these processes. Some of the processes are quite mature, e.g., pyrolysis, but others are very new.

Chapter 4 – Raw materials

There are a variety of potential processes for chemical recycling such as:

- Hydrolysis.
- Aminolysis.
- Ammonolysis.
- Methanolysis.
- Glycolysis.

All these processes use depolymerising agents to break the polymer chains and involve complex chemical engineering. The depolymerisation process works best for step-growth (condensation) polymers, e.g., PU, PA, PLA, PC, PHA and PEF, and thermosets but is less efficient for chain-growth (addition) polymers, e.g., PE and PP.

- **Tip** – PS is an addition polymer but depolymerisation works well for PS.

Chemical recycling technologies have been developed for single plastic inputs such as: PET (chemical, thermal and biological), PE (all types – chemical and thermal), PP (chemical and thermal) and PS (and EPS) and new technologies are opening the capability for mixed waste streams.

- **Tip** – Chemical recycling partially decontaminates the basic monomer but some additives can be trapped in the output.
- **Tip** – There are many players developing chemical recycling technologies. Not all will succeed. This is an industry that is still at the pilot stage.

Thermochemical recycling

Thermochemical recycling (also sometimes termed 'thermal cracking' or 'thermolysis') is the decomposition of a molecule under heat. The two main processes are pyrolysis (heating without oxygen) and gasification (heating with limited oxygen). In thermochemical recycling, not only is the main polymer backbone broken but the monomers are also broken and the output is a mixed petrochemical feedstock, e.g., wax, oil and gas, that can be used to replace conventional petrochemical or biomass feedstocks in the production of new polymers or as fuels. These are very energy intensive processes but both can deal with mixed plastic waste, i.e., no sorting is required.

Feeding the output products of thermochemical recycling back into the petrochemical production cycle is recycling but using the products as fuel is sometimes regarded as not being true recycling.

Pyrolysis

Pyrolysis is carried out by heating of the input material in the absence of oxygen and, as with any chemical process there are several types of pyrolysis, e.g., thermal pyrolysis, catalytic pyrolysis and co-pyrolysis. The temperature used for the process varies with the exact process and is in the range 400-1000°C. Catalysts can be used to both lower the temperature and to influence the output materials (depending on the exact process).

Gasification

Gasification of plastics waste differs from pyrolysis in that the process allows limited amounts of oxygen and the resulting partial oxidation produces a gas mixture of CO, CO_2, H_2 and CH_4 and a solid residue (tars) from any additives to the input plastics. The resulting gases ('Syngas') can either be used directly as a fuel or treated as a petrochemical feedstock.

- **Tip** – Thermochemical recycling is vastly different to incineration where the amount of available oxygen is effectively unlimited and full combustion is both sought and generally achieved.

Is this the answer to recycling?

There is increasing activity in the area of chemical recycling and it is being promoted by some as 'the answer to recycling' or even 'a new ecosystem for plastics'.

The reality is that, whichever process is chosen, the capital costs, the process economics and the process losses do not yet compete with mechanical recycling (see Section 4.6) for commodity plastics. However, for mixed plastics waste of 'other' plastics these processes potentially offer a more sustainable option than incineration with energy recovery (see Section 4.11) or landfill (see Section 4.12).

- **Tip** – There is still much work to be done to clarify the LCA for the various processes and to solve some of the issues regarding contamination of the input plastics and the environmental impact of any by-product outputs.
- **Tip** – Any polymer produced by chemical recycling will inevitably have a higher carbon footprint than the original material – this is the way that thermodynamics works.

If the outputs from thermochemical recycling are simply burnt as fuel then is that recycling and what about the circular economy?

Chemical recycling has the potential to be complementary to other waste treatment processes to raise recycling rates.

The current status is that the industry should seek first to improve mechanical recycling rates for the most widely recycled plastics, i.e., PET, PE-HD and PP, before seeking to use chemical recycling for mixed plastic waste.

4.11 Energy recovery – energy from waste (EfW)

It can't all be recycled

Mechanical or chemical recycling can capture large amounts of plastic waste and should always be the first choice. However, there will always be some plastics waste that is either too contaminated or in such low quantities that mechanical or chemical recycling are not feasible. At this stage the waste, generally commingled waste from municipal solid waste (MSW) collection, can be either landfilled (see Section 4.12), simply incinerated or incinerated with energy recovery. This is termed Energy from Waste (EfW) or 'quaternary recycling'.

Interest in EfW is partially driven by external forces. In the past, many countries exported waste plastics to South East Asia. These were sorted for recyclable materials and the rest sent to landfill or poorly disposed of. Restrictions on importing waste in these countries has driven interest in alternative methods of treating MSW.

It has already had a good life

Most petrochemical use is 'single use', i.e., $\approx$ 87% of the production of oil and gas is used only once for the embodied energy. The $\approx$ 4-6% of petrochemicals used to produce plastics has a useful life as product and EfW allows the material to then be used as an energy source. EfW reduces the use of other petrochemicals for energy production and allows the collection of metals that are not otherwise recovered.

For these plastics, EfW can be a resource-efficient solution when compared to landfilling or to enforced recycling.

• **Tip** – Incineration without energy recovery is not a preferred option due to the loss of the embodied energy.

Incineration is widely used in Europe for MSW where incineration rates are in the region of $\approx$ 40% and rising, although there are regional variations, e.g., Switzerland uses EfW for $\approx$ 70% of MSW, and Greece uses virtually none. In Europe, landfill is rapidly being replaced with EfW and the recovered energy is used for power generation and district heating.

Incineration for EfW also varies widely around the world and is largely driven by the availability of suitable landfill sites, e.g., the USA does not have a significant EfW capacity.

Calorific values

EfW can be carried out for any MSW that has a high calorific value. MSW not only contains plastics ($\approx$ 10%) but also contains biomass, paper, food waste and other non-recyclables. A general mix of plastics will have a higher calorific value than coal, wood or paper and the average MSW will be just below that of coal, wood or paper. This means that an EfW incinerator needs no additional fuel.

• **Tip** – Some EfW plants have had to reduce the burn rate to account for the increased calorific value of MSW from plastics.

The process

The EfW process (shown in the diagram below) burns waste at $\approx$ 850°C to produce heat and steam. The steam is used to drive a turbine for electricity and can also be used for district heating. This is similar to CHP (see Section 6.24) except that fuel is MSW.

Modern EfW plants are very different to older style incinerators which were primarily designed to reduce the mass of the waste. They are designed to recover not only the embodied energy but also to recover metals and other non-combustibles. These plants are a major long-term investment and are

Critics of EfW have termed this 'skyfill' on the basis that the material is diverted from landfill but the combustion products are released to the atmosphere. This ignores the fact that plastics use $\approx$ 4-6% of the world's petrochemicals and that most petrochemicals ($\approx$ 87%) are directly burnt for energy, heating or transport and only have one life.

Plastics products have had at least two lives.

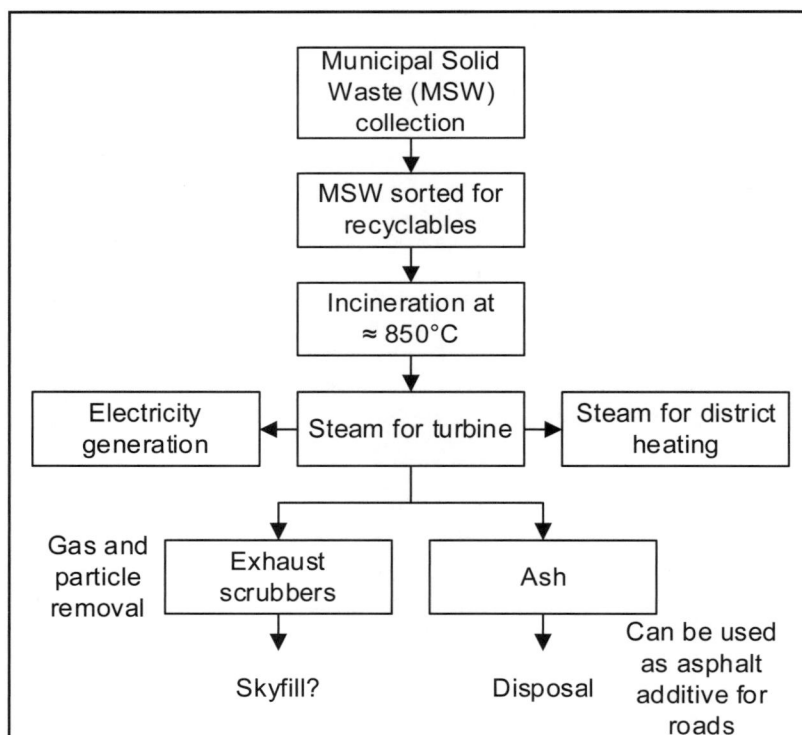

The EfW process

EfW is not without outputs. It does not completely reduce the input material, there are emissions to atmosphere (CO_2, NO_x and SO_x) and solid residuals such as char and ash. EfW recovers the embodied energy to replace other resources that may have been used.

not plastics processing sites, they are the realm of large waste management companies. A modern EfW site will need > 200,000 tonnes of MSW/year to be profitable and will have a 20-30 year life with a long payback time.

A significant issue for EfW is the efficiency of the process relative to conventional gas-fired electricity generation (see Section 4.12). The current efficiency rates are in the region of 15-30%. This is low but is increasing rapidly.

- **Tip** – For plastics companies there is an opportunity to sell mixed waste (after removing all the recyclables) to EfW sites. This may require waste sorting but sites should already be doing this (see Chapter 9).
- **Tip** – Critics of EfW contend that all materials should be recycled and re-used. Whilst there is bound to be some sympathy for this argument, the reality is that not all materials can be recycled and/or re-used.

Emissions and other outputs

The largest emission is CO_2 (between 250-600 kg CO_2/tonne of waste) but there are also SO_x and NO_x emissions. The combustion of any organic matter, e.g., wood, coal, oil and plastics, will produce dioxins such as CDD and CDF. Dioxins are a group of chemicals that are persistent and harmful to health.

Emissions and other outputs of EfW are very highly regulated and modern incineration technology with scrubbers and flue gas treatment has dramatically reduced the emissions of heavy metals and dioxins. Data from Germany, the USA. and the UK reveals that the use of filters has reduced the emission of heavy metals by 90% and dioxins by more than 99%[1].

EPA data shows that EfW emissions decreased 95.5% from 14.0 kg TEQ in 1987 to 0.6 kg in 2012 and that EfW plants in the USA were only responsible for 0.54% of the controlled dioxin emissions, and 0.09% of all dioxin emissions from controlled and open burning sources[2].

In the UK, Public Health England states "PHE's risk assessment remains that modern, well-run and regulated municipal waste incinerators are not a significant risk to public health. While it is not possible to rule out adverse health effects from these incinerators completely, any potential effect for people living close by is likely to be very small."[3]

EfW can reduce waste volume by > 90% but the residual char and ash must be disposed of carefully due to the increased concentration of heavy metals and other contaminants. This material has been successfully used in roads as an asphalt additive for roads and separation techniques are being developed to recover precious and rare earth metals.

- **Tip** – Dioxins are also produced in forest fires and backyard incinerators which are not regulated and are major sources.

Prepare for energy recovery

Designers and specifiers of plastics products can help prepare for energy recovery by selecting the right plastics:

- Most thermoplastics are suitable for EfW unless they contain chlorine, fluorine or other halogens, e.g., PVC and PTFE, or if they contain brominated flame retardants. These are more difficult to incinerate and may need additional flue gas treatment (scrubbers) to be safely incinerated. Suitable plastics for EfW are PP, PE (all types), PET, ABS, PS (all types).
- Thermoplastics that are highly filled with combustible fillers, e.g., wood-plastic composites, are suitable for EfW.
- Thermoplastics that are highly filled with non-combustible fillers, e.g., glass fibre, talc, $CaCO_3$ and TiO_2, can be used for EfW but will increase the residual ash content.
- Highly cross-linked polymers, e.g., epoxies, thermosets, BMC and SMC are either less suitable for EfW or not suitable at all.

Cement kilns?

It is also possible to use MSW to produce solid recovered fuel (SRF) to fire cement kilns. These operate at $\approx 1,400°C$ and the high calorific value of plastics makes them an ideal fuel substitute for coal or other fuels but they must be fitted with flue gas scrubbers.

Between 1987 and 2012, dioxin emissions from EfW incinerators in the USA were reduced by 96% and today represent only 0.09% of all dioxin emissions.

For contaminated or low volume plastics which are not suitable for mechanical recycling, energy recovery can be a resource-efficient solution compared to landfill or enforced recycling.

- 1. de Titto & Savino. 2019. 'Environmental and health risks related to waste incineration'. Waste Management & Research. 37. October.
- 2. Dwyer and Themelis. 2015, 'Inventory of U.S. 2012 dioxin emissions to atmosphere'. Waste Management, 46, December.
- 3. Public Health England, 2019 'PHE statement on modern municipal waste incinerators study'. www.gov.uk.

The electricity generated by the incineration of a single plastic bag is enough to power a 60 W light bulb for 10 minutes.

4.12　Landfill – storage for the future?

This is where I came in

In 1972, the local dairy in my home town of Geelong introduced PE-HD milk bottles. This many not sound radical today but at the time there was no recycling infrastructure and the used bottles were simply sent to landfill. I was concerned about this waste of resources but the dairy owner was adamant that this was the 'wave of the future' – even though we were already running out of holes for landfill. As a result, I began working with environmental groups in Melbourne looking at the difficulties of MSW disposal in Melbourne.

The past 49 years have shown that we were both right, PE-HD milk bottles were the wave of the future and landfill is an increasingly important issue.

The trend in landfills

Landfill is still used around the world with a greater or lesser degree of control, i.e., countries with poor waste management systems tend to use large amounts of landfill and have poor landfill processes. In Europe, landfill is used in roughly inverse proportion to incineration (see Section 4.11) and landfill rates are an average of $\approx 25\%$ and decreasing although there are regional variations, e.g., Switzerland uses no landfill for MSW and Greece uses landfill for $> \approx 75\%$ of MSW.

Landfill is also widely used in the USA due to the availability of land and the USA currently has > 3,000 landfill sites.

The global trend is that the amount of MSW being sent to landfill is decreasing as recycling and EfW incineration are increasingly being used for MSW disposal. The driver for this is largely policy decisions, e.g., landfill taxes, at government levels but the case is not a clear cut as it may appear.

The modern landfill

The modern landfill bears little relation to the simple holes in the ground, e.g., abandoned quarries etc., that were previously used. It is a complex engineered structure designed to store MSW safely, to recover value from the any gases generated, e.g., CH_4, to treat any leachate from the MSW and has sophisticated controls to monitor the local environment. A cross section of a sealed modern landfill is shown on the right and this shows the methods used to control the structure.

A modern landfill is designed to keep water and oxygen out of the structure and to prevent leachate escaping from the structure. Any leachate formed in the structure is captured by perforated drainage pipes and pumped out to be collected and treated before release.

- **Tip** – Not all landfills are 'modern' and there are still many 'old style' landfills in use where leachate control is poor (or non-existent) and there is no energy recovery from the CH_4 generated, i.e., it is simply flared off.

Landfill as an energy source

The collection and combustion of the CH_4 generated in a landfill is a vital part of the modern landfill, this can be used to generate energy but it is equally important not to let the CH_4 escape to atmosphere. CH_4 has a Global Warming Potential (GWP) of 28 compared to a GWP for CO_2 of 1. Capturing

A modern landfill is a major capital project costing > £2 million.

It is not simply a 'hole in the ground', whatever it may look like

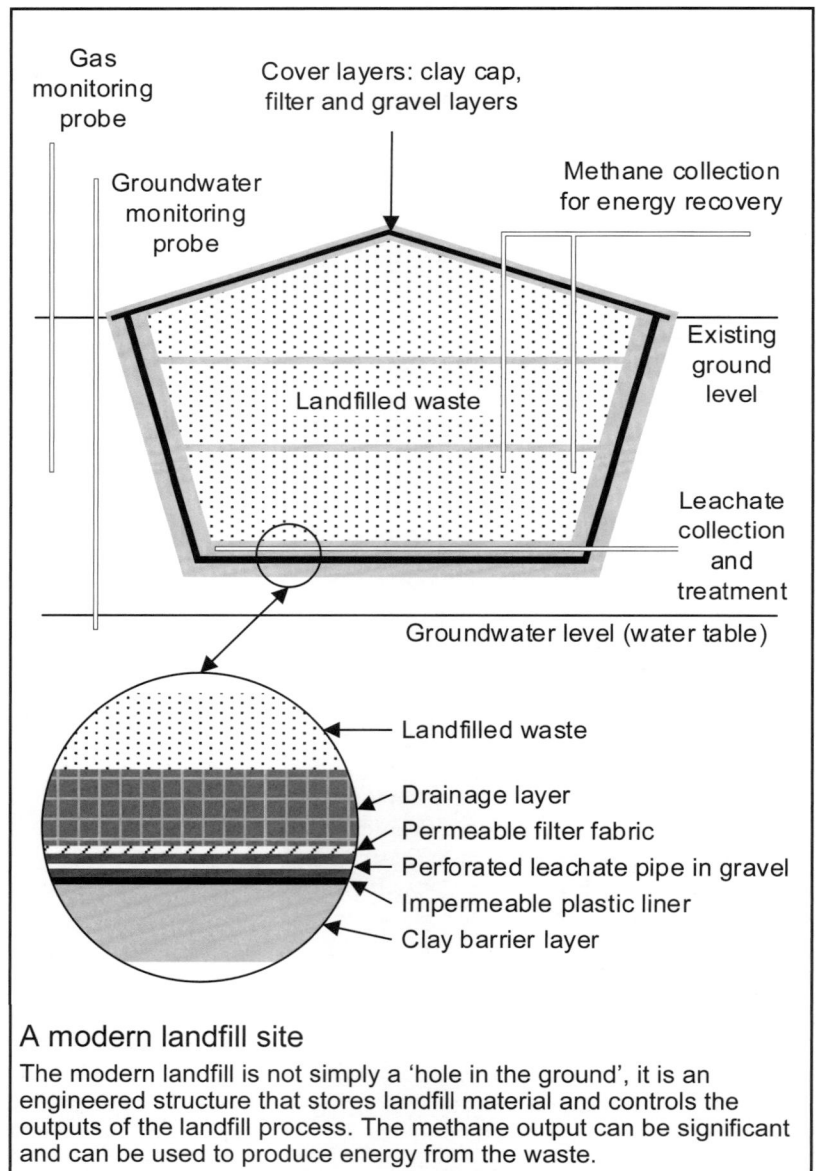

A modern landfill site

The modern landfill is not simply a 'hole in the ground', it is an engineered structure that stores landfill material and controls the outputs of the landfill process. The methane output can be significant and can be used to produce energy from the waste.

CH_4 and recovering the energy is necessary on two levels but a landfill site is generally a net contributor to climate change rather than a benefit.

- **Tip** – Unfortunately, energy recovery from CH_4 is not achieved at most landfill sites and much more needs to be done to encourage this.

Issues with landfill

Even modern landfills are not without potential issues such as:

- Leachate leaks to groundwater due to failure of the clay or PE-HD plastic liner. These can be from diffusion through the clay liner and aging of the liner. This is monitored using a groundwater probes but rectification in the event of failure is difficult.
- Clogging or blocking of the leachate collection system with silt or bacteria.
- Water ingress into the structure due to failure of the cover layers as a result of animals and plant or tree growth.

Plastics in landfill

Depending on recycling and EfW rates, the plastics content in landfilled MSW can be between 5 and 25%. In most countries with good waste management these plastics will be those that are not currently recycled or are highly contaminated but these can still represent a significant proportion of the total plastics waste stream, e.g., $\approx$ 30% of the total plastic waste generated.

Degradation in a landfill

Degradation in a modern landfill is anaerobic (without oxygen), there is little water present (due to the clay cover layer and the design of the structure) and no sunlight for photodegradation. This means that degradation of most of the waste will be very slow and even waste foodstuff can take years to degrade, e.g., landfill archaeologists or 'garbologists' (some people have the best jobs) have found recognisable food over 25-years old and readable newspapers over 50-years old.

Plastics in an anaerobic landfill, even if they are biodegradable or photodegradable, will take many years to degrade. The exact times depend on many factors but a PET bottle will take at least 500 years and probably much longer. There may be some biological or chemical degradation due to H_2S in the leachate but this will still take many years.

A benefit to landfill?

Plastics waste in a landfill is largely inert, does not contribute to odour generation and can act as a stabiliser for the structure. Additives to the base polymer are also

tightly bound in the plastic and there is a low probability of any substantial leaching of additives from the plastic.

- **Tip** – This is a good thing now that the EU has prevented the recycling of PVC-U containing lead stabilisers and all of this will now end up in landfill.

Plastics also act as a 'carbon sink' in landfills, i.e., they have had a useful life and the carbon used in their production is now fully locked up and stored, as opposed to being burnt and releasing the sequestered carbon as CO_2.

Storage for the future

Keith Freegard (www.freegard.net)[1] has suggested that landfill is the best place for plastics that cannot be mechanically recycled. The argument is:

- The efficiency of an EfW plant is $\approx$ 15-30% (see Section 4.11).
- The efficiency of a Combined Cycle Gas Turbine (CCGT) is $\approx$ 50-60%.
- A CCGT can therefore produce the same amount of electricity as an EfW plant but with only 40% of the CO_2 emissions.
- Even if the EfW plant has integrated heat recovery (which can be produced using a standard boiler at > 90% efficiency) then a CCGT can therefore produce the same amount of electricity as an EfW plant but with only 65% of the CO_2 emissions.
- It is therefore more logical to use the most efficient means (CCGT or renewables) to generate electricity and to use landfill as a long-term 'carbon sink' or that can be mined in the future when new processes make this more attractive and resource efficient.

There is a compelling logic to this argument that is not often recognised. Is the modern landfill really 'storage for the future'?

Mechanical recycling is still the preferred option for plastics products.

Making products out of the right materials and recycling them is always better.

There is already considerable interest in landfill mining and even 'enhanced landfill mining' to recover materials, e.g., Waste to Materials (WtM), and energy (EfW).

Things are getting interesting and waste may be, in future, regarded as 'something we don't need just now' rather than 'something we don't need ever again'.

Freegard also notes (private communication) that as separate food waste collection grows then 'normal' MSW contains much less moisture and organic waste.

This makes the landfill almost inert with little material breakdown or leaching and an even better storage unit for the future.

It is not simply modern landfills that we need to be concerned with. There are hundreds of thousands of old landfill sites around the world (perhaps greater than 1,000,000) and these are already repositories of billions of tonnes of waste that could be mined in the future.

- 1. Freegard, 2018. 'What's best for low grade plastics?', axiongroup.co.uk/news/whats-best-low-grade-plastics.

4.13   Materials recovery – where are you now?

Use the molecules wisely

How plastics processors use and treat raw materials is an important measure of their sustainability and it is growing in importance with the concept of the circular economy. As an industry, we must make sure that every piece of plastic we buy makes it into a product and that the products sent out of our factories are captured in a waste management system at the end of their life to be recycled back into new products.

Not all of this process is within the industry's control (see Chapter 10) but processors can help by maximising material utilisation and providing a market for the materials that are recycled at the end of life. This is not simply an aspiration for sustainability but is also a cost control measure.

Legislation is increasingly making Enhanced Producer Responsibility part of the landscape and this brings cost advantages to using recycled materials or, more correctly, cost disadvantages to using virgin materials.

Processors need to start making smart materials choices to minimise impacts and costs in the future.

Completing the chart

This chart is completed and assessed as for the previous charts.

Recovering and using the plastics we already have in the system is rapidly moving from 'good' to 'essential'.

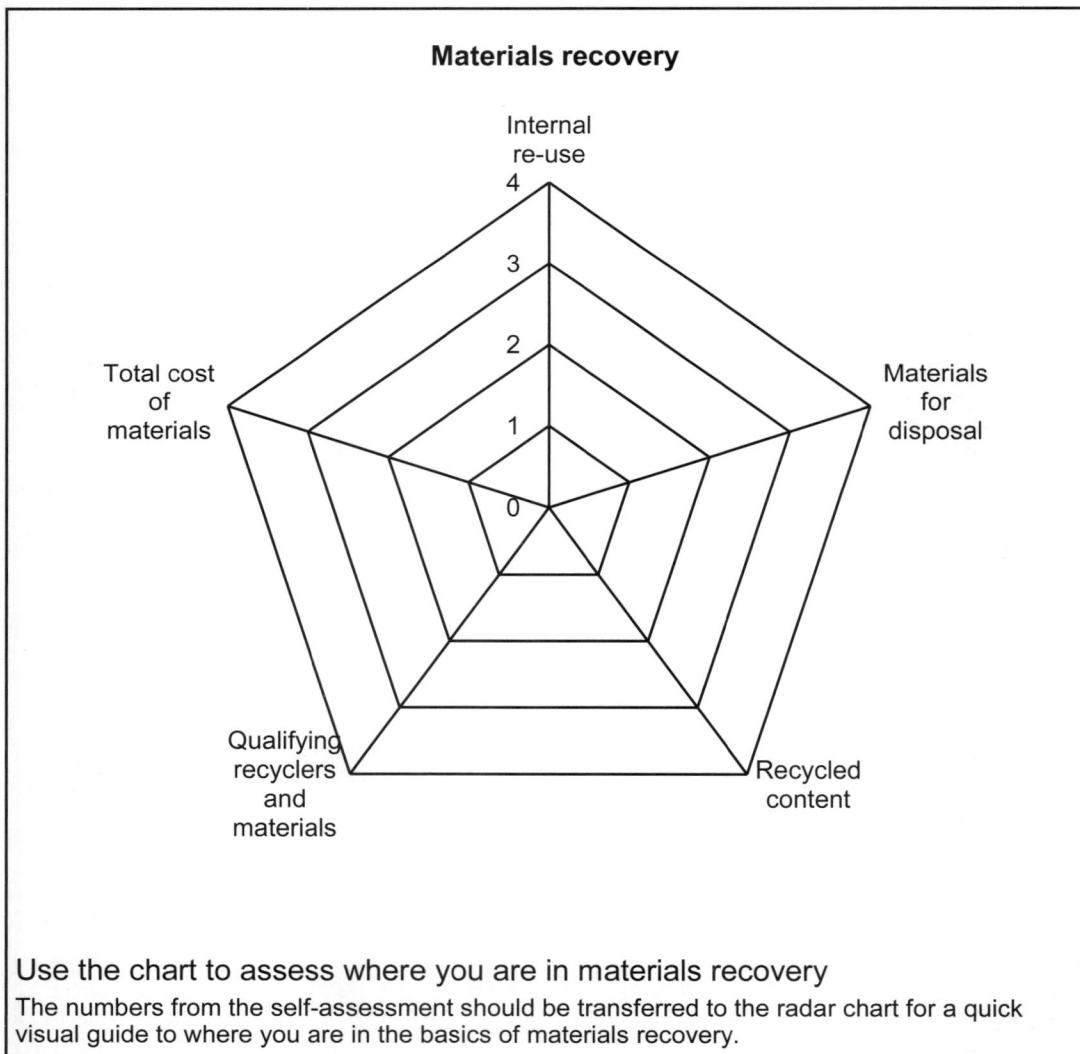

Materials recovery

Use the chart to assess where you are in materials recovery

The numbers from the self-assessment should be transferred to the radar chart for a quick visual guide to where you are in the basics of materials recovery.

I am old enough to remember when you paid people to take your waste plastic away.

How times have changed.

Level	Internal re-use	Materials for disposal	Recycled content	Qualifying materials	Total cost of materials
			Materials recovery		
4	All possible materials re-used internally.	No landfill waste stream at all. All materials, including plastics, recycled by registered & approved recyclers.	Use of external recycled content is maximised for all products (> 50% of materials used are externally recycled materials).	Certified recyclers used for all materials. Full traceability available for all materials. Data sheets & MSDS available for all materials. Recycled content declared.	Excellent understanding of compliance costs & these are managed well. Total cost of materials (purchase price + compliance costs) used for cost of materials calculation.
3	Internal re-use of materials is high (> 75% of available materials are internally re-used).	Good sorting of plastics from materials being sent for disposal. No plastic materials in landfill waste stream. No plastics materials sent for disposal.	Good use of externally recycled content (> 30% of materials used are externally recycled materials).	Certified recyclers used for most materials. Traceability available for most materials. Data sheets & MSDS available for most materials. Recycled content declared.	Good understanding of compliance costs. Compliance treated as a manageable cost. Cost of materials includes compliance costs for most products.
2	Internal re-use of materials is good (> 50% of available materials are internally re-used).	Average sorting of plastics from materials being sent for disposal. Small amounts of plastic in landfill waste stream.	Moderate use of externally recycled content (> 15% of materials used are externally recycled materials).	Certified recyclers used for some materials. Traceability available for some materials. Data sheets & MSDS available for some materials. Recycled content not declared.	Moderate understanding of compliance costs. Compliance treated as a fixed cost. Cost of materials includes compliance costs for few products.
1	Internal re-use of materials is low (< 25% of available materials are internally re-used).	Poor sorting of plastics from materials being sent for disposal. Significant amounts of plastic in landfill waste stream.	Minimal use of externally recycled content (< 5% of materials used are externally recycled materials).	Recyclers locally approved but no certification. Traceability not available. Data sheets & MSDS available for some materials. Recycled content not declared.	Poor understanding of compliance costs. Compliance treated as a fixed cost. Cost of materials for products is based only on purchase price.
0	No internal re-use of materials.	No sorting of materials being sent for disposal & large amounts of plastic in waste in landfill waste stream.	No externally recycled materials used in products.	Recyclers locally approved but no certification. Traceability poor. Poor materials definition. Recycled content not relevant to product (cost reduction only).	No consideration of compliance costs. Cost of materials for products is based only on the purchase price.
Score					

4.14 Bio-based plastics – raw materials

Can we have some clarity here?

The terms bio-based plastics and biodegradable plastics are often confused, for clarity and avoiding long definitions:

- 'Bio-based plastic' describes the source of the material, i.e., where the feedstock comes from. It is about raw materials. Note that we have used the term 'bio-based plastic' and not the more confusing term 'bioplastic'.

- 'Biodegradable plastic' describes the sink of the material, i.e., where the material goes to. It is about 'end-of-life'.

The terms 'bioplastic', 'biodegradable plastic' and 'bio-based plastic' have been misused so often and sometimes so wilfully that it appears that 'confusion marketing' is being used.

It is where it comes from, not where it goes to

In the initial development of the plastics industry, almost all plastics were bio-based, e.g., horn, Parkesine, Celluloid, Galalith and Bakelite. This all changed in the 1930's when the petrochemical industry developed the technology to use oil and gas-based source materials and the cost advantages drove the industry to use these. Bio-based materials have undergone a resurgence on a sustainability basis and, whilst they are not yet produced in substantial volumes (they are < 1% of the total plastics market), the market is growing due to legislative and consumer pressure and they represent a potential route to improving the sustainability of the industry.

There are various definitions of bio-based plastics but essentially, bio-based plastics are made partially or fully from biological materials rather than from conventional petrochemicals.

We use the phrase 'partially or fully' because, as ever, there is a spectrum of materials, some can be fully bio-based, e.g., PLA, some are fully fossil-based, e.g., conventional PVC, and some are partly bio-based, e.g., PVC manufactured using bio-based ethylene or PUR manufactured using bio-based polyols. Bio-based plastics are not a binary choice. Care is needed when evaluating claims such as 'bio-based' and 'environmentally friendly'.

Sources

There are many sources of biomass and these are all renewable, i.e., they can be renewed within 1-2 years and do not involve depletion of non-living (abiotic) resources. The sources include almost all the types of biomass that can be imagined. Typical biomass sources are:

- Agricultural sources, e.g., crops grown for biomass such as:
  - Corn, wheat, potato starch – in fact any available starch containing crop.
  - Sugar cane – where the sugar is used to produce ethanol and then PE or other plastics.
- Vegetable oils and other food waste.
- Forestry and farming waste, e.g., wood, straw and other waste biomass.

Even though the raw material source is biomass, bio-based plastics are still produced in large-scale chemical plants. This is not a retreat to some rustic agrarian economy.

Material types

Bio-based describes the source of the raw material but it does not describe the properties of the polymer produced. It is possible to use biomass to produce conventional polymers such as PE, PP, PET, PA and some polyesters. These are bio-based materials but will have the same properties as their fossil-based counterparts and are termed 'drop-in' bio-based plastics.

It is also possible to use biomass to produce a range of materials that have improved performance in terms of biodegradability and the major current types are:

- Starch and starch derivatives.
- Polylactic acid (PLA).

Bio-based plastics may, or may not be, biodegradable. Biodegradable materials are discussed in Section 4.15.

The 'generations' of bio-based sources are:

- First generation – agricultural crops, e.g., sugar cane.
- Second generation – cellulosic crops and residues/waste products, e.g., wood, straw and waste biomass.
- Third generation – non-traditional organisms, e.g., algae.

The sustainability of a bio-based source will vary with the source.

Ethylene is ethylene, it doesn't matter, or know, where it comes from.

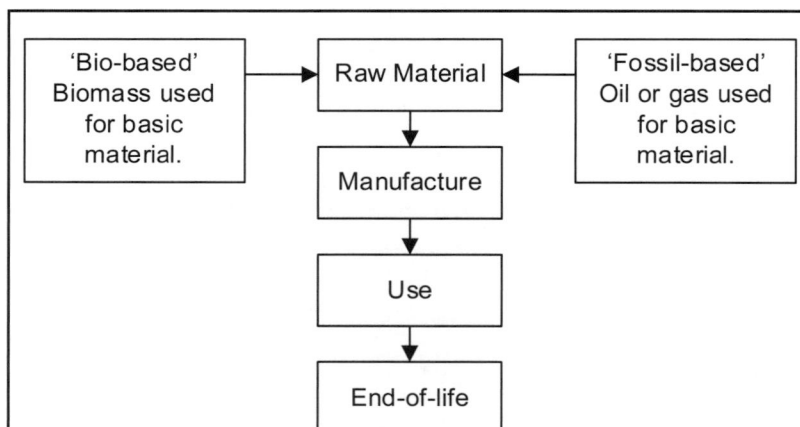

```
┌─────────────┐      ┌─────────────┐      ┌─────────────┐
│ 'Bio-based' │─────▶│Raw Material │◀─────│'Fossil-based'│
│ Biomass used│      └─────────────┘      │ Oil or gas used│
│  for basic  │             │             │   for basic   │
│  material.  │             ▼             │   material.   │
└─────────────┘      ┌─────────────┐      └─────────────┘
                     │ Manufacture │
                     └─────────────┘
                            │
                            ▼
                     ┌─────────────┐
                     │     Use     │
                     └─────────────┘
                            │
                            ▼
                     ┌─────────────┐
                     │ End-of-life │
                     └─────────────┘
```

Where it comes from

Bio-based describes the source of some or all of the raw material. Raw materials can come from fossil-based materials such as oil and gas or from biomass raw materials such as plant matter. This has nothing to do with the end-of-life phase of the material.

- Polyhydroxyalkanoates, e.g., PHA, PHB and PHBV – these are made using microorganisms and fermenting the biomass.
- Cellulose and some derivatives.
- Protein (soy) based materials.
- Polyethylene furanoate (PEF) – a bio-based replacement for PET (with significantly better properties).
- Aliphatic- and aliphatic-aromatic copolyesters, e.g., PBAT.
- **Tip** – Except for the 'drop-in' types (which have exactly the same properties), bio-based plastics tend to have properties in the same range as commodity plastics. There are few engineering and performance bio-based plastics, although they are being developed, and PA 11 has always been a bio-based plastic (manufactured from castor oil).
- **Tip** – Simply replacing a conventional plastic in a part with a bio-plastic (except for a 'drop-in') and expecting it to perform the same is not a smart thing to do. The data sheet doesn't tell you everything.

Are they better?

Bio-based plastics are sometimes presented as being more sustainable and the future of plastics processing but, as with all aspects of sustainability, the picture is more nuanced than that:

- Natural resources – bio-based plastics do not deplete non-renewable sources (although many of the precursors for plastics production are the waste products of oil refining that were once simply flared off).
- Carbon footprint – bio-based plastics have a reduced carbon footprint and the materials sequester CO_2 from the atmosphere and reduce the carbon footprint of the material. This means that even if the materials are incinerated then there is no net gain in CO_2 to the atmosphere, i.e., bio-based materials are carbon neutral.
- Other environmental effects – where the source material is specifically grown for biomass then there are inevitable additional side effects to their production such as:
    - Increased water use – this is particularly important for areas of the world which may be water stressed (see Section 8.14).
    - Increased fertiliser use.
    - Increased transport emissions.

The LCA case for bio-based plastics is not as positive as it might appear at first sight.

- Reduction in land for food production –

where the source material is specifically grown for biomass then there is a reduction in the amount of land available for food, particularly in agrarian economies where the cash crop may replace food crops.

- Biodegradability – some, but not all, bio-based plastics are biodegradable (see Section 4.15).
- Cost – the cost of bio-based polymers is always higher than the cost of fossil-based polymers but this may be due the youth of the industry and the optimised production of fossil-based plastics. In the long-term, the cost of bio-based polymers may be more stable than fossil-based polymers where costs can change due to oil price effects.
- Processing machinery – bio-based plastics can be processed on current machines, e.g., injection moulding, extrusion (film and profile), EBM and thermoforming, with modified processing requirements (except for the 'drop-in' materials).

Bio-based plastics have a definite place in the plastics processing industry provided they are used wisely and sensitively.

Testing

It is possible to determine the biomass content by radiocarbon analysis to determine how much Carbon-14 is present in the material, 'old' carbon from petrochemical sources will have no Carbon-14 present whilst 'new' carbon from biomass sources will have Carbon-14 present. Testing[1,2] for the presence and amount of Carbon-14 allows the amount of bio-based material to be calculated to validate a claim of using bio-based materials.

- **Tip** – The bio-based content (%) is the ratio of the mass of bio-based carbon in the product/the total mass of carbon in the product.

There are a range of programmes around the world encouraging the use of bio-based materials.

One of the largest is the USDA 'BioPreferred®' programme which has mandatory purchasing requirements for federal agencies and their contractors and a voluntary labelling initiative for bio-based products. This covers not just plastics but a whole range of products and is a sign of the direction of movement. See www.biopreferred.gov for a list of the products.

'Drop-in' bio-based plastics are totally compatible with plastics recycling systems.

The other bio-based plastics are usually fully compatible with plastics recycling systems but need NIR sorting to avoid contaminating other plastics streams, e.g., PLA in PET.

The bio-based plastics industry is developing and investment is growing to meet demand.

However, they currently have less than 1% market share and the growth rate is not much greater than the growth rate of conventional plastics.

That is a lot of noise from a small part of the market.

- 1. ASTM D6866: 'Standard test methods for determining the bio-based content of solid, liquid, and gaseous samples using radiocarbon analysis'.
- 2. EN 15440: 2011: 'Testing for solid recovered fuels'.

4.15   Biodegradable plastics – end-of-life

It is where it goes to, not where it comes from

'Biodegradable plastic' tells you one of the exit routes at the end of life. It is about where the material goes to and is not related to where it comes from. The issue is that the term 'biodegradable' is also so often misused that it appears to be 'confusion marketing' and 'greenwashing'.

Plastics can exit the system in many ways (see Section 4.5) and the degradation exits are shown in the diagram on the right. These are.

- Degradable – all plastics degrade, i.e., they suffer from chemical cleavage of the polymer chain and eventually break down into smaller particles, the driver for this can be from UV light or other mechanisms. This degradability is not the result of any microorganisms, it is a chemical process and for most plastics it can take a long time unless the plastic is stabilised with anti-oxidants etc.

- Biodegradable – this is a degradation process that involves biological activity, especially enzymes. In this process, the polymer chain is broken and metabolised by microorganisms.

- Oxo-degradable – this is a degradation process that uses chemical additives to induce chain scission and later degradation.

Biodegradability of a plastic is not determined by the source of the feedstock, i.e., fossil based or bio-based, but by the structure of the polymer chains.

- **Tip** – Not all biodegradable plastics are bio-based, some fossil-based plastics are biodegradable, e.g., PBAT.

- **Tip** – Not all bio-based plastics are biodegradable, some bio-based plastics are not bio-degradable, e.g., PA 11 and all of the 'drop-in' bio-based plastics.

- **Tip** – Biodegradability is not the solution to ocean waste or littering. There are very few plastics that biodegrade in the ocean or in simple soil conditions.

Biodegradable plastics

Biodegradation is the mineralization of organic structures by micro-organisms (bacteria or enzymes). Depending on the process conditions, i.e., with or without oxygen present, biodegradation can be divided into composting (aerobic – with oxygen) or biomethanation (anaerobic – without oxygen).

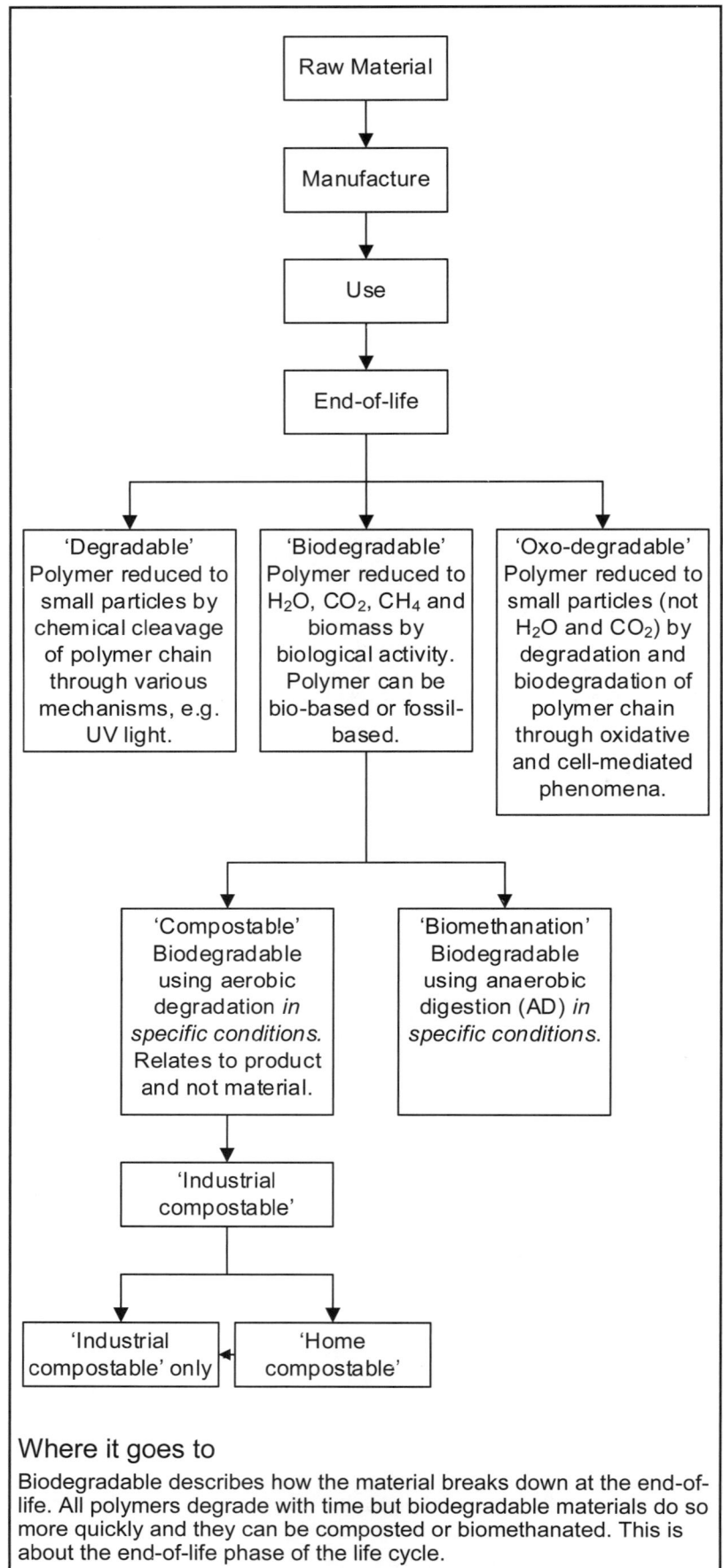

```
Raw Material
      |
      v
  Manufacture
      |
      v
     Use
      |
      v
  End-of-life
```

'Degradable' Polymer reduced to small particles by chemical cleavage of polymer chain through various mechanisms, e.g. UV light.

'Biodegradable' Polymer reduced to H_2O, CO_2, CH_4 and biomass by biological activity. Polymer can be bio-based or fossil-based.

'Oxo-degradable' Polymer reduced to small particles (not H_2O and CO_2) by degradation and biodegradation of polymer chain through oxidative and cell-mediated phenomena.

'Compostable' Biodegradable using aerobic degradation *in specific conditions*. Relates to product and not material.

'Biomethanation' Biodegradable using anaerobic digestion (AD) *in specific conditions*.

'Industrial compostable'

'Industrial compostable' only

'Home compostable'

Where it goes to

Biodegradable describes how the material breaks down at the end-of-life. All polymers degrade with time but biodegradable materials do so more quickly and they can be composted or biomethanated. This is about the end-of-life phase of the life cycle.

The general mineralisation process is:
Polymers $\rightarrow$ Oligomers $\rightarrow$ Monomers $\rightarrow$ Biochemicals $\rightarrow$ Minerals (CO_2, CH_4, H_2O) and biomass.

- For composting, the outputs are CO_2, H_2O and biomass.
- For biomethanation, the outputs are CH_4, CO_2, H_2O and biomass.

Compostable plastics

In most cases, the term 'compostable plastics' refers to industrial composting, this is an aerobic process using microorganisms to break organic material down to CO_2, H_2O and biomass. It is carried out under controlled conditions and should leave no waste or toxic residues.

To be termed 'compostable', a material should meet the requirements of ISO 17088, EN 13432 (for packaging), EN 14995 (for general plastics) or ASTM D6400.

- **Tip** – These basic standards are supported by a variety of test methods and labelling requirements.

The standards are very similar in their requirements (although the times vary slightly):

- During composting, the materials should have an oxygen level of > 10%, be kept at a temperature of 60-75°C and be turned biweekly.
- To show disintegration: after ≈ 12 weeks, no more than 10% of the original dry mass shall remain after sieving through a 2 mm sieve.
- To show biodegradation: after ≈ 26 weeks, more than 90% of the original organic carbon shall be converted to CO_2, i.e., 90% mineralization.
- The resulting compost shall be able to support plant growth, e.g., summer barley and cress.
- **Tip** – Not all biodegradable materials are compostable according to the standards. The standards have time limits for degradation and some materials will biodegrade but not within the time limits.
- **Tip** – The requirement for 90% mineralization within six months is based on the biodegradation of cellulose which achieves 90% degradation in six months.

Home compostable

Home composting has lower temperatures but higher time scales than industrial composting. This means that an industrially compostable material may not be suitable for home composting (although a home compostable material should be suitable for industrial composting). There are several standards for home compostable materials,

e.g., France, Italy and Austria, but there is no current internationally recognised standard.

Recycling of biodegradable plastics

Biodegradable plastics can be recycled using any of the methods discussed earlier in this chapter. If they are to be mechanically recycled then they should be separated by NIR sorting and recycled separately. They should not be processed with other plastics to avoid contaminating the recyclate stream.

Landfills are anaerobic and biodegradable and compostable materials should not be sent to landfill unless there is a methane collection system present.

Anaerobic digestion (Biomethanation)

Biomethanation is an anaerobic (without oxygen) process that uses microorganisms to break organic material down to CH_4, CO_2, H_2O and biomass. It is carried out by anaerobic digestion (AD) and the CH_4 can be captured and used as a biofuel. Some compostable materials can be treated by AD but the operators of AD plants do not like to have plastics in the digester because of sorting issues and tend to remove all plastics from the input material (including compostable food caddy liners and compostable packaging).

ISO has published a range of test methods for AD of plastics, e.g., ISO 14853 and ISO 15985, which vary depending on the conditions in the AD. The equivalent ASTM test method is ASTM D5511.

Oxo-degradable

Oxo-degradable materials contain additives which initiate degradation under specific conditions. The degradation process is initiated by the additives to give chain scission (oxo-degradation) and subsequent degradation is by bacteria or fungi as for standard plastics. Oxo-degradable plastics do not currently conform to EN 13432/ASTM D6400 as they take longer to degrade than allowed.

There is considerable controversy over the use of oxo-degradable plastics and their effect on the recycling chain. It is believed that introducing oxo-degradable into the recyclate stream will contaminate the complete stream. The EU (2019/904) prohibits the introduction of products containing oxo-degradable plastics as of 2020.

Biodegradability or compostability do not necessarily mean soil degradability and are not a license to litter.

Some polymers will biodegrade but the timescales do not allow them to be considered to be compostable or suitable for biomethanation

A special type of biodegradable plastics are the 'bioabsorbable' plastics which can be used for medical applications and which break down inside the body.

"It is ironic that many of the challenges of sustainability faced by polymer scientists of today arise from the polymer scientists of yesterday doing their job too well. Our predecessors developed innovative routes to materials that are, perhaps, too robust, last too long, and come from resources that are too inexpensive. A key responsibility of the polymer community in the second century is to continue innovating through creative chemistry while not forgetting the lessons of the past."

Brent Sumerlin

The Next 100 Years of Polymer Science. Macromolecular Chemistry and Physics, Wiley-VCH Verlag, 2020, 221 (16), pp.2000216.

Key tips

- Reducing materials use will improve sustainability.
- Use a Materials Team to reduce the materials use costs.
- Involve the suppliers and work with them.
- Identify the real use to decide where to start.
- Go for the big users first – they are likely to give the biggest opportunities for big reductions.
- Justify materials use or remove it.
- Compare current products with competitors' equivalents for ideas.
- Set aggressive targets in terms of materials cost reductions.
- Look at standard materials to get volume discounts.
- Reduce packaging and handling and ask for the money off the purchase price.
- Use suppliers to help to design the products to reduce materials content.
- Use sales and design staff to light-weight products at the design stage.
- Use other design techniques and tools to reduce the materials content of existing and new products.
- Use new technology to reduce the materials content of existing and new products.
- Consider retooling to reduce materials use.
- Recovering material is more sustainable than using new material and there are many ways of recovering material or its embodied energy.
- Mechanical recycling can be used for > 75% of the plastics in MSW and there are many processes that can produce standard or food-contact grades.
- The supply chain for recycled materials is complicated but is ripe for consolidation due to the pressures from legislation, the waste processors and the processors.
- The market for recycled materials is growing as a result of legislative and consumer pressure.
- The qualification of recyclers and recycled materials is becoming increasingly important as a method of providing process validation and traceability.
- Chemical recycling takes the polymer back to either polymer, monomer or to basic feedstocks.
- Chemical recycling is currently a candidate technology for recycling polymers when the basic electrical distribution system is decarbonized.
- Incineration with energy recovery is used to recover the embodied energy of mixed plastic waste. This has a lower efficiency that gas-fired systems.
- Modern landfills are an economical and safe method of storing plastics for the future.
- The term 'bio-based' plastics describes where the material comes from and not where it is going to.
- Bio-based plastics have a wide range of raw materials but all are biologically based.
- The term 'biodegradable plastic describes where the material is going to and not where it comes from.
- Not all biodegradable plastics are bio-based, some fossil-based plastics are biodegradable.
- Not all bio-based plastics are biodegradable, some bio-based plastics are not bio-degradable.
- 'Compostable' is not the same thing as 'home compostable'. The compostability specifications are written for industrial composting.
- Oxo-degradable plastics are probably not going to be a future technology.

Biodegradable plastics are still plastics, despite what some marketeers would have you believe.

Chapter 5

Manufacturing – general

Sustainable companies reduce resource use in all areas and, in plastics processing, the manufacturing area is a major user of all types of resources.

Many people think that sustainability costs money but this is untrue. When sustainability is considered holistically, i.e., in the three key areas of environmental, social and economic sustainability then most companies can achieve sustainability at little or no cost. This is especially true for aspects such as energy and water use where improving sustainability will lead to significant cost reductions. It is also true that getting the operational aspects of plastics processing right will reduce costs whilst improving sustainability credentials.

Optimising production to increase materials productivity, improve quality and reduce external impacts through good procurement and distribution are tasks that have faced Operations and Production Managers for many years. These are not new tasks and it is only in recent years that these traditional tasks are being seen as also relevant to the sustainability agenda. Production Managers need to see these not as 'new' tasks but simply as redefining the 'old' tasks that have always faced them.

Sustainability is not something that is only the responsibility of the Sustainability Manager, just as quality is not only the responsibility of the Quality Manager. Quality is everyone's responsibility and, similarly, sustainability is everyone's responsibility.

The important point is that, if we get it right, we can improve profits at the same time as improving our sustainability credentials. This can be in many diverse areas and this chapter covers most of the general manufacturing aspects of sustainability such as reducing materials waste and losses to improve materials yield, improving quality management to reduce product losses due to poor quality, improving

procurement to positively impact and improve social sustainability and distribution to reduce transport impacts.

Some of the most important and 'traditional' aspects of manufacturing sustainability are covered separately in other chapters, e.g., energy (Chapter 6), water (Chapter 8) and waste (Chapter 9). These topics are not considered in this chapter.

Manufacturing professionals can link operational, environmental and social improvements to achieve a synergy that benefits everyone and also achieves a sustainable future.

We just have to start doing it.

When it happens, everybody will not only feel good about it but will be contributing to the sustainability of the industry.

5.1　Measuring raw material losses

Assessing performance

Raw materials are the largest cost for every plastics processor but many processors do not effectively measure and control their use. To find out the current costs and performance, use the tables on the right to calculate the specific costs. The information needed should be easy to obtain from:

- Accounts records of purchased material and invoices for contract recycling, waste disposal, etc.
- Production records to find out how much plastic is used, rather than how much is ordered and delivered.
- Waste transfer notes to find out how much solid waste has left the site. If wastes are not segregated then estimate the waste plastics percentage.

FTY and MBY

First time yield – FTY %

FTY is the weight of good production divided by the total throughput of the process (including regrind). FTY is a benchmark for similar sites and processes, it measures how much is produced right first time. For most plastics processors, FTY will be ≈ 95% (higher is better).

FTY can be increased by reducing recycling in the process or by improving MBY (or both).

Mass balance yield – MBY %

MBY is the weight of good production divided by the actual weight of virgin material used in the process. MBY is always equal to or higher than FTY. MBY is a benchmark for similar sites and processes. For most plastics processors, MBY will range from 99% down to 30% or less (higher is better).

MBY can only be increased by converting more raw material into finished product.

The goals

After calculating FTY and MBY, it is possible to start looking for the losses that are shown by the difference between the two numbers. These can be because the data is incomplete or wrong or materials are being lost from the system. Start by:

- Checking that the process losses agree with the quantities disposed of.
- Measuring the waste figures by two or more different methods and reconciling the answers.

The material must go somewhere!

Calculating the COW for plastics

The total cost of waste (COW) for raw materials is the sum of the FTY cost and the MBY cost.

FTY cost

The FTY cost is calculated from the FTY and the materials records (see box below) and the cost of running the process. This is the cost of processing the material to the point where it is lost to the system.

MBY cost

The MBY cost is calculated from the MBY and the annual cost of materials (see box below). It can also be calculated from the sum of the MBYs for specific materials.

Other costs

There will be other general waste costs associated with plastics waste (see Table 4 on the opposite page) and these can be added to the FTY and MBY costs to give the total cost of plastics waste.

In the UK, the Envirowise Programme produced extensive information on waste minimisation but this is no longer available.

Get an archived copy of EG 252 – Benchmarking waste in plastics processing from: www.tangram.co.uk.

If materials are being lost to the system then this is an environmental impact and costs money.

If you don't measure it then you can't manage it.

First Time Yield (FTY%)

- $= \dfrac{\text{Weight of good production}}{\text{Virgin material used + plastic reground on site (through the nozzles)}}$
- Expressed as %.

Mass Balance Yield (MBY%)

- $= \dfrac{\text{Weight of good production}}{\text{Weight of virgin material used}}$
- Expressed as %.

The goals

- FTY = MBY
- MBY = 100%

The total cost of polymer waste (COW) is

- COW = FTY cost + MBY cost + other costs (see Table 4)

where:

- FTY cost = (100 − FTY) × annual cost of the process
- MBY cost = (100 − MBY) × annual cost of materials

Calculating the losses

These calculation sheets for the amount of polymer wasted require only basic, easily available information.

Plastics use (Table 1)

Identify the three main plastics, group the rest under 'Others' and complete Table 1.

Cost of plastics waste (Table 2)

Complete Table 2 to determine the cost of wasted plastics.

Waste/rejects reground on site: The cost of regrinding is about 5% of the plastics cost. This includes rejects, trimmings, etc., which are reground and fed back into the process. Some of these wastes may not be measured and may be hidden.

To contract recycler: If a contractor is used to regrind this can be cost-effective, but it is useful to examine the full financial case. Transport costs will always add cost.

Loss in value of plastics sold as scrap: Sending plastics for scrap will reduce the value by at least 50%. Any income from scrap represents, at best, a corresponding loss of revenue to the same sum. Fill in the amount received for your scrap as the value lost will be at least this.

Sent for final disposal: This may include items such as purgings which require specialist regrinding, or items which have become contaminated with oil or dust. It may also include materials generated in small quantities.

If waste costs are high, then change the disposal route to maximise the value, e.g., use a contract regrinder or regrind in-house rather than selling as scrap.

First Time Yield (plastics only) (Table 3)

Calculate the FTY plastics yield. Is the site better or worse than the general industry average of ≈ 95%?

This may seem acceptable but it is not just the cost of the plastics that is lost or recycled. There will also be associated losses in direct and additional labour, overhead costs such as energy, consumables and the simple costs of recycling. A steady FTY of ≈ 95% therefore costs ≈ 1% of turnover in the short term, and in the long-term costs ≈ 3% of turnover. First Time Yield waste has a large and direct impact on profitability!

Cost of general waste disposal (Table 4)

Calculate the cost of general waste from this table.

Table 1: Plastics use

Plastic Type	Amount used (tonnes/year)	Cost (£/tonne)	Annual cost (£)
1.			
2.			
3.			
Other			
Total plastic use			

Table 2: Cost of plastic waste

Plastic Waste Route	Amount (tonnes/year)	Cost (£/tonne)	Annual cost (£)
Waste/rejects reground on site			
To contract recycler			
Loss in value of plastic sold as scrap			
Sent for final disposal			
Total plastic waste			

Table 3: First Time Yield (plastic only)

First time yield (plastic) = $\dfrac{\text{Weight of good production}}{\text{Virgin material used + plastic reground on site}}$	%

Table 4: Cost of general plastic waste disposal

General Waste Route	Amount (tonnes/year)	Cost (£/tonne)	Annual cost (£)
Disposal charges, e.g., skip lifts			
Less income from segregated waste			
Total general waste (tonnes)			

5.2 Minimising raw material losses

How do you compare?

Measuring and tracking FTY and MBY are key techniques in identifying and minimising raw material waste and sites should be using these values internally to track their performance. For plastics processors, the FTY is generally the best value to track and improving the FTY will almost always lead directly to improvements in the MBY.

Typical values for the FTY in plastics processing are given in the table on the right and these values vary with the process for obvious reasons, e.g., the lack of sprues and runners in extrusion, the presence of tops and tails in extrusion blow moulding and the presence of web waste in thermoforming. Despite the process differences, there are also large differences between the best and the worst performers in the industry and this will inevitably be reflected in the relative environmental and economic sustainability.

- **Tip** – Measuring and tracking FTY needs to become a key performance indicator for plastics processing sites.

Improving FTY means making sure that every pellet that enters the factory leaves as part of a saleable product and we need to work in all areas of the site to ensure that this happens.

Measuring and quality

Measurement and attention to detail are essential in improving FTY and the actions are:

- Calculate and record FTY by polymer, site and machine on a regular basis and track any variations found.
- Calculate and record MBY by polymer, site and machine on a regular basis and track any variations found.
- **Tip** – Calculating and recording MBY by machine can be difficult if the material used is calculated by stock difference, i.e., if there is a central silo and direct machine distribution, it may not be possible to calculate MBY by machine.
- Check that MBY is close to 100% and that FTY is close to MBY.
- Track reject rates by process and/or machine in real time to allow quick reaction to any changes in performance.

Effective quality management will reduce scrap and defective products at all process stages and improve FTY (see Section 5.5).

Materials handling

Good materials handling and housekeeping are essential at all process stages to prevent material escape or loss and the actions are:

- Closely supervise polymer deliveries to reduce the risk of spills and waste.
- For large volume materials, use silos and conveying systems to remove any waste during handling and to reduce packaging disposal costs.
- Regularly inspect and immediately fix any leaks in conveying systems.
- For smaller volume materials, use octobins instead of bags to reduce potential material loss and packaging waste and ensure that octobins are covered to prevent material contamination.
- Avoid the use of bags and manual loading of hoppers.

> Avoid contaminating waste materials in any way as this lowers its value by at least 50%.

> Raw materials use has improved enormously in plastics processing during my time in the industry. Most companies are now aware of the costs involved but there is always more to do.

FTY for plastics processes[1]			
Process	Best	Worst	Average
Injection moulding	96-99%	75-80%	94%
Extrusion	96-99%	85-90%	96%
Extrusion Blow Moulding[*]	75%	55%	70%
All processes	99%	75%	95%
[*]Estimated			

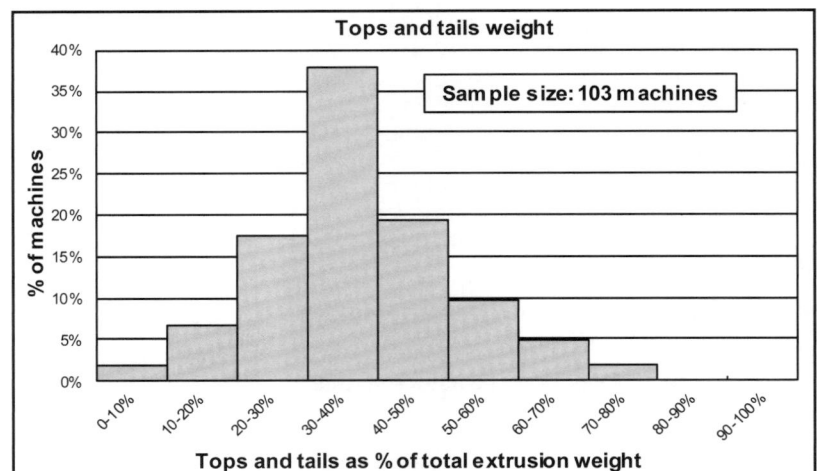

Tops and tails weight

Sample size: 103 machines

% of machines vs Tops and tails as % of total extrusion weight

Tops and tails in extrusion blow moulding

The average tops and tails produced in EBM is ≈ 35%. All this material is heated, processed and then reground to go back into the process. No material is lost but all the energy, time and capacity are lost and the FTY is very low for the process (see Section 6.14).

- Set hoppers and feeders/loaders to operate 'on demand'.
- Provide operators with systems, tools and training to prevent spills during delivery and to prevent the escape of any pellets (see Section 5.4).
- Carefully control all regrind to reduce the possibility of contaminating virgin material.

Processing

Processing is often where the losses occur and good housekeeping is again an essential part of reducing material losses and increasing FTY. The main actions are:

- Use design (see Chapter 3) to reduce the amount and type of plastics used to allow effective use of regrind and to reduce waste from sprues and runners.
- Minimise process scrap by using quality management techniques such as Statistical Process Control (SPC) to detect when a process is out-of-control and requires action to continue producing good product (see Section 5.7, Section 5.8 and Kent[2]).
- Optimise production planning to reduce colour and tool changes to reduce product and material losses at changeovers.
- Use automated handling to minimise damage to products.
- Use gravimetric dosing for additives to control levels and avoid overdosing.
- Protect all materials and products from contamination. This includes purging waste, rejected product, sprues and runners, skeletal waste, edge trim or saw off-cuts. If any of these materials 'touch the floor' or are contaminated in any way then they are not suitable for simple regrinding and re-use. Contamination will reduce the value of the material by at least 50% and potentially make it only suitable for disposal.

Regrinding

Regrinding and regrind are dealt with in Section 5.3.

Waste management

Material that is contaminated, multi-coloured, too large for internal regrinding, e.g., head waste, or in small volumes may not be suitable for regrinding and feeding back into the original process. This must be disposed of and the actions are:

- Optimise waste segregation and external recycling to reduce or eliminate waste sent to landfill.
- Use contract recyclers to recycle raw materials. The value may be small but it is better than landfill.

- Examine the contents of the skips at regular intervals to identify any issues or opportunities.

- 1. GG 376: 'Benchmarking report on waste in plastics processing', Envirowise, 2000.
- 2. Kent, R.J. 2016. 'Quality management in plastics processing', Elsevier.

If you are generating more regrind than can be fed back into the process then the issue is not how to increase the regrind level but how to reduce the amount of regrind generated.

When I started work with one company I asked "What is the reject level?"

The reply was "Zero, we regrind it all and feed it back into the product".

I fear they missed the point that it was best not to produce the regrind in the first place.

5.3   Regrinding (internal re-use)

Internal re-use

Whilst it is preferable to not to generate material for regrinding in the first place, it is inevitable that regrind will be generated for most processes. Regrind material can come from:

- Sprues and runners (injection moulding).
- Edge trim (film and sheet extrusion)
- Tops and tails (blow moulding).
- Start-up scrap and out-of-specification product (all processing methods).

Regrind and the need to both account for it and treat it is fundamental in plastics processing.

- **Tip** – Even more fundamental for economic operation is to generate as little material for regrinding as possible.

After material to be reground has been generated then it is important to regrind and re-use it as effectively as possible to maintain the value of the material and to allow it to be fed back into the original process. The actions are:

- Monitor scrap and regrind levels produced by machine and/or process for early identification of concerns.
- Minimise start-up scrap by good setting and management control.
- Maximise in-house regrinding to reduce contract recycling which has higher environmental impact and costs.
- Maintain the integrity and cleanliness of all regrind to maximise utility and value.
- Feed regrind directly back to the original process wherever possible.
- Establish upper limits of allowable regrind addition based on process or customer limitations.
- Control machine-side and central regrinders to only operate when needed. They will use energy and cost money when they are simply idling with no material being reground (see box on the right). For a longer discussion of the energy aspects of regrinding in plastics processing see Section 5.52 of Kent, R.J. 2018. 'Energy management in plastics processing', Elsevier (Third edition).
- Get the right size of regrinder for the job, a large regrinder will have large losses and even higher idling costs.
- **Tip** – Regranulation of purgings and head waste has always been a issue but the 'Purging Recovery System' from Maguire (www.maguire.com) uses a unique system

to shave purgings down to a size where they can be reground. Think of a power plane mounted upside down and being run over the purging. This is a great system and worth a look to recover purgings from any process.

Regrind management is important

Tracking the flow and value of regrind is crucial because of the high raw material value and the increasing cost and embodied energy put into the product as it goes through the process. It is strongly recommended that the generation and flow of regrind at every site is tracked by a separate process flow chart (see Sections 2.5 and 8.3). This is a method of quantifying the material flows in a process and can be used to find any material leakages from the system.

- **Tip** – If we are going to make sure that every pellet ends up in a product then we have to know exactly where each pellet goes at the site.
- **Tip** – Regrind is not 'free' material, the material may be recycled but all the embodied processing costs, e.g., energy, are lost as waste and it costs real money to recycle the material.
- **Tip** – Waste polymer sent to an external recycler loses the purchase cost of the polymer and all of the embodied processing costs. These costs dwarf any money received from the recycler for the material yet some companies insist on including this

Remember: Every pellet in a product!

If regrind is not fed directly back into the process then look after the regrind materials: Separate and code them according to material, colour and grade. This makes re-using regrind easier and if it is not re-used then it makes it more valuable to a recycler.

Adopt a Pareto approach to regrind: Find the top 20% of the processes, products or people who generate regrind and solve these issues. This will reduce regrind production by up to 80%.

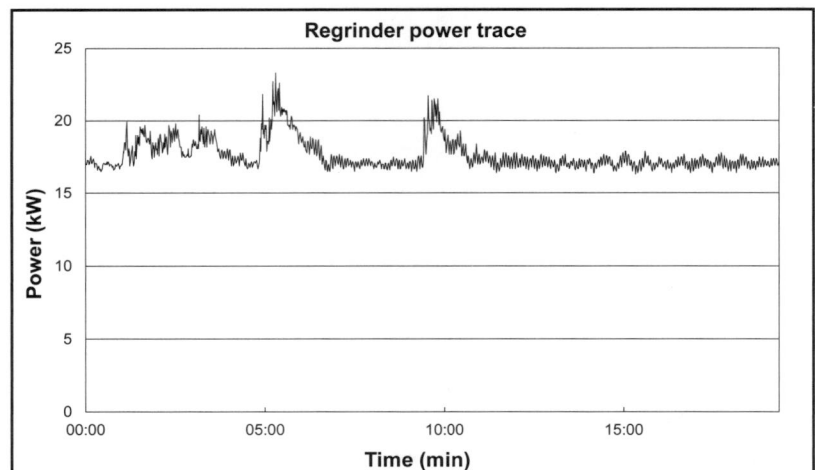

Regrinder power trace

Power trace for a regrinder

This regrinder draws an average power of 17.6 kW and costs ≈ £1.76/hour to operate but for most of the time it is not actually doing anything other than use energy. Regrinding recovers material but it is not free. Increasing the FTY saves money at many levels.

as 'income'. It would be more accurate to record the 'value lost' as a result of generating the waste.

- **Tip** – The amount of regrind is a sensitive measure of the effectiveness of a plastics processing operation.

Regrinding is not recycling

Some processors mistakenly claim to be 'recycling' when what they are doing is regrinding and re-using material that has already been through the process, i.e., the material has never left the site boundaries and is re-used in the original process. This material is therefore not recycled material because it is treated in a closed-loop.

It may be classed as Post-Industrial Regrind (PIR) but not as recycled material.

ISO 14021:2016 'Environmental labels and declarations – Self-declared environmental claims (Type II environmental labelling)' is very clear that only pre-consumer and post-consumer material may be considered as recycled content. The definitions of these two types of material makes this clear:

- "Pre-consumer material – material diverted from the waste stream during a manufacturing process. Excluded is reutilization of materials such as rework, regrind or scrap generated in a process and capable of being reclaimed within the same process that generated it".

- "Post-consumer material – material generated by households or by commercial, industrial and institutional facilities in their role as end-users of the product which can no longer be used for its intended purpose. This includes returns of material from the distribution chain".

The recovery and use of regrind in plastics processing can therefore be referred to as internal re-use of PIR but cannot be declared as being recycled content. The logic of this is clear – take the example of tops and tails in blow moulding, these will be ≈ 35% of the mass of the blow moulding (see Section 5.2) and are collected, reground and fed back into the process without ever leaving the closed-loop of the process. It would be irrational for every blow moulder to try to declare that their recycled content was > 35% when the material had never left the process boundary.

Using PIR from another process and, ideally, an external source may be classed as recycling. In terms of ISO 14021:2016 this material can be classed as pre-consumer material because it has not yet reached the consumer. We used to buy PVC 'jazz' from wire coaters, dose it with carbon black and use it for the inner core of garden hose. This was good for the application, low cost and reduced the amount of waste sent to landfill.

This was diverting material from the waste stream and was thus recycling of pre-consumer material.

- **Tip** – If regrind is produced from the waste of one process, reground and used in another process then it may be considered as recycled content. Ideally, the regrind material would have been diverted from the waste stream and have left the site boundary.

Regrind should be clearly identified, marked and held on the stock control system as with any other raw material.

This is vital for accurate financial reporting.

Regrind not only needs management time and energy for treatment, it also represents lost production time and opportunity.

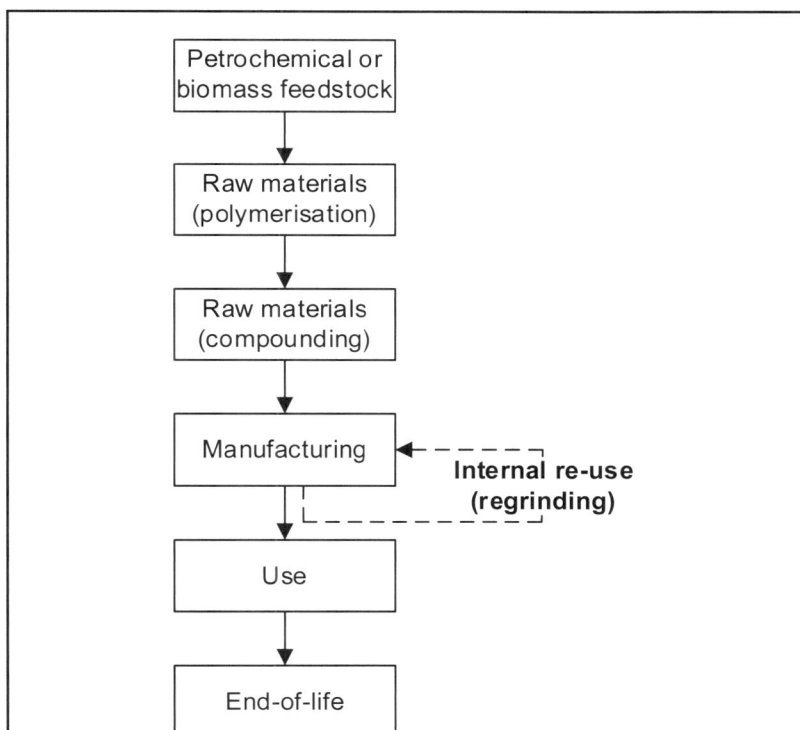

Internal re-use (regrinding) is not recycling

Whilst the use of regrind is necessary for economic operation of a site, it is definitely not recycling and should never be referred to as such. It is internal re-use of PIR and, although it may make the industry look good, it is ultimately 'greenwashing'.

5.4   Pellet control

This is important

One of the biggest issues for plastics processors around the world is the debate around plastics in the oceans. Whilst the industry can legitimately claim not to be responsible for the presence of litter in the oceans, the industry cannot absolve itself of responsibility for the presence of pellets, flakes or powder. These can be directly related to the industry as they have never been anywhere near the consumer.

In 1991, as a response to evidence of plastics pellets in waterways, the Society of the Plastics Industry (SPI), now the Plastics Industry Association (PLASTICS), started Operation Clean Sweep® (OCS) in the USA. In 2004 the American Chemical Council (ACC) joined as a partner and the two organisations have jointly administered the OCS programme since then. The programme was rolled out world-wide in 2011 when royalty-free licences were offered to other international plastics organisations and the programme is now administered by local trade associations in over 50 countries throughout the world.

In 2019, Plastics Europe, which runs OCS in Europe, made adherence to OCS principles a requirement for all member companies rather than a voluntary commitment and aims to have all member companies externally audited by 2025 – the pressure is increasing and this is not a short-term issue.

OCS is concerned with product stewardship and aims to ensure that everybody handling pellets, flakes or powder uses good housekeeping to prevent materials loss to the environment and to contain materials in the event of losses. The aim is to achieve zero pellet and material loss to the environment.

Implementing the OCS programme has multiple benefits throughout a company and some of these are:

- Improved sustainability credentials.
- Improved water quality by preventing pellet, flake or powder loss to the environment.
- Improved compliance with legislation and reduced exposure to fines and other sanctions.
- Improved staff safety, e.g., pellets are a trip/slip hazard.
- Improved housekeeping.
- Improved local reputation as an ethical and sustainable operation.

- Improved profits by ensuring that every pellet entering the site leaves as a product.

The good thing is that there is a wealth of free information on implementing OCS and every plastics processor should sign up to OCS and take action to implement OCS.

- **Tip** – Operation Clean Sweep has produced a manual for OCS and this can be downloaded at www.opcleansweep.org/resources/. There are also additional checklists on a variety of topics available.

  In many cases these resources are also available from the local OCS partners who promote OCS (in various languages) and there are also local initiatives for implementing OCS.

It is all along the supply chain

A unique feature of OCS is that it is not simply the plastics processing industry, it is relevant along the complete supply chain and covers every process where pellets, flake or powder can escape to the environment. OCS covers:

- Raw material producers.

Plastic losses
It doesn't matter if it is pellets, flake, or powder. In all these cases, the plastic was lost to the system and in places where they could easily be washed into drains, watercourses or rivers. This is not only financially irresponsible; it is also totally unacceptable for any site.

This is important:

You must sign up for Operation Clean Sweep and you must take action.

There are no excuses.

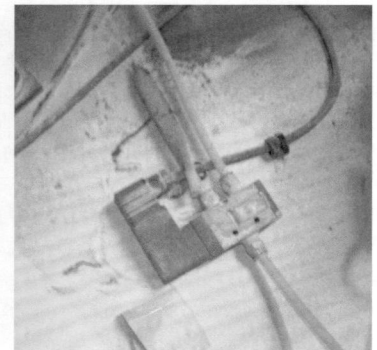

- Storage and warehousing facilities.
- Logistics, transport companies and facilities:
  - Ports and shipping terminals.
  - Shipping companies.
  - Railways.
  - Contract transport companies
- Plastics processors.
- Recyclers.

This is not simply a processor issue; plastics materials can escape to the environment in many ways and, as with many things, it is often at the interfaces where materials are transferred or moved that escapes occur.

- **Tip** – Look for areas where materials are being transferred or moved from containers, e.g., bulk transport to silos or other containers. These are generally the highest risk areas for escape.

Avoid, contain, clean up, recycle/dispose

The OCS process follows a hierarchy of: avoid spills, contain spills, clean up spills and finally recycle/dispose of any contaminated material. The OCS manual has a wealth of information and guidance on implementing OCS and it is not proposed to repeat this here.

For plastics processors the main actions are:

Avoid

Avoiding spills is the highest priority, a spill that is avoided is one that requires no further action. Areas to examine are:

- Delivery of materials – delivery spills are possible whatever the delivery method. In all cases the systems and processes should be examined to ensure that spills are avoided. The issues will vary with the delivery method but in most cases, the solutions are obvious and easily implemented.
- Internal spills can occur at materials handling systems, machines and other material transfer points or where materials are not supplied in closed systems.
- **Tip** – Prevention is always better (and cheaper) than cure.

Contain

If a spill does occur then it should be contained as soon as possible to prevent any material reaching the drains and then watercourses. Containment can be internal, i.e., in the building, as a first line of defence or external, i.e., around the factory area, as a backup to the internal containment.

- **Tip** – Never disperse spills, concentrate them for easier clean up.

- **Tip** – As someone who has spent most of his life in plastics factories, the trip/slip/slide hazard of pellets on the floor cannot be overstated. Contain spills and keep staff safe.

As a last barrier, all drains should be fitted with suitable traps and filters to prevent any pellets escaping to the drainage system.

Clean up

Clean up should take place as soon as possible after a spill to prevent material dispersing around the site.

- **Tip** – Provide clean up materials and resources at likely spill locations and make sure they are used promptly.
- **Tip** – Train staff in the operation of cleaning equipment and make sure that they use it at the first possible opportunity.

Recycle/dispose

Any material collected in clean-up operations is likely to be contaminated and should be recycled rather than being sent to landfill.

Audits are important

A management commitment to OCS is important and the most effective tool to encourage improvement is the use of regular site audits. These should act to improve operations and should never be seen as a 'search for the guilty party'.

- **Tip** – A monthly audit in the initial stages can be reduced to a quarterly, or yearly, audit as progress is made and the OCS ethos is embedded in the company.

There is no excuse

Preventing pellet loss is an opportunity to drive the sustainability message into the company operations and into the wider community. There is absolutely no excuse for not signing up to OCS ('taking the pledge') and achieving zero pellet loss from every site. It is easy to sign up, easy to do and is a real benefit to the company.

The industry calls them 'pellets' but they are often referred to as 'nurdles'. Nobody is sure where this name came from but it is now in common use.

'Surfers against Sewerage' also calls them, somewhat more emotively, 'mermaids tears'.

The name doesn't matter, they shouldn't be there.

In the UK, the British Standards Institution, with the assistance of the industry and the Scottish Government, has developed PAS 510: 2021 for the management of pellet handling and prevention of leakage to the environment.

This is not the same as a British Standard but it gives good guidance on preventing pellet loss.

It is available free from BSI at (shop.bsigroup.com/).

Powder is much more difficult to avoid, contain and clean up than pellets and flakes. Sites compounding materials such as PVC dry powder blends will often have PVC dust in the air. Controls and maintenance of dust collection equipment need to be higher at these sites.

5.5 Quality management and sustainability

The ultimate cost

Dr. W. E. Deming stated that 'Cutting costs without improvement in quality is futile' and, as in many things, he is right. Equally, the route to improved quality almost always results in reduced environmental impacts and costs across a company.

Reducing the environmental impact of products counts for nothing if the 'total product' produced is not of acceptable quality. Resource efficient product design, well-sourced materials and efficient production all contribute to sustainability but are wasted if the product does not fulfil the primary function and is destined for recycling or disposal soon after it is despatched to the customer. However, this quality must cover the total product and not be simply the physical product that the customer receives. The total product includes the whole range of services and contacts with the customer. The physical product itself is of great importance but it is not the total product.

This has long been recognised in quality management standards. The early issues of ISO 9001 were written around 'quality' as a property of the product and the standard was primarily aimed at manufacturing companies. As the concept of total quality grew, companies as diverse as solicitors and travel agents applied for and gained ISO 9001 approval. The concept of quality goes beyond the simple product and ISO 9001 now focuses on total quality as perceived by the customer.

- **Tip** – ISO 9001 is an MSS similar to those described in Chapter 2 but everybody should already know that.

ISO 9001 recognises that quality is not simply the responsibility of the production area or any one group or person – it is everyone's responsibility. A broader view is necessary and this is the concept of 'Total Quality' in all aspects of customer contact.

- **Tip** – Quality also covers sustainability as part of the social and economic responsibilities of a company.

The quality building blocks

The drive for improved product quality, reduced variability, improved reliability and reduced cost really has little to do with public relations and marketing (although these are valuable spin-offs). It is about the simple issue of surviving and prospering in a world where excellence is only a 'ticket to the game' and not a guarantee of winning.

The building blocks for Total Quality are:

Management commitment

The management of the company must be committed to the concept of quality. This involves both leading the way and being the facilitators for quality improvement. The management must provide:

- The quality policy.
- The resources to get the job done.
- The organisation and framework for success.
- The training in the necessary skills.
- The delegated authority to allow the people to get the job done.

Quality tools

These are the tools used to achieve Total Quality: e.g., Pareto analysis, cause and effect diagrams, flow charts and statistical process control (SPC). The tools are not only used for quality improvement but for all process improvement.

Quality improvement

This is the discipline of never accepting that things are good enough, it is the search for never-ending quality improvement. It involves problem identification, problem solution, identifying causes and not symptoms as well as a continual effort to reduce quality costs.

Quality systems

These are the systems that make it all work and provide control, communication, continuity and confidence. The systems requirements are generally fulfilled by the

Does your quality system work?

The lack of understanding of the essential quality building blocks is the reason that some companies complain about the lack of effectiveness of their quality system. The presence of a quality system to ISO 9001 or any other standard will not automatically reduce costs or improve quality.

If the other building blocks are absent the quality system is seriously weakened from the start.

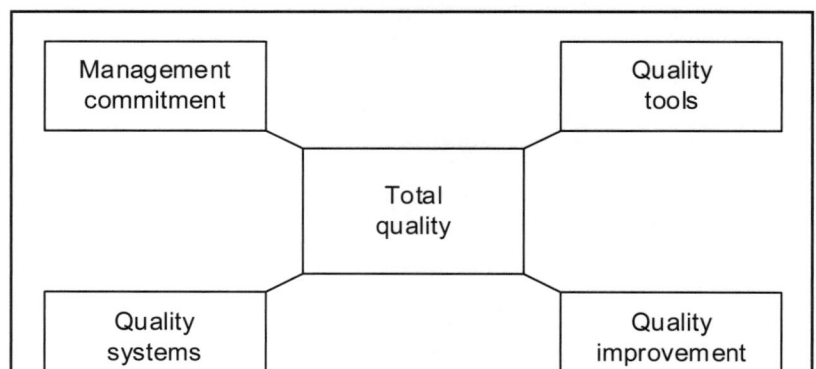

Management commitment		Quality tools
	Total quality	
Quality systems		Quality improvement

The building blocks of the Total Quality Model

It is not enough to have some of the building blocks present, all must be present for TQM to succeed. The importance of management commitment cannot be overstated, without good management then neither quality nor sustainability management will be effective.

ISO 9001 system.

All these building blocks must be present to succeed in the search for 'Total Quality'. If any of the components are not present then attempts to reduce costs through quality improvement will fail.

Total quality management

TQM is more than simply a philosophy or a tool. It requires a change in organisation and company culture. TQM focuses the organisation on continual improvement by regarding everything in the company (not just manufacturing) as a process. All processes must be improved using scientific methods and decisions based on facts. The goal of this focus is perfection as assessed by the customer in all areas. The driving force is everyone, whether as an individual or as part of a team.

To foster this culture change, Dr W.E. Deming gives 14 points as obligations for management (or the facilitators). Whilst the Western mind may not agree with all the 14 points, there is more than a grain of truth in most of them.

The reverse is also true and Deming also identified the 5 deadly sins that plague Western management. He regards these sins as the root cause of the decline of Western manufacturing over the last 50 years. How many of those reading this are guilty of at least one of these deadly sins?

- **Tip** – Deming's 14 points and 5 deadly sins were devised for improving quality management but have an equal resonance in improving sustainability management.

- **Tip** – Training and empowering staff are part of the labour practices aspects of social responsibility (see Section 11.5).

- **Tip** – Producing high quality products is part of the consumer issues aspect of social responsibility (see Section 11.8).

Deming's guidelines set the scene for the culture change required before TQM can become effective and only by accepting the need for this culture change can the other components be put into place. Having a quality management system to ISO 9001 does not by itself improve quality but simply gives one of the tools necessary to achieve quality.

The basic requirements for a TQM implementation programme are:

- Good quality management systems for quality assurance, quality control and quality improvement (they are different) .

- Control of quality at the point of production by providing tools, training and responsibility.

- Control of supplier's quality.

- Data recording and action on the records for both production details and the cost of quality.

Quality is no longer a 'winner' it is only a ticket to the game.

For a longer discussion of quality management in plastics processing see: Kent, R.J. 2016. 'Quality management in plastics processing', Elsevier.

Deming's 14 points and 5 sins of quality management

The 14 Points

1. Consistency of purpose.
2. The new philosophy.
3. Cease mass inspection.
4. End 'lowest tender' contracts.
5. Constantly improve systems.
6. Institute training.
7. Institute leadership.
8. Drive out fear.
9. Break down barriers.
10. Eliminate exhortations.
11. Eliminate targets.
12. Permit pride of workmanship.
13. Encourage education.
14. Top management's commitment.

The 5 Deadly Sins

1. Lack of constancy of purpose.
2. Emphasis on short-term profits.
3. Evaluation by performance, merit rating or annual performance.
4. Mobility of management.
5. Running a company on visible figures alone.

After Deming (The Deming Institute, deming.org)

5.6   Quality costs/quality savings

The cost of poor quality

One of the basic concepts of quality is that it is better and cheaper to prevent defects than to detect them after you have made them. If defects are prevented by an efficient system then waste is reduced, sustainability is improved and costs are lowered. Quality is not an abstract theory but a vital tool in cost and sustainability management for any company.

The process of defect detection allows faulty and incorrect products to be manufactured and paid for before they are detected. The more effective process of defect prevention prevents this happening and saves money.

The cost of quality is not just the waste and cost of inspection and scrap materials; it is the total cost of not getting the product 'right first time'. Quality costs are not always due to bad production but can also be created by initial specifications or customer expectations that are unrealistic with respect to the actual production capabilities (see Section 5.7).

The cost of poor quality for the average plastics processing company is estimated to be between 5-25% of turnover. At an average of 15% of turnover this is more than the profit of most plastics processing companies. A high proportion of these costs are avoidable and yet few companies have seriously tried to identify or reduce these avoidable costs. If the avoidable costs were to be reduced by 50% it could mean an increase of 50% in profits. A simple calculation for most companies will show the possible magnitude of the costs and yet in most cases these cost savings and sustainability improvements are ignored.

The true quality costs

Quality costs may be divided into three separate areas, i.e., prevention, appraisal and failure (PAF). In most companies, only 5% of costs are spent on prevention and 95% of quality costs are expended on failure and appraisal. Failure and appraisal costs add nothing to the product value and are total wastes. Increasing the money spent on prevention can reduce the overall quality cost by between 30 and 50%. It may well be the most highly geared management action that a company can make.

The approaches of detection or prevention have very different results and these are shown in the diagram on the right. In these terms it hardly makes sense to purchase cheap materials because the cost only returns via another route – cheaper in the short term may be more expensive in the long term. This change of emphasis from failure and appraisal to prevention requires a change in current ideas about quality management and control activities. At present these are seen as a cost to the company but by changing the emphasis they can be seen as a gain to the company by increasing quality (and hence sales) whilst at the same time decreasing the overall product cost.

Most companies are shocked to find that their true quality costs are far higher than simply the cost of their Quality Control Department. Even rough figures will show that the cost of quality is a significant proportion of turnover.

Prevention, appraisal and failure

A typical breakdown of the sorts of costs that are considered to be quality costs is:

Failure costs:

- Internal – scrap material, labour overhead, sorting, selective production and downtime.
- External – faults and complaints, investigations, interest on unpaid invoices and product recall costs (transport, paperwork, etc.).
- Intangible – lost sales through bad

Overselling?

A major cause of 'quality issues' is overselling of the product in the first place. If the customer is promised features or services that you cannot deliver then they will always feel dissatisfied with your service.

Sales staff desperate to get an order can often set you up to fail by overselling the product in the first place.

Over-promise and under-deliver and you will always have issues.

Under-promise and over-deliver and you will always have satisfied customers.

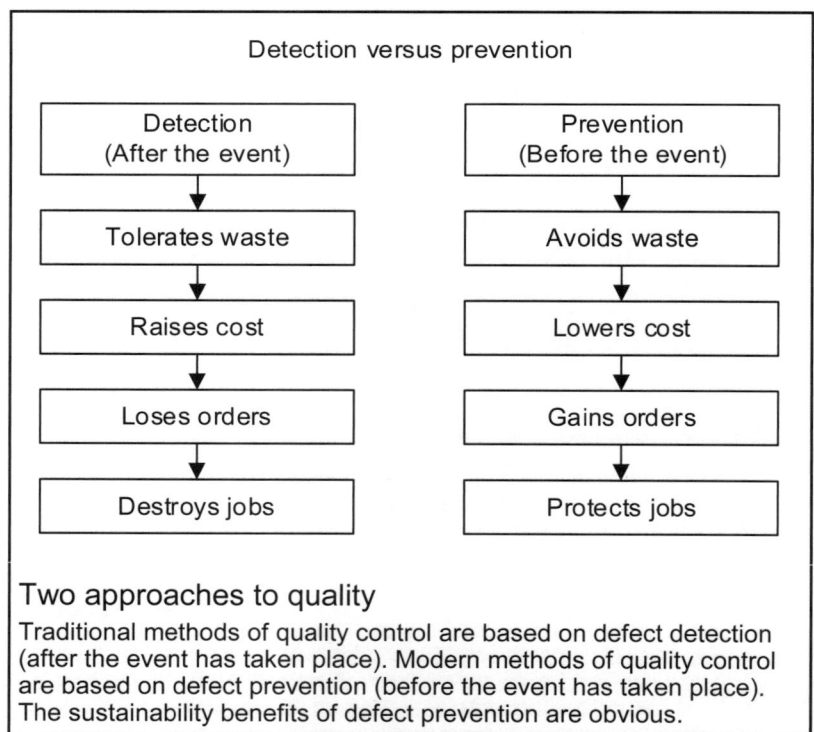

Detection versus prevention

Detection (After the event)	Prevention (Before the event)
Tolerates waste	Avoids waste
Raises cost	Lowers cost
Loses orders	Gains orders
Destroys jobs	Protects jobs

Two approaches to quality

Traditional methods of quality control are based on defect detection (after the event has taken place). Modern methods of quality control are based on defect prevention (before the event has taken place). The sustainability benefits of defect prevention are obvious.

reputation and production delays.

Appraisal costs:

- Incoming, in-process and final inspection.
- Appraisal test equipment.
- General quality control overheads.
- Cost of inspectors' wages.

Prevention costs:

- Quality plans.
- Sourcing via quality suppliers.
- Realistic design tolerances.
- Housekeeping.
- Packaging.
- Training of personnel in quality.
- Statistical process control (see Section 5.8).

Getting the information

There are no current national or international standards for the cost of quality. There was a British Standard (BS 6143) but this was withdrawn some years ago and not replaced. There are several models for data collection and these are variations on the PAF model based on the work of Feigenbaum, Juran and Crosby. Each of the models has positives and negatives:

- If the need is simply for cost of quality data then my preference is for the PAF type of model because of the easy integration with existing costing and accounting systems.
- If the cost of quality is simply part of a move to world-class manufacturing then Activity Based Costing provides a set of tools that have benefits across the whole company and not only in quality.

Some years ago, we used the simple PAF to collect these costs for an in-house moulding company. We used the standard management accounts and set up a report based on their standard cost allocation codes. After the numbers were complete and the magnitude of the cost was revealed there was no longer any issue with investment approval for quality management and the 'cost of quality' report became a regular management report at Board level.

After a quantified view of the costs is available, it is a long-term process to transfer costs from the appraisal and failure categories to the prevention category and to reduce the overall magnitude of the costs. The cost of quality report and other non-financial numbers provide a focus for these activities. Significant causes can be identified using tools such as Pareto analysis and 'cause and effect' diagrams (see Section 9.4), and tools such as statistical process control (see Section 5.8) can be used to prevent defects and reallocate resources.

Sustainability?

Reducing the cost of quality is a key area in cost and sustainability management because of the large financial implications but more importantly, quality will be the one thing that decides if a company survives or not.

Sustainability is about economic sustainability as much as about social and environmental sustainability.

Get some rough figures first – do not seek great accuracy at the early stages but simply try to find out where you are spending money on poor quality.

These areas can be targeted for rapid improvements using the appropriate tools.

For a longer discussion of quality management in plastics processing see: Kent, R.J. 2016. 'Quality management in plastics processing', Elsevier.

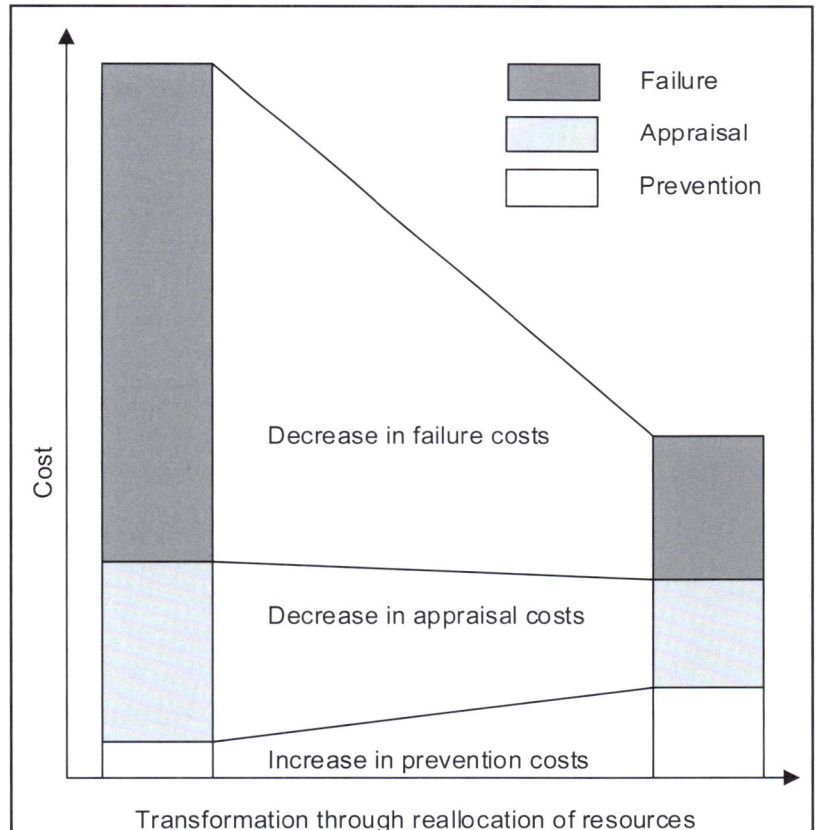

Transformation through reallocation of resources

Total cost reduction by reallocation of resources

The total cost of quality in most companies is 5-25% of sales. Allocating resources to prevention instead of inspecting products and rectifying failed products leads to a decrease in overall costs that is transferred straight to the profit of the company.

5.7　Statistical process control to reduce waste – capability studies

Can we make it?

Moving from detection of poor quality to prevention reduces material use and costs. An essential part of this is understanding if it is possible to make the product and this requires a capability study for the process. This is a method of analysing if a process is capable of producing to the tolerances specified. It enables tolerances to be based on statistical evidence rather than on feel and precedence.

If enough samples are taken and if the cell size is decreased then a conventional histogram will often begin to look like the 'normal distribution' or bell-curve (shown on the upper right). This normal distribution describes how many things vary but most importantly it shows the results of many processes.

It is physically impossible to produce every successive part from a process to exactly the same dimensions, but if enough measurements are taken then the normal curve will often begin to appear. The greatest number of parts is near the centre of the curve with small numbers of parts being produced over the edges.

This distribution is predictable and can be fully described by just two numbers:

- Mean – this is the average of all the individual values. It is the centre of the distribution and gives the 'where' value. It is written as $\overline{X}$.

- Standard deviation – this is the 'spread' of the values and is related to the variability of the process. It is calculated by a simple formula, (it is even marked on many calculators) and is written as σ.

The standard deviation is such that:

- The limits $\overline{X}$ +/- σ contain 68.26% of the samples.

- The limits $\overline{X}$ +/- 2σ contain 95.44% of the samples.

- The limits $\overline{X}$ +/- 3σ contains 99.73% of all the samples.

If the mean is 10 and the standard deviation is 1.0, then 68.26% of the samples will have a value in the range of 10 +/- 1 (between 9 and 11) and 95.44% of the samples will have a value in the range of 10 +/- 2 (between 8 and 12).

The value +/- 3 σ or 6σ is a special value because we know that 99.73% of all the results will lie within this band. This is termed the 'process spread' or process variability.

Tolerances

Spread (C_p)

The spread relative to a given tolerance band is described by C_p where:

$$C_p = \frac{\text{Specified Tolerance}}{6\sigma}$$

If $C_p = 1$ then the normal distribution will only just fit within the tolerance limits but any movement of the spread or location is likely to produce out-of-tolerance parts.

The limit for acceptable production is a minimum C_p value of 1.33 to allow for some movement in the spread of the process. This means:

- If $C_p > 1.33$ then the process is capable of reliably producing in-tolerance parts with small movement in the spread.

- If $C_p < 1.33$ then the process is not capable of reliably producing in-tolerance parts and even small movements in the spread will produce out-of-tolerance parts.

The UK Envirowise Programme produced excellent information on process control but this is no longer available. Get archived copies of:

GG223 – Preventing waste in production: industry examples.

GG224 – Preventing waste in production: practical methods for process control

from: www.tangram.co.uk.

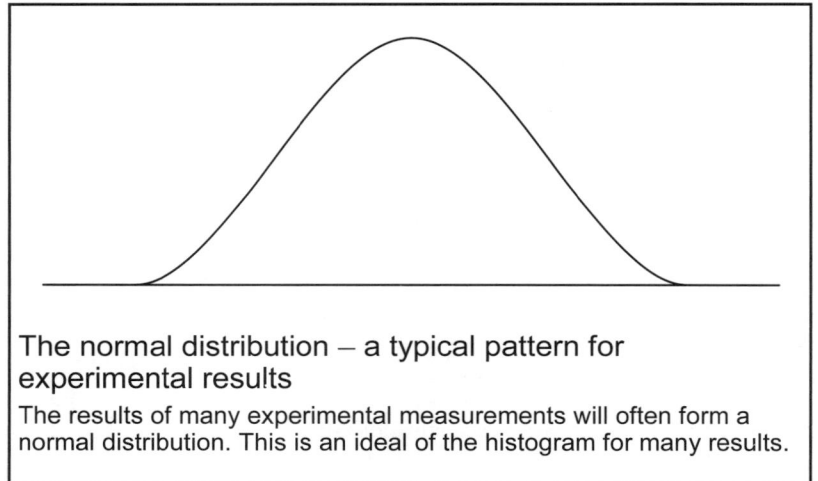

The normal distribution – a typical pattern for experimental results

The results of many experimental measurements will often form a normal distribution. This is an ideal of the histogram for many results.

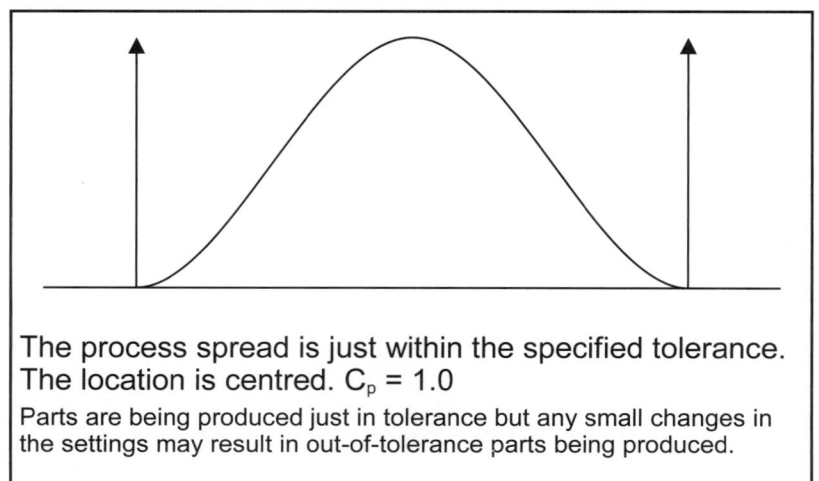

The process spread is just within the specified tolerance. The location is centred. $C_p = 1.0$

Parts are being produced just in tolerance but any small changes in the settings may result in out-of-tolerance parts being produced.

Location (C_{pk})

The location relative to a given tolerance band is described by finding the smaller of two values, Z_{upper} and Z_{lower}:

$$Z_{upper} = \frac{\text{Upper Tolerance Limit - Mean}}{\sigma}$$

and

$$Z_{lower} = \frac{\text{Mean - Lower Tolerance Limit}}{\sigma}$$

The smaller of Z_{upper} and Z_{lower} is called Z_{min} and is converted to the C_{pk} value by:

$$C_{pk} = Z_{min}/3$$

The limit for acceptable production is a minimum C_{pk} value of 1.33 to allow for some movement in the location of the process. This means:

- If $C_{pk} > 1.33$ then the process is reasonable well centred in the tolerance band and is capable of reliably producing in-tolerance parts with some small movements in the location.

- If $C_{pk} < 1.33$ then the process is not capable of reliably producing in-tolerance parts and even small movements in location will produce out-of-tolerance parts.

Z values can also be used with standard statistical tables to provide an estimate of how much of the production will be out of specification at the upper and lower limits.

Using the information

A capability study will quickly and easily give information about a process and what you can expect to achieve from it. Typical uses are:

Process studies

Capability studies will tell you if a machine is operating properly and if you can ever expect to get good results from it.

Process setting

Never set a process based on a single random sample. Depending on the location of the sample in the distribution the process adjustment can be low or high but is rarely correct. This gives see-sawing of the process in and out of control. Never rely on a single sample to set or adjust a process.

Tolerance setting

Capability studies provide the essential information for the setting of realistic and achievable tolerances.

Machine purchasing

Capability studies tell you if a machine is capable of the claims made by manufacturers. If a supplier tells you that a saw will cut to +/- 0.1 mm then C_p and C_{pk} values will reveal if the machine is capable of this (both should be > 1.33).

Tooling acceptance

Tooling is often accepted based on small numbers of measurements. Capability studies reveal the true performance of tooling before acceptance.

More information

Capability studies answer the question 'Can we make it OK?' and this naturally leads on to the question 'Are we making it OK?' To answer this question, SPC is needed (see Section 5.8).

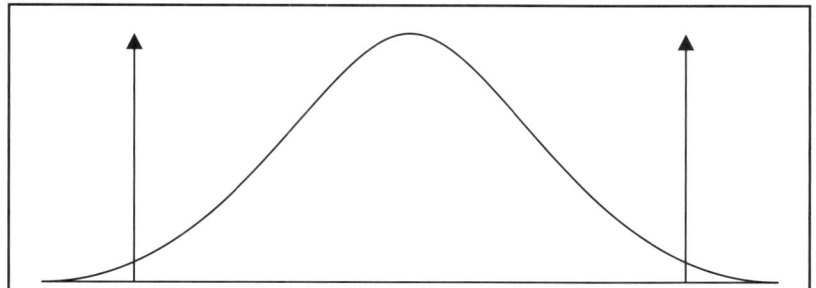

The process spread is greater than the tolerances. $C_p < 1.0$ and $C_{pk} < 1.0$

This is a process that can never produce to the tolerances. Out-of-tolerance parts will always be produced at the upper and lower tolerances. The process is not capable.

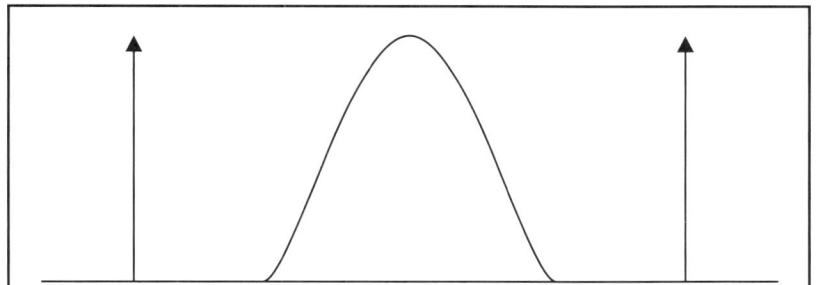

The process spread is much less than the tolerances and the location is good. $C_p > 1.33$ and $C_{pk} > 1.33$

It is easy to produce in-tolerance parts provided the location is kept near the centre of the tolerance band. This is an excellent process that is easy to control.

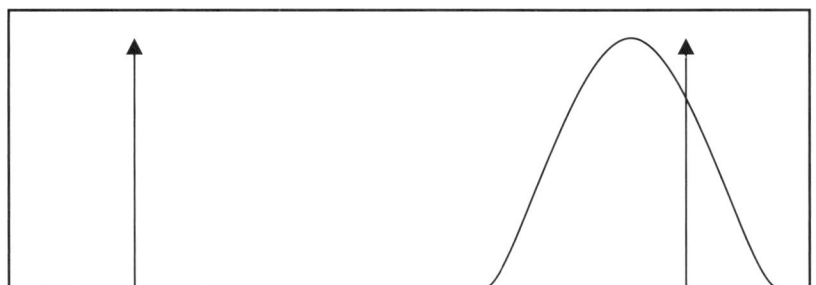

The process spread is much less than the tolerances but the location is poor. $C_p > 1.33$ but $C_{pk} < 1.33$

Out-of-tolerance parts are being produced. The machine needs adjustment to bring the location back to the centre of the tolerance band.

5.8 Statistical process control to reduce waste – control charts

The benefits of process control

Statistical process control (SPC) is used by operators for ongoing control of a process. It helps a process to perform consistently and predictably and gives better quality, fewer rejects and reduced losses at lower cost. More importantly SPC gives the entire company a common language for discussing process performance and process improvement.

Statistical process control enables companies to distinguish between the two types of causes of product variation:

- Common causes – these are the causes that will always be present because of tooling and machinery variations. They have many small sources, are predictable and permanent. They generally require management action to remove or reduce them.

- Special causes – these are the causes that have major sources, are unpredictable and irregular. They can generally be reduced at a local level.

Control charts have been in existence for many years and offer an easy way to improve product quality and reduce rejects throughout the business. They 'feed forward' to control the process before defects occur. Despite the statistical basis they are remarkably easy to use and operators enjoy the improved control and reduced defects that real control brings.

As an example, for injection moulding the easiest variable to control is the product weight. This has been used to successfully control complex products and achieve spectacular results in terms of improved process consistency and improved product quality – it does not have to be complex to work.

Setting the limits

The initial control chart parameters ($\overline{\overline{X}}$, $\overline{R}$ and the relevant UCL and LCL) are set using the following simple rules:

- Select the subgroup size (number of samples/ measurement interval). Use 5 as a default.

- Select the measurement interval. Use hourly as a default.

- Calculate $\overline{X}$ and R for each subgroup where $\overline{X}$ is the average of the measured values and R is the difference between the highest and lowest values.

- Plot $\overline{X}$ and R for the subgroup on the control chart.

- After measuring 20 subgroups, calculate $\overline{\overline{X}}$ and $\overline{R}$ ($\overline{\overline{X}}$ is the average of the averages and $\overline{R}$ is the average of the ranges for the subgroups measured).

- Calculate the control limits from:

  - $UCL_R = D_4 \times \overline{R}$
  - $LCL_R = D_3 \times \overline{R}$
  - $UCL_{\overline{X}} = \overline{\overline{X}} + A_2 \times \overline{R}$
  - $LCL_{\overline{X}} = \overline{\overline{X}} - A_2 \times \overline{R}$

- The constants A_2, D_3 and D_4 depend on the chosen subgroup size and can be found in Kent[1], Price[2] or Oakland[3].

- Draw the control limits on the control chart and start to get control of the process.

Completing the chart

The control chart should be regularly completed by the operator as follows:

- Measure five samples for each interval and calculate $\overline{X}$ and R.

- Plot these on the control chart and join to the previous point with a straight line.

> Control charts let you look forward rather than backward.
>
> Prevention is better than detection.

> Is process control and other waste linked to the time of day, shift or other time factors?

> Control the process and the product automatically follows.

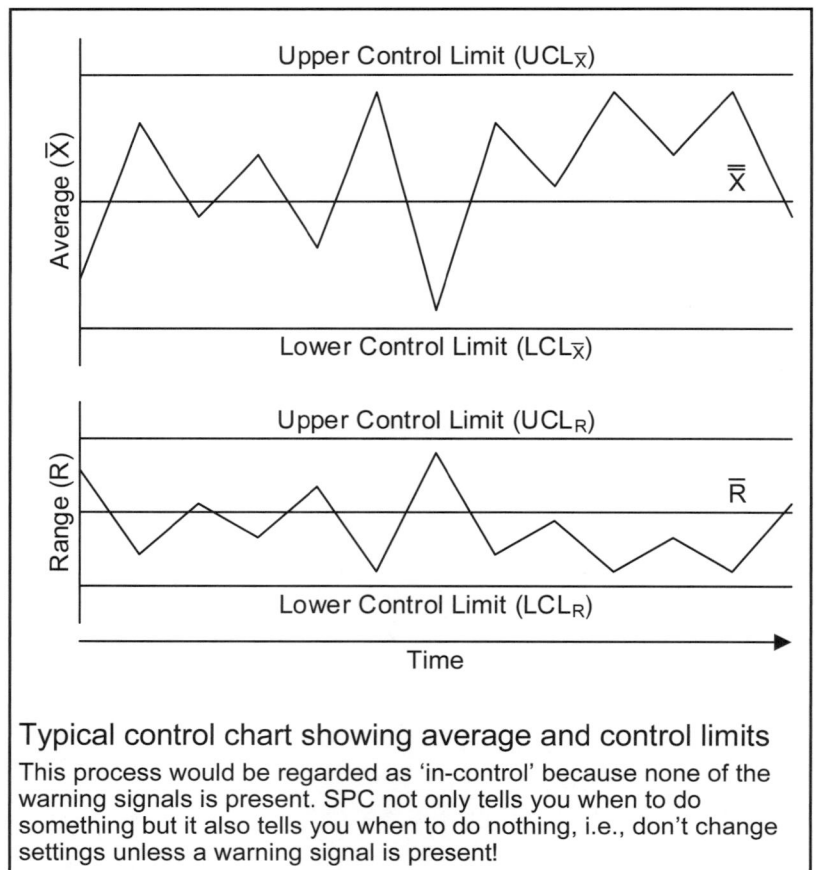

Typical control chart showing average and control limits

This process would be regarded as 'in-control' because none of the warning signals is present. SPC not only tells you when to do something but it also tells you when to do nothing, i.e., don't change settings unless a warning signal is present!

• Sign the chart with time and date.

Operators must regard the chart as a working document. They should mark it, write on it and note any changes in operator, materials or conditions. Above all they should do nothing unless the chart tells them to do something.

Action points

The alarms for action are:

• Any points outside the control limits.

• Seven points above or below $\overline{\overline{X}}$ or $\overline{R}$.

• Seven intervals going up or down.

• Two thirds of the points should lie in the middle one third of the control limits.

• One third of the points should lie in the outer two thirds of the control limits.

The first step is to check the R chart for alarms – if these are in order then the $\overline{X}$ chart is checked. The alarms apply to both the $\overline{X}$ and R plots.

If any alarm condition is seen then the operator should call for assistance.

Taking action

One of the huge benefits of control charts is knowing when to do nothing. Unless the charts indicate action then no action should be taken. A simple edict that action without an alarm is a disciplinary offence will remove the 'fiddling' that wastes so much time in plastics processing.

If an alarm is triggered then management action is required to find out what has changed.

Other control charts

The 'control chart for variables' described above is only one type of control chart. For visual defects an attribute control chart is used but this is beyond the scope of this publication (see Kent[1]).

The future

If you are processing plastics and are not using SPC then you will really never know if your process is in control and will constantly be hoping that your processes will produce the right results. Start to look at SPC now to improve processes and reduce costs.

• 1. Kent, R.J. 2016. 'Quality management in plastics processing', Elsevier.

• 2. Price, F. 1986, 'Right First Time', Gower Publishing.

• 3. Oakland, J.S. 2007, 'Statistical Process Control', Butterworth-Heinemann.

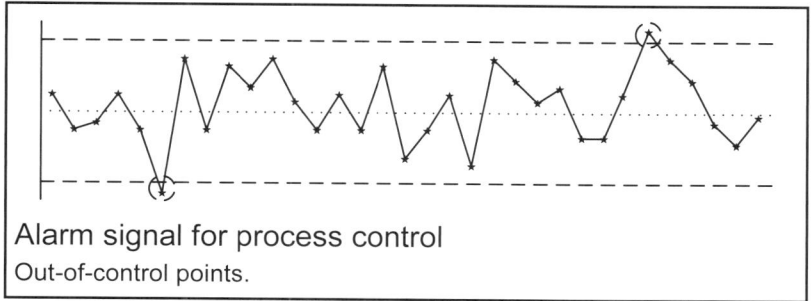

Alarm signal for process control
Out-of-control points.

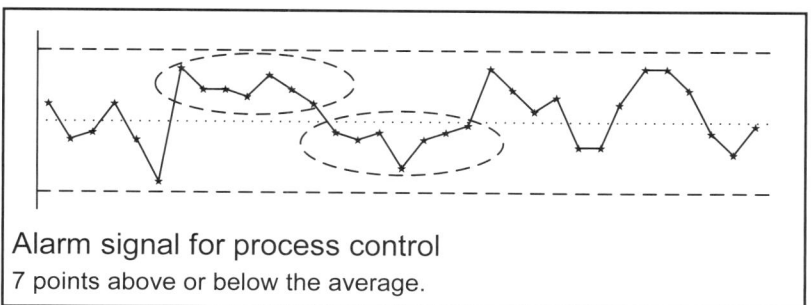

Alarm signal for process control
7 points above or below the average.

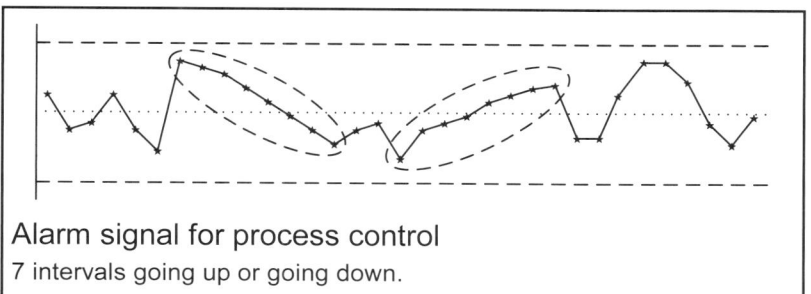

Alarm signal for process control
7 intervals going up or going down.

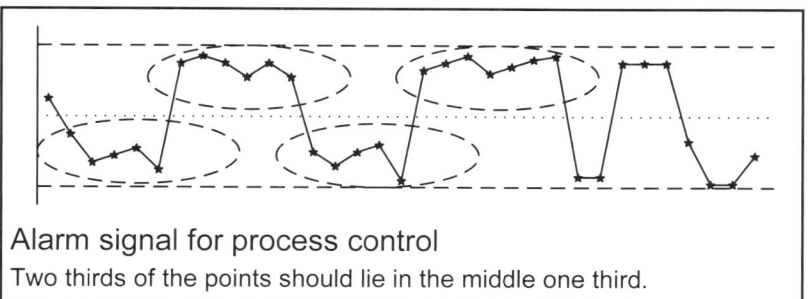

Alarm signal for process control
Two thirds of the points should lie in the middle one third.

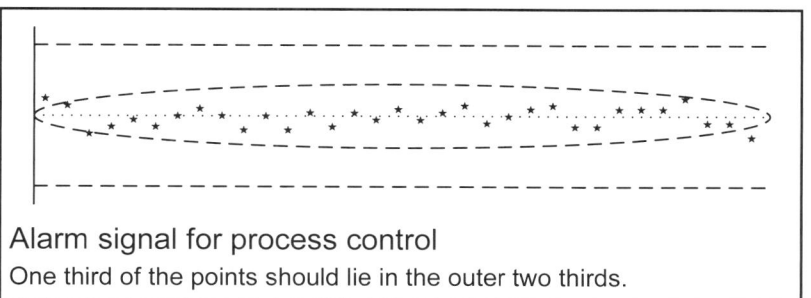

Alarm signal for process control
One third of the points should lie in the outer two thirds.

5.9 Waste and non-value activities

What are waste activities?

Sustainability requires that we effectively use all resources and waste is 'the expenditure of resources that do not add value to the product at least equal to the cost of the resources expended'. Chapter 9 covers waste in detail but this section looks at the common wastes in manufacturing processes and why these should be eliminated or reduced.

Reducing waste in manufacturing translates into improved performance and economic sustainability. All aspects of the process, from order processing to production, need to be investigated to reduce waste, this is not simply a manufacturing issue.

What are the wastes?

The seven wastes defined by Taiichi Ono (Toyota's Chief Engineer) are:

Overproduction

Making a product with no current sales:

- Ties up capital.
- Uses space.
- Reduces delivery performance.
- Occupies bottleneck machines.

Waiting

Machines waiting for goods, maintenance, product or other action:

- Wastes time.
- Reduces throughput.
- Uses space for idle products.

Transportation

Moving products:

- Increases cycle time.
- Increases Work In Progress (WIP).
- Creates waiting waste.

Process

Using inefficient processes:

- Increases cycle times.
- Increases WIP.

Stock

Stock hides other wastes and:

- Ties up capital.
- Increases stock losses due to damage.
- Uses space.

Motion

Movement that does not create added value:

- Increases cycle time.

ACTIVITY	ADDING VALUE	WASTE
Moving		✓
Storing/Waiting		✓
Processing	✓	
Over-production		✓
Counting		✓
Inspecting		✓
Scrapping		✓
Re-working		✓
Assembling	✓	
Sorting		✓

How much time do you spend on waste activities?

Classification of factory activities into value-adding processes and waste processes. Measuring the time spent on each of these processes in the production area gives a measure of the time wasted in the production area.

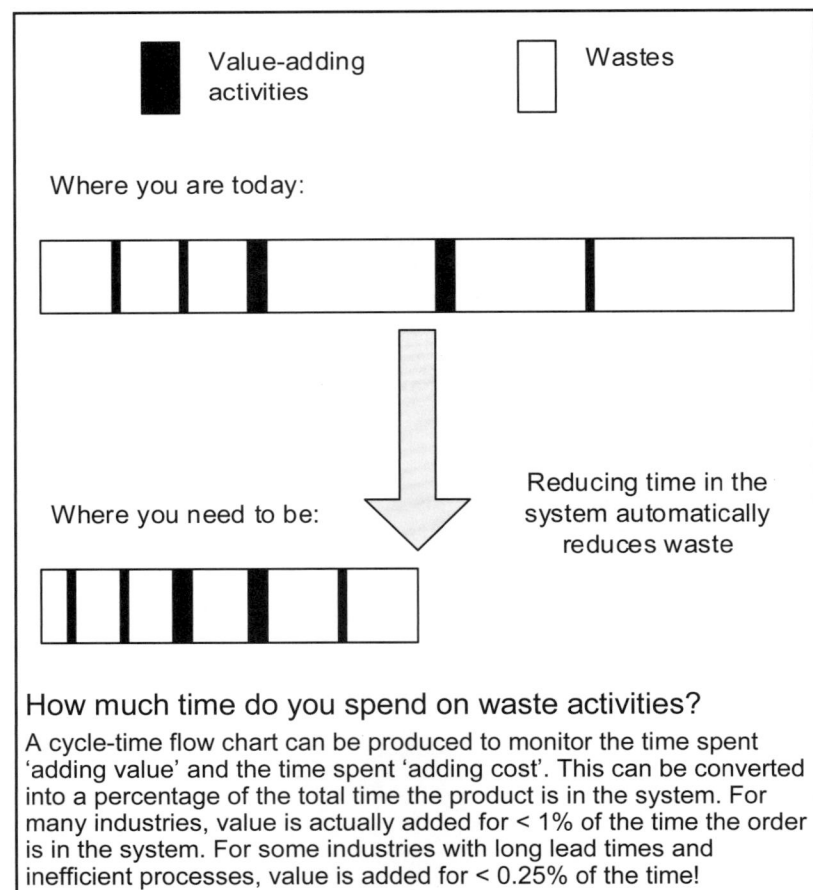

■ Value-adding activities   □ Wastes

Where you are today:

Where you need to be:

Reducing time in the system automatically reduces waste

How much time do you spend on waste activities?

A cycle-time flow chart can be produced to monitor the time spent 'adding value' and the time spent 'adding cost'. This can be converted into a percentage of the total time the product is in the system. For many industries, value is actually added for < 1% of the time the order is in the system. For some industries with long lead times and inefficient processes, value is added for < 0.25% of the time!

- Causes product damage.
- Wastes employee effort.
- Needs investment in space and machines.

Defects

- Create all the other wastes and add unnecessary cost to the product.

Quantify the waste activities

Quantifying waste is critical to success and the first step is to find areas or processes that waste resources, of any description, and then to eliminate or redesign the process to reduce the waste.

In many firms the material in the manufacturing system is only having value added to it for < 1% of the total time it spends in the system. The rest of the time, it is WIP or inventory, and in the real world, inventory is a liability rather than an asset.

Paperwork processes

A prime, but often neglected, target for waste reduction is the paperwork or control systems of the company. In some companies it takes less time to make the product than it does to complete the paperwork. These systems need to be redesigned to catch up with the production system. This is the essence of Business Process Re-engineering (BPR), which applies production techniques such as Just-in-Time (JIT) to office as well as to production.

The greatest waste of all

This is the waste of not using all of the talent and ideas that we have in the company. We ignore our employees' ideas and treat them as a 'body for hire'. Truly the greatest waste of all.

Find waste

We have all seen signs in factories saying 'Stop Waste' or 'Eliminate Waste'. These treat workers and staff as if they wouldn't stop waste if they saw it. The biggest issue is not that we don't stop waste but that we accept it as a normal part of the system and don't even think of it as waste.

We should change the signs to read 'Find Waste' to challenge all staff to find an area of wasted effort each week and to eliminate it on the spot. It is not actually that difficult if the emphasis is changed.

	Time (Minutes)
Elapsed time order is in the order processing system	
Elapsed time order is in the production system	
Total cost adding time	= X

Actual time taken to <u>process</u> the order. (Ignore waiting times and transport times)	Process 1:
	Process 2:
	Process 3:
	Process 4:
	Process 5:
Actual time taken to <u>produce</u> the order. (Ignore waiting times and transport times)	Process 1:
	Process 2:
	Process 3:
	Process 4:
	Process 5:
Total value adding time	= Y

Adding value time/ Adding cost time (%)	= X/Y(%)

Adding value – adding cost

Use the worksheet above to quickly calculate how much of the time you are adding value and how much of the time you are adding cost. Anything above 5% is very good. Anything above 10% and you probably haven't collected the information correctly. Reducing the cost-adding time will improve process throughput and customer responsiveness. It will also release cash back into the business.

5.10 Procurement for sustainability

Buying right

The introduction of a Materials Team (see Section 4.1) to reduce materials use in products and production is not intended to take over from procurement but to assist and strengthen the procurement role. As they have done in the past, procurement must continue to manage and develop suppliers effectively but they also have an important role in managing and achieving sustainability.

- **Tip** – Procurement and purchasing are often used interchangeably but they are different. Many companies focus on a reactive transactional approach (purchasing) when they should be developing a pro-active relationship approach (procurement). Sustainable procurement can help drive this transition.

The cost of the raw materials in typical plastics products will vary between 45% of the total cost for a 'technical' product and 80% for a 'mass produced' product but for the average processor, direct materials will be ≈ 55% of the cost of the product and the cost breakdown will be as shown in the diagram on the upper right. Managing the materials supply chain is a vital part of achieving sustainability.

What is sustainable procurement?

Sustainable procurement is the process of:

- Meeting the company's needs for goods and services.
- Achieving value for money on a whole-life basis.
- Achieving the most positive environmental, social and economic impacts at all phases of the product life cycle.

Sustainable procurement is not just about the purchases, it is also about the suppliers. Buying sustainable products from suppliers who do not conform to standards for environmental, social and economic performance is not sustainable procurement (see diagram on the lower right).

- **Tip** – Supplier social performance carries a huge reputational risk. To check this, search for 'Nike sweatshops': Nike have spent a lot time and money trying to improve social performance in sub-contractors but the issue still keeps coming back to haunt them.

Sustainable procurement does not simply consider the purchase price but also:

- The acquisition, use and end-of-life costs (the total cost of ownership).
- The risks and opportunities of ownership.
- The external environmental and social costs and benefits.

This is much broader than the simple 'lowest cost' purchasing approach that has traditionally been the case.

Sustainable procurement is, like most of sustainability issues and management systems (see Chapter 2), a process based on reducing risk and improving performance. This is a process of continual improvement where companies get aim to get better with

Companies implementing sustainable procurement can improve their reputation, develop better relationships with suppliers and still deliver cost savings.

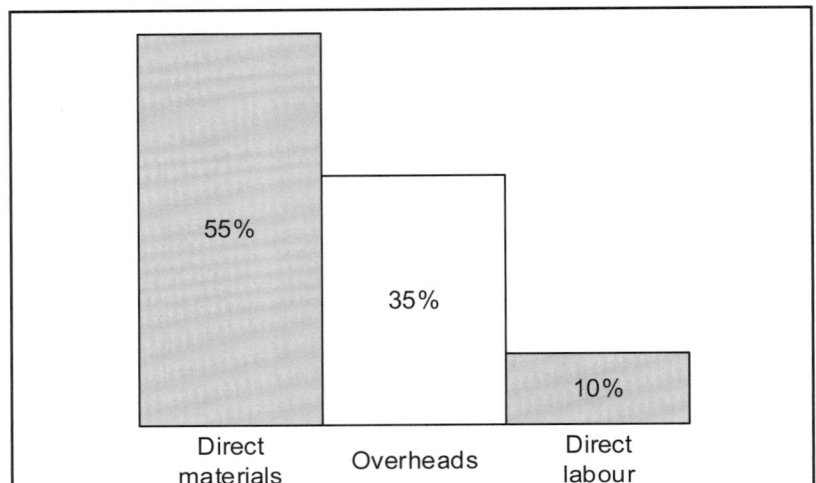

Most of the spend is on raw materials

Plastics processors spend ≈ 55% of their total costs on plastics raw materials and some of their spend on overheads will be on other raw materials, e.g., cleaning. Getting procurement right can reduce impacts and improve sustainability credentials.

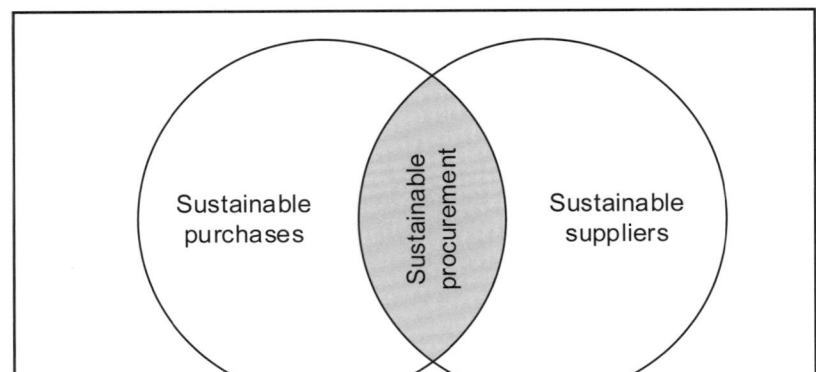

It is not just about the purchases

Sustainable procurement is not simply about sustainable purchases and products, it is just as much about dealing with sustainable suppliers who meet the all the social, economic and environmental sustainability requirements.

time by continually working with suppliers, customers and stakeholders.

- **Tip** – There is no finish line in sustainability issues.

Why sustainable procurement?

There are many reasons for implementing sustainable procurement and improving management of the supply chain. Some of these are:

- Consumer expectation and concern – there is a rising consumer expectation that companies are responsible for the activities of their supply chain and consumers are increasingly concerned about the ethics of companies. Failures in the supply chain can easily, and rapidly, lead to reputational and brand value damage.

- Financial markets expectations and concerns – it is not only consumers who have rising expectations, financial markets are also increasingly concerned about supply chains and ethics. Failures in the supply chain can have financing implications and financial institutions are increasingly concerned about sustainability issues.

- Customer pressure – customers are aware of the consumer and financial risks from failures in the supply chains. They are pushing sustainable procurement down the supply chain to reduce their exposure to environmental and social risks.

- Non-compliance risks – the rising amount of legislation, e.g., RoHS, REACH, WEEE and ELV, means that companies can be at direct financial risk from failures in their supply chain. Using sustainable procurement will reduce exposure to regulatory risks.

- Growing markets – sustainability is increasingly becoming a strategy for successful companies and they are searching for sustainable products and supply chains.

- Published targets – sustainable procurement is fundamental to meeting any published targets or commitments for improved sustainability.

Sustainable procurement is necessary because companies are only as sustainable as their supply chains and simply looking at 'internal' sustainability is no longer sufficient.

The benefits of sustainable procurement

Sustainable procurement should not be thought of purely as a 'risk avoidance' measure. In many areas sustainable procurement can bring direct institutional and financial benefits, some of these are:

- Strategic supply risks – sustainable procurement can be used to identify strategic risks in continuity and cost. Action can then be taken to avoid disruption and improve the security of supply. This is becoming increasingly important in some areas.

- Supply chain efficiency – sustainable procurement can reveal opportunities to improve collaboration with key suppliers in the supply chain and to reduce costs by working with the key suppliers.

- **Tip** – Implementing sustainable procurement is an opportunity for the procurement function to show leadership in a key strategic area of the business.

- Access to technology – improved supplier relationships can improve access to innovation and lower cost technologies and products. This can reduce costs and increase productivity through resource efficiency (see Section 3.3) and waste minimisation (see Chapter 9).

- Cost savings – using sustainable procurement can produce direct cost savings by reducing the external spend on raw materials. This means working with suppliers to:
  - Simplify specifications and make them performance-based.
  - Improve processes.
  - Minimise waste.

- **Tip** – Sustainable procurement doesn't have to increase costs. It can help with cost savings.

Without sustainable procurement, today's savings can easily be tomorrow's costs.

Companies need to look at:
- What their products are made from.
- Where they came from.
- Who made them.

For a longer discussion of supplier management and development, see Kent, R.J. 2018. 'Cost management in plastics processing', Elsevier.

Sustainable procurement can reduce negative environmental and social impacts at the same time as improving business value and resilience.

5.11   Implementing sustainable procurement

What to do now?

Implementing sustainable procurement is not a race and companies should not try to do everything at once. The process should begin by understanding the company's business and sustainability objectives, policies and priorities and then aligning the procurement process with these.

- **Tip** – The key to implementing sustainable procurement is to identify which issues you can take responsibility for and which you want to prioritise.
- **Tip** – This section focuses on sustainable suppliers rather than on sustainable purchases.

Map the supply chain

The first thing to do is to map the supply chain to gain an understanding of the suppliers. This can be as simple as generating a list of the main suppliers along with the relative spend with each supplier or using a process flow chart (see Section 9.3) to identify the main inputs to the process.

It can also be useful to develop a spend profile for the supplier, i.e., is it small/large and is it regular/irregular? Focusing on large/regular suppliers can reduce work at the start.

- **Tip** – Don't forget that this is not simply plastics purchases, i.e., it is all purchases.

Identify potential issues and risks

Sustainable procurement uses the seven core subjects of ISO 26000 (see Chapter 11) as a framework for assessing issues and the impact. The diagram on the opposite page shows how the core subjects can be used to assess and rate issues and impacts for a range of purchased goods and services. Not all the issues will be relevant for all suppliers but all purchasing categories should be assessed to a common framework.

The risks will also not be the same for every supplier in a purchasing category. The actual risk will depend on:

- The product and the application, e.g., food contact, toys.
- The relevant legislation, e.g., ROHS, REACH, etc.
- The specific supplier, e.g., spend, location and historical performance.
- The production processes and materials.
- The strategic importance of the purchased goods to the company.

Assessing the risks by supplier and product category allows the priority work areas to be identified and dealt with first.

Review current practices and suppliers

This is the ideal time to review the current procurement practices. This should include:

- Reviewing current contracts.
- Reviewing supplier standards and codes of conduct.

Specifications for sustainability

Implementing sustainable procurement is also the ideal time to review current specifications to optimise these for sustainability. This can include category specific initiatives but should include reference to the concept of Life Cycle Assessment (see Section 1.5).

- **Tip** – Look for cost savings/reductions that can be used to fund sustainable procurement.
- **Tip** – Sustainability criteria should also be written into all new contracts and renewals.

Supplier assessment

Suppliers are traditionally assessed in terms of quality/performance and sustainability adds to the complexity of supplier assessment (see Section 5.13). Initial assessment can be by self-assessment questionnaires, supplier visits and audits, independent NGOs or existing approvals, e.g., ISO 14001 and ISO 50001.

- **Tip** – Supplier assessments should cover all aspects of sustainable procurement but in the initial stages it may be easier to focus on the key issues with the highest impact.
- **Tip** – Sustainable procurement covers environmental, social and economic issues. Do not focus simply on environmental issues.

Check with key stakeholders

Sustainable procurement is never done in isolation and it is always driven by what the customer wants and achieved by what the suppliers can deliver. An open dialogue with the key stakeholders during the process will deliver the best results. Remember to "under-promise and over-deliver".

Develop a policy and strategy

At this stage, it is possible to fully develop the general sustainable procurement policy

When looking at the major suppliers, it is worthwhile asking them about their suppliers too.

Do not lose sight of the fact that sustainable procurement can be used to reduce costs as well as reduce risks.

The process can be used to re-boot procurement and save money.

Communication is the key to success. Talking with customers about what they want and suppliers about what they can deliver helps to set achievable targes and to reach them.

and process that will be rolled out to the supply chain. This may require different approaches based on the strategic priority and risks for each specific issue but the result should be a company strategy to deliver sustainable procurement.

Implement the policy and strategy

Implementation of the policy and strategy should start with a public statement of the policy to all suppliers (even those not directly affected in the early stages). Implementation should also consider issues such as:

- Compliance management – how compliance is to be managed in the future.
- Supplier management – how suppliers will be motivated to engage in sustainability (see sidebar) and how sustainability will be built into the existing supplier assessment.
- **Tip** – In some cases, e.g., large plastics suppliers, the supplier may well be in advance of the processor in sustainability. Why not engage with them and get them to help you?

Reporting

No programme is complete without some type of reporting. This requires:

- Setting the metrics that are going to be used.
- Setting the goals to be achieved (linked to the overall company sustainability goals).
- Setting the performance reporting system (see Chapter 12) with reference to goals, standards, traceability and legal requirements.

Motivate suppliers towards sustainable procurement by initiatives such as:

- Giving them more business.
- Training and guidance, e.g., newsletters, guidance publications and conferences.
- Rewards and recognition.
- Greater access to your site and people, including access to the results of your efforts.
- Establishing joint initiatives to improve sustainability.

	Raw materials (plastics)	Raw materials (packaging)	Raw materials (other)	Sub-contracted services	Logistics
Governance (ISO 26000: Clause 6.2)					
Human rights (ISO 26000: Clause 6.3)			Working conditions. Fair wages. No forced labour.	Working conditions. Fair wages. No forced labour.	Working conditions. Fair wages. No forced labour.
Labour practices (ISO 26000: Clause 6.4)	Healthy and safe work place. Staff development.	Healthy and safe work place. Staff development.	Healthy and safe work place. Staff development.		Healthy and safe work place.
Environment (ISO 26000: Clause 6.5)	Energy efficiency. Resource depletion. Granule escape.	Waste minimisation. Recycling.	Energy efficiency.	Cleaning and general materials safety.	Reduce CO_2 emissions. Improve efficiency.
Fair operating practices (ISO 26000: Clause 6.6)	No bribery or cartels in pricing. Fair competition.	Fair competition. Property rights (incl. intellectual).	No bribery or cartels in pricing. Fair competition.		No bribery or cartels in pricing. Fair competition.
Consumer issues (ISO 26000: Clause 6.7)	Materials data. Labelling.	Honest marketing. Accurate labelling. Food safety.	Materials data. Sourcing.		
Community involvement (ISO 26000: Clause 6.8)	Local suppliers. Local community initiatives.	Local suppliers. Local community initiatives.		Local suppliers.	Local suppliers.

High impact	Medium impact	Low/No impact

An example of the core subjects of ISO 20400 related to the major purchased goods and services

ISO 20400 uses the seven core subjects of ISO 26000 (see Chapter 11) as the core subjects for sustainable procurement. The core subjects can be related to the major categories of purchased goods and services and rated according to the potential impact. Not all the issues will be relevant for all suppliers and the table should be modified to reflect the relative potential impact. Start work on the highest impact areas first.

5.12   Local suppliers

It is not just about food

There is a rise in the awareness of local suppliers and 'food miles' in the food supply chain and many consumers are choosing local food in preference to food that has been transported large distances. This conversation is not as clear cut for plastics processing but still needs to take place with respect to all procurement. Sites should consider sourcing from local suppliers where this is possible.

One issue with local suppliers is the difficulty defining what is 'local'. Is it defined by country borders, by continental borders or by distance? A local supplier in the USA could be 200 miles away, but in Europe this distance could involve four separate countries where the languages and culture are very different and would not be regarded as local. The reality is that 'local' is whatever you want it to be and it depends on the nature and scope of the purchases.

Even for food, where the argument for local sourcing is clearer, there is often no easy answer, e.g., is food grown naturally in Spain and transported to the UK worse than food grown in the UK which needs heated greenhouses and fertilisers?

In most situations, there is a need for a quick LCA to assess the real benefits of local suppliers.

The rise of the global economy has made procurement from around the world far easier and this sometimes confuses the issue, e.g., if I purchase an office chair from a local supplier then the chair could well have been made anywhere in the world and could have travelled thousands of miles to the supplier.

There is still a natural desire to use local suppliers and using them will not only reduce transport emissions but will also help develop the local community and the company's ties to the community.

- **Tip** – Using local suppliers may reduce purchases in non-local developing countries and adversely affect their development.
- **Tip** – The global economy is not inevitable and rising trade barriers across the world may make a local supplier base an essential for security of supply.

The benefits of local suppliers

Using local suppliers can have many benefits and some of these are:

- Lower transport emissions and costs – local sourcing reduces transport emissions for the supplier to get goods to the site. Transport emissions for suppliers will appear in their carbon footprint (see Section 7.2) but this does not mean that processors should not consider the effect of their choices wherever the impact occurs. Reduced transport distances can also reduce transport costs.

- Increased supplier diversity – local sourcing increases supplier diversity and increases community involvement and development for the social aspects of sustainability (see Section 11.9).

- Improved supply chain transparency – local sourcing makes supplier assessment for sustainable procurement easier and makes it easier to audit and control suppliers (see Section 5.13).

- Reduced exposure to exchange rates and trade agreements – local suppliers will be invoiced and paid in the local currency. This removes any exposure to fluctuations in exchange rates and makes costs certain. The rise of the global economy may have increased material flows across the world but these are still dependent on trade agreements which can change and introduce uncertainty in tariffs and supply.

- Improved supplier dedication and service – local suppliers will tend to have improved dedication and customer service in a common language.

- Improved supplier control – local sourcing improves supplier control by making personal inspection of facilities and projects easier. This can improve the supplier's quality and responsiveness to changes.

- **Tip** – We once had an international supplier who provided regular reports on tooling progress. Suspecting something was amiss when they declared that the tooling was 90% complete, we made an unannounced flight to the supplier and found that they hadn't even started making the tooling. The process would have been much easier to control if they were local.

- Improved innovation in the supply chain – local suppliers will tend to be SMEs and support from other SMEs improves innovation and cooperation in the supply chain.

The debits of local suppliers

Working with local suppliers is not without potential issues and some of these are:

"Think globally, act locally"

A good slogan but the benefits must be evaluated and clear.

Apply the same sustainable procurement criteria to local suppliers as you do to all other suppliers (see Section 5.13).

- Technical capabilities – local suppliers will tend to be SMEs and this can mean that their technical capabilities are not as advanced as larger international suppliers. The decision on technical capability should never be over-ridden by a desire for local suppliers.

- Cost – small local suppliers may be more expensive than large international suppliers. Sites should assess the financial implication of local sourcing.

- Product range – local suppliers may be more flexible but may not have the range of products available to large international suppliers. Sites should assess any impact of a reduced product range on the local sourcing decision.

There will inevitably be products and services where local suppliers do not have the capabilities for the product or service. Companies should never allow the desire to use local suppliers override the needs of the business.

When to use local suppliers

There are many areas where choosing local suppliers is simply good business and most companies will intuitively make the right choice. Procurement of products and services such as general office supplies, cleaning services, maintenance services and other general services are natural choices for very local suppliers. Procurement for more specialised products and services will necessarily be less local.

- **Tip** – It is relatively easy to identify the areas with the most potential for local sourcing but you have to know who the local suppliers are to use them. This sounds obvious but some companies have no idea of the local resources.

For most plastics processing sites, the major spend will be raw materials and these are almost never local suppliers. The amount of procurement spend that can be directed to local suppliers is therefore limited but companies should still support local suppliers.

- **Tip** – The choice is yours, but never make the sourcing choice based on locality alone.

Taking action

In many cases, companies do not use local suppliers, even when they exist, because they have never tried to attract them or made efforts to engage with them.

Actions to improve the use of local suppliers are:

- Break down internal barriers to local sourcing.

- Actively encourage tenders from SMEs and local suppliers.

- Break contracts and requirements down into smaller lots to encourage small local suppliers.

- Advertise locally for products and services that can be sourced locally.

- Use your web site to encourage local suppliers to become suppliers.

- Manage the process to bring local suppliers into the business as partners.

Making a positive effort to increase the number of local suppliers and the amount of work given to them will reveal opportunities to improve the local community and sustainability.

Using local suppliers will increase the number of people who look upon the company as a resource for the community and as a good partner in the community (see Section 11.9).

A quick scan for local suppliers

Mapping the supply chain (see Section 5.11) will allow a quick scan of the supplier base. This can be done by simply running a report from the purchase ledger.

- List the supplier name, the total value of purchases for the year, the approximate distance travelled and whether this could be classed as local or not.

- The highest spend will be on raw materials and the opportunities will be further down the list.

- A quick scan of the list will reveal anomalies in terms of local suppliers.

- Look for the suppliers who are a long distance away and consider if there are local suppliers who can provide the same (or better) service.

5.13   Supplier assessment for sustainability

It is more than internal

Supplier assessment traditionally focuses on the ability of the supplier to provide the required goods and services but supplier assessment should also look outside of the actual product. This means looking at the broader aspects of the product and company such as the sustainability performance (social, economic and environmental). This is not simply about the sustainability of the product; it is also about the sustainability of the supplier. It is about how they meet their social and legislative obligations.

Materials data

Suppliers must ensure that raw materials do not contain any prohibited product, material or substance. In Europe, this is controlled by the REACH (Registration, Evaluation, Authorisation and restriction of Chemicals) system. In the USA, it is controlled by the American Toxic Substance Control Act (TSCA).

Suppliers should ensure that their products do not contain a 'critical or hazardous' substance and that suitable controls are in place.

Materials Safety Data Sheets (MSDS)

These have various names around the world but all serve as a method of proving compliance with REACH (or equivalent). Suppliers should be prepared to supply MSDS for any substance supplied.

An MSDS will provide vital information on:

- Identification of the substance/mixture and of the company/undertaking.
- Hazard identification.
- Composition/information on ingredients.
- First-aid measures.
- Fire-fighting measures.
- Accidental release measures.
- Handling and storage.
- Exposure controls/personal protection.
- Physical and chemical properties.
- Stability and reactivity.
- Toxicological information.
- Ecological information.
- Disposal considerations.
- Transport information.
- Regulatory information.
- **Tip** - Suppliers to the automotive industry also need to comply with the requirements of the European End of Life Vehicles (ELV) directive.
- **Tip** - Suppliers to the automotive industry may need to use the International Material Data System, which is the automobile industry's material data system.

Health and safety

All suppliers should ensure health and safety risks are reduced and that operations are carried out to the relevant regulations, approved codes of practice and industry best practice. All suppliers should show a commitment to health and safety and maintain effective policies and procedures.

- **Tip** - Ask about training (manual handling, fire safety, emergency response, etc.), risk assessment, personal protective equipment, improvement plans, accident incidence and their monitoring.
- **Tip** - Health and safety issues can shut down a supplier and prevent them from delivering to you so this is not simply about

Sustainability makes a real difference in terms of knowing whether you are dealing with a serious supplier or not.

Asking the questions will tell you a lot about a supplier's attitudes.

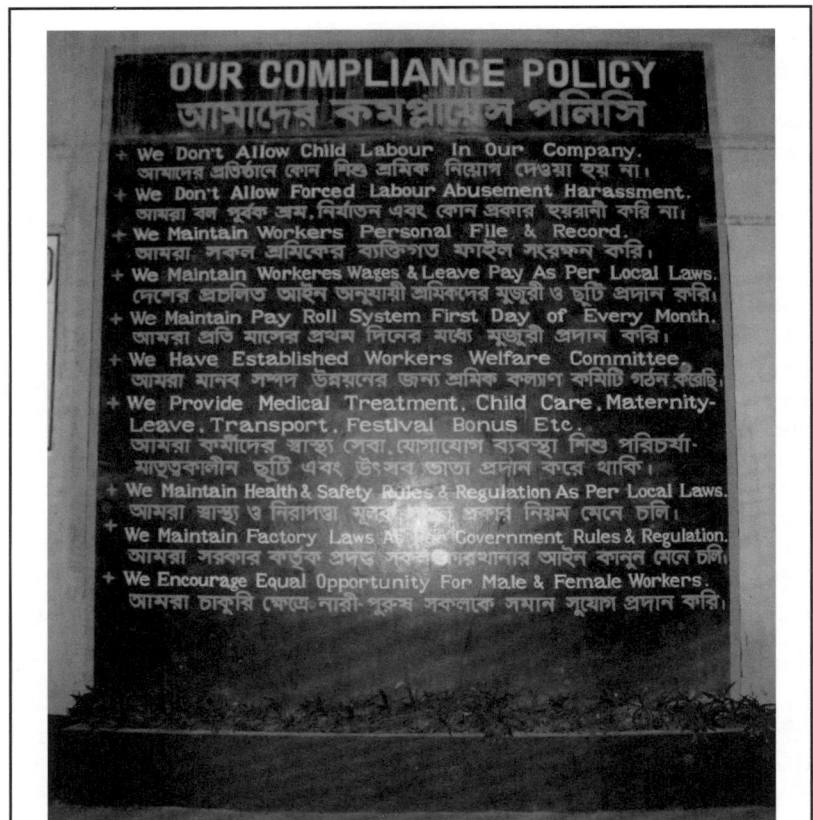

This Bangladeshi company is proud

The need for checking compliance may be higher in some countries than in others but this Bangladeshi company proudly states their compliance and displays it at their site. They also actually do all the things that they say they do. We checked!

them. After you have had a fire at a tool maker and lost 4 months on a project you take this sort of thing seriously.

Social

The social activities of a supplier are also important and all suppliers should:

- Obey the laws and regulations in all the countries in which they operate and/or sell.

- Not use child labour, i.e., those under the age of 16 or the country's legal minimum age, whichever is higher, and comply with the International Labour Organisation's provisions for the health, safety and morality of young people aged between 15 and 18.

- Not use forced or compulsory labour, i.e., any work which is forced upon any person under the threat of a penalty and which the person has not entered of his/her own free will.

- Ensure that employees understand their rights to payment of fair wages, overtime and retention of identity documents, etc.

- Ensure that working hours comply with the country's laws & regulations and international conventions, e.g., overtime work should be voluntary and paid.

- Comply with all applicable laws and regulations, including those relating to minimum wages, overtime hours and benefits as well as paying employees in a fair and timely manner.

- Not discriminate against any worker based on race, colour, age, gender, sexual orientation, ethnicity, disability, religion, political affiliation, union membership, national origin, social origin, or marital status.

- Respect the right of workers to associate freely, form and join worker's organisations, seek representation and bargain collectively, as allowed by the applicable laws and regulations.

Economic practices

Fair business practices are an essential to building a relationship with suppliers and are a good indication of how they will treat not only their customers but also their suppliers. All suppliers should:

- Have processes and procedures in place to prevent corruption, bribery and extortion. They should not, directly or indirectly, offer, give, demand or accept any bribe to obtain or retain business.

- Have internal controls, ethics and compliance programmes to prevent and detect bribery, e.g., financial and accounting procedures to ensure transparent and accurate accounts.

- **Tip** - In days gone by some suppliers offered substantial 'gifts' or 'reward programmes' to purchasing departments for business. This is no longer acceptable and everybody should know this (see Section 11.7).

- Have processes and procedures in place to ensure fair competition, e.g., no cartels.

- Have processes and procedures in place to identify, reveal and avoid conflicts of interest.

Environment

Environmental issues are of growing importance for any company and all suppliers should:

- Comply with all relevant environmental laws and regulations.

- Have and publish an environmental policy.

- Take all possible measures to protect the environment and minimise the effect of their products on the environment during the whole of the product life cycle.

- **Tip** - Great suppliers will have an Environmental Management System that meets ISO 14001 (see Section 2.6).

- Calculate their carbon footprint and take all possible measures to reduce this.

- **Tip** - Great suppliers will have an Energy Management System that meets ISO 50001 (see Section 2.10).

Suppliers to the food industry may already have met this in the Walmart™ 'Sustainability Index' where suppliers have to answer 16 questions on sustainability. They are serious about the sustainability efforts of their suppliers.

If you think that 'business ethics' is an oxymoron along the lines of 'military intelligence' then perhaps you have bought the wrong book.

Ethics is fundamental to business operation. It is not a county to the North-East of London, that is 'Essex'.

A good source for things to look for in the social, economic and environmental areas of sustainability is ISO 26000 (see Chapter 11).

This covers social responsibility for plastics processors but is equally applicable for assessing suppliers.

For a longer discussion of supplier assessment and quality improvement, see: Kent, R.J. 2016. 'Quality management in plastics processing', Elsevier.

5.14 Sustainable procurement – ISO 20400

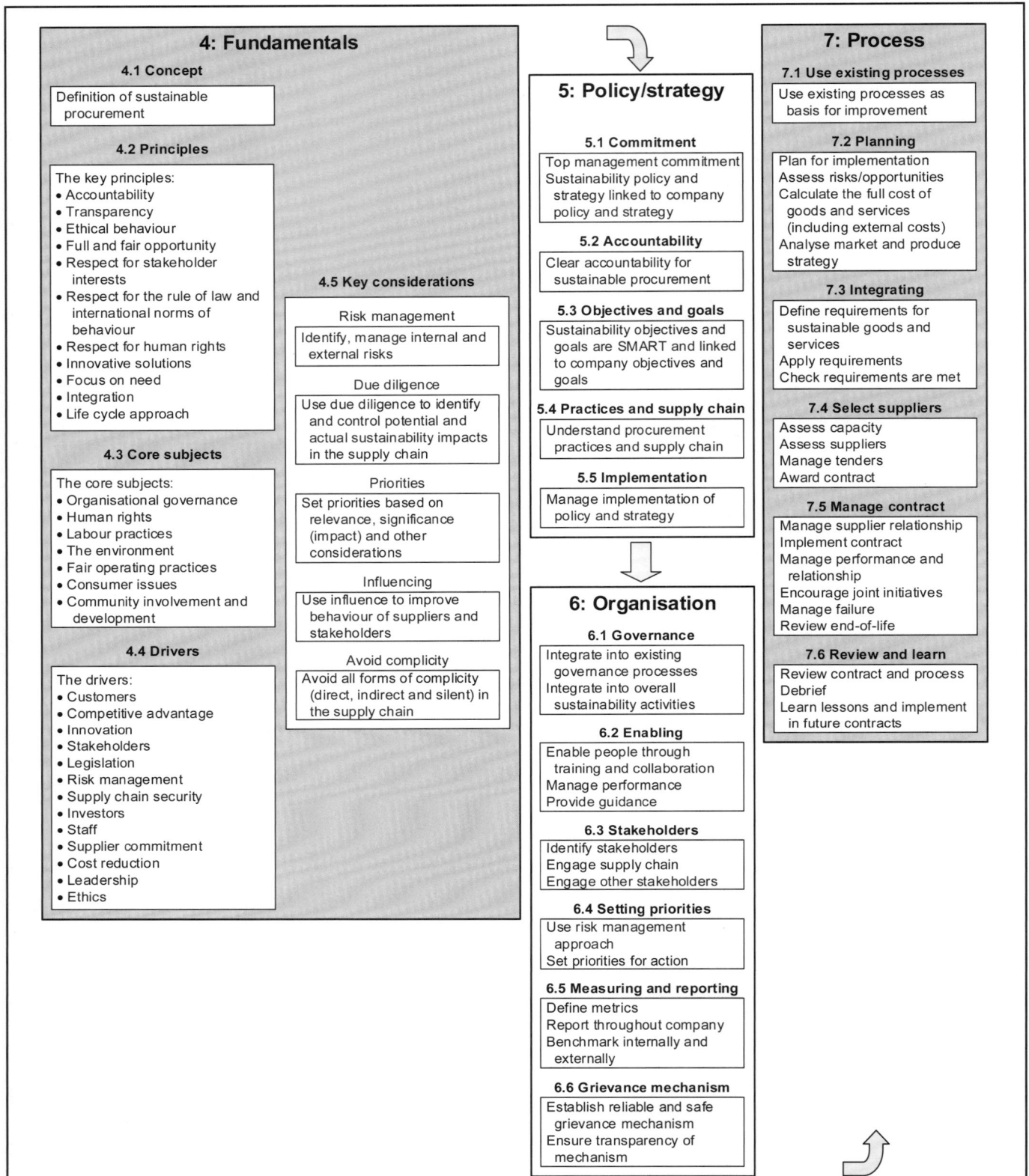

4: Fundamentals

4.1 Concept

Definition of sustainable procurement

4.2 Principles

The key principles:
- Accountability
- Transparency
- Ethical behaviour
- Full and fair opportunity
- Respect for stakeholder interests
- Respect for the rule of law and international norms of behaviour
- Respect for human rights
- Innovative solutions
- Focus on need
- Integration
- Life cycle approach

4.3 Core subjects

The core subjects:
- Organisational governance
- Human rights
- Labour practices
- The environment
- Fair operating practices
- Consumer issues
- Community involvement and development

4.4 Drivers

The drivers:
- Customers
- Competitive advantage
- Innovation
- Stakeholders
- Legislation
- Risk management
- Supply chain security
- Investors
- Staff
- Supplier commitment
- Cost reduction
- Leadership
- Ethics

4.5 Key considerations

Risk management

Identify, manage internal and external risks

Due diligence

Use due diligence to identify and control potential and actual sustainability impacts in the supply chain

Priorities

Set priorities based on relevance, significance (impact) and other considerations

Influencing

Use influence to improve behaviour of suppliers and stakeholders

Avoid complicity

Avoid all forms of complicity (direct, indirect and silent) in the supply chain

5: Policy/strategy

5.1 Commitment

Top management commitment
Sustainability policy and strategy linked to company policy and strategy

5.2 Accountability

Clear accountability for sustainable procurement

5.3 Objectives and goals

Sustainability objectives and goals are SMART and linked to company objectives and goals

5.4 Practices and supply chain

Understand procurement practices and supply chain

5.5 Implementation

Manage implementation of policy and strategy

6: Organisation

6.1 Governance

Integrate into existing governance processes
Integrate into overall sustainability activities

6.2 Enabling

Enable people through training and collaboration
Manage performance
Provide guidance

6.3 Stakeholders

Identify stakeholders
Engage supply chain
Engage other stakeholders

6.4 Setting priorities

Use risk management approach
Set priorities for action

6.5 Measuring and reporting

Define metrics
Report throughout company
Benchmark internally and externally

6.6 Grievance mechanism

Establish reliable and safe grievance mechanism
Ensure transparency of mechanism

7: Process

7.1 Use existing processes

Use existing processes as basis for improvement

7.2 Planning

Plan for implementation
Assess risks/opportunities
Calculate the full cost of goods and services (including external costs)
Analyse market and produce strategy

7.3 Integrating

Define requirements for sustainable goods and services
Apply requirements
Check requirements are met

7.4 Select suppliers

Assess capacity
Assess suppliers
Manage tenders
Award contract

7.5 Manage contract

Manage supplier relationship
Implement contract
Manage performance and relationship
Encourage joint initiatives
Manage failure
Review end-of-life

7.6 Review and learn

Review contract and process
Debrief
Learn lessons and implement in future contracts

An overview of the ISO 20400 guidance

ISO 20400 provides excellent guidance on how to define, implement and manage sustainable procurement and this diagram shows the main recommended activities. The structure of the standard is logical and it provides considerable guidance in choosing how to approach and implement sustainable procurement. The standard does not conform to Annex L (see Section 2.1) because it is a 'guidance' standard and not a 'requirements' standard.

Great guidance

Sustainable procurement has naturally made it into the standards world and ISO issued ISO 20400 "Sustainable procurement – Guidance" in 2017 to provide guidance in the area of sustainable procurement. This follows the outline of the previous BS 8903: 2010 and can be applied to any organisation or stakeholder affected by procurement processes. It provides an understanding of what sustainable procurement is and how sustainability affects a range of procurement issues but best of all it provides good guidance on implementing sustainable procurement and driving this down the supply chain.

- **Tip** – ISO 20400 is one of the better ISO standards, it is clear, concise, well-written and gives loads of good ideas for implementing ISO 20400.
- **Tip** – ISO 20400 does not replace any other legislation or requirements. It is guidance for sustainable procurement.
- **Tip** – Sometime in the future you will be asked about this by your customers. Getting ready now and implementing sustainable procurement will prepare you for that phone call.

An outline of the structure of the standard and what it expects is shown on the opposite page and this shows the four main elements of the standard.

Fundamentals

This is designed to be read by everybody concerned with sustainable procurement and covers:

- The concept of sustainable procurement.
- The key principles.
- The core subjects – these are the same as the core subjects of ISO 26000 (see Chapter 11) and these are also linked to the GRI standards for reporting (see Chapter 12).
- **Tip** – Sustainable procurement to ISO 20400 fits easily into the rest of the sustainability agenda.
- The drivers for sustainable procurement – these will vary by company but some will be present in every company.
- The key considerations – these provide a general framework for good practice in sustainable procurement.

Policy and strategy

This is designed for top management and covers how sustainable procurement can be implemented at the company level via top management commitment and the setting of procurement policies and strategies that align with the overall company policies and strategies.

- **Tip** – Unless sustainable procurement policies and goals are aligned with the company policies and goals then it won't happen.

Organisation

This is designed for procurement managers who are responsible for organising the procurement function and establishing the framework for sustainable procurement. It covers the organisational requirements to control and manage procurement, e.g., setting priorities for action (see Section 5.11).

- **Tip** – Getting the organisation and conditions right is essential for sustainable procurement.

Process

This is designed for the people who are responsible for carrying out the procurement and covers the actual management and operation of sustainable procurement. This is the largest section of the standard and gives guidance on all areas from planning to supplier assessment (see Section 5.12) and managing sustainable contracts.

Performance needs to be managed and the standard focuses on the specification and the initial contract conditions to provide the basis for managing the supplier relationship.

- **Tip** – Whilst designed for the actual procurement process, this is, in my opinion, the most important part of the standard.

It is guidance

Just as ISO 26000 is a 'guidance' standard (see Section 11.2), ISO 20400 is a 'guidance' standard, i.e., it is not a requirements standard like ISO 14001. This means an organisation cannot be certified for compliance but can be evaluated and/or advised by a competent third party. A company can only say that it "follows the guidance of ISO 20400".

- **Tip** – If a company says it is 'certified to ISO 20400' then maybe it tells you something about the company (and not in a good way).

Spot what is wrong in this press release:

"ASUS today announced that the company received the ISO 20400:2017 certification for responsible practices in sustainable procurement from SGS, a leading global inspection, verification and certification firm".

www.asus.com/ News/ nxpsitxmff8lwhpe 15/09/2020

A great source of information, checklists, videos, self-assessment tools and everything you ever needed to know about implementing ISO 20400 is www.iso20400.org/.

5.15 Sustainable procurement – where are you now?

Good procurement reduces impacts

The high value of raw materials in relation to turnover makes procurement a key area in reducing the sustainability impact of plastics processors. There is a high potential reputational risk from poor procurement and easy wins can be made in this area.

Sustainable procurement also raises the profile of procurement professionals and gives them an added skill to help the company prosper in social, environmental and economic terms. Their skills in integrating suppliers into the Materials Team, helping them to improve sustainability performance and ensuring that purchased goods and services have low impacts can ensure that the actual price paid is both fair and reasonable.

Procurement is not simply about prices, it is also about managing the product life-cycle, getting specifications and contracts defined so that the supplier has a chance to reduce both impacts and prices.

Completing the chart

This chart is completed and assessed as for the previous charts.

Invest in the development of the 'soft' purchasing skills to add value to the process rather than to simply drive prices lower.

This is a skilled profession and should be treated as such. If you want simple 'buyers' then be prepared to pay the cost and accept the inevitable reputational and other risks.

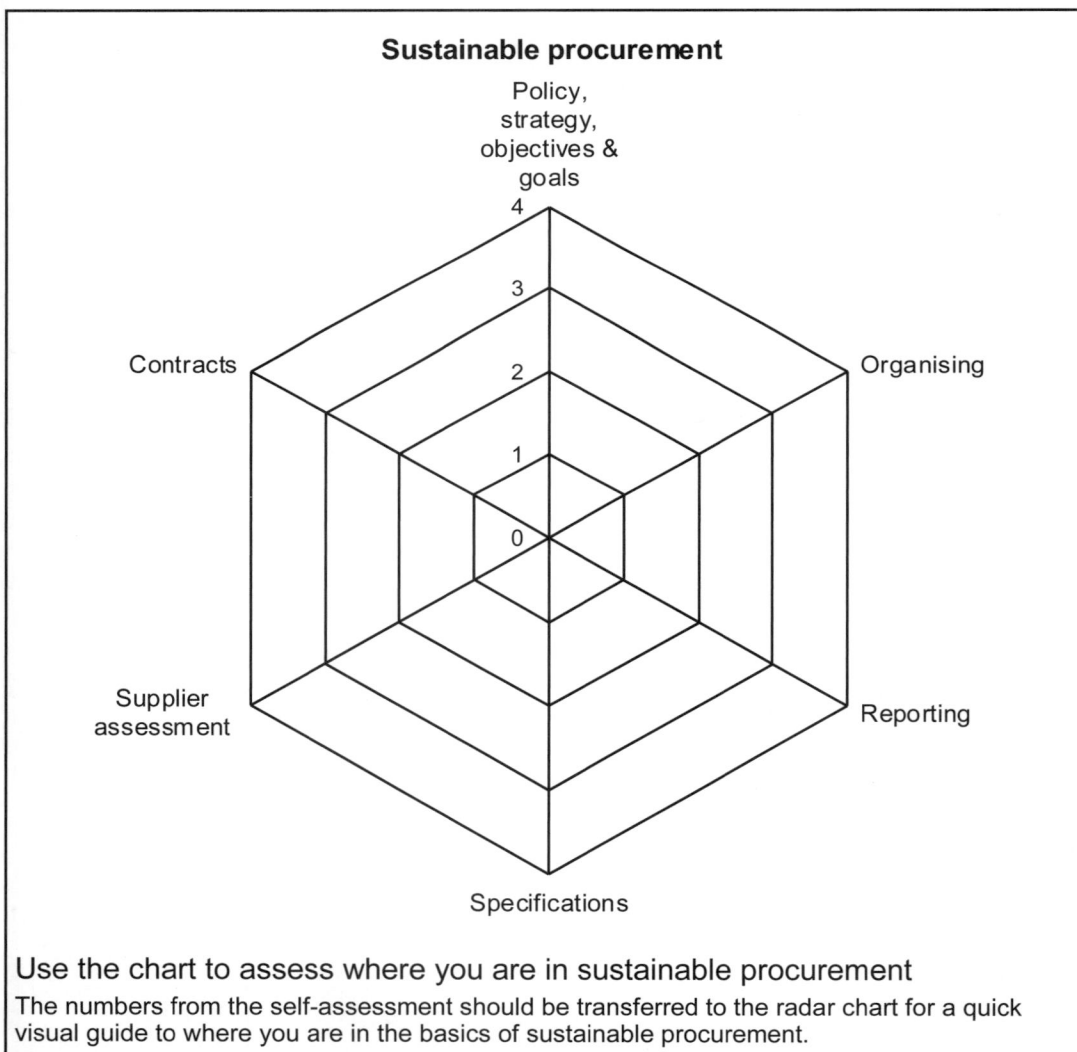

Sustainable procurement

Use the chart to assess where you are in sustainable procurement

The numbers from the self-assessment should be transferred to the radar chart for a quick visual guide to where you are in the basics of sustainable procurement.

Capitalism may be short sighted but it is not blind.

Every customer needs good suppliers and must be prepared to pay for their continued existence.

			Sustainable procurement			
Level	Policy, strategy & goals	Organising	Reporting	Specifications	Supplier assessment	Contracts
4	Policy, strategy, objectives & goals available. Consistent with company policy, strategy, objectives & goals.	Procurement integrated into sustainability activities. All suppliers involved in process. Priorities set for improvement.	Good metrics consistently used. Reported widely. Internal & external benchmarks used.	Clear & concise specifications set with supplier to use supplier's skills. Specification includes sustainability requirements.	Supplier assessment for all sustainability issues, e.g., environmental, social & economic. Reputational risk minimised.	Cost based on life-cycle costing. Supplier treated as partner & relationship managed well. Contract review looks at overall performance.
3	Policy & strategy available but no objectives or goals set. Policy &strategy are consistent with company policy & strategy.	Procurement integrated into sustainability activities. Most suppliers involved in process. Some priorities set for improvement.	Good metrics consistently used. Internal reporting only. Internal benchmarks only.	Good specifications but little use of supplier's skills. Specification includes sustainability requirements.	Supplier assessment for most sustainability issues, e.g., social & environmental. Low reputational risk.	Cost based on total cost of ownership. Supplier treated as partner & relationship managed well. Contract review looks at overall performance.
2	Objectives & goals set but no policy or strategy available. Objectives & goals are consistent with company objectives & goals.	Some integration into sustainability activities. Some suppliers involved in process. Some priorities set for improvement.	Some consistent metrics used. Internal reporting only. Internal benchmarks only.	Specifications exist but considerable room for improvement in use of supplier's skills. No consideration of sustainability issues.	Supplier assessment for limited sustainability issues, e.g., social only. Moderate reputational risk.	Purchase cost assessment only. Supplier treated as partner with good relations. Contract review looks at overall performance.
1	Policy, strategy, objectives & goals available. Inconsistent with company policy, strategy, objectives & goals.	Little integration into sustainability activities. Suppliers not involved in process. Few priorities set for improvement.	Some metrics but inconsistent use. No reporting. No benchmarks used.	Poor & ambiguous specifications arbitrarily imposed on suppliers. No consideration of sustainability issues.	Supplier assessment with no focus on sustainability issues. High reputational risk.	Purchase cost assessment only. Adversarial contract handling with poor relations. Poor contract review.
0	No policy, strategy, objectives or goals available.	Procurement seen as separate to sustainability. No engagement with supply chain. No priorities for improvement set.	No metrics available. No reporting used. No benchmarks used.	Specifications are non-existent or vague. Frequent disputes with suppliers over standards & no consideration of sustainability issues.	Supplier assessment not carried out. Purchase cost assessment only. Very high reputational risk.	Purchase cost assessment only. Poor contract handling process, supplier is the 'enemy'. No contract review process.
Score						

5.16 Sustainable distribution – the method

The transport scene

At the global level, overall transport emissions are 25-30% of total CO_2 emissions but are higher in terms of total CO_2e (see Section 7.1) due to higher NO_x and SO_x emissions from diesel vehicles used for freight and public transport. In terms of oil use, transport use is higher due to the lack of gas use in transport applications. Transport emissions can be divided according to the type of transport and this is shown in the diagram on the lower right. This shows that the highest proportion of transport emissions are the result of private vehicles ($\approx$ 42%) and that freight vehicles represent $\approx$ 22% of transport emissions.

- **Tip** – Product transport is also responsible for emissions in other transport methods, e.g., light commercial vehicles, rail and air transport, but most plastics products are transported by freight vehicles.

Improving the impact of the transport sector is vital for sustainability and general transport technology is improving rapidly, e.g., electric vehicles, but this is primarily restricted to private vehicle developments. This does not mean that the industry can ignore the sustainability benefits of improving the efficiency of product distribution and there are many simple actions that can be taken.

Transport emissions

For most plastics processing companies, transport will be $\approx$ 10% of the carbon footprint (see Section 7.4). A carbon footprint component of $\approx$ 10% may appear relatively small but this is largely due to the very high energy use in plastics processing rather than to any specific good practice in the industry. Another factor is the relative proximity of most processors to their customers – plastics processing sites tend to be small and close to their customers and this also reduces the transport emissions and costs.

- **Tip** – If customers collect products then they are NOT included as part of the carbon footprint for the site. However, customer collection is relatively rare in the sector.

It is unusual for plastics processors to have their own transport fleet and most will use contract transport for customer deliveries (even if it is 'badged' with the processor's name). This means that most of the carbon footprint from transport will appear as Scope 3 emissions.

Transport methods and emissions

The relative transport emissions for a range of transport methods are shown in the diagram on the opposite page. This is expressed in terms of kg CO_2e/tonne.km, i.e., the amount of CO_2e emitted from transporting a tonne of product for 1 km and is based on the data given in DBE&IS 2020 (see Section 7.2).

- **Tip** – CO_2e is particularly important for much of the transport used for plastics products due to the high proportion of diesel-based transport used. CO_2e includes not simply the CO_2 emitted by the transport but also the NO_x and SO_x emissions.

- **Tip** – Processors need to understand the emissions of the various potential transport methods so that they can choose the most sustainable method of product transport.

Heavy goods vehicles (HGV)

HGVs are the most common transport method for plastics products throughout the world. The carbon impact of distribution by HGV is assessed as part of the processor's Scope 3 carbon footprint but the widespread use of contract transport means that the processor is not fully in control of the condition of the vehicle used, the route taken or methods used to maximise the backhaul capacity of the network. Companies should

The emissions from HGVs have decreased from 0.129 kg CO_2e/tonne.km in 2011 to 0.107 kg CO_2e/tonne.km in 2020. This is a 17.5% improvement due to vehicle improvements.

Monitor and target HGV utilisation variables such as:
- Tonne-kilometre/vehicle/year.
- Vehicle load factor.
- Space utilisation.
- Empty running.

Simple M&T measures can provide an insight into measures to reduce transport emissions.

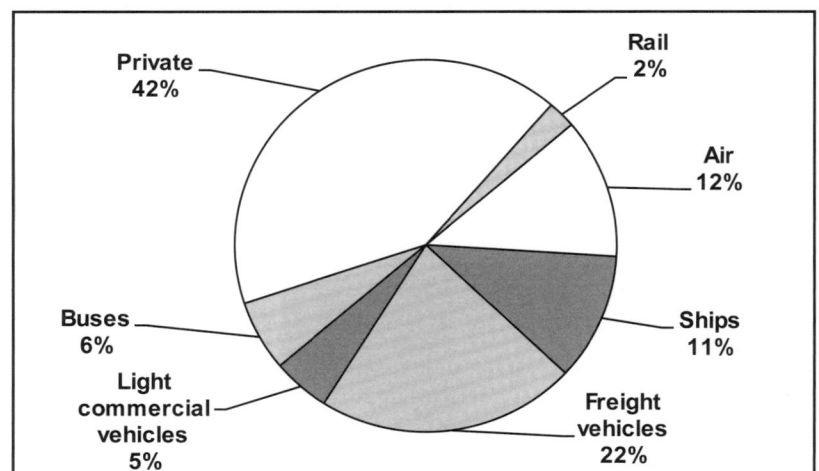

Transport emissions (CO_2) by sector for 2019

The highest proportion of transport emissions are from private vehicles with freight emissions only 22% of the total. Most plastics products will be transported by land freight.

Source: Energy Technology Perspectives, www.iea.org/reports/energy-technology-perspectives-2020

work with their transport contractors to improve the efficiency of HGV transport in areas such as:

- Improving routing and scheduling to reduce the transport distance.
- Improving vehicle aerodynamics to reduce air resistance losses.
- Limiting the speed of vehicles to reduce air resistance losses.
- Reducing tyre losses.
- Improving vehicle maintenance.
- Improving management controls.
- Maximising the backhaul capacity to avoid empty running.

These actions can reduce transport emissions by over 20% by both reducing the distance travelled and improving vehicle efficiency.

Vans

Small vans are not widely used to transport plastics products except for limited deliveries to customers and this is fortunate as a 3.5 tonne van (the most common small delivery vehicle) has a carbon intensity (in tonne.km) nearly 6 times that of a HGV.

- **Tip** – Avoid the use of dedicated small vans for product transport unless required for urgent local deliveries.
- **Tip** – Sites should check if small vans are used and take action to consolidate loads internally or to use freight companies who can consolidate loads to reduce emissions.

Air freight

Air freight is the most carbon intensive transport method and it is also the most expensive transport method. The cost is the primary reason that air freight is rarely used for commodity plastics products. However, it is sometimes used for low-volume, high-value products where the processor is located some distance from the customer, e.g., medical components.

- **Tip** – Air transport should only be used for very low-volume, high-value products where urgent delivery is required and there is no alternative transport method.
- **Tip** – Sites should eliminate or minimise the use of air freight by encouraging customers to order sufficiently in advance to allow alternative delivery methods.

Rail

Rail is a very sustainable transport method and has low emissions but it is not widely used for plastic product transport (although some processors in the USA accept bulk polymer deliveries via the rail network). Rail transport is primarily aimed at high-volume and heavy products such as steel or similar

products. Some extrusion companies, e.g., pipes, occasionally use rail transport for long distance delivery where the schedule allows but even in these cases the majority of product is transported by HGVs.

Ship

Ship transport via containers has low carbon emissions but is not widely used in the plastics processing industry even though it is suitable for inter-continental product transport where the demand can be accurately predicted.

- **Tip** – Rail and ship transport both have very low emissions in terms of kg CO_2e/ tonne.km, but do not forget that most plastics products will also need road transport by HGV to rail terminals or docks and road transport by HGV for the 'last mile' to the customer.

The future

Transport emissions are critical for sustainability although HGVs will probably continue to be the main distribution method. New technology such as electric vehicles is unlikely to have an impact at the HGV scale but hydrogen powered HGVs represent a possible route for the future.

Improved vehicle loading can be achieved by measures such as:

- Increasing the backhaul (return) loading.
- Increasing the limits on carrying capacity. Most plastics products will reach the maximum volume before they reach the maximum load.
- Using space-efficient systems and packaging (see Section 5.17).
- Implementing transport-efficient order cycles, i.e., delivering on a nominated day.
- Sharing vehicle capacity.

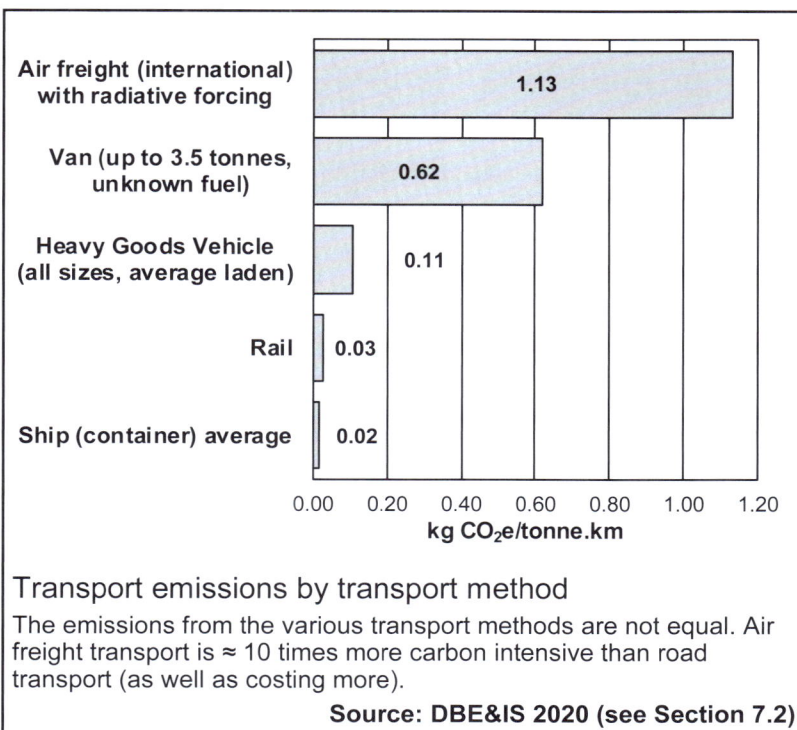

Transport emissions by transport method

The emissions from the various transport methods are not equal. Air freight transport is ≈ 10 times more carbon intensive than road transport (as well as costing more).

Source: DBE&IS 2020 (see Section 7.2)

5.17 Sustainable distribution – improving vehicle use

Pack it in and reduce the impact

In most cases the distribution method (see Section 5.16) will be determined primarily by the customer demands although processors need to understand the relative impacts of each distribution method to be able to discuss these with the customer and potentially change the method to reduce the impacts.

Most plastics products are transported using HGVs and this makes improving vehicle use the most effective method that plastics processors can use to improve the sustainability of their distribution supply chain. These actions will not only reduce emissions and resource use but will also reduce distribution costs.

- **Tip** – This section should be read in conjunction with Sections 8.6 to 8.9 which deal with minimising waste in packaging, which can also contribute to sustainable distribution.

- **Tip** – Whilst this section is focused on improving sustainability in distribution from the processor to the customer, the concepts can also be used with suppliers to reduce their impacts. Do not lose the opportunity to use the lessons learnt to help them reduce their distribution impacts.

The programme to improve vehicle use

Vehicle use can be improved by a structured programme as shown in the diagram on the right.

Stage 1 – Minimise the demand

Minimising the demand is the first stage in the programme because it is not logical to attempt to optimise the supply (use) until the demand is minimised.

Step 1: Reduce the distance travelled

Advances in vehicle telematics (an inter-disciplinary field covering vehicles and optimising their use) mean that vehicle routes can be controlled and programmed to minimise the distance travelled and the time idling due to traffic issues.

- **Tip** – Most people now use 'satnavs' (or GPS in some parts of the world) to plan our driving routes to minimise the distance travelled or the driving time. Telematics is simply an extension of this technology.

Telematics takes the idea of a satnav further by optimising the potential for load sharing

with other companies or backhaul to reduce empty running (see below). Reducing the distance travelled will make an immediate impact on the sustainability of distribution.

- **Tip** – Work with transport contractors to use telematics to schedule and route vehicles efficiently and be prepared to adjust delivery schedules to so they are transport-efficient.

In some sectors of the industry, e.g., PET and HD-PE bottles, the introduction of 'hole-in-the-wall' sites has effectively reduced the distance travelled to zero. This has eliminated the impact, a far better result than reducing it.

- Tip – Innovation can reduce the distance travelled and provide other logistical benefits.

Step 2: Reduce the amount transported

Reducing the amount transported (in mass terms) is not generally possible for plastics processors due to the light-weight nature of the product.

- **Tip** – The light-weight nature of plastics products may make 'load sharing' with other companies possible.

Stage 2 – Optimise the supply

After the demand is minimised then work can begin on optimising the supply of the service.

Sustainable distribution covers all aspects of the process for getting products from the manufacturer to the customer. It is about packaging design, storage, transport methods and collecting returnable packaging.

Global trading has increased the number of tonne.km faster than GDP and this is seen even inside areas such as the EU or the USA where HGVs are the main transport method. This leads to rising transport impacts in sustainability terms.

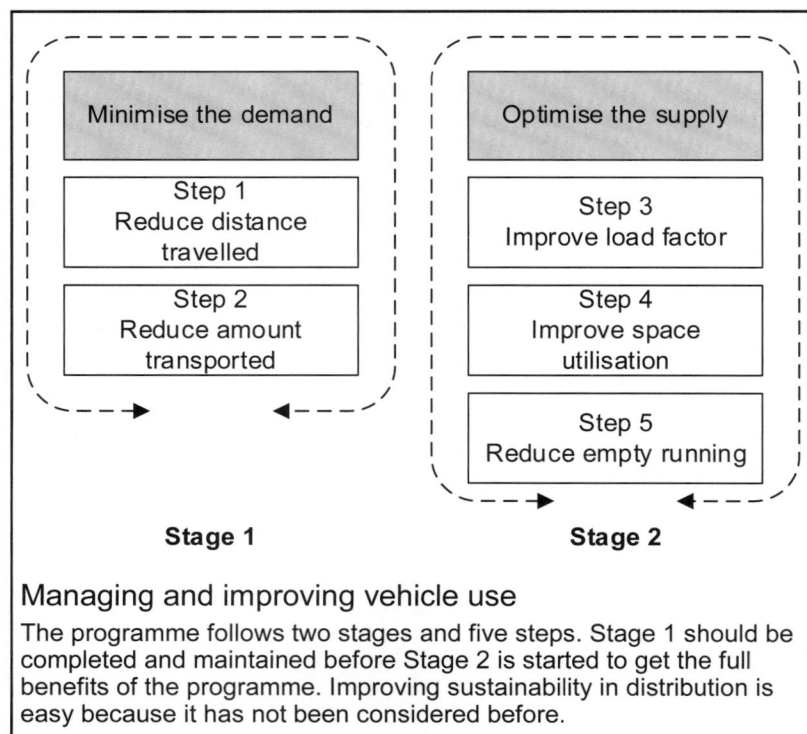

Stage 1

Minimise the demand

Step 1
Reduce distance travelled

Step 2
Reduce amount transported

Stage 2

Optimise the supply

Step 3
Improve load factor

Step 4
Improve space utilisation

Step 5
Reduce empty running

Managing and improving vehicle use

The programme follows two stages and five steps. Stage 1 should be completed and maintained before Stage 2 is started to get the full benefits of the programme. Improving sustainability in distribution is easy because it has not been considered before.

Step 3: Improve the load factor

The vehicle load factor is the ratio of the actual weight of the load to the load that could have been carried if the vehicle were to be fully loaded. For most plastics products the light-weight nature of the products means that the weight of the load will rarely approach the load carrying capacity. This will be so even if the vehicle is volumetrically 'full'.

Step 4: Improve space utilisation

This is a key factor for plastics products and most vehicles will become 'full' in volume terms before any weight limits have been reached. In many cases, this will not mean that the vehicle is actually volumetrically full, simply that no more products can be loaded. This may be due to:

- Poor use of the floor area of the vehicle.
- Limits on the stacking height of the existing packaging.
- A failure to use the full height of the vehicle.

Fully utilising the whole space of the vehicle can dramatically increase loading and reduce the number of trips to get the same amount of product to the customer.

- **Tip** – Check the volumetric loading of all vehicles to ensure that the vehicle is full and that the space utilisation is acceptable.
- **Tip** – Consider adjusting pack sizes, e.g., using modular heights, or specifications to allow greater stacking heights and to increase volumetric loading. This may mean increasing the quality of the transit packaging or changing the transit packaging method, e.g., cages with flexible packs instead of cardboard packaging. If returnable packaging is used then it will, of necessity, be sturdier than single-trip packaging and allow greater stack heights.
- **Tip** – Consider alternative loading methods for products, e.g., can bulk loading be used instead of cartons?
- **Tip** – If the product is not subject to transit damage then consider moving to flexible packaging to increase the packing density.
- **Tip** – Consider if the product can be safely 'nested' to reduce the volume being transported and increase space utilisation.
- **Tip** – Any of the methods chosen to increase space utilisation must still provide adequate protection for the product.

Optimizing transit packaging materials and their design can considerably increase space utilisation and increase load stability.

Working with the transport contractor in this area can help to deliver the maximum amount of product delivered for each kilometre travelled to reduce CO_2 emission and costs.

Step 5: Reduce empty running

Empty running of vehicles produces emissions for no productive or economic benefit and can be up to 30% of the distance travelled by vehicles. Decreasing empty running will not only provide economic benefits but also environmental benefits. This makes it an ideal opportunity for both companies and freight contractors to work together for mutual benefits. Companies should work with their freight contractor to:

- Collect reclaimed packaging for re-use (see Section 9.8).
- Collect any unsold or end-of-life products for recycling or re-use.
- Allowing vehicle capacity sharing during returns – in the 1980's our profile delivery vehicles collected used oil on their return trips for treatment and use as a fuel in the Ffestiniog Railway steam trains. Reduced emissions, recycling and a social benefit all in one.

Set some targets

Typical targets could be:

- Reduced vehicle mileage – set a target for % decrease in miles/employee/year or miles/kg of product/year.
- Reduced fuel use – set a target for % decrease in litres/employee/year.
- Distribution fuel efficiency for goods – set a target in % decrease in litres/tonne-km.

Using resource efficient design

Resource efficient design (see Chapter 3) should also consider distribution and storage. Designers should look at:

- The method and length of the distribution chain – can the product be made closer to the point of use, e.g., 'hole-in-the-wall, and can it be made in response to demand, i.e., minimum stock?
- The type of packaging currently used – can returnable and re-usable packaging be used instead of single trip disposable packaging? (see Section 9.6).

Designers should ask the dispatch department, drivers and customers for areas and ideas to reduce packaging use, implement re-usable packaging or packaging with a lower environmental impacts and costs (see Section 9.6).

The International Road Transport Union (www.iru.org) has good resources on improving vehicle use through driver training.

"Unlike passengers, who usually return to their starting point, most freight travels only in one direction."

Alan McKinnon

Empty running can be up to 30% of the distance travelled by vehicles.

5.18   Greening the office

Small things

Every plastics processor will have some office functions and these will naturally impact on the sustainability of the company. For a large commercial office, the office functions are very important and the major area for sustainability action. For a plastics processor these are less important because the impact of office functions will be small in relation to the impacts of the production area. Despite this, the psychological importance of greening the office should not be underestimated. Sustainability is a company-wide issue and involving the office functions is essential to gain acceptance.

- **Tip** – The office will be one of the first things that staff focus on even if the impact is small. Accept this and start work in the offices to improve sustainability.

Management

The managers responsible for implementing sustainability programmes will be based in offices and it is appropriate that they should set an example and contribute to sustainability. They need to show commitment and to promote activities to improve sustainability.

Energy

Most of the energy used in offices will be for heating/cooling, ventilation and lighting (≈ 75%) and these are considered in Section 6.19. The remaining energy use (≈ 25%) will be in general office equipment, computers (including servers) and printers/copiers. A typical office computer will draw between 50 and 250 Watt depending on the type of computer and monitor. At 250 Watt, this is a total cost of ≈ £50/year/computer.

Most of this equipment is rarely switched off after it is switched on in the morning but there are a range of simple measures that can be taken to reduce energy use:

- Make sure that all office equipment is switched off rather than left in standby mode. Electrical appliances in standby mode use up to 50% of their operating power.
- Select 'Energy Star'-rated equipment at purchase to ensure that the whole life energy costs are reduced.
- Set up energy-saving features such as stand-by to reduce energy use by up to 95%. Most features are easily set up and once done will operate for the life of the machine.

- Monitor out-of-hours use and be sure to switch off all equipment at the end of the day. The argument that computers can fail if frequently switched on and off is false.
- Screen savers were designed to prevent an image burning itself into CRT monitors – they do not save energy and are not generally required with modern flat screen monitors. Disable them.
- Turn monitors off if they are not going to be used for > 10 minutes but set monitors to go into stand-by after a set period in case users forget to turn monitors off.
- Set computers to power down hard discs if not used for a set period. Simple settings can make a big difference.
- Fit 7-day timers to photocopiers to stop them being left on when not being used. Printers are 'personal' and people tend to turn them off. Photocopiers are 'communal' – nobody is responsible, so nobody turns them off.
- Use 7-day timers for drinks vending machines and water machines with heaters (coffee machines, etc.). These generally have a small heater inside to keep water near the use temperature and

A green office sets the scene for sustainability and if the staff can see that the management is 'walking the walk' then they will be much more likely to buy in to the overall process.

Don't think that email is 'carbon-free', every email uses energy in computers and servers.

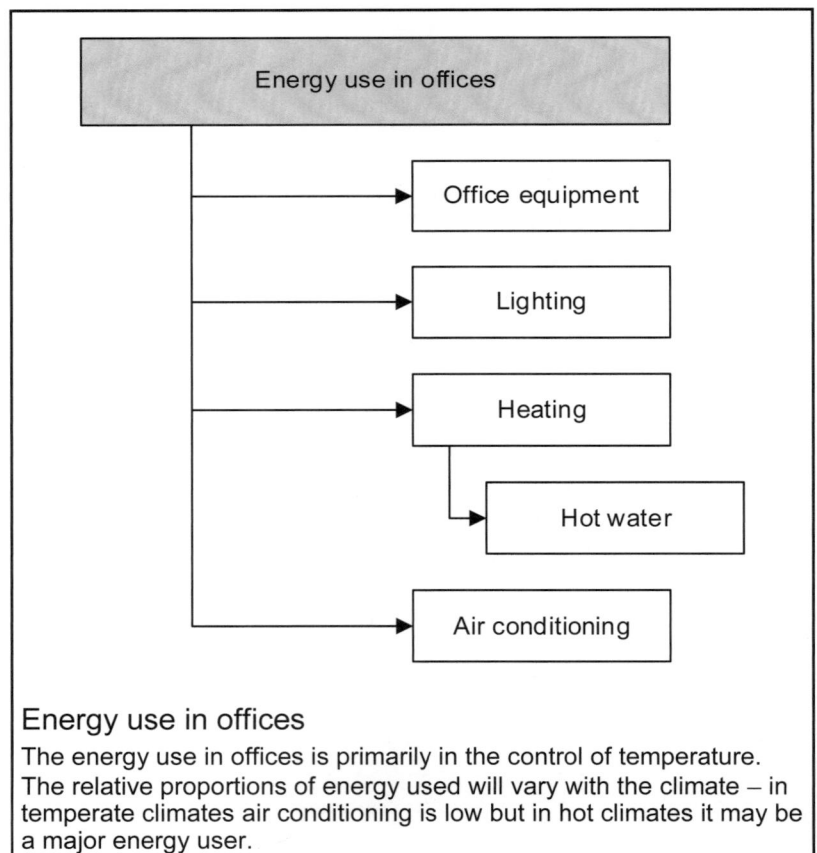

Energy use in offices

Office equipment

Lighting

Heating

Hot water

Air conditioning

Energy use in offices

The energy use in offices is primarily in the control of temperature. The relative proportions of energy used will vary with the climate – in temperate climates air conditioning is low but in hot climates it may be a major energy user.

7-day timers can save up to 75% of the energy used by these machines.

Water

General water use is covered in Chapter 8 and water use in facilities, e.g., offices, is covered in Section 7.9.

Follow the water management programme to reduce water use throughout the site and not simply in offices.

- **Tip** – If bottled water is provided for staff, then a water cooler connected directly to the mains supply can be much cheaper and just as effective.
- **Tip** – If tea and coffee are provided then avoid the use of single-serve portions of milk and sugar (but keep hygiene standards high).

Paper and stationery supplies

Paper and stationery supplies are often regarded as minor but paper use is still high in some companies even if it is steadily reducing as more companies move to a 'paperless' office.

Tangram Technology has been 'paperless' since 2010:

- No documents are kept except as computer files.
- Emails are not printed (ever) – this avoids that annoying note on the bottom of emails saying "Please do not print this email unless you have to".
- Confidential documents are shredded and recycled.
- General documents are simply recycled.

The major paper use for 2020 was in printing a draft copy of this book for manual editing.

- **Tip** – Only purchase paper with a high recycled content (up to 100% is available) and aim for 100% post-consumer waste (as opposed to pre-consumer waste which is not the same thing).
- **Tip** – Try to purchase paper which is chlorine free. There is often confusion between elemental chlorine free (ECF) and totally chlorine free (TCF) – contact your paper supplier for clarification.
- **Tip** – Use 80 gsm paper as standard to reduce paper use.
- **Tip** – All paper should be collected and recycled by competent recyclers.
- **Tip** – Avoid printing of marketing materials that are date or event specific to reduce wastage.

Paper is not the only stationery material and most offices will have huge quantities of pens and other items hiding in individual desks.

- **Tip** – Centralise supplies, i.e., a stationery cupboard, to reduce stocks and waste due to old age.
- **Tip** – If possible, purchase 're-manufactured' printer cartridges and consumables. This will not only save money but also encourage the recycling of consumables.

Office waste

General waste is the waste that is generated at each of the office desks. This can be reduced by removing individual waste bins and fitting an 'office recycling centre' where everybody recycles everything. This will encourage people to segregate waste for recycling and allow identification and action for materials that are not being recycled.

- **Tip** – Check the 'non-recyclables' bin regularly to ensure that correct sorting is taking place.
- **Tip** – If outside contractors are used to cater for events and meetings then try to avoid the use of single-serve plates or packaging (and recycle everything).

WEEE waste

Disposal of computers and other electronic goods falls under the WEEE regulations (see Section 3.14) and must be disposed of properly and by a contractor who has the appropriate authorisation. There is a duty of care associated with the disposal of WEEE waste.

- **Tip** – Work with your waste contractor to separate WEEE waste and ensure that it is handled correctly.

It can be done

It is easily possible to reduce the sustainability impact of offices and most office staff are ready and willing to help to do this. They simply need the right structures in place and the encouragement to do so. Greening the office can contribute to a company's sustainability effort and help in the social aspects of sustainability by showing that everybody can contribute to sustainability.

Some sites have replaced single-use cups with ceramic cups and claim to be 'green' but the real impact of a ceramic cup is in the heating of the water to clean it and a plastic cup can easily be recycled. Do not become confused by the first impact – remember the law of unintended consequences.

Nothing helps as much as measurement. Start to measure things like:

- Energy used by the office /month (read or fit a sub-meter for the office areas.
- Amount of paper used/month.
- Measure the amount of 'non-recyclables' and set a target of zero.

Nothing motivates people more than measurement and a target.

Transport
The methods used for staff to get to and from work can also be part of 'greening the office' but this is a very long-term plan and needs very careful thought.

Key tips

- Every pellet entering a site should leave the site as part of a product.
- Raw materials (plastic) waste is a key performance indicator for plastics processing companies.
- Companies need to know their First Time Yield and their Mass Balance Yield to assess performance.
- Calculate the First Time Yield and the Mass Balance Yield.
- Increasing First Time Yield and Mass Balance Yield improves sustainability by maximising the use of resources and reducing plastic disposed of to recycling.
- Re-using plastic via regrinding saves the material but the energy, processing time, labour and other inputs are lost forever. It is not free.
- Regrind management is important, track the flow of regrind and make sure that it is kept clean and uncontaminated to maximise the value.
- Regrinding is not recycling and should never be declared as such.
- Pellet control is important and every company should be signed up to Operation Clean Sweep and implement pellet controls.
- Pellet losses should be avoided, contained, cleaned up and disposed of carefully.
- Product quality is important in sustainability.
- Collect Quality Cost information to see where to improve.
- Set up plans for reducing the cost of quality.
- Get rid of inspectors. Make operators responsible for their own work and the quality of that work.
- Remove Quality Control (after the event) and substitute Quality Assurance (before the event) by using SPC or other techniques.
- Use process control and capability studies to check that the process is capable of producing the product to specification.
- Use process control and control charts to check that the product is being produced to specification.
- Identify the value-adding and waste processes. Take the acid test – walk around the factory and check where people are being efficient at wasting money. Don't make it efficient – eliminate it!
- Sustainable procurement is not only about sustainable purchases but also about using sustainable suppliers.
- Supplier sustainability failures, particularly in the social area, have a high reputational risk factor.
- Companies need a sustainable purchasing policy and strategy and to implement sustainable purchasing and assess areas where they are at risk.
- Using local suppliers can reduce transport emissions, increase supplier diversity, improve supply chain transparency and improve social sustainability.
- Supplier assessment for sustainability is a key to implementing sustainable purchasing.
- ISO 20400 provides excellent guidance on implementing sustainable purchasing but it is a 'guidance' document' and it is not possible to be certified for compliance to a guidance standard.
- Transport is a major source of carbon emissions and companies need to choose product transport methods carefully to reduce emissions.
- Most plastics products are transported by HGV and companies need to examine transport requirements to minimise the distance travelled, reduce the amount transported, improve load factors and space utilisation and reduce empty running.
- Offices may be small in resource use but they can be improved with little effort and it will involve the office staff in the process.

Chapter 6

Manufacturing: energy

Reducing energy use in the manufacturing phase is not only beneficial in sustainability and environmental terms but has a direct and quantifiable benefit in cost terms. This is a double benefit – turning the bottom line black whilst turning the company green is good for both the company's finances and the company's reputation.

This chapter looks at manufacturing energy use as part of the overall sustainability agenda but is largely based on the contents of a previous book on the topic of cost management[1]. This is entirely appropriate because for many companies energy management is a real and pressing business issue as well as a sustainability issue.

In high labour cost countries, energy costs generally represent the third-largest variable cost (after materials and direct labour) and, in low labour cost countries, the cost of energy is generally the second-largest variable cost (after materials). This means that energy management is not simply a 'sustainability' issue, a 'green' issue or a 'carbon management' issue. It is a real cost management issue and, in many companies, it is the last 'undiscovered frontier' of cost management. For companies producing 'commodity' items it can be a simple survival issue but, then again, that is also part of the general sustainability agenda.

The importance of energy costs, the amount of material and the potential for cost savings led to the publication of a separate book on this topic[2] in 2008. To reflect the growing importance of energy management, new editions of this book has expanded by more than 50% in the last 10 years and cover energy management to a far greater depth than is possible here where the focus is on the broader importance of sustainability. However, most of the basic points are common and are repeated here because of their importance and relevance. Readers who want a more in-depth treatment are directed to the specialist text on this topic.

Note 1: In plastics processing the dominant energy source is electricity and other fuels such as gas and oil are primarily used for heating. When the term 'energy' is used in the subsequent pages it should be presumed to be referring to electricity unless otherwise noted.

Note 2: Where cost calculations are presented then these are based on a nominal cost of £0.10/kWh for electricity.

- 1. Kent, R.J. 2018. 'Cost management in plastics processing', Elsevier.
- 2. Kent, R.J. 2018. 'Energy management in plastics processing', Elsevier.

6.1 Improving sustainability and saving money by reducing energy use

The magnitude of the costs

Energy is increasing in importance as a critical cost and critical element of sustainability. Every plastics processor knows this but remarkably few have actually taken significant action to reduce energy use and costs. Many have taken some 'token' actions in energy management and then got back to their efforts in reducing labour costs.

For the majority of plastics processors, the cost of energy is in the region of 6-8% of the turnover and, in many cases, this is approximately equal to the cost of direct labour and the profit of the site. For processors in low-margin sectors, such as in packaging or automotive parts, the cost of energy can be much greater than the profit margin. Savings in energy management can have a dramatic effect on profitability for most plastics processing sites.

Equally, energy use is a major element of the carbon footprint and, for most sites, the direct use of electricity will account for $\approx$ 75% of the site carbon footprint (see Section 7.4).

The magnitude of the savings

The potential savings from good energy management are in the region of 30% of the current energy spend and in the region of 27% of the site carbon footprint for most plastics processing sites.

These values vary with the status of the site but they will naturally be higher for 'novice' sites, i.e., those that have taken little action in the past and those that have little experience in energy management. In some cases, energy savings of up to 50% and carbon footprint savings of up to 37.5% have been identified, although savings of this magnitude are rare.

These savings can be delivered virtually irrespective of the industry sector or process used. It is not the case that any particular plastics process wastes more energy than another. The process appears to make little difference in the potential savings – it is the management that makes the difference. These savings are possible though simple actions and improvements in management, maintenance and investment.

Management actions

Recognising that the rules have changed and that we need to manage energy use as much as we manage direct labour can produce energy savings of up to 10% and related carbon footprint savings.

This involves simple techniques such as monitoring and targeting (see Section 6.5) and implementing management control systems. These are all effectively 'no-cost' actions.

Maintenance actions

Maintenance actions are those where the payback is < 1 year (irrespective of the amount required to achieve the saving). These actions would typically come from the revenue budget rather than the capital expenditure budget due to the rapid payback.

Simple quick-fix actions such as controlling the use of services both in the process and in the building services can easily produce energy savings of up to 10% and related carbon footprint savings.

This includes small investments in technologies such as variable speed drive (VSD) control of air compressors, cooling tower fans, pumps for chilled and cooling water and air handling fans.

Capital investments

Capital investments are when the payback is > 1 year (irrespective of the amount required to achieve the saving). These investments would typically come from the capital expenditure budget rather than the revenue budget due to the longer payback.

Investment in energy-efficient processing technologies that reduce energy use in the process and, just as importantly, in the effective management of these, can again

> Energy costs can easily reach 8% of turnover. Do you spend this proportion of your time managing the cost of energy?
> Do you spend any time at all?

> Why am I not impressed by statements such as "We have reduced our energy use by 5% over the past 3 years".
> This should be done in 3 months!

> Low-cost energy efficiency measures can improve profits significantly.

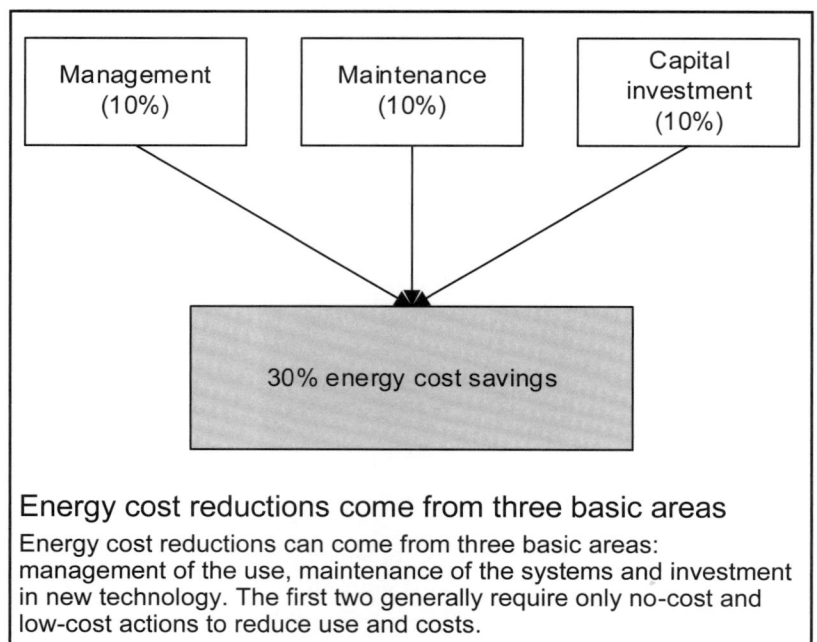

Management (10%)   Maintenance (10%)   Capital investment (10%)

30% energy cost savings

Energy cost reductions come from three basic areas
Energy cost reductions can come from three basic areas: management of the use, maintenance of the systems and investment in new technology. The first two generally require only no-cost and low-cost actions to reduce use and costs.

deliver energy savings of up to 10% and related carbon footprint savings.

This includes investment in projects with longer paybacks such as completely new machines and technology.

The payback

Savings can be delivered through a balanced combination of no-cost, low-cost and investment actions. Some actions have very rapid paybacks (< 2 months) but the average payback for investment in energy management is, in our experience, in the region of 6-9 months. This is true even when the payback is calculated using nominal costs for internal management efforts using existing resources.

These returns make investment in energy management extremely attractive from a purely financial point of view and very rewarding from a sustainability point of view. Not many capital investment projects achieve a payback of less than 1 year and continue to deliver the benefits virtually indefinitely. Despite this, many sites do not seem to accept or encourage investment in energy management because of the misguided view that energy costs are fixed and uncontrollable. Capital investment proposals are still primarily presented based on direct labour reductions and rarely put forward purely based on energy use reduction.

The source of the current costs

Few sites are able to accurately allocate their energy consumption in terms of where energy is used despite the fact that this is a relatively easy task (see Sections 6.2 and 6.9). A basic understanding of the reasons for energy use will show where a site should spend time and effort in energy use reduction efforts. Efforts at reduction should normally be relative to the size of the use and/or the ease of implementation.

The general approximate energy cost distribution for most plastics processing sites is shown on the right. The exact percentages for each individual site will depend on the process used, e.g., compressed air use is normally higher at blow moulding sites, cooling is normally higher at extrusion sites and sites with a large amount of assembly work will also generally use more compressed air. Despite these local variations, the ratios will be approximately correct for the majority of sites.

- **Tip** – A simple walk-around to identify the major energy use areas will often identify hidden use that the site has never really considered before.

- **Tip** – Many amateur energy surveyors (or those not familiar with the plastics

industry) get excited about lighting and heating. This is a sure sign that they have a lot to learn. The main energy use in plastics processing is in the machinery and services (92%). Lighting, heating and offices are minor costs (8%).

- **Tip** – If your energy-saving actions to date have been confined to 'replacing fluorescent or high-bay lighting with LEDs' then you haven't understood where the costs really are.

Compare your site energy use to the values shown below. They won't be that different.

Now compare your efforts to where the costs really are.

If they are different then you need to change your efforts.

Energy is an asset to be managed to generate revenue and competitive advantage but many companies do not have an energy strategy with specific targets, actions or budgets.

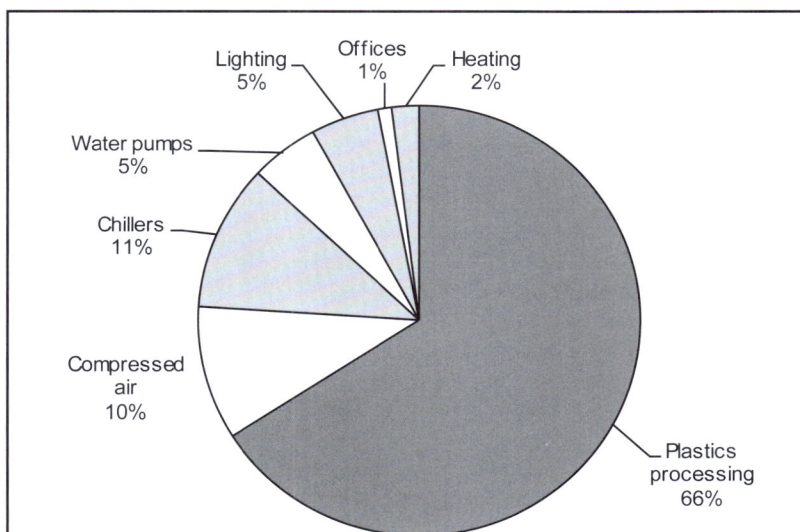

Energy use distribution in plastics processing

The main energy use and cost is in the plastics processing machinery and services (92%). Lighting, heating and offices are minor costs (8%). It pays to get a sense of perspective and to focus on the big users so that you can make the biggest savings.

6.2 The vital questions

Reducing energy costs – the first steps

The plastics processing industry generally regards energy as an overhead and as a fixed cost. In many companies the energy bill is sent by the supplier, passed by the Accounts Department and automatically paid 14 days later with no assessment or critical examination. Even when it is passed to the Production Department it is considered an overhead and is likewise signed off with little critical examination.

Energy costs are regarded as fixed and uncontrollable and are always somebody else's problem.

This is untrue, energy is both a variable and a controllable cost (see Section 6.4).

The vital questions

Before it is possible to reduce energy costs it is first necessary to understand where, when, why and how energy is being used. Additionally, it is necessary to understand what and how you are paying for energy. This information provides the signposts for low-cost and rapid improvements.

Where are you using energy?

The main energy users in plastics processing are motors and drives, heaters, compressed air, cooling systems and lighting systems. A simple site energy use map, such as that shown in the diagram on the upper right (but in more detail), will show where energy is being used. This is generally easily prepared by the site electrician from his existing knowledge.

The energy use map can be combined with actual use data to produce a site energy spreadsheet (shown in the table on the lower right). This can be produced by recording the nominal motor or heater sizes, lights, etc. and using the actual duty %'s to assess the actual operating load (in kW). Simply applying the operating hours/year will allow an assessment of the total kWh/year consumed by the equipment or service. The total of these can be compared to the overall billing data as a check on the assumptions. It should be possible to get agreement within ± 5% with little difficulty. This process will locate the key areas for monitoring and improvement.

If there is only a single meter for the site, it may be cost effective to introduce sub-metering to obtain further information on the areas of high energy use. The introduction of sub-metering enables

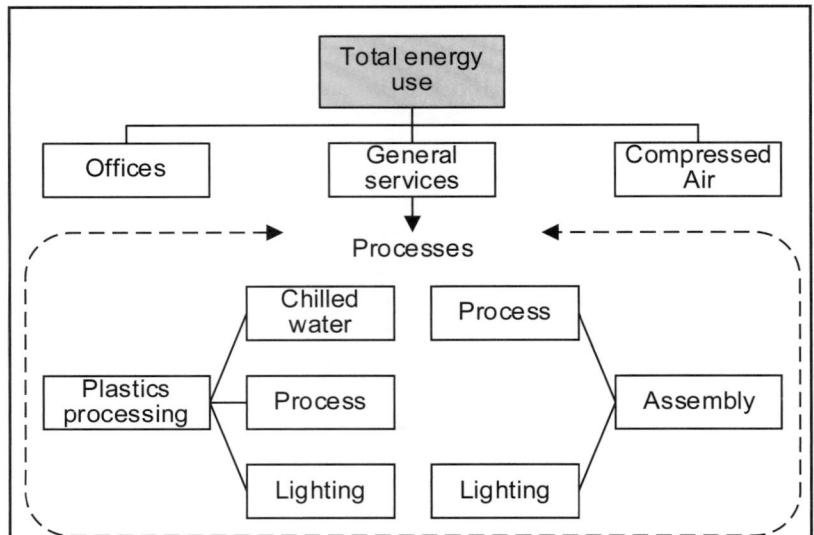

Where? Site energy use map

A simple site energy use map (obviously more detailed than this example) will reveal areas of high energy use and allow targeting of improvement efforts in the most rewarding areas. The relative use values need only be estimates at this stage.

	Nominal (kW)	Duty (%)	Load (kW)	Hours	kWh/ year	Total (kWh/year)
Services						
Compressed air						
Fixed speed compressor	50	100%	50	8736	436,800	
Driers	11	100%	11	8736	96,096	**532,896**
Cooling						
Chiller 1	80	100%	80	8736	698,880	
System pumps (3 x 7 kW)	21	100%	21	8736	183,456	
Process pumps (2 x 7 kW)	14	100%	14	8736	122,304	**1,004,640**
Materials handling						
Vacuum pumps (1 x 15 kW)	15	75%	11	8736	98,280	**98280.0**
A/C Plant						
Offices	30	25%	8	8736	65,520	**65,520**
Services sub-total						**1,701,336**
Main process						
IMM						
1 Main motor	55.9	45%	25.17	6659	167,602	
Heaters & MTC (5 x 5 kW)	25.0	10%	2.50		16,648	**184,250**
2 Main motor	89.5	45%	40.27	7209	290,274	
Heaters & MTC (5 x 5 kW)	25.0	10%	2.50	8736	18,022	**308,296**
20 Main motor	74.6	45%	33.56	2651	88,945	
Heaters & MTC (5 x 5 kW)	25.0	10%	2.50		6,627	**95,571**
Main process sub-total (all machines)						**3,423,655**
Site						
Lighting						
Main production areas	45	100%	45	8736	393,120	
Offices	22.5	40%	9	3000	27,000	**420,120**
General equipment						
Main power	45	45%	20.25	8736	176,904	**176,904**
Site sub-total						**597,024**
Global total						**5,722,015**

Where? Site energy spreadsheet

The site energy spreadsheet lists the nominal sizes of motors and loads, assesses the % duty and uses the operational hours to generate a predicted kWh/year use. This can be compared to the actual billing data as a reality check.

calculation of the cost of energy for each operation and identification of the areas of high energy use – a key factor in reducing energy costs.

When are you using energy?

The time at which energy is used is important in cost terms (it is generally cheaper at night) and the total demand (in kWh) plotted versus time will give invaluable information on how to reduce the time-dependent energy costs. A sample of typical interval data is shown on the right.

• **Tip** – Energy Lens from Bizee (www.energylens.com) is an excellent programme for analysing interval data.

The upper chart shows the daily electricity use over a 9-month period and it is easy to see a progression from 5 days/week to 7 days/week operation. It is also possible to zoom in on a specific period and this has been done for a 1-week period to show the effect of the weekend closure. This shows a high use on Sunday even though the site was not operating. Zooming in on the Sunday data shows the site starts up at around 18:00 in Sunday – despite the fact that production did not begin until 06:00 in Monday. The site was warming up all the machines on Sunday to be ready for operation on Monday morning. It takes around 2 hours for an injection moulding machine to be ready-to-run from cold and the site was wasting at least 10 hours of energy use for no production. This was changed very simply and quickly.

The data for such plots should be freely available from your supply company – they monitor the use as part of the charging mechanism. Interval data are 1/2-hourly for the UK but can be 10- or 15-minute data in other parts of the world.

When reviewing interval data, look for:

• Unusual daily peak variations.

• Unusual patterns during the day.

• Electricity use when there is no production, particularly over weekends and holidays.

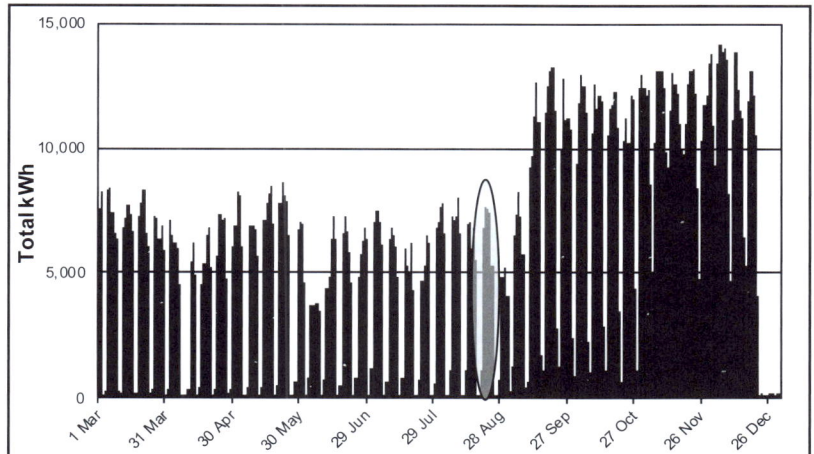

Daily demand over 9 months – Data can be viewed over any convenient period.

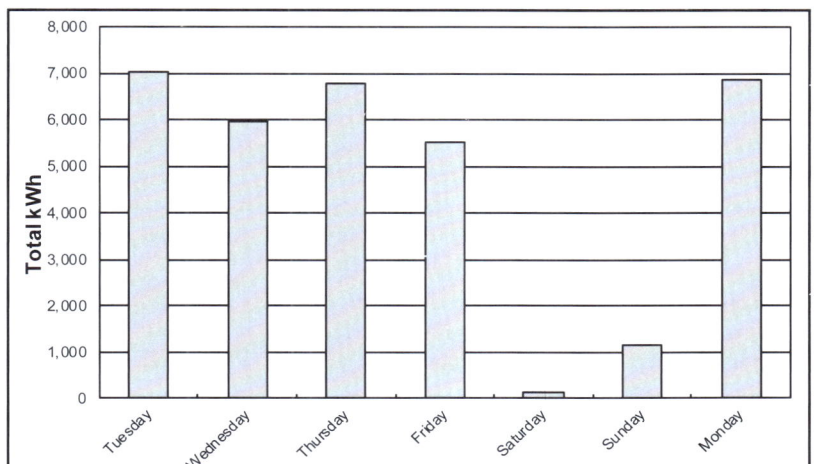

Daily demand over 1 week – Data can be viewed daily over a single week.

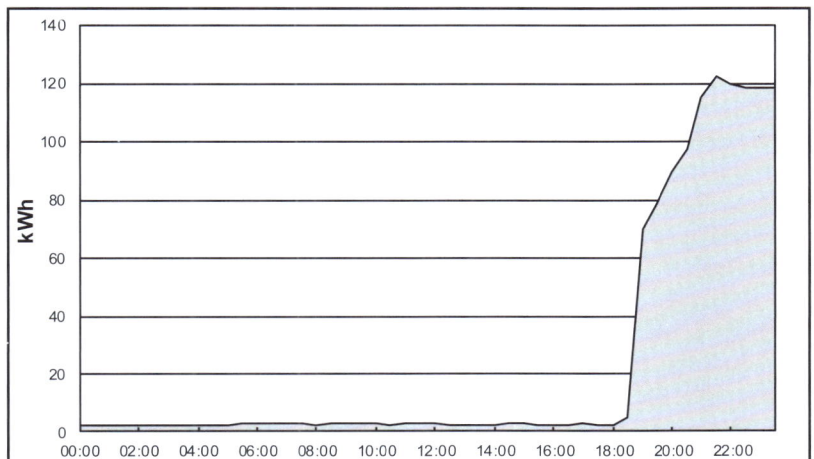

Hourly demand over 1 day – Data can be viewed hourly over a single day. The start-up of the site is clearly seen.

Plotting site energy use using interval data

Freely available interval data allow energy use to be seen on a yearly, weekly and daily basis. The data show that the site was switching on the machines at 18:00 on Sunday so that they were ready at 06:00 on Monday. A practice that was soon changed.

6.3 More vital questions

How much energy are you using?

Simple interval data will also allow sites to determine how much energy they are using in any given period by simply adding up the consumption for the required intervals. This can act both as a check for the billing data (see below) and as an essential part of the performance assessment process (see Section 6.4).

For plastics processing sites, the amount of energy used is normally directly related to the amount of plastic processed (see Section 6.4) but many managers see the amount of energy used, and hence the cost, as being fixed and uncontrollable. They fail to realise that energy is a variable and controllable cost. Recognising that energy is a variable cost is a major step for many management teams and simply recording how much energy is used each week or month is part of the process of changing this misconception.

- **Tip** – Sub-metering is very useful to see the relationship between activity and energy use. Be prepared to pay 2% of the annual electricity bill for a good sub-metering system.

Why are you using energy?

Energy should be used only to provide essential services or to process plastic for good product. Despite this, many sites have only a vague idea of why or where they are using energy and simple tools such as the site energy map (see Section 6.2) are rarely used.

Finding out why the site is using energy will often reveal a wide range of possible steps for reducing energy use.

There are two components to the total energy use at any plastics processing site. These are:

- The base load – this is the energy used simply to have the site open and ready for production. In this state, a site will use electricity for heating, lighting, compressors (mainly to feed leaks in the system), cooling (mainly to remove parasitic heat gain), pumps (for unnecessary flow of water), idling motors/machines and other non-productive activities. Reducing the base load is a sure way to make savings because this is energy that is not related to production.

- The process load – this is the energy used for productive plastics processing and is related to the process being used and the

---

Key tips for reducing the cost of electricity

The terms used will vary between countries and suppliers but look for the following terms and units to find some easy wins in reducing the cost of electricity:

Available capacity, contract demand, maximum power requirement (kW or kVA) – this is the maximum power that a site can draw from the supply without tripping the area circuit breakers. If a site exceeds the available capacity then penalty charges may be applied. In the worst case, the area trips will operate causing an interruption to supply for the *area* and very significant costs to the site.

If the capacity is available, it is normally free to increase the available capacity but this will also normally increase the monthly fixed costs. If no capacity is available and the supplier has to upgrade the distribution system then this is very expensive.

If the available capacity is set too high then it is also generally possible to reduce the capacity and save considerable costs.

Setting the available capacity is a strategic issue for management.

Energy management can minimise any additional demand from increasing production volumes and avoid the cost of increasing the available capacity. Reduce the cost by:

- Matching the available capacity to the actual requirements and getting the available capacity right for the site.

- Staggering start-ups to avoid exceeding the available capacity.

Maximum demand (kVA or kW) – this is the maximum power drawn at the supply voltage for the billing period (even if only for 1 minute). Reduce the cost by:

- Staggering start-ups of machinery to reduce initial loads.

- Giving machinery time to stabilise before starting new processes.

- Peak demand lopping (see sidebar on the right).

Power factor (PF or cos φ) – electrical machines induce a phase shift between the supply voltage and current if there is a high reactive impedance (inductance or capacitance). Electricity suppliers don't like low power factors because they need high network capacities for low consumption charges. Low power factors can cause problems with running the distribution network and suppliers may charge for reactive power (kVArh) if the power factor is < 0.95. Improve the power factor by:

- Running electric motors efficiently to get power factors close to 1.

- Using PF correction equipment to improve the PF.

Load factor (LF) – this is a measure of the number of hours/day that the user draws from the supply. A 9-hour single shift working pattern gives a load factor of 9/24, i.e., 37.5%. Variable 'peaky' aggregate electricity demand forces the supplier to have standby capacity that runs only on peak demands. The supplier has the same fixed costs maintaining a distribution system that is used for 9 hours/day as for one that is used for 24 hours/day, but less consumption revenue to offset them. Reduce the cost by:

- Running for greater than a single shift.

- Carrying out some operations outside the main shift pattern, e.g., regrinding.

---

amount of material being processed. Every process has a different energy intensity and it is not possible to compare different processes because of this.

Understanding the relationship between the base and process loads is important to understanding the correct methods of setting targets and assessing performance.

What are you paying for?

Reading the supplier's electricity bill is a key skill that needs to be developed at every site. It is estimated that 5% of electricity bills actually contain errors (although this has greatly decreased with the introduction of automated meter reading) but most sites do not actually understand the charges listed on their electricity bill. This is not a surprise because electricity bills can have up to 12 lines of detailed charges, are often poorly laid out and can be very confusing.

A close survey of the bill will often reveal areas for potential savings, sometimes actions as simple as changing the tariff or reducing the available capacity can reduce costs at little or no cost. The key tips for reducing the cost of electricity are shown in the box on the left and the importance of the power factor is explained in the box on the right.

- **Tip** – If nobody at the site understands how to read the electricity bill then you are at the mercy of your supplier. Learn how to read the bill. If necessary, get the supplier to explain all the charges on the bill and suggest methods to reduce them if possible. Understanding the energy charges is a key management task.

- **Tip** – Read the meters (manually if there is no independent metering system) at the start of each month and compare this with the billing data.

- **Tip** – Establish a simple spreadsheet to cross-check every line item on the bill for consistency (month to month) and with the actual use from the meters and the production activity.

- **Tip** – Fixed costs will often be as much as 30% of the total energy bill. There are some easy wins to be made in looking at these sections of the bill.

How are you paying for your energy?

Most sites have few controls on payment of the energy bill despite the fact that it is normally between 6 and 8% of the turnover. Section 6.5 deals with performance assessment for electricity use and this forms a natural basis for setting electricity use targets. Every site must critically examine the process of paying the bill and make sure that the people who use the energy are

responsible for explaining any variation in the bill and particularly explaining any variation from the target electricity use.

'Peak demand lopping' can be very effective in reducing short peaks in the maximum demand.

This automatically shuts off low-priority activities to avoid exceeding the maximum demand for short periods.

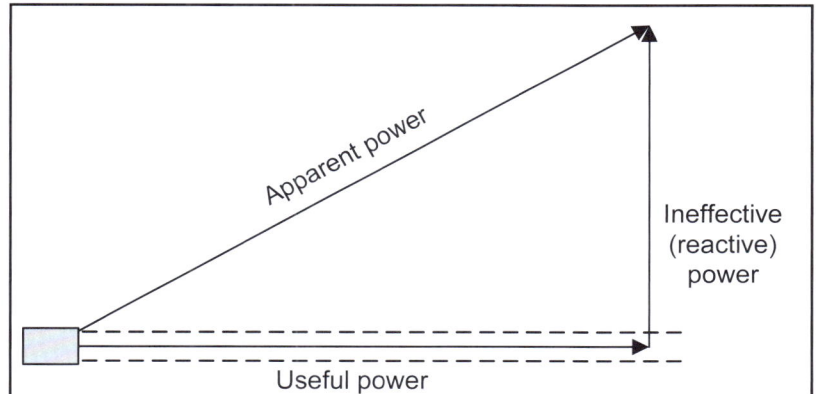

Mechanical analogy for power factor

Imagine pulling a heavy load along a set of tracks. You are not able to pull it directly along the tracks but having to pull from one side of the tracks.

The useful power is that required to move the load down the tracks. The apparent power is the power you actually need – this includes the 'wasted' power that is not really used to move the load down the tracks but is expended by simply pulling against the side of the tracks.

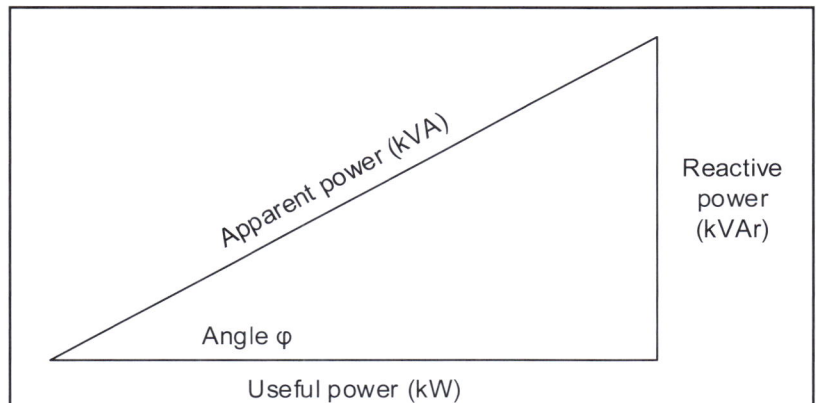

Power factor = Useful power/apparent power = cos φ

Power factor in electrical supply

The useful power (kW) actually required for the load, the apparent power (kVA) is that drawn from the supply and the reactive power (kVAr) is the useless or wasted power. The value of cos φ is the amount of the apparent power doing useful work.

6.4   Internal site benchmarking

Getting the data and presenting the results

At any plastics processing site it is possible to determine the base and process loads for the site from simple energy use and production volume data.

The total site base and process loads can be quickly estimated using the following method:

- Record the production volume (in kg) for a number of months and record the energy use (in kWh) for the same period. This is shown on the right for a sample injection moulding site but can be done for any forming process.

- Plot the energy use (in kWh) versus production level (in kg) in a simple scatter chart. A graph using the sample data is shown on the lower right with a linear line of best fit plotted for the points and the equation of the line of best fit shown.

- The intersection of the line of best fit with the 'kWh' axis indicates the 'base load' for the site. This is the energy use when no effective production is taking place but machinery and services are available. This is the base cost of operating the site.

- The slope of the line of best fit is the 'process load' and shows the average energy being used to process each kilogram of polymer.

The equation of the line of best fit for this data is:

kWh = 1.5751 × Production volume + 152,440   R² = 0.9397

The good R^2 value (0.9397) indicates that the data set is very consistent with the line of best fit. The correlation coefficient (R^2) can vary between 0 (no correlation between the two variables) and 1 (perfect correlation between the two variables). A high R^2 (> 0.7) value indicates good correlation of the data and gives good confidence that the equation describes the actual relationship. If the R^2 value is < 0.70 then there is generally a problem with the data, e.g., the data collection periods for the energy use and the production volume are not the same, or the site energy management is poor and inconsistent. This can also indicate poor overall management of the site.

The equation of the line of best fit is the Performance Characteristic Line (PCL).

The energy cost to the sample site consists of a base load (the intercept of the line of best fit) of approximately 152,440 kWh/month

Month	Energy use (kWh)	Production volume (kg)
January	425,643	182,421
February	463,772	197,897
March	504,675	248,742
April	437,307	204,228
May	492,613	212,716
June	518,940	225,239
July	532,322	217,864
August	469,029	207,615
September	676,008	347,845
October	711,119	343,468
November	671,962	311,174
December	409,526	147,378

Sample energy use and production volume data for an injection moulding site

Simple measurements can provide a remarkable insight into the energy use at a site. These 24 numbers form a 'fingerprint' of the site energy use – it is unique to this site.

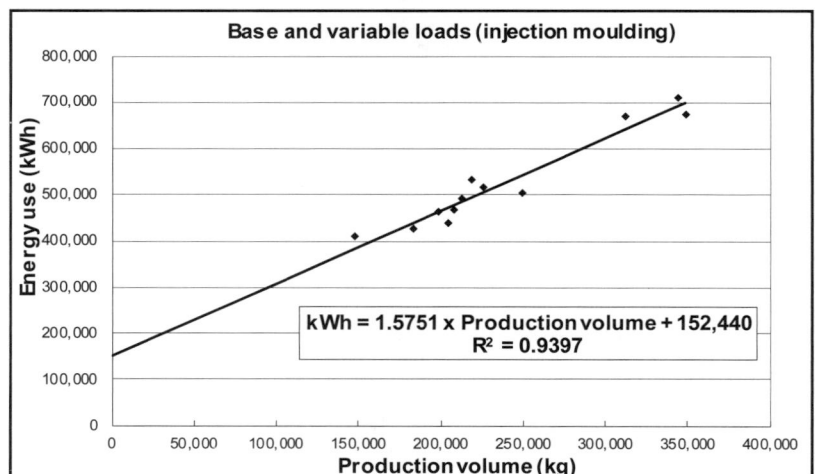

Scatter chart of energy use and production volume data with simple line of best fit trend for the data

The sample data can be plotted as a scatter chart to reveal the Performance Characteristic Line of the site. The equation of the best fit line gives both the base load and the process load for the site.

and a process load (the slope of the line of best fit) of 1.5751 kWh/kg of plastic processed.

It is possible to record the energy use/ production volumes over other periods, such as weeks and this gives faster data collection, faster feedback and quicker resolution of any concerns.

The base and variable loads

The base load information from the sample site implies that even if no production is taking place then the site will consume approximately 152,440 kWh/ month. At an energy cost of £0.10p/kWh, the total cost of this base load is approximately £15,244/ month or £182,928/ year. This base load is approximately 30% of the monthly cost of energy to the site – this is regarded as average for the plastics processing industry where base loads can range from 20 to 40%.

The base load is the energy 'overhead' and is primarily due to machinery being left on with no production, or services being left operational with no productive output, e.g., compressed air leaks, parasitic heat gain in cooling water piping, lights on, conveyors running with no production, etc.

Reductions in the base load (translating the complete trend line downwards) can generally be made without affecting production rates, quality or operations. These savings are very profitable to carry out because the base load is largely a dead weight on the site that is unrelated to the actual production output.

The process load information from the sample site shows that for each kilogram of plastic processed, the site uses approximately 1.5751 kWh. The process load shows how efficient the site is at plastics processing.

Reductions in the process load (reducing the slope of the graph) indicate improved process efficiency or machine utilisation. These are often more difficult to achieve.

Whilst separating the base and process loads may appear easy in theory, in practice it is often more difficult because many loads have both a fixed and a variable element.

Plastics processors need this type of information to enable correct targeting of energy use improvements for both the base and the process load. The actions required in each case are very different.

Process dependency

The method is naturally process independent but the results are highly process dependent, particularly in terms of the process load. This is shown in the box below where similar simple process data are shown for an extrusion site.

Injection moulding has a higher process energy requirement than extrusion and therefore it is to be expected that the process loads will be significantly different. The sites have broadly similar base loads but the extrusion site has a process load of 0.4467 kWh/kg compared to the injection moulding site, which has a process load of 1.5751 kWh/ kg.

Alternative production measures

The use of production volumes in 'kilograms' is sometimes not possible at sites that regard themselves as being in 'medical products' or automotive products' rather than in 'plastics processing'. In this case, the production volume in 'parts' can be used instead of production volume in 'kilograms' and provided there is a reasonably consistent mix of part size, the use of parts as a variable still allows assessment of the base load and process load.

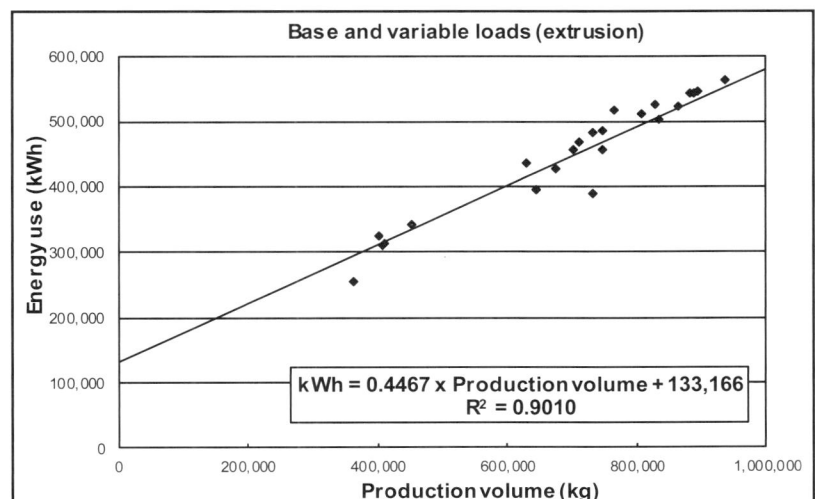

Base and variable loads (extrusion)

kWh = 0.4467 x Production volume + 133,166
R^2 = 0.9010

Base and process loads for extrusion

The base and process loads can be found for any plastics forming process but the process load will change depending on the energy intensity of the basic process being used. Extrusion is less energy intensive than injection moulding.

6.5 Performance assessment and forecasting

Performance assessment

The simple energy use and production volume data and the derived equation for the Performance Characteristic Line (PCL) can be used to assess site performance on a monthly basis (or on a weekly basis if the data and equation are weekly).

The equation of the line of best fit for the injection moulding site discussed in Section 6.4 was:

kWh = 1.5751 × Production volume + 152,440

This equation can be used to assess energy use for a given production volume in a month.

If the production volume is 200,000 kg, then the predicted energy use will be:

kWh = 1.5751 × 200,000 + 152,440

or

kWh = 467,460 kWh

Therefore, the predicted energy use for a production volume of 200,000 kg in the month is 467,460 kWh and predicted energy cost is £46,746.

This simple approach enables the production of a performance table (shown on the right) for performance assessment and prediction of the monthly energy cost to the company.

Energy use can no longer be regarded as a fixed cost. The PCL shows clearly that it is a variable cost that has a direct relation to the production volume at the site.

Site managers now have a direct and simple method of assessing energy use performance and assigning production accountability by the following method:

- Set up a simple spreadsheet to calculate the kWh to be used for a given production volume. This uses the PCL to generate the predicted kWh to be used for a given production volume.
- Determine the volume of material processed in the past month and calculate the predicted energy use.
- Determine the actual energy use for the past month.
- Compare the predicted energy use to the actual energy use.
- If the actual energy use is less than the predicted energy use then the site performed better than it has done historically – find out what the site did right and do more of it.
- If the actual energy use is more than the predicted energy use then the site

performed worse than it has done historically – find out what the site did wrong and do less of it.

The PCL is an invaluable tool to set targets in terms of energy use for a given production volume. These targets can be used for performance assessment based on real production volume and internal energy benchmarks generated from the historical site performance.

If the site has sub-metering then a PCL can be generated for each sub-metered area to allocate responsibility to the area managers. As a general rule, when using the PCL to drive improvements and cost reductions it is most useful to use weekly data for energy use and production volume. This gives faster feedback to the production managers and allows greater control of the improvement process.

It is important to realise that the PCL does not, of itself, drive improvement or provide any external benchmarking. The PCL is based on the historical performance and only assesses the site against previous achieved performance.

The PCL is a measure of performance but only against internal historical performance. It does not provide an external benchmark.

Energy is a variable and a controllable cost.

Production volume (kilograms in month)	kWh (in month)	£/month
0	152,440	£15,244
50,000	231,195	£23,120
100,000	309,950	£30,995
150,000	388,705	£38,871
200,000	467,460	£46,746
250,000	546,215	£54,622
300,000	624,970	£62,497
350,000	703,725	£70,373

kWh = 1.5751 x Production volume + 152,440
Energy cost calculated at £0.10/kWh

Using the PCL to assess monthly energy

The PCL gives a simple method of assessing monthly site performance and targeting areas of excessive use and excessive cost. Energy use is no longer an uncontrolled and unknowable variable – it is directly related to the production volume.

Predicting costs

The equation for the PCL can also be used for energy budgeting and the prediction of energy use based on the predicted sales volumes for the month/year.

The sales volumes (in kg or in parts) can be taken from the sales forecasts and used as shown in the table on the right.

We now have a tool for the accurate prediction of the future energy use and cost of the site based simply on the historical energy use of the site corrected for production volume and the current energy prices – much more useful than any current method available.

The pitfalls of simple kWh/kg

Most companies take a simple approach to energy efficiency and calculate a single Specific Energy Consumption (SEC) in terms of 'kWh/kg' each month as an assessment method. They calculate this from the kilograms processed in the month and simply divide by the kWh used in the month. This provides a good snapshot of energy performance but can be misleading if production volumes change significantly over the period.

The monthly SEC type of measurement is affected by both the monthly production volume and the base load. Increasing the production volume will automatically reduce the SEC because the base load will be amortised over a greater production volume and lead to the impression that energy efficiency is improving.

Companies therefore must be careful in assessing energy efficiency changes by simply comparing SEC values; these can be affected by simple changes in production volume rather than real changes in the energy efficiency of the process. Obviously, this will be less significant where the base load is low in comparison to the process loads but the simple number can often be misleading.

This is not an issue when production volumes are rising, there is a decreasing SEC and the production managers can accept congratulations for doing nothing at all. When production volumes are decreasing and the SEC is increasing despite their efforts, then they are less happy to accept the criticism.

Month	Forecast production volume (kg)	Forecast energy use (kWh)	Forecast energy cost (£)
January	150,000	388,705	£38,871
February	200,000	467,460	£46,746
March	250,000	546,215	£54,622
April	240,000	530,464	£53,046
May	235,000	522,589	£52,259
June	225,000	506,838	£50,684
July	235,000	522,589	£52,259
August	248,000	543,065	£54,306
September	267,000	572,992	£57,299
October	287,000	604,494	£60,449
November	210,000	483,211	£48,321
December	160,000	404,456	£40,446
TOTAL	2,707,000	6,093,076	£609,308

kWh = 1.5751 x Production volume + 152,440 (monthly)
Energy cost calculated at £0.10/kWh

Budgeting for future energy use

The PCL can be used with the sales forecasts (however accurate) to produce an energy use forecast based on the real site characteristics. The only variable is the energy cost and this can be estimated from the current contract and future projection.

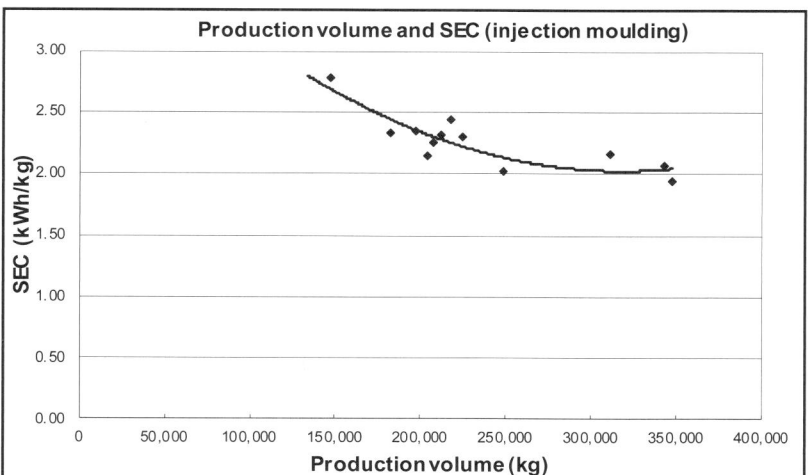

Production volume and SEC

Changing production volumes will directly affect the SEC because the base load will automatically be amortised over a greater process load. The simple SEC value is dependent on production volume and is an unreliable metric.

6.6   External benchmarking by site

External benchmarking

The simple methods and analysis described previously allow internal benchmarking against historical performance but do not provide the essential external reference to drive real improvements in performance. This needs external benchmarking.

Process dependency

Plastics processes are not equally energy intensive and each process needs a specific external benchmarking reference. It is impossible to compare an injection moulding site with an extrusion site. The process energy intensities are very different.

Production rate dependency

Average site SEC data (including the base load) is available from two sources[1,2] and this is shown in the table on the right. The two data sets are relatively consistent in terms of the overall average site SEC but the average SEC takes no account of the production rate. Most plastics processing sites have high fixed base loads and this results in a high rate dependency as the fixed loads are amortised into the variable loads (see Section 6.5).

The average site SEC is therefore of no use for external benchmarking because the overall output rate matters and it is essential that any benchmarking corrects for this production rate dependency.

- **Tip** – The average site SEC (in kWh/kg) is always higher than the process load (in kWh/kg) from the PCL because the base load is included in the site SEC, whereas the process load only considers the processing energy use.

Processes

Injection moulding

An operating curve for SEC (kWh/kg) versus production rate for injection moulding has been generated from data from 171 injection moulding sites throughout the world[2].

This is shown on the right and it is now possible for an injection moulding site to be externally benchmarked based on the production rate. This is done by calculating the global production rate (in kg/h/machine) over a full year of operations and using the operating curve to calculate a benchmark SEC for the production rate. This can be compared to the actual site SEC which is calculated from the total electricity use of the site over the full year and material processed over the full year.

Several points should be noted:

- Defining the number of machines in operation over the full year may require some assessment of 'part-machines' if production is not continuous over all machines.

- The benchmark SEC is a best-fit value and some companies have a much lower site SEC for the injection moulding process, particularly at lower production rates.

- Achieving the benchmark SEC is not a sign of good practice, only a sign of average practice.

- The machines used at any site will be of various sizes but for this analysis the average consumption is assumed. This assumption does not appear to introduce any large anomalies.

- The polymer processed would be expected to have some effect but the available data show that this has little effect on the overall assessment.

Comparing site SEC values without taking production rate into account is meaningless – it is comparing apples with pears.

Process	Average site SEC (kWh/kg)	
	EURecipe[1]	Tangram data[2]
Injection moulding	3.118	3.133
Profile extrusion	1.506	1.316
Extrusion blow moulding	N/A	2.229
Average site SEC for various processes		

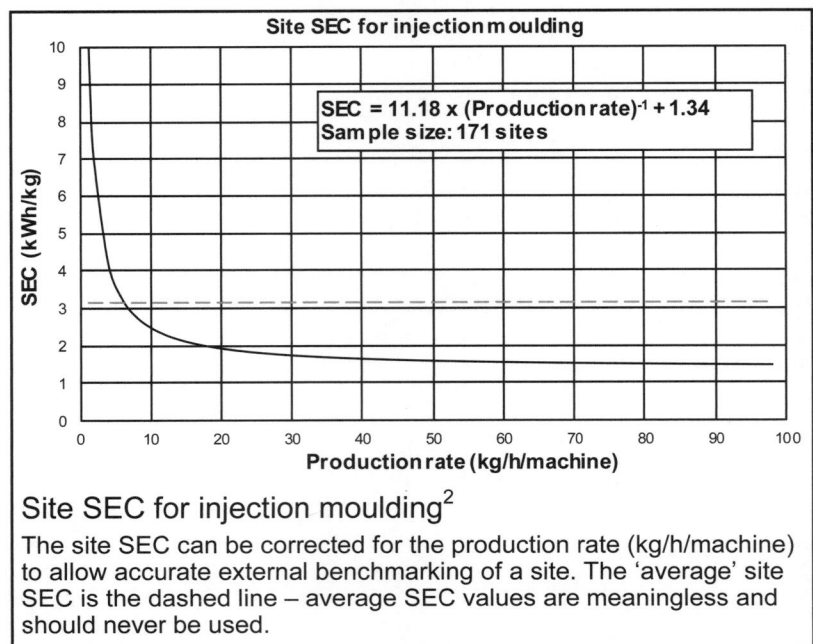

Site SEC for injection moulding

$$SEC = 11.18 \times (\text{Production rate})^{-1} + 1.34$$
Sample size: 171 sites

Site SEC for injection moulding[2]

The site SEC can be corrected for the production rate (kg/h/machine) to allow accurate external benchmarking of a site. The 'average' site SEC is the dashed line – average SEC values are meaningless and should never be used.

Chapter 6 – Manufacturing: energy

Extrusion

A similar analysis is possible for extrusion and this has been done for 49 extrusion sites throughout the world to produce an equivalent operating curve for extrusion[2]. Extruders can use this to benchmark themselves against established practice and other similar sites at the same production rate. Again, it is emphasised that achieving the benchmark SEC is not a sign of good practice, only a sign of average practice.

- **Tip** – Extrusion is a very efficient process with low fixed loads. This is the reason that the curve goes essentially flat after about 200 kg/h/machine. The fixed loads have been almost fully amortised into the process loads by this stage.

Extrusion blow moulding

A similar analysis is possible for extrusion blow moulding and this has been done for 34 extrusion blow moulding sites throughout the world to produce an equivalent operating curve for extrusion blow moulding[2].

Other processes

Data sets for other processes (injection blow moulding, thermoforming and rotational moulding) are available[2] but the data sets are smaller in size and, whilst it is possible to provide a benchmark, the degree of confidence is lower.

Energy efficiency versus production rate

Despite any reservations regarding machine size, machine utilisation and polymer type (and the variations that these will inevitably introduce) the operating curves illustrate an important point – improving energy efficiency in plastics processing in no way contradicts improving processing output.

Unlike cars where you drive slowly to increase energy efficiency, in plastics processing the harder you push the site the better the energy efficiency. This is due to the high fixed loads of sites. Increasing output amortises the fixed loads over greater outputs and improves the overall energy efficiency.

Summary

The simple operating curves shown allow sites to carry out external site benchmarking corrected for production rate. This provides companies with the essential driving force for change – if the site is above the operating curve then it is possible to set targets for cost reduction based on external data.

There is no conflict of interest

Being green can also be profitable. Running a plastics processing site harder improves the energy efficiency of the site.

- 1. European Benchmarking Survey of Energy Consumption and Adoption of Best Practice' – EURecipe, 2005, www.eurecipe.com.
- 2. Tangram Technology Ltd.: Internal data from 171 injection moulding, 49 extrusion and 34 extrusion blow moulding sites.

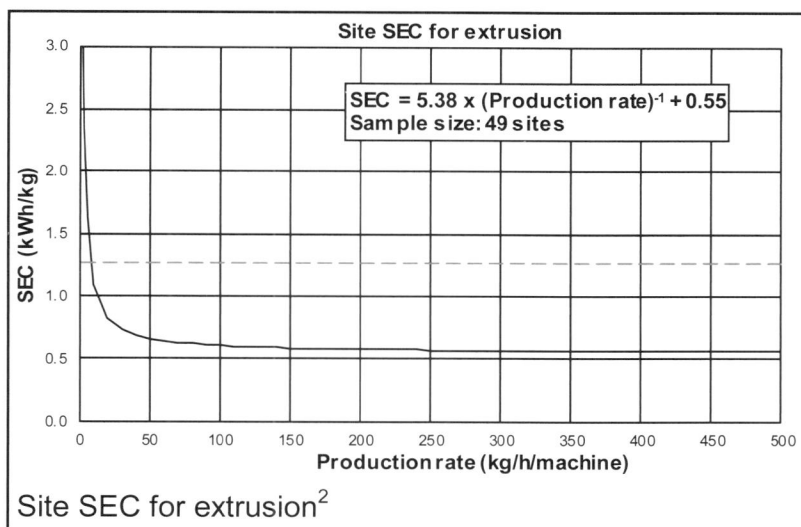

Site SEC for extrusion

$$SEC = 5.38 \times (\text{Production rate})^{-1} + 0.55$$
Sample size: 49 sites

Site SEC for extrusion[2]

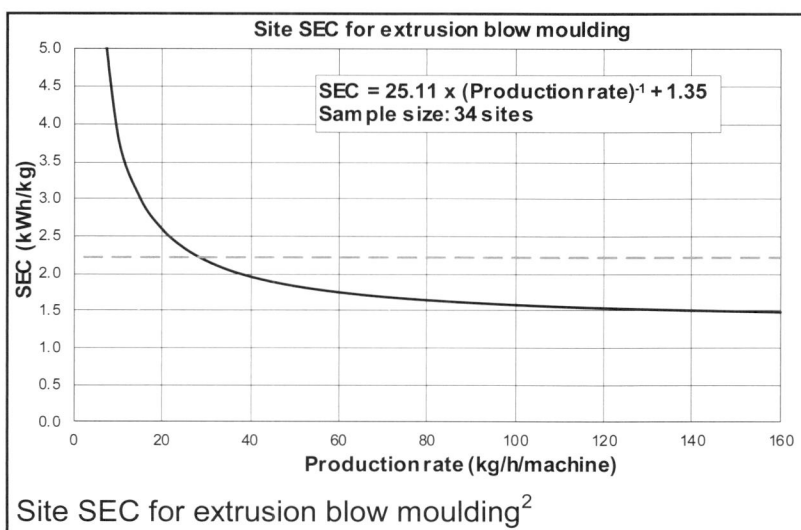

Site SEC for extrusion blow moulding

$$SEC = 25.11 \times (\text{Production rate})^{-1} + 1.35$$
Sample size: 34 sites

Site SEC for extrusion blow moulding[2]

6.7 External benchmarking by machine

The heart of the process

The analysis of a site SEC corrected for production rate provides a vital external global benchmark for a site but many sites would also like to be able to benchmark individual machines. This will enable them to determine which machine is the most energy-efficient under given operating conditions and to investigate machine settings to reduce energy consumption through improved machine setting.

Process information

Machine energy consumption is available from a variety of sources but this is generally applicable for only a limited number of machines and is not generally 'production'-oriented data. It is generally laboratory data which examines a limited range of machines. Tangram Technology has carried out detailed energy measurements on 395 injection moulding machines, 94 extruders and 99 extrusion blow moulding machines throughout the world. These machines were all production machines producing commercial products in a variety of materials and were from a variety of manufacturers. The quality of the setting of the machines is naturally variable but in most cases the settings were acceptable.

Process dependency

It will come as no surprise that plastics processing machinery is not equally energy intensive. Therefore, each distinct process again requires a specific external benchmarking reference but sufficient data are really only currently available for injection moulding, extrusion and extrusion blow moulding. Fortunately, these processes account for the majority of the materials processed and are the most relevant to the bulk of the industry machines.

Machine energy use (which includes any machine base load and, in some cases, a small ancillary load) was measured, converted to an SEC (kWh/kg) and then correlated to the production rate (kg/h) for each process using best-fit lines.

The data for each process is very different and this reinforces the strong process dependency of energy use in plastics processing, i.e., extrusion is significantly less energy intensive at the machine level than injection moulding and extrusion blow moulding is located in the middle due to the additional machine operations in extrusion blow moulding.

Production rate dependency

As for site SEC values (see Section 6.6), it is obvious that the use of a simple average SEC for any process is irrelevant due to the effect production rate has on the apparent SEC of the process. This is again due to the amortisation of the base loads inherent in the process.

Processes

Injection moulding

Data from the analysis of 395 injection moulding machines have been used to produce the operating curves shown below for both hydraulic and all-electric injection moulding machines[1]. These operating curves correct the predicted SEC for variations in production rate for injection moulding machines.

Note: Hybrid machines have been included with the all-electric machines in the data.

It is now possible to benchmark individual injection moulding machine/tool combinations against other typical machines/tool combinations. This can be done by:

- Calculating the average machine production rate (kg/h).
- Finding the predicted machine SEC from the operating curve at the specific production rate using the relevant equation (kWh/kg).
- Calculating the actual machine SEC from the total electricity use (kWh) and the total

> Comparing machine SEC values without taking production rate into account is meaningless – it is comparing apples with pears.

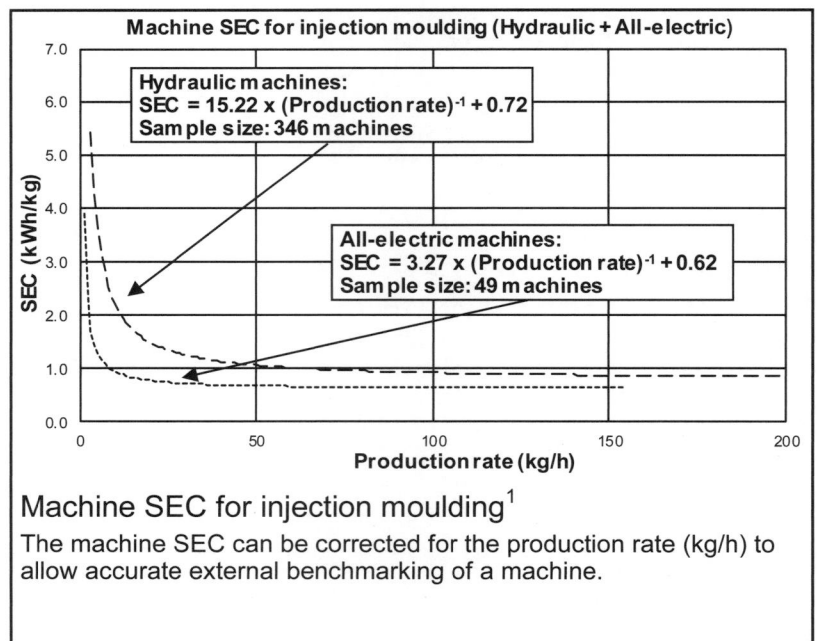

Machine SEC for injection moulding (Hydraulic + All-electric)

Hydraulic machines:
SEC = 15.22 x (Production rate)$^{-1}$ + 0.72
Sample size: 346 machines

All-electric machines:
SEC = 3.27 x (Production rate)$^{-1}$ + 0.62
Sample size: 49 machines

SEC (kWh/kg) vs Production rate (kg/h)

Machine SEC for injection moulding[1]
The machine SEC can be corrected for the production rate (kg/h) to allow accurate external benchmarking of a machine.

material processed (kg).

- Comparing the actual machine SEC at the given production rate with the machine operating curve SEC benchmark.

Sites can now benchmark their machine/tool energy performance relative to the performance of the sample machine/tool combinations for the two main types of injection moulding machine.

Several points should be noted with regard to this analysis:

- The benchmark machine SEC is an average value and some machine/tool combinations have a considerably lower machine SEC, particularly in the area of lower production rates.
- Achieving the benchmark machine SEC is not a sign of good practice, only a sign of average practice.
- The machines used at any site will vary in size/clamp force but the graph is consistent without any reference to the absolute machine size.

Extrusion

Similar data for 94 extruders has been used to produce an equivalent operating curve that can be used to benchmark extruders against other machines with similar output rates[1].

Extrusion blow moulding

Similar data for 99 extrusion blow moulding machines have been used to produce an equivalent operating curve that can be used to benchmark extrusion blow moulding machines against other machines with similar output rates[1].

Other processes

Data sets for other processes (injection blow moulding, thermoforming and rotational moulding) are available but are much smaller and whilst it is possible to provide a benchmark, the degree of confidence is lower[1].

Energy efficiency versus production rate

As with sites, the harder you push plastics processing machines the better the overall energy efficiency of the overall process. This is due to the high fixed loads of operating most plastics processing machinery. Increasing output amortises the fixed loads over greater variable process loads and improves the overall apparent energy efficiency.

There is no conflict of interest

Being green can also be profitable. As with sites, at the machine level, running plastics processing machinery harder improves the energy efficiency of the machine.

- 1. Tangram Technology Ltd.: Internal data from 395 injection moulding machines, 94 extruders and 99 extrusion blow moulding machines.

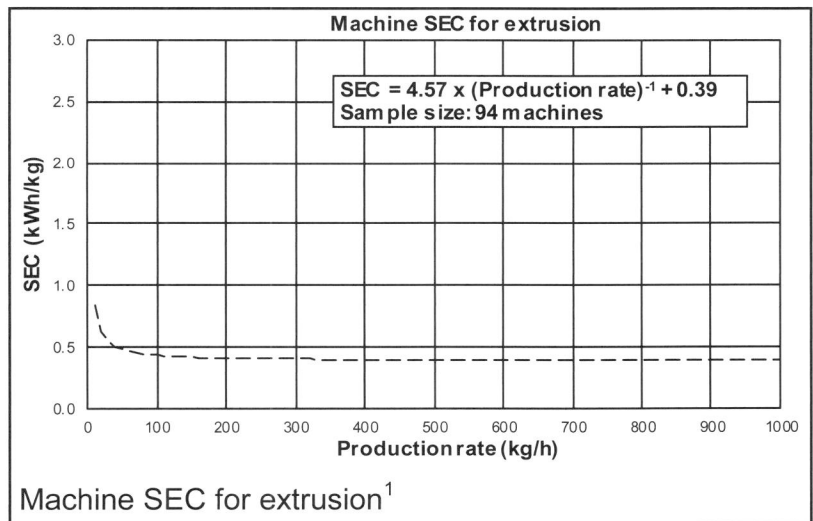

Machine SEC for extrusion

SEC = 4.57 x (Production rate)$^{-1}$ + 0.39
Sample size: 94 machines

Machine SEC for extrusion[1]

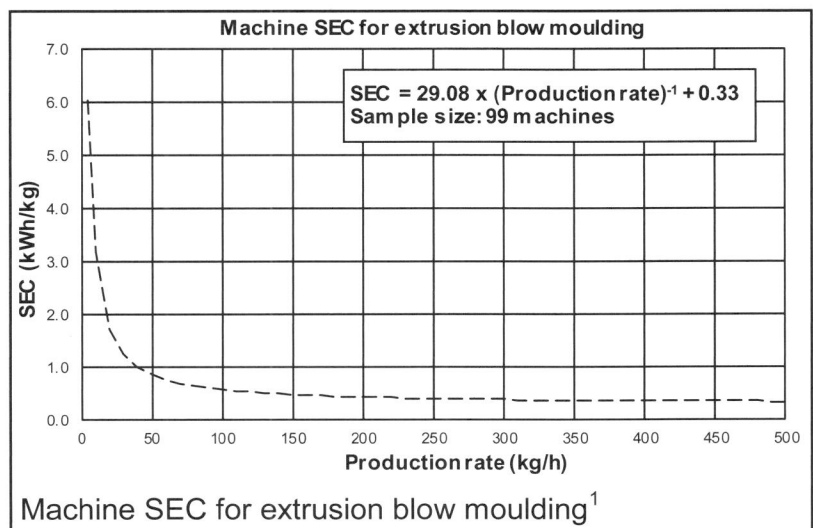

Machine SEC for extrusion blow moulding

SEC = 29.08 x (Production rate)$^{-1}$ + 0.33
Sample size: 99 machines

Machine SEC for extrusion blow moulding[1]

6.8   Integrating energy into the accounts

Accounting – a key function

The potential energy savings from simple actions are known to be large but when sites attempt to reduce energy use the achieved reductions are often far smaller than those known to be possible. One of the major reasons is that energy management is seen simply as a 'technology fix' that has little to do with the financial aspects of the site. The activities are seen as involving the technical and production areas but are not rewarding enough for the other staff to get involved. Energy management is somehow seen as a 'sideshow' or as a 'minority sport'.

Nothing could be further from the truth and some of the most effective energy management efforts can come directly from the accounting function. So why is it that the accounting function pays little attention to energy efficiency and the positive benefits?

The magnitude of the costs

The approximate magnitude of the cost of energy in plastics processing is shown on the right. The exact breakdown will vary with the site considered but for many products the cost of energy is already the same magnitude as the cost of direct labour.

For mass-produced volume parts the energy cost represents around 5.8% of the product cost and for complex technical parts it represents around 5.3% of the product cost. Good energy management can reduce these costs by up to 30% and therefore reduce the overall product cost by up to 1.5%. If the profit margin is low then the cost of energy is almost certainly higher than the profits of the site. A reduction in internal costs such as the cost of energy will translate into a significant increase in profits for no extra sales. This is typically 25-30% but can be up to 50% in some cases.

Any internal low-risk activity that can raise profits by up to 50% should certainly attract attention at any site. Many sites do not hesitate to spend money on trying to increase sales but fail to see the benefits of spending money on increasing profits by increasing energy efficiency.

The efforts, risks and rewards of improving profits via energy management are entirely internal and within the control of the site. Increasing profits by 50% through the conventional external approaches means increasing sales by 50% with the associated concerns of increasing production capacity and increasing risk.

The standard procedures

At many sites there are few controls on the energy spend. The energy bills are received by the accounts department, regarded as a fixed cost and paid. The people who control the expenditure (the Production function) rarely see the bills and in any case also regard them as part of the 'cost of doing business'.

This separation of authority and responsibility makes management and control of the energy spend nearly impossible and reduces the effectiveness of any efforts to reduce the costs. The spend to reduce the costs is allocated to the maintenance department but the benefits of the spend are not seen by that department and even more rarely appreciated by anybody in management. When an activity receives all the costs but none of the benefits or recognition then it is not surprising that little gets done.

Section 6.4 covered the variable nature of energy use where the PCL gives a relationship of the form:

Total energy use = Process load × Production volume + Base load

Accountants can add value to a business by using their analytical skills to improve energy efficiency.

Nobody needs approval to spend money on energy.

Everybody needs approval to spend money on saving energy.

Does anybody see a fundamental disconnect here?

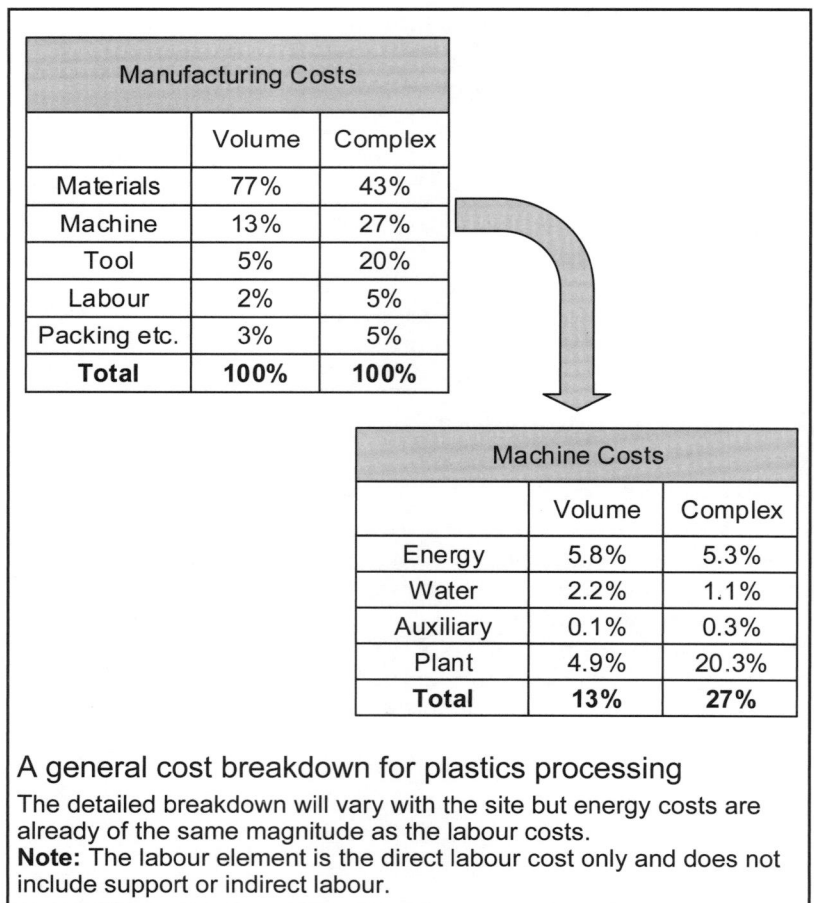

Manufacturing Costs		
	Volume	Complex
Materials	77%	43%
Machine	13%	27%
Tool	5%	20%
Labour	2%	5%
Packing etc.	3%	5%
Total	**100%**	**100%**

Machine Costs		
	Volume	Complex
Energy	5.8%	5.3%
Water	2.2%	1.1%
Auxiliary	0.1%	0.3%
Plant	4.9%	20.3%
Total	**13%**	**27%**

A general cost breakdown for plastics processing

The detailed breakdown will vary with the site but energy costs are already of the same magnitude as the labour costs.
Note: The labour element is the direct labour cost only and does not include support or indirect labour.

Accountants are familiar with calculating the total costs for a site from the fixed and variable components, where:

Total cost = Production volume × unit variable cost + Fixed costs

Therefore, the use of a similar approach for energy management should hold no surprises for most accountants.

- **Tip** – I have said this before and been taken to task by some of my colleagues: 'Accountants can be your friends'. Use their undoubted skills with numbers and analysis to embed energy into the management accounts as a variable cost.

Measuring to manage

These ideas allow energy management to be integrated into the accounts as with any other cost element and accounting for energy management can be treated in the same way as other items in the accounts systems. The overall aim is to achieve cost-effective energy management and integration of energy reporting into the accounting function which allows energy to take its place on the management agenda as a part of the normal management of the site.

In spite of this, the majority of sites have no active energy management programme or even reporting system. They consider that it is not central to their core business and are unaware of the potential for improvement or the substantial returns that can be made through small investments in energy efficiency. This needs to change to improve profitability.

To integrate energy into the accounting function there is a need to establish the measures that will be used to assess performance and this requires monitoring and targeting.

Monitoring and targeting (M&T)

In energy management, the concept of M&T is used to focus attention on energy consumption and the identification of cost reduction opportunities with attractive returns on investment.

M&T is the collection, interpretation and reporting of information on energy use. It measures and maintains performance and locates opportunities for reducing energy consumption and cost.

At most sites, the information needed for initial M&T can be taken from the existing Management Information System (MIS) and a large proportion of the benefits can be achieved by simple analysis of existing information. Basic historic data and a spreadsheet can be used to set up a simple system to start formal M&T.

Gathering data will not, in itself, provide results. The potential benefits of energy efficiency cannot be achieved by collecting large amounts of data or preparing lengthy reports. Data are meaningless without careful analysis, and reports are useless if they are not targeted at people with the authority and the will to act.

For any site, the amount of M&T needs to be appropriate to the energy spend. Effective M&T may need improved metering capability for accurate cost allocation. The decision is not whether to install meters to allow a breakdown of the use and the costs, but how many meters and where to put them.

Data are not the same thing as information. Data are simply a collection of numbers whereas information provides the basis for management action.

Sites can be awash with data and still have little real information.

6.9   Measuring energy costs

Data collection

Measuring costs always involves the collection of data but there is no point in spending more money to collect data than can be saved by the useful application of that data. It is always necessary to critically assess the cost–benefit balance for data collection. It is also important to recognise that good-quality data are not necessarily the same thing as highly accurate data. The need is for the minimum amount of data necessary to produce the relevant information.

Most of the core production data are probably already being gathered for cost and production control purposes. These data can often be used, with minimal changes, as part of an energy monitoring and targeting (M&T) system and sharing the data may only require simple modifications to enable a basic but effective M&T system.

Production data are either directly related to amounts, e.g., weight, volume, number of items, or do not directly relate to amounts, e.g., density, moisture content. Amount-related data are 'additive' and information for a week can be obtained by adding the daily data. Data not related to amounts are 'non-additive' and can sometimes be difficult to summarise and use. It may be difficult to establish an effective M&T system without recording some 'non-additive' data, e.g., it may be difficult to assess energy use when there are large variations in polymer moisture content and the drying process uses a significant amount of energy.

Other data may require specific collection for M&T, e.g., regular meter readings, and this may be collected either manually or automatically, e.g., the interval data available from the electricity supplier.

The basic data

Energy accounting falls into three categories:

Consumption data

This is the most basic data of all and is collected via the site metering system. There is a common misconception that M&T requires the installation of large numbers of meters. The use of sub-meters enables accurate allocation of costs but substantial progress on M&T can often be made with only a single meter using techniques such as the simple PCL approach (see Section 6.4). For medium to larger sites, where the cost of sub-metering is small in relation to the

amount of energy used, it is often economic to install limited sub-metering and a possible arrangement is shown below. This simple arrangement allows the process energy requirements (generally variable) to be separated from the building and utilities requirements (generally more fixed) but is dependent on the layout of the distribution boards.

Cost data

Only by expressing energy data in cost terms can integration into the accounting function take place. Cost data come from the supplier bills and must be part of the M&T system but they are not the most important in terms of actual management.

Driver data

These are data about the factors that influence energy consumption and can be divided into 'activity' and 'condition' drivers:

- Activity drivers are where the company activity influences energy consumption. For most plastics processing sites, the main 'activity' driver is the number of kilograms processed. Activity driver data come from internal sources, e.g., production data.

- Condition drivers are those where the consumption is not affected by the activity but by external conditions. For most companies, the main condition driver is the

> The prime function of the energy information system is the support of the overall strategy of the organisation.

> Be prepared to pay 1-2% of the total annual electricity bill for an effective sub-metering system.

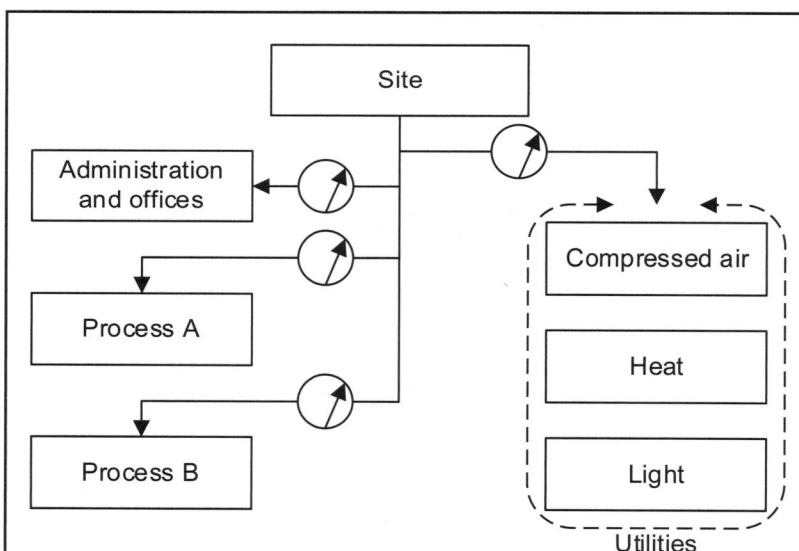

Typical energy use and metering diagram

Recording energy use simply from the main meter does not give all the information needed. Sub-metering allows a site to allocate costs and responsibility based on real energy use. Responsibility, assessment and recording drive real improvement.

weather (which influences the heating or cooling loads). Condition driver data come from external sources, e.g., degree day data.

Converting data into information

Data are useless unless they are converted into information; the simple presentation of data is not sufficient to account for energy efficiency. Producing information requires basic skills in data analysis and typical analytical techniques are:

- Performance Characteristic Line (PCL) – these are essential in the production of simple relationships between the drivers and the energy consumption. They aid prediction and allow an understanding of how various drivers affect consumption.
- Specific Energy Consumption – the SEC can be used to benchmark processing efficiency provided it is corrected for production rate. It is generally better to use the PCL approach to compare the predicted and actual use.
- Trend lines – these show the trend in energy use over time. They can be produced using 'moving averages' to remove 'noise' in the data but do not generally drive improvement.
- Energy profiles – these use supplier's interval data to show changes over short periods. They are particularly useful in looking at consumption over weekend periods when no production is taking place.
- CUSUM – CUmulative SUM of variance from standard performance charts are one of the most powerful methods of identifying and quantifying the impact of changes in energy use. They can be used to identify the time of changes in performance and are an invaluable tool.
- Comparisons against current and past energy performance and variances – whilst frequently used in accounting, they do not identify areas for improvement or drive improvement.

Assigning the costs

Cost reductions only happen if the person who controls the use of the resource is made directly responsible for them. After data have been converted into real information it is necessary to attempt to assign the costs to create ownership.

When monthly accounts are prepared for any business the operational costs are directly assigned to individuals who are responsible for the performance of that section of the operation. Energy costs should be treated in the monthly accounts package in the same manner as other operational costs.

Methods of assigning accountability vary with companies but typical methods are:

- Energy Accountable Centres – these make departmental managers accountable for the energy costs of their department. They allocate energy costs to departments and require them to operate within the allocated budget and to achieve agreed targets. This needs the ability to measure energy consumption by area.
- Quality-centred M&T – is based on the existing quality and environmental management systems (ISO 9001, ISO 14001) and shares information with these. Applying energy accounting within existing management structures means that it is less likely to be marginalised.
- Activity-based costing (ABC) – in traditional ABC costs are added by 'activities', where some activities add value and others do not. ABC identifies activities that add more cost than value, this can be extended to cover activities such as energy that have traditionally been included in overheads.

Planning the way energy management is organised is a senior executive responsibility which will determine the success or failure of the energy policy …

… this goes well beyond simply appointing an energy manager.

Degree days

Degree days are a method of measuring how hot or cold it is over a given period.

The best source for degree day data and some excellent explanations of the use of the data is www.degreedays. net.

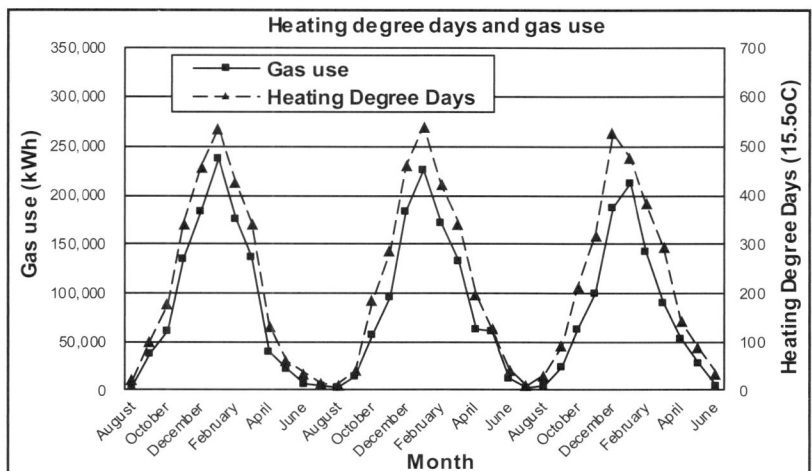

Heating degree days and gas use

Weather is a condition driver and the amount of cold weather affects heating energy use

A simple plot of heating energy use (in this case gas) versus the number of heating degree days in the month shows the strong correlation between the weather and gas use.

6.10   The site energy survey

The initial site energy survey

An initial site energy survey is the starting point for all improvement plans. The objective of the initial site energy survey is to gain an overview of the general site energy use. It is a walk around the site with an 'energy management' hat on. This will help to identify some of the rapid no-cost or low-cost improvements that can be made to reduce costs.

The survey should be carried out as soon as possible – if energy is being wasted now, it is costing money now. The main areas of energy use in plastics processing are in motors, heating, cooling and the provision of other site services such as compressed air. Use these pages as a guide during the walk-around to find areas of high or unnecessary energy use.

Carrying out a site survey

Take an unannounced walk around the site at around mid-shift. If there is no night shift it can also be profitable to take a walk around the factory when there is no production being carried out. This will identify the areas of base load use such as compressors left switched on and machinery idling with no production.

The questions to ask

Typical questions to ask during the initial site survey are:

- Which areas have the largest electrical load? Look for the largest machines, they will most likely also have the largest motors and create the largest loads when they are used.

- Is the thermal insulation, if present, on all the machines in good condition? If there is no insulation then why is it not present?

- If it is hot and you are paying to heat it, then it probably needs insulation.

- If it is cold and you are paying to cool it, then it probably needs insulation.

- Look for signs of machines that are not in production but have motors or ancillary equipment running. Typical examples are conveyors, pumps, granulators, fans, machine heaters. All of these can be fitted with simple on-demand switches to stop the operation when there is no product or requirement for operation.

- Look for machines being kept idling before the next production run. Is there a shut-down procedure to specify the longest time that machines are to be left idling?

- Look for motors that are left running when not doing productive work.

- Why are the motors the size they are? Would a smaller motor be more efficient and cost less to run?

- Which cooling water pumps (and chillers) and vacuum pumps are still running when there is no production?

- Are chillers being used when they are not necessary?

- Is the airflow from fans being throttled back with dampers to get the correct flow rate and could variable speed drives be used instead?

- Look for areas of energy use where no productive work is being done and yet machines are running and using energy. This is particularly important after the factory has stopped for the day or week.

- Look for water, air or steam leaks. These should be obvious in most cases (you should be able to see or hear them) and in many cases represent not only lost energy but in the case of steam leaks they are also a potential health and safety concern.

- Where can you hear steam and compressed air leaks? The hissing noise you hear from leaks is costing real money. If there is no

> Low-cost energy efficiency measures can improve profits significantly.

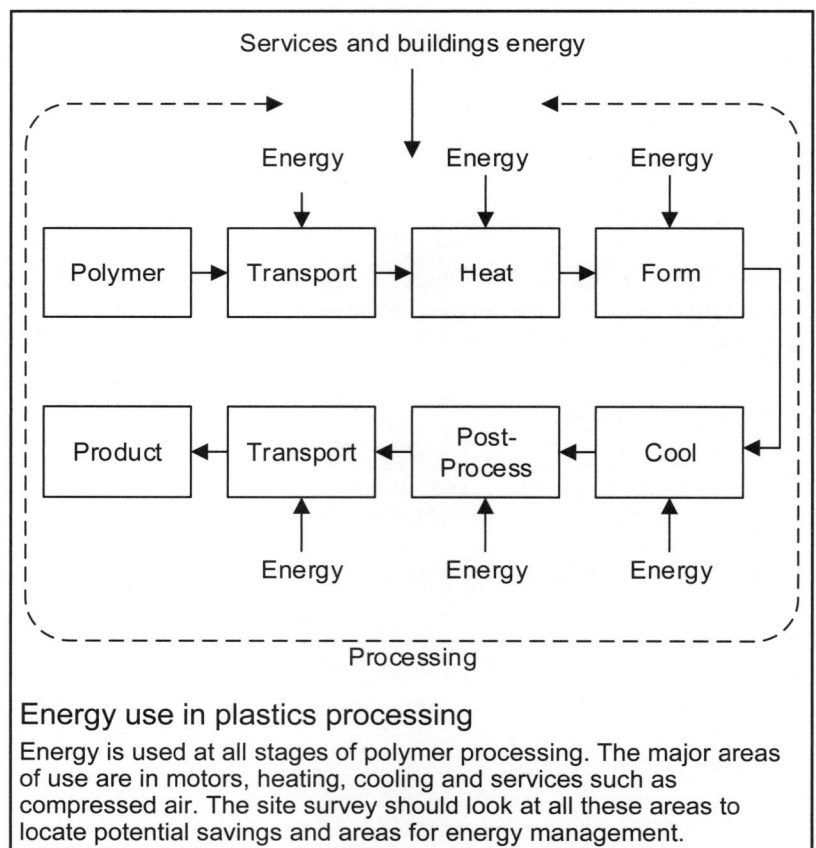

Services and buildings energy

Polymer → Transport → Heat → Form
Energy ↓ (Transport, Heat, Form)

Product ← Transport ← Post-Process ← Cool
Energy ↑ (Transport, Post-Process, Cool)

Processing

Energy use in plastics processing

Energy is used at all stages of polymer processing. The major areas of use are in motors, heating, cooling and services such as compressed air. The site survey should look at all these areas to locate potential savings and areas for energy management.

production being carried out then why is the compressed air system still running?

- Is compressed air being used for expensive applications where other cheaper methods can be used, e.g., cleaning or drying?
- Does the compressed air pressure need to be so high, or the vacuum so low?
- Is the lighting dirty or broken?
- Can natural daylight be used to reduce the need for artificial lighting?
- What are the good, simple maintenance measures that can be adopted to reduce energy use?
- Are 'accepted' practices wasting energy? Can they be modified at no cost at all?
- Are there clear setting instructions for all machines and products and are they implemented when a machine is set up?

The answers to questions such as these will generate a series of actions to take in order to reduce site energy use.

Turning the survey into energy savings

A site energy survey is useless unless action is taken as a result of the findings.

- Use the survey to estimate the excess energy use of the site. It is generally in the region of 30% so be brave, but never over-estimate. If you cannot find this much then consider getting help from an experienced energy auditor.
- The energy distribution map and spreadsheet (see Section 6.2) will give the overall site energy use and the location of the largest energy consumers.
- Arrange for the site electrician to measure the total factory electrical load and to calculate the costs involved.
- Get further details on energy-saving equipment and procedures for the plastics processing industry from the companion book[1].
- Use the survey and information to identify operating practices or machinery that cost money and can be changed for little or no cost.
- Send the results of the survey to the Managing Director and Production Director with a fully costed action plan.
- Carry out the action plan to reduce the energy costs for the site.

Raising energy awareness

As with any exercise, the cost savings and sustainability benefits possible from energy efficiency will only be achieved if there is a real management commitment to reducing the costs, improving sustainability and carrying out the actions identified. This commitment is best ensured by developing a company energy policy that is as much a part of the overall company operations as the quality policy.

The energy policy can form part of a broader company environmental/sustainability policy but must be formally adopted with top-level management commitment. The energy policy should also be the responsibility of a designated person who has clear responsibility for energy matters. An essential part of the policy is that there should be regular communication (both formal and informal) with the major energy users. These major users should also be held accountable for the energy use that they control and the energy use should be monitored and targets should be set.

The quantified savings from the actions implemented as a result of the site survey and energy policy should be promoted within the company and used to create a favourable climate for the investment assessment of other energy reduction initiatives.

Energy efficiency is an integral part of sustainability as well as cost management and can provide a competitive advantage in any market. The initial site energy survey is the starting in gaining an advantage for the site.

ISO 50001 (see Section 2.10) provides a really good framework for implementing the results of an initial survey BUT does not provide the projects necessary for reducing energy use.

- 1. Kent, R.J. 2018. 'Energy management in plastics processing', Elsevier.

6.11   Injection moulding

Machine operation

Over 90% of the energy costs in injection moulding are due to electricity use. This makes energy management a key cost management area for moulders.

For hydraulic machines, in the complete cycle, only 5-10% of the total energy used is actually input to the polymer, the other 90-95% is used simply to operate the machine and large savings can be made without affecting the product in any way.

Machines

As with most machines, the initial cost of a moulding machine will be less than the cost of energy used during its lifetime but the energy cost will be even more for machines that are not energy efficient. Although they may cost more initially, energy-efficient machines will save money in the long term – an important factor when customers are beginning to expect price decreases through the lifetime of a product. In this market it is important to use a 'whole life costing' approach when purchasing new machines and to include the energy costs in these calculations. Machinery suppliers are aware of these changes in the market and the new-generation machines often have improved energy efficiency. In some instances, this can reduce product costs by over 3%. Where the basic machine is not energy efficient, many machinery suppliers can provide additional equipment to reduce energy consumption. This will increase initial costs but produce long-term savings.

Getting the right machine for the job is vital for energy efficiency and the machine capacity should be closely matched to the product. Using large machines for small products is inherently wasteful. Large oversized motors at part-load are less efficient than small motors at full-load. As with most machines, moulding machines are most efficient near their design load and total machine energy efficiency decreases as the operating conditions move further away from the original design conditions.

- **Tip** – Check that all jobs are on the appropriate machines.

'All-electric' machines are an energy-efficient moulding solution and can both reduce energy use and make computer control easier and more direct (see Section 6.12).

In hydraulic machines the system needs to provide peak power for a very limited time and the hydraulic system is overrated for most of the time. The use of accumulators for rapid hydraulic energy release can allow significant reductions in the size of the hydraulic system.

Electric motors account for over 80% of the electricity used in moulding machines and the highly variable loads in the moulding cycle can give low power factor values, e.g., in the region of 0.7, this is low and increases costs. PF correction equipment can easily be retrofitted to increase the PF to greater than 0.95 and this generally has a payback of less than one year.

Controlling the start-up sequence of machines can reduce energy costs with few other effects. Attempting to start multiple machines at the same time will increase the MD (see Section 6.3) and the cost of energy.

- **Tip** – Fit a warning device to the MD meter to sound or flash when the MD approaches the allowable limit.

- **Tip** – Plan and control the start-up sequence so that not all machines are heating up at the same time to limit the MD.

Heat transfer to the barrel is improved by pre-seating the heating element to the barrel and by using flexible metal bearing compounds. Thermal efficiency can also be improved by barrel insulation. This has a rapid payback (generally under one year) and improves other areas such as health and safety and barrel temperature fluctuations due to air currents.

Machines use energy even when idling and the amount varies with the machine type but can range from 52–97.5% of the full

> Good practice is inexpensive and reduces all costs – not just energy costs.

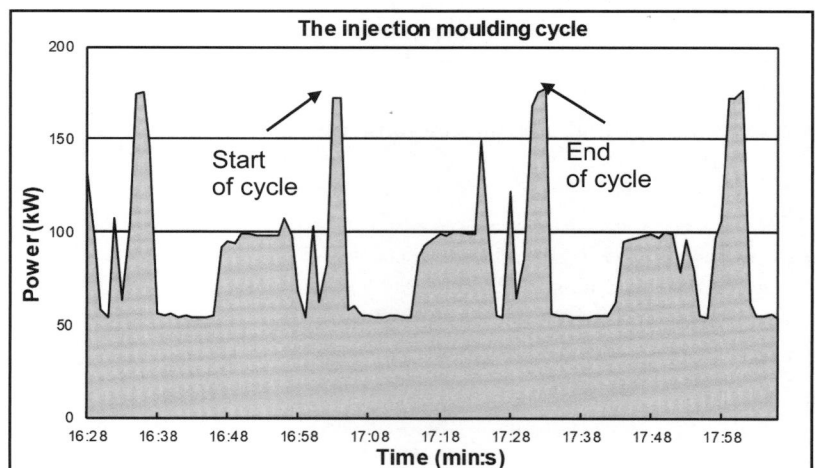

The injection moulding cycle

Start of cycle

End of cycle

Power drawn in the injection moulding cycle

The power drawn during the injection moulding cycle fluctuates regularly through the cycle and it is possible to see the machine movements that draw power and use energy. Each plot is a unique signature of the machine, tool and settings.

moulding energy consumption. An idling machine is not 'free' it is costing large amounts of money. Idle periods of greater than 20–45 minutes may make it cheaper to switch off and restart.

- **Tip** – Define an 'idling' mode for all machines – heaters reduced, hydraulics off, compressed air off, conveyors stopped.
- **Tip** – Switch off barrel heaters and cooling fans between runs.
- **Tip** – Stop cooling water circulating through idle tooling.
- **Tip** – Stop supplying compressed air to idle machines.

Preventative maintenance such as de-aeration of the oil system and maintenance of the controls will reduce energy costs.

- **Tip** – Monitor machine energy use to identify deterioration of the machine.
- **Tip** – Increased maintenance can lead to significant energy savings.

Moulds

Product cooling time is often more than 50% of the total cycle time. Good cooling will reduce cycle times and energy use – a double benefit. Air in the cooling system reduces cooling efficiency and degassed and pressurised systems will reduce cycle times and energy use. Cooling systems are often set and forgotten – check that cooling water is at the maximum possible temperature and minimum quality and is efficiently treated and distributed (see Section 6.17).

Excessive tool change times will waste energy even if the machine is idling. Rapid set up of tooling reduces energy use and improves overall factory effectiveness.

- **Tip** – Plan tool changes into production schedules and use rapid set-up methods.

Ancillaries and services

Ancillaries use energy in electric motors and in utilities consumption. In a highly automated factory, the ancillary energy demand can even be comparable to the main machine energy demand. The main opportunities are in minimising the demand for utilities because the motors are generally small and run intermittently so it is not often cost effective to retrofit more efficient motors or controls.

- **Tip** – Specifying energy efficiency during design of handling and ancillaries will give rapid payback on any additional costs involved.
- **Tip** – Design handling systems to operate 'on-demand' only.

Granulation and scrap recovery also use large amounts of energy and can raise costs considerably if they are carried out at the

wrong time or if the regranulator continues to idle with no product. Carrying out granulation at night will reduce costs.

Heat recovered from hydraulic systems and chiller units through heat exchangers can be used to provide space heating for offices and other areas with payback times of around six months.

- **Tip** – Look for opportunities to recover heat and re-use energy. It has been paid for so why not use it?

Management

Optimising machine settings reduces the electrical energy needed. Get machines set right, record the settings and do not change them unless absolutely necessary.

The goal

Management is at the heart of energy efficiency. Without good management, neither energy efficiency nor any other change in operating practices will be effective.

Energy-efficient injection moulding is good moulding practice.

It is inexpensive and reduces all costs – not just energy costs.

For more information on energy saving in injection moulding, see:

Kent, R.J. 2018. 'Energy management in plastics processing', Elsevier.

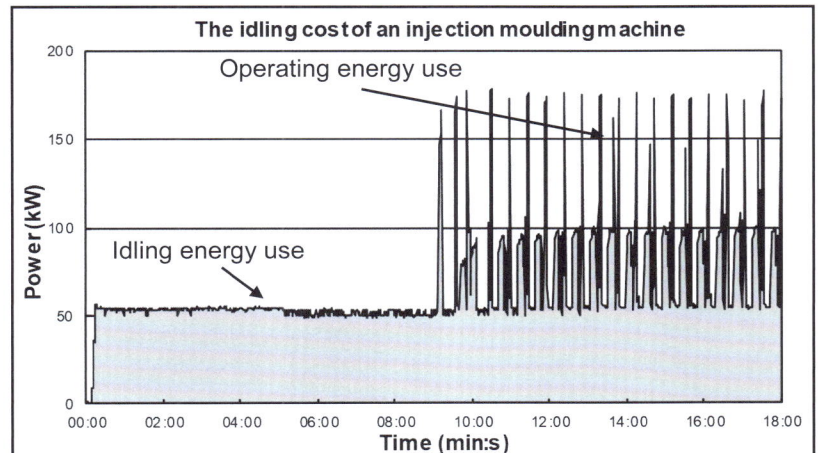

The idling cost of an injection moulding machine

Idling cost of a hydraulic injection moulding machine

The idling power drawn by a stopped hydraulic injection moulding machine will be between 52 and 97.5% of the running power. This is a considerable cost and hydraulic motors should be stopped as soon as possible.

6.12 All-electric injection moulding machines

A mature technology

All-electric injection moulding machines have seen a rapid rise in application in many parts of the world, primarily because of their energy efficiency.

Traditionally the cost of energy has represented between 4 and 5% of the cost of a moulding but rising energy costs are increasing this considerably (see Section 6.8). The cost of energy used in producing a moulding can easily be the difference between profit and loss for the job and represent the profit that is achieved in the business.

Early all-electric machines had a significant purchase cost differential compared to conventional hydraulic machines (in the order of 50%) but, as with any new technology, this differential has decreased rapidly and continues to decrease. As noted earlier, the initial purchase should not be the deciding factor and the important cost is the 'whole life cost' of the machine (initial cost + operating costs).

Machine size

All-electric machines are also rapidly increasing in size and clamp force. When first introduced, the maximum clamp pressure available was approximately 30 tonnes. This has now increased to over 1,000 tonnes and continues to rise with technology improvements in transmissions and motor design. A decrease in the inertia of servo motors has also allowed faster reaction times during the injection phase and higher speeds during the clamping phase.

The benefits

All-electric machines have many benefits that are independent of the specific manufacturer and these are:

Energy savings

All-electric injection machines have the potential to reduce the energy use in injection moulding by between 30 and 60% depending on the particular moulding and the all-electric machine being used. Energy profiles through the moulding cycle show that energy is saved during all the phases of the moulding cycle.

Controlled trials carried out by suppliers show significant energy savings across a broad range of materials (from PS to PC). The energy savings can be achieved even if the cycle time is kept at that required for the conventional machine.

In most cases, decreased cycle times are also achievable by carrying out operations in parallel (such as clamping and injection and opening and ejection) to reduce energy consumption and increase productivity. Using all-electric machines and optimised cycle times maximises the energy saving and productivity of the machine.

On hydraulic machines, the hydraulic system provides peak power for a very short part of the cycle and is overrated for much of the time. Accumulators can be used as storage for rapid hydraulic energy release to reduce the energy consumption but the hydraulic system is generally overrated. In contrast, all-electric machines use only the power needed and at the time it is needed.

The use of all-electric machines also eliminates the need for hydraulic heating/cooling and both the equipment and the energy use associated with this.

Operations

Removing the hydraulic system from the machine is one of the major effects with all-electric machines and this removes a significant variable from the process. This has a multitude of benefits:

- No hydraulic system present, means no requirement for hydraulic oil to be stocked, provided, filtered, changed or disposed of – all operations that take time, cost money and use energy.
- There is no hydraulic oil present to contaminate the operations area or the environment.
- No hydraulic system means no waiting for the hydraulic oil temperature to stabilise and quicker start-ups.

All-electric or 'hybrid'?

Some manufacturers use a 'hybrid' technology combining both electric and hydraulic operation for specific applications. This allows moulders to benefit from the advantages of both electric and hydraulic operations.

'Hybrid' machines might not achieve all of the benefits listed.

All-electric machines show energy efficiency during all phases of the injection cycle.

Application	Typical recorded energy saving
Medical product (inhaler)	58%
Medical product component	60% in PS (53% in PC – with same mould conditions)
Automotive product (connector)	62%
Household product (shower panel)	55%
Cap stack tool	Between 28% and 64%
Garden product (flower pot)	40%

Typical recorded energy savings for all-electric machines in a variety of applications

The improved performance and positional accuracy of servo drives gives greatly improved process control and a process that is easier to set up, easier to adjust and calibrate and more stable in series production. Typical all-electric machines can control machine movements and shot weights up to ten times better than hydraulic machines. This accurate control of the machine prevents processing too much material, optimises the amount of material used and the energy used to process it, as well as significantly reducing rejects and improving reproducibility.

All-electric machines are directly driven – the motor directly controls the machine movements – unlike a hydraulic machine where the drive from the motor is indirect and via the hydraulic system. The reduced system inertia (no valves to open or close) makes operations quicker, more direct and more controllable. This can give cycle time reductions of up to 30% without any degradation of product quality. This also reduces the energy use and costs.

Secondary operations

Hydraulic control of cores on existing moulds that are currently powered from the machine's hydraulic system is easily possible on all-electric machines by the use of a small hydraulic power pack. This allows existing tooling to be used on all-electric machines with no modification.

Installation

Installation of all-electric machines is generally easier and cheaper than the installation of hydraulic machines because plumbing, cooling and filling of hydraulic oil systems is not necessary.

Maintenance

One of the benefits of all-electric machines is the reduced maintenance load of the machines. Hydraulic systems account for a large proportion of the maintenance requirements of hydraulic machines and the removal of the hydraulic system significantly reduces the maintenance load of the machine:

- There are no consumables, e.g., oil and filters, and maintenance stockholding is greatly simplified and reduced.
- No hydraulic system means no need for cleaning and servicing and no oil leaks.
- The reduced number of operating parts means fewer parts to service and replace. Servicing is also reduced in complexity, time and cost but may require more highly qualified service technicians.

Overall, all-electric machines have a reduced risk of failure and can be more easily used in

'lights out' operations than conventional machines.

Costs

All-electric machines can have a significant benefit in overall cost terms for moulders. The reduction in the energy use is a key factor and whilst energy savings of 30% are very common, these can rise to 60% for specific products when cycle time reductions are taken into account.

The improved cycle times, increased reproducibility and precision results in improved productivity, reduced production capacity requirements, and can lead to significant overall cost reductions.

All electric machines have a much-reduced heat output because more of the energy used is applied directly to the process, in clean rooms this will give a reduced load on the air-conditioning and filtration plant and a reduction in energy costs.

All-electric machines can control machine settings more accurately to the set point.

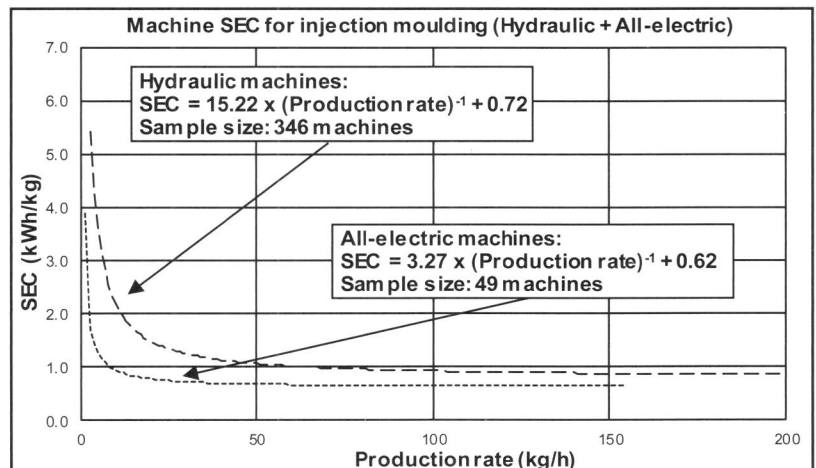

Machine SEC for injection moulding (Hydraulic + All-electric)

Hydraulic machines:
SEC = 15.22 x (Production rate)$^{-1}$ + 0.72
Sample size: 346 machines

All-electric machines:
SEC = 3.27 x (Production rate)$^{-1}$ + 0.62
Sample size: 49 machines

SEC (kWh/kg) vs Production rate (kg/h)

The energy savings from all-electric machines

All-electric and hybrid machines can save up to 60% of the energy depending on the production rate. The savings are less at higher production rates because of the decreased effect of the base load.

6.13   Extrusion

The key process

Extrusion is not only a final forming process for products but is also an intermediate process for most other processing techniques such as injection moulding, blow moulding and film blowing. The cost-effective operation of extrusion screws is therefore essential to much of the plastics processing industry.

The process is highly dependent on electricity and most of the energy used is directly related to operation of the extruder screw. For profile extrusion 50% of the total energy is used to drive the extruder and the remaining energy is used for items such as ancillaries and utilities (see box on the right). Most of the thermal energy involved in plasticising and heating the polymer comes from the shearing of the polymer whilst it is being moved by the extruder screw. As with many other plastics processing techniques, industry surveys show that a typical company should be able to reduce energy use by up to 30% without major capital outlay.

The extruder

The initial cost of energy-efficient extruders may be higher but they will give rapid returns on the extra investment. High-efficiency AC motors and VSDs (see Section 6.15) are now almost standard equipment on new machines and are also cost effective when replacing motors and drives. New style Permanent Magnet (PM) motors are also very effective in extruders.

Whatever the age of the machine, it is essential to get the right extruder for the job and the screw diameter and design should be checked to make sure they are right for the polymer and product. Extruders run most efficiently (not only in energy terms) when operating at the design conditions. The extruder should be set to run at the maximum design speed, as this is usually the most efficient speed. The screw speed should be controlled to give an extrusion rate as close to the maximum as possible and still produce good product.

- **Tip** – Using large extruders for small profiles wastes energy and costs money.

Extruder motors run most efficiently close to their design output and the electric motor should be sized and controlled to match the torque needed by the screw.

Optimising the extruder speed in this way maximises the heat from mechanical work and minimises the amount of electrical energy needed to heat the plastic.

- **Tip** – If the downstream equipment does not limit the output, the SEC (kWh/kg) of an extruder can be decreased by nearly 50% by doubling the rotational speed of the extruder.

Good extrusion demands that the polymer is kept at the optimum processing temperature whilst at the same time prevented from overheating. Depending on the material, this 'processing window' is small and overheating from shearing is common unless accurate temperature control is present. Accurate temperature control will not only produce good product but will also minimise energy costs.

- **Tip** – Check extruder controls to make sure that the heating and cooling are working together and not competing.

- **Tip** – In most cases, insulation is not recommended for extruder barrels. Most of the heat input comes from shear heating and this is the most efficient heating method. If barrel insulation is used then take care to avoid overheating and a runaway process.

- **Tip** – Insulation downstream of the extruder screw tips is strongly recommended for most applications. This can be on die heads (see box on the lower right), on transfer pipes from secondary to primary extruders, on oil transfer piping and on any other area where electrical heating is applied and there is little shear heating. In areas such as these, insulation has a payback of around 12 months and

Extrusion is a key forming process and is integral to many other processes.

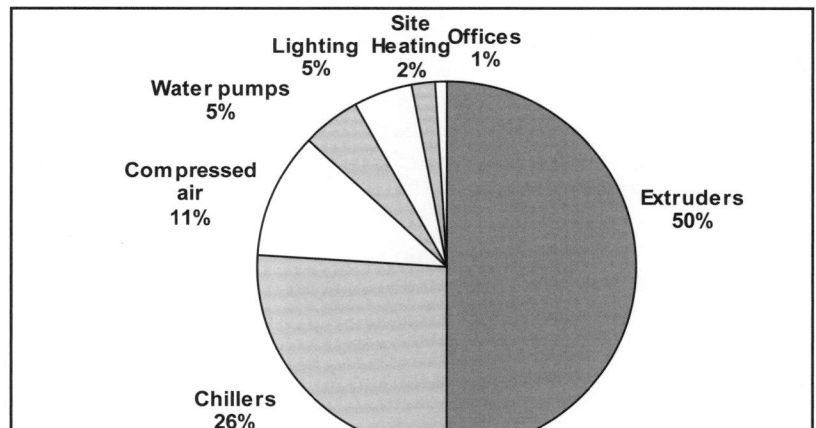

Energy use at a typical profile extrusion site

The extruder drives use the most energy (50%) but chillers, water pumps and compressed air use nearly as much (37%). Minimising the demand for these three key elements will reduce energy use and costs in extrusion.

Chapter 6 – Manufacturing: energy

also reduces health and safety issues.

'Standby' operation of extruders will not have a large energy use in the main motor (it will be switched off) but can use significant amounts of energy in utilities through barrel heaters, cooling water, calibration vacuum and lights.

- **Tip** – Find the minimum 'standby' settings and set up routines to always leave machines in this condition when they are not producing.
- **Tip** – Turn off barrel heaters and cooling fans between runs when the time between runs is sufficient and after barrels have been purged (if required).
- **Tip** – Turn off cooling water on idle machines where possible.
- **Tip** – Turn off vacuum generation on any vacuum calibrators when production stops.

The energy use in an extruder is also a sensitive measure of the condition of the extruder and can be used as a diagnostic tool. Increasing energy consumption is an early warning of deterioration of the machine/screw condition and the need for maintenance of the machine.

- **Tip** – Increasing the frequency of maintenance involves effort and cost but will lead to significant energy savings.

Setting up a Total Productive Maintenance (TPM) programme involves some additional effort and costs but will lead to significant energy savings in processing and will also keep machines and systems in the best condition.

The ancillaries

The largest opportunity for energy saving in ancillaries is in minimising the demand for utilities.

- **Tip** – Specifying energy-efficient ancillaries will give a rapid payback on any additional costs involved.

A necessary first step in energy-efficient extrusion is to get the main extruder set correctly, if the extruder is at the optimum conditions the need for downstream cooling and calibration will be minimised.

For utilities, the approach should always be to 'minimise the demand and then optimise the supply' (see Section 6.15).

- **Tip** – Find the maximum acceptable extrudate temperature after cooling and set the maximum cooling water temperature to achieve this. Do not overcool the product.
- **Tip** – Check that cooling water is not circulating through idle calibrators, e.g., fit controls to isolate the water if the main motor is not working.
- **Tip** – Check that cooling water is treated,

chilled and distributed efficiently.

- **Tip** – Check that compressed air is not supplied to idle machines, e.g., fit controls to isolate compressed air if the main motor is not working.
- **Tip** – Check that compressed air is generated and distributed efficiently at the minimum pressure.
- **Tip** – Check that the vacuum supply is the minimum needed and that it is generated and distributed efficiently.
- **Tip** – If vacuum calibration is used then ensure that the calibration boxes are adequately sealed to prevent vacuum leakage.
- **Tip** – Check that the vacuum supply is switched off when it is not needed, e.g., fit controls to isolate vacuum supply if the main motor is not working.
- **Tip** – If replacing electrical motors then match the size to the actual demand and fit energy-efficient motors.

For more information on energy saving in extrusion, see of:
Kent, R.J. 2018. 'Energy management in plastics processing', Elsevier.

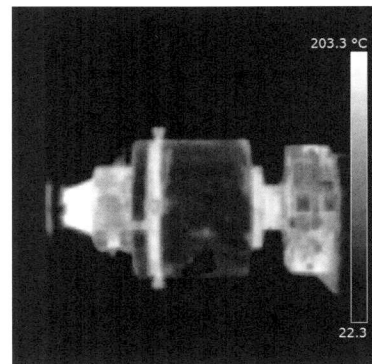

Thermograph of hot sheet extrusion die head

The extrudate is at 203°C but large areas of the uninsulated die head are > 200°C and would benefit from insulation to reduce heat losses to the atmosphere and to stabilise the process. Exposed hot surfaces also have health and safety implications.

6.14   Extrusion blow moulding

Machine

The major component of energy use is the extruder area of the blow moulder and this typically uses over 55% of the total energy supplied to the machine with machine motions using over 28%. As with other processes, energy-efficient machines may be initially more expensive but will have lower long-term operating costs and will show a rapid payback of the extra investment.

For blow moulding, the use of all-electric machines is an energy efficient option because these machines remove the inevitable energy losses at the electro-hydraulic interface. These EBM machines are now becoming more widely available and substantially reduce the process energy costs.

Whichever type of basic machine is used, good process parameter control gives more efficient operation and reduces the cost of operation in all areas, not simply in terms of energy efficiency. Process controller improvements can give good paybacks and it is often worthwhile investigating in upgrades to machines. Improved process control will give controlled, accurate and minimised wall thickness and parison length and will improve energy efficiency and materials use.

The energy use in blow moulding is extremely dependent on the cycle time and the process should be set to use just enough energy to complete each process stage. Energy costs can be reduced by reducing the heating time, cooling time and other cycle stages. This will improve process efficiency and reduce costs.

Blow moulding machines generally use only small amounts of externally applied heat (most is generated mechanically) and heat transfer from barrel heaters can be maximised and evenly distributed by good seating of the heaters to the barrel and the use of conductive metal compounds.

- **Tip** – Energy use may be reduced by barrel insulation but this should be checked by monitoring the machine to prevent overheating. Insulation can also reduce start-up times and improve the health and safety concerns of hot barrels. Barrel insulation generally has a pay-back of in the region of 1 year.

- **Tip** – Machine temperatures should naturally be set at the minimum temperature the polymer actually needs to reduce the need for external heating.

- **Tip** – There is often poor insulation of the 'cold' side of extrusion blow moulding, i.e., the cooled moulds are rarely insulated and this increases the load on the chillers through parasitic heat gain.

Tops and tails

Depending on the product and design, the tops and tails can be up to 80% of the total extruded parison weight. The material in tops and tails is almost always recovered and re-used but the time and energy lost in the first (and subsequent) process passes is lost forever (see Section 5.1).

The industry average for tops and tails is in the region of 35% but best practice is as low as 5% (although this may not always be possible due to product designs or materials handling constraints).

- **Tip** – Calculate the amount of tops and tails being produced by each machine and tool combination. Be prepared for a surprise!

- **Tip** – Do not be confused by the low volume of the tops and tails versus the high volume of the actual product. Tops and tails are often solid whereas the product is largely air.

- **Tip** – Specify the allowable weight of tops and tails and monitor this regularly. Tops and tails must be part of the control measurements for the process during production.

- **Tip** – Do not allow setters to hand a machine over to production unless the

Large tops and tails represent a loss of productive capacity to a site.

Top and tails must be minimised to reduce energy use and improve productivity.

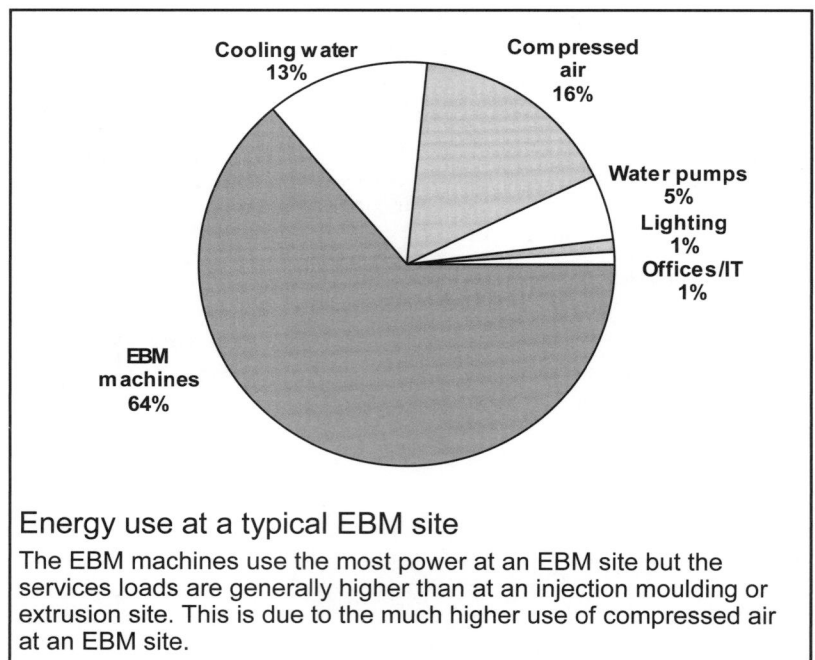

Energy use at a typical EBM site

The EBM machines use the most power at an EBM site but the services loads are generally higher than at an injection moulding or extrusion site. This is due to the much higher use of compressed air at an EBM site.

amount of tops and tails is within specification.

- **Tip** – Improved control of the parison and final product size will improve energy and process efficiency.

Regranulation of tops and tails can be done off-line (at night) to minimise energy costs, but the first step is to minimise the production of tops and tails – reduce and then re-use.

When a machine is not producing for a short time it is often not practical to shut down the extruder but temporarily shutting down the hydraulic systems can give considerable energy savings. Between runs, the barrel heaters and cooling fans should be turned off to reduce energy wastage.

Start-up procedures should be established to bring the energy demands online at the best (and latest) possible time, i.e., heaters, hydraulics and finally the extruder drive. Similarly, shutdown procedures should be developed to switch off the energy-intensive areas of the machine as soon as possible.

- **Tip** – Develop start-up and shut-down procedures to save energy and time.

Ancillaries

The formation of the parison must be complete before the outside surface chills and stops surface texture formation. This places large transient demands on the compressed air system. The compressed air pressure for blowing should be adjusted to be just sufficient to form the parison before chilling is complete but it can then be reduced to simply hold the parison against the mould surface.

- **Tip** – Excessive air pressures for blowing or holding wastes energy. Reduce the pressures used to the minimum needed.

- **Tip** – Investigate the use of accumulators to cope with high transient demands.

Most of the heat put into the material during the softening stage must be removed before the product can be released from the die. The product cooling time is about 50% of the cycle time and minimising the melt temperature will save energy in both heating and cooling as well as reducing the cycle time.

- **Tip** – Setters may raise temperatures or increase cooling times to get a job running. Check the settings and reduce these to the minimum actually required.

- **Tip** – Establish the optimum settings and ensure that these are used at all times.

Chillers use large amounts of energy in operation and the process efficiency of chilling affects both the time taken and the energy used in the process. Water has a

better cooling efficiency than air and bubbles in the cooling water will decrease the efficiency of the cooling (see Section 6.17).

- **Tip** – Seal, degas and pressurise the cooling water system.

Cooling is most efficient with good contact between the parison and mould and this should be kept by holding the pressure at the minimum required during the cooling stage.

Hydraulic systems for mould closing should be matched to the demand (blowing pressure × projected area) to reduce the energy needed and the hydraulic oil should be de-aerated on a regular basis to improve the efficiency of the hydraulic system. The hydraulic fluid should also be kept at a steady temperature to improve process control and prolong the life of the oil.

Top and tail management in EBM is critical to both energy use and productivity.

It is not accurate to think that because all the tops and tails are 'recycled' that there is no waste in the process.

For more information on energy saving in extrusion blow moulding, see:

Kent, R.J. 2018. 'Energy management in plastics processing', Elsevier.

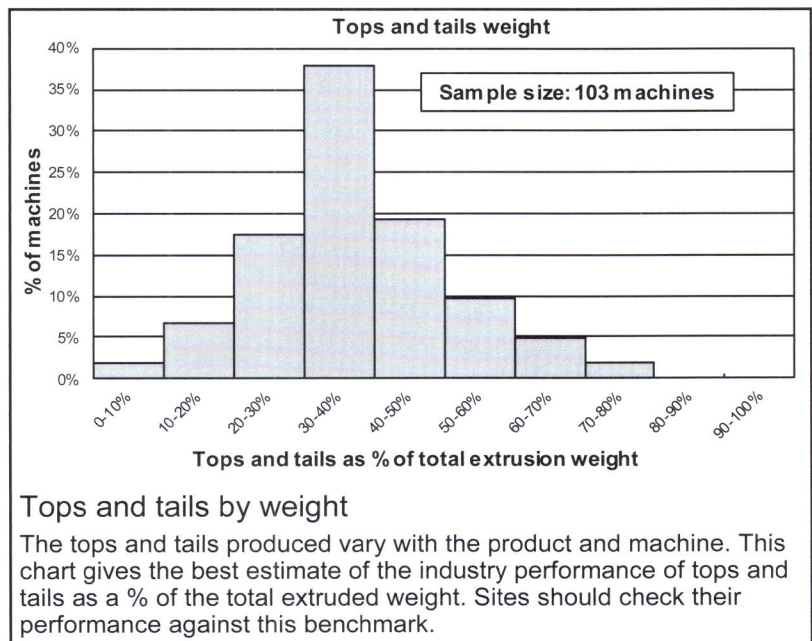

Tops and tails weight

Sample size: 103 machines

% of machines

Tops and tails as % of total extrusion weight

Tops and tails by weight

The tops and tails produced vary with the product and machine. This chart gives the best estimate of the industry performance of tops and tails as a % of the total extruded weight. Sites should check their performance against this benchmark.

6.15 Motors

The motor management policy

The greater importance of running costs over the initial purchase price means that companies need to change the way they look at motors. Traditionally decisions have been made by the electrician on the spot and are governed by the need to keep production running. A motor management policy sets the framework for the decisions based on the 'whole life cost' of the motor where all purchase, maintenance, repair and operating costs are considered. The changes with the development of Variable Speed Drives (VSD) and High Efficiency Motors (HEM) mean that, in order to reduce costs, companies should establish a motor management policy for the purchase and operation of motors. This policy should include guidelines on

- Repair and replacement based on lifetime costing.
- The specification of HEMs for all new purchases (see below).

When new motors are required, the benefits of opting for HEMs are obvious, i.e., they are up to 6% more efficient than standard motors. However, the failure of an existing motor raises the question of whether the motor should be repaired or replaced. Rewinding a failed motor may initially appear to be a cost-effective action but rewinding can reduce energy efficiency by up to 1% and may not be the most economical long-term action.

A motor management policy will provide the rules for making the best financial decision and save costs in the long term.

The motor management programme

All services should be examined within a common framework of 'minimise the demand and then optimise the supply'. For motors, this is the 'motor management programme' that provides a structure for reduced energy use and costs.

Stage 1 – Minimise the demand

Step 1: Turn the motor off

Operating a motor that is not required is a total waste of energy and money. It is generally easy to find motors operating when not needed and doing no useful work. You simply have to look for them.

There are many low-cost methods for turning motors off and the method chosen depends on the application and the potential savings.

- **Tip** – Link downstream equipment to the main machine so that it is turned off when not needed.

The main barrier is a lack of management attention and a failure to realise that operating motors needlessly costs money.

Step 2: Reduce transmission losses

Transmission losses are often a major component of the overall losses in the motor system and can easily exceed the losses in the motor itself. Improving the transmission method reduces the overall load on the motor and can be used to optimise the motor size. Transmission losses will occur with gearboxes, belts and chain drives. Reducing transmission losses is primarily about asking: Are we maintaining the system correctly and is there an alternative method of driving the system?

- **Tip** – For belt drive systems, reduce transmission losses by replacing conventional V-belts with cogged belts. These use the same pulleys but need less maintenance, run cooler, last longer and are around 4% more efficient than standard V-belts.

Step 3: Reduce the driven load

A significant number of motors do not actually perform a useful task. In some cases, the task was once relevant but has now been removed but nobody thought to

> The energy cost of running a motor can exceed the purchase cost after just 1,000 hours.

> Set up a 'motor register', this is a list of all the motors on a site (> 3 kW) with details of the rating, frame size and an estimate of the run hours. This list should include:
> - The main machine motors.
> - The services on the machines, e.g., vacuum pumps and regrinders.
> - The main services motors, e.g., pumps, compressors, chillers and fans.
> - Identify 'mission critical' motors and the necessary spares to be carried.

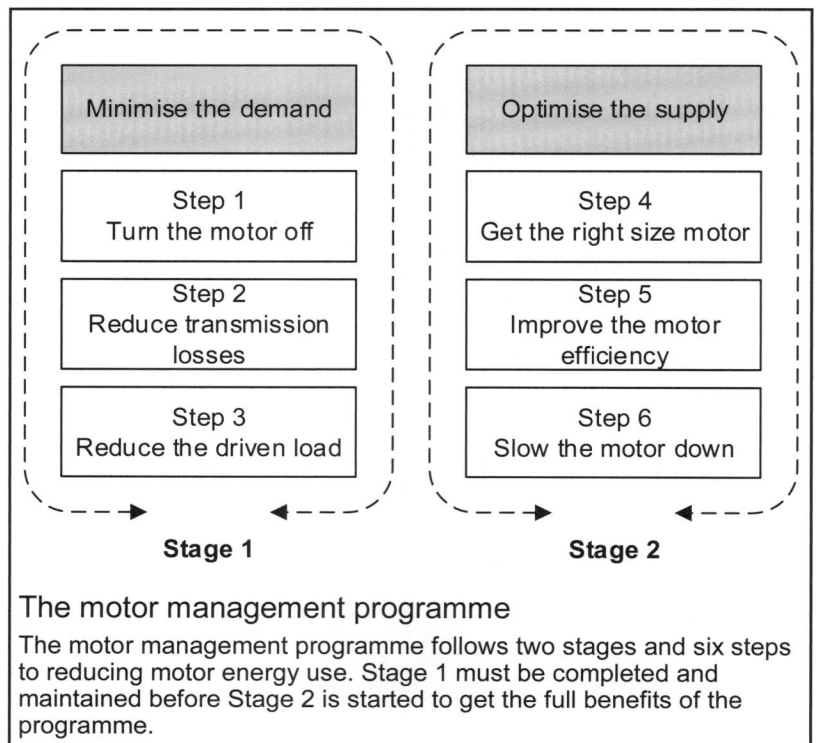

Minimise the demand		Optimise the supply
Step 1 Turn the motor off		**Step 4** Get the right size motor
Step 2 Reduce transmission losses		**Step 5** Improve the motor efficiency
Step 3 Reduce the driven load		**Step 6** Slow the motor down
Stage 1		**Stage 2**

The motor management programme

The motor management programme follows two stages and six steps to reducing motor energy use. Stage 1 must be completed and maintained before Stage 2 is started to get the full benefits of the programme.

stop the motor – after all it doesn't cost anything does it? In other cases, the task was never relevant but again the motor was free so why not use one.

Eliminating the load is simply a matter of asking: What does this motor do and do we need to do it? Ask this often enough and you will find motors that are not doing useful work and that can be eliminated.

Reducing the load at source is simply a matter of asking: What does this motor do and can we reduce the load? Ask this often enough and you will find loads that can be reduced with little effort.

Stage 2 – Optimise the supply

Step 4: Get the right size motor

Motors are most efficient when their load equals, or is slightly greater than, their rated capacity and are highly inefficient when operating at a small proportion of the rated load. Fitting a large motor for a small load will cost money for the length of the motor life – bigger is not better and size does matter.

Despite this, it is rare to find a motor that is correctly sized for the application – during the design process the sizes are almost always 'rounded up' and 'safety factors' are applied. The result is a motor that is often twice as big as needed and that runs inefficiently for the whole of its life (until it fails and is replaced by an even bigger motor 'just in case').

- **Tip** – It is strongly recommended that expert advice on motor sizing is sought to reduce costs.
- **Tip** – Where motors can be accurately predicted to run at < 33% of the rated output it is possible to reconfigure the motor from Delta to Star connection. This can produce savings of up to 10%.

Step 5: Improve the motor efficiency

The cost premium for High Efficiency Motors (HEM) is small and easily offset by the energy cost savings that result from their use. HEMs achieve efficiency levels of up to 6% more than standard motors and have a peak efficiency at 75% of load, thus reducing both energy costs and oversizing issues. A 6% efficiency gain may not sound much, but a £500 motor uses approximately £50,000 worth of energy over its year life and a 6% saving is £3,000 – the equivalent to six free motors.

Step 6: Slow the motor down

The rotational speed of an AC motor is nominally constant and is fixed by the number of poles and the supply frequency. As a result, many pumps and fans are driven at a constant speed, even though the demand varies considerably.

Almost all pumps and fans in the plastics processing industry run at constant speed when the actual demand is variable, i.e., there is not a constant need for cooling. The best and most cost-effective way of meeting a variable demand is to fit a Variable Speed Drive (VSD) to the motor. Fitting a VSD allows the speed of an AC motor to be varied and the pump output or pressure can be accurately matched to the variable demand. Thanks to the magic of the 'Cube Law' even small reductions in motor speed can produce large energy savings.

VSDs can be applied to fans, water pumps, air compressors and almost any application where the load varies with time.

- **Tip** – Investigate the use of VSDs for all pumps and fans to reduce energy costs.

For constant loads, the use of a correctly sized motor is still the best option and the motor policy should take this into account.

The ratings for HEMs:
- IE 4 (super premium efficiency), no current NEMA equivalent.
- IE3 (premium efficiency) = NEMA Premium efficiency.
- IE2 (high efficiency) = NEMA Energy efficient = EFF1.
- IE1 (standard efficiency) = NEMA Standard efficiency = EFF2.

The 'whole life cost' of a motor is often over 100 times the purchase cost.

VSDs represent an outstanding opportunity to reduce energy use, improve process control and save money.

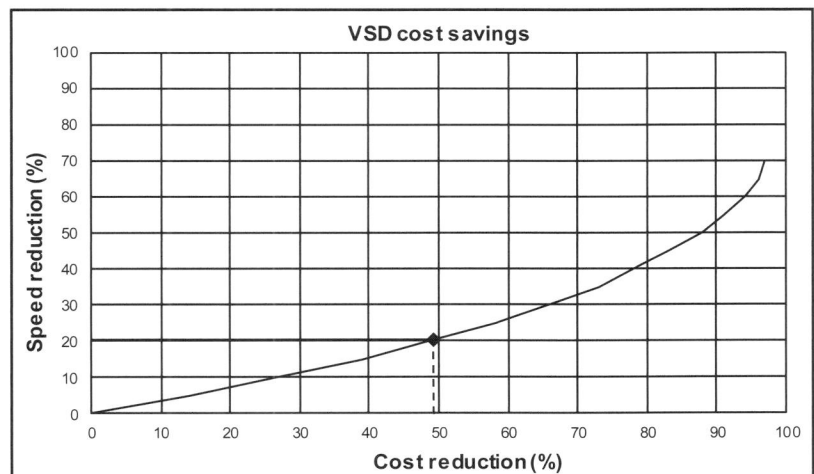

VSD cost savings

Speed reduction (%) vs *Cost reduction (%)*

The effectiveness of VSDs

Even small decreases in rotational speed can lead to large decreases in energy use. Reducing the speed by 20% will reduce the energy used by almost 50% and reducing the speed by 30% will reduce energy use by 66%.

6.16 Compressed air

The invisible cost

Compressed air is a convenient and often essential utility, but it is very expensive to produce. Most of the energy used to compress air is turned into heat and then lost to the atmosphere. At the point of use, compressed air costs more than ten times the equivalent quantity of electrical power, i.e., an equivalent cost of around £1/kWh if the energy cost is £0.10/kWh. At this cost, it should never be wasted and should only be used when absolutely necessary. Despite this, compressed air is treated in most factories as though it were free. Air also needs to be treated to remove moisture, oil and dirt, and the higher the quality required, the greater the energy consumed by the treatment system.

The diagram on the upper right shows the typical cost of a compressor over a 10-year period. In a typical 24/7 operation, a 100-kW compressor will use energy worth approximately £84,000 per year, assuming the cost of electricity to be £0.10/kWh. At these cost levels, an energy-efficient system is highly cost effective, even if it costs slightly more.

The cost of compressed air makes it an expensive resource and the way to achieve the best savings is to minimise the demand and then to optimise the supply. Up to 30% savings can be made by simple, no-cost good housekeeping measures such as making end users aware of the cost of compressed air and enlisting their help in reporting leaks.

The compressed air management programme

The 'compressed air management programme' provides the structure for reducing energy use and costs in compressed air systems.

Stage 1 – Minimise the demand

Step 1: Reduce leakage

A significant amount of energy is wasted through leakage of compressed air. Typical leak rates in industry are between 20 and 40%, i.e., up to 40% of the generating power is wasted in feeding leaks in the distribution system. A 3 mm hole in a system at 7 bar will cost ≈ £1,800/year.

A walk around to listen to the sound of the compressed air escaping from the system is the same as listening to the sound of money escaping from the business.

Leak surveys and good maintenance can

produce dramatic cost reductions and, in some cases, leak reporting and repair has enabled companies to shut down some compressors for all or most of their operating time.

- **Tip** – Estimate the losses due to leakage from the graph on the lower right and set targets for demand reduction.
- **Tip** – A walk-round survey, with leaks tagged and repaired as soon as possible, will greatly reduce leakage. The only tools

Compressed air is an expensive resource. It is not free.

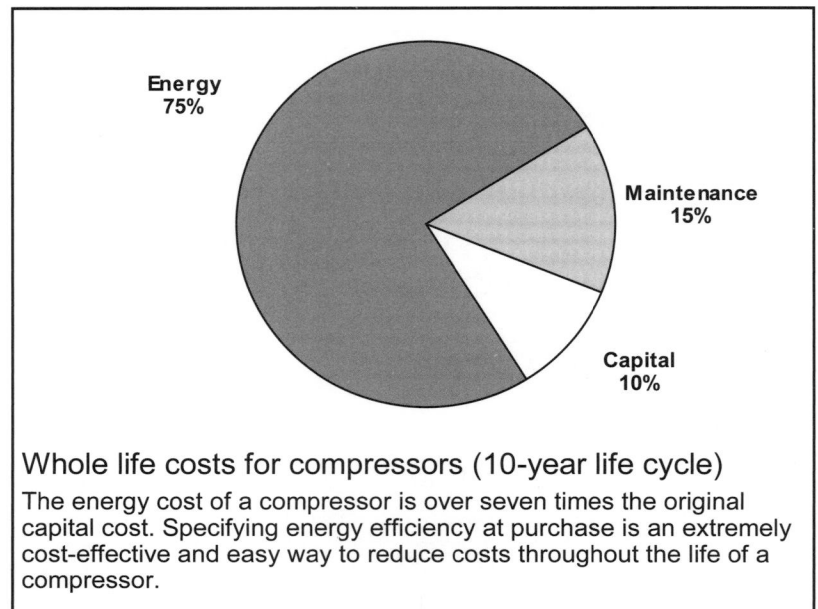

Whole life costs for compressors (10-year life cycle)
The energy cost of a compressor is over seven times the original capital cost. Specifying energy efficiency at purchase is an extremely cost-effective and easy way to reduce costs throughout the life of a compressor.

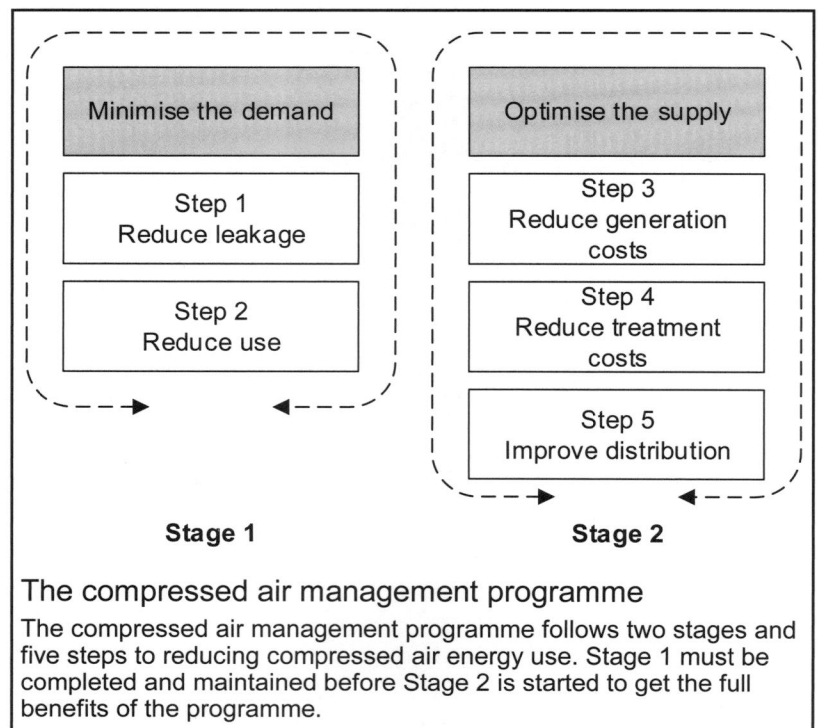

The compressed air management programme
The compressed air management programme follows two stages and five steps to reducing compressed air energy use. Stage 1 must be completed and maintained before Stage 2 is started to get the full benefits of the programme.

needed are a good sense of hearing, some water, some detergent and a brush.

- **Tip** – Isolate redundant pipe work, this is often a source of leaks.

Step 2: Reduce use

Compressed air is often wasted because everyone assumes it is cheap. Check every application to see if it is essential or simply convenient. If you want to estimate the cost of any application using open compressed air nozzles then use the chart on the lower right.

- **Tip** – Stop the use of compressed air for ventilation or cooling – fans are cheaper and more effective.
- **Tip** – Fit high-efficiency air nozzles – the payback can be as short as 4 months.
- **Tip** – Consider the use of electric tools instead of compressed air tools.
- **Tip** – Never use compressed air to move granules or products.

Stage 2 – Optimise the supply

Step 3: Reduce generation costs

The higher the air pressure, the more expensive it – twice the pressure uses four times as much energy. In many cases, the real pressure needed is less than the pressure being supplied. In some cases, a machine is rated as needing a 7-bar supply but pressure reducers are fitted inside the machine. Find out what the real needs are.

- **Tip** – Check that compressed air is not being generated at a higher pressure than needed.
- **Tip** – Switch off compressors during non-productive hours. They are often only feeding leaks or creating them.
- **Tip** – Check that compressors are not left idling when not needed – they can draw up to 80% of full power when 'off-load'.
- **Tip** – Position compressor air inlets outside if possible – cold air is easier to compress.
- **Tip** – If there is a machine or area that needs compressed air for longer than the rest of the factory, consider zoning or a dedicated compressor.
- **Tip** – Investigate electronic sequencing to minimise compressors going on- and off-load without good reason.
- **Tip** – Maintain the system – missing a maintenance check increases costs.

Step 4: Reduce treatment costs

- **Tip** – Treat the bulk of air to the minimum quality necessary, e.g., 40-micron filters. Five-micron filters will increase purchase cost, replacement frequency and the pressure drop.

- **Tip** – Test filters regularly to make sure that the pressure drop does not exceed 0.4 bar – if the pressure drop is > 0.4 bar, replace the filters. The cost of the power to overcome a pressure drop > 0.4 bar is usually greater than the cost of a filter.
- **Tip** – Manual condensate traps are often left open and act as leaks. Fit electronic traps to replace any manual condensate traps.

Step 5: Improve distribution

The longer the compressed air pipeline, the greater the pressure loss over the pipeline and the greater the cost generating the compressed air to feed the system.

- **Tip** – Make sure that pipe work is not undersized, this increases the resistance to air flow and causes unnecessary pressure drops.
- **Tip** – Use a ring main. Air can then converge from two directions. This reduces the pressure drop and makes changes to the system easier.
- **Tip** – Avoid sharp corners and elbows in pipe work, they cause turbulence and high pressure drops.

The limit of an audible leak is in the region of £200/year.

If a site is quiet then any leak you can hear will be costing a minimum of £200/year.

That 'sssssssss' noise you hear is not a normal operating condition, it is energy and money being wasted.

If compressed air were hydraulic fluid, we would never accept the leaks found in a typical factory.

During one energy survey we discovered that the compressor system was doing nothing but feed the leaks.

The system had been installed, was maintained and was being paid for but the compressed air was not actually being used for anything. The only consumer of compressed air was the leakage of the system itself!

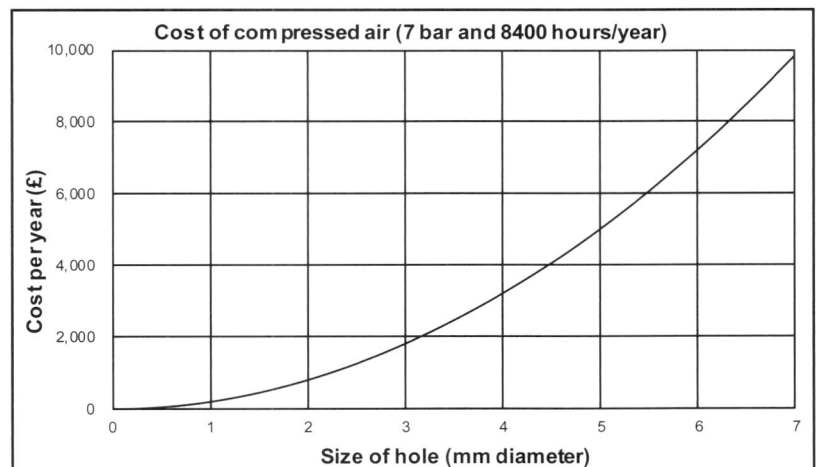

Cost of compressed air (7 bar and 8400 hours/year)

The cost of compressed air leakage and use

The cost of a compressed air leak or use for a hole at a pressure of 7 bar and for 8400 operating hours/year. A 3 mm diameter leak or use costs £1,800/year. A simple bowl feeder can cost up to £18,000/year depending on the number of jets used.

6.17　Cooling water

Another invisible cost

Plastics processing uses large amounts of energy to heat materials and to form these into products. In every case there is also a need to remove this heat to complete the process. This means that the provision of a reliable and consistent source of cooling is essential for fast and repeatable processes.

There is a great deal of emphasis in the industry on energy-efficient heating and processing but there is much less emphasis on energy-efficient cooling – a process that uses large amounts of energy and that has huge opportunities for energy-efficiency improvements.

- **Tip** – Cooling plant is generally ignored unless there is a problem. Regular analysis of performance data will quickly detect any losses in efficiency.

The chilled and cooling water management programme

The 'cooling water management programme' provides the structure for reduced energy use and costs.

Stage 1 – Minimise the demand

Step 1: Reduce heat gains

Eliminating or reducing ineffective heat gains ('parasitic loads') can have a significant impact on the running costs of any cooling installation and identifying the loads and reducing these is critical to improving energy efficiency.

- **Tip** – Fit adequate insulation to minimise 'parasitic loads'. Heat gains can occur on any long run of chilled water piping where the pipe is inadequately insulated. This is not as critical for cooling water piping.

- **Tip** – Only supply cooling where it is needed and isolate cooling circuits when they are not needed.

Step 2: Increase temperatures

Overcooling is a large factor in excessive energy use. Setting the temperature to the maximum appropriate to the process will reduce the energy consumed.

- **Tip** – Use the maximum possible chilled water temperature; a 1°C rise in the supply temperature from a chiller will reduce the energy required by about 3%.

Stage 2 – Optimise the supply

Step 3: Reduce cooling costs

Every conventional chiller is a compressor that pumps refrigerant; for every 100 kW of cooling capacity, it will draw about 30 kW. Even a small plastics processing site can need a 200-kW chiller, with an operating cost of over £50,000/year. As with any energy-intensive system, the total cost of ownership is far greater than the initial purchase cost. For a typical chiller operating full time then over a 10-year life cycle the energy costs will be 90% of the total cost and the initial capital cost will be only 9% of the total cost. However, simple measures can often improve the energy efficiency of chillers significantly.

- **Tip** – Scroll and screw compressors are more efficient and can replace existing chillers.

- **Tip** – Avoid running chillers at low loads.

- **Tip** – Use large evaporators and condensers and avoid direct expansion evaporators if possible.

- **Tip** – Where part loads are required, e.g., when using free cooling, then consider multi-compressor chillers which perform better under part loads.

- **Tip** – Optimise existing systems. Use the most suitable refrigerant and optimise the system for high part-load and winter efficiency. This is particularly important when additional chillers have been added to the system after the original design and installation.

- **Tip** – Balance pumps and chillers and match them to the normal load (with controls to match any variable loads).

A huge and hidden energy cost

Cooling and refrigeration plant use approximately 11% of all the energy consumed in manufacturing.

When was the last time anybody in your factory looked at the cooling system?

There is often confusion about chilled and cooling water. To clarify:

- 'Chilled water' is at <15°C and is generally used for moulds and tooling.

- 'Cooling water' is at >20°C and is generally used for machine cooling.

There are normally separate water circuits for chilled and cooling water.

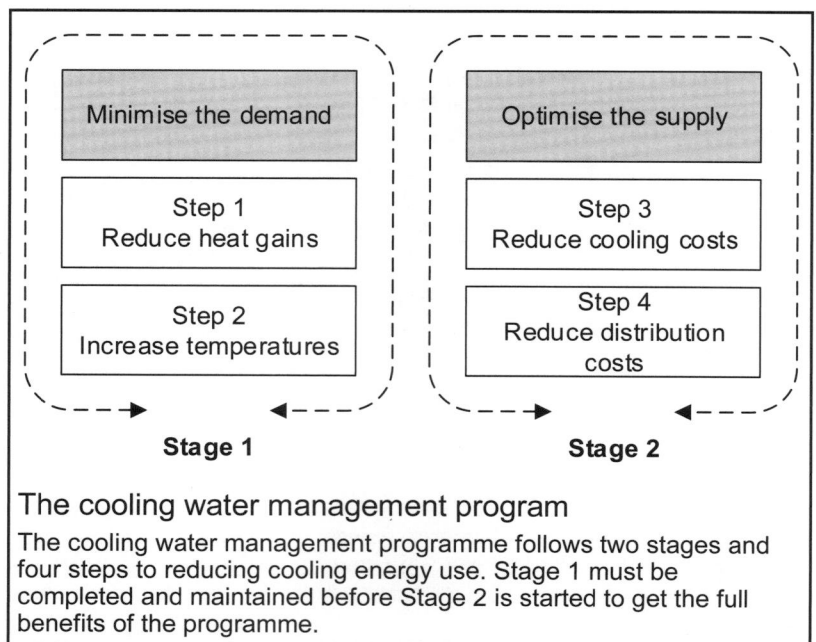

The cooling water management program

The cooling water management programme follows two stages and four steps to reducing cooling energy use. Stage 1 must be completed and maintained before Stage 2 is started to get the full benefits of the programme.

- **Tip** – Keep chillers well ventilated to provide good airflow over the condensers.
- **Tip** – Use heat recovery to provide energy for space heating and hot water.

Chillers often operate at low efficiency due to a lack of routine maintenance.

- **Tip** – Service chillers regularly and keep records of plant conditions.
- **Tip** – Clean evaporators, free coolers and heat exchanger surfaces regularly to maintain high efficiency.
- **Tip** – Check flow/return temperatures and system flow rates to verify these are correct and optimised. This can act as an early warning of possible degradation in compressor efficiency.
- **Tip** – Ensure that chillers have the correct charge of the correct refrigerant.
- **Tip** – Design moulds, cooling baths or spray tanks to provide good heat transfer from the plastic to the cooling water.
- **Tip** – Set all system components to turn off automatically when not in use.

Standard chilled water systems do not take full advantage of cold ambient weather conditions and constantly use energy to provide cooling. The flow temperatures used in plastics processing mean that 'free cooling' can significantly reduce energy costs. Free cooling pre-cools the return water from the process to reduce chiller loads and energy use.

If the ambient temperature falls to 1°C or more below the return water temperature, then the return water from the process is diverted through the free cooler. The more the ambient temperature is below the return water temperature, the greater the free cooling effect. It is normally possible to switch off the main chiller when the ambient temperature is 3°C below the flow water temperature.

- **Tip** – Chiller systems with new or retrofitted free cooling circuits can show large reductions in operating costs.
- **Tip** – Chiller systems with free cooling circuits have shorter compressor running times, lower maintenance costs and extended chiller life. The size of the overall chiller package should be capable of providing the total cooling load to cope with periods when free cooling is not operational.

Free coolers are available for capacities as low as 5 kW with no effective upper limit as units can be linked together to provide greater cooling capacity. They can be supplied as standard equipment for new installations or retrofitted to existing cooling systems to improve their energy efficiency. The typical payback period for a free cooler

system in temperate climates is less than 2 years, for cold climates it is considerably less.

Cooling towers are used for cooling water at many sites and can effectively provide low-cost cooling water even when the cost of treatment for Legionella is included.

- **Tip** – Use VSDs to control the fan motors on cooling towers. Control the motor speed with feedback from the sump to slow the fan down when the sump water is cold enough.

Step 4: Reduce distribution costs

Pumping chilled and cooling water is expensive, especially if fixed-speed pumps are used. Pumps are ideal applications for VSDs (see Section 6.15) and can be controlled by either the pressure or the return temperature.

- **Tip** – Use VSDs for all process water pumps and control the pump speed with feedback from the return water temperature or pressure. This also gives more consistent water temperatures and better process control.
- **Tip** – Check that pipe work and pumps are sized correctly for current demands.

Greenhouse gases

Old chillers using R22 as a refrigerant are less efficient than the new generation of chillers. Chillers using R22 are also subject to increasing legal controls.

New refrigerants (to replace older refrigerants that are now prohibited) are more efficient and can reduce chiller operating costs by between 12 and 30%.

Whichever refrigerant is being used the refrigerant loss should be monitored and any leaks repaired. Under EU Regulations it is required that all refrigeration equipment containing more than 3 kg of gas must be leak tested at least annually.

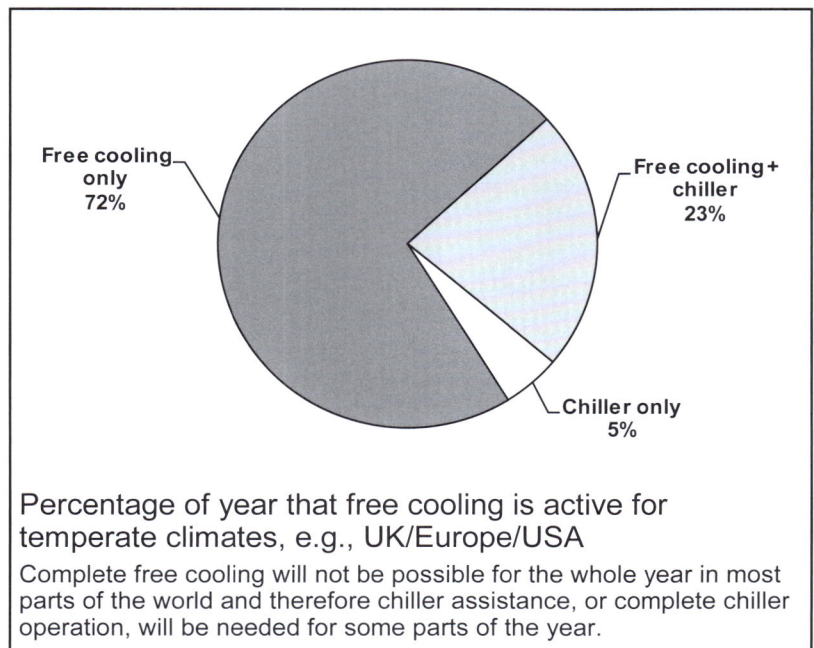

Free cooling only 72%

Free cooling+ chiller 23%

Chiller only 5%

Percentage of year that free cooling is active for temperate climates, e.g., UK/Europe/USA

Complete free cooling will not be possible for the whole year in most parts of the world and therefore chiller assistance, or complete chiller operation, will be needed for some parts of the year.

6.18   Polymer drying

A processing essential?

Drying uses large amounts of energy but is necessary for processing hygroscopic polymers (i.e., those that absorb water). If hygroscopic polymers are not dried to the correct level then the moisture will be converted to steam during processing and create surface marks or weaknesses in the moulding.

However, simple measures can achieve significant energy savings during drying.

- **Tip** – Drying is a hidden cost. Find the time taken, the optimum temperature and the energy used.
- **Tip** – Drying systems always have 'hidden' motors for blowers, etc., correct specification and motor management will reduce costs.

The drying management programme

The 'drying management programme' provides the structure for reduced energy use and costs.

Stage 1 – Minimise the demand

Step 1: Dry the right materials

Hygroscopic polymers will absorb moisture to an equilibrium level in normal atmosphere and must be dried before processing.

Non-hygroscopic materials should not require drying but if the storage is poor then they may carry surface moisture that should be removed before processing.

- **Tip** – Drying of non-hygroscopic materials is 'insurance drying' and should be avoided.

Step 2: Store materials correctly

Good storage of non-hygroscopic materials will remove the need for insurance drying and good storage of hygroscopic materials can reduce the moisture level and the length of the drying cycle.

- **Tip** – Good storage of materials in a warm, dry environment will reduce the moisture content before drying.
- **Tip** – After drying, any material should be kept in a sealed container or be conveyed by a sealed system to prevent it absorbing moisture from the air after drying.

Step 3: Improve control systems

Drying is difficult to control for optimum performance but improved control systems and modelling technology are now available to control and optimise the process to reduce energy use and to avoid overheating of materials. These can be retrofitted to existing dryers to reduce energy and improve material quality.

- **Tip** – Computer controls and predictive technology can be used to reduce drying times and temperatures.

Most drying is based on simple rules such as 'Dry for 6 hours at 80°C' with little or no consideration of the fact that what is really wanted is a consistent final moisture content of the polymer. Improved controls can significantly reduce drying times and produce a more consistent product.

- **Tip** – Reducing cycle times in warm dry weather will reduce energy use.

Stage 2 – Optimise the supply

Step 4: Reduce drying costs

Oven and hot air drying both have high energy costs and high batch-to-batch variations. They are not recommended for drying except for the removal of surface moisture from very cold non-hygroscopic polymers that have surface condensation.

Desiccant dryers are the industry standard and work by passing normal air through desiccant beads to produce dry air, which is then heated and passed through the drying hopper containing the polymer. The warm dry air removes moisture from the granules and the wet cooler air is recycled back to the dryer for further drying and use.

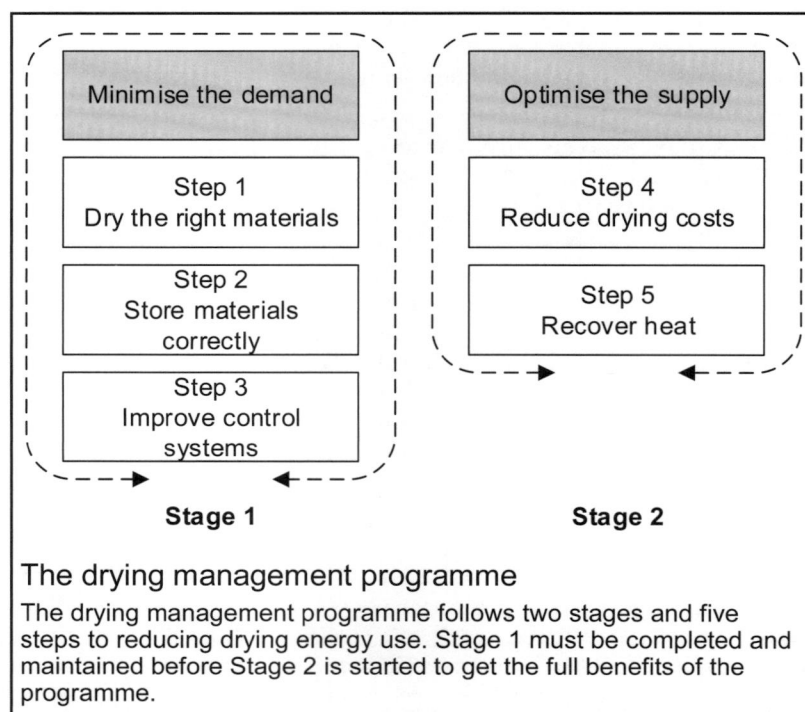

The drying management programme

The drying management programme follows two stages and five steps to reducing drying energy use. Stage 1 must be completed and maintained before Stage 2 is started to get the full benefits of the programme.

A typical desiccant dryer uses several desiccant canisters that are cycled through the drying and regeneration stages. Carousel drying uses a rotating wheel that is impregnated with desiccant crystals and which continuously rotates to pass the desiccant through the adsorption, regeneration and cooling cycles.

- **Tip** – Units that have automatic desiccant regeneration controlled by dew point sensors or preferably by material moisture content are more consistent.

- **Tip** – The lower the dew point of the air supplied, the quicker the drying time – but this needs to be balanced against the frequency of regeneration and the energy used for this.

- **Tip** – Small, spherical desiccant sieves give faster drying, better reactivation and greater adsorption.

- **Tip** – High reactivation temperatures improve reactivation and give greater adsorption in use.

- **Tip** – Optimise cycle times for the desiccant during drying to avoid overloading the desiccant and thus reducing process efficiency.

- **Tip** – Reactivated desiccant must be cooled after reactivation and a cooling stage (either passive or active) should be present before use. Ideally, this should not use ambient moisture-containing air.

- **Tip** – Design desiccant drying systems to be 'closed loop' to exclude ambient air and obtain the lowest dew point from the process.

- **Tip** – Desiccant systems need to be correctly sized for the demand.

Low-pressure drying (LPD) uses a vacuum applied to the dryer cabinet to accelerate drying. The vacuum reduces the boiling point of water from 100°C to around 56°C, and water vapour is driven out of the granules even at low temperatures. LPD reduces drying times by up to 85% and can reduce energy use by 50-80%. It is suited to machine-side drying of materials and rapid material changes. The short drying time enables a rapid start-up, and the smaller batches of material reduce the clean-down and changeover times.

Infrared drying uses infrared radiation to directly heat the polymer granules. The energy applied creates internal heating though molecular oscillation. This internal heat drives moisture out of the material into a stream of cool ambient air that removes it from the process. Infrared drying is particularly suitable for drying reprocessed PET material because it can combine the processes of recrystallisation and drying into a single-pass operation.

Step 5: Recover waste heat

In a conventional desiccant dryer only $\approx 35\%$ of the total input energy is used to heat and dry the polymer. The rest of the input energy is lost before or after drying the polymer.

These dryers do not recover the heat lost from the dryer during the process and often incur cooling costs. The latest machines use integral heat exchangers to recover heat from the exhaust air and recycle this back to pre-heat the cooler dried air from the desiccant dryer. This process can improve the heat balance so that up to $\approx 55\%$ of the input energy is used to actually dry the polymer. This almost doubles the efficiency of the system and significantly reduces energy use and costs.

- **Tip** – Look for dryers that recycle the heat from the drying process. These have reduced energy costs.

Absorb and Adsorb

Absorption = The penetration of a substance into the body of another.

Adsorption = The taking up of one substance at the surface of another.

The difference is very well defined, a desiccant takes water out of the air to dry it and adsorbs it – there is no penetration of the water removed from the air into the body of the desiccant so the desiccant does not absorb the moisture. All the moisture is on the surface of the desiccant.

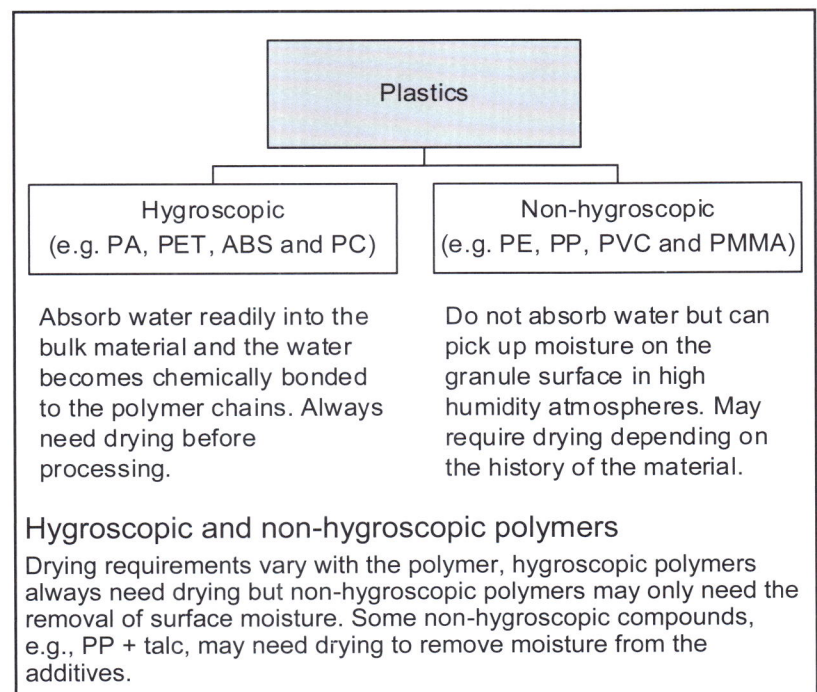

Plastics	
Hygroscopic (e.g. PA, PET, ABS and PC)	**Non-hygroscopic** (e.g. PE, PP, PVC and PMMA)
Absorb water readily into the bulk material and the water becomes chemically bonded to the polymer chains. Always need drying before processing.	Do not absorb water but can pick up moisture on the granule surface in high humidity atmospheres. May require drying depending on the history of the material.

Hygroscopic and non-hygroscopic polymers

Drying requirements vary with the polymer, hygroscopic polymers always need drying but non-hygroscopic polymers may only need the removal of surface moisture. Some non-hygroscopic compounds, e.g., PP + talc, may need drying to remove moisture from the additives.

6.19   Buildings

Buildings

Process-related energy use is the largest factor for plastics processors and buildings-related energy use is far lower and normally in the region of 7%. However, buildings are an easy area to make energy savings because any changes have no impact on production. In most cases, a simple site survey can reduce the site energy costs considerably.

For the plastics processing industry, recent years have seen vast improvements in factory buildings and working conditions. This upgrading of conditions has produced significant improvements in all-round site efficiency, and this has resulted in a general reduction in energy use. However, large opportunities still remain for energy savings in areas such as lighting, space heating and general hot water supplies.

Many plastics processes generate excess heat, e.g., processing machinery and compressors, and it is worth investigating if this excess heat can be used for other purposes, such as space heating on colder days or local heating or preferably energy recycling through a heat exchanger.

Note: Processes that involve solvents will require local exhaust ventilation.

The starting point is an audit of the buildings and systems. This is similar to the general energy walk-around but concentrates on the building elements.

Heating

Existing buildings

- **Tip** – Reducing heating load is the top priority, so the first step is to prevent unnecessary heat loss by making buildings as air-tight as possible. Draught-proofing doors and windows is cheap but effective.

- **Tip** – Replacement double glazing, or at the very least secondary glazing, will reduce heat loss and make offices more comfortable as well as extending the comfort zones of the office.

- **Tip** – Ensure that doors and windows are effectively sealed and maintained properly.

- **Tip** – Automatic fast-acting roller shutters save energy losses at external access doors which are used with forklifts and other mechanised access.

- **Tip** – High ceilings increase heating costs. Investigate the use of false ceilings, or destratification fans to blow hot air from the roof space down to the working area.

- **Tip** – Restrict the areas to be heated by using partitions or local systems to control the key areas. Do not ventilate or heat the whole building space for a few small areas.

- **Tip** – Do not heat areas where you have windows or outside doors open.

- **Tip** – Do not heat lightly occupied stores or warehouses when you are only trying to prevent excessive dampness.

- **Tip** – Set the heating level as required by the area. Typical settings should be:
  - Offices: 19°C
  - Factory areas: 16°C
  - Stores areas: 10°C to 12°C

- **Tip** – Reducing the heating temperature by 1°C will reduce the heating cost by about 8%.

- **Tip** – Make sure that thermostats are located and set correctly and preferably cannot be changed by staff. For larger sites, Building Energy Management Systems control energy costs without relying on staff.

- **Tip** – Set time controllers to match the occupancy patterns of the building.

- **Tip** – Insulate supply pipes to radiators.

- **Tip** – Permanently seal unused doors and windows.

- **Tip** – Prepare a company energy policy for buildings and ensure that it is followed.

New buildings/refurbishment

- **Tip** – Ensure building insulation and fabric meet the current best practice for the country. High standards of thermal

> Building energy costs are not a significant percentage of the total energy costs for most plastics processors.

> Improving building energy efficiency also improves staff comfort and work output.

Type	Acceptable		Unsatisfactory	
	Electricity (kWh/m²/y)	Gas (kWh/m²/y)	Electricity (kWh/m²/y)	Gas (kWh/m²/y)
Light industry	<43	<175	>70	>300
Storage and distribution	<29	<135	>43	>185
Office in factory	<72	<150	>100	>225
Separate office building	<95	<120	>110	>200

Approximate energy benchmarks for a range of buildings

The values are typical energy benchmarks for industrial buildings and offices located in them. These values are for temperate climates and need to be adjusted for hot or cold climates as well as for occupancy and local climate conditions. They are a guide only.

efficiency in buildings are now common and these standards will almost certainly rise in the future.

- **Tip** – Double glazing can both reduce heat loss and improve comfort. Modern low-e glass, inert gas filling of sealed units and advanced systems are even more effective than standard double glazing.
- **Tip** – Condensing boilers are the best option for new or replacement small hot water systems.

Lighting

Although it is only a relatively small part of the overall energy use, lighting systems offer easily demonstrable opportunities to save energy. Pay attention to areas with:

- High or continuous lighting levels and no or low occupancy. Use occupancy sensors or time switches to turn off lighting when areas are not occupied.
- Fluorescent tubes at high levels without reflectors. The use of reflectors increases light levels which may mean that the number of fittings can be reduced.

In lighting, simple measures can save money easily and a good lighting system can be a permanent energy-saving feature.

A key point is that lighting should be considered in terms of 'ambient' lighting for general movement and processes and 'task' lighting for specific tasks such as inspection. Many sites have relatively high levels of what is 'task' lighting when what is really needed is 'ambient' lighting.

Survey all areas of the site to set the appropriate task and ambient lighting needs (see table on the right).

- **Tip** – Many major lamp manufacturers also offer advice and contract consultancy on lighting. Use the help that is available for free to save energy.
- **Tip** – Replacing standard tungsten bulbs with LEDs saves money and gives a very clean white light. They use only 25% of the energy of tungsten bulbs and last about 10 times longer. The reduced maintenance costs, especially for lights in high-bay fitments, can easily fund the extra purchase costs.
- **Tip** – High-frequency tri-phosphor T5 tubes should always be installed when replacing or refurbishing systems where good colour is needed.
- **Tip** – High-pressure metal halide or high-pressure sodium lighting for warehouses or site lighting can be economically replaced with 'screw-in' LEDs that give excellent light levels.
- **Tip** – Lighting switched on in the morning will rarely be switched off until the evening

(if ever) – whatever the changes in light levels in the intervening period.

- **Tip** – Use natural daylight where possible and keep skylights clean to reduce the amount of artificial lighting needed.

Management

The key to reducing energy costs in buildings is the same as in many other areas. Good management can reduce costs through no-cost measures and good low-cost investment can give further reductions in costs, generally with rapid payback periods.

Producing a 'Lighting Map' of the site to show the position and size of lighting and the position/extent of the controls will reveal where it is possible to reduce costs.

The 'Lighting Map' should be used to create an 'Action List' that can be carried out by electricians.

Lighting is a very visible sign of a company's commitment to energy management. The savings are easily achieved and it sends a real message to the staff that the company is serious about energy management.

Don't forget that IT is not energy free. Servers (both local and in the cloud) use energy and many are simply sitting there, using energy and waiting for somebody to try to access an email sent in 2010. Look at data retention policies and practices to take redundant servers off-line or to delete data that is no longer required.

Location/Activity	Lux
Packing work, passages	150–300
Offices and computer work stations	300–500
Visual work at production line	300–750
Inspection work	750–1,500
Small parts assembly line	1,000–2,000

Approximate light levels required for various activities

The lighting level required depends on the activity and the task lighting needs should never be confused with ambient lighting needs. Good use of task lighting can reduce lighting costs considerably. Note that there may be legal limits for lighting levels!

6.20　Financial energy management – where are you now?

Without money it won't happen

Energy management is the same as any other project or process – starve the process of adequate and appropriate investment and it will fail. All projects, even nominally no-cost and low-cost projects need investment in staff time and much progress can be made in these areas.

Eventually, the process will exhaust the no-cost and low-cost projects and the process will require financial investment of some magnitude and this must be justified before progress can be made.

Energy management does not require preferential funding. Most energy management projects can easily meet the standard investment hurdles and analysis that are in place at most sites. The main concern is that energy management receives the appropriate level of funding for the benefits that it can deliver.

Completing the chart

This chart is completed and assessed as for the previous charts.

Areas of disagreement may show that there is something happening at the site that is not well known, i.e., the Finance function is actually keeping records but is not telling the Production function that they are doing so.

Energy: financial management

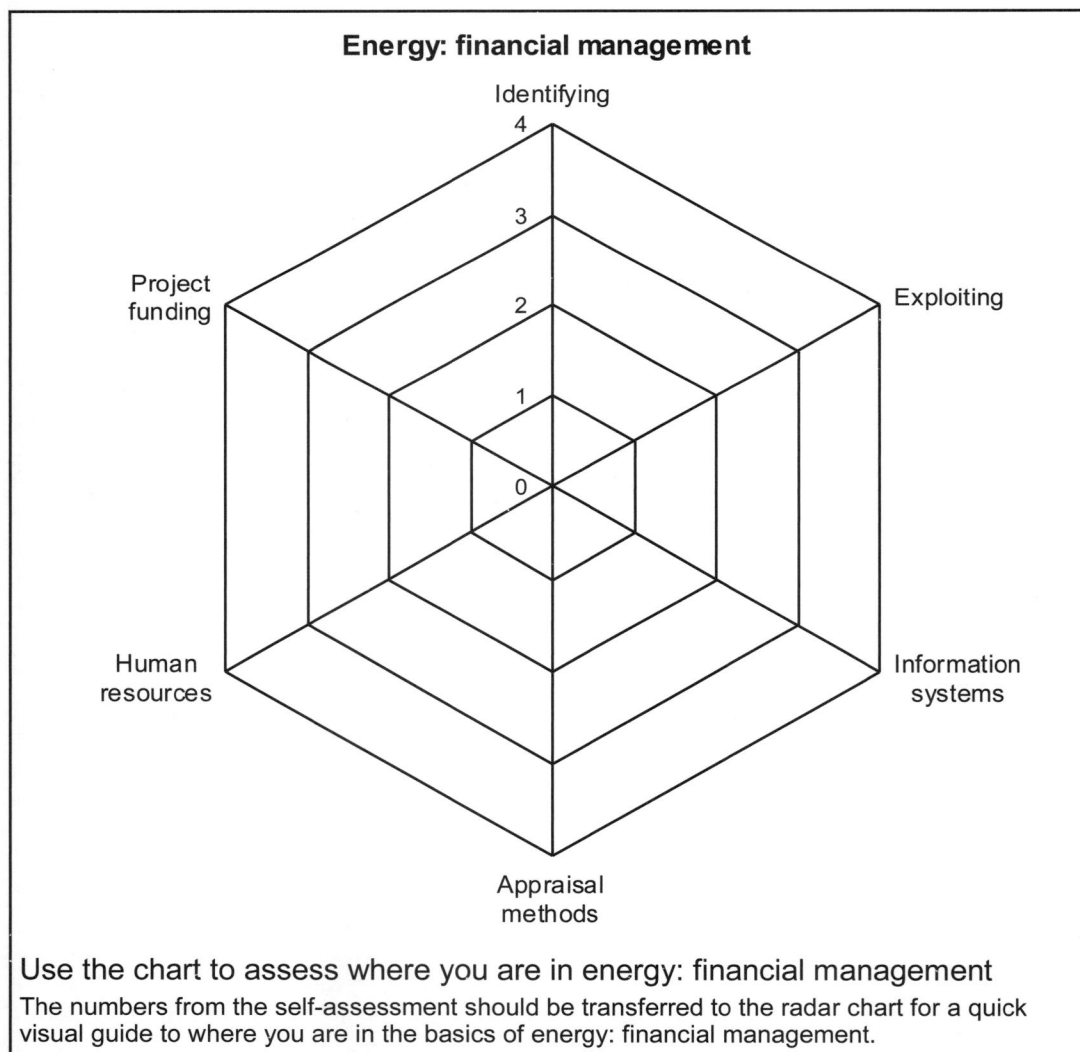

Use the chart to assess where you are in energy: financial management

The numbers from the self-assessment should be transferred to the radar chart for a quick visual guide to where you are in the basics of energy: financial management.

Level	Identifying	Exploiting	Information systems	Appraisal methods	Human resources	Project funding
Energy: financial management						
4	Detailed energy surveys regularly updated. Opportunities already costed & ready to proceed.	Formal requirement to identify the most energy-efficient option. Decisions made on the basis of life cycle costs.	Full management information system enabling identification of past savings & further opportunities for investment.	Full discounting methods using internal rate of return & ranking priority projects as part of an ongoing investment strategy.	Board take a proactive approach to long-term investment as part of a detailed environmental strategy in full support of the energy team.	Projects compete equally with other areas. Full account taken of indirect benefits, e.g., marketing opportunities, environmental factors.
3	Energy surveys conducted for areas likely to yield largest savings.	Energy staff required to comment on all projects. Energy efficiency options often approved but no account is taken of life cycle costs.	Promising proposals are presented to decision-makers but insufficient information, e.g., sensitivity or risk analysis, results in delays or rejections.	Discounting methods using the organisation's specified discount rates.	Energy manager presents well-argued cases to decision makers.	Projects compete for capital along with other business opportunities, but have to meet more stringent requirements for return on investment.
2	Regular energy monitoring / analysis used to identify possible areas for saving.	Energy staff notified of all proposals that affect energy usage. Proposals for energy savings are at risk when capital costs are reduced.	Adequate management information available, but not in the correct format or easily accessed.	Undiscounted appraisal methods, e.g., gross return on capital.	Occasional proposals to decision makers by energy managers with limited success & only marginal interest from decision makers.	Energy projects not formally considered for funding, except for very short-term returns.
1	Informal ad hoc energy walkabouts conducted by staff with checklists to identify energy saving measures.	Energy staff use informal contacts to identify projects where energy efficiency can be improved at marginal cost.	Insufficient information to demonstrate whether previous investment has been worthwhile.	Simple payback criteria are applied. No account taken of lifetime of the investment.	Responsibility unclear & those involved lack resources to identify projects & prepare proposals.	Funding only available from revenue on low risk projects with paybacks of less than one year.
0	No mechanism or resources to identify energy-saving opportunities.	Energy efficiency not considered in new-build, refurbishment or plant replacement decisions.	Little or no information available to develop a case for funding.	No method used irrespective of the attractiveness of a project.	No-one in organisation promoting investment in energy efficiency.	No funding available for energy projects. No funding in the past.
Score						

6.21 Technical energy management – where are you now?

The plant is the thing

The distribution of energy use in plastics processing is very different to that in an office, the major energy users are the services and the plastics processing machinery and this is where the efforts must be concentrated.

This requires good technical knowledge of the services and processes used and good technical management of the processing itself.

This chart tries to provide an assessment of these technical aspects of energy management.

Even when the majority of the operational plant was not originally designed with energy efficiency in mind there are many simple actions that can be taken to improve the energy efficiency of existing plant. These range from good maintenance action, where simple low-cost tasks, such as the alignment of motor drives, can easily reduce energy use

for existing plant through to involving the operators to reduce energy use.

Completing the chart

This chart is completed and assessed as for the previous charts.

Agreement by various people across the site shows consistency – a starting point for improvement.

Energy: technical management

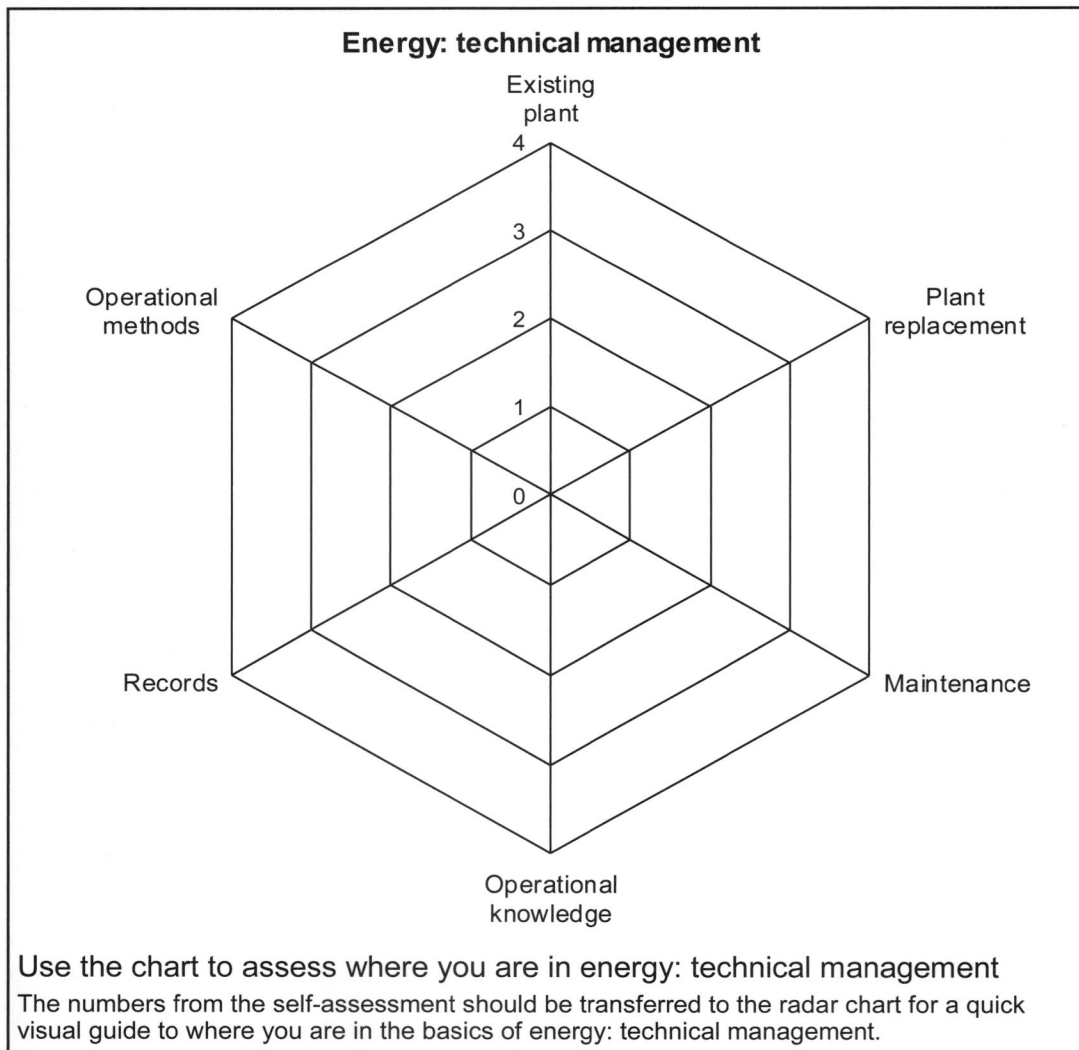

Use the chart to assess where you are in energy: technical management

The numbers from the self-assessment should be transferred to the radar chart for a quick visual guide to where you are in the basics of energy: technical management.

Operators often know how to reduce energy use by changing operational methods. Despite this, their knowledge is routinely ignored in favour of the less relevant knowledge of someone in an office.

An opportunity lost and a huge pity.

				Energy: technical management		
Level	Existing plant	Plant replacement	Maintenance	Operational knowledge	Records	Operational methods
4	Majority of existing equipment uses best practice energy efficient features, is correctly commissioned & well maintained.	Equipment chosen is the most appropriate for application. Life cycle costs & energy efficiency are major factors in selection.	Maintenance is based on needs, with condition appraisal used for all equipment & fabric elements affecting energy efficiency. Results acted upon.	Staff know how their actions affect energy efficiency & take positive steps to minimise energy use. Staff have targeted training in energy issues.	Detailed descriptions of systems, plant control & operation. Detailed schedules of all plant, instrumentation & controls.	Operational methods & settings for energy efficiency well defined & implemented. Full utilisation of feedback from monitoring.
3	Equipment & plant is appropriately selected, energy efficient, commissioned for low energy consumption & well maintained.	Equipment is appropriate for application with energy efficiency considered. Life cycle costs & energy efficiency are evaluated.	Regular surveys carried out on equipment & fabric elements affecting energy efficiency. Action undertaken for most defects identified.	Staff are aware of how they affect energy use & take all good housekeeping measures to save energy. Training on a regular basis.	Detailed descriptions of plant control & operation, & outline systems. Reasonable schedules of all plant, instrumentation & controls.	Operational methods & settings for energy efficiency poorly defined & implemented. Informal use of information from monitoring.
2	Most equipment is not specifically energy efficient, but either was commissioned or is being regularly maintained for low energy consumption.	Equipment selected to be fit for purpose, bearing in mind likely life cycle costs & energy efficiency factors.	Condition surveys carried out regularly on all equipment & fabric elements affecting energy efficiency. Remedial work constrained by budgets.	Most good housekeeping practices are adhered to in an attempt to reduce energy usage. Occasional energy efficiency training received.	Basic descriptions of plant control & operation. Basic plant, instrumentation & control schedules for most control systems.	Targets set against realistic budgets, & maintained through financial procedures.
1	Equipment is not energy efficient, but has been commissioned for economy & undergoes periodic maintenance.	Power efficiency data on products obtained as part of selection process.	Condition surveys carried out occasionally, prompted by plant failure or safety considerations. Remedial work only carried out on major defects.	Energy-saving techniques are only adopted where they can be easily accommodated within traditional working practices.	Minimal or poor plant control & operation. Plant instrumentation & control schedules for only some of the plant & control systems.	Targets set by default through budget setting procedures.
0	Energy performance has not been considered during the procurement, commissioning or maintenance of existing plant & equipment.	No consideration of energy efficiency in product selection.	No regular surveys or maintenance carried out.	No consideration is given to energy efficiency during working operations.	None available.	No targets set.
Score						

6.22 Energy awareness and information – where are you now?

Knowledge is the key

As with any new area there is a need to both specify what people are going to do and to ensure that they are aware of their responsibilities.

One of the keys to energy management is 'show results to get resources' and there is a need for clear reporting of successes in energy management both to get resources and to motivate the team. Equally there is a need to provide all staff with training and development opportunities. A training course on variable speed drives may appear a luxury but if it saves real money then it is a good investment in both the staff and the company.

Energy management is a rapidly developing field and there are very few people with experience or understanding of this area – keep staff well trained and up-to-date with the latest market developments.

Completing the chart

This chart is completed and assessed as for the previous charts.

Low scores are not bad but simply show areas with improvement potential.

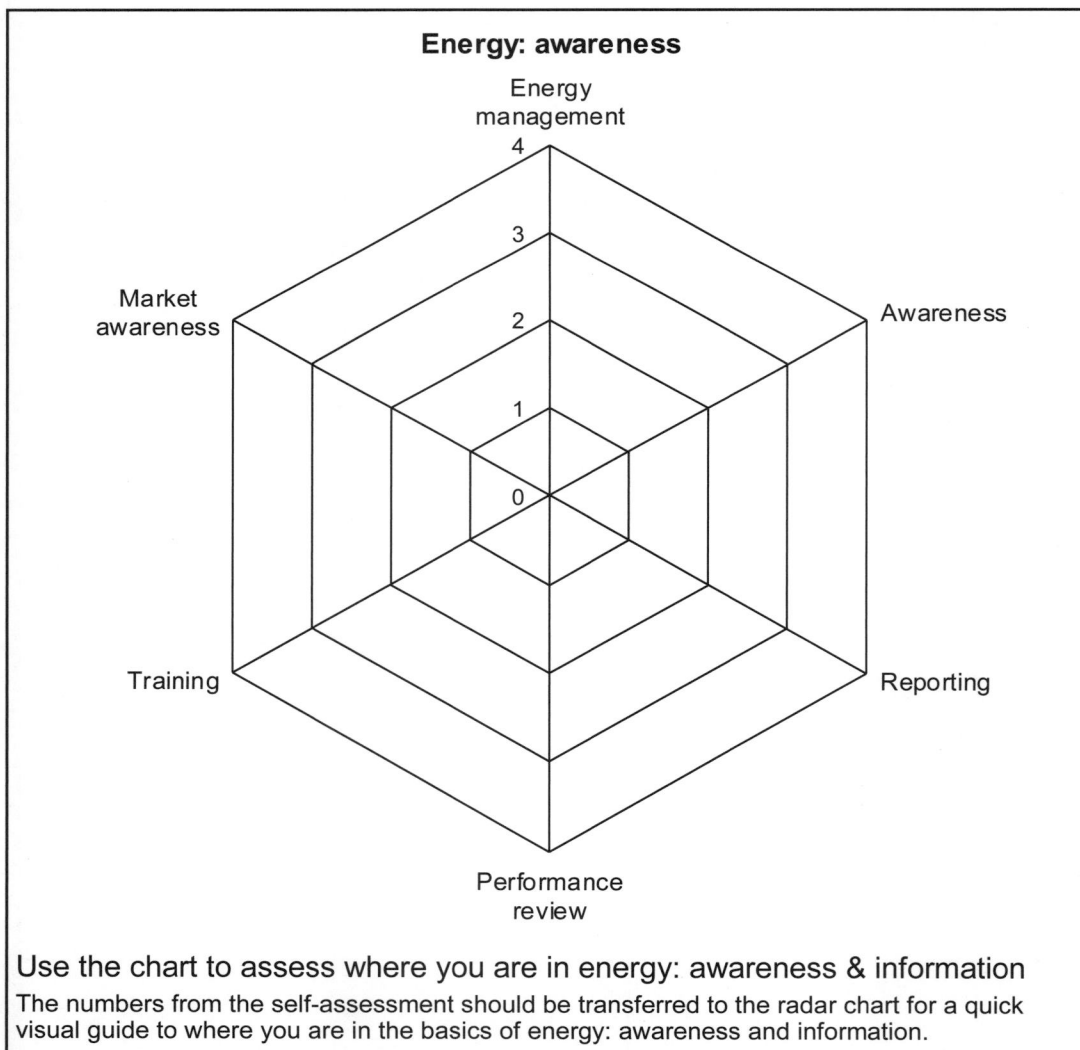

Energy: awareness

If you don't know that you are doing it wrong then how can you hope to improve?

Use the chart to assess where you are in energy: awareness & information
The numbers from the self-assessment should be transferred to the radar chart for a quick visual guide to where you are in the basics of energy: awareness and information.

If you think training is expensive then consider the cost of ignorance.

				Energy: awareness		
Level	Energy management	Awareness	Reporting	Performance review	Training	Market awareness
4	Lists of responsibilities & their assignment exist & are comprehensive & regularly reviewed. All staff have responsibilities.	Energy efficiency status regularly given to all staff. Full use made of publicity. All methods used to promote new measures for saving energy.	Wide reporting of current status compared with best practice, on regular basis & for a range of audiences. Full support to public statements.	Progress regularly reviewed. Performance compared against internal & external benchmarks. Ideas actively sought.	Training properly resourced for technical & premises staff. Active technical library. All staff have access to an energy efficiency library.	Keep abreast of technological developments by monitoring of trade journals, literature & other sources on issues affecting energy efficiency.
3	Lists of responsibilities & their assignment exist for key energy staff & all departments.	Energy efficiency status presented to all staff at least annually. Occasional but widespread publicity to promote energy saving.	Status reports issued annually to shareholders & staff. Impartial reporting of performance to staff & departments on a regular basis.	Frequent energy efficiency reviews using monitored consumption & cost data. Analysis is regular, wide-ranging but ritualistic.	Some professional development for technical staff. Some staff are aware of & have access to an energy efficiency library.	Regular studies carried out on trade journals, literature & other sources to assess current developments impacting on energy efficiency.
2	Some staff & departments have written responsibilities.	Energy performance presented to staff on a regular basis. Occasional use of publicity to promote energy saving.	Occasional issue of energy efficiency status reports. Concentrates on good news.	Occasional technical energy efficiency reviews. Regular cost checks with exception reporting. Analysis of limited scope.	Technical & premises staff development by professional & technical journals. Occasional initiatives to train staff in energy efficiency.	Trade journals, literature & other sources scanned on an ad-hoc basis for information on developments relating to energy efficiency.
1	Unwritten set of responsibility assignments.	Energy performance occasionally reported & known to very few staff. Energy-saving measures are rarely promoted.	Reports only issued if prompted by a business need. Most reports will contain only good news.	Energy review activity based on revenue costs. Limited exception reporting only.	Few staff have knowledge of energy efficiency techniques & facts. Little training in energy efficiency for staff.	Trade journals, literature & other sources studied for energy implications when a purchase is imminent.
0	No evidence of assignment of energy efficiency tasks & duties.	No staff have explicit responsibilities or duties.	No reporting.	No monitoring activity to underpin review processes	Staff have little, if any, knowledge of energy efficiency. No attempt to inform staff of techniques & benefits of energy efficiency.	Energy efficiency not a consideration when keeping up to date on products or technology.
Score						

6.23 On-site energy generation – solar and wind

Renewables

Some sites have taken action to use solar or wind-power to generate some of their own electricity in an attempt to reduce energy costs. The key word here is 'some' and for almost all plastics processing sites the use of on-site solar or wind will never be able to meet more than a small fraction of the demand. Solar and wind-power installation levels have greatly increased across the world in the last decade but, in most industrialised countries, solar and wind-power do not have sufficient power density to provide a viable alternative to other methods of generation.[1]

Solar

In assessing the viability of solar panels for a plastics processing site we will assume that the panels are to be mounted at the site. This immediately presents a concern as most sites are compact and do not have sufficient space for a large array (even if it were to be active all of the time). This means that for most sites, any solar module installation would be roof-mounted.

Typical outputs

Solar panels are rated in terms of the peak electricity output (kWp) which is assessed by a constant irradiance of $1,000/W/m^2$ in the plane of the panel. This is not the same thing as the actual electricity output as this will vary widely with the panel technology, the location, the installation conditions and other factors. For solar panels installed in the UK, the annual electricity output will be $\approx$ 100-180 kWh/m^2/year but this is highly variable.

For a large plastics processing site some typical numbers are:

- Total roof area: 24,000 m^2.
- Usable roof area allowing for orientation and other effects: 75% = 18,000 m^2.
- Solar output: 140 kWh/m^2/year.
- Total solar output: 2,520,000 kWh.
- Actual electricity used: 14,000,000 kWh.
- Solar supply: 15.75% of requirements.

Typical solar outputs (kWh/m^2) for an excellent day and a poor day are shown on the upper right and the output by day for a year is shown on the lower right. These graphs highlight the main issues with solar generation, it will reduce energy use only when the sun is shining and this is unpredictable except in broad terms (more in summer and none at night).

- **Tip** – If a site is operating 24/5 and has no large-scale battery storage to store excess electricity generated then these values will decrease and only 11% of the annual use will be met by the solar panels.
- **Tip** – These values will increase for sites in countries which receive higher annual solar radiation, i.e., nearer the equator.

Payback

The payback on solar panels around the world is heavily dependent on government support via grants, rebates and feed-in tariffs (where the generation is subsidised by

> Do not be confused or misled by the use of the words 'installed capacity' for solar panels, this is always much more that the actual output of the panels.

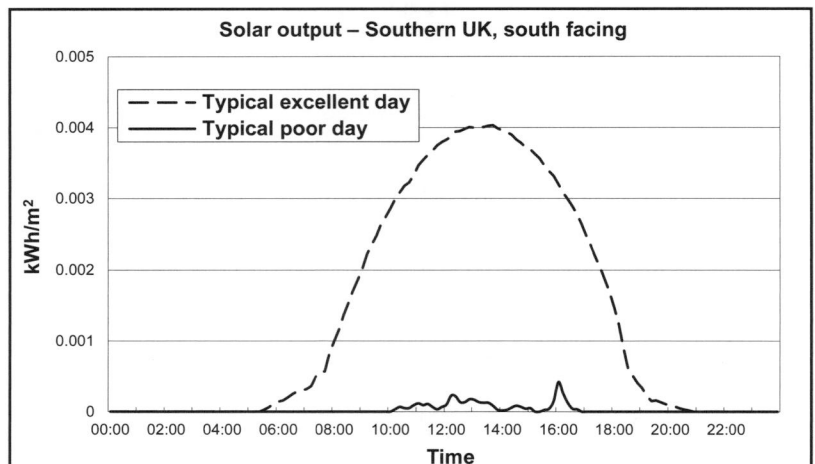

Solar output – Southern UK, south facing

Solar output for excellent and poor days

The difference between excellent (summer and a cloudless sky all day) and poor (winter and overcast most of the day) is dramatic. In winter, the days are also shorter and the opportunity to gather energy is much shorter.

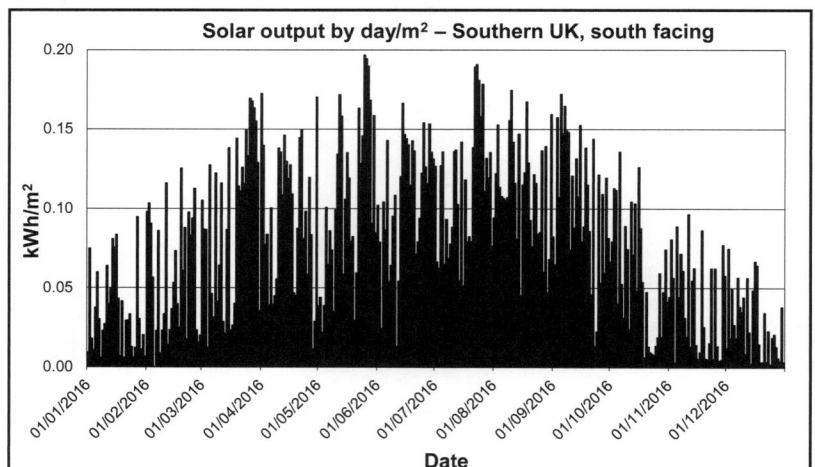

Solar output by day/m^2 – Southern UK, south facing

Solar output by day for a complete year

This is the actual solar output (kWh/m^2) over a complete year. The difference in the generation output between summer and winter is stark. Obviously solar panels will be more beneficial in sunnier climates.

the government). These were initially very high to encourage the growth of the solar industry but have been dramatically reduced as the industry has grown larger.

Depending on the location and the government support available the payback time is in the region of 7-9 years. This is normally well above the investment criteria threshold for plastics processing sites.

Summary

Installing solar panels may well make a site feel good and look good in the annual sustainability report but there are many projects that are easier to carry out and are more financially attractive.

Wind

As for solar power, we will assume that wind turbines are to be mounted at the site and this again presents the issue of the size of the site and the turbines and the amount of energy that can be generated by wind turbines. It is highly unlikely that any suitable wind turbine could be roof-mounted and a site therefore needs a large site area to make any meaningful difference in electricity generation terms.

- **Tip** – In most parts of the world, off-shore wind is more consistent and stronger than on-shore wind. Do not be confused by the generation yields for off-shore wind.

Outputs

For the UK, it is estimated that the power density for on-shore wind is ≈ 2 W/m^2 of land area used.[1] This assumes an average wind speed of 6 m/s in the UK (although this is much less for heavily populated areas of the UK) and typical windmill heights of 50 metres spaced at 5d where 'd' is the diameter of the wind turbine[1].

If the site considered for solar energy (above) had a free area suitable for wind turbines equal to twice the roof area, i.e., 48,000 m^2 then installing wind turbines would give a nominal output of 96 kW and over a year would generate 840,960 kWh or 6% of the annual requirements. Not many sites have this much free space or are located in areas where the average wind speed is as high as 6 m/s over the year.

- **Tip** – A major issue with wind turbines is the necessity for planning permission and this may be difficult to get for a substantial wind turbine farm.

- **Tip** – As for solar systems, if a site is operating only 24/5 and has no large-scale battery storage to store excess electricity generated then these values will decrease and only 4.2% of the annual requirements will be met by the wind turbines.

Payback

The payback on wind turbines is heavily dependent on government support but this is much less for on-shore generation and many countries make it difficult to install on-shore through planning constraints.

Depending on the location and the government support/constraints the payback time is in the region of 5-6 years. This is normally well above the investment criteria threshold for plastics processing sites even if the physical space is available.

Summary

Installing wind turbines may also make a site look and feel good but there are many projects that are easier to carry out and are more financially attractive.

Storage

It might be argued that storage could change the arguments for solar or wind-power but the current available output levels mean that most sites would use all the power available as it was generated.

Do we do it?

For most sites the use of solar or wind-power for on-site generation of electricity is not sufficiently large or financially attractive to make this a consideration. For some private companies, long-term investments of this type may be financially justified but these will be rare.

Do not be confused or misled by the use of the words 'peak capacity' for wind turbines, this is always much more that the actual output of the turbine.

Some sites with access to large areas of land close to the site are using this for solar, wind or anaerobic digestion plants. The power generated is then fed to the site and used to meet some of the demand (and reduce costs).

Innovative thinking is needed to make local generation viable but it can be done to reduce the costs, although the payback will still be high.

- 1. MacKay, D.J.C. 2008. 'Sustainable Energy – without the hot air'. UIT. Available for purchase or free download at www.withouthotair.com.

6.24 On-site energy generation – combined heat and power and tri-generation

Going off-grid

Many people will be familiar with the concept of co-generation or combined heat and power (CHP), where a gas-fired turbine/generator is used to provide both electricity and heat (see diagram of the energy flow in CHP on the lower right). This technology is widely used in buildings and district heating schemes where both the electricity and heat generated can be used effectively. The generated electricity can run the lighting, services and machines and the waste heat can be used for heating or hot water.

Whilst this is useful for buildings, most of the plastics processing industry needs cooling or chilled water more than it needs heat and CHP is not widely used in plastics processing.

Tri-generation

Tri-generation or combined cooling, heat and power (CCHP) is similar to CHP in most respects but also different in how the waste heat is used. In CCHP, a gas turbine is used to produce electricity, a linked boiler is used to produce steam from the waste heat and the resulting steam is used either for heating or for the production of chilled water using absorption cooling (see diagram of the energy flow in CCHP on the far right).

All plastics processing sites use large amounts of chilled and cooling water for the process. Chilled and cooling water is normally provided by electrically powered chillers or cooling towers and these are very expensive to operate, particularly chillers (see Section 6.17).

CCHP can generate power, heat (steam) and also chilled water to allow a site with access to good gas supplies to go off-grid for almost all the energy requirements. It is also relatively immune to power outages, although it does need a reliable gas supply. In addition, removing the cooling demand from a site's electricity demand will dramatically reduce the electricity use by ≈ 11% by reducing chiller use.

- **Tip** – It is the production of cooling and chilled water that holds the key for sites to use almost all of the energy stored in the much cheaper raw gas input fuel.

- **Tip** – It may seem strange that steam can be used to produce cold and chilled water but the technology has a long history, is well developed and uses absorption cooling where steam is used instead of the expansion valve in a chiller.

It can be partial

CCHP does not have to meet the complete cooling or electrical needs of a site and can be sized to meet:

- The site electrical load – this would take the site 'off-grid' in terms of electricity but may result in excess heat or cooling production depending on the exact balance.

  Note: The electricity output does not need to be the same as the current demand because of reduced chiller loading and would be ≈ 11% less.

- The site heating load – for most sites this would still require a grid connection but with reduced demand. The system would meet all the heating needs in cold weather but only partially meet the cooling and electricity demand.

- The site cooling load – for most sites this would still require a grid connection but with reduced demand. The system would meet all the cooling needs in hot weather and most likely all the heating needs in cold weather but only partially meet the electricity demand.

Whichever size was chosen, with suitable controls a CCHP system for temperate climates would be set up as follows:

- Hot seasons – the CCHP system is used to generate the maximum amount of cooling to take the maximum amount of load from

CCHP is not a new technology and is well known in Europe, the USA and Asia. Perhaps the future is bright for tri-generation for a multitude of reasons?

Absorption cooling can also use an ammonia-water cycle where ammonia is the refrigerant and water is the absorber

Absorption cooling systems have a low COP (however defined) but the low cost of the fuel, i.e., waste heat, can make systems very cost-effective.

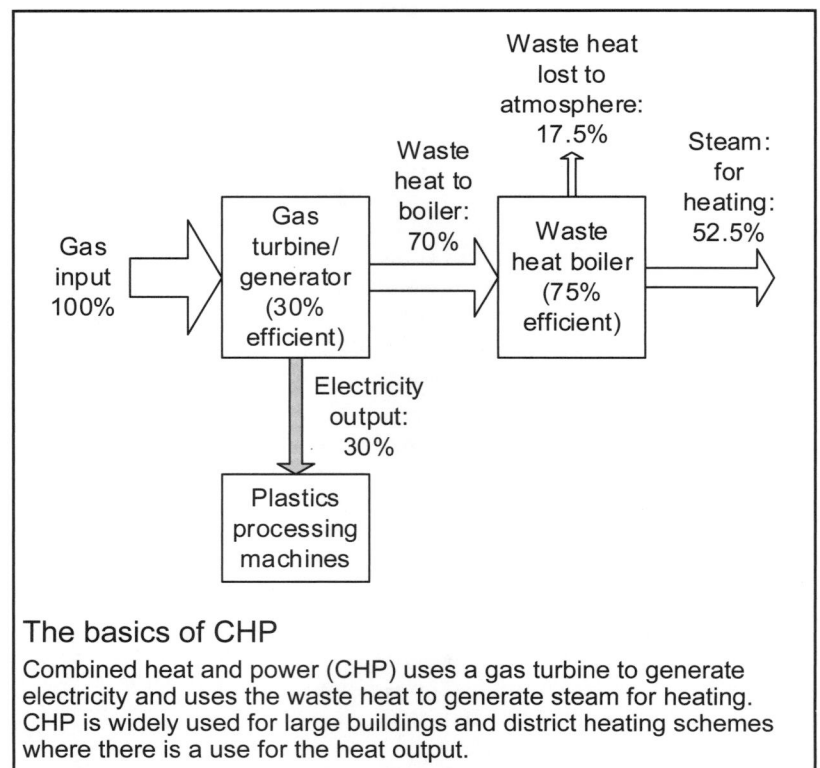

The basics of CHP

Combined heat and power (CHP) uses a gas turbine to generate electricity and uses the waste heat to generate steam for heating. CHP is widely used for large buildings and district heating schemes where there is a use for the heat output.

the chillers. The system is not used for any heating.

- Cold seasons – the CCHP system is used to generate the maximum amount of heating to reduce heating gas use. Any remaining heat is used for cooling but the majority of the cooling load is met by air blast chillers (see Section 6.17) and any excess cooling load by chillers.
- Intermediate seasons – the CCHP system is flexed between heating and cooling depending on the heating and cooling loads.

What do we need to change?

A CCHP system does not need changes to existing chilled or cooling water systems and can be set up as follows:

- The CCHP system generates chilled water via absorption cooling and distributes this to the existing chilled and cooling water tanks.
- Existing chillers remain in place but are set to switch on at high temperatures in the chilled water sump ($\approx 20°C$), i.e., in the event the CCHP system cannot generate enough chilled water.
- Existing process pumps remain in place to distribute chilled water from the sump to the process and are VSD-controlled (see Section 6.17).
- Existing system pumps, e.g., to towers and chillers, remain in place but are turned off and controlled to meet the cooling demand.

What are the savings?

The capital costs for CCHP with absorption cooling are currently high but falling rapidly. CCHP has advantages in terms of overall energy costs (gas is much cheaper than electricity) and insulating a site from power outages.

- **Tip** – For sites with an unreliable grid supply then the benefits of CCHP may be more than simply the energy aspects.
- **Tip** – Find out more about tri-generation and the benefits of going off-grid by carrying out a quick feasibility study to clarify the economics of CCHP.

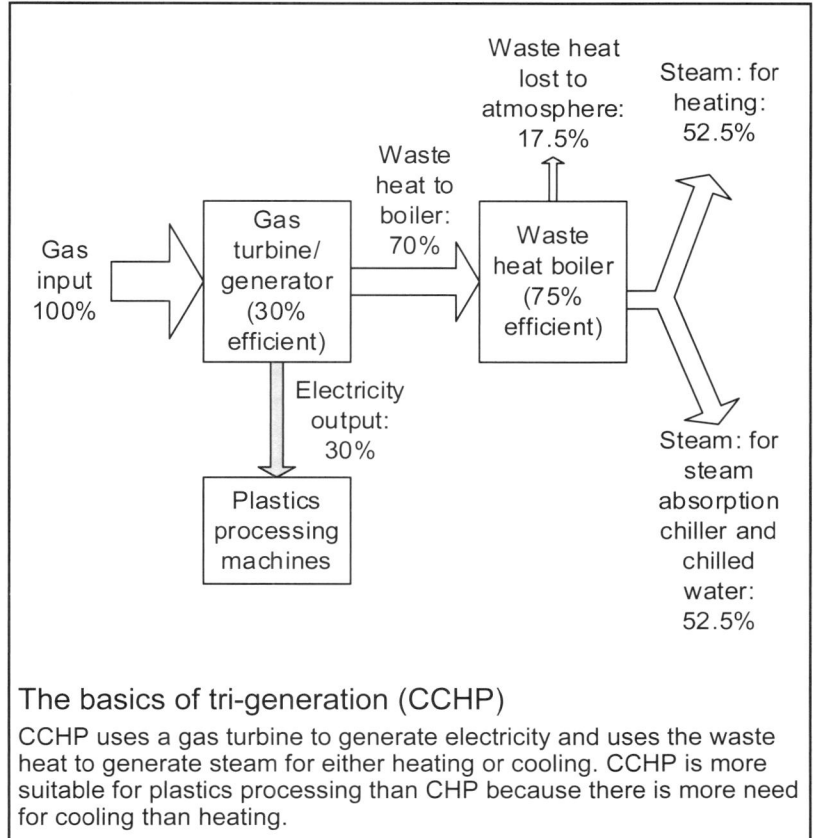

The basics of tri-generation (CCHP)

CCHP uses a gas turbine to generate electricity and uses the waste heat to generate steam for either heating or cooling. CCHP is more suitable for plastics processing than CHP because there is more need for cooling than heating.

How does absorption cooling work?

It may seem counter-intuitive to be able to use heat for cooling but gas-operated refrigerators have a long history (although they are no longer common).

Cooling using absorption cooling is similar to traditional vapour compression cooling except that the compressor is replaced by a chemical cycle taking place between an absorber, a pump, and a regenerator. Absorption cooling dissolves a vapour in a liquid (the absorbent), pumps the solution to a higher pressure in the regenerator and then uses heat to evaporate the refrigerant vapour out of the solution.

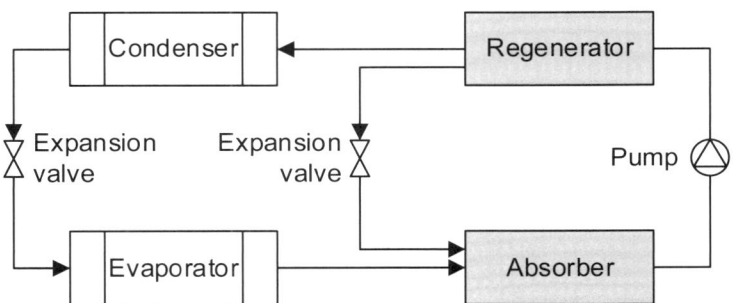

Basic absorption cooling cycle

The most common absorption cycle uses water as the refrigerant and lithium bromide (LiBr) as the absorber. These systems can be single, double or triple stage, where increasing the number of stages increases the efficiency but also the cost.

The COP of an absorption chiller is in the region of 0.65 (single effect) -1.2 (double-effect) and the capital cost is higher than that of a conventional chiller. Despite this, if the heat source is very low cost, then the cost of providing process cooling can be very low compared to conventional chillers.

Key tips

- Find out what the bills were last month and examine them for agreement and accuracy with the current supply contract.
- Take action to reduce the cost of energy.
- Establish an energy management plan for the future.
- Investigate competitive pricing for energy supplies.
- Investigate 'peak demand lopping' by internal generation.
- Get the interval data for electricity and look for the use patterns.
- Establish controls on the use of electricity.
- Monitor use and calculate the base load and the energy used per kg of processed plastic.
- Locate the large energy users in the company and consider sub-metering in these areas.
- Do not use kWh/kg as a monthly performance metric. It may be appropriate for yearly measurement but it will vary too much with monthly production volume.
- Benchmark the site and the machines to understand where you are in the rankings.
- Integrate energy into the accounting package so that it is visible to management.
- Carry out an energy walk-around to identify prominent areas of wasted energy.
- Look for machines on stand-by that are not due to be operated for some time.
- Establish start-up and shut-down procedures for all machines.
- Shut down utilities to machines as the machine is shut down.
- Optimise machine temperature settings. Plastic is heated up to process it and then cooled down – energy can be wasted at both ends of the cycle.
- Examine ancillaries (conveyors, etc.) as well as the main machines – they can use up more energy in total.
- Plan machine start ups to minimise the maximum demand.
- Use preventative maintenance to reduce energy use.
- Set up a company motor management policy to only use high-efficiency motors and specify these for all applications.
- Set up a motor management policy to avoid rewinding motors except where absolutely necessary.
- Get the right size motor for the application and implement VSDs wherever possible.
- A 3 mm hole in a 7 bar compressed air line costs about £1,800 per year.
- Energy consumption varies as the square of pressure so increasing the supply pressure from 2 to 4 bar requires four times as much energy.
- Compressed air use can often be reduced by 30% by simple management measures. Minimise the demand for compressed air as much as possible.
- Insulate chilled water piping and isolate unused water circuits.
- Increase chilled water temperatures to the maximum possible.
- Only dry materials that actually need it.
- Examine the lighting throughout the company and establish the most energy-efficient solution. Consult lighting suppliers for a lighting audit.
- Check heating and ventilation systems for settings and lock staff out of these to prevent tampering.
- Fit 'rapid' doors and entrances to reduce heat losses.
- Concentrate the minds – set the heating so that it doesn't operate if doors are open.

Chapter 7

Manufacturing: carbon footprinting

The rising recognition of the effect of atmospheric CO_2 on climate change and global warming has driven efforts to reduce anthropogenic (man-made) CO_2 emissions. In order to reduce emissions, the first step is to quantify them and this is the function of carbon footprinting.

Carbon footprinting is simply a method of quantifying the carbon impact of processes, products and activities in terms of total equivalent CO_2 emissions (CO_2e). This allows an assessment of the emissions and their impact. This has already been referred to for transport emissions (see Section 5.16) but in this section we will look at carbon footprinting at both the site and the product level.

Carbon footprinting is a technique that is rising in importance around the world and customers are beginning to ask for carbon footprint values as part of the purchasing process. Companies need to be prepared for the now seemingly inevitable question about their carbon footprint.

The importance of knowing and understanding the ramifications of the carbon footprint are also being driven by increasing numbers of countries around the world which are setting targets to be 'net zero' in terms of carbon emissions. This is a often by a specified date, e.g., 2050, which is so far in the future that the people who set the targets and make the promises will no longer be around to be held accountable. Whilst these promises do not quite fit into the category of greenwashing, in some cases they verge on the practice of 'virtue signalling'. The idea of 'net zero' is almost in danger of being 'weaponised' as countries try to outdo each other in terms of the dates promised.

This is particularly the case when the concept of 'net zero' is still developing and when there are a number of unresolved issues in the concept. At the country level, it appears relatively easy to calculate and define a net zero framework but the concept of embodied carbon in imported or exported products is still very much under development. At the product level, there is no generally agreed definition of a 'low carbon product' and how this is calculated. Some definitions look only at the carbon emissions resulting from production of the product whereas others attempt to include carbon emissions over the complete product life cycle.

There was a temptation to include net zero in the bulk of the text but, as this is a practical workbook, we will concentrate on the things that are known and can be achieved by plastics processors. This means we will focus on what the carbon footprint includes, how to calculate it and how to present it.

7.1 The basics of carbon footprinting

Site or product?

A carbon footprint can refer to either a specific site or to a complete product and these are very different:

- A site carbon footprint is only for the specific site and does not include any embodied energy of the raw materials – this can be calculated by the raw materials suppliers from their individual site carbon footprints.
- A product carbon footprint is the total of all the carbon footprints for the complete supply chain (cradle to grave) of a product and includes all the embodied energy of the raw materials.

Most plastics processors will be primarily concerned with their site carbon footprint so that this can be provided to the customer to feed into the total product carbon footprint.

What is a carbon footprint?

A carbon footprint is simply the total of all the carbon emissions that result either directly from a site or as a result of the site's activities. This is expressed in terms of the total CO_2 emissions. In some cases, the actual gases emitted are not simply CO_2 but include other greenhouse gases (GHG), e.g., methane (CH_4) or nitrous oxide (N_2O). These emissions are converted to CO_2 equivalents (CO_2e) to allow a total equivalent CO_2e to be calculated.

There are a range of reporting methods, some web based, some specific to a country or area, some designed for industry and some designed for domestic/consumer use. Most of the methods are based on the GHG Protocol (www.ghgprotocol.org) which is a standard format for classifying and reporting emissions.

Many of the rules for carbon footprinting are similar to accounting rules and are designed so that emissions can be added up across the supply chain without double-counting. At the site level, we can ignore these and simply calculate the emissions within a site's control.

Emission types (scopes)

Emissions are grouped by the control that a site has over them:

Scope 1: Direct emissions

These are emissions that a site directly causes and controls. In this case, the CO_2e is actually emitted either at the site or by an asset that the site controls, e.g., gas burnt on a site directly emits CO_2.

Scope 2: Indirect emissions from imported utilities

These are emissions from purchased electricity or other utilities such as imported heat or steam. In this case, the CO_2e is emitted at some distance from the site by the power station that generated the electricity, heat or steam.

Scope 3: Indirect emissions

These are emissions that a site causes to occur but where it does not control the asset, e.g., transport in vehicles owned by other organisations such as air travel.

For a basic carbon footprint, the focus is on the larger common emissions from Scope 1 and Scope 2 which can be easily and consistently measured. Some standards regard coverage of Scope 3 emissions as optional depending on the scale of the site and data availability.

Emission sources

The table below gives a list of common emission sources listed by scope. A major area of confusion is whether to allocate an emission as Scope 1 or Scope 3. The key issue is whether the site has control of the asset or not. Actual ownership is less

One of our clients paid a consulting company $42,000 to provide a carbon footprint – it took three consultants 3 weeks.

We had the energy data (as part of the site survey) and carried out the other data gathering on the back of a place mat over a steak and a good beer.

Our numbers differed by only 0.5%, i.e., well within the potential error of the calculation.

It is not that difficult and the numbers are relatively easy to either estimate or calculate.

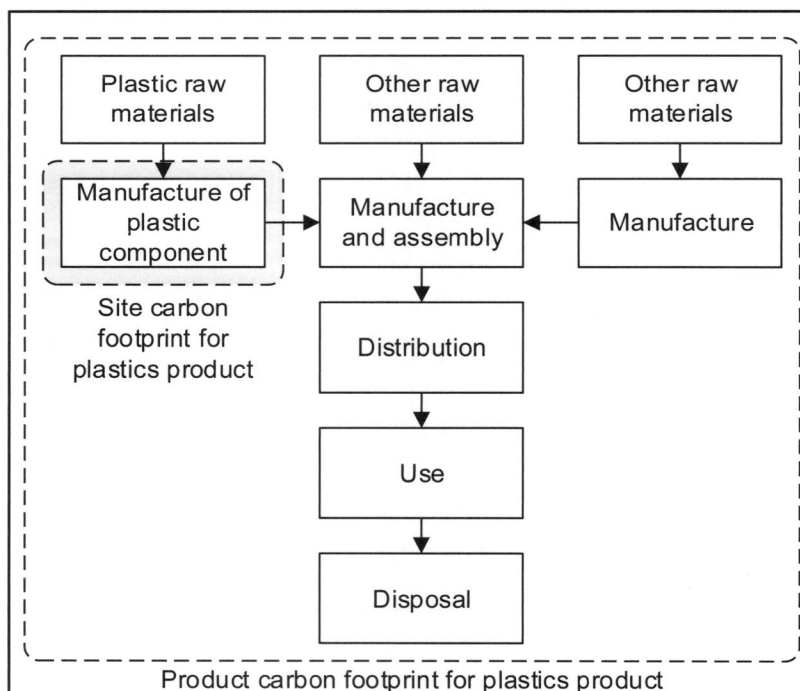

Product carbon footprint for plastics product

The boundaries for site and product carbon footprints

The site carbon footprint for most plastics processing sites will only be part of the total product carbon footprint. The contributions from each of the product life cycle stages can be added up to give the product carbon footprint.

important than control, e.g., emissions from a vehicle which is either owned or is on a long-term lease contract are included in Scope 1 but emissions from a vehicle which is not owned by the site are included in Scope 3.

Getting the data

To create a carbon footprint, 'activity data' such as distance travelled, litres of fuel used or electricity used must be converted into equivalent CO_2e emissions. The best, and most respected, set of conversion factors is available free from the UK government (see sidebar). This spreadsheet gives the values to be used to make conversions and provides step-by-step guidance on how to use the factors.

- **Tip:** These conversion factors are regularly updated. Be sure to use the correct year.

The main factor that is not given in this spreadsheet (except for UK readers) is the conversion factor for electricity used. This varies widely with the country depending on the mix of generation for the country.

- **Tip** – Get the local conversion factor for consumption from the supplier.

Preparing an initial carbon footprint will necessarily involve estimations and approximations because some of the data will not be easily available. After a site has prepared the initial carbon footprint the data collection should become part of the normal operations. It is much easier and more accurate if the data are collected continuously.

- **Tip** – Sites should regularly record and refine their carbon emissions to quantify the carbon footprint.

Reporting the data

Carbon footprints should be reported for a complete year in terms as 'tonnes of CO_2e'. The year chosen can be either the company reporting year or a calendar year but consistency is important. The reporting period should be both declared in the footprint and consistent.

- **Tip** – Consider reporting the carbon footprint via the Carbon Disclosure Project (www.cdproject.net).

Make energy savings look good

Saving energy (see Chapter 6) not only reduces costs but also reduces the carbon footprint of a site and product. This can have significant benefits both in public relations and in conforming to a range of government regulations appearing around the world.

Carbon footprinting requires good data collection but is not difficult.

Carbon footprinting is a sub-set of a full LCA (see Section 1.5) because it only focuses on the greenhouse gas emissions.

One of the best and simplest explanations of how to measure and report is the free UK government publication 'Guidance on how to measure and report your greenhouse gas emissions'.

The emission factors are given in another UK government publication: 'Greenhouse gas reporting conversion factors'.

Scope 1	Scope 2	Scope 3
Direct emissions (site has control of the asset)	Indirect emissions from imported utilities	Indirect emissions (site does not have control of the asset)
■ Gas (process or heating). ■ Oil (process or heating). ■ Bottled liquid or gaseous fuels (e.g., LPG for fork lift trucks). ■ Other fossil fuel. ■ Owned or leased cars, buses, trucks or other vehicles. ■ Process emissions. ■ Refrigerant emissions (e.g., replacement of losses due to leakage). ■ Other direct emissions.	■ Emissions from purchase of electricity. ■ Emissions from import of heat or steam.	■ Employee business travel – personal car. ■ Employee business travel – train, bus and other means. ■ Employee business travel – plane. ■ Employee business travel – rental car. ■ Employee business travel – taxi. ■ Employee commuting. ■ Water. ■ Product transport – where the company does not own the vehicle. ■ Waste disposal/recycling.

The scopes for carbon footprinting

The three main divisions of the emissions are used to assess the total emissions from a site. The objective is to be consistent in collecting information. Much of the published guidance is on how to report the available data so that it can be compared with other sites and avoids issues such as 'double counting' in the production of a site carbon footprint. These conventions can then be used to build up a full product carbon footprint.

7.2 Site carbon footprinting – Scope 1

Direct emissions (controlled assets)

Scope 1 covers the direct emissions at a site, i.e., the emissions that a site directly causes and controls.

This can be divided into various emission sources and these will be covered individually for a typical plastics processing site. A typical report format is shown at the bottom of the opposite page. Some sites will not have all the emissions discussed.

Purchased fuels

The CO_2e for fuels purchased for use on site can be calculated using the tables in the 'Fuels' tab of DBE&IS (see box on the right for details of DBE&IS). This is simply a matter of deciding on the relevant measure (tonnes, litres or kWh) and using the conversion factor to calculate the CO_2e resulting from this factor.

These values are relatively independent of the application and location. The results are a function of basic chemistry and assume full combustion of the fuel.

Gas (heating and process)

Gas when purchased via a mains feed will be invoiced either by mass (tonnes), by volume (litres) or by energy (gross calorific value or net calorific value).

The gas supply for the year should be available from the billing data and this can be easily converted into CO_2e using DBE&IS. The only difficulty is when the supplier quotes 'kWh' (see upper sidebar). In most cases the kWh quoted will be the gross CV but this should be clarified with the supplier to ensure that the correct factor is used.

- **Tip** – Check that you are using the correct table in DBE&IS. There is a large difference between gross CV and net CV.

Fuel oil (heating and process)

As with gas, fuel oil can be purchased in a variety of ways but in most cases, this will be in litres. The amount of fuel oil can be converted to CO_2e using DBE&IS.

Fork lift truck gas

Where fork lift trucks (FLTs) are gas-powered the gas will generally be LPG and will be delivered by bottle. The use can be calculated from the litres delivered in the year. If the data are not available then a reasonable use of a FLT will be ≈ 6,000 litres/year.

Obviously, if electric FLTs are used then this can be ignored.

Other fossil fuel

DBE&IS lists a wide range of fuel types and can be used to convert almost any fuel use into CO_2e.

Owned cars

Owned cars includes all company cars provided for use by employees. It does not matter if the car is leased or owned. The important issue is the control of the vehicle. If the site has control of the vehicle then it is counted here.

DBE&IS has conversion factors for a wide range of vehicle types. The input data are in terms of distance travelled (in miles or km).

- **Tip** – The only time I haven't been able to find a relevant conversion factor was in Brazil where the vehicles used ethanol as fuel.
- **Tip** – If data are available in terms of the amount of fuel used then the 'Fuels' tab will give more accurate results.

Owned trucks or vehicles

Some sites will lease or own trucks, vans, buses or other vehicles and the data for these are available in DBE&IS for both passenger and goods vehicles (see the 'Delivery vehicles' tab).

As with owned cars, there are a variety of options and it is simply a matter of choosing the correct activity measure and conversion factor.

Calorific value of gas

Gross calorific value (gross CV) is the calorific value under laboratory conditions and will be higher than the net calorific value (net CV) which is the useful calorific value that will be obtained in the real world.

Suppliers will usually quote gross CV but this should be checked.

The absolute value of CO_2e is generally reported in tonnes as tCO_2e. This is the metric ton or metric tonne and is equal to 1,000 kg.

Conversion factors for emissions

There are a variety of sources for data on emissions such as those published by the GHG Protocol and various organisations around the world.

The most reliable, consistent and comprehensive set of data is that published by the UK government ('Greenhouse gas reporting – Conversion factors'). This is produced by the Department for Business, Energy & Industrial Strategy and is updated yearly. At the time of writing, the latest edition is 02 June 2021.

The conversion factors are supported by a Methodology Paper ('2021 Government Greenhouse Gas Conversion Factors for Company Reporting: Methodology Paper for Conversion Factors') that gives background data on how the factors were calculated and how they should be used.

All conversion factors used in this chapter are taken from the 2021 edition of this document. This document includes not only the CO_2 emissions but also allowances for CH_4 and N_2O emissions. For convenience we will refer to the 2021 version of this document as 'DBE&IS'.

Readers are advised to check for updates and extensions to DBE&IS.

- **Tip** – If company owned trucks are used to collect goods from suppliers or to deliver to customers then this is where the emissions are collected. If suppliers deliver goods or customers collect them then they are NOT included as part of the site carbon footprint.

Process emissions

Few plastics processing operations have actual process emissions and this is not normally relevant. If, however, the site uses a foaming process and vents a gas to atmosphere then the Global Warming Potential (if any) of the gas should be found and inserted in this area.

Refrigerant emissions

Most sites will have chillers, compressed air dryers and A/C units. These will all contain refrigerant gases and most of these are GHGs. These will all leak slight amounts of refrigerant gas to the atmosphere and these emissions should be calculated.

Method 1

The most accurate method is to record the amount and type of GHG used to service the equipment each year. This will also be the amount and type of refrigerant lost to the atmosphere and can be used with the factors in DBE&IS to give a CO_2e value.

Method 2

The easiest initial method is to produce a 'GHG List' for the site. This is a list of the location, application, volume and type of each GHG. The total volume of each type of

GHG present at the site can then be calculated. The loss of refrigerant can be calculated by assuming a loss factor for each item of equipment. DBE&IS gives full details of this method and typical loss factors.

- **Tip** – If in doubt about leakage then apply a leakage factor of 8% for industrial chiller systems.

The GHG leakage (in kg) for each refrigerant type can then be used with the factors in DBE&IS to give a CO_2e value.

- **Tip** – Whichever method is used the losses should be calculated over a whole year.

Other direct emissions

This is not normally relevant for most plastics processing sites.

The GHG List can also serve as part of the control process to remove GHGs from the site.

There are separate factors for GHG emissions for installation and disposal. These should be used when installing or disposing of equipment.

Emission type	Emission source	Emission		
		Tonne CO_2e	%*	Notes
Scope 1: Direct emissions (controlled assets)	Gas (heating and process)	2.85	0.08%	Billing data.
	Fuel oil (heating and process)	0	0.00%	N/A.
	Fork lift truck gas	0	0.00%	Electric FLTs.
	Other fossil fuel	0	0.00%	N/A.
	Owned cars	20.25	0.60%	Distance travelled.
	Owned trucks or vehicles	0	0.00%	N/A.
	Process emissions	0	0.00%	N/A.
	Refrigerant emissions	3.84	0.11%	Refrigerant data.
	Other direct emissions	0	0.00%	N/A.
	Sub-Total	**26.94**	**0.79%**	

Example of reporting of Scope 1 emissions

This is a typical format for reporting Scope 1 emissions (all conversion factors taken from DBE&IS). Data for fuel use are taken from the billing or delivery data. Data for cars and trucks are taken from the 'distance travelled' DBE&IS tables.

* This is the percentage of the total emissions from all three scopes, i.e., in this example the total Scope 1 emissions are 0.79% of the total of all three scopes.

7.3   Site carbon footprinting – Scope 2

Indirect emissions (imported utilities)

Scope 2 covers indirect emissions from imported utilities and the most common of these is the purchase and use of electricity. Some sites also purchase steam or heat from external sources and this is also covered under Scope 2.

The data needed are simply the total number of kWh used in the year and the relevant carbon intensity factor for the supply country. The number of kWh used in the year should be easily available from the electricity bills and the interval data (see Section 6.2).

Emissions from electricity use

The generation of electricity will inevitably result in CO_2e emissions but the size of these emissions depends on the generation method. Nuclear, thermal or hydroelectric generation result in low CO_2e emissions, whereas coal, gas and other fossil fuels result in high CO_2e emissions. Each country will have a mix of generation for grid electricity that will therefore affect the total emissions. This is also true at the supplier level, i.e., a supplier can have a different emission factor to the country as a whole if their generation profile is significantly different to the country profile.

For example, a country generating a large proportion of grid electricity from nuclear sources or renewable energy will have a lower CO_2e emissions factor than a country generating a large proportion of grid electricity from coal or other carbon-based fuels. The factors for any specific country will change with time as the generation pattern changes.

- **Tip** – Always use the latest data for emissions factors to account for changing generation patterns.

Generation, transmission and consumption emissions

Grid electricity is generated at the power station and this results in emissions, but the transmission and distribution of the electricity to the user also results in losses, i.e., transmission and distribution losses. The emission factor at the point of consumption is therefore higher than the emission factor at the point of generation and

Emission factor (consumed) =
Emission factor (generated) +
Emission factor (losses).

Country	kg CO_2/kWh
South Africa	0.8573
India	0.8291
Australia	0.8136
Poland	0.7739
Indonesia	0.7584
Saudi Arabia	0.7529
People's Rep. of China	0.7525
Malaysia	0.7218
Czech Republic	0.5919
South Korea	0.5518
Japan	0.5294
Thailand	0.5249
USA	0.4985
Germany	0.4718
Turkey	0.4644
UK	0.4622
Egypt	0.4542
Russian Federation	0.4498
Mexico	0.4484
Italy	0.399
Netherlands	0.399
Pakistan	0.3945
Spain	0.2891
Belgium	0.1894
Canada	0.164
Brazil	0.0693
France	0.0586
Iceland	0.0002

Scope 2 carbon intensity by country (2015)

The carbon intensity (kg CO_2e/kWh) varies widely across the world depending on the generation method. These data include emissions from all Scopes and is the 5-year rolling average.

Source: DBE&IS 2015.

- **Tip** – Generation emissions are recorded under Scope 2 but loss emissions are recorded under Scope 3 (see Section 7.4).

Supplier data

The most accurate value for the emission factor will be from the specific supplier because this will reflect their generation profile. Most suppliers should be able to supply this on request.

- **Tip** – If requesting data from suppliers then be sure to get the generation emission factor and check if it is CO_2e data and not simply CO_2 data. If it is only CO_2 data then note this in any reporting.

- **Tip** – The carbon intensity for any supplier will change with time as their generation profile changes. The best value to use is either the most current year or a moving average for the last 5 years.

Country data

If the supplier cannot supply a generation emission factor then it is possible to use a country emission factor but this will not be as accurate.

Until 2015, country emission data for a wide range of countries were given in DBE&IS and the table on the left gives the 2015 consumption emission factors for selected countries (DBE&IS 2015) and shows the range of emission factors around the world.

The countries with the highest emission factors, e.g., South Africa, India, Australia and Poland, mainly use coal for electricity generation. The countries with low emission factors, e.g., Brazil and France, use low-carbon electricity generation methods such as hydro or nuclear.

After 2015, DBE&IS stopped giving this information but this is now available for purchase (at great expense) from the International Energy Agency (IEA).

- **Tip** – Always use the emission factor for electricity generation and not for consumption in Scope 2. The transport and distribution losses in the distribution system are covered in Scope 3 (see Section 7.4).

US regional data

In the USA, the size and diversity of the electricity generation capacity means that the global single country value from the DCBE&IS is not really relevant for most of the country. For a more accurate approach it is possible to use the eGRID data produced by the US Environmental Protection Agency. The USA is divided into 26 sub-regions for which detailed emissions data are available. The latest version of the data was released in January 2020 and is for 2018. This site also has an excellent Frequently Asked Questions section and summary tables for quick reference.

- **Tip** – The US eGRID data are sometimes difficult to work with because the sub-region boundaries do not align with physical state boundaries. Try to get supplier data before using eGRID.

Emissions from imported heat or steam

Importing heat or steam from local generators is rare in plastics processing. If heat or steam is imported then DBE&IS provides UK data but if the site is located outside the UK then the emission factor should be calculated by the generator of the heat or steam.

The imported utilities, i.e., electricity, will almost always be the largest part of the carbon footprint for plastics processors.

It is essential to get the right data for this Scope.

Emission type	Emission source	Emission		
		Tonne CO_2e	%*	Notes
Scope 2: Indirect emissions (imported power)	Electricity	2,484.57	73.11%	Supplier CO_2e data.
	Sub-Total	**2,484.57**	**73.11%**	
Example of reporting of Scope 2 emissions This is a typical format for reporting Scope 2 emissions. The data for actual fuel use are taken from the billing or interval data. The emissions factor is taken from supplier data. * This is the percentage of the total emissions from all three scopes, i.e., in this example the total Scope 2 emissions are 73.11% of the total of all three scopes.				

7.4   Site carbon footprinting – Scope 3

Direct emissions (not controlled assets)

Scope 3 covers the direct emissions that a site uses but does not control. As an example: An employee has to fly to visit a customer. The flight has direct emissions but the site does not control the asset (the aeroplane) and is not responsible for the complete emissions of the flight. The site therefore takes a share of these emissions based on the aeroplane's emissions.

Data for Scope 3 emissions are sometimes difficult to get and the errors can be high. In many cases, Scope 3 emissions are not mandatory and not calculated and this is unfortunate because it is usually possible to estimate them quickly using available information.

For most plastics processing sites, Scope 3 emissions will be < 10% of the total emissions. Even large estimation errors will not affect the total emissions greatly but including Scope 3 emissions provides a better estimate of the total carbon footprint.

Scope 3 can be divided into various emission sources and these will be covered individually for a typical plastics processing site. A typical report format is shown on the lower right on the opposite page. Some sites will not have all the emissions discussed.

Transport and distribution losses (T&D)

Transport and distribution losses for electricity are recorded in Scope 3. As of 2018, DBE&IS no longer gives T&D values for other countries but this is also available for purchase from the IEA. These factors can be used with the electricity consumption (kWh) to calculate the emissions from T&D losses.

Well-to-tank losses (WTT)

The production of electricity involves well-to-tank losses. These are the emissions from the extraction, refining and transportation of fuels before they are used in electricity generation. WTT loses are still given in DBE&IS 2020 for many countries and these can be used with the electricity consumption (kWh) to calculate the emissions from WTT losses.

Employee business travel

Business travel has several elements:

Private car

If employees use their private cars for company business then this is included here. This may be calculated from company payments for private car use with the emission factor for the car type taken from DBE&IS. If in doubt, then use the 'average car/unknown fuel' factor.

- **Tip** – A quick method is to estimate how many people use their car for business and then to estimate the average distance travelled/year.

Rail/bus

Most sites make little use of rail or bus but these can be calculated from DBE&IS.

Flights

Emission factors for air travel vary for short- and long-haul travel and also for the various travel classes, i.e., first class has higher emissions than economy class. DBE&IS gives the relevant emission factors for both short and long haul and for a range of classes.

The total emissions can be calculated from travel records and the distances travelled by each flight. The numbers needed here are the number of passenger km travelled/year in the relevant categories.

- **Tip** – The approximate distance travelled for flights can be found on the Internet.
- **Tip** – A quick method is to estimate how many people fly for business and then to estimate their typical journey.

Scope 3 calculations will always have greater errors than other scopes due to the diverse nature of the components.

This is no excuse for not estimating the magnitude of the Scope 3 emissions. There are quick methods for estimating many of the Scope 3 effects.

Flight emissions can be calculated using 'with radiative forcing' or 'without radiative forcing'.

Radiative forcing is a measure of the additional environmental impact of aviation from emissions of NO_x and water vapour when emitted at high altitude.

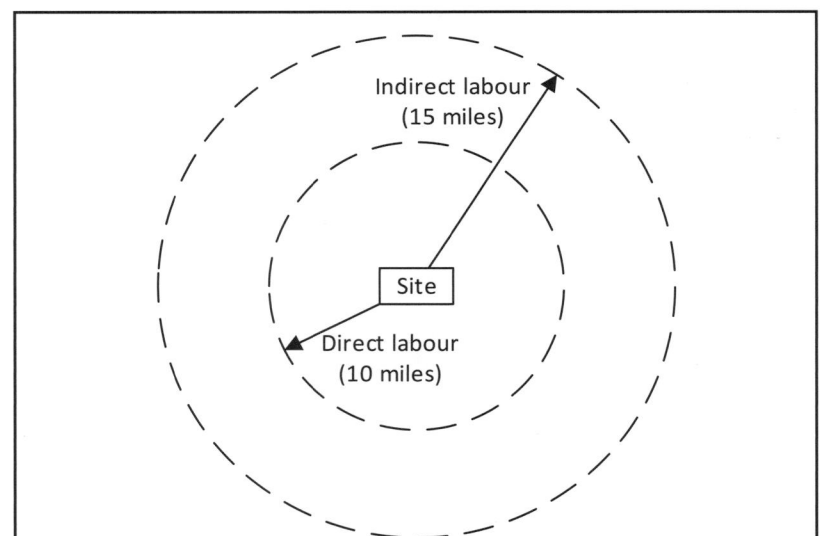

Estimating commuting mileage

The quickest way to estimate commuting mileage is to identify the average distance travelled by direct labour and by indirect labour. The indirect labour will travel further because they are better paid.

Rental cars/taxis

This can be accurately found from travel records (employee expense claims) and using DBE&IS. If in doubt then use the 'average car/unknown fuel' factor.

- **Tip** – Rental cars and taxis are mainly used for travel at the start and end of flights. A quick method is to use the flight numbers (see above) and to allocate some taxi and rental car use to each flight.

Employee commuting

This can be difficult to calculate accurately. Some sites use staff surveys but the quickest method is to assume that each employee travels an average distance to the site. The average distance will be greater for indirect labour than for direct labour. This average distance can be used with the number of days worked/year and the 'average car/ unknown fuel' factor.

- **Tip** – When car sharing is used then the values should be decreased in proportion to the amount of sharing.

- **Tip** – If commuting is by train or bus then DBE&IS can also be used.

Product transport

In most cases, product transport is by contractors who control the vehicle. DBE&IS has tables for all types of transport from trucks to air freight.

When calculating the product transport emissions by road there is a choice of:

- Calculating emissions as if the complete vehicle was used, i.e., all the emissions are taken by the site. In this case the data needed are the number of loads and the average distance to the customer.

- Calculating the emissions as if only part of the vehicle was use, i.e., the site only takes a part of the emissions. In this case the data needed are the number of tonne.km. This is the distance travelled (km) by each tonne of product.

Sites should assess their transport method and use the appropriate tables.

- **Tip** – A quick method is to estimate the average distance to the customer for each transport method and the amount of product transported by this method. This gives a rapid estimate of the emissions from product transport.

- **Tip** – If the customer collects the products then the carbon emissions are allocated to them and not to the site.

DBE&IS also gives emission factors for water supply and treatment.

These can be used to calculate the emissions due to water supply and treatment.

The 'Travelmath' web site is good for finding driving and flight distances.

DBE&IS also gives emission factors for material use and waste.

These can be used to calculate the emissions due to basic material use and to waste disposal and recycling.

Emission type	Emission source	Emission		
		Tonne CO_2e	%*	Notes
Scope 3: Direct emissions (not controlled assets)	T&D losses	212.44	6.25	DBE&IS data.
	WTT losses (generation/T&D)	376.36	11.07%	DBE&IS data.
	Employee travel – own car	0	0.00%	Estimated.
	Employee travel – air transport	4.34	0.13%	Estimated.
	Employee travel – taxi	1.73	0.05%	Estimated.
	Employee travel – rental car	0.37	0.01%	Estimated.
	Employee travel – commuting	10.78	0.32%	Estimated.
	Product transport	274.20	8.07%	Estimated.
	Water use	0.38	0.01%	Estimated.
	Waste disposal	6.41	0.19%	Estimated.
	Sub-Total	**855.78**	**25.41%**	

Example of reporting of Scope 3 emissions

This is a typical format for reporting Scope 3 emissions. The data can be based on estimates if no accurate data are available. The relevant emission factors are taken from DBE&IS.

* This is the percentage of the total emissions from all three scopes, i.e., in this example the total Scope 3 emissions are 25.41% of the total of all three scopes.

7.5   Site carbon footprinting – putting the scopes together

The site carbon footprint

Calculating the effects of the various scopes (see Sections 7.2 to 7.4) allows an estimate of the various elements of the total site carbon footprint. These are put together into the complete site carbon footprint on the opposite page and for this example site the total CO_2e emissions are estimated at 3,749.54 tonnes of CO_2e/year.

There is naturally an error associated with this value and this is normally in the region of ± 2%. The methods used to gather the data will affect this error value and this may increase if estimates are used extensively.

- **Tip** – It is possible to exclude certain elements of the carbon footprint if they are < 5% of the total.

It is also useful for most sites to look at the relative effects of the various scopes and this is shown in the pie chart on the lower right. This distribution of the carbon emissions is typical for a plastics processing site where the emission factor for electricity is in the region of 0.3-0.6 kg CO_2e/kWh, i.e., most of Europe and the USA. The reason for this is the high use of electricity in plastics processing as a source of energy.

Assessing the carbon footprint

The absolute value of the site carbon footprint in tonnes/year is useful as a reporting measure but fluctuating production volume will obviously affect this value because of the relationship between production volume and electricity use. It is therefore common to report the carbon footprint in terms of a relative measure such as 'tonne of CO_2e/tonne of product' or 'tonne of CO_2e/product'.

These relative measures can provide a measure of progress towards carbon reduction targets but they will also suffer from the same problems as using kWh/kg as a metric (see Section 6.5). For any site there will be a 'base carbon load' from heating, lighting and other fixed loads and a 'process carbon load' from production-related loads. This is analogous to the base and process loads discussed in Section 6.4. Despite this, if the assessment is over a period of at least 12 months then the effect becomes small.

Comparing carbon footprints between sites located in different countries is possible but the large variations in emission factors between countries (see Section 7.8) will inevitably make this comparison more dependent on the country emission factor than on how the site operates.

Reporting

Calculating a site carbon footprint is not the end of the story. The carbon footprint provides a convenient metric for the following actions:

- Reporting progress in carbon reduction to staff, customers, stakeholders and external bodies such as government – this can be via the annual report or via notice boards/ staff newsletters.

- Setting targets for carbon reduction via an energy management programme as electricity is key to the carbon footprint.

- Providing customers with data for incorporation into product carbon footprints (see Section 7.6). This is becoming more important as major customers demand these data for their own products.

- **Tip** – One of the most respected ways of externally reporting carbon footprints and progress in carbon reduction is via the Carbon Disclosure Project (www.cdp.net). This is an open access database of freely disclosed carbon data from many of the world's major companies).

For sources that are less than 1% of the total it is sensible to use estimates rather than spend resources calculating the data for accurate calculation.

The aim is to have good data for ≈ 95% of the emissions.

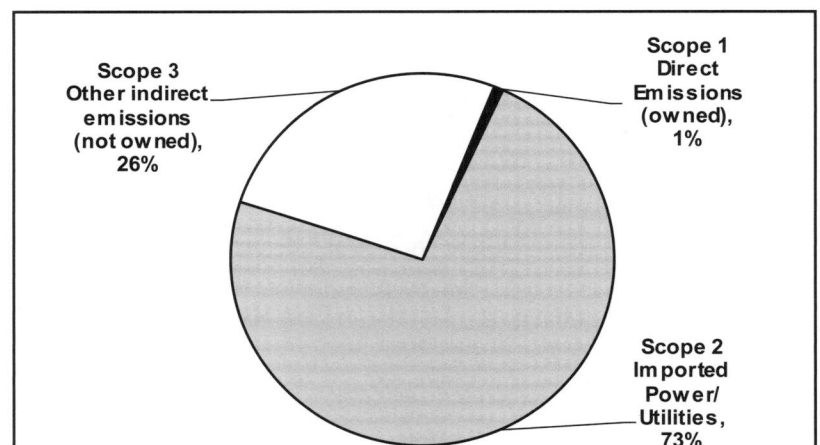

The carbon footprint by scope

As with many plastics processing sites, Scope 2 is the largest contributor to the carbon footprint with the Scope 1 and Scope 3 emissions being much lower. This is simply a reflection of the heavy dependence of the industry on electricity as a source of energy.

Period: 2020 Calendar Year				
Emission type	Emission source	Emission		
		Tonne CO_2e	%	Notes
Scope 1: Direct emissions (controlled assets)	Gas (heating and process)	2.85	0.08%	Billing data.
	Fuel oil (heating and process)	0	0.00%	N/A.
	Fork lift truck gas	0	0.00%	Electric FLTs.
	Other fossil fuel	0	0.00%	N/A.
	Owned cars	20.25	0.60%	Distance travelled.
	Owned trucks or vehicles	0	0.00%	N/A.
	Process emissions	0	0.00%	N/A.
	Refrigerant emissions	3.84	0.11%	Refrigerant data.
	Other direct emissions	0	0.00%	N/A.
	Sub-Total	**26.94**	**0.79%**	
Scope 2: Indirect emissions (imported power)	Electricity	2,484.57	73.11%	Supplier CO_2e data.
	Sub-Total	**2,484.57**	**73.11%**	
Scope 3: Indirect emissions (not controlled assets)	T&D losses	212.44	6.25	DBE&IS data.
	WTT losses (generation/T&D)	376.36	11.07%	DBE&IS data.
	Employee travel – own car	0	0.00%	Estimated.
	Employee travel – air transport	4.34	0.13%	Estimated.
	Employee travel – taxi	1.73	0.05%	Estimated.
	Employee travel – rental car	0.37	0.01%	Estimated.
	Employee travel – commuting	10.78	0.32%	Estimated.
	Product transport	274.20	8.07%	Estimated.
	Water use	0.38	0.01%	Estimated.
	Waste disposal	6.41	0.19%	Estimated.
	Sub-Total	**855.78**	**25.41%**	
Total		**3,749.54**	**100%**	

The complete site carbon footprint (all scopes)

This is a typical format for reporting a complete site carbon footprint. It is important to specify the period of the calculation.

This is an example calculation to show the format of the report only.

7.6　Product carbon footprinting

The complete life cycle

A product carbon footprint (PCF) is designed to allow companies and consumers to understand the environmental impact of a product over the complete life cycle from the raw materials to disposal and/or recycling.

The theory is that this will allow companies to measure, monitor and manage energy use, set targets for emission reductions, reduce costs and communicate with stakeholders.

The PCF concept has generated many schemes and systems and there are currently over 20 schemes around the world that are used to measure PCFs. Whilst the methods are all similar, they can vary in the details.

The two most robust current methods for developing a PCF are:

- 'PAS 2050: 2011' – developed in the UK by BSI and the UK Government.
- 'Product Life Cycle Accounting and Reporting Standard' (2011) – developed by the GHG Protocol, WBCSD and WRI.

Their methodology is very similar and they are both freely available (see sidebar on the lower right). They were essentially developed together and are broadly compatible, although they vary in some details of the requirements, primarily in the areas of recording and public reporting. A similar and compatible approach was also used to develop ISO 14067:2018 'Greenhouse gases – Carbon footprint of products – Requirements and guidelines for quantification and communication'.

Steps to calculating a PCF

Whichever scheme is chosen, the steps to calculating a PCF are broadly similar. These are:

Step 1 – Build a process map

The first step in defining the PCF is to create a process map of the product life cycle. A basic example process map is shown on the right and when fully completed this gives details of all the materials, activities and processes in the complete product lifecycle. These may be very different for nominally similar products.

Step 2 – Assess boundaries and determine priorities (materiality)

The process map is used to assess the boundaries (the limits of the life cycle that are being considered) and to calculate an initial overview PCF to locate the 'hot spots' of high emissions. At this stage the 'materiality' is also considered – this is to minimise the effort of data collection. If a single source contributes < 1% of the total footprint then it can be excluded provided that the total of all the exclusions is not > 5% of the total PCF. This allows an initial high-level PCF calculation to establish high emission areas where data collection accuracy is important.

Step 3 – Collect data

Two type of data are needed for a PCF:

- Activity data – these are the measurable materials and energy used in each of the life cycle stages. Ideally these would be data derived from meters or invoices.
- Emission factors – these are the GHG emissions associated with each unit of the activity data. These can come from databases such as DBE&IS or other accredited databases.

Step 4 – Calculate the footprint

The activity data are multiplied by the emission factors (as for the site carbon footprint) to calculate the actual emissions for each stage of the life cycle.

Step 5 – Check uncertainty

Uncertainty is critical in producing a robust PCF, i.e., if the uncertainty is high then the PCF is not robust. The GHG Product Standard specifically requires a statement of the uncertainty in the report but PAS 2050 notes this only as guidance to use data that will reduce uncertainty.

Step 6 – Verify the footprint

The PCF should be verified and this can be by simple self-certification, by a third party or by an accredited verification company. The choice depends on costs and how the PCF is going to be used.

- **Tip** – For most plastics processors, a PCF will be unnecessary but processors should be aware that their site carbon footprint data may be requested by customers as part of their PCF.

There is not complete agreement on the benefits, efficiency or even accuracy of product carbon footprinting. There are considerable criticisms of the methodology and application of the results.

Most PCFs are suitable for comparing the impact of changes to materials, processes, etc. over time but do not provide the level of detail or accuracy necessary for product comparison.

The best (and easiest to understand) documents on product carbon footprinting are:

- PAS 2050:2011 'Specification for the assessment of the life cycle greenhouse gas emissions of goods and services'. Available free from BSI: shop.bsigroup.com/ en/forms/PASs/ PAS-2050.
- 'Product Life Cycle Accounting and Reporting Standard'. Available free from GHG Protocol: ghgprotocol.org.

Process map for a plastic vacuum flask

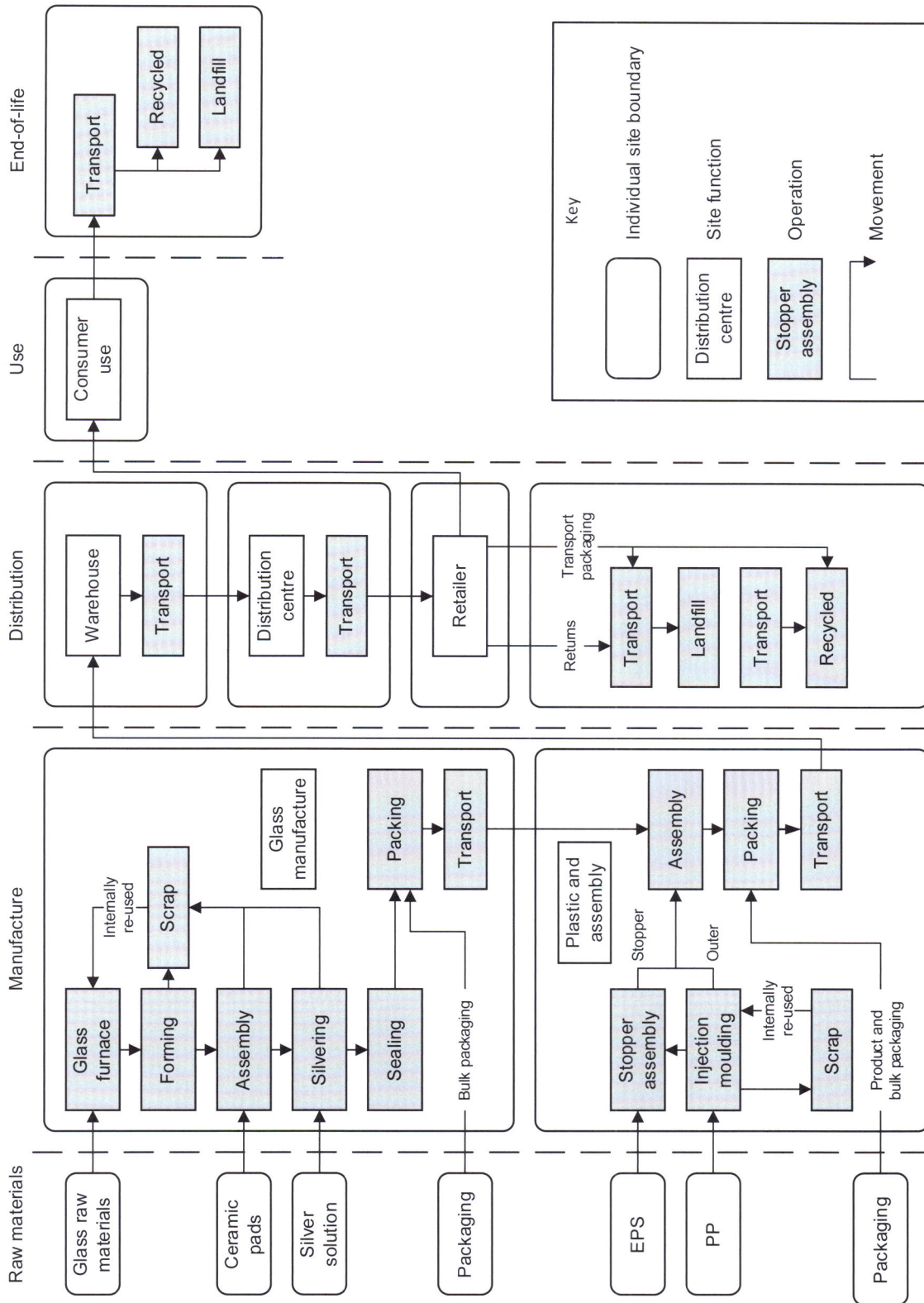

The basic process map for the life cycle of a plastic vacuum flask

The process map is the start of creating a PCF. It defines the basic processes and enables the emissions 'hot spots' to be identified. If the map covers the complete life cycle then it is 'cradle to cradle' or 'cradle to grave', if it stops at distribution then it is 'cradle to gate'.

7.7   Product carbon footprints

Cradle-to-cradle or cradle-to-gate?

Most plastics processors do not make final products but product carbon footprints have been created for everything from potato crisps through to cars. For final products, the complete life cycle gives a 'cradle-to-cradle' carbon footprint but in some cases a partial carbon footprint or 'cradle-to-gate' that does not include the use or end-of-life phases is used. This may be because the manufacturer has no knowledge or control over these phases but the use of a cradle-to-gate inventory should be both disclosed and justified.

- **Tip** – It is possible to use 'standard user' profiles to model the use phase and these can be gathered from user surveys or industry averages.

We have attempted to carry out some 'cradle-to-cradle' carbon footprints based on some broad model assumptions and four of these are shown on the opposite page. These are not externally validated and are presented simply to show some typical results of product carbon footprinting.

- **Tip** – The use and end-of-life designations for these products are covered in more detail in Chapter 10.

The important point is that for long-life products, the use phase is the dominant phase for carbon emissions and that for short-life products the raw materials and manufacturing phases are the important phases. Whilst this may not come as a surprise, it can help the industry to focus on improvement efforts.

- **Tip** – The model carbon footprints also use some assumptions on the amount of recycled material and the effect this has on the total carbon footprint. We have modelled the effect of mechanical recycling on the raw materials carbon footprint and this shows an $\approx 80\%$ reduction in raw materials carbon footprint can be achieved by using mechanical recycling.

- **Tip** – For a great, funny and totally illuminating read, we suggest "How Bad are Bananas? The carbon footprint of everything" by Mike Berners-Lee[1]. He covers the carbon footprint of everything from a text message (yes, it has a carbon footprint) to a war with diversions through a bottle of wine and a heart bypass.

Where do I start?

There are already an enormous amount of data and databases available with standard processes already input and, in many cases, it is possible to use these to quickly create a carbon footprint. The ones that we use are:

- CCalc2 (www.ccalc.org.uk) – this was produced by the University of Manchester and models life cycles following ISO 14044:2006 and PAS 2050:2011. The latest version has some excellent case studies that can be modified for many plastic products as well as good access to a free database (CCalC database).
- OpenLCA (www.openlca.org) – this is 'open source' software run by Green Delta in Berlin with a combination of free and 'for purchase' databases.

Most plastics processors will find CCalC2 sufficient and easy to use.

Is it worth it?

The benefit of a product carbon footprint is that it tells us where we want to spend our efforts to reduce the carbon footprint of the product. Finding the pressure points of the complete product system allows industry to know where to work to reduce the carbon footprint. Knowledge, even imperfect knowledge at the start, is power in reducing the impact and improving sustainability.

Reducing the 'use' phase energy use in windows is why I spent about 6 years developing and implementing an energy rating system for windows in the UK (www.bfrc.org).

It is important to note that a carbon footprint does not show how much energy can be saved by using plastics products, this can only be done by comparing the carbon footprints for competing products.

In this case, selecting the boundary conditions must be done with great care to avoid prejudicing the validity of the results.

Product carbon footprints tell you the carbon (actually CO_2e) of a product. They do not tell you about the carbon avoided by the use of the product.

- 1. "How Bar are Bananas – The carbon footprint of everything", Mike Berners-Lee, Profile Books, 2020.

Single use coffee cup lid (very short-life)

A single use coffee cup lid has an average use life of < 1 day and uses virtually no energy at all in the use phase, i.e., it is a container, although it does prevent leakage and spills. This means that ≈ 96% of the carbon footprint contribution is in the raw materials and manufacturing phases.

End-of-life has a very low carbon impact because this type of product will not generally be captured by the MSW stream and become landfill, i.e., it is an 'on-the-go' product and ends up in a mixed waste stream. This makes lightweighting and efficient manufacturing the key to reducing the carbon footprint of very short-life products such as coffee cup lids.

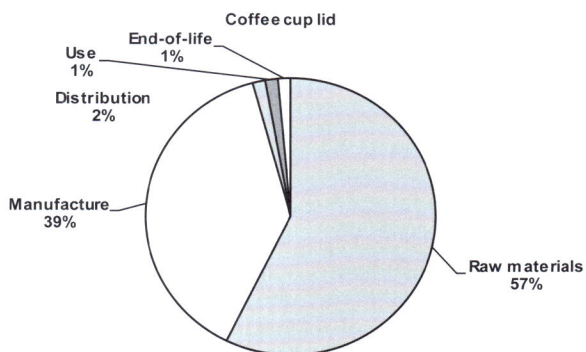

Coffee cup lid

- End-of-life 1%
- Use 1%
- Distribution 2%
- Manufacture 39%
- Raw materials 57%

Food packaging (short-life)

Typical food and drink packaging will have a use life of < 2 years because the products are required to provide functional product protection, e.g., barrier layers, hygiene or transport. The main carbon footprint contributions are from raw materials and manufacture (≈ 88%).

The end-of-life contribution is higher for this type of product because they will be captured in the MSW stream for recycling and this has an impact on the carbon footprint.

Lightweighting and efficient manufacturing are still important but these products also need good functional analysis to ensure that the plastic is used wisely.

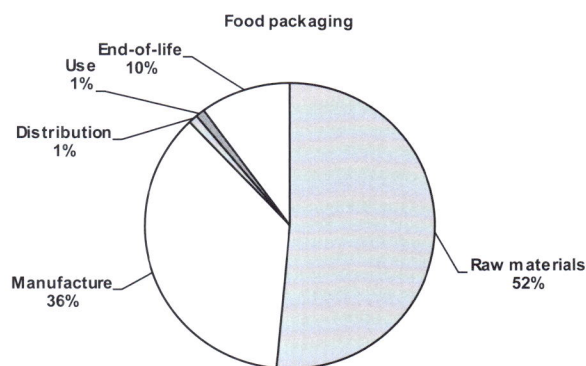

Food packaging

- End-of-life 10%
- Use 1%
- Distribution 1%
- Manufacture 36%
- Raw materials 52%

Car part (medium-life)

Cars have a use life of ≈ 2-15 years and largest component of the carbon footprint is in the use phase. The rise in the amount of plastics use in cars has driven down the weight of cars and improved their energy efficiency. The use phase is not as high as in some products because the plastic parts are only part of the overall vehicle. End-of-life is also increased because of the more complicated refuse stream.

For car parts, reducing the raw materials carbon footprint by lightweighting and improved design not only reduces the raw materials contribution but also further reduces the use contribution.

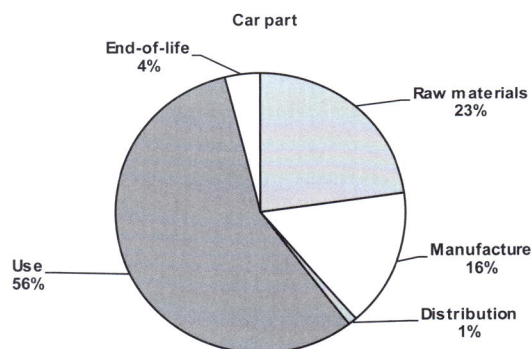

Car part

- End-of-life 4%
- Raw materials 23%
- Manufacture 16%
- Distribution 1%
- Use 56%

PVC-U window profile/window (long-life)

PVC-U windows have a use life of ≈ 25-50 years and the largest component of the carbon footprint is in the use phase. Windows are 'appliances' that use/lose energy throughout their lifetime. In fact, windows use/lose so much energy that the use phase dwarfs the other phases.

Any action taken to reduce the use phase energy use will be beneficial not only to the consumer but also to the environment even if it uses more material or costs more to produce.

Other long-life products, e.g., plastic pipes will be even better in carbon footprint terms because they do not use any energy in the use phase.

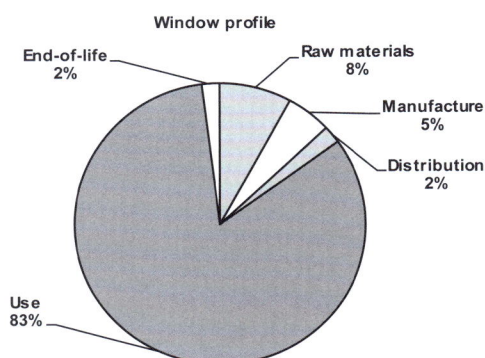

Window profile

- End-of-life 2%
- Raw materials 8%
- Manufacture 5%
- Distribution 2%
- Use 83%

7.8 Country plastics processing carbon footprints

The global level

Section 6.27 gave the relative carbon intensity for electricity generation in a variety of countries and Section 6.28 discussed the effect of transport and distribution (T&D) losses and well-to-tank (WTT) losses. Under the standard carbon footprinting conventions, these carbon emissions are reported separately in Scopes 2 and 3 but they can be combined to give a country total carbon emission for electricity consumed that will include all the emissions.

Tangram Technology has developed a simple model which:

- Takes the total processed plastics production for a country.[1]
- Divides this into the various types of plastics based on general world-wide trends for plastics production.
- Divides each type of plastic into the various production methods used to process the material based on the general world-wide trends for plastics processing.
- Totals each type of plastic used for each process to give a total volume of processed plastic by process.
- Assigns an average site process energy to each specific process (kWh/kg).
- Calculates the total process energy (kWh) for each specific process.
- Calculates the total country energy use for all plastics processing methods (kWh).
- Uses the total country energy use for plastics processing and the country total carbon emissions as a result of electricity use to estimate the total country CO_2e emissions as a result of plastics processing.
- Uses an average electricity cost of £0.10/kWh to estimate the total energy cost (in £) for the country.

The results of this process for 26 countries are shown in the table on the right.

- **Tip** – These calculations do not include the embodied energy of the plastics processed, i.e., the energy used to produce the actual plastic.
- **Tip** – These calculations do not include any consideration of the product carbon footprint, i.e., they are only for the plastics processing operation.

Error estimation

This process is based on a series of estimations and inevitably contains errors. We have compared the model results with other similar estimates and the results are within ± 7% in all cases. This is regarded as an acceptable error given the assumptions necessary to make the estimates, i.e., not every country will have the same proportion of plastics processed and not all processing will use energy with the same carbon intensity as the general country carbon intensity.

Therefore, these estimates are provided for guidance only.

Country plastics processing carbon footprints

The top three countries, in terms of processed volume, electricity used, carbon emissions resulting from plastic processing and energy cost are China, USA and India.

For the 26 countries considered, the total electricity used is estimated to be ≈ 531,682 GWh, the total cost is estimated to be ≈ £53 billion and the total carbon emissions are estimated to be ≈ 400 million tonnes. The table only lists the major plastics processing countries but this is estimated to capture ≈ 95% of the total world-wide volume of plastics processed.

Note: The data is not fully aligned in dates, i.e., the 2015 carbon intensity for generation has been used for each country, the 2019 transport and distribution losses have been used and the 2017 data for plastics consumption has been used. This is because of the time lag in getting reliable data. We have used the most recent available data for each component of the calculation.

Plastics processing is an energy-intensive operation wherever it is carried out.

- 1. Processed volume values from 'Plastics Resin Production and Consumption in 63 Countries Worldwide', 2019, Euromap (www.euromap.org).

	Generation (kg CO_2e/ kWh) 2015	T&D + WTT (kg CO_2e/ kWh) 2019	Total (kg CO_2e/ kWh)	Processed volume[1] (tonnes) 2017	Electricity used (GWh)	CO_2e (Mtonnes)	Energy cost (Million £)
Belgium	0.1894	0.0357	0.2251	2,313,000	5,017	1.13	502
Brazil	0.0693	0.0246	0.0939	7,081,000	15,360	1.44	1,536
Canada	0.1640	0.0397	0.2038	3,210,000	6,963	1.42	696
China	0.7525	0.1688	0.9212	99,497,000	215,822	198.82	21,582
Czech Republic	0.5919	0.1337	0.7256	1,366,000	2,963	2.15	296
Egypt	0.4542	0.1368	0.5910	1,773,000	3,846	2.27	385
France	0.0586	0.0158	0.0744	4,853,000	10,527	0.78	1,053
Germany	0.4718	0.0950	0.5669	8,716,000	18,906	10.72	1,891
India	0.8291	0.3232	1.1523	15,928,000	34,550	39.81	3,455
Indonesia	0.7584	0.2016	0.9600	5,019,000	10,887	10.45	1,089
Italy	0.3990	0.0904	0.4894	6,649,000	14,423	7.06	1,442
Japan	0.5294	0.1036	0.6330	9,394,000	20,377	12.90	2,038
Malaysia	0.7218	0.1603	0.8820	2,530,000	5,488	4.84	549
Mexico	0.4484	0.1617	0.6100	6,298,000	13,661	8.33	1,366
Netherlands	0.3990	0.0843	0.4832	1,800,000	3,904	1.89	390
Pakistan	0.3945	0.1489	0.5434	1,626,000	3,527	1.92	353
Poland	0.7739	0.1908	0.9647	3,533,000	7,664	7.39	766
Russia	0.4498	0.1393	0.5892	5,920,000	12,841	7.57	1,284
Saudi Arabia	0.7529	0.2029	0.9558	2,921,000	6,336	6.06	634
South Africa	0.8573	0.2302	1.0874	1,491,000	3,234	3.52	323
Spain	0.2891	0.0763	0.3654	3,407,000	7,390	2.70	739
Thailand	0.5249	0.1115	0.6364	5,006,000	10,859	6.91	1,086
Turkey	0.4644	0.1650	0.6294	6,984,000	15,149	9.54	1,515
UK	0.4622	0.0768	0.5390	3,422,000	7,423	4.00	742
USA	0.4985	0.1117	0.6101	34,376,000	74,566	45.50	7,457
Total					531,682	399.10	53,168

Country carbon footprints of major plastics processing countries

The relative total electricity carbon intensity for a country can be combined with the production volume for the country and the various processing methods to give a total country carbon footprint for the plastics processing industry in that country.

Note: This is for the processing only and does not include any embodied carbon in the basic polymer production.

7.9   Carbon footprinting – where are you now?

Assessing the impact

Carbon footprinting assesses the impact that a site or organisation has on the atmosphere and is a performance metric that is growing in importance. External organisations are increasingly asking suppliers for access to carbon footprint calculations and every site should be assessing this impact.

Good energy management for plastics processing companies will not only reduce the amount of energy used and the cost of this but will also reduce the carbon footprint. Companies may embark on energy management primarily for the cost benefits but calculating and monitoring the carbon footprint will also reveal the benefits to society of good energy management.

As with the score charts shown earlier, this is a self-assessment exercise to allow a site to benchmark their current status in terms of carbon footprinting.

Completing the chart

This chart is completed and assessed as for those presented previously.

Most of the data needed for carbon footprinting will be generated automatically as part of energy management.

There is very little extra work involved.

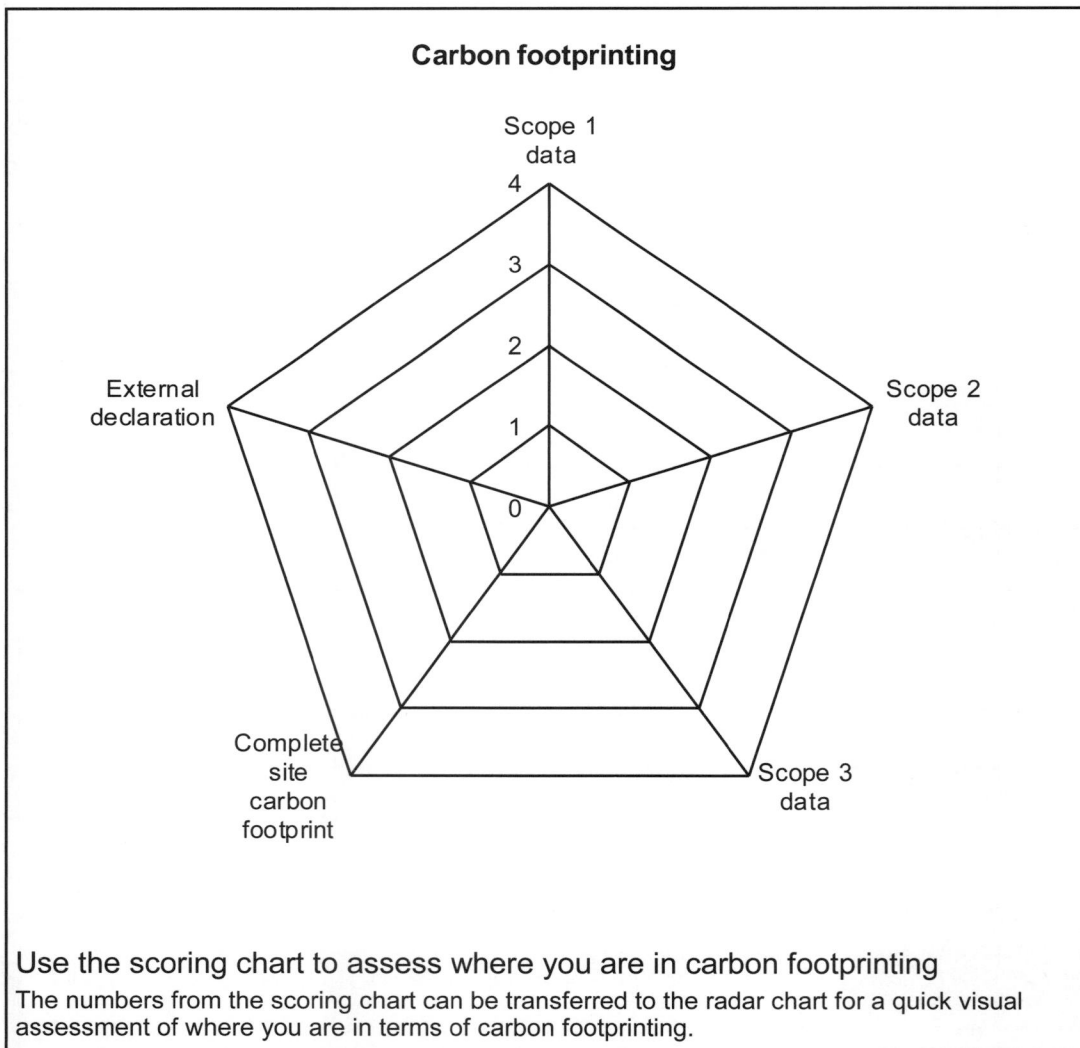

Carbon footprinting

Use the scoring chart to assess where you are in carbon footprinting

The numbers from the scoring chart can be transferred to the radar chart for a quick visual assessment of where you are in terms of carbon footprinting.

Calculate your carbon footprint now so that it is ready for when your customers ask for it.

Being pro-active is far better than being reactive.

Carbon footprinting

Level	Scope 1 data	Scope 2 data	Scope 3 data	Complete site carbon footprint	External declaration
4	All relevant Scope 1 data collected on a monthly basis using existing accounting systems for greater accuracy.	Scope 2 emissions from electricity calculated using supplier's current specific carbon intensity for generation.	All relevant Scope 3 data collected on a regular basis using existing accounting systems for greater accuracy.	All relevant data for Scopes 1 to 3 combined on a monthly basis using existing accounting systems for greater accuracy.	Full external declaration of organisation carbon footprint for Scopes 1 to 3.
3	All relevant Scope 1 data collected on an annual basis using existing accounting systems.	Scope 2 emissions from electricity calculated using area or region carbon intensity for generation.	All relevant Scope 3 data collected on an annual basis using existing accounting systems.	All relevant data for Scopes 1 to 3 combined on an annual basis using existing accounting systems.	Full external declaration of site carbon footprint for Scopes 1 to 3.
2	All relevant Scope 1 data estimated on an annual basis.	Scope 2 emissions from electricity calculated using general country carbon intensity for generation.	All relevant Scope 3 data estimated on an annual basis.	All relevant data for Scopes 1 to 3 combined on an annual basis using good estimates for a number of factors.	Full external declaration of organisation carbon footprint for Scopes 1 & 2.
1	Some relevant Scope 1 data not calculated at all.	Scope 2 emissions from electricity calculated using unvalidated carbon intensity factor for generation.	Some relevant Scope 3 data not calculated at all.	Scope 1 & 2 data combined for partial carbon footprint but no Scope 3 data estimated or included.	Full external declaration of site carbon footprint for Scopes 1 & 2.
0	No calculation of Scope 1 data.	No calculation of Scope 2 data.	No calculation of Scope 3 data.	No complete site carbon footprint prepared.	No external declaration of organisation or site carbon footprint.
Score					

Key tips

- Carbon footprinting needs good data for an accurate calculation.
- Data for needs to be collected regularly to increase accuracy.
- Calculate a provisional site carbon footprint using the available accurate data and estimate the other data.
- Estimate the potential errors in the provisional site carbon footprint.
- Start to gather reliable data on the Scope 1 emissions such as:,
  - Gas use.
  - Fork lift truck gas (if used).
  - Mileage or fuel purchase for owned cars and trucks.
  - GHG gas use.
- The energy data (see Chapter 6) can provide reliable information for the calculation of Scope 2 emissions.
- The calculation of Scope 3 emissions is the most difficult area for most companies.
- Start to gather reliable data on the Scope 3 emissions. This will always be less accurate than Scope 1 and Scope 2 emissions because of the sources of the data.
- Scope 3 data to be collected is:
  - Electricity transport and distribution losses.
  - Employee business travel by transport mode.
  - Employee commuting details
  - Product transport methods.
  - Recycling details.
  - Water use (sources and sinks).
- Publicise the site carbon footprint to the staff – they will be interested.
- Declare the site carbon footprint externally and report regularly.
- Carry out some simple product carbon footprints (cradle-to-gate) to be ready for when you are asked by your customers.

Chapter 8

Manufacturing: water

Earth has about 1.4 billion cubic kilometres of water but only about 3% (42 million cubic kilometres) of this is fresh water. The remaining 97% is salt water. Of the 3% fresh water, most of the water is frozen as ice, glaciers and snow (2%) or in groundwater (1%). Only approximately 0.01% of the total is available as surface water, i.e., rivers, lakes and swamps.

Clean and fresh water is one of the most undervalued natural resources of all and water quantity and supply directly affects economic growth, food supply and health care throughout the world. Without plentiful supplies of clean fresh water then most countries will face severe problems and limits to peaceful growth. Global water security is not simply about individual production sites and countries, it also impacts on neighbouring countries. How the world's supplies of fresh water are shared out and used may become a potential flashpoint in terms of global security. Plastics processors need to consider this because sustainability is about the whole world and not simply about a single country.

Despite the limited amount of fresh water in the world, the demand for water is growing throughout the world not only for agriculture but also for industry. For example, the water footprint (see Section 8.14) of 1 kg of a typical plastic is approximately 183 litres. This sounds high but the water footprint of 1 kg of cotton is approximately 11,000 litres. To make it even worse, the water footprint of a standard bottle of white wine is 551 litres and for a typical car is in the region of 400,000 litres. Obviously, wine and cars are not to be taken together.

Most countries have not only an actual water demand but also a 'virtual' water footprint. This is in terms of the water footprint that is imported in the form of food, products and energy that need water for production and transportation. For example, in the UK, more than 66% of the total water footprint is estimated to be 'imported' from countries that may already be 'water stressed'. This means that countries that are not water stressed will be contributing to water stresses elsewhere in the world and impacting on global scarcity. To paraphrase Arthur C. Clarke: "There is enough water to go around but it is not evenly distributed".

Water is a precious natural resource that needs to be used sustainably to improve water security and reduce costs.

The absolute cost of water to most plastics processing companies is not as high as the cost of energy but sustainable plastics processing companies still need to manage and minimise their water use to reduce their environmental impact.

Adopting a systematic approach to water use reduction can typically achieve a 20–50% reduction in the amount of water used for most sites. This can give significant savings in water supply costs and in wastewater and effluent disposal costs.

8.1 Water use in plastics processing

The essential fluid

Water is the essential fluid in plastics processing and, as with many processes, modern plastics processing would not be possible without using water. An outline if the areas where water is used in plastics processing is shown on the right. In most of the common processes, plastics processing does not use particularly large amounts of water for the actual processing and the systems are mostly closed circuits. The industry is therefore very good in this regard when compared to other industries such as the food and pharmaceuticals industries. In most cases the amount of direct and indirect water (see Section 8.14) used is relatively small.

However, water is used extensively as a heat transfer medium for cooling and, in some cases, for heating because water has a higher capacity for heat removal than air even if it is more energy intensive to process and distribute.

The cost of water as opposed to the cost of cooling or heating it is therefore relatively minor compared to many other costs but it is still significant and can be reduced.

Cooling services

Cooling is an essential part of plastics processing (see Section 6.17) and most processes use extensive amounts of chilled and cooling water for product or machine cooling. The energy used for this cooling is in the region of 10–15% of the total energy used at a site.

When chillers are used, the system will be a closed system to prevent heat gains and the amount of water actually used is often quite small provided the system does not have substantial leaks.

- **Tip** – If chilled water systems containing glycol have leaks then they should be sealed as soon as possible to avoid potential groundwater contamination.

When cooling towers or free coolers with adiabatic cooling (water spray) are used the system will be open and the cooling effect is primarily provided by evaporative cooling. In these cases, the evaporative, blowdown and drift losses can be significant particularly for cooling towers that are in continuous use in warm climates (see Section 8.7).

In the absence of substantial leaks in the system, the storage and distribution losses in cooling services are generally quite low and very visible if present.

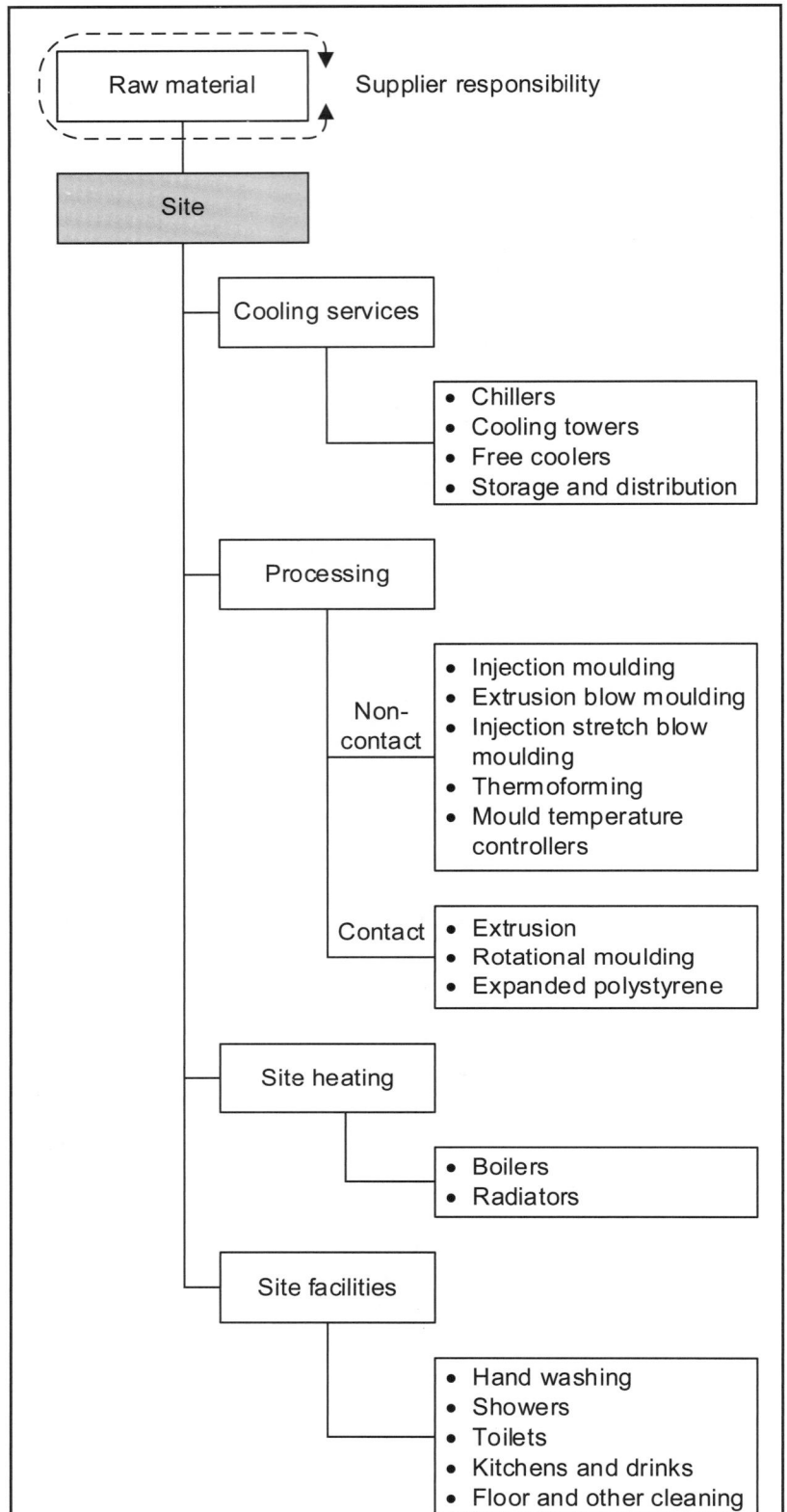

Where water is used in plastics processing

Water is a vital resource for successful plastics processing. It is used throughout the business for a variety of reasons. Without water for cooling, processing, heating and sanitary applications, modern plastics processing would not be possible.

Processing

Cooling water systems at the point of the cooling demand can be either:

- Non-contact processes – where the cooling or chilled water is in a closed system and does not come into contact with the product. Injection moulding is a typical non-contact process; chilled water is contained inside the mould and machine cooling water is contained inside the heat exchanger.

  Where mould temperature controllers (MTC) are used, these can use either water or oil as the heating medium. Older MTCs using an open direct method can have high water use from evaporation and other losses.

- Contact processes – where the cooling water comes into contact with the product as in extrusion or into contact with the mould as in water spraying of rotational moulds to cool the mould.

  Expanded polystyrene (EPS) production is a contact process but in this case the contact is via the steam used for the pre-blow and blow processes.

Site heating

Site heating via radiators is not generally used for production areas due to the heat rejected from processing and warehouses tend to use alternative methods due to the amount of space to be heated. This means that water-based site heating is normally restricted to office areas (see Section 8.9). This generally uses a closed system and water use mainly consists of make-up water in the event of leaks.

Site facilities

For many plastics processing sites, the site facilities will be the largest component of the water used. Water is used for:

- Hand washing – water use will be particularly high at sites processing food contact products where strict hygiene regulations require effective hand washing before entering any production area.

- Toilets – these are a major water user especially if the urinals and toilets are old and poorly regulated (see Section 8.9).

- Showers (where used) – these are high water users and a typical 5–8 minute shower will use approximately 65 litres. Not all sites provide showers but those that do will have higher water use.

- Kitchens and drinks – at sites with an on-site kitchen, the kitchen use will be significant for food cleaning and preparation. Every site will also provide water for hot or cold drinks and this is another area of water use.

- **Tip** – The average direct daily water use per person in the UK is about 142 litres for home use and, whilst there is no available data for industrial use, it is estimated that each person at a plastics processing site will use about 25–40 litres per day and perhaps more if there is an on-site kitchen.

A single dripping tap can easily cost £7.50/year.

It all adds up

Initially, most people would think that water use in plastics processing is minimal but the various areas of water use all add up. Most sites could do more to use this valuable resource more effectively and to reduce not only initial use but also effluent discharges to the drains.

The water industry is energy intensive and any plan to save energy should also include measures to save water and limit water waste.

Some initial questions:
- How much water is the site using?
- Where is the site using water?
- What is the site using water for?
- Is the site paying too much in wastewater charges?
- Have there been any attempts to reduce water use?
- Is there more action that can be taken?
- What has the competition done?

The next questions:
- What is the water meter size?
- Is the meter the correct size?
- What is the water pressure?
- Can the pressure be reduced without affecting operations?
- Is the water supply frost protected?
- Is the meter located in a dry environment and easily read?
- Is there leakage at or near the water meter?
- How often is the meter read?
- Is the meter a pulse system (for automatic reading)?
- Is the meter reading checked against the bill?
- Is water use monitored on site?
- If so, how is it monitored?
- Is water use measured against any benchmark?
- Is there an alarm system for excessive water use?
- Are drawings for the main water system available?
- Are there water systems on the site other than the mains water, i.e., boreholes, wells, rainwater recovery?
- Are all valves maintained and regularly checked for leakage?
- Has there been a recent leak detection survey?

8.2   Sources and sinks

It is mostly in the pipes

Establishing the water sources and sinks for a site is a first step in understanding, controlling and reducing water use:

- A water source defines where the water comes from. At most sites, the primary water source is the water main to the site.
- A water sink defines where the water goes to. At most sites, the primary water sink is the sewer system via the site drains.

Defining the sources and sinks at a site is the start of understanding and reducing the impact of water use and an essential part of the water balance (see Section 8.3).

Sources

The main water sources are shown in the diagram on the right. Not all sources will be present.

Water mains

Most of the water used in plastics processing is 'potable water' from the mains. 'Potable' water has been treated to a defined standard and is considered safe for human consumption. These standards vary around the world but in most cases the standards will be higher, or substantially different, to those required for process use. Other water sources may be lower in quality but still suitable for industrial use with minimal treatment.

The method of charging for potable water varies throughout the world but in many areas the fixed cost is based on the size of the meter and the size of the water main (see box on opposite page). For sites with a low or reduced water use and a large initial meter, replacing the existing main meter with a smaller more appropriate meter can give dramatic reductions in fixed costs.

- **Tip** – Check that the size of the meter is correct for the site's needs and demand profile. The size of the meter should be marked on the bill.
- **Tip** – Each building (and preferably the complete site) should have only a single supply point and meter, so that there is only one fixed charge.
- **Tip** – Where possible, the meter should have a pulsed output to allow connection to a building management system (see Section 8.4).

Rainwater

Rainwater is a clean source of water but water harvested from roofs and other areas should be regarded as non-potable due to dust or other potential contamination. Rainwater should be either used for non-potable applications, e.g., toilet flushing, or filtered and treated for process applications.

Surface water runoff

Any surface water runoff from adjacent land should be regarded as non-potable. In some cases, this can be filtered and treated for hygiene or process applications.

- **Tip** – If surface water enters a site's drains then it may charged as wastewater. Ensure that the site is not paying for surface water runoff from a neighbour's land.

Rivers and lakes

Water from rivers and lakes should be regarded as non-potable. In some cases, this can be filtered and treated for hygiene or process applications and then returned to the source provided it has been segregated from the process.

Boreholes

Water extracted from boreholes will generally be quite clean and, as for water from rivers and lakes, can be filtered and

The introduction of high-quality water treatment in the UK in the 1880s led to a 15-year increase in life expectancy over the next 4 decades and reduced deaths and damage to children from diarrhoeal diseases by more than 20%.

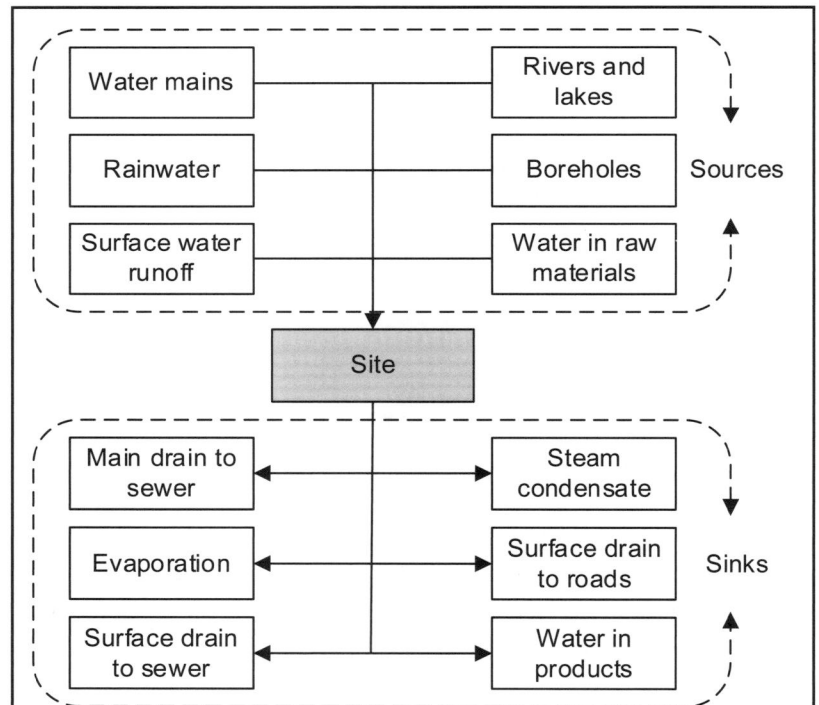

Sources and sinks for water use

Water is supplied to the business via sources and leaves the business via sinks. Understanding the sinks and sources for a site is the key to creating a water balance (see Section 8.3) which is a vital resource for reducing water use.

treated for various applications and then returned to the source.

- **Tip** – The removal and use of water from rivers, lakes or aquifers will generally require an 'abstraction licence' and is prohibited in some areas.
- **Tip** – If water from rivers, lakes or aquifers is used for process cooling and returned to the source then physical segregation from any treated process water is essential (see Section 8.12).

Water in raw materials

Water can enter the site in raw materials although this is not common in plastics processing.

Sinks

The main water sinks are shown in the diagram on the left. Not all sinks will be present.

Main drain to sewer

The main drain to the sewer will be the largest water sink but not all the water entering a site will leave via the drains. Depending on the processes used at the site, up to 50% of the source water will not leave via the main drain. This is called 'non-return to sewer' and the supplier will make an allowance for this in the charges to account for evaporation losses and underground leaks.

Identifying other water sinks can be used to increase the non-return to sewer allowance and decrease charges.

- **Tip** – Check the 'non-return to the sewer' allowance for the site. This varies depending on the location and activities.
- **Tip** – Getting the correct size meter is also important for the disposal charges because fixed wastewater charges can vary with meter size.

Evaporation

Depending on the site operations, evaporation can be a large water sink. Cooling towers, hot water and steam systems will all lose water to the atmosphere and reduce the amount of water discharged to the drains.

Surface drain to sewer

Surface water from rainfall on roofs, car parks and other areas will generally be discharged to the drains.

Steam condensate

Steam is not used in many forming processes but where it is, e.g., EPS, rubber processing etc., the losses from steam condensate can be large and will reduce the amount of water discharged to the drains.

Surface drains to roads

In some cases, surface water will be discharged to roads and/or adjacent land and not to the drains.

Water in products

Water can leave the site in products although this is not common in plastics processing.

At this stage we will not consider the embodied water content (see Section 8.14 on the Water Footprint).

Wastewater can be defined as 'any water that has been adversely affected in quality by man-made activities arising from domestic residences, commercial properties, industry, and/or agriculture'. It can cover a range of contaminants and concentrations.

The cost of water supply:

The fixed charge

Most water supply companies will have a fixed charge in the water bill. In many parts of the world this will be based on the meter size and this is determined by the internal diameter of the meter to reflect the size of the water main used to deliver water to the site. The increase in the fixed charge is not linear: A 20 mm main feed will cost ≈ £200/year in fixed supply charges but a 200 mm main feed will cost ≈ £120,000/year in fixed supply charges.

The variable charge

The variable cost of water is the charge for each cubic meter (m^3) of water used. This will generally be 'stepped' so that the first 100,000 m^3 of water supplied costs more than the next 100,000 m^3 of water supplied. In the UK this cost is ≈ £1.00/m^3 for the first 100,000 m^3.

The cost of water disposal

The fixed charge

As for the cost of supply, there will typically be a fixed charge based on the meter size because this reflects the size of the sewers needed for the wastewater. The increase in the fixed charge is again not linear: A 20 mm main feed will cost ≈ £200/year in fixed disposal charges but a 200 mm main feed will cost ≈ £100,000/year in fixed disposal charges.

The variable charge for wastewater

The variable cost of water is the charge for each cubic meter (m^3) of wastewater disposed of. This will generally be constant for each cubic metre and in the UK this cost is ≈ £1.50/m^3.

The variable charge for trade effluent

Trade effluent is wastewater produced on industrial sites (excluding domestic sewerage which is defined as wastewater). Trade effluent charges do not use a simple volume charge but generally use a formula which varies with the volume, chemical oxygen demand (COD) and suspended solids in the effluent (see Section 8.13).

8.3   The water balance

An essential part

A water balance is the sum of the water inputs and outputs from the site (the sources and sinks from Section 7.2). It should also note how much, where, when and how water is being used at a site.

A water balance is an essential tool to:

- Reveal excessive, unknown or unauthorised use and potential areas for savings.

- Reveal unknown or unauthorised discharges to the sewer and thus not only save money but also reduce potential fines and legal action.

- Reveal leaks and quantify them.

As with an energy balance (see Section 6.2), the water balance enables the best water saving opportunities to be identified and projected cost savings to be calculated. This is essential in targeting the best projects.

- **Tip** – The water balance is the first step in reducing water use and costs through water management. It reviews current practice and is an initial water mass balance at the site.

- **Tip** – Start with the simplest diagram possible and add data as necessary to fill in the gaps. Use 'sources and sinks' information (see Section 8.2) as a first step.

- **Tip** – The WRAP publication "Tracking Water Use to Cut Costs"[1] provides more detail and useful information on how to carry out a water balance.

How much water is used?

The first step in the water balance is to establish how much water is being used at a site. The input water use is generally relatively easy to find and track from the main water meter and other meters from sources such as boreholes.

- **Tip** – Do not trust the water bill for use data. This may have 'estimated' readings that are not related to actual use. Find the water meter, learn to read it and read it daily over a few weeks to get an accurate reading.

- **Tip** – If the existing meters are 'manual' types then get them replaced with pulse meters to allow Automatic Meter Reading (AMR) for real time and accurate data.

Where is water used?

Mapping the water distribution system is essential to finding where water is used at a site. Use a site plan to track the water pipes and the drains at the site and to create a map of where they go and their relative size.

- **Tip** – A 'walk-around' to map the site can often reveal instant savings that nobody has ever thought of before.

- **Tip** – Although not essential in the first stages, it is wise to map all the water flows at the site, i.e., map the chilled, cooling and heating water flows. This will allow the water balance to be updated at a later stage to include all water flows at the site.

A water map of a site gives a first view of where water is used but a full water balance needs actual or estimated numbers. Allocating use to the various areas can initially be done by estimation (see diagram below) but it is far better to meter the significant use areas identified by the initial water map.

- **Tip** – Use the water map to find the significant use areas and install meters (preferably pulse meters) to get accurate data for current volumes and costs and for monitoring and targeting (see Section 8.4).

- **Tip** – Seal and isolate unused spurs in the system at the earliest possible point to reduce the potential for leaks or poor use.

- **Tip** – The diagram in Section 7.1 can be used as a guide for places to look for water use.

The water balance is an essential tool in minimising water use and costs.

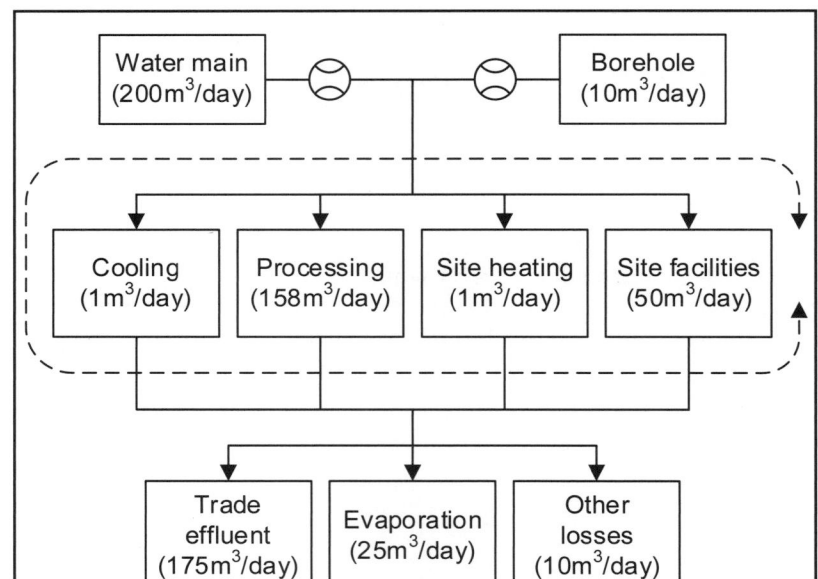

An initial water balance

An initial water balance gives the basics from analysis of the sources and sinks (see Section 8.2) but should be extended to include all the internal flows, i.e., water that flows around the chilled, cooling and heating water systems, when additional data is available.

Where does wastewater go?

Quantifying and locating the output water is often more difficult than quantifying the input water because at most sites there will be multiple wastewater and sewer discharge points. Site plans should be used to track where wastewater and effluent goes. All drainage systems should be identified as:

- Sewers (toilets etc.)
- Trade effluent
- Surface water
- Combined drainage systems.

The type, direction of flow and approximate volume flow for each system should be marked on the plans and transferred to the water balance. Methods for estimating effluent flows are given in the WRAP publication[1].

When is water used?

As for energy use, the 'when' of water use is also vital to understanding and reducing use. If a site operates 24/5 then any use whilst not producing, sometimes known as the 'night use', can provide vital information.

Any use when the site is not operating is potentially due to leaks after the meter that can be sealed to reduce use and costs.

- **Tip** – A site in Scotland carried out a water balance and looked at the 'night use'. They found that there was significant use when no operations were taking place and as a result found a large leak between two of the buildings on the site. This had been present for at least 18 years but because the water bill was 'constant' then nobody was concerned. The leak was sealed and the site saved around £5,000/year in water and wastewater charges.

Does it balance?

After the initial water balance has been created, this should be checked against the water bills. An initial water balance should aim to capture > 95% of the water entering and leaving the site. If there is a significant mismatch between the inputs and outputs then this can indicate:

- Errors at the input meters.
- Leaks in the system after the input meter.
- Errors in estimating the use at various area.
- Errors in the output meters.

All of these areas are opportunities to improve accuracy, reduce use and reduce costs.

Improving the details

The initial water balance provides the direction for future work and is a tool for improvement. The initial data gathered can always be improved by adding the details of water and heat flows in closed systems such as chilled and cooling water systems but this should only be done after the basic balance is completed.

As with any tool, the water balance must be used but it provides the essential data for critical examination of the actual use of water at a site.

Tracking water use and improving processes can easily lead to a 30% reduction in water use and costs.

- 1. "Tracking Water Use to Cut Costs", WRAP, 2013, www.wrap.org.uk.

If the water balance shows more than ≈ 20% error there is a high probability of a significant leak.

Data for the water balance

Much of the data for the initial water balance will already be available, Look for data such as:

- Water bills – look also for the meter details as these will often be on the bill.
- Abstraction licenses, costs and any volume restrictions.
- Internal water treatment systems and costs (for all water systems).
- Site plans for the location of the water distribution system, drains and any installed meters. These should be treated with caution as they will rarely be updated and can easily be wrong.
- Site plans for the location of the chilled, cooling and heating systems and temperature/flow details.
- Trade effluent and sewerage bills.
- Effluent treatment costs for trade effluent before discharge.
- Effluent removal costs for any tanker removal of effluent.
- Site plans for any effluent treatment plants.

Depending on the site there may be even more data available and it is worthwhile getting it all into a single place before starting the water balance.

Convert the data into information

Gathering data is itself a worthless exercise. The benefit comes when the data is converted into information that can be used to guide management action. Use the data to:

- Assess the total volume of water used at the site.
- Assess the total cost of water (purchase, treatment and disposal) to the site.
- Assess the cost-effectiveness of proposed actions.
- Drive the water management programme (see Section 8.5).

8.4   Water monitoring and targeting

Data driven improvement

The water balance provides good basic information for a site but improvement is really driven by monitoring and targeting (M&T). Monitoring automatically raises staff awareness of water use and will often reveal simple actions to reduce water use, especially if no action has been taken in the past. Targeting, using meters and standards, provides data for further improvement by highlighting any deviation from expected use. Both actions are essential in reducing water use and costs.

The initial issues

M&T and indeed good water management, may prove initially difficult due to a range of issues, the most common are:

Lack of top-level commitment

Top-level commitment is often hard to achieve but the water balance and the associated costs can help to focus attention on the potential savings (see Section 8.3). Top-level commitment means that capex requests for improvements are more likely to be approved.

Lack of knowledge

There is often a lack of knowledge about where water is used and the attitude is "as long as it comes out of the tap and goes away again then all is OK". The initial water balance and associated cost data (see Section 8.3) will highlight where water is used and its importance in the site's operations.

Equally, there is often a lack of billing data and a lack of understanding of how to read the bill. In fact, many sites do not understand the difference between supply and discharge costs.

- **Tip** – Learning to read the water bill is an essential task.

Lack of data

Water billing data is rarely as good as energy billing data and many sites will have a combination of:

- Non-automated meter readings, i.e., manual readings taken at sporadic and irregular intervals.
- Significant numbers of 'estimated' readings due to the need for manual reading.
- Poorly located, difficult to read and 'invisible' meters.

These factors make it difficult to establish an M&T programme because of the 'invisibility' of use locations or patterns,

particularly when the site is unoccupied. At most sites the lack of data is a critical issue.

- **Tip** – It is strongly recommended that all sites have recording meters with a pulsed output that can be remotely read and which can provide time-based data similar to that seen for electricity (see Section 6.2). This allows water use to be measured and tracked continuously to give early warning of excessive use.

All sites should check and validate the water billing data instead of simply filing it and paying it. There are often overcharges through incorrect billing for supply and discharge and simple analysis of the historical data can reveal these.

For companies with multiple sites there are often issues with billing for sites and locations that are no longer occupied or with fixed costs being levied for sites and locations that no longer use water, e.g., warehouse operations.

- **Tip** – Create a register of operational sites to look for inconsistencies in bills and to target sites for further investigation.
- **Tip** – Look particularly for sites where the use pattern has changed significantly. These can be opportunities for simple meter replacement to reduce fixed costs.

Lack of M&T standards

Whilst there are some approximate benchmarks for use in office buildings, there is a real lack of similar data for plastics processing sites. Most sites will have to rely on internal benchmarking to provide guidance.

Initial monitoring

Initial M&T efforts should concentrate on establishing the baseline water use and the water balance is essential for this (see Section 7.3). This provides the starting point for M&T – only by knowing where you are starting from can you assess if you are making progress and if actions have been effective.

Whilst automatic meters are strongly recommended, significant initial data can be gained simply by reading the main water meter at 1-hour intervals during a standard working day and at the start and finish of the weekend (or the other non-production hours). This should be done over several days and weekends to increase the data available and to provide a better view of the use. These meter readings can then be plotted on a simple time graph to show

Simply monitoring and targeting water use can easily result in year-on-year savings of > 5%.

Check the night time use when no water is being used to help monitor leakage.

Where manual water meters are used then they should be located where they are visible and easily read.

The simple presence of a water meter can act as an incentive to reduce water use.

demand through the day and weekend.

This simple process will identify periods of high use during working days and the non-production use (the 'night use') when the site is not producing. The purpose is to set the use baseline and to detect the cause of any period of high use.

The amount of water used in non-production hours should be minimal and should consist only of the final top-up of storage tanks and potentially a small amount for urinals (see Section 8.9).

- **Tip** – As a guide, non-production use should be < 35 litres/100 people/day. If it is more than this then first check the measurement and then find the leaks or hidden uses.

Better data makes monitoring more effective and the best solution is to install water meters in the different services and process areas to continuously monitor use (similar to electricity sub-metering). The water balance can be used to locate the best meter positions, i.e., the largest users, so that the data can be used to target areas with inconsistent or inefficient performance and to set progressively tighter consumption targets.

- **Tip** – Water management should be part of a complete system for the overall control of all utilities.
- **Tip** – Do not be surprised if water use drops whilst trying to find the baseline even with no action. The simple act of measurement changes things.
- **Tip** – As with any meter, water meters can fail. This may not be a cost issue (although they will eventually find out and the back-charges may be substantial). A regular check of the meter by opening some taps when there is no other use will show if the meter is working at low flow rates.

Targeting

After establishing a baseline value, it is possible to move on to targeting performance improvements and the choice of performance metric is important. For water use there may be more than one driver:

- The number of staff and the size of the buildings will drive facilities use.
- The external temperature will drive evaporative losses, e.g., cooling towers and adiabatic cooling.
- The production volume will drive production related use.

The multiple drivers at a site therefore make targeting complicated and, in the first instance, the best method is to use techniques similar to those used for electricity, i.e., the PCL approach discussed in Section 6.4. This allows the

'fixed' (facilities) and 'variable' (process) uses to be separated and treated separately, as they should be in any case.

- **Tip** – As for electricity targeting, the use of a simple 'm³/kg' is not recommended due to the high base loads created by facilities use.
- **Tip** – The use of the PCL approach also allows budgeting for the future (see Section 6.4) and integration of water use into the accounts (see Section 6.8).
- **Tip** – The use of sub-metering provides more data and, if available, the most relevant driver can be used for targeting.

Reporting

The results of M&T need to be reported to get the best out of the system and reporting works best if the results successes are distributed widely in a simple and easy to understand format. This allows staff to be part of the process of water use reduction.

- **Tip** – It is important to balance the cost of any programme against the savings that are achieved and reporting water use results may not need to be as regular as reporting of energy use results.

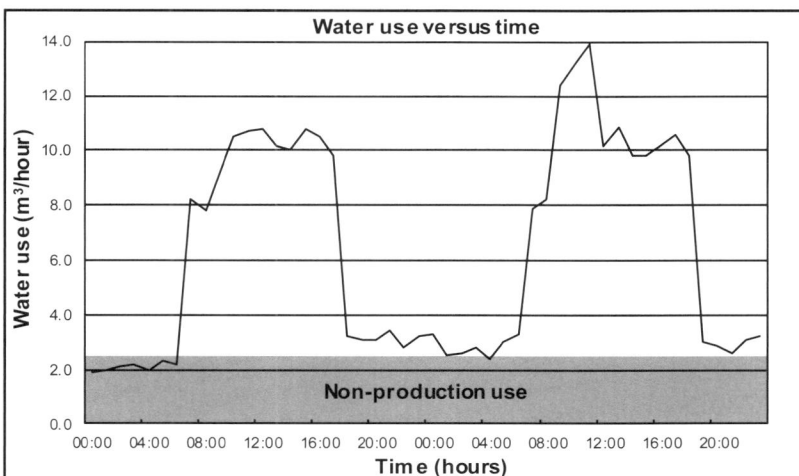

Simple meter readings can give an insight into water use

Simply recording water use from the main meter can give insights into water use. This is shown by the results over two days of manual reading. The non-production use is ≈ 2.4 m³/hour (is this leakage?) and what caused the 'spike' at ≈ 12:00 on the second day?

8.5   The water management programme

The key to success is knowledge

The preceding sections of this Chapter have mainly been concerned with gathering information. During this process it is typical that improvement opportunities will be identified but the real benefits come through a structured water management programme that systematically attacks the use and cost of water at a site – this is a continual improvement programme rather than a single exercise.

The programme

An outline of the water management programme is shown on the right and this shows the three stages and nine steps of the programme. For the average site, a water management programme can produce savings of > 50% over a period of 3 years by inexpensive measures because, in most cases, the use of water has been neglected as a minor priority.

Stage 1: Minimise the demand

Minimising the demand is the first stage in the programme and should be completed before moving on to the other stages. It is not logical to optimise the supply for excessive demand based on excessive leakage or poor use. Reducing leakage and poor use will also automatically reduce discharge costs by reducing the total volume of water discharged to the sewers.

Step 1: Reduce leakage

Water leakage is a total waste to the system and must be minimised as the first step. Leaks will always be present in poorly maintained systems and systems that have had no leakage survey in the past 3 years will almost certainly have leaks. This is literally 'money down the drain' and reducing leaks is an essential first step.

Leakage rates in the water system (before the site) can range from small (5-10%) to high (> 50%) depending on the country and the condition of the distribution network. A similar range of values are present at plastics processing sites although it is likely to be higher at sites with old systems.

Step 2: Reduce use

After leakage has been controlled it is logical to investigate the actual use of water at the site to determine if the use is justified and if it can be reduced by changes to the process.

Step 3: Use alternative sources

At most sites the majority of water used is potable water (see Section 8.2). Establishing the actual purity/quality requirements of the water and using alternative sources of appropriate quality can reduce the demand for highly (and costly) treated water from the mains.

Stage 2: Optimise the supply

After the demand is minimised it is then possible to optimise the supply to deliver the water at the required level.

Step 4: Water recovery

Water recovery is the on-site recovery of water for the same purpose as the original use. Water recovery can be by treating wastewater or by capturing water vapour

Water is similar to any other service, i.e., reduce the demand and then optimise the supply but must be extended to reduce effects and cost of disposal.

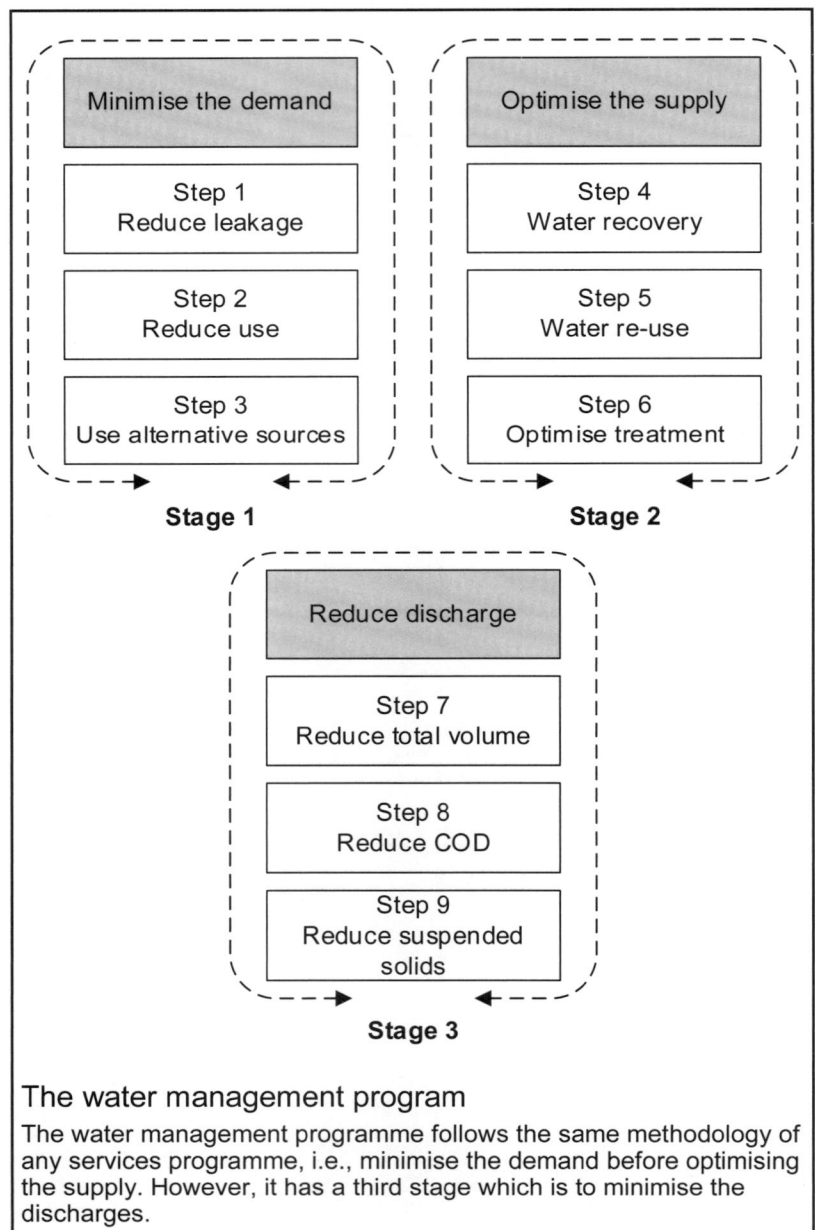

Stage 1

| Minimise the demand |
| Step 1 Reduce leakage |
| Step 2 Reduce use |
| Step 3 Use alternative sources |

Stage 2

| Optimise the supply |
| Step 4 Water recovery |
| Step 5 Water re-use |
| Step 6 Optimise treatment |

Stage 3

| Reduce discharge |
| Step 7 Reduce total volume |
| Step 8 Reduce COD |
| Step 9 Reduce suspended solids |

The water management program

The water management programme follows the same methodology of any services programme, i.e., minimise the demand before optimising the supply. However, it has a third stage which is to minimise the discharges.

from evaporation processes. Both processes extend the utility of input water to the site and reduce use and costs.

Step 5: Water re-use

Water re-use is the re-use of water already used for one purpose for another purpose where the level of contamination from the first process does not affect the second process.

Note: Rainwater harvesting may be considered as an alternative source or re-use because of the change in location. At this stage we will consider it as an alternative source.

Step 6: Optimise treatment

Treatment costs, even of potable water for process use, are often a substantial but hidden cost. These costs should be minimised by only treating water to the minimum quality required for the process. This can include water treatment for cooling towers where treatment is required for Legionella prevention. In this case, the potential cost of a Legionella event means that treatment costs are a secondary concern.

Stage 3: Reduce discharges

Discharge costs to the mains via drains, sewers and other effluent streams can easily be as much, if not more, than the supply costs. Discharges also need to be as clean as possible to reduce the potential for contamination of the output stream. Discharges may all be charged as for domestic discharges (sewage) or may be calculated separately depending on the composition of the discharge.

In the UK the cost of disposal for trade effluent, based on the composition, is commonly calculated by the Mogden formula (see Section 8.13) which has three main components; the amount of suspended solids, the chemical oxygen demand and the total volume. Authorities around the world use similar mechanisms but vary in the detailed calculation and charging method.

Whatever the detailed calculation method used, the discharge charges are always lower if the discharge is cleaner and easier to treat.

Step 7: Reduce total volume

The total volume of effluent charged for is made up of:

- foul sewage.
- trade effluent.
- surface water drainage.
- highway drainage.

Minimising the total output volume will be an automatic outcome of the first two stages of the programme which will minimise the input volume but care needs to be taken that this does not increase the suspended solids or COD at the same time.

Step 8: Reduce chemical oxygen demand (COD)

The chemical oxygen demand (COD) measures how much oxygen is required to consume the organic components of an effluent stream during treatment. Effluent from food processing sites will normally have a high COD but plastics processing sites will normally have a low COD due to the lack of organic components for biological oxidation.

Step 9: Reduce suspended solids

The amount of suspended solids in the effluent stream is an indicator of water quality and indicates how 'cloudy' the water is. Decreasing the amount of suspended solids has a direct effect on the cost of treating the water.

The first steps

The first steps in water management are:

- Buy-in – gain buy-in from senior management with responsibility for facilities, finance and operations.
- Make someone responsible – appoint a member of staff to track water use and water minimisation opportunities.
- Educate staff about water efficiency – train staff and add water to job responsibilities, preferably link these to a water policy statement.
- Set a budget – after the initial information has been collected, and before the programme is started, it will be possible to gain an oversight of the potential projects and paybacks. At this stage a preliminary budget needs to be established for the programme. Many of the actions in the water programme are no- or low-cost and will have a good payback but some projects will need capex and a preliminary budget should plan for these at the outset.

In most cases the energy used to treat water for the process and the site (heating or cooling) is in the region of 10-15% of the total energy costs of a plastics processing site.

Sewerage: the system of pipes and infrastructure carrying sewage or surface water.

Sewage: the combination of water and solid waste transported by the sewerage system.

Do not rush – this is a marathon and not a sprint.

8.6 Minimise the demand – reduce leakage

Pure waste

Leaks in the water distribution system, especially where the leak runs direct to the drain are a waste of water and money. Despite this, most sites have not carried out a water leak assessment in the past 3 years and have no idea of the amount of leakage at the site. Leaks can occur from underground pipes and tanks or from internal spillages, leaks and overflows but in all cases are a waste.

Quantifying leakage

Leakage can be quantified by two independent methods:

- Using the water balance (see Section 8.3) where the 'other losses' required to balance the inputs and outputs will provide an estimate of the leakage.

- Measuring water use during non-production hours (see Section 6.4) will give an alternative estimate of the leakage, although this may not give the full value if the site shut-down is very good.

- **Tip** – A quick check for non-production leakage is to listen to or watch the meter during non-production hours. Any noise of flowing water or movement of the meter indicates leakage in the distribution system.

There will be errors in both approaches but both will give an estimate. It is recommended that both values be calculated and compared.

- **Tip** – The balance between the cost of checking for leaks and the cost of wasted water must be assessed.

These methods will detect existing leaks but new leaks can also be found by using the M&T data (see Section 8.4). If the main meter or sub-meters show a significant increase then this may be due to a leak. In this case, sub-meters can be used to isolate the potential leak to a specific area.

Supply leaks

Leaks on the supply side of the meter are the responsibility of the supplier but leaks after the meter are the responsibility of the site. Underground pipes after the meter are one of the largest sources of leaks as they are hidden and cannot be seen from the surface.

Some leaks can be detected by examining the flow at drainage manholes. Check for increased flow between manholes to detect water leaks into the effluent stream or decreased flow between manholes to detect water leaks from broken pipes. Obviously, this only works if the leaks are to or from the pipes and not if the leak is to or from the ground.

Finding supply leaks

If a supply leak is suspected, it is worthwhile using the water supplier or a specialist contractor to find the leaks. These use a range of techniques such as acoustic detection, thermal imaging and tracer gases to find underground or other hidden leaks.

- **Tip** – Damp patches on paths or roads may indicate leaks if they are close to the supply pipes.

- **Tip** – Excessively green plants or lawns may indicate leaks if they are close to the supply pipes.

- **Tip** – Water running from any gutter on a dry day may indicate water leaks.

Services and process leaks

Services and process leaks are generally more visible than underground leaks especially if the leak is external to the process and the water ends up on the site floor. 'Overflows' are less visible as the leak will be contained in the system and can run direct to the drainage system with no control or measurement. These are mainly due to poor control of the process.

Leaks and overflows can be caused by:

- Physically damaged pipes, connections, flanges, tanks or other system components.

- Corroded pipes, connections, flanges, tanks or other system components.

- Poorly set or worn control valves.

> After three years with no water leak checks, approximately one-third of all premises will have water leaks that need sealing.

> In this section we are not talking about poor use of water, this is covered in subsequent sections, but about leaks in the system where water is lost to the system with no control.

Leakage rate	Yearly loss (m^3)	Supply cost (£)	Effluent cost (£)	Total cost (£)
Two drops/ second	9.5	9.5	14.25	23.75
Drops into stream	31	31	46.5	77.5
2 mm	146	146	219	365
3 mm	336	336	504	840
5 mm	528	528	792	1,320

The yearly cost of leaks

It is possible to estimate the cost of leaks by the size of the water stream. This table is based on variable costs of £1.00/m^3 for supply and £1.50/m^3 for discharge. It does not include any of the fixed costs as a result of any leaks.

• Poorly set or flooded tank control floats.

In cases where the distribution system is insulated (hot or cold) then any leak under the insulation will be difficult to find. Even if the leaks are small and the water cost is small, then the effective cost will be much greater because waterlogged insulation is ineffective and energy will be lost as a result of the leak.

• **Tip** – A thermal camera will often quickly reveal these leaks.

Steam leaks

Steam leaks are a special case of process leaks because of the high cost of generating and distributing steam. Steam has a very high energy content even if it does not contain a great deal of water. This is due to the high latent heat of vaporisation of water.

Steam leaks not only waste considerable amounts of water and energy, they can also present a significant health and safety risk.

Leaks in steam systems are quite often easily visible and this makes leak detection much easier.

• **Tip** – Early detection of leaks from valves and traps can reduce overall maintenance costs.

• **Tip** – Reduce losses from steam systems by regularly checking, reporting and repairing leaks.

• **Tip** – Working with steam is dangerous and should only be carried out by qualified specialists.

Finding process leaks

It is almost inevitable that there will be process leaks at any site. The first action should be part of the water balance (see Section 8.3) where the simple act of tracking the water system should include noting any leaks or areas of excessive use that are not process related. During the water balance walk-around:

• Check all piping and system components for water leaks.

• Check every process that uses water for water leaks.

• Listen for drips or unexpected water flows (particularly overflows).

Facilities leaks

Most of the facilities leaks and overflows will occur in washrooms, showers and catering areas. As for process leaks, the main sources of leaks will be:

• Worn or perished tap washers.

• Corroded pipes, connections, flanges, tanks and other system components.

• Poorly set or worn control valves (primarily in cisterns).

• Poorly set or flooded control floats (primarily in cisterns).

• **Tip** – Check that overflows are visible.

• **Tip** – Reduce losses by regularly checking, reporting and repairing leaks.

• **Tip** – Install easy to use shut-off valves.

Finding facilities leaks

As with process leaks, the water balance walk-around (see Section 8.3) should be used to scan for facilities leaks. During the water balance walk-around:

• Check all piping and system components for water leaks.

• Check every facility that uses water for water leaks.

• Listen for drips or unexpected water flows (particularly overflows in cisterns).

Toilet areas are a prime source of leaks from areas such as urinals with no flush controls and poor overflows or toilets leaking and continuously dripping.

• **Tip** – Toilets can be easily checked with dye to reveal leaks.

Leaks and losses are at least 5% of the input.

Protect against cold weather-related leaks by insulating pipes to prevent frost or freezing damage.

Additive leaks?
If clean rainwater is not being used at the site (see Section 8.10) then check to see if the water can be diverted away from the drains to reduce the amount of trade effluent.

It may be possible to divert clean water away from the site onto adjacent land to reduce trade effluent costs.

Water pressure

High mains water pressure will cause leaks or make leaks worse as well as increasing water use and increasing the maintenance load of the distribution system.

Mains water systems typically operate at 2-4 bar but there is rarely an upper limit to the mains pressure and in most cases the delivery pressure is uncontrolled. It is possible to use Pressure Reducing Valves (PRVs) to both reduce and control the pressure at the incoming mains or at points around the distribution system. These will reduce the effects of excessive pressure in the system, protect the system and deliver a controlled system pressure of 1.5-6 bar under variable flow conditions. PRVs are freely available and cost £30-£300 each (depending on the diameter required).

When considering a PRV then:

■ Choose a variable PRV with an integral pressure gauge.

■ Check all equipment to ensure that the reduced operating pressure is suitable.

8.7 Minimise the demand – reduce use in services

We are already good at this

Water is used extensively in the plastics industry for both services and processes, primarily as a heat transfer medium. Most of the use in services is with closed systems and the industry uses relatively small amounts of water in comparison to many other industries.

Cooling towers

Evaporative cooling towers are often used for cooling of processing machines, chillers and compressors (see Section 6.17) although their use is decreasing due to concerns with Legionella and the rising use of closed-circuit air blast coolers which reduce water and treatment chemical losses. Despite this, there are still many cooling towers used in the industry and these use considerable amounts of water. Cooling towers need the addition of 'make-up' water to replace that lost to evaporation, drift and blowdown.

Evaporation losses

Evaporation losses depend directly on the cooling load on the cooling tower and can be minimised by reducing the cooling load on the tower. In most cases, this is difficult because the cooling load is fixed by the process requirements. It is possible to use heat exchangers to recover heat from the warm process return water and reduce the water temperature but the heat is relatively low-grade and this is not often used.

Drift losses

Drift losses are caused by droplets of water being caught in the air stream and escaping from the tower system. This loss can be reduced by the use of drift eliminators which are simply baffles that force the exiting air to suddenly change direction. This separates the water droplets from the air and puts them back into the system.

- **Tip** – Fit and maintain drift eliminators to cooling towers to reduce drift losses.
- **Tip** – Where automatic blowdown is used (see below) then drift losses are effectively blowdown and real losses are small. It is often better to install automatic blowdown control than to improve drift control.

Blowdown losses

Evaporation of water from a cooling tower removes the water but obviously leaves behind any dissolved solids. Blowdown is needed to stop the build-up of dissolved solids in the water, tower packing and other system components that will reduce the

system's cooling efficiency. Blowdown is simply the removal of some of the system water with high levels of dissolved solids and replacing this with mains water (make-up water) with a lower level of dissolved solids.

Blowdown can be either:

- Periodic/intermittent with a defined volume of water being removed and replaced.
- Continuous using an automatic blowdown control which measures the total dissolved solids (TDS) in the water and adjusts the blowdown rate to keep the TDS level below the maximum allowable level.

Automatic blowdown is preferred because it controls the TDS levels in the system whilst minimising the loss of water and treatment chemicals.

One measure of the amount of TDS in the system is the 'cycles of concentration'. This is a measure of the amount of TDS in the system relative to the amount of TDS in the make-up water, i.e., a cycle of concentration value of 3 indicates that the system contains 3 times the TDS of the make-up water. The optimum cycle of concentration for a cooling tower is in the region 3-5:

- If the cycle of concentration is high (>5) then blowdown is low, less water is being lost and treatment chemical use is low but TDS build-up can start to affect system efficiency.
- If the cycle of concentration is low (<2) then blowdown is high, more water is being lost and treatment chemical use is high.
- **Tip** – The amount of water used by a cooling tower will also depend on the air temperature and relative humidity.
- **Tip** – Cooling towers will be dosed with chemicals to prevent the growth of Legionella and excessive blowdown will use more chemicals to keep the concentration levels correct.
- **Tip** – If cooling towers are used then be sure to claim for a reduction in wastewater charges. The evaporation means that some to the water supplied to the site will not be discharged to the sewers but will be released as evaporation (see Section 8.13).

Boilers

Steam is not used in many processes but it is essential for processes such as EPS and rubber processing. As with cooling towers, the feed water for boilers contains dissolved solids and the evaporation process to produce steam will concentrate these solids

> Cooling tower water is often contaminated by dust and cooling towers are very efficient 'dust collectors'. Be sure that the water is clean enough for the process.

in the boiler and system. This means it is necessary to control the TDS level in steam boilers and if this is not done correctly then it will not only waste hot water and fuel but can also lead to boiler failure. Blowdown and TDS control is essential for effective boiler operation.

Blowdown can be either:

- Periodic/intermittent with a defined volume of water being removed from the bottom of the boiler to capture any settled sludge. The amount of blowdown can be based on the level in the gauge glass or if the TDS values are available then the % blowdown can be calculated from:

$$\% \text{ blowdown} = \frac{S_f}{S_b - S_f} \times 100$$

where:

S_f = TDS level of feedwater in ppm.

S_b = desired TDS level in boiler in ppm.

- Continuous, using an automatic blowdown control from a feed at the nominal water level. This will remove suspended solids in the water but will not remove settled sludge at the bottom of the boiler.

Boiler blowdown is more critical than for a cooling tower and it is strongly recommended that automatic blowdown control based on the water properties, e.g., conductivity, is used.

- **Tip** – When using automatic blowdown, it is highly recommended to use a periodic blowdown from the bottom of the boiler to remove any settled sludge.

- **Tip** – Blowdown should be minimised to give the correct water conditions. Excessive blowdown wastes both water and energy. In many cases blowdown represents a significant heat loss from the boiler and is second only to the heat losses from flue gases. Good control systems can save large amounts of money.

- **Tip** – Heat recovery from blowdown water is often very economical due to the high-grade heat being produced. Heat recovery rates of > 50% can often be achieved. Calculate the annual heat (energy) loss in water being discharged to the sewers and consider heat recovery.

- **Tip** – Blowdown water should be cooled to < 43°C before discharge to the drains and this may also need additional cold water and/or treatment costs.

Vacuum pumps

Liquid ring vacuum pumps are common in some processes and these need a continuous flow of cold water to seal the rotors and provide good vacuum performance. In some cases, this is achieved using 'once-through'

mains water at a rate of ≈ 1 litre/sec. This is extremely wasteful and costly. The two potential solutions are:

- Use a recirculation system with a heat exchanger to cool the water before it is recirculated to the vacuum pump, e.g., use cooling tower water to cool the recirculated water via a heat exchanger. This may also mean installing filters for the recirculated water to remove any entrained particles.

- Replace the liquid ring vacuum pump with a standard dry vacuum pump. This is the preferred option but may not always be possible.

- **Tip** – If liquid ring vacuum pumps are essential and must use 'once-through' mains water then a shut-off valve should be fitted to stop pump operation and water flow when the vacuum is not needed.

- **Tip** – Fit liquid ring vacuum pumps with VSDs to ensure that the pump is only generating the required level of vacuum and keep the liquid ring temperature low to allow the pump to operate at the highest efficiency.

Washdown using hose pipes or sprays

Washdown using simple hoses and sprays will use large amounts of water. Any washdown operation should use trigger-operated spray guns that shut off when the trigger is released. These significantly reduce water consumption by not only reducing the flow by up to 80% but also produce a more concentrated jet and have an automatic shut off when the trigger is released.

- **Tip** – All hose pipes should be fitted with trigger-operated jets.

Process water use is a key area for plastics processors but do not ignore the potential for large savings in washrooms where systems such as low-flush toilets can be an easy win.

Closed systems should, by their design, not use any water. If regular topping up is needed then the system is leaking (see Section 8.6).

Shut off water pumps when they are not performing useful work to save not only energy but also the cost of any water treatment.

8.8 Minimise the demand – reduce use in processes

The good news

Process water use is also primarily for heat transfer purposes. For most processes, the use is in closed systems and the actual water use is quite low but some processing methods use semi-open systems for heat transfer or cleaning, e.g., extrusion, recycling (washing and extrusion). In addition, rotational moulding uses water for direct evaporative cooling, e.g., spray cooling of moulds.

In these cases, process use (losses) can be significant and have the potential to be reduced through process changes.

- **Tip** – Sometimes it is possible to reduce water demand by simply changing the way the existing equipment is operated, rather than replacing or modifying it.
- **Tip** – To reduce leakage losses, pipes should be frequently inspected and any leaks identified promptly repaired.

Closed systems

Injection moulding, IBM, ISBM and EBM all use water for cooling (moulds and machines) in closed systems and water use is generally minimal. Process water use is mainly due to losses at moulds during mould changes and even this tends to be small. Any use/leakage from the cooling water system at the mould area will be very obvious from watermarks on the product and fixed quickly.

Extrusion

Profile extrusion uses water for cooling, either in older style full immersion calibration baths or spray cooling in the newer style tables (see photograph A on the opposite page).

Where vacuum calibration is used the calibration table must be sealed to retain the vacuum and this also helps to retain water in the cooling system.

Where open calibrators are used there is often considerable wastage of water to the site floor and area.

- **Tip** – Fit spray covers to all open calibration baths to reduce water losses, keep the area water free, reduce evaporative losses and prevent dust from getting into the system.

Compounding is a special case of extrusion and the final compound is generally extruded into an open water bath (see B on the opposite page). The water is then cooled in a separate system or allowed to naturally cool through heat losses to the site. Whilst

the water in the bath can become quite warm, the temperature is limited by water evaporation and as the temperature increases so do the losses through evaporation.

- **Tip** – Fit covers to water baths to recover condensate and feed it back into the system.

Mechanical recycling

Mechanical recycling is one of the most water intensive plastics processing operations. The recycling process uses large amounts of water for initial rinsing, wet grinding, friction washing (for label removal) and final rinsing before extrusion of the final recyclate (see Section 4.6).

This process will use ≈ 1-1.5 litre of water for every 1 kg of recycled plastic processed. In some, more advanced, processes, the water is automatically filtered and recirculated in the process.

Techniques used to filter recirculated water include membrane filtration, ultra-filtration and reverse osmosis. Whichever technique is used, there will be substantial amounts of solid waste to be disposed of due to label fragments and other contaminants. This waste will almost certainly contain water.

- **Tip** – Increasing the number of stages in a washing operation can reduce water use.
- **Tip** – Check for areas where low-volume, high-pressure water (nozzles) is better than high-volume, low-pressure rinsing or flushing.
- **Tip** – Cyclone or centrifugal drying can be used to both dry flakes and recover rinse water.
- **Tip** – Recycling lines should always include water treatment capability to optimise water use and minimise effluent discharge.

> Shut off water pumps when they are not performing useful work to not only save energy but also the cost of any water treatment.

Questions for water use:

- Is the water use necessary?
- Is process using too much water?
- Can the water use be reduced?
- Can a lower quality water be used?
- Can the water be re-used or recovered for use elsewhere?

Questions for wastewater:

- Why are we producing this wastewater/effluent?
- Are we losing/using clean water as part of the process?
- Are any discharges authorised and legal?
- Can the wastewater be re-used or recovered?

- **Tip** – Check strainers and filters regularly to ensure that they do not become clogged. Clogged filters cause losses in pressure and reduced effectiveness.
- **Tip** – Reducing effluent discharge is not simply about diluting the discharge it is about removing the contaminants.

Extrusion of the final recyclate (see photographs B and C on this page) should follow the rules given for extrusion (see above).

Rotational moulding

Rotational moulding often uses water sprays on the mould cooling fans to increase the cooling rate (see photograph D on this page). The evaporation of the water assists cooling but there is often no consideration of the cost of the water used. Rotational moulding sites using water spray cooling should minimise the water use by ensuring that the water used is all being evaporated, i.e., if there is a significant amount of water on the floor around the moulds then the spray feed is excessive and the flow control valves should be adjusted to minimise the water use.

- **Tip** – If water sprays are used then check that the cooling phase of the process is actually the dominant process phase and that spray cooling is really necessary.
- **Tip** – Rotational moulding sites using water spray cooling should check that the evaporated water is not being included in their discharge to sewer costs.

General checks

Whichever process is used there are some general checks that should be made to minimise process water use. These are:

- Hoses and taps – check for hoses and taps that are left on and ensure that they are turned off when not needed or fit trigger sprays (see Section 8.7).
- Redundant branches – remove redundant and unused branches and streamline piping systems.
- Reduce evaporation losses – reduce evaporation, energy losses and the potential for dust contamination by installing covers on all tanks.
- Once-through water – avoid the use of once-through water at all costs. Investigate if a once-through system be converted to a circulating system and revise the water distribution system to allow re-use of process water wherever possible.
- Cleaning in Place (CiP) technology – CiP can reduce water use and maintenance costs. CiP allows the internal surface of pipes and vessels to be cleaned automatically without the need for manual cleaning or disassembly.

Process water use is a key area for plastics processors but do not ignore the potential for large savings in washrooms where systems such as low-flush toilets can be an easy win.

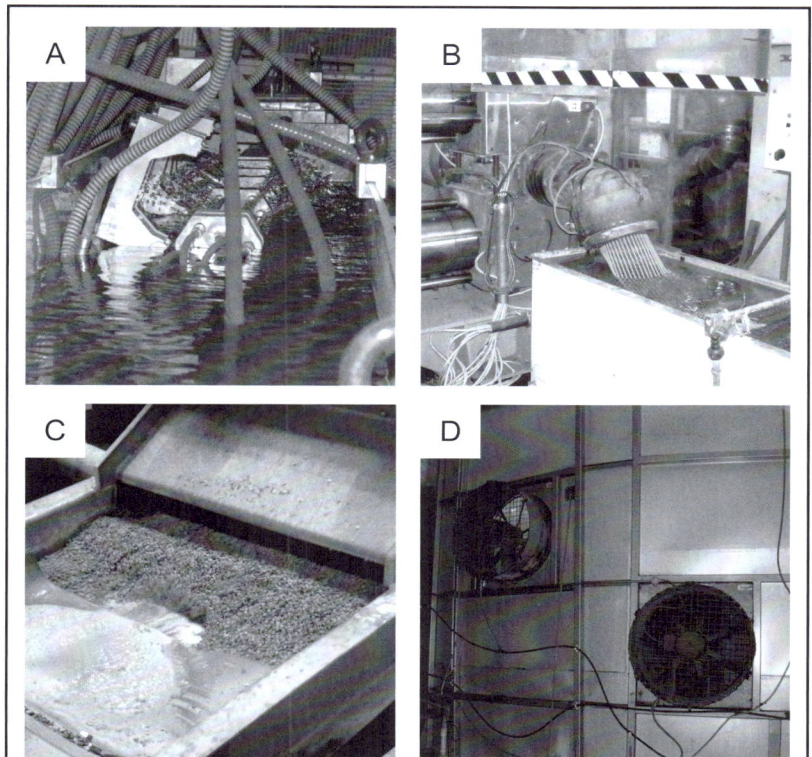

Water use in extrusion and rotational moulding

Extrusion (profiles and recycling) uses water in open circuits and rotational moulding sometimes uses water spray in the cooling phase to speed up the process. In both cases, the evaporation losses can be high.

Water pinch analysis

Water pinch analysis is similar to conventional pinch analysis used for chemical processes. In chemical processes, pinch analysis is used to calculate the theoretical minimum energy target for a process and then how to achieve this through optimised heat recovery and process operating conditions.

Water pinch is a relatively recent technique for analysing water networks and reducing water costs for processes. It uses a graphical design method and complex mathematics to identify and optimise the best water re-use, regeneration and effluent treatment opportunities.

Water pinch is best used for new system design to adjust processes to maximise water re-use and recycling.

8.9   Minimise the demand – reduce use in facilities and heating

It is probably the largest demand

For the majority of plastics processing sites, the water use in facilities and heating will be the largest water use. This is fortunate because there are many basic (and cost effective) water-saving devices for facilities.

Facilities

The best opportunities to reduce use in facilities will be in the washrooms and toilets (the largest use will be in toilet flushing). Staff awareness of good housekeeping practices will help reduce use in all facilities.

Sinks and hand washing

Taps

If older-style rotating or lever taps are still used then these can result in large losses when not turned off properly. A small 3 mm stream from a tap will cost ≈ £840/year (see Section 8.6) and the returns from simple improvements can be very good. Whilst it is unlikely that any single tap will be left on for the whole year, at a site with many taps the losses can mount up quickly especially if simple maintenance such as regular washer replacement is not carried out.

- **Tip** – Simple maintenance of taps will reduce drips and leakage.

The use of 'push-to operate' or 'percussion' taps that require user pressure to turn on and which turn off automatically after a pre-set time will reduce use and remove any possibility of taps being left open or dripping.

Percussion taps can also be fitted with adjustable flow restrictor valves to reduce the water flow and further reduce use. If the water flow is reduced then it is possible to use aerator style nozzles to mix air with the water flow to retain the perceived flow rate but to use less water.

- **Tip** – Replacement and upgrading of taps should be made part of a long-term water management programme.
- **Tip** – Reduce water use by reducing the water pressure in areas that do not need high pressure.
- **Tip** – Fit percussion taps to reduce water use on taps and drinking water supplies.
- **Tip** – Fit flow restrictor valves to feed pipes to reduce water flow.
- **Tip** – Fitting smaller basins will reduce use but check with users first.

- **Tip** – Reducing water use should never compromise site hygiene requirements. Test solutions before wide implementation.

Hot water losses

The cost of dripping or leaking taps is even higher if the tap is for hot water, in this case the cost of the energy to heat the water is much higher than the cost of the water itself. Hot water pipes should be insulated and well separated from cold water pipes to reduce the time to get hot water from the tap and to reduce parasitic heat gains.

- **Tip** – Insulate hot water pipes and separate hot and cold water pipes.
- **Tip** – Check for hot water tank overflows and prevent or minimise these.
- **Tip** – Check the hot water heating set point. Adjust this to the minimum required temperature and lock it off to prevent tampering.

Toilets

Flush controls

Most older style toilet cisterns will be fitted with a single flush mechanism that delivers the same amount of water irrespective of whether the waste is solid or liquid. Modern toilet cisterns use dual flush systems ('two-button' cisterns) that adjust the amount of water depending in the type of waste.

- **Tip** – Upgrading an old toilet (≈ 13 litres/flush) to a new dual-flush toilet (≈ 6 litres/flush) can save up to 7,000 litres/person/year.

Many toilets at sites are old and have high flush volumes (up to 13 litres/flush) compared to new toilets that have flush volumes of < 6 litres/flush.

An easy win?

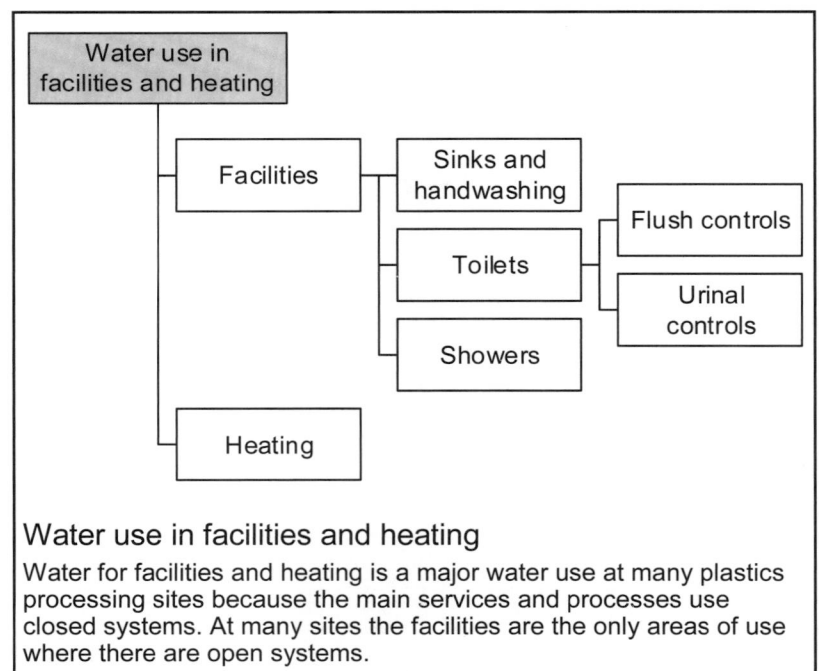

Water use in facilities and heating

Water for facilities and heating is a major water use at many plastics processing sites because the main services and processes use closed systems. At many sites the facilities are the only areas of use where there are open systems.

- **Tip** – The replacement and upgrading of single flush cisterns with dual flush cisterns should be made part of a long-term water management programme.

Where replacing and upgrading existing single flush cisterns is not cost-effective then the simplest alternative is often the use of a Cistern Volume Adjuster (CVA). A CVA reduces the amount of water used in the flush, typically by 1-2 litres/flush but by up to 6 litres/flush (depending on the cistern and type of CVA used). This is not as much as can be saved using a dual-flush toilet but is much cheaper. Most CVAs are simple bags of water or water containers that are placed in the cistern. In the UK, some water companies will supply these free of charge but they only cost $\approx$ £2/CVA.

- **Tip** – CVAs should not be used with cisterns of < 7 litre capacity to ensure that the flush is adequate.
- **Tip** – You can buy commercial CVAs, e.g., hippo-the-watersaver.co.uk, or you can simply use a litre bottle of water in the cistern (as long as it doesn't affect the flow). It will do the same job. An even simpler alternative is to adjust the float control to stop the cistern from filling completely. This is a 2-minute job using a screwdriver or hand adjustment.

Overflows from faulty float controls or washers in header tanks or cisterns can easily be invisible and discharge directly to the cistern or the sewer. A simple overflow fault in a toilet cistern can discharge $\approx$ 90 m³ of potable water per year and cost more than £200/year.

- **Tip** – Check cistern float controls to ensure that the cistern is not leaking to the cistern or the sewer.

Urinal controls

Older style urinal controls will typically flush every few minutes depending on the time taken to fill the urinal cistern and the system settings. If this is uncontrolled then urinals will continue to flush even when there is no use. Based on a 20-minute continuous use cycle and a 13-litre flush, this is $\approx$ 341 m³ of water/ year and a cost of $\approx$ £850/year for each urinal. For sites not operating 24/7 then the waste can be up to 30% or £250/year. This will be even higher if the urinal is set to flush at a shorter interval.

Urinal controls such as Passive Infrared (PIR) sensors can detect the presence of people to flush only when necessary and match water use to actual use. PIR controls can reduce water use by up to 75% and have payback times of < 3 months.

- **Tip** – Fit PIR controls to all urinals to save water and money.

- **Tip** – PIR push buttons are an alternative control method for both toilets and urinals. These have a PIR and are a contactless method of flushing toilets and urinals.
- **Tip** – Manual push button controls on urinals is also used but hygiene demands make these a less desirable alternative to automatic methods.

Waterless urinals are a relatively recent development but remove water use entirely. There are several different technologies, e.g., microbiological, liquid barrier and valve barrier, but all are effective at removing water use and preventing odours from urinals. Most of these will also use some type of regular flush or treatment to prevent microbial growth.

Showers

At sites where showers are provided, these can use large amounts of hot water. The most effective solution is the use of percussion controls or to use PIRs to automatically shut off the flow after a specified time or when no occupancy is detected.

As for taps, the use of pressure regulators and aerators can reduce use but retain the user experience.

- **Tip** – Install flow (pressure) regulators to showers.
- **Tip** – Install aerator heads to showers.

Heating

Where commercial boilers are used then the boiler blowdown (see Section 8.7) should be optimised to prevent excessive use in blowdown.

Where domestic style boilers are used then modern condensing boilers are much more effective than older style boilers.

- **Tip** – Consider replacing old hot water boilers with high efficiency units.

A site with 100 staff using old style cistern toilets can save £500/year simply by fitting cistern volume adjusters.

Kitchens and drinks

Relatively few plastics processing sites have full-scale kitchens but most will have break rooms and drink facilities.

Treat the taps in these areas as for taps for sinks and hand washing.

8.10 Minimise the demand – use alternative sources

Does it have to be potable water?

As noted in Section 8.2, the majority of water used at plastics processing sites is potable water from the mains. However, many services, processes and facilities uses do not require high-quality potable water and suitably managed alternative sources can be used to improve sustainability and reduce both supply and discharge costs.

Why use alternative sources?

Using alternative water sources to supply non-potable demand can be highly effective in reducing demand but there are additional benefits such as:

- Cost reduction – alternative sources are almost always cheaper to use than mains supply.

- Security of supply – alternative sources can be more secure than mains supply and provide back-up in the event of disruption.

- Reduced environmental impact – alternative sources reduce the carbon impact of mains water delivery and discharge.

Reducing potable water use simply means asking: "Can we hygienically and effectively use alternative sources in an activity?"

Alternative sources

Rainwater harvesting

Rainwater harvesting is the use of rainwater (rather than potable water) for non-potable applications, e.g., toilet flushing, or it can be filtered and treated for process applications. In many parts of the world, rainwater and/or snow melt captured from roofs is a plentiful source of water with a low pH and low hardness (see box on the right). The low pH makes rainwater more corrosive to pipes than standard mains supplied water. The low hardness (low level of dissolved solids) is due to the fact that the water has not been in contact with any minerals to dissolve them.

The low hardness of rainwater means that it can be treated and used at high cycles of concentration (see Section 8.7) in cooling towers and boilers to reduce the use of treatment chemicals and blowdown water use.

Water from rainwater harvesting systems should be filtered to remove any suspended solids picked up from the roof and the system should prevent any backflow into the potable water system.

Surface water runoff

Surface water runoff from adjacent land is similar to rainwater harvesting although it may contain more suspended solids. It should be regarded as non-potable and treated as water from rainwater harvesting.

Rivers and lakes

Water from rivers and lakes can also be used for non-potable applications. The water will have a slightly higher pH than rainwater (slightly alkaline) and a slightly higher, but not excessive, hardness.

River and lake water can be filtered and used for applications such as toilet flushing or filtered and treated for use in chilled and cooling water systems, e.g., cooling towers. As for rainwater, the low hardness of river and lake water means it can be used at high cycles of concentration in cooling towers.

- **Tip** – If water from rivers, lakes or aquifers is used for process cooling and returned to the source then physical segregation from any treated process water is essential. This can use simple water to water heat exchangers but keep the pressure of the source water higher than that of the process water to ensure that any leaks in the system are safe.

Boreholes

Water extracted from boreholes will generally be quite clean but have a higher pH (alkalinity) and much higher hardness than rainwater or other sources. This water will also need filtering as for other

Rainwater harvesting systems should be designed to avoid introducing *Legionella* or other health hazards and to separate non-potable water from potable water.

EN 16941 covers the design, installation, water quality, maintenance, and risk management of rainwater harvesting systems.

Measures of water

- pH – this is a measure of the acidity or alkalinity of the water on a logarithmic scale of 1-14 where < 7 is acidic, 7 is neutral, and > 7 is alkaline. Most potable water will be in the range 6.8-8.

- Hardness – this is a measure of the concentration of Ca^{2+} and Mg^{2+} ions in the water where 'soft' water has fewer dissolved minerals and 'hard' water has more dissolved minerals. Predicting the effect of hardness is not simple as it depends not only on the amount of dissolved minerals but also on the pH and the water temperature. There are several measures, e.g., Langelier Saturation Index, Ryznar Stability Index and Puckorius Scaling Index etc. and all attempt to describe the relationship between hardness and scale deposition.

- Suspended solids – suspended solids are different to dissolved solids and are typically the dust and dirt that will be present in many alternative sources. They are measured in terms of mg/L.

- Microbiological matter – most alternative sources of water should be assumed to contain microbiological matter of some description. Whilst much of this is harmless, some can cause disease and many can form slime or algae in water systems. Suitable water treatment is recommended in all cases.

alternative water sources but its use in cooling water systems will be limited by the low cycles of concentrations and the higher use of chemicals such as water softeners and pH modifiers (acids).

Using alternative sources

Alternative water sources can dramatically reduce demand (and costs) when used appropriately. Typical uses are:

- Toilet flushing and handwashing – there are generally no restrictions on using alternative sources in toilets and urinals provided the water is adequately filtered. The wastewater will enter the standard sewer for treatment. The only issue is that, if well water is used, then there could be a build-up of scale if toilets or urinals suffer from drips or leaks.

- Showers – water captured from alternative sources is not generally suitable for staff showers.

- Watering of plants and lawns – water from alternative sources is suitable for plants and lawns with no additional treatment.

- **Tip** – Non-potable water systems should always be clearly signposted as 'Not drinking water' to prevent staff from drinking the water.

- Open process cooling systems – these are systems where evaporation can take place to remove water from the system that is replaced from the alternative source, e.g., cooling towers. In these cases, the use of alternative sources is generally acceptable provided the water is filtered and chemically treated to prevent microbial growth. This is routinely done for cooling towers by biocides and does not represent an additional step. Borehole water is not generally recommended for cooling towers due the higher requirement for chemical treatment.

  The use of open cooling tower water to cool compressors, chillers, machine hydraulics etc. is possible because these mainly use larger bore 'shell and tube' type heat exchangers that can be acid washed if they need cleaning.

  The use of open cooling tower water to cool moulds, 'plate' or 'plate and frame' is not generally recommended because the smaller bores make fouling from any deposits more likely.

- **Tip** – The use of open cooling systems for extrusion water baths should be carefully examined due to the possibility of deposits on the extrusion after exiting the calibration system. Such systems should be well filtered and strained as part of the flow.

- Closed process cooling systems – the lack of evaporation in a closed system means that the water losses (and make-up) should only be during mould changes and should be minimal. This means that these systems will not concentrate dissolved minerals and appropriately treated water from alternative sources can be used in the system.

- **Tip** – It is possible to use cooling towers for closed process systems by using cooling tower water and a heat exchanger. The cooling tower water is then separated from the process water by the heat exchanger and there is no risk of scale build-up on the process side of the system.

Is it worth it?

The use of alternative water sources can easily reduce water use and bring sustainability benefits to a site but alternative sources need to be well-managed to ensure that they meet process requirements, hygiene requirements, legislative requirements and preserve the quality of the source water.

- **Tip** – Use the water programme to identify how and where alternative sources can be used as a replacement for mains water.

Water stress is defined as an annual supply of less than 1,700 m^3/person, water scarcity is defined as an annual supply of less than 1,000 m^3/person and absolute scarcity is defined as an annual supply of less than 500 m^3/person.

Water scarcity is felt strongly by more than 1 billion people in 43 countries.

8.11 Optimise the supply – recovery and re-use

What is the difference?

Water efficiency at a site can be optimised through both recovery and re-use. The two processes are fundamentally different:

- Water recovery – this is the on-site recovery of water at the site for the same process. The water is recovered into the same process after suitable treatment.

- Water re-use – this is the re-use of water from one part of the site at another part of the site for a different service, process or facility. This may or may not involve treatment of the water after the first use.

- **Tip** – The detailed water balance (see Section 8.3) should be used to map the flow of water through the site to identify water that can potentially be recovered or re-used.

The processes, although very similar, extend the use of any water input to the site to reduce use, reduce costs and improve sustainability.

Water recovery

Water recovery itself may also be divided into two distinct types of operation:

Wastewater recovery

This is treating the wastewater from a process and recovering this, after suitable treatment, back to the original process (see diagram on the right).

Recovered water always requires treatment (regeneration) to remove contaminants from the process and to prevent contaminant build-up in the system. Regeneration can be carried out by processes such as filtration, reverse osmosis, ion exchange, stream-stripping or carbon adsorption. This will reduce the overall amount of water used, the amount of water discharged to the sewers and the contaminant load of the discharged wastewater (see Section 8.13).

At most plastics processing sites, the purchase or operating cost of any regeneration process, other than simple filtration or straining, will make these technologies not cost effective.

- **Tip** – Most regeneration processes, other than simple filtration or straining, will be costly to implement for plastics processors and have long pay-back times. These processes will generally only be viable if required by local regulations.

Condensate recovery

This is capturing condensate from steam or other systems that generate hot water and recovering this back to the system that generated the condensate (see diagram below).

In most cases, condensate recovery does not require any regeneration unless it has been contaminated during capture because it is condensed steam and is essentially distilled water.

- **Tip** – Check all water sent to the drains and main sewer for opportunities to recover the water and decrease both water use and discharges.

- **Tip** – Check the treatment required for any recovered water before it is fed back into the process.

Rainwater harvesting (see Section 8.10) can also be considered as water re-use, even if you don't pay for it, and is a good way to reduce potable water use.

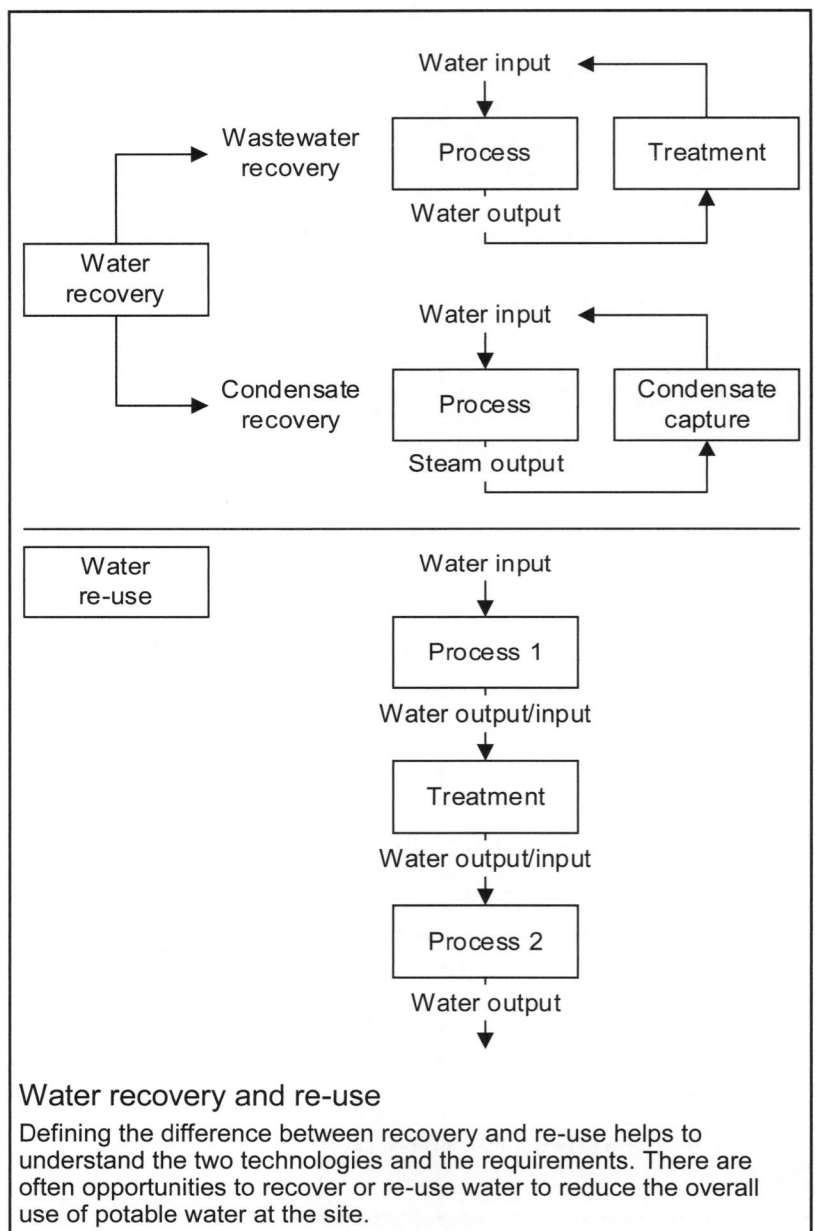

Water recovery and re-use

Defining the difference between recovery and re-use helps to understand the two technologies and the requirements. There are often opportunities to recover or re-use water to reduce the overall use of potable water at the site.

Water re-use

Re-using water at a site may also be considered to be a type of recovery, although strictly speaking it is a 'downcycling'. In this case, any contaminants introduced by the first process are either removed or are already at such a level that they do not affect the efficiency of the second process (see diagram on the left).

Water re-use without treatment reduces the overall amount of mains water used and the amount of discharged wastewater but does not reduce the overall contaminant load in the discharge water. It will reduce the concentration, i.e., the same load but less volume of water.

Sometimes treatment (regeneration) is needed before re-use to remove contaminants which would otherwise prevent re-use. Similar techniques such as those used for wastewater recovery are available (see above). However, as for wastewater recovery, the cost of any process, other than simple filtration, will generally make these technologies not cost effective for plastics processors.

- **Tip** – Check all water sent to the drains and main sewer for opportunities to re-use the water and decrease water use.
- **Tip** – Check if any re-used water needs treatment before re-use.

Opportunities for recovery and re-use

Despite the high costs of treatment for recovering or re-using water, there are several opportunities for optimising the supply by simply filtering. Process washing, process cooling and pump sealing systems do not always need potable mains water. In these cases, filtering and treating with UV light to remove the majority (99.9%) of waterborne microorganisms will allow the re-use of slightly contaminated, but compatible, wastewater or rainwater harvested from roofs. If wastewater is to be treated then sites should:

- Re-use rinse water from cleaning operations, (check any product quality implications).
- Re-use water from handwashing and taps (filtered and UV treated) for toilet flushing. This may need small holding tanks to top up the system if the re-used water is not sufficient in volume.
- Re-use uncontaminated cooling water, this will need careful consideration if the water contains glycol (see Section 8.12) or biocides.
- Install sufficient holding tanks for the re-use system.

- Check strainers and filters regularly to prevent pressure losses from blocked filters.
- Investigate if recovered and/or re-used water is suitable for applications such as toilets, gardens, etc. to reduce the amount of water discharged to the sewer.

Dilution is not the solution

Minimise emissions at source. Do not dilute them down to an apparently 'acceptable' level.

This only increases the demand.

Strainer or filter?

The main difference between a strainer and a filter is the size of the particle that is trapped and the ability of a strainer to re-use the filter.

A filter uses a single-use filter medium to filter small particles from the water flow. The particle size filtered depends on the filter medium.

A strainer (basket or Y-type) uses a reusable metal plate or screen to remove larger particles from the water flow. Cleaning a strainer is simply a matter of opening the strainer and allowing the water flow to clean the plate or screen. The size of the holes in the metal plate or screen can be varied to vary the size of particle captured.

- Is rainwater recovered at the site?
- If yes, are gutters/ collectors checked for correct operation, i.e., no blockages?
- Is waste water measured and monitored?

8.12 Optimise the supply – optimise treatment

This is for every site

Whether the water used is recycled, re-used or from the mains, it will generally need some type of treatment before it is used at a plastics processing site. This water treatment is necessary to control microbiological growth, corrosion, scale formation (fouling), freezing and to reduce blowdown rates (see Section 8.7) and water use (see Section 8.8).

Good water treatment also reduces maintenance costs and improves heat transfer efficiency by reducing or preventing scale formation.

- **Tip** – Even a very small layer of scale will significantly reduce heat transfer in moulds, boilers or chillers, e.g., 0.25 mm of scale will reduce heat transfer efficiency by ≈ 10%.

- **Tip** – Scale is very hard to remove once it has formed on heat transfer surfaces. It is much easier and effective to treat water to prevent scale formation in the first place than to remove scale after it has formed.

Water treatment is essential for cooling towers, chilled water systems and boilers and optimised water treatment uses less chemicals and still allows effective water use. In addition, water treatment is not free and the operating costs for treatment before use or for recycling or re-use include:

- Consumable chemicals costs.

- Effluent discharge costs.

- Energy costs for system operation.

Water treatment is very important but many sites do not give this the attention that is deserves and pay the price later in increased maintenance and other costs.

Treatment may be divided into two types:

- External treatment – this is where the water is treated before it enters the system, e.g., removal of minerals from the feedwater to prevent scaling.

- Internal treatment – this is where the water is treated whilst it is in the system, i.e., where the chemicals are directly added to the process water.

The treatment may be automatic or manual:

- Automatic systems can either simply treat the water the same, irrespective of the actual properties, or can measure the water properties (see below) and treat the water to the required level.

- Manual systems can also either simply treat the water the same, e.g., regular

manual dosing, or can use manual measurement of the water properties to assess dosing levels.

- **Tip** – Automatic systems controlled using a measure of the water properties are preferred as these will minimise chemical/ system use and treatment costs.

Typical treatments in plastics processing

Some of the most common treatment methods used in plastics processing are:

- Biocides – these are used to control Legionella and other microbial growth in water systems. Chlorine (Cl) or Chlorine dioxide (ClO_2) are used as disinfectants to remove biofilm and kill bacteria, spores and viruses.

- Water softening – this is needed to remove Ca^{2+} and Mg^{2+} from input water to prevent the formation of scale in pipes, heat exchangers, boilers and other surfaces. The main method used in plastics processing is ion exchange which uses common salt (NaCl) to replace the Ca^{2+} and Mg^{2+} ions with Na to prevent the build-up of scale. The Ca^{2+} and Mg^{2+} ions are collected and then regularly flushed away.

- **Tip** – One disadvantage of water softening is that if all the Ca^{2+} and Mg^{2+} ions are removed then the water becomes corrosive and can affect the systems. Automatic control of water softening, e.g., Testomat from Lenntech, is strongly recommended for water softening systems to maximise the effectiveness of treatment.

- UV treatment – this is an alternative to

Measurement and close control of water treatment is essential to minimise water use, chemical use and treatment costs.

For all you ever wanted to know about almost every type of water treatment and the science behind them then go to www.lenntech.com.

A treasure trove of information that goes well beyond a simple sales web site.

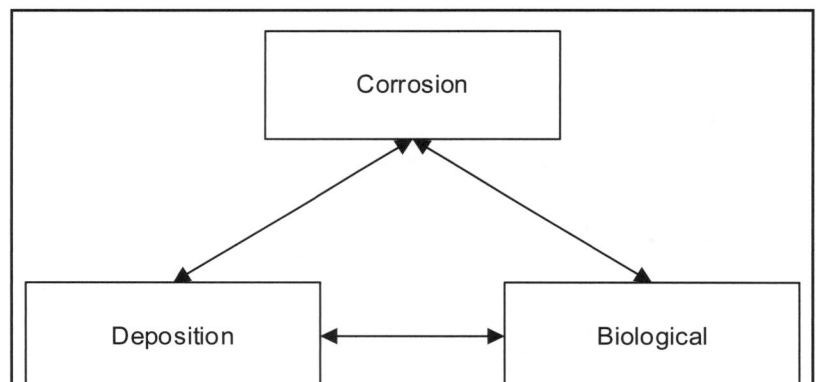

The treatment triangle

Water treatment seeks to control corrosion, deposition and biological growth. Corrosion is perhaps the biggest issue as corrosion products will reduce heat transfer and encourage microbial growth but control of biological products is also essential.

biocides for some applications and is more environmentally friendly than chemical treatments. It is, however, an 'end-of-pipe' application and has no residual effectiveness, i.e., it is useful in the recycling and re-use of water but not so good for open systems such as cooling towers.

- Straining and filtering – straining and filtering (see Section 8.11) are necessary to remove particles that may cause damage to pipes, pumps or other system components.
- Glycol – glycol is added to chilled water systems in cold climates to provide freeze protection to the system and the dose used is increased in colder areas.
- Corrosion protection – corrosion protection treatments, e.g., nitrite, molybdate, etc., are necessary to reduce corrosion which will not only affect system components but also reduce heat transfer and promote biological growth.

Cooling tower water treatment

For sites using open systems, such as cooling towers, the minimum treatment required is:

- Biocide treatment is essential for Legionella control.
- Blowdown is essential for TDS control and softening may also be necessary to improve TDS control.
- Strainers are necessary to remove larger pieces of scale or fouling.

The effectiveness of biocide treatment should be checked on a weekly basis with dip slides and every 3-months with a Legionella specific laboratory test.

- **Tip** – Dip slides will give an overview of the 'health' of the water in the system, i.e., the amount of bacteria in the system, but they are not specific to Legionella. This is why there should be Legionella specific testing every 3-months.
- **Tip** – Legionella testing should always conform to the local regulations. They will vary with the location of the site.
- **Tip** – Take advice from a water specialist for any additional treatment that may be needed to meet any local regulations.

Chilled water treatment

Despite being primarily closed systems, chilled water systems are still subject to corrosion, scale and microbial growth that require treatment. For sites using closed systems, the minimum treatment required is:

- Softening of make-up water to prevent scale build in the system,
- Corrosion protection to prevent damage to system components.

- Biocide treatment to prevent microbial growth.
- Glycol treatment for freeze protection in cold climates. Glycol should not be added unless necessary for freeze protection as it is not as good as water at transferring heat and also increases the viscosity of the water. Glycol at > 2% will provide some degree of protection against bacterial growth but at 0.1-1% can actually promote microbial growth.
- **Tip** – Some glycol products include a corrosion inhibitor but this will never be as good as proper corrosion protection.

Boiler water treatment

Steam boiler water should always be treated and conditioned before use to reduce both TDS levels and corrosion. If this is not carried out then 'carryover' is possible. This is where solid, liquid or vapour contaminants leave the boiler with the steam and enter the system. Carryover can cause corrosion, reduce heat transfer, damage the system components and cause water hammer. Carryover can be due to mechanical factors, e.g., operating the boiler in excess of the design load or by suddenly increasing the load, but the main reasons are often chemical and a result of high TDS levels, high pH levels (alkalinity) or the presence of oils or soaps in the system.

- **Tip** – Carryover can be controlled by correct treatment and using a water treatment specialist is strongly recommended.
- **Tip** – The use of blowdown to control TDS levels in boilers is covered in Section 7.7 and it is recommended that automatic blowdown control is used on all boilers to minimise the water and energy losses that result from blowdown.
- **Tip** – Most equipment and chemical suppliers will train staff in water treatment. They are the experts and should be used to provide expert assistance and staff training in water treatment.

8.13 Reduce discharge – reduce volume, reduce COD and reduce suspended solids

The second charge

In most areas of the world the assessment of the cost of discharging trade effluent is different to the assessment of the cost of discharging standard sewage. In the UK, trade effluent charges are largely based on the Mogden formula, named after the Mogden Sewage Treatment site in West London (see box on the right for the actual formula and variables). This is the second charge for water use, i.e., the charge for disposing of it.

The Mogden formula uses three basic factors in the calculation:

- Volume – how much trade effluent is being discharged to the sewers, i.e., the volume in m^3.
- Cleanliness – how dirty the discharged trade effluent is compared to standard sewerage, i.e., the chemical oxygen demand (COD).
- Solids – how much suspended solids the trade effluent contains compared to standard sewage, i.e., the suspended solids value.

Reducing any of these factors reduces the impact that the trade effluent has on the treatment system and the cost of treating the effluent.

Understand effluent flows

Whilst some sites will understand the flow of water coming onto the site, most will have little idea of the effluent flows of water from the site. The water balance (see Section 8.3) needs to establish the location and volume of the various flows, e.g., effluent, surface water and foul sewer water, and for effluent flows the location of the sampling points and any meters present.

- **Tip** – Reducing discharges needs data on where and how the discharges flow from the site.

Reduce volume (V)

Reducing water use will automatically reduce trade effluent charges. Minimising the demand (see Sections 7.6 to 7.9) by reducing use will obviously reduce the discharge volume. Using alternative sources (see Section 8.10) may reduce the discharge volume but this depends on what happens to the water from the alternative source after use, i.e., if it is clean and discharged back to groundwater then there will be no extra charge but if it is discharged via the sewers then there will be an extra charge.

Optimising the supply by recycling and re-use (see Section 8.11) will reduce the total discharge volume.

In many cases, the real issue is whether the site is being charged for the correct amount of effluent being discharged. This is because the treatment company will generally assume that the amount of effluent discharged from a site is the same as the metered amount of water supplied.

If input water is not returned to the sewer, for any reason, then sites will be paying too much for the discharge.

Typical reasons for not returning water to the sewer are:

- Process losses – where water is used in the services or process and is not discharged to the sewers, e.g., water lost to evaporation in cooling towers, water lost to evaporation in spray cooling of rotational moulds or water lost to evaporation in any other process.
- Product losses – where products containing water leave the site. This is unlikely for plastics processors.
- Facilities losses (domestic sewage volume adjustment) – where water is used in sewerage then an allowance for this is possible in trade effluent volumes.
- Leaks – where a site has leakage (see Section 8.6) that does not return to the sewers.

Where sites do not return all of their input water to the sewers then they should not be charged for this and can apply to the supplier for a rebate on effluent charges.

If you live in London then you use the work of Joseph Bazalgette every day and should thank him for it. He was the engineer responsible for the central London sewerage network and saved millions of lives through an 82-mile sewer network that virtually eliminated cholera and decreased typhus and typhoid.

Bazalgette and other sewer engineers made the modern city possible.

Trade effluent

This is any liquid (with or without solids) produced in the course of any trade or industry at trade premises but does not include domestic sewage.

The Mogden formula

Trade effluent charges in the UK are not a simple volume charge based on m^3 but are based on the Mogden formula, which is:

Trade Effluent Charge = R + V + Bv + [B x (Ot/Os)] + [S x (St/Ss)]

where:

- R = Reception and conveyance charge (pence/m^3).
- V = Volumetric treatment charge (pence/m^3).
- Bv = Volume charge for biological treatment, if needed (pence/m^3).
- B = Biological oxidation charge (pence/m^3).
- Ot = Chemical oxygen demand of trade effluent (mg/l).
- Os = Chemical oxygen demand of settled sewage (mg/l).
- S = Unit cost for sludge disposal (pence/$metre^3$).
- St = Solids value of trade effluent (mg/l).
- Ss = Solids value of settled sewage (mg/l).

Similar formulas will be used in other parts of the world although the detail may vary and the charges obviously will.

- **Tip** – Sites should be prepared to provide records (inputs and outputs) or calculations to justify rebates.
- **Tip** – Reducing the volume of trade effluent may have the unintended effect of increasing the overall COD and suspended solids. This is a delicate balance.
- **Tip** – Surface water should never be diverted to the foul sewer or trade effluent flow. It should preferably go to grassed areas, gravelled areas, soakaways, a Sustainable Urban Drainage System (SUDS) or to the public surface drainage sewers. If it is sent to the trade effluent flow then the site could be charged twice for the discharge, i.e., for the surface water and for the trade effluent.

Reduce COD (Ot)

The Ot value is given by a chemical test to measure how much oxygen is needed to consume the organic component of the trade effluent. The majority of the effluent discharge from plastics processing sites will be from cooling water or boiler blowdown and these will have a low COD and hence a low Ot.

If the site also processes food, i.e., an integrated site, then the COD (Ot) can be high and it may be worthwhile investing in processes to reduce the COD. This can be done by:

- Precipitation – this uses coagulants and flocculants to form large clumps of sludge for sedimentation.
- Bacteria – this uses bacteria (aerobic or anaerobic) to break down organic compounds.
- Oxidation – this uses oxidisers, e.g., hydrogen peroxide (H_2O_2), to reduce the COD.
- Fenton's Reagent – this is an oxidation reaction that uses hydrogen peroxide (H_2O_2) with Iron Sulphate ($FeSO_4$) to oxidise the contaminants.
- Advanced Oxidation Process – this is an advancement on Fenton's reagent and uses Fenton's reagent with ozone injection to rapidly oxidise contaminants.
- Filtration and activated carbon – this uses filtration and activated carbon to remove and absorb contaminants.

In most cases, reducing the Ot will not be necessary for plastics processors.

Reduce suspended solids (St)

The St value is given by a test that measures the solids content of the trade effluent. It indicates how 'cloudy' the water is, i.e., a very cloudy discharge will have a high St value.

Effluent containing large quantities of suspended organic and inorganic material should be treated by screening, filtration or settling/flotation prior to discharge.

At most plastics processing sites, the primary contaminants in trade effluent will be oils and greases and the most common point of introduction of these is the compressed air system. An oil-water separator is therefore an essential component of the compressed air system. These separate the oil and water to allow the St of the water to be reduced. The clean water can then be discharged as trade effluent and the oil sludge recovered or disposed of.

- **Tip** – Oil-water separators can be used for general trade effluent if there are significant quantities of oil in the effluent. They can separate mineral oil, synthetic oil, polyglycol, and other lubricants.
- **Tip** – Examine the trade effluent being discharged. If it is very cloudy and the St costs are high then screening, filtration or settling can be effective in improving the quality of the trade effluent discharged and in reducing costs.

Check the waste stream being monitored at each sampling point to ensure that this is representative of the complete site discharges.

Dilution is not the solution

Minimise emissions at source. Do not dilute them down to an apparently 'acceptable' level.

8.14 The water footprint

A new (and developing) concept

The water footprint is used to measure the water impact of a process, product or country. The concept of the water footprint was devised in 2002 by Arjen Hoekstra and in 2008 Hoekstra helped to establish the Water Footprint Network (waterfootprint.org). In 2011 Hoekstra et al produced 'The Water Footprint Assessment Manual[1]' which gives the basis for water footprint assessment and these concepts were incorporated in ISO 14046:2014 'Environmental management. Water footprint. Principles, requirements and guidelines.' produced as part of the ISO Life Cycle Assessment series of standards. Whilst ISO 14046 gives the requirements for compliance with the standard, the Water Footprint Assessment Manual gives far more of the background to water footprinting and is recommended as a first resource.

There is a temptation to think that the water footprint and the carbon footprint (see Section 7.1) are analogous. They are both part of the family of footprint concepts and they are very similar in measuring impact (although they measure different things). However, there are also significant differences between the two, perhaps the biggest being the consideration of location and time in the water footprint, i.e., water use is a 'local' issue whereas carbon emissions are considered to be a 'global' issue.

The types of water

The water footprint considers water in terms of both the direct and indirect impact, i.e., water used at a site and also water used indirectly by the raw materials and services supplied to a site. It also considers the location and time of the water use. There are three types of water use in a water footprint but there is a further division into 'consumption use' or 'pollution use'. These divisions are shown on the right and are:

- Green water – Water from rain that does not run off but is stored in the upper soil layers and is used by plants or evaporates. This is the main type of water used for agriculture, horticulture and forestry.

The green water footprint is the volume of rainwater consumed during crop or plant growth. It includes the water used in evaporation or transpiration (evapotranspiration) as well as water included in the crop or plant. Green water use is considered as water consumption.

- Blue water – water from rain that is stored in rivers, lakes and groundwater. This is the main type of water used for domestic and industrial purposes and for agricultural irrigation.

The blue water footprint is the volume of water used and then evaporated or included in a product. It includes water from a source that is returned to another catchment or the sea but does not include water that is returned to the same source. Blue water use is considered as water consumption.

- Grey water – water needed to dilute pollutants in any output water to an agreed quality standard.

The grey water footprint is the volume of water that is required to dilute any pollutants so that the water quality is better than an agreed quality standard. The grey water footprint is an indicator of the water pollution associated with a product over the full supply chain. Grey water use is a measure of water pollution.

The water footprint is a simple volume measure of water consumption and pollution. It does not measure the local impact of the consumption and pollution as these are local, regional or national issues.

How is the footprint calculated?

The water footprint methodology is designed to be flexible, it can be used to calculate the water footprint of a process, product, person, business, business sector, country or even the world.

> "The water footprint is a measure of humanity's appropriation of fresh water in volumes of water consumed and/or polluted."
>
> **Water Footprint Network**

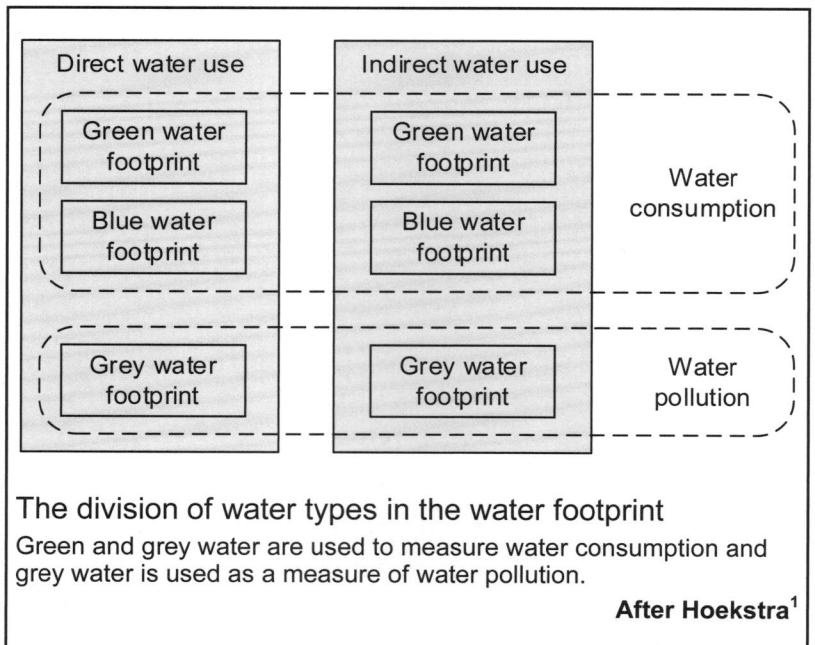

The division of water types in the water footprint

Green and grey water are used to measure water consumption and grey water is used as a measure of water pollution.

After Hoekstra[1]

- **Tip** – The flexibility of water footprinting means that the goals, system boundaries and purpose of the water footprint need to be carefully specified.

The water footprint is also related to time and space, i.e., the water footprint of rice grown in the USA is very different to the water footprint of rice grown in South-East Asia because of the differences in local conditions. This means that the response to the final water footprint may be very different depending on the location, this is very different to the location independence of carbon footprinting.

The water footprint is built up from the process step level. The footprint for each process step (including the supply chain) is calculated and these are combined for all the process steps in a product to give a product water footprint.

The relevant product footprints can then be combined to give a site or company footprint for all the products produced at the site or by the company. Similarly, a consumer footprint can be produced by combining the footprints of all the products consumed, including imported products, e.g., food and consumer goods.

Sector or national footprints can be built up by combining the relevant consumer footprints and area footprints (regional, national or international) which can be calculated from the total of the processes taking place in the area.

- **Tip** – There is a difference between a national footprint calculated by combining the consumer footprints and the national footprint calculated from the processes taking place in the area. This is because of the 'virtual water' in products that are exported and imported, e.g., rice eaten in the UK will contain virtual water that was used for its production in another country. Developed countries therefore import 'virtual water' in the form of products.

- **Tip** – The time relationship of the water footprint means that a product footprint should be in the form of volume/mass, e.g., litre/kg, but a consumer footprint should be in the form of volume/time, e.g., litre/year.

As with carbon footprinting, there are very specific rules for the calculation of the water footprint to avoid 'double counting' of any of the inputs.

In the case of a plastics processing site, there will be four general components of the water footprint, these are:

- Internal product use, i.e., process load.
- Internal overhead use, i.e., base load.
- External (supply chain) product use.
- External (supply chain) overhead use.

How is the footprint used?

The water footprint does not itself specify the actions to be taken after the calculation. This is decided by the aims of the footprint. For many sites, the typical aims should be:

Green water

For plastics processing sites, the green water element of the water footprint is not relevant.

Blue water

For plastics processing sites, the aim should be to reduce the blue water footprint to zero through full recycling and zero losses through evaporation.

- **Tip** – Water recovery and re-use (see Section 8.11) mean that the water footprint will decrease.

Grey water

For plastics processing sites, the aim should be to reduce the grey water footprint to zero with full recycling, heat capture from effluent and suitable treatment to reduce discharges to zero.

Using the water management programme

A water management programme as set out in this chapter will provide a clear route to reducing the water footprint. The water footprinting process provides good data to support the water management programme and a clear method of tracking progress in water management and sustainability (see Section 12.9).

The total division of fresh water use in the world:
- Agriculture ≈ 70%.
- Industry ≈ 19%.
- Domestic ≈ 11%.

The water footprint process, by breaking use down into process steps, provides a method of identifying 'hotspots' of water use and thus the areas for action and improvement.

For most sites, the external supply chain footprint will be larger than the internal footprint.

The water footprint process will highlight external hotspots as well as internal ones.

- 1. The Water Footprint Assessment Manual, Water Footprint Network, 2011. Available free from https://waterfootprint.org/en/resources/publications/water-footprint-assessment-manual.

8.15 Water management – where are you now?

Use it wisely

Water management is often neglected in plastics processing because the main processes do not generally use much water. Despite this, good water management can improve a site's sustainability and reduce costs. The benefits are also often easily and quickly achieved because it is not an area that many plastics processors have concentrated on before.

The water management programme sets out a series of actions to enable a site to assess water use, to identify the areas of excessive leakage or use and provides the tools and techniques to minimise these.

Water is unique amongst the services where the standard process is to reduce the demand and then to optimise the supply. With water it is also necessary to reduce the effluent discharges to minimise the environmental impact and costs.

Completing the chart

This chart is completed and assessed as for the previous charts.

Low scores are not bad but simply show areas with improvement potential.

The barriers to water efficiency:

- Lack of commitment – staff need to be motivated and committed. Involving the relevant stakeholders, e.g., internal and external staff, contractors, suppliers and members of the general public, will ensure improvement programmes are effective.

- Lack of understanding – most staff are not aware of the cost of water to the site. As with compressed air, it is seen as free.

- Low priority – there are always limited staff resources to identify and implement water efficiency measures. Water is often seen as less important than energy (and we have enough trouble implementing energy efficiency).

Many sites will find it hard to justify the time, funding and commitment for water efficiency, but this does not mean that it is not worthwhile.

Water management

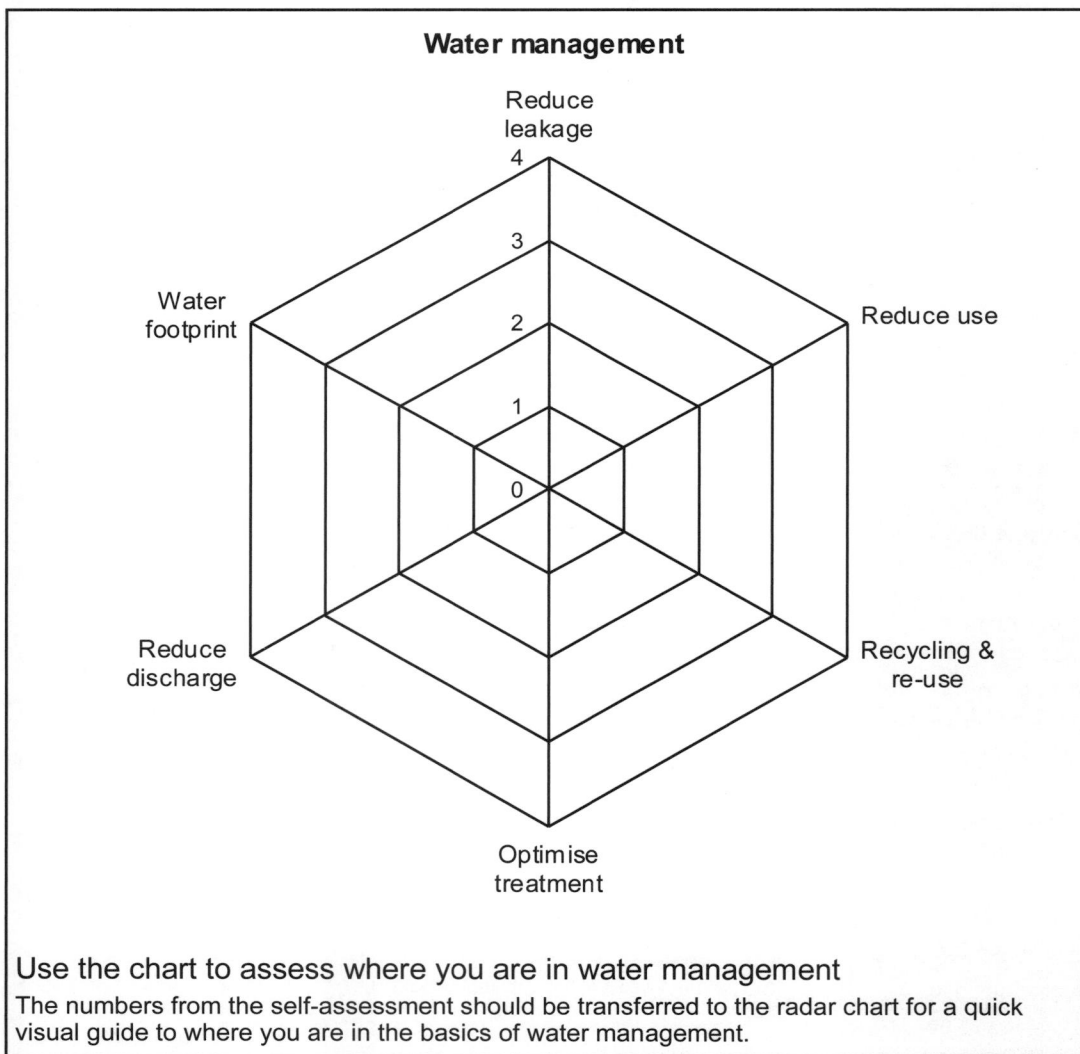

Radar chart with axes: Reduce leakage, Reduce use, Recycling & re-use, Optimise treatment, Reduce discharge, Water footprint. Scale 0 to 4.

Use the chart to assess where you are in water management

The numbers from the self-assessment should be transferred to the radar chart for a quick visual guide to where you are in the basics of water management.

	Water management					
Level	Reduce leakage	Reduce use	Recycling & re-use	Optimise treatment	Reduce discharge	Water footprint
4	Inspection & remedial action carried out for all areas within last 6 months.	Water use reduced to practical minimum in all areas.	All potential recycling & re-use opportunities investigated & implemented. Discharges minimised.	Water treatment fully optimised & automatically controlled to reduce treatment to practical & regulatory minimum. External input to process.	Discharges well controlled to minimise volume, COD & suspended solids.	Full water footprint completed for direct & indirect blue & grey water use.
3	Inspection & remedial action carried out for all areas within last 12 months.	Water use minimised in processes.	Water re-use carried out where water does not require any treatment.	Water treatment manually controlled for all systems. External input to process.	Discharges well controlled to minimise volume only.	Water footprint completed for direct (internal) blue & grey water use.
2	Inspection & remedial action carried out for all areas within last 2 years.	Water use minimised in services.	Limited water recovery & re-use carried out.	Water treatment manually controlled for open systems but poor or no control on closed systems. External input to process.	Discharges currently uncontrolled & within permits. Discharge reduction considered but no action taken.	Water footprint completed for direct (internal) blue water use only.
1	Inspection & remedial action for facilities only carried out within last 2 years.	Water use minimised in facilities & heating.	Minimal water recovery & water re-use.	Water treatment excessive & uncontrolled. No external input to process.	Uncontrolled discharges (within permits) & no consideration of reducing discharges.	Some knowledge of water footprint concept & implications.
0	No inspection or remedial action carried out for any area in previous 5 years.	No effort made to reduce water use in any area.	No water recovery or re-use	Water treatment inadequate with potential for breach of regulations, health risks or damage to systems.	Uncontrolled discharges & potentially breaching discharge permits.	No knowledge of water footprint concept.
Score						

Key tips

- Start by asking the obvious questions about how much, where, when, what and why.
- Map the sources and sinks for the site to understand where the water is coming from and where it is going to.
- Carry out the initial water balance quickly from knowledge of the sources and sinks.
- Try to capture > 95% of the water entering and leaving the site to have confidence that the major flows have been captured.
- Add detail to the water balance for the critical areas and map the internal water flows for the site.
- Get the water balance to agree within ± 5%
- Use meter readings to gain an understanding of water use when the site is not operating. This will give an estimate of leakage at the site.
- Use meter readings to understand the production related use of water and relate this to production volumes and time.
- Set up the water management programme for a structured approach to reducing water use.
- Leakage may well be the largest user at some sites and minimising leakage is the first task.
- Investigate pressure reducing valves to reduce leaks and maintenance costs.
- Reducing use is often simple because it has been ignored in the past.
- Look carefully at the services, blowdown use in cooling towers and steam systems. These are often large users and poorly controlled (if at all).
- Many processes use little water but those that do use water should be investigated for easy wins in water use reduction.
- Facilities (washing, toilets etc.) offer very easy wins and the technology is ready for implementation at most sites. Advanced (and simple) flush controls and urinal controls can easily and significantly reduce water use.
- Alternative sources are not always used effectively or even considered. Take a walk around the site and look for potential sources that can be accessed (they may need permissions) for water supply. The effort will be worth it.
- Water recovery and re-use can reduce the amount of water entering the site and the discharges from the site.
- Water treatment is needed not simply for health reasons but also to protect the water system from deposition, scale and biological growth.
- Treatment systems should be automatic to reduce chemical use and meet discharge requirements.
- Reducing discharge volume is the easiest way to reduce the effect of discharges but not if it significantly increases the COD or the suspended solids content.
- A water footprint calculation is an excellent method of driving water management, it can provide a method of assessing the water impact of a site better than simply calculating the extraction requirements of the site.

Chapter 9

Manufacturing: waste minimisation

Sustainability is about using the materials and services that we have in the most effective way and this naturally involves minimising the amount of waste that is produced in our processes. Waste minimisation is a very broad subject and can include wasted effort (see Section 5.9), wasted energy, wasted water and wasted materials.

This chapter deals with general waste in plastics processing and the methods available to reduce this. It does not deal with waste in terms of energy, water, material re-use by in-house regrinding or recycling and value recovery after use; these are dealt with elsewhere in this book.

I first started working in waste minimisation for the plastics processing industry in 1999 with the Envirowise programme to reduce the environmental impact of UK industry. At that stage we were focused on reducing waste, in the broadest possible terms, to improve not only environmental impact but also to improve the profitability of industry. This was because reducing waste is not only good for the environment but also very profitable for companies who reduce waste. There are benefits in being 'green' that go far beyond the 'greenwashing' we sometimes see.

The Envirowise programme was subsumed into WRAP (Waste and Resources Action Programme) in around 2005 and most of the extremely valuable published documents were lost or filed so deep in the WRAP site that they are virtually lost. This is a huge pity but many of these documents are still available at www.tangram.co.uk.

The resource efficiency training modules ("Resource efficiency: cut costs in plastics processing") is a series of practical workbooks for the plastics processing industry that provide training templates and presentations for directors, managers, supervisors and operators. These are particularly valuable for companies who want a pre-prepared series of training packages.

Resource efficiency is at the heart of sustainability and all sectors of the industry need to strive to "make more with less and to create less waste". This is not about waste in the sense of Post-Consumer Waste but is about avoidable waste inside a plastics processing site.

This is about reducing the environmental impact and outputs of the process and is applicable to every plastics processing site for both the environmental benefits and for the cost benefits.

9.1 Waste minimisation

The business reasons

Wasted effort in production and wasted energy through inefficient operations are both unnecessary costs but waste of physical resources is equally costly and can often easily be reduced. Waste, in terms of materials alone, is estimated to cost approximately 5% of the total turnover for the average plastics processing company. In most companies, the cost of waste can easily be reduced by at least 20% (1% of turnover) through the implementation of a simple, but formal, waste minimisation programme. This uses simple tools and techniques to reduce environmental impact, improve sustainability, reduce waste costs and raise profits. Even when investment is required, the payback periods are generally short and the returns high.

- **Tip** – Most companies will deny that their cost of waste is approximately 5% of turnover but most of the cost of waste is indirect and hidden (see Section 9.5).

Waste costs real money

Waste is often ignored and is always assumed to be somebody else's problem. However, waste costs real money and this comes directly off the company profits. The box on the right will give an initial estimate of the basic and total cost of waste. At a net margin of around 10%, a reduction in waste costs by 20% is the equivalent to increasing sales by 10%, i.e., internal efforts spent minimising waste can produce equivalent benefits to substantially increased sales. Most plastics processing companies would regard a sales increase of 10% as very desirable yet the same benefits can be achieved totally internally and with no risk at all. The only risk is that the competition does it and you do not.

The true costs are hidden

The numbers from the box on the right show that there is a difference between the visible and the true cost of waste. Waste costs are either direct or indirect. Direct costs are visible and include waste collection and disposal costs. Despite the fact that they are visible these costs are still largely ignored by most companies.

The bulk of the true waste costs are indirect and hidden even from the accountants (see Section 9.5). Most companies have developed special words and phrases as a code for waste and a list of the most common code words is given in the box on the far right.

How many of these words do you use in your company to disguise waste as a part of normal operations? Have you even invented new ones?

These words make the waste appear an inevitable part of the operations. It is not. These hidden wastes make up the largest portion of the total waste costs in any business and include:

- Excess raw material costs.
- Effluent generation.
- Excess packaging use.
- Excess and wasted factory and office consumables.
- Wasted time and effort.

These costs are either not recorded or are hidden in the accounts and not shown as separate items. Despite this, they exist even for efficient companies. They arise whether you like it or not, and they are significant whether you realise it or not. After simple waste assessments, many companies have found that their waste costs were more than 20 times higher than the initial estimates.

Responding to stakeholders

Waste minimisation shows how effectively and efficiently you control operations.

Waste is not simply the contents of the skips – it is much more than that (although a bit of 'dumpster diving' can show some remarkable results).

Do not insult people by putting up posters saying:

STOP WASTE!

The greatest waste is the waste we don't see and the posters should read:

FIND WASTE!

After waste is found then it is generally easy to stop.

The potential benefits

Calculate the potential savings based on raw materials losses:

Amount of main raw material used last year, e.g., tonnes.	A
Amount of product produced last year, e.g., parts.	B
Amount of main raw material/unit of product, e.g., polymer/part.	C
Quantity of main raw material sold in parts last year = B x C.	D
Wasted main raw material = (A – D).	E
Purchase cost of main raw material.	F
Cost of wasted raw material = (E x F).	£

These calculations only show the visible purchase cost for wasted raw material. The true and total cost will also include wasted production costs, labour, storage, etc. Consideration of all areas of waste will give a much higher figure.

Calculate your potential savings based on a cost reduction of 1% of turnover:

Turnover last year	£
Potential savings (1% of turnover)	£

Customers, employees and investors all have a growing interest and knowledge of sustainability and environmental performance:

- Customers and other stakeholders are increasingly asking for evidence of good environmental performance (see Section 12.9). Waste minimisation is a fundamental part of this commitment and is a key part of environmental management.
- Investors want the highest possible return on capital and high dividend growth, banks want to see efficient use of borrowed capital. Waste minimisation can help to deliver both requirements by improving resource use and minimising costs.
- Employees know where materials and resources are being wasted and can see the cost benefits that will make the company more competitive and safeguard their future. For companies involved in waste minimisation, increased employee satisfaction was among the top five benefits of the process.

Good investment returns

Cost-effective waste minimisation is also a valuable investment that pays dividends in reduced costs for any company. Large savings can be made from small capital investments. Money spent on waste minimisation is a sound investment and waste minimisation can significantly reduce costs for plastics processors.

The legal requirements

Companies, and the key directors and managers, can face stiff penalties for failing to comply with environmental legislation, which gets tougher year on year. An effective waste minimisation programme:

- Can be used to prove conformance with existing laws.
- Helps to reduce the cost of conformance.
- Can reduce rapid, disruptive and expensive changes due to changes in legislation.
- Can be used as part of the EMS (see Section 2.6) to show continual improvement.

The essentials of waste

Waste minimisation may seem difficult to start but the essentials are:

- In any system, what goes in must come out.
- Every waste has a source, a destination and a justification (even if it is wrong).
- Understanding the sources and destinations makes it possible to improve control, efficiency, quality and sustainability.

- Waste can always be reduced but only if action is taken.

Cut waste, and you will reduce environmental impact, improve sustainability and boost profits.

Money saved from waste minimisation goes straight to the bottom line.

The code words for waste in plastics processing

These are some of the code words used for waste in plastics processing. Do you use any of these? A rose by any other name ....

Allowance	Overproduction
Batch growth	Overspec
By-product	Overweight
Cancelled orders	Packaging
Contaminated solids	Process loss
Conveyor loss	Purgings
Credits	Quality samples
Customer returns	Reaction loss
Damage	Recoverable loss
Defects	Reel ends
Deposit loss	Reel strippings
Dipstick error	Regrind
Dirty solvent	Rejects
Doubles	Residue
Downgrade	Resort
Drainings	Returns
Dregs	Rework
Drool	Rubbish
Dross	Runners
Dumped	Samples
Dust	Scrap
Edge trim	Start-up allowance
Effluent	Second quality
Evaporation	Seconds
Excess stock	Shortages
Extraction	Shrinkage
Flash	Side run
Garbage	Slow-moving stock
Giveaway	Snot
Gob	Spare mix
Head waste	Sprues
Hidden losses	Stock loss
Inspection loss	Sub-standard
Invisible loss	Surplus
Leakage	Swarf
Make-ready	Sweepings
Make-ups	Tails
Material variance	Tops
Mixing over-run	Trash
Non-conformances	Trials
Obsolete stock	Underdelivery
Offcuts	Unrecoverable loss
Out-of-shelf life	Use allowance
Out of specification	Use variance
Overcount	Workaway
Overdelivery	Yield loss
Overfill	+ more that you have invented
Overissue	for your company

9.2　The waste minimisation programme

It is similar to society

Waste minimisation inside a plastics processing site can be related to the more general efforts of society to improve sustainability by minimising waste but the system boundaries for site waste minimisation are drawn around a specific site rather than the complete society. However, it is useful to use the same type of model for site-level waste minimisation to retain consistency with society's efforts.

The EU waste hierarchy

The EU introduced the Waste Framework Directive in 1975 but this was updated and revised in 2008 (Directive 2008/98/EC). The directive states that the first objective of a waste policy should be to "minimise the negative effects of the generation and management of waste on human health and the environment".

The directive uses the waste hierarchy as a tool for waste policy to reduce resource use and improve resource efficiency. The waste hierarchy (shown in the box on the right) provides the priorities for waste prevention and management, legislation and policy.

Prevention

Waste prevention is the most sustainable option. Eliminating or reducing the amount of waste materials will result in the lowest environmental impact and cost.

The focus should be on:

• Eliminating the waste stream.

• Reducing the amount of materials consumed and waste generated.

Prepare for re-use

Re-use is next in the hierarchy and refers to the continued use of products or materials for their original function. At the product level, re-use involves reduced processing, i.e., checking, cleaning and/or repairing. In plastics processing, this is an effective option for wastes such as packaging (see Section 9.6).

Recycle

Recycling includes any operation where waste is reprocessed into products or materials for the original or other purposes. It includes the reprocessing of organic material but does not include energy recovery.

Segregating waste streams, e.g., cardboard, paper, plastics and metals, allows them to be processed by specialist recycling companies.

Recovery

This is where the waste serves a purpose by replacing other materials which would otherwise have been used to fulfil the function, or waste being prepared to fulfil that function, in the plant or in the wider economy. This includes processes such as:

• Recycling/reclamation of organic substances not used as solvents. This includes gasification and pyrolysis, i.e., chemical recycling (see Section 4.10).

• Solvent reclamation/regeneration (see Section 4.10).

• Recovering the embodied energy, i.e., Energy from Waste (EfW) (see Section 4.11).

Disposal

Disposal is the least favoured option and the primary methods in plastics processing are landfill (see Section 4.12) and incineration (with no energy recovery or using an incinerator with a low efficiency).

Disposal is also often the most expensive option due to landfill and other taxes.

A waste minimisation programme

At the site level, the EU waste hierarchy does not always provide the focus needed for waste minimisation. Waste minimisation can be viewed using the same structure used

> The Waste Directive states that costs should be allocated in such a way as to reflect the real costs to the environment of the generation and management of waste.

> The polluter-pays principle is a guiding principle at European and international levels.
>
> The waste producer and the waste holder should manage the waste in a way that guarantees a high level of protection of the environment and human health.

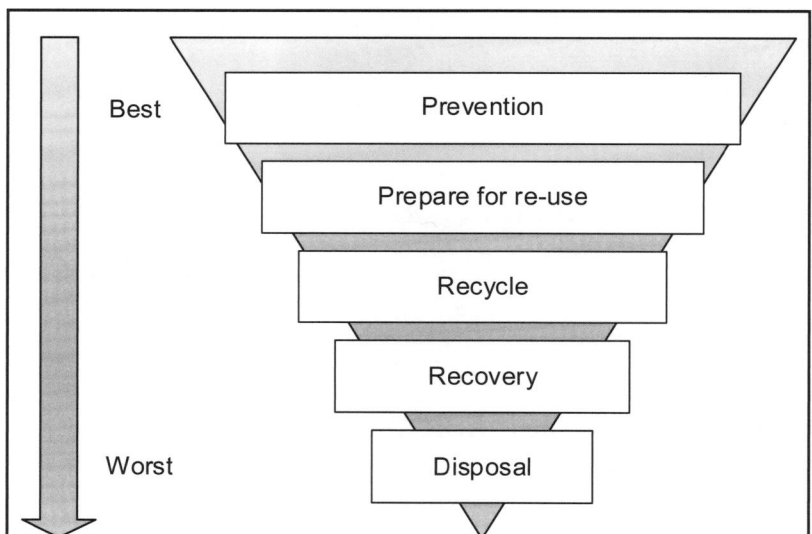

The waste hierarchy

The waste hierarchy provides a framework for evaluating the options in life-cycle thinking. The best option is at the top and the worst is at the bottom. The European waste hierarchy shown above is referred to in the Waste Framework Directive (Directive 2008/98/EC).

for services and water, i.e., 'minimise the demand and then optimise the supply', but modify this to 'minimise the demand and then optimise the output'.

Stage 1: Minimise the demand

Minimising the demand is the first stage in the programme, it is not logical to optimise the output based on excessive demand.

Step 1: Eliminate the use

Eliminating the use of a material completely removes the environmental impact and costs. This is the first option to be explored.

Step 2: Reduce use

Reducing the use of a material or process does not totally remove the environmental impact and associated costs but where the material must be used then this can be very effective. This is the second option to be explored.

Step 3: Re-use

This is about using materials or products more than once for the purpose that they were originally intended, e.g., reusable transit packaging or containers inside the site or between the site and customers or suppliers.

For plastics processors, this can also include the use of internal regrind if the regrind has not left the site boundaries and is re-used for the original process. Typical sources could be:

- Tops and tails (blow moulding).
- Sprues and runners (injection moulding).
- Edge trim (sheet extrusion)
- Start-up scrap and out-of-specification product (all processing methods).

All of these are treated in a closed-loop and can be classed as Post-Industrial Regrind (PIR). They do not count as recycled content (see ISO 14021:2016).

- **Tip** – This Chapter does not deal with re-use of plastic materials (see Section 5.3).

Stage 2: Optimise the output

After demand is minimised, then optimise the output to minimise the impacts and costs.

Step 4: Recycle

Recycling is using material that has left the site boundaries to fulfil the same or a new function, e.g., it has left the closed-loop of the re-use system. This includes sending paper, carboard, metal or other materials to recyclers.

For plastics processors this could include:

- Using PIR from another process and an external source. In terms of ISO

14021:2016 this material can be classed as Pre-Consumer Material because it has not yet reached the consumer (see Section 5.3).

- Using Post-Consumer Material. This is waste generated by the end user and includes material returned from the distribution chain (ISO 14021:2016).
- **Tip** – This Chapter does not deal with materials recovery and recycling of plastic materials (see Section 4.5).

Step 5: Recover value

Recovery can be split into 2 options:

- Recovering, reclaiming or regenerating the material (see Section 4.10).
- Recovering the embodied energy using the material as a fuel (see Section 4.11).
- **Tip** – See Chapter 4 for details of sustainable raw materials.

Step 6: Disposal

This is the last resort after all the other options have been exhausted.

Reduce, re-use and recycle – the standard symbol.

The greatest waste is the waste we don't see.

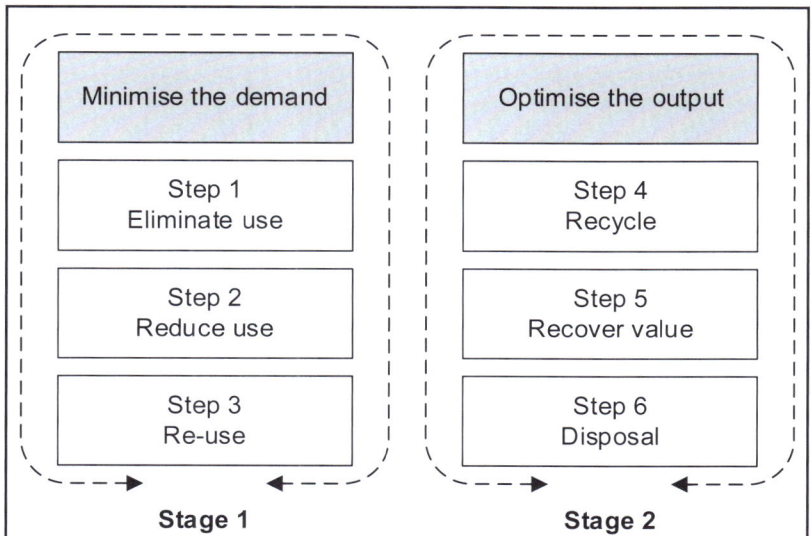

```
┌─────────────────────────┐  ┌─────────────────────────┐
│   Minimise the demand   │  │    Optimise the output  │
├─────────────────────────┤  ├─────────────────────────┤
│         Step 1          │  │         Step 4          │
│       Eliminate use     │  │         Recycle         │
├─────────────────────────┤  ├─────────────────────────┤
│         Step 2          │  │         Step 5          │
│        Reduce use       │  │      Recover value      │
├─────────────────────────┤  ├─────────────────────────┤
│         Step 3          │  │         Step 6          │
│         Re-use          │  │        Disposal         │
└─────────────────────────┘  └─────────────────────────┘
        Stage 1                      Stage 2
```

The waste minimisation programme

It is also possible to show waste minimisation in a similar format to that used for other services. The standard 'minimise the demand and then optimise the supply' can be easily changed to 'minimise the demand and then optimise the output' to show a consistent approach.

9.3 Waste minimisation tools – the process flow chart

Waste minimisation tools

The key tools in waste minimisation are the 'process flow chart' and the 'waste tracking model' (see Section 9.4). These are fundamental in finding both 'quick starts' and in developing a systematic approach for long-term sustainability savings. This makes it worthwhile examining each of these in detail.

The overview process flow chart

A process flow chart is a simple visual representation of the main processes that take place at the site. It can be created rapidly for the basic processes to give an overview of the full production process. This is the same type of process chart as used for environmental management (see Section 2.5) and this can be re-used and extended for waste minimisation.

- **Tip** – Process charts are an essential tool in waste minimisation; they help to understand processes and reduce costs (as well as complexity).

Each step in the process flow chart adds value to the product but also incurs costs from the labour, materials and utilities consumed during the process step. There is generally only one major input (from the previous process) and one major output (to the next process) and these should be easy to identify. The overview process flow chart should be created for the main process steps to show the main inputs and outputs. A typical example of an overview chart is shown on the right.

Every process step also creates a variety of wastes that have environmental impacts and cost money. At this stage, do not try too hard to put an accurate cost on the wastes – this can come later (see Section 9.4). The main objective is to try to identify, in broad terms, the amount and types of waste that are being generated at each process step. For plastics processing, it is important to include the services, e.g., compressed air, and to map the flow of regrind at the site.

It is important to get the overview process flow chart correct in the first instance because this will act as the road map for detailed cost collection and future waste minimisation efforts.

- **Tip** – The 'other inputs' are often the easiest places to look for waste. Companies tend to focus on the major inputs and outputs and ignore all the others as 'overheads'.

- **Tip** – The true cost of waste includes the cost of wasted resources and rejects at each step in the process. The cost of rejects should include the value added to the material by the time it is rejected and this increases through the process, i.e., rejects cost more the further down the process they are.

- **Tip** – The process flow chart can also include office processes, e.g., sales and marketing, R&D and finance, where these functions generate waste. These functions can be treated as per services (see below).

The detailed process flow chart

After the overview process chart has been completed, it is possible to create a separate chart for each process step. This will be more

> This flow chart is based on the process but an alternative method is to start with a physical map of the site and to mark on the map the main inputs, outputs and wastes.

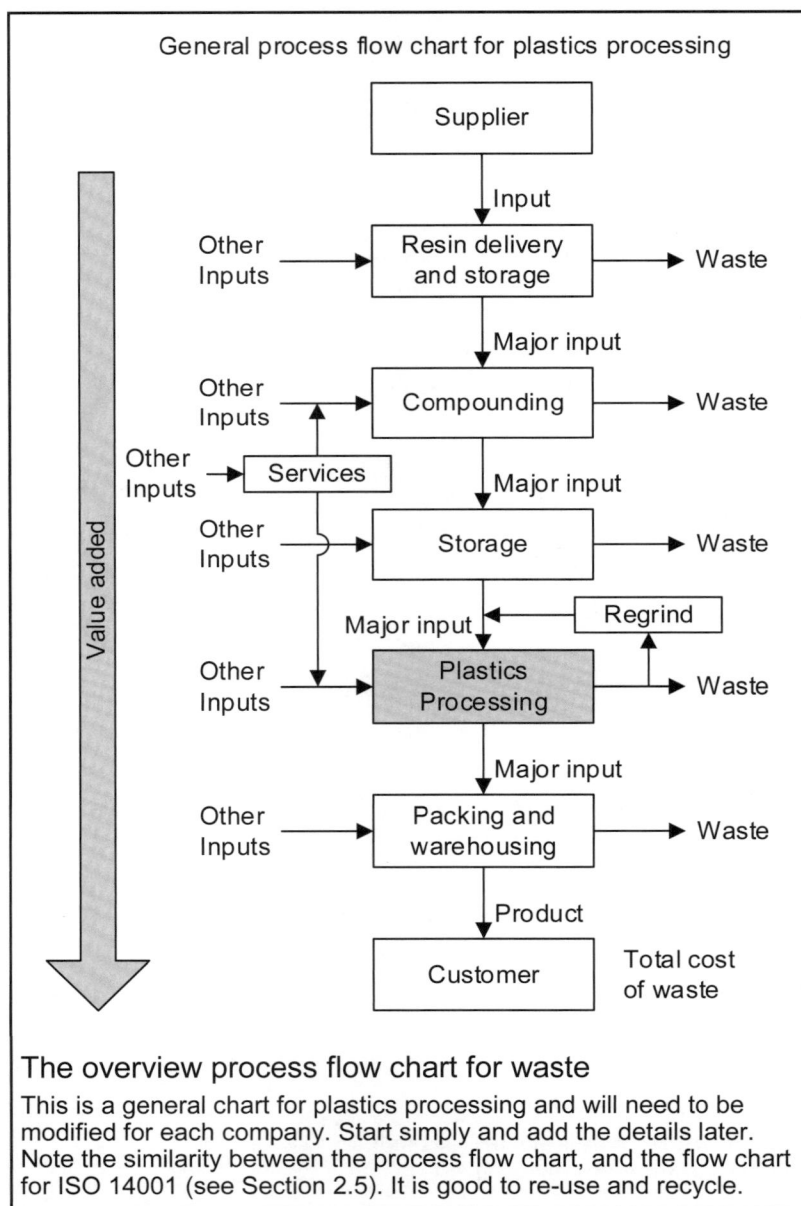

General process flow chart for plastics processing

The overview process flow chart for waste

This is a general chart for plastics processing and will need to be modified for each company. Start simply and add the details later. Note the similarity between the process flow chart, and the flow chart for ISO 14001 (see Section 2.5). It is good to re-use and recycle.

detailed and in addition to the main inputs and outputs will include many of the small inputs (some of these are wastes and can be eliminated) and the many small outputs (generally these are wastes and should be eliminated at source). A typical example of a detailed overview chart is shown on the lower right.

Create a detailed chart for each intermediate process to show the detailed inputs and outputs, concentrating on the possible input and output wastes. This detailed chart acts as a map for future waste minimisation work.

At this stage, the wastes can start to be quantified in terms of weight, volume or any other convenient measure. This does not need to be financial at this stage (see Section 9.4).

- **Tip** – When all the input and output wastes from the intermediate processes have been allocated it is useful to check the results by investigating the waste outputs from the site (from the skips or other waste disposal routes). Cross check these wastes against the wastes already allocated to intermediate processes.

- **Tip** – It will sometimes be difficult to allocate a waste to a specific process. These wastes can be aggregated over the complete process if dividing them up becomes difficult.

- **Tip** – Overweight production is often the highest waste in plastics processing and is due to poor process control (see Section 5.8). Overweight products 'give away' material and add cost.

Services

Services are one area where allocating waste can be very difficult, e.g., compressed air will be supplied to the complete site and leaks will occur all over the site. In most cases, central services can be regarded as a 'process step' and a detailed process flow chart generated for each service. This should concentrate on the area where the service is generated rather than on the area where the service is used. This is because many of the services wastes will already have been covered in terms of energy (Chapter 6) or water (Chapter 8).

- **Tip** – Services wastes have been included in the detailed process chart shown on the right for the sake of completeness.

- **Tip** – Do not forget to include the services in the process flow chart. These are often ignored but can be sources of both environmental impacts and costs, e.g., refrigerant leakage from chillers, compressed air dryers or A/C units and heating where the outputs will include GHGs.

Regrind

The use of regrind in plastics processing is common (see Section 5.3). Tracking the flow and value of regrind is part of waste minimisation and is considered in that section.

The 'reality check'

When you have done all this then go and stand by the machines to do a 'reality check' and make sure that you have collected all the wastes.

The UK Envirowise Programme produced information on waste minimisation (I wrote some of the plastics publications) and some of this is available from WRAP (www.wrap.org.uk).

Search for:

ET 219 – Waste Mapping: Your Route to More Profit.

Don't try searching the WRAP site because it is not listed!

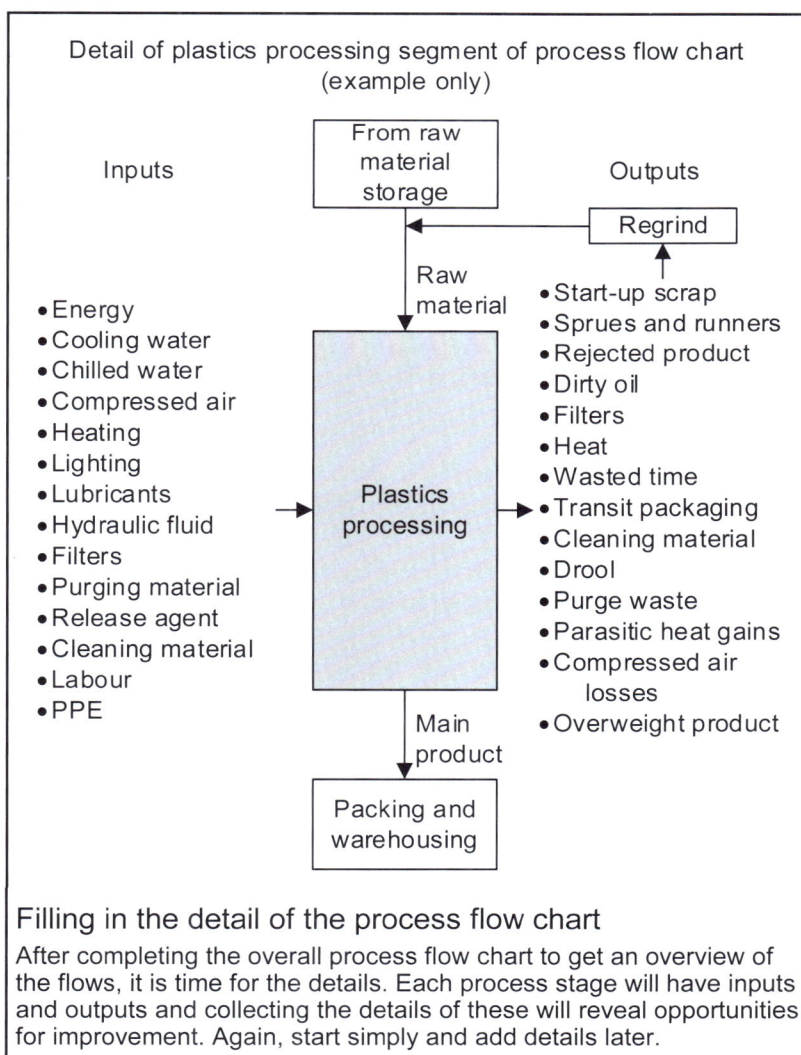

Detail of plastics processing segment of process flow chart (example only)

Inputs
- Energy
- Cooling water
- Chilled water
- Compressed air
- Heating
- Lighting
- Lubricants
- Hydraulic fluid
- Filters
- Purging material
- Release agent
- Cleaning material
- Labour
- PPE

From raw material storage → Raw material → Plastics processing → Main product → Packing and warehousing

Regrind

Outputs
- Start-up scrap
- Sprues and runners
- Rejected product
- Dirty oil
- Filters
- Heat
- Wasted time
- Transit packaging
- Cleaning material
- Drool
- Purge waste
- Parasitic heat gains
- Compressed air losses
- Overweight product

Filling in the detail of the process flow chart

After completing the overall process flow chart to get an overview of the flows, it is time for the details. Each process stage will have inputs and outputs and collecting the details of these will reveal opportunities for improvement. Again, start simply and add details later.

9.4 Waste minimisation tools - the waste tracking model

Putting numbers to the waste

The process flow chart helps to identify the key waste areas but driving waste minimisation needs numbers and measurements. These do not all have to be financial measurements at the start, in fact, it is better to start with 'non-financial' measurements and fill in the values later.

The waste tracking model

The information from the detailed process flow chart (see Section 9.3) for each process step can easily transferred to the waste tracking model for a better estimation of the true cost of waste at that step. A sample waste tracking model sheet is shown on the opposite page and this includes the inputs from the previous step and the outputs to the subsequent process step.

A similar waste tracking sheet should be created for each process step and for areas such as services and offices where these have been created.

It is best to get some data on the volume of waste in physical terms and to estimate or calculate the full financials later for the biggest wastes. When gathering information, people can often tell you waste values such as 'about 100 boxes per week' even though they have no knowledge about the actual financial implications. The physical data can then be used to estimate the financial implications of the waste identified.

- **Tip** – When estimating the costs, do not forget to include the time and effort involved in handling or dealing with the waste.

The waste tracking model gives a much more accurate estimation of the cost of waste for each process step and it also identifies where real savings can be made.

The combined waste tracking models for all the process steps will provide an initial estimate of the total cost of waste to the business (see Section 9.5).

- **Tip** – Combine the details from the various waste tracking sheets to see if there are any discrepancies in overall values, i.e., between identified and total actual water use, and raw material and energy consumption.

- **Tip** – Dig deeper if there are any major discrepancies. They may be a major cost and a major savings opportunity!

Prioritising action

The waste tracking model is also an 'opportunity sheet' for each process step. The decision about which opportunities to pursue can be based on the quantity of waste, the environmental impacts, the cost or the easiest projects to complete. Whichever method is used, the next step is to get some of the staff together to generate ideas for waste minimisation and to prioritise action to complete the projects (see Section 9.13).

- **Tip** – It should take less than one day to identify several projects with the potential to make immediate savings and put them in order of priority.

- **Tip** – Take photographs of waste. Nothing drives home the message more than a photograph.

The cost of waste isn't the cost of the skip, it is more the cost of the things in it and what it cost to make these things.

5 Whys

This is asking 'why does this happen' at least five times. This method tries to avoid looking at symptoms and looking for the root causes.

If presented with an oil leak on a machine it is not enough to clean it up. Using '5 Whys' can reveal that the real cause was that the machine was not subject to regular maintenance.

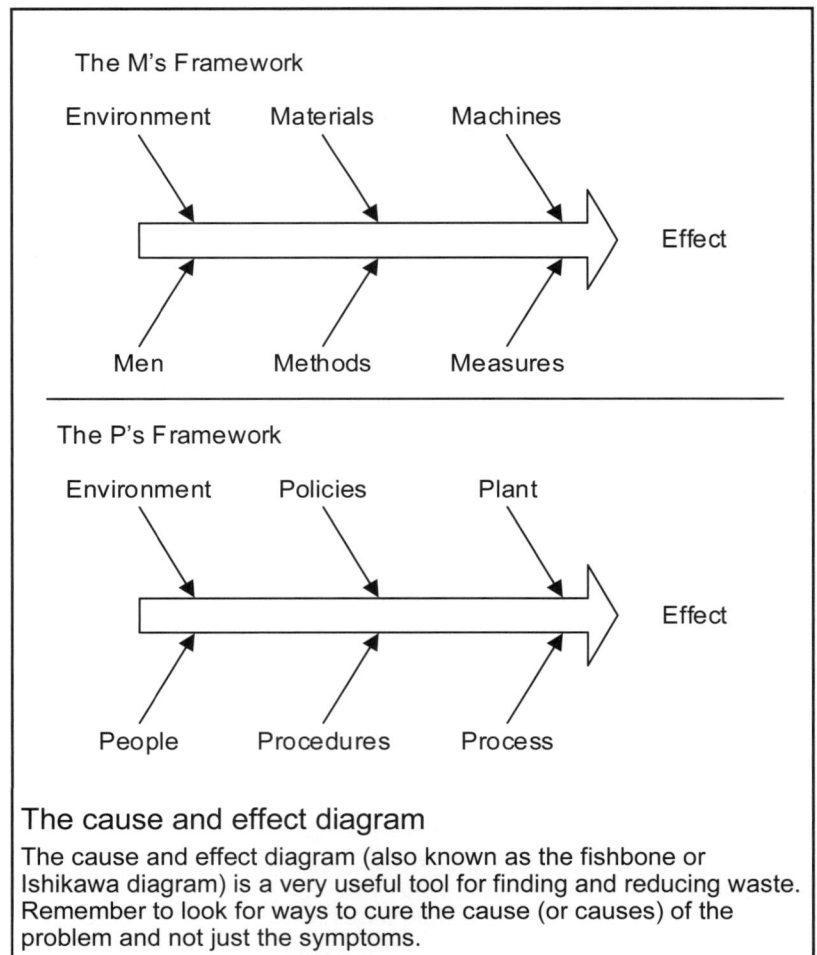

The cause and effect diagram

The cause and effect diagram (also known as the fishbone or Ishikawa diagram) is a very useful tool for finding and reducing waste. Remember to look for ways to cure the cause (or causes) of the problem and not just the symptoms.

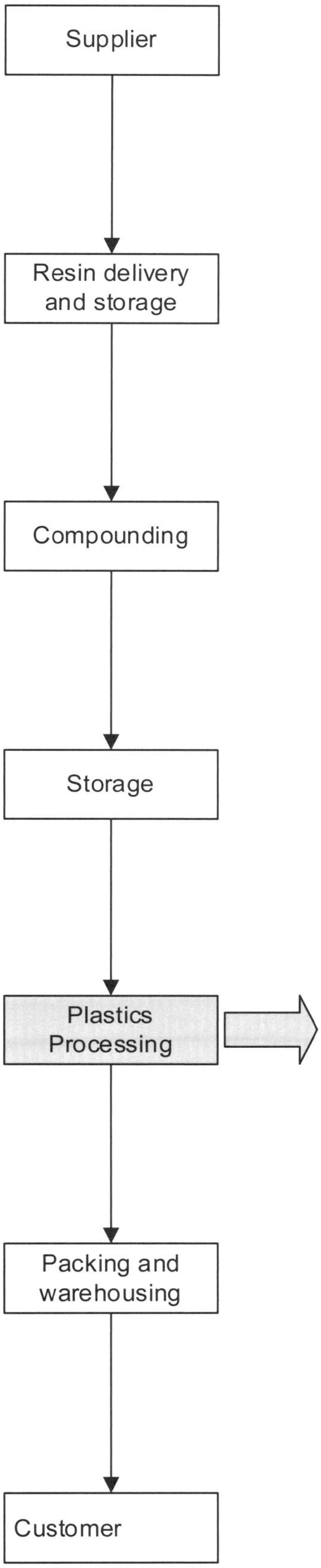

The flow chart (left column)

- Supplier
- Resin delivery and storage
- Compounding
- Storage
- Plastics Processing
- Packing and warehousing
- Customer

The waste tracking model

Name:	Date:	Sheet: of

Process description:

Supplier/input:

Resource/ Material/Utility	Estimated quantity wasted per month	Monthly Cost (include purchase and disposal)	Current waste reduction activities
Start-up scrap			
Sprues and runners			
Rejected product			
Dirty oil			
Filters			
Heat			
Wasted time			
Transit packaging			
Cleaning material			
Drool			
Purge waste			

Total

Customer:
or next process

General waste tracking model for plastics processing

The process flow chart identifies the wastes at each step of the complete process and the waste tracking model starts to put quantities and values on each of the individual wastes for each step. The waste tracking model provides an overview of the potential opportunities for each process step and a total of the potential cost savings for the process step. Combining the waste tracking models for the complete site reveals the potential benefits of a site waste minimisation programme.

9.5 Accounting for waste

Get waste into the accounts

It is entirely possible to implement a good sustainability initiative and waste minimisation programme without using accounts data but good accounts data shows where you are now and how fast you are getting to where you want to be. Enlisting the help of the accountants can pay great dividends because of their ability to highlight discrepancies in data and to present the information to top management. In many cases, the top management are also accountants, so get used to talking to them .

Accounting for waste will involve both management and financial accountants:

- Management accountants will be interested in the collection and presentation of cost data and investment appraisal.
- Financial accountants will be interested in the potential liabilities from putting products on the market and the 'polluter pays' principle.

Sustainability and waste do not respect professional boundaries.

Direct and indirect costs

Unfortunately, when waste is discussed then most people will think of the direct costs such as the skip charges. However, as with quality, most of the costs of waste are indirect and hidden. The concept of the quality 'iceberg' is just as true for waste minimisation and the waste iceberg is shown in the diagram on the right. What we see as waste is only the tip of the iceberg and the rest of the impacts and costs are hidden in the accounts as code words (see Section 9.1) or as general overheads. The hidden costs are much higher than the obvious costs and the total cost of waste is usually about 20 times the first estimate.

Finding and reporting these hidden costs needs the skills of good accountants who are interested in measuring, allocating and controlling costs. Without an understanding of the cost of waste the environmental cost of products and services will not be allocated properly and prices will not be set at the right level.

- **Tip** – The Pareto principle (the 80:20 rule) works for waste minimisation as well:
  - 80% of the waste volume will come from 20% of the products.
  - 80% of the waste impacts will come from 20% of the products (they may not be the same ones).

- 80% of the waste costs will come from 20% of the products (they may not be the same ones).

Make sure that the costs are being allocated accurately and reflect reality.

- **Tip** – Obsolete product is sometimes not scrapped in the forlorn hope that it will be accepted later (after the Christmas party?). It never is, but with a bit of luck we can push it into the next financial year.

Getting the numbers

For most companies, all the information needed for a comprehensive waste report is already being collected but simply not being collated and attributed correctly. The

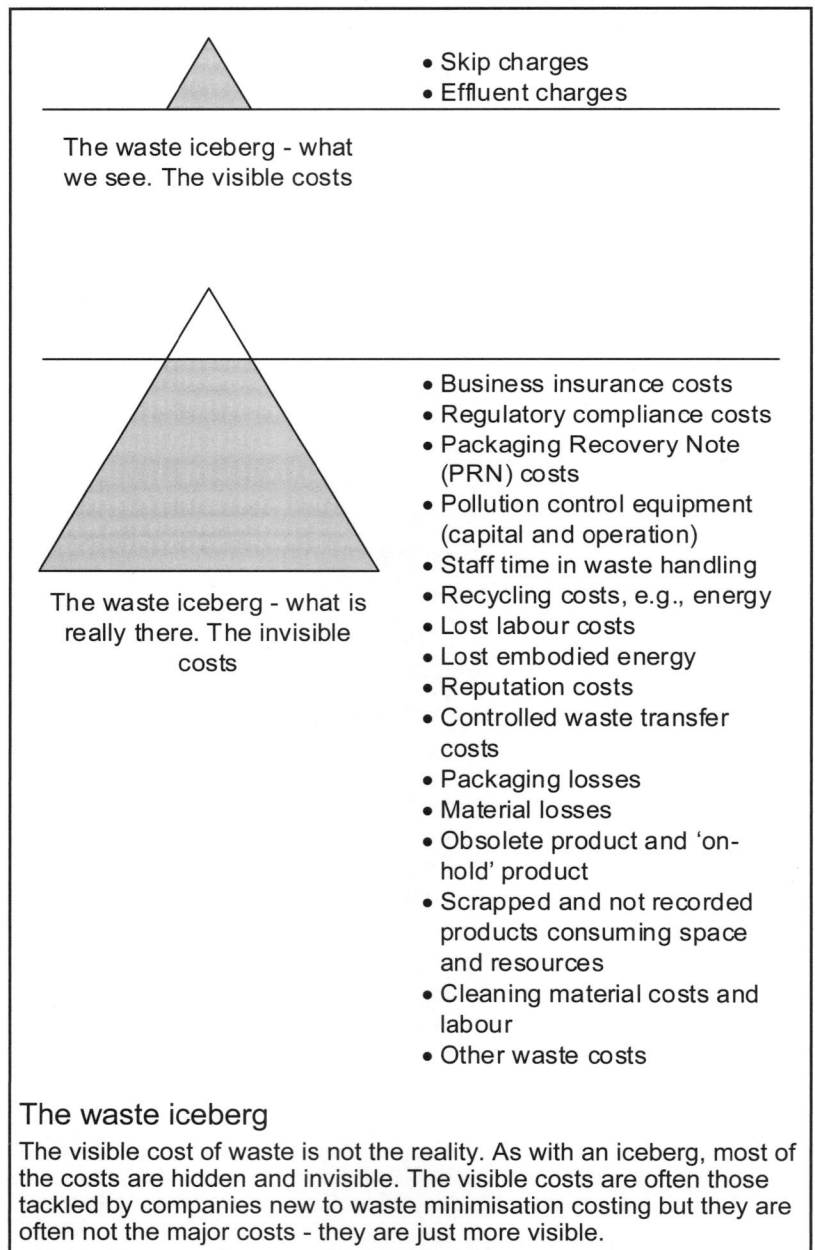

The waste iceberg - what we see. The visible costs

- Skip charges
- Effluent charges

The waste iceberg - what is really there. The invisible costs

- Business insurance costs
- Regulatory compliance costs
- Packaging Recovery Note (PRN) costs
- Pollution control equipment (capital and operation)
- Staff time in waste handling
- Recycling costs, e.g., energy
- Lost labour costs
- Lost embodied energy
- Reputation costs
- Controlled waste transfer costs
- Packaging losses
- Material losses
- Obsolete product and 'on-hold' product
- Scrapped and not recorded products consuming space and resources
- Cleaning material costs and labour
- Other waste costs

The waste iceberg

The visible cost of waste is not the reality. As with an iceberg, most of the costs are hidden and invisible. The visible costs are often those tackled by companies new to waste minimisation costing but they are often not the major costs - they are just more visible.

> Resource efficiency and waste minimisation are everybody's job – accountants cannot hide.

> Record information only once and do this at the point of origin.

indirect and hidden waste costs are simply the part of the waste iceberg that we do not see. There is inevitably valuable information in invoices, purchase ledgers, stock control data, PRN information, consumables use data, production waste records (make sure that they are treating the waste correctly) and a myriad of other data that is already being collected.

If the company has a good set of management accounts and a good management accountant then this can sometimes be a simple process: Sit down with the management accountant and go through the cost codes for each department to identify the direct and indirect costs of waste. This process will give a first pass estimate of the true cost of waste.

The identified costs for each cost code can then be allocated to:

- A general waste report (see below) for a company-wide cost of waste report.
- Specific products or processes to improve costing and accountability.
- Prevention, Appraisal and Failure (PAF) categories as for the traditional Cost of Quality report (see Section 5.6), where prevention costs are the costs of preventing waste (usually < 5% of the costs), appraisal costs are the costs of finding out that the waste was created and failure costs are the costs of cleaning up after the waste was created (usually appraisal and failure are > 95% of the costs).
- **Tip** – At this stage, it is worthwhile taking the advice of the accountants on the most effective presentation method, you might be pleasantly surprised at their enthusiasm.

When collecting the costs, it is important that some of the more intangible costs are included. These are costs such as:

- The cost of labour in scrapped product.
- The cost of consumables in scrapped products.
- The cost of utilities in scrapped products.
- **Tip** – All of these costs will increase as the product moves through the value chain.

These are the truly hidden costs and may need some approximations in the first stages but it is worthwhile trying to estimate these to get close to the true cost of waste (which is not the number you first thought of).

- **Tip** – Get the accountants on board as part of the process of reducing the impact and cost of waste. They know the numbers and can work with them to improve operations.
- **Tip** – When getting the numbers, the accountants should start with the costs and then work back to the physical quantities. This is the reverse of the

method used for the waste tracking model (see Section 9.4) and can act as a cross-check on the numbers.

The waste cost report

The output of the accountants' work should be a waste cost report to concisely report the progress made in waste minimisation. This should report:

- Raw materials use.
- Raw material waste separated by disposal route, e.g., recycled, incinerated or landfilled.
- Energy use (if there is no separate report for energy).
- Water use and effluent disposal (if there is no separate report for water).
- Packaging waste.
- Solvent waste.
- Oil waste.
- Special waste (including emissions to air, e.g., VOCs).

The report should give the quantity of waste, the cost of the material, the cost of the disposal and metrics to assess improvements over time (see Chapter 12).

The waste report is a fundamental output of working with the accounts and will drive improvements in sustainability by reducing waste at all levels.

- **Tip** – The waste report should be linked to the EMS to show continual improvement and should use the EMS data where applicable (see Section 2.5).

Investment appraisal

Waste minimisation is important for investment appraisal of projects to improve sustainability, new cleaner products and process improvements to reduce environmental impacts. Only by understanding the current and future true cost of waste can an effective life cycle cost be calculated to minimise impacts and costs.

Optimise waste segregation and recycling to minimise the amount of waste requiring disposal.

You will never capture everything and not everything is worth capturing.

9.6   Minimising packaging waste – the basics

There is more of it than you think

Packaging can be divided into 3 basic types:

- Primary (sales) packaging – this is the packaging that most consumers see, it is the packaging that is around the goods or products at the point of purchase, e.g., a bottle, a plastic bag or a thermoformed clam-shell.

- Secondary (collation or grouping) packaging – this is the packaging which groups several products together until the point of sale or the use in the supply chain, e.g., a box or container in which products are supplied to the next step in the supply chain. This type of packaging is not generally seen by the consumer and is kept within the supply chain.

- Tertiary (transit) packaging – this is the packaging which allows handling and transport of several grouped secondary packaged items, e.g., the pallet carrying several boxes or containers and any shrink wrapping or strapping used to keep them on the pallet. This type of packaging is not generally seen by the consumer and is kept within the supply chain.

Plastic packaging is a major and visible issue for the world in terms of use and disposal. However, this visible 'sales packaging' is not what is being considered in this section. Although many plastics processors produce primary packaging, the design and production of this packaging is a separate topic. The aim of these section is to minimise the amount of secondary and tertiary packaging used by plastics processors.

Secondary and tertiary packaging are the boxes, plastic bags, containers and all the other packaging used to contain, protect, handle and deliver goods from the processor to the next step in the supply chain.

What does it cost?

The typical cost of secondary and tertiary packaging to a plastics processor can vary between 1 and 15% of turnover depending on the specific operations. This is simply the purchase cost of the packaging and is not the total cost of the packaging. The total cost of the packaging is much higher and some of the additional packaging costs are shown in the box on the right. The total cost of packaging can easily be double the initial purchase cost, i.e., 2 to 30% of turnover.

Most companies have no clear idea of their total cost of packaging and in many cases no idea of even the initial purchase cost of packaging. This is because these costs are simply accepted and included either as part of the product cost or hidden away in the 'overheads' category. This lack of visibility means that there is rarely any focus on how packaging is used and the total impact in environmental or cost terms. The procurement function is focused on reducing the cost of packaging but rarely has the skills or remit to reduce the total amount of packaging used.

A good packaging waste minimisation programme will not only reduce impact on the environment but also reduce the total product cost.

- **Tip** – Compliance with legislation can be very time consuming. Don't forget to try to estimate the time and costs of compliance as part of the cost of packaging.

- **Tip** – Minimising the amount of packaging also benefits the customer because they will pay less for the products and have less packaging waste to dispose of.

The packaging flow chart

A first step in reducing packaging use and costs is to map the flow of packaging through the site. The diagram on the lower right shows a packaging flow chart from the packaging inputs to the packaging outputs. Sites should start with a simple flow chart

Packaging management is a systematic approach that allows the most efficient use of packaging and packaging materials to reduce costs and waste without compromising product protection, handling and storage.

Do not forget to look at the distribution chain for products. Packaging can affect the supply chain as well (see Section 5.17).

Minimising packaging use reduces the environmental impacts and the costs at source.

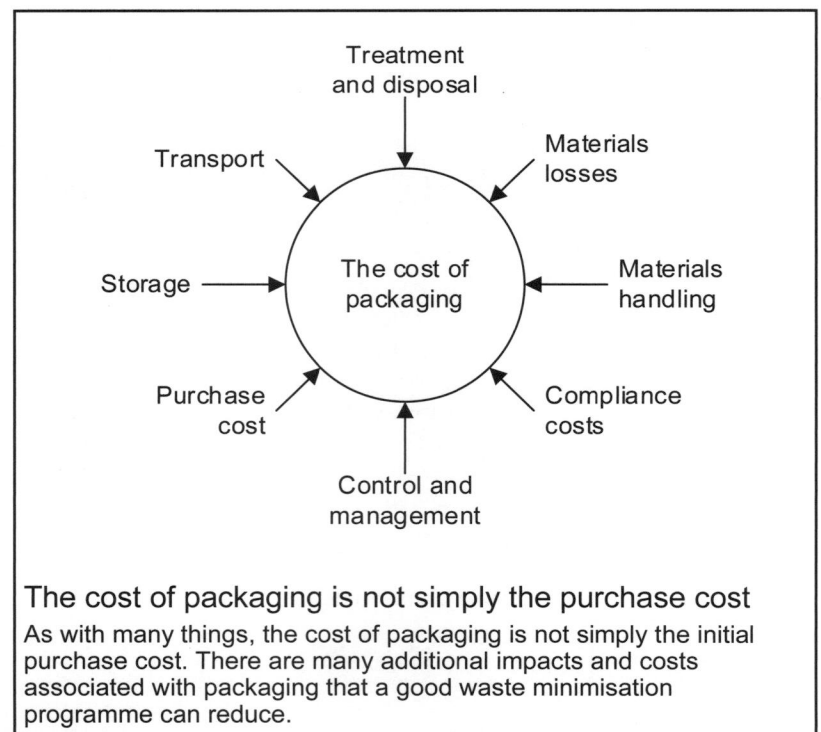

The cost of packaging is not simply the purchase cost

As with many things, the cost of packaging is not simply the initial purchase cost. There are many additional impacts and costs associated with packaging that a good waste minimisation programme can reduce.

and map the supply and use of packaging around the site.

Inputs

The packaging inputs to a site will be:

- Supplier packaging – this is the packaging around the goods that the site buys. It includes all the bags, sacks, pallets and other secondary or tertiary packaging that the suppliers use to get their goods to the site. Some of this will be for production goods or components, i.e., regular purchases, but some will be for irregular purchases.
- Product packaging (new) – this is the new packaging that will be used to protect goods supplied to customers.
- Packaging returned from customers – if the site has implemented a re-usable packaging programme (see Section 9.8) then the packaging returned by customers is included as an input.

Outputs

The packaging outputs from a site will be:

- Product packaging – this is the most visible packaging and should include not just the secondary packaging but also the tertiary packaging such as pallets, shrink wrapping and plastic bags used inside cardboard boxes.
- Packaging returned to supplier – if the site has an agreement to return packaging to suppliers, e.g., octobins, then this is an output from the site.
- Internal transit packaging – some sites use internal transit packaging when a product is produced, stored and later used as part of an assembly. This will generally be re-used multiple times but will eventually fail and become waste packaging for recycling/ or disposal.
- Waste packaging for recycling or disposal – this is packaging that has reached the end of its life, whether non-returnable packaging from suppliers, losses from damaged product packaging or old internal transit packaging.
- **Tip** – Spending time understanding the flow of packaging flows at a site is never wasted because the flow chart starts to show the obvious areas for improvement.

Recording

The packaging flow chart should reveal some of the hidden packaging use but it becomes a more powerful tool if the chart includes the actual amounts of packaging. The data can be expressed in mass values, money values or any convenient value for the site.

It may be easiest to start with the cost accountants and the cost codes to get a first

pass at the values (see Section 9.5) and to later refine these values and improve their accuracy.

- **Tip** – Try to put actual numbers on the flow chart so that the size of the issue becomes clear and quantified.

Packaging technologists can help

Packaging technologists are very skilled at understanding the use (and abuse) of packaging in both retail and industry. They can add real value in understanding and reducing packaging use. Unfortunately, they are not often used in plastics processing and it is unlikely that many processors will have a packaging technologist on the staff.

- **Tip** – If the costs from the packaging flow chart justify it, then consider employing a consultant packaging technologist (with the relevant experience) to review packaging use in the company.

Packaging must be considered for the whole of the manufacturing and distribution system; it should not be dealt with in isolation as this can result in product damage.

Most secondary and tertiary packaging is not really 'designed'. In most cases it is specified and bought the same as it has been in the past because it is not thought of as important.

Less packaging = lower environmental impact = lower costs.

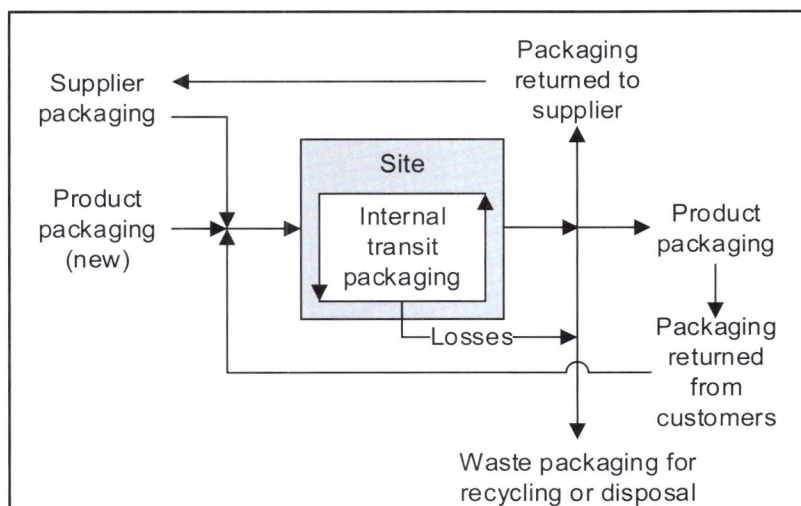

The packaging flow chart

Charting the flow of packaging in the supply chain will reveal areas for improvement. This is similar to the process flow chart (see Section 9.3) but focused on packaging. It should be possible to 'balance' the quantities of packaging used in the flow chart.

9.7   Minimising packaging waste – eliminate or reduce

The first and second steps

Eliminating or reducing consumption at source is the first step but it is essential that the type and quantity of packaging should be matched to the level and type of protection needed, e.g., cleanliness, fragility. Packaging should never be thought of in isolation, it is part of a system designed to get the product to the customer in the correct condition. This does not mean accepting what is currently used but considering how best to achieve this through changes in product design, improved handling, bulk delivery or through changes in on-site handling and distribution methods.

Supplier packaging

The decisions on supplier packaging will largely be made by the supplier on your behalf but these decisions are not waste- or cost-free and it is always worthwhile reviewing incoming packaging requirements and suitability. Select the five largest suppliers to the site and, with the supplier, start to redesign their packaging to minimise handling, transport and product damage.

- **Tip** – Start to talk to suppliers about eliminating or reducing packaging. They probably want to do it too but have never been asked or asked the question.

- **Tip** – If supplier packaging is not returned or re-used then segregate the waste for recycling or disposal (see Section 9.9).

Customer packaging

Product packaging should be an integral part of the product design process (see Section 3.7) and not an afterthought. The packaging choices made during design should provide protection and ease of handling during production and distribution. This means understanding:

- The product 'fit' into the packaging.
- How the packaging works in production.
- How the packaging is stored and distributed.
- If the packaging materials are fit-for-purpose and have minimum environmental impact.
- If it is possible to standardise the packaging to reduce stock and increase order size.
- If the packaging will be returned or re-used by the customer (see Section 9.8).

Over packaging gives unnecessary layers, additional handling, waste and increases

costs. We have all had products delivered that were grossly 'over packed', is this also the case with the products delivered to our customers?

- **Tip** – Start to talk to customers about eliminating or reducing packaging. They probably want to do it too but have never been asked or asked the question. They could well turn out to be the greatest supporters.

Packaging to eliminate or reduce

The variety of packaging makes it difficult to specify the packaging that can be eliminated or reduced but consider the following:

Bags and sacks

Bulk deliveries direct to silos or returnable octoboxes are preferable to bags and sacks for raw materials. They reduce the potential for spills and have little residual packaging.

Drums and IBCs

Steel drums can be replaced by plastics drums which are lighter, long-lasting and recyclable and, for some materials, IBCs (Intermediate Bulk Containers) are a good compromise between drums and tankers.

Boxes and cartons

Boxes and cartons are extensively used for small items. Sites need to ask:

- Are they necessary if shrink-wrap can give the same protection but is only 30% of the weight?
- Are they of the optimum design and are the flutes oriented correctly to provide strength in the needed direction?
- Is the box space fully used with minimum head space? (see Section 5.17)
- If plastic bags are used to prevent the product touching the cardboard then is the film gauge minimised?

Good packaging design means considering the format of the container, the collation unit, the size and shape of container and the type and grade of material used.

Assess replacing virgin packaging materials with recycled materials. This cuts waste to landfill, preserves finite resources and helps to stimulate the market for reclaimed materials.

Even if you are not obligated under packaging waste regulation to report packaging use then a packaging waste minimisation programme can help both sustainability and profits.

It doesn't matter how big or small your company is, the benefits can still be significant.

Hole-in-the-wall

The industry has already made great strides in eliminating packaging using 'hole in the wall' systems. These are factories (EBM milk bottles or PET drinks bottles) which are co-located with the foodstuff producer (milk or drink). They produce the packaging and feed it through a 'hole-in-the-wall' directly to bulk storage or the filling line. There is no secondary or tertiary packaging in the supply chain. The container is produced and sent to the filler with no human or packaging contact. There is still secondary and tertiary packaging between the filler and the consumer but no packaging inside the supply chain. Packaging eliminated, transport eliminated and sustainability increased.

Filler material

If boxes are not completely full then filler materials may be used. Sites need to ask:

- Is it possible to reduce or eliminate fillers by changing the box design?

If separators are used then orient them correctly to provide strength in the needed direction (opposite to the box itself) and they should be easily re-used or recycled.

Box closures

If boxes are used then they must be closed and sealed and sites need to ask:

- Is it possible to avoid using staples? Staples make boxes difficult to open without damage and reduce re-use or recycling potential.
- Is it possible to avoid using adhesive tapes (or to minimise the size)? Adhesive tape can also make boxes difficult to open without damage and reduce re-use or recycling potential.
- If tape must be used then is it being used effectively?
- Is it possible to use strapping to close the box and prevent damage on opening?

Pallets

If pallets are used then sites need to ask:

- Is the pallet to the standard ISO footprint for easy re-use by the customer or return?
- Is the pallet of the right quality for the application, i.e., single trip or multi-trip?
- Is the load stability compromised by non-standard box sizes?
- Is load stability retained by using stretch-wrap?
  - Would strapping be better?
  - Would layer separators stabilise the load more effectively?
- Is it possible to eliminate pallets entirely?

Shrink wrap

Where shrink wrap is used then sites need to ask:

- Is the shrink wrap needed and effective?
- Can the shrink wrap thickness be reduced, e.g., from 40µ to 20 µm, and still provide protection and stability?
- Can the amount of shrink wrap be reduced?

Collation trays

Collation trays may be used for internal transit or shipping to the customer and sites need to ask:

- Are collation trays necessary and how much do they cost?
- Can shrink wrap can be used instead?

- Can the format or quantities be increased to decrease packaging use?

Variation and standardisation

Sites need to ask:

- Is it possible to standardise the outer packaging and to use liners for different products?
- Is standardisation causing issues with pack volumes and wasted space?

Work practices

Supplier packaging can only be returned for re-use if it is in good condition. Sites should:

- Assess unpacking difficulty, if it is difficult and damages the packaging (or the contents) then inform the supplier.
- Treat packaging carefully, if damaged then the potential for re-use decreases greatly.

The benefits

Eliminating or reducing packaging can have great benefits by reducing use, environmental impacts and costs. Sites should examine all packaging for the potential to eliminate or reduce it.

Keep records of packaging received. It is often a legal requirement and it starts to highlight the issue of packaging waste minimisation.

Keep records of packaging re-used. This can also help with legal requirements and reporting.

Involve suppliers and customers in discussions about how to reduce packaging.

Discussing packaging elimination or reduction with suppliers and customers is often the quickest way to start to process of eliminating or reducing packaging.

This is a team effort and the benefits go both ways.

Eliminate or reduce

Some of the packaging items to look at for elimination or reduction are:

- Adhesive tapes
- Air-tight sealing films including shrink and stretch-wrap
- Box lines (plastic)
- Clips and staples
- Containers, boxes, cartons and crates
- Cores, reels or tubes
- Disposable cups, trays, plates, dishes etc.
- Drums and IBCs
- Edge protectors
- Envelopes
- Fillings, e.g., paper, polystyrene and air bags
- Glue containers
- Kraft paper tape
- Inks
- Labels or wing tickets

- Layer pads or padding
- Non-hazardous process chemical containers
- Oxygen absorbers or silica gel
- Pallets, trays or slip sheets
- Pins
- Rubber bands
- Ribbons
- Sacks, bags or sachets
- Screw-topped bottles
- Seals
- Separable lids or caps
- Separators, dividers or interleaves
- Shrink film
- Strapping
- String, thread or twine
- Top frames
- Wrapping materials

9.8 Minimising packaging waste – re-use

The third step

If it is not possible to eliminate or reduce packaging then re-using the packaging either in a closed loop or at some other part of the supply chain can be an effective method of reducing environmental impacts, i.e., re-usable packaging can be in a dedicated closed loop from the site to the customer and back or for re-use by the customer for onward delivery with the savings shared.

Closed-loop systems are the most economical and effective in reducing environmental impact and dedicated re-use systems are increasingly common. These are particularly effective for short distances but are less effective for overseas or one-way trips where lightweighting is a better solution.

Sites should assess the opportunities for re-use of their existing packaging and ensure that any packaging materials sent to customers is re-usable.

- **Tip** – Not all packaging is suitable for re-use and existing packaging needs to examined to see if it has the potential for re-use and if it can be redesigned to make it suitable for re-use.

- **Tip** – Re-usable packaging is only beneficial if it is re-used, the initial impacts and cost of re-usable packaging will probably be more than for single use and the number of 'trips' is critical in justifying re-usable packaging.

Design for re-use

In many cases, packaging can be redesigned to make it suitable for re-use in terms of functionality and operations. The rules are:

- Design simple and standardised outer packaging to maximise flexibility in use.

- Design packaging that can be multipurpose, e.g., PVC-U profile stillages (see box on the right) are used throughout the supply chain, stillages for film rolls can also be used throughout the supply chain and secondary collation packaging can be designed to serve as display packaging in stores.

- Design packaging to stack, collapse or nest when empty to minimise storage space and transport costs, e.g., PE-HD corrugated tote boxes can be folded flat to have minimal volume on the return journey.

- Design packaging for easy cleaning and maintenance, e.g., keep it simple and avoid dead space or areas that are difficult to clean.

- Design packaging to include labelling and identification features, e.g., logos and other relevant information such as instructions for use. These should be simple and easy to read for the life of the packaging.

- Design packaging to reduce manual handling and health and safety concerns, e.g., packaging that can easily be handled by fork lift trucks and transported at the customer's site.

- **Tip** – As with any packaging issue, packaging technologists and designers can often help. They will have experience in designing re-usable packaging.

Modifying existing packaging

In some cases, rather than design or source new packaging, it is possible to modify existing packaging to be re-used. This usually requires adding components to improve the strength of the packaging or to reduce damage and contamination of the packaging. Simple modifications can be:

- Increasing the thickness or gauge of the packaging to increase the service life.

- Adding protection (physical or finishes) to reduce damage to the packaging.

> **Think laterally**
> It may be possible to sell or give away packaging arriving at your site to encourage re-use.
> It may even be possible to buy used packaging for tertiary use.

> Keep all packaging clean and uncontaminated and segregate at source to encourage re-use.

Re-usable PVC-U profile stillages

The PVC-U window profile industry re-uses packaging extensively. All window profiles are packed directly into stillages at the end of the extruder. These are used for warehousing at the site, transport to the customer and by the customer as bulk storage and fed directly from the stillage to the saws at the start of the production line. The stillages are collected by the delivery lorry and returned to the extruder. This is a major logistics exercise and the stillages are quite costly but the whole distribution chain works around the tertiary packaging and the value it adds is enormous.

This works because of the closed-loop nature of the packaging chain.

- Using the existing packaging with new re-usable packaging providing protection.

Modifying existing packaging is generally regarded as being less effective than designing new containers dedicated to re-use.

- **Tip** – Whether packaging is designed or modified for re-use, it is important to provide customers with information about the packaging and the materials used. This will help them to handle, re-use and eventually recycle the materials. Providing this information is often a legislative requirement.

- **Tip** – Do not think only about re-usable packaging in terms of a closed loop, it may be possible for the customer to re-use the packaging in their supply chain after it has been delivered to them. This is already done for articles such as pallets and is not only a waste reduction but can also be an income stream.

Managing re-use

Re-usable packaging has many benefits but it is not cost free and needs more management than single-trip packaging, i.e., this is about managing the distribution and logistics process. Before introducing re-usable packaging, a company must assess the costs and plan the logistics of the re-use cycle. This requires determining:

- Who will manage the logistics, i.e., is it going to be the supplier, the customer or a logistics company?

- Who is responsible for returning the packaging, i.e., the supplier or the customer?

- How the packaging is going to be returned or backhauled, i.e., is it going to be collected and returned during routine deliveries?

- How return is going to be enforced, e.g., deposits, invoice at cost if not returned etc?

- **Tip** – When using stillages for PVC profile we found that the customers were using them for all sorts of things and had to 'encourage' them to return them.

- Is individual packaging going to be tracked to allow enforcement?

- How is the packaging going to be checked and cleaned before re-use?

- How are the benefits of re-usable packaging going to be shared, i.e., will there be lower prices for products supplied in re-usable packaging?

Whether considering re-usable packaging for suppliers or customers, there must be internal and external consultation to answer these and similar questions and to ensure that the concept is viable and has widespread support. This consultation must involve the customer to clarify the issues, create ownership and to provide the right incentives to make the system work.

- **Tip** – Think about all levels of packaging for re-usability. This is not just for secondary packaging, e.g., will plastic pallets last longer and be more economical than conventional wooden pallets?

- **Tip** – A logistics company may be able to improve operational performance with a properly structured contract.

- **Tip** – Even 'single trip' packaging can often be suitable for several trips before recycling if it is suitably checked and maintained before re-use.

Working with suppliers

Suppliers are often enthusiastic about re-using packaging, especially if it can be returned with new deliveries and reduce costs.

Select the five largest suppliers and ask them about the potential for re-usable packaging that can benefit both parties.

Tip – Start to talk to suppliers about re-using packaging. They probably want to do it too but have never asked the question.

Working with customers

As well as suppliers, customers are often enthusiastic about re-using packaging, especially if it can be returned with new deliveries and reduce costs. For some customers this can be an additional positive feature if the packaging can be labelled to demonstrate their 'green' credentials.

Working with customers and arranging for the re-use of packaging can benefit both parties and build stronger relationships as it becomes part of future contracts and policies.

- **Tip** – Start to talk to customers about re-using packaging. They probably want to do it too but have never asked the question.

- **Tip** – Identify the materials used on the packaging itself, to help your customers to re-use or recycle.

Re-usable packaging is used extensively in the automotive industry where it serves not only as packaging but also as 'just-in-time' storage delivered direct to the production line.

Re-usable packaging is particularly suitable for high-value items that are being used on the customer's production line. It provides protection, stock control and can be integrated into the production line.

Do you re-use any packaging at all? If it all comes in and either goes out to customers or recycling then you have missed the point of the circular economy.

Discussing packaging re-use with suppliers and customers is often the quickest way to start to process of re-using packaging.

This is a team effort and the benefits go both ways.

9.9 Minimising packaging waste – recycling, recovery and disposal

The last steps

If packaging waste cannot be re-used then it has already lost most of its value. Recycling, recovery and disposal are used to try to minimise the environmental impact of the material and to minimise the economic loss that has already occurred. Whilst recycling and recovery may lead to a notional income stream, it is a fraction of the original cost and value of the packaging and is more 'damage limitation' than a true income stream.

It is worthwhile considering all packaging waste as a potential source of income. You will never get the full value back but you can minimise the loss.

Segregate it and keep it clean

As with re-use, the first step in recycling, recovery and disposal is to make sure the waste is both segregated and kept clean. Segregation of packaging waste can significantly reduce the cost of recycling and disposal and, in the UK, can also reduce the cost of Packaging Recovery Notes (PRNs).

Segregation needs good identification of the material and keeping waste clean demands discipline in waste handling. Mixed material packaging will be more difficult to segregate. Suppliers should be encouraged to use single material packaging or, if mixed materials must be used, then should use methods of easily separating the various materials for recycling (this also applies to labels).

Suppliers should also be required, as part of the purchase contract, to provide details of the packaging materials used to allow easy segregation and recycling. Sites should follow these instructions to minimise the amount of general packaging waste generated.

- **Tip** – Packaging segregation should be as close to the source as possible, i.e., remove packaging on delivery and segregate there. It is much easier at source than later.

- **Tip** – Contact suppliers if there is not enough information to segregate easily and get them to mark it on the packaging.

- **Tip** – Think about your own packaging. If you have issues with recycling mixed waste then consider the effect of your packaging on your customers. Design your packaging to be easy to recycle, e.g., made from a single material.

- **Tip** – Think about your own packaging. Is it clearly and correctly marked to help your customers recycle it?

Flatten, bale, shred and crush

Whether the packaging waste is destined for recycling or disposal, volume reduction is an important issue and many waste handling firms base their charges on volume. Sites should take every opportunity to:

- Flatten cardboard boxes to reduce space used and keep them segregated.

- Use a baler, compactor or shredder to reduce the volume of segregated waste before recycling or disposal. A typical baler or shredder can reduce waste volume by 80% as well as making handling much easier.

Reducing the volume of waste may not reduce the environmental impact but it will reduce the number of collections needed and the cost of waste disposal.

- **Tip** – 'Dumpster diving' (also known as 'looking in the skips') can be a quick check for adequate segregation. It will also reveal if things are being thrown away that can be re-used.

Recycling

Recycling is only used after the potential for re-use has been exhausted and will preserve the value of the raw material. Recycling is well developed in many parts of the world and segregated waste can easily be recycled. Instead of using a large skip and unsegregated waste, sites should use at least three smaller skips for the most common recycling streams, i.e., cardboard, metals, glass and plastic (head waste etc. that cannot be recycled in-house). It is probably also worthwhile to have a separate collection for waste paper (shredded for security), a skip should not be necessary – if it is then you are generating too much waste paper.

Choosing the best type of skip can help with recycling. Traditional open skips are not always the best solution for segregating and recycling, it is possible to use covered front-end-loader skips (FEL), rear-end-loader skips (REL) or large 'wheelie bins' that can be emptied directly into the collecting vehicle to reduce handling. This is often cleaner and cheaper than removing and replacing traditional skips.

Some recyclers will offer a 'mixed' recycling contract where paper, cardboard, plastic bottles and small metal items, e.g., drinks and food cans, can all be placed in a single container and sorted at the recycler's Materials Recovery Facility (MRF) but this will inevitably cost more to recycle than segregated waste.

Waste and packaging regulations vary around the world.

Get to know what the regulations are in your area and comply. Most countries have quite severe penalties for breaches of the regulations.

Start a discussion about waste packaging with the staff.

They know the operations and will have suggestions for improving the handling of waste packaging.

This also increases their involvement in minimising waste.

Some recyclers will also offer to complete the Duty of Care/Transfer notes required by legislation instead of the site having to do this.

- **Tip** – Before recycling, consider if the waste can be re-used by another local company via a waste exchange scheme.

- **Tip** – Colour code skip areas and clearly label the areas with the type of material being collected. Do not allow contamination of the waste stream to devalue the material.

- **Tip** – It is always worthwhile shopping around to check that the site has the best recycling contractors as recycling is highly legislated in most countries. Always use a licensed contractor and check their license. The environmental impact, penalties and reputational damage for failing to get it right are severe.

Recovery

When the potential for recycling has been exhausted then it is possible to consider energy recovery as a method of diverting materials from landfill. Most packaging waste has a high calorific value and if there is sufficient waste then it may be viable to consider EfW and use an incinerator/boiler or CHP unit (see Section 6.24) to recover the embodied energy of the material.

This will recover some of the embodied energy but the efficiency of waste incinerators is generally low and this must be considered in the calculations. EfW produces only a fraction of the energy that can be saved by recycling and the resource is destroyed forever, i.e., it is always better to recycle. The use of an incinerator may also involve additional requirements in terms of local environmental regulations.

Incineration should not be regarded as moving towards 'zero waste' (however it is defined) as there is still residual ash that must be disposed of.

- **Tip** – If there is sufficient waste to make an incinerator or CHP unit viable then perhaps it is better to review the work done on reduction, re-use and recycling.

- **Tip** – It may be possible to collect waste from other sites in the area (not necessarily plastics processors) to increase throughput and improve the economics.

- **Tip** – It may be possible to dispose of wastes by giving them to other companies for recycling or recovery.

Disposal

Disposal is the last, and most costly, resort for dealing with packaging waste. If a site has implemented all the previous actions, then it is possible that there will be no need

for disposal of packaging waste. If there is any residual packaging waste then this must be disposed of with the least environmental impact and cost.

This means that skips must be used effectively, i.e., use the largest and most appropriate skips possible and use baling, crushing or compacting to minimise uplifts.

- **Tip** – It is always worthwhile checking that the site has the best waste disposal contract but waste disposal is heavily legislated in most countries. Always use a licensed waste contractor and check their license. The environmental impact, penalties and reputational damage for failing to get it right are severe.

- **Tip** – Always visit the waste contractor's site. It provides confidence that they are doing the right thing and can generate ideas about what else the contractor can do for you.

Rotate skip use so that only one of each type is in operation and, when this is full, close it and open a new one.

Make sure that the only skips lifted are full ones to reduce costs.

If in doubt about recycling, recovery and disposal, contact your waste recycler/disposal company. They are the experts in this.

If they are not (or if they are not willing to talk) then get a new recycler/disposal company.

9.10　Minimising oil and hydraulic fluid waste

An essential component

Every plastics processing site will use oils for lubrication and most will also use hydraulic fluids for power transmission. These are different functions (see box on the right) but the waste minimisation actions are very similar and from this point forward we will refer to them generically as 'oils'. A good oil waste minimisation programme can lead to savings of between 10-40% in use and better handling of oils at the end of their life.

Oil management

Understanding the current situation is the key to minimising oil use and managing the waste minimisation programme. Most sites will have a wide range of oils present and many of these will be specified by different supplier numbers but will be almost identical in functional and operational properties. The existence of the two different systems for oil specification (see box on the right) based on different test temperatures also makes management difficult for non-specialists, e.g., ISO VG 220 is the equivalent of SAE engine grade 50 and SAE gear oil 90 but this is not clear to the non-specialist and it is possible that all three grades are kept at a site.

This means that sites should seriously consider using a single supplier or agent for all oils and working with the supplier to rationalise the oil holdings. As with most stocks, the Pareto Principle will apply and 80% of the use will be in 20% of the stocks. Using a single supplier provides:

- The potential for rationalisation and consolidation in the number of oils held, i.e., elimination of duplicate oils. Oils can be selected based on the viscosity grade required (ISO or SAE) because the supplier is the expert and will know about grade equivalence.

- Improved stock management with oils replaced and managed to avoid 'use by' dates and wasted oil.

- Improved stock levels and monitoring of oil use.

A reduction in the number of oils held will:

- Lead to improvements in housekeeping.

- Allow larger containers and less waste, e.g., IBCs can be used instead of drums to almost eliminate bottom waste in drums.

- Improve storage and handling, e.g., First In First Out (FIFO).

- A reduction in manual handling to get at the container or the oil.

- Improve the choice of oils, i.e., fewer oils mean less choice, less mistakes and reduced waste.

Using a good supplier can allow a site to contract out oil management and reduce environmental impact and costs.

- **Tip** – Oil should never be allowed anywhere near the drains. The penalties for emissions are justifiably severe. All new and used oils and hydraulic fluids should be kept in bunded areas to prevent accidental loss to the drains.

- **Tip** – Emergency spill kits should be easily accessible and clearly labelled. In some sites, the machines are mounted inside steel bunds (a metal 'tray' about 3 cm deep)

Reducing oil waste is a 'cost avoided' rather than a specific saving.

High oil use on a machine indicates bigger issues than waste minimisation, i.e., maintenance issues that may lead to machine failure.

Oils and hydraulic fluids – a first primer

Most sites will use both oils and hydraulic fluids and these have different functions:

- Oils – these are primarily used for lubrication. e.g., the oil in the sump of a car is there to lubricate the metal-to-metal contact in the engine, it is not used to transmit power.

- Hydraulic fluids – these are used primarily for power transmission, e.g., the hydraulic fluid in a hydraulic injection moulding machine is there to transmit power from the pump to the operating parts of the machine, it also lubricates the moving parts but this is not the primary function.

Most sites will also use greases – these are also primarily for lubrication but have a much lower viscosity and are 'stickier'. They can not only lubricate but also act as a barrier.

Both oils and hydraulic fluids can be mineral oils (primarily manufactured using distillation of hydrocarbons) or synthetic oils (which also primarily use hydrocarbons but these are modified rather than simply distilled). In either case, the main property that is used to specify the material is the viscosity.

Viscosity is the resistance of a fluid to deformation by shear stress and in non-scientific terms it is a measure of the 'thickness' of the fluid:

- High-viscosity fluids are 'thick' fluids, e.g., honey or treacle.

- Low-viscosity fluids are 'thin' fluids, e.g., water or petrol.

For oil and hydraulic fluids, the viscosity measured is the 'kinematic viscosity' which is the ratio of the dynamic viscosity to the density of the fluid. For most fluids, and particularly for hydraulic fluids, the viscosity will decrease with increasing temperature, i.e., the fluid will get thinner and flow more easily.

The measure of how much the viscosity will change with temperature is the viscosity index (VI) which is an arbitrary measure of the change in viscosity with temperature of a fluid:

- A low VI means a high change in viscosity with temperature.

- A high VI means a low change in viscosity with temperature.

One issue with specifying oils and hydraulic fluids is that there are two competing specifications for oil although both are based on the viscosity. These are the ISO grade system which is based on the viscosity at 40°C and the SAE grade system which is based on the viscosity at 100°C (or at 0°C if it is a W for winter grade). There is no direct comparison between the two systems (viscosity changes with temperature) although there are many 'conversion charts' available.

---

that captures any fluid leak. This is good for preventing escape to the drains but should not be treated as an excuse to ignore leaks on the machines.

- **Tip** – We have been to sites where they never change their hydraulic fluid. This is simply poor management. Oil and fluids degrade with age and use. Regular testing and replacement are essential.

Reducing oil use

Reducing the number of oils and grades present at a site, whilst staying within the manufacturer's guidance, will reduce oil losses from bottom waste and make the complete site more efficient in oil use.

Sites need to carry out an oil survey which documents:

- Oil use by type and grade in litres/year or similar.
- The application or applications of the oil.
- The location and access of the filling ports for each application.

This is the basic document for rationalising and consolidating the oil stocks held at the site.

- **Tip** – Clearly mark all machines with the type of oil to be used beside the filling port and check access for the filling port. This should be updated after rationalisation of the oil stocks.

Many sites will replace the oil at regular intervals without knowing if the oil needs replacement. This is even though most suppliers offer low-cost oil testing services to analyse if an oil needs replacement. Testing and condition monitoring can extend maintenance periods if the oil is acceptable for continued use. It can also tell a site when oil needs to be replaced to prevent damage to the machine.

- **Tip** – The frequency and type of oil testing to be carried out should be added to the oil survey for each application.
- **Tip** – Maintain equipment regularly to prevent oil leaks. Use a drip tray where oil leaks still occur but preferably fix the leak!

Re-using oil

During normal use, oil will pick up contaminants such as solid particles, sludge, varnish, water and become aerated. Leaving these in the oil system can lead to damage to machines and oil oxidation that requires the oil to be replaced more often. Machines may be fitted with conventional oil filters but these will not remove most of these contaminants. Cleaning with advanced systems, e.g., www.triple-rrr.com, will often allow cleaned oils to be re-used for their original purpose but, if in doubt, they should be tested by the supplier.

Recycling oil

It is often possible to recycle oil by either laundering or by re-refining but these are specialist processes that need to be undertaken by an off-site recycler. Using a single supplier means that the supplier can undertake to recycle the oils where possible.

Disposal

Oils that have reached the end of their life at the site should be kept separate from oil that is to be used in production machines to prevent accidental use. This storage should not be in the oil store but should also be a bunded area.

Different grades and types of oils should never be mixed and drums should be clearly labelled to show the contents.

- **Tip** – Disposal is a specialist activity that can normally be provided by the oil supplier who can remove old oil when delivering new oil.

High quality filtration and deaeration can be off-line to periodically remove contaminants or in-line to continuously remove contaminants.

Bunds

- Bunds should be a minimum of 110% of the largest bulk storage tank or 25% of the total product stored in the bunded area (whichever is the greatest.
- Regularly check the condition of drums and IBCs in the bunded areas for corrosion and/or leaks.
- Regularly check that the bund itself can provide the required isolation.

Good practice for oil storage:

- All oils have an allocated space, are in that space and are visibly and clearly labelled.
- Similar oils are grouped together.
- Stock control is based on FIFO to give stock rotation and 'use by' date control.
- All oil issued is recorded.
- All drums are stored horizontally.
- All drums or IBCs have taps or pumps fitted.
- All drums or IBCs have drip trays under taps or pumps.
- Where oil is to be transported in drums then drum transport devices are available.
- Oil storage areas are bunded and segregated from other areas.
- Spill kits are available and in good condition.
- Spills and leaks are promptly cleaned.
- Old and potentially waste oil has a separate bunded storage area to segregate it from good stock.
- Used spill kit materials are stored with old and potentially waste materials.

9.11 Minimising solvent waste

It doesn't just evaporate

Not many plastics processing sites use significant amounts of solvents but those that do will find that they are both expensive and difficult to dispose of. Solvents such as acetone, hexane, ethanol, methyl ethyl ketone (MEK) and trichloroethylene (TCE) are generically known as Volatile Organic Compounds (VOCs) and these will all evaporate at room temperature and can escape to the atmosphere. VOCs are recognised as having health and safety and environmental impacts and are strongly regulated throughout the world.

In plastics processing, most sites will use very small amounts of solvents (primarily for degreasing) but some sites will use large amounts for painting, printing or in some adhesives. For any site using large amounts of solvents, it is essential to map the solvent use at the site (see box on the right) to establish where solvents are entering the site, where they are being used and how they are leaving the site (controlled or otherwise).

- **Tip** – It may be necessary to break solvent use down to individual processes to have the detail needed for effective action.
- **Tip** – Solvents are expensive to purchase and expensive to dispose of, good solvent management can also reduce costs.

Eliminate or reduce

Eliminating solvent use by process or material changes is the obvious first step as this removes any need to deal with solvents and VOCs.

In the case of paints and printing, solvents can be eliminated by moving from solvent-based paints to either water-based paints or solid UV-curable inks.

Water-based paints have been used successfully for automotive applications for many years and this not only reduces solvent use and VOCs but also allows for easier cleaning of paint guns and reduced paint loss caused by evaporation/drying out of paint before it is fully used.

Solid UV-curable inks have also been used for many years and these not only reduce solvent use and VOCs for printing on plastic parts, e.g., tubes and containers, but can also significantly reduce the plant size for printing.

In the case of metal cleaning and degreasing for insert and outsert moulding, MEK and TCE have successfully been replaced with aqueous systems at many sites.

There may be some extra costs in replacing solvents but these will often be more than balanced by reduced health and safety issues and environmental charges.

- **Tip** – Solvents can often be eliminated by using detergent, warm water and regular cleaning.
- **Tip** – Citrus based (non-VOC) cleaners can be used to eliminate VOCs from print equipment cleaning.

In many cases, eliminating solvent use is easier, in process terms, than reducing it. Elimination involves a new process and potentially high capital investment but will remove all the environmental issues and costs associated with solvent use.

Reducing solvent use is often more difficult because of the original process design. Improvements will therefore tend to be smaller and more incremental than transformational, i.e., they will be centred around small process or management improvements and the environmental impacts and costs will remain high.

Improving solvent management to reduce use will focus on good housekeeping issues such as:

- Improving container size to reduce leakage and use.
- Improving storage conditions.
- Improving and monitoring transfer and transport mechanisms to minimize leakage and evaporation.

> Solvents are hazardous and their use is strongly controlled throughout the world.
>
> Disposal is also difficult as they are generally classed as hazardous waste.

> Not all VOCs are man-made, many VOCs are naturally occurring compounds and most plant and animal scents and odours are VOCs.
>
> All smells, good and bad, are the result of molecules interacting with your sensory system, i.e., all smells are molecules and many are VOCs.

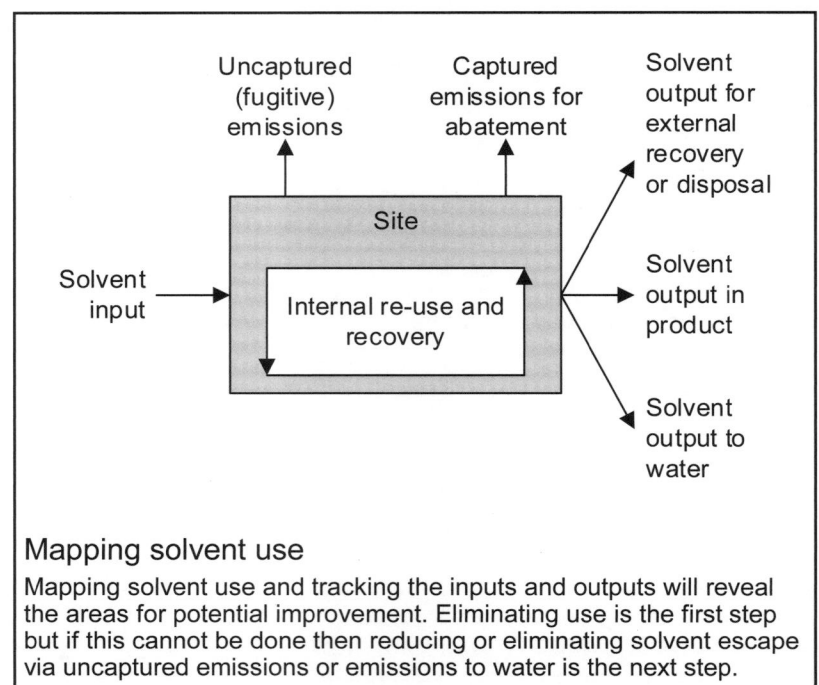

Mapping solvent use

Mapping solvent use and tracking the inputs and outputs will reveal the areas for potential improvement. Eliminating use is the first step but if this cannot be done then reducing or eliminating solvent escape via uncaptured emissions or emissions to water is the next step.

- **Tip** – If solvents are stored in drums with breather valves, then consider using conservation valves to reduce the losses.
- **Tip** – Waste mapping should include solvents and, where used, they should be on the process flow chart (see Section 9.3).

Process improvements to reduce use should focus on reviewing the process design and operation for potential process improvements to minimise VOC use and fugitive emissions.

Re-use and recovery

Solvent re-use, recovery and disposal are complex topics that are beyond the scope of this book and are only briefly reviewed here.

It is often possible to recover solvents for re-use by recovery processes (see box on the upper right). Recovery reduces the loss of solvents to the atmosphere, the need to operate abatement plants and the purchase of new solvents. The choice of the appropriate technique (absorption, adsorption, condensation or membrane) depends on factors such as:

- The properties of the air stream that contains the solvent.
- The solvent being recovered.
- The process using the solvent.

Recovered solvents can be re-used in-house if needed or sold back to the manufacturer for re-use in the original products containing the VOC.

- **Tip** – If solvent is collected for off-site recovery or disposal then keep the waste separate to allow accounting for the waste flow.

Disposal

Disposal can take place on site using one of the available destructive techniques and an abatement plant (see photo on the lower right). These break the VOCs down into CO_2, and H_2O, they are expensive to install and expensive to operate even if heat recovery is used to reduce the operating cost.

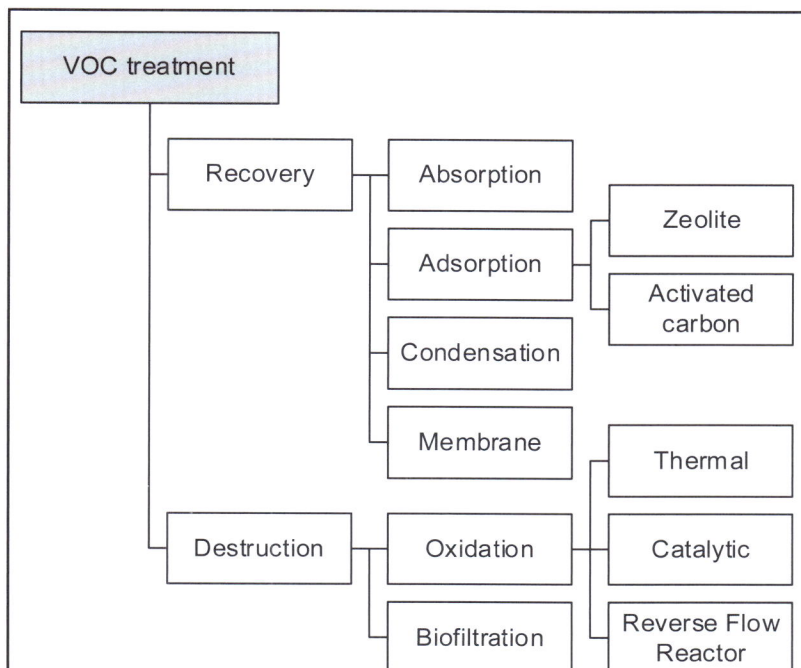

VOC abatement techniques

If VOCs cannot be removed from the process then there are a range of VOC abatement techniques. Recovery is preferred but is not always possible. It depends on the solvents used, the concentration of the solvents and the flow rate of the solvent-laden air.

A medium-scale thermal oxidation VOC abatement plant

A thermal abatement plant for VOC destruction is rarely a small-scale piece of equipment. These are large and energy intensive, even if fitted with thermal recovery. The major electrical loads are from fans and blowers.

9.12　The site waste survey

Walk-around

The first step in waste minimisation is to find the waste in your business. The best tool for this is to carry out a quick site waste survey. This is to gain an overview of the processes and identify some rapid no-cost or low-cost improvements that can save money and reduce costs. After the survey is complete other methods can be used to find some of the hidden wastes.

The survey is carried out in the same way as an energy survey (see Section 6.10) and should be carried out as soon as possible – waste is happening now and it is costing money now.

No-cost and low-cost actions

Eliminate, reduce, re-use, recycle

In waste minimisation the waste 'hierarchy' (see Section 9.2) is an important decision-making tool: first attempt to eliminate the source of waste, then reduce the amount, then re-use any waste that does arise, then recycle the waste and only when these have been eliminated should waste be disposed of. During the site survey always look for opportunities to eliminate the waste rather than simply accepting it.

Materials

- Avoid pellet spills by improving storage and polymer handling techniques (see Section 5.4).
- Record polymer utilisation wherever possible and track any variations.
- Monitor how much polymer has to be reground and how much is returned from your contract recycler (if you use one).
- Find out where and why waste polymer is being generated by your process. Getting it 'Right First Time' is the easiest way to decrease waste.
- Review product design (see Section 4.2). Could less polymer be used? Could waste polymer, e.g., in sprues, be reduced?
- Minimise the need for polymer recovery, regrinding and re-use. Apart from the additional processing, transport and administration costs, converting the recovered polymer into saleable product occupies process time that could be used to make more product (see Section 5.3).
- Plan production to minimise changeover losses.
- Establish total material loss over a given period. Compare this with the utilisation

rate to find the relative importance of process and material handling losses.

Packaging

- Re-use any packaging for your products, where appropriate.
- Discuss ways of minimising packaging use with both suppliers and customers.
- Packaging from suppliers is never free – it is in the product price, find ways to reduce the packaging that you 'accidentally' buy from your suppliers.

Oils and hydraulic fluids

- Review hydraulic oil purchase, storage, handling and disposal procedures.
- Consider the benefits of installing bypass filters in all hydraulic equipment.

Other measures

- Ensure machines are suitable for the processes being carried out, are set up for the minimum polymer and energy consumption, and are maintained regularly.
- Ensure employees are trained and understand the effects of their actions. Employees are vital to the success of waste segregation and minimisation.
- Don't forget to look in the skips – it is the best place to find real waste and is an excellent starting point.
- Measure the amount of polymer used on each machine, how much is reground and how much is sent off-site for reprocessing or disposal.

Fast starts

The site survey should have identified some obvious areas for improvement. Waste is obvious when looked for. It is now possible to make some 'fast starts' to reduce waste and costs. Fast starts should be simple, no-cost and effective. Some of these will have been found in the site survey. Record the starting position and publicise improvements to both motivate employees and start to gain commitment for the initiative from senior management.

Reducing waste by no-cost and low-cost measures will significantly decrease environmental impact and increase profits.

"Zero Waste: The conservation of all resources by means of responsible production, consumption, re-use, and recovery of products, packaging, and materials without burning and with no discharges to land, water, or air that threaten the environment or human health."

Zero Waste International Alliance (www.zwia.org) Updated 20/12/2018

"Zero landfill is not zero waste

Any term that includes "zero" in it must achieve at least 90% diversion from landfills, incinerators and the environment, and commit to a goal of reducing the amount of materials discarded, and any discards going to thermal processes as part of a continuous improvement system to zero."

Zero Waste International Alliance (www.zwia.org)

Opportunity checklist

Department type: Incoming materials.

Areas: Loading docks, pipelines, receiving areas.

- ☐ Packaging/containers.
- ☐ Off-spec deliveries.
- ☐ Damaged containers.
- ☐ Spill residue.
- ☐ Cleaning rags, etc.
- ☐ Pallets (non-returnable).
- ☐ Gloves, overalls, etc.
- ☐ Lining paper.

Department type: Storage (raw materials, parts, final products).

Areas: Tanks, silos, warehouse, drum storage, yards, storerooms.

- ☐ Tank bottoms.
- ☐ Off-spec materials.
- ☐ Excess materials.
- ☐ Damaged containers.
- ☐ Empty containers.
- ☐ Leaks from pumps/valves/pipes.
- ☐ Out-of-date materials.
- ☐ No-longer-used materials.
- ☐ Damaged products.
- ☐ Any of the other code words for waste (see Section 9.1).

Department type: Production.

Areas: Moulding, extruding, blowing, forming, coating, machining.

- ☐ Cooling water.
- ☐ Wash water.
- ☐ Solvents evaporating.
- ☐ Still bottoms in tanks.
- ☐ Off-spec product rejects.
- ☐ Catalysts.
- ☐ Empty containers.
- ☐ Sweepings.
- ☐ Ductwork clear out.
- ☐ Additives.
- ☐ Oil.

- ☐ Rinse water.
- ☐ Excess materials.
- ☐ Filters.
- ☐ Leaks from tanks/pipes/valves.
- ☐ Spill residue.
- ☐ Swarf/off-cuts.
- ☐ Sludge.
- ☐ Head waste.
- ☐ Start-up scrap.
- ☐ Sprues and runners.
- ☐ Packaging of final goods.
- ☐ Any of the other code words for waste (see Section 9.1).

Department type: Support services.

Areas: Laboratories, maintenance shops, garages, offices.

- ☐ Chemicals.
- ☐ Samples & containers.
- ☐ Solvents.
- ☐ Cleaning agents.
- ☐ Degreasing sludge.
- ☐ Sand blasting waste.
- ☐ Lubricating oil & grease.
- ☐ Scrap metal.
- ☐ Caustics.
- ☐ Filters.
- ☐ Acids.
- ☐ Batteries.
- ☐ Office paper, etc.

Department type: Water.

Areas: Processes, toilets, kitchens.

- ☐ Urinals flushing continuously.
- ☐ Underground leaks.
- ☐ Taps left running.
- ☐ Wasteful wash-downs.

Department type: Other

Areas: Consumables.

- ☐ Detergents.
- ☐ Overalls.
- ☐ Other used PPE.

9.13 Managing waste minimisation

Making 'fast starts'

The first task in making 'fast starts' is to identify the priority areas for action. Implementing some quick and cheap cost-saving measures will provide evidence of the benefits of waste minimisation.

Gather available information

Key actions:

- Carry out a site waste survey (see Section 9.12).
- Write down the quantities and direct costs of the 'wastes' that you can see.
- Identify the major sources of waste, such as packaging, lubricants, energy, water and rework.
- Don't worry if information is not always available. Make 'best' estimates or take simple measurements.
- **Tip** – Take photographs of waste and where it is produced. These will show how much waste there is, and help for comparison with future improvements.

Useful further actions

- When estimating costs, remember to go beyond the obvious, e.g., wasted material. Estimate the consequential costs of wasted process time, etc.
- Try to identify the main areas and quantities of energy, water and raw materials use. Compare these values with the total use. If there are major discrepancies, try to find out why. The unexplained use of energy, water or raw material may be one of the biggest sources of waste!

To make sure all the wastes have been identified use a process flow sheet (see Section 9.3) and go through it with key staff.

Identifying priorities

- Find the major sources of waste.
- Identify the priority areas. These may be the largest quantities, e.g., effluents or solid waste to landfill or the highest net costs, e.g., disposal costs, energy consumption, raw material wastage.
- Talk to the staff involved in the activity producing the waste to understand why it is produced. Is it because no-one has seriously considered there is an issue or because it is an established practice which is no longer relevant?
- Use a waste reduction team or other staff to come up with ideas for preventing major

wastes. Simply asking staff for ideas can often be very useful. Estimate the savings you will achieve from the best ideas.

- Focus on a few major areas with the largest financial savings and where there are practical ideas for changes.
- Fill out a waste minimisation opportunity worksheet (see example on the right) to record the projects chosen.
- **Tip** – It should be possible to identify potential actions for 'fast start' savings and put them in order of priority in under a day.

Making the first savings

To make your first savings:

- Make an action plan.
- Agree who is doing what and when.
- Involve the 'front line' staff controlling operations that produce waste to define the aims and priorities, as well as to allocate the responsibilities.
- Set the plan in motion, and review progress against the plan's aims.

Measuring the savings

To demonstrate savings, it is necessary to measure:

- Waste production, e.g., skips/month.
- Raw material use, e.g., orders/month.
- Utility use, e.g., what was the last bill?

As part of the action plan, make sure that simple measuring systems are used. These should be cost-effective and appropriate. Plan measurements to check progress and include regular checks in the plan. Use simple information such as:

- Stock control information.
- Installed meters for energy and water.
- Separating different types of important solid waste to measure waste simply by weight or volume.
- Counting waste containers – helpful for less important wastes, but not as effective as weighing.

Achieving more savings

Progress reviews will provide evidence that waste minimisation is worth the commitment and effort.

- **Tip** – Take photographs to record changes.

Use this evidence to convince management and employees that the waste reduction programme is improving sustainability and extend the programme.

In the UK, the Envirowise Programme produced extensive information on waste minimisation but this is no longer available.

Get an archived copy of "ET30 – Finding hidden profit – 200 practical tips for reducing waste" from: www.tangram.co.uk.

This has a selection of great ideas for simple waste reduction measures.

The cost reduction benefits of minimising waste come only from action.

Start work on your own action plan:

- Establish a firm board-level commitment for waste minimisation.
- Appoint a part-time waste minimisation 'Champion' to establish the true cost of waste and to motivate the workforce.
- Produce regular financial one-line reports on the cost of waste collection and disposal and on the total cost of waste.

Good communication is essential to success. Involve people in reducing waste and tell them about the successes.

- Report success to management.
- Market the process.

Management support

The success of any waste minimisation programme depends on the active support of the Managing Director and other senior managers. Senior management should:

- Demonstrate visible leadership.
- Encourage employee participation.
- Set clear waste minimisation targets.
- Promote a company environmental affairs policy.

Choosing a coordinator

A waste minimisation coordinator is needed because reducing waste in one area of the business may require action in other areas. The coordinator needs three essential qualities:

- Management authority or direct access to senior management.
- Enthusiasm and ability to motivate people.
- A working knowledge of waste management or a willingness to learn.

The waste minimisation coordinator is not normally a full-time job and could be the Managing Director, Production or Quality Manager.

The main responsibilities are:

- Co-ordinating waste segregation and measurement.
- Identifying opportunities to prevent waste.
- Locating priority action areas.
- Setting up waste reduction teams.
- Allocating 'ownership of waste'.
- Raising waste reduction awareness.
- Creating monitoring systems for feedback to managers and the workforce – this is essential for success.
- Working with suppliers to identify areas for materials reduction or recovery.

This may sound a lot, but the role is to coordinate and facilitate. It will be the waste minimisation team, or teams, who actually achieve the results.

Prepared by:		Date		Rev:
Site:		No.:		Page:

Description: Current and proposed practice
Current:

Proposed:

Quantification		Payback	
Baseline consumption			
Unit of measure		Zero cost	
Unit cost		< 1 year	
Saving		1 - 2 years	
% saving		2 -3 years	
Measured		> 3 years	
Estimated		Improved service	
Material cost saving		Improved quality	
Manpower cost saving		Environmental	
Total cost saving		Other criteria	
Capital cost			
Operational cost			
Total cost			

Notes:

Input category				Notes
Raw material		Operating costs		
Consumables		Fossil fuels		
Paper & packaging				
Water				
Electricity				

Output category		Notes
Off-site disposal		
Liquid effluent		
Air emissions		
Incineration		
Degraded product		

Project status		Timeframe		Project type	
Implemented		0 - 6 months		Product	
Feasible		7 - 12 months		Input	
Not feasible		1 - 2 years		Technology	
Study underway		2 -5 years		Procedure	
Study not started		Never		Housekeeping	

Waste minimisation opportunity worksheet

Use a simple form to explain and decide on which projects to carry out. Make it easy to tick the boxes and get the project started. After the project is approved then use a one-page project plan to plan the project.

9.14 Waste minimisation – where are you now?

Waste minimisation

Waste is a major factor in sustainability, reducing the amount of waste at any site will improve sustainability credentials and reduce impact on the environment. It is also an overhead that is rarely treated with any seriousness by most management teams.

Waste is seen as 'what is in the skips' and not in the broader sense of anything that does not add value to the process or the product.

Simple but organised action to reduce waste can reduce operating costs by 10% and the company can become 'greener' by reducing waste and discharges to the environment. Waste of materials is very similar to a waste of energy. They are both the result of management failing to notice that the world has changed and that reducing the cost of direct labour is no longer the only key component of the cost of operations.

Failing to have a plan to reduce the cost of waste is not only bad for sustainability but also financially bad for most companies.

Completing the chart

This chart is completed and assessed as for the previous charts.

A simple waste 'walk-around' will reveal many areas where the current systems tolerate waste as part of the normal operations of the business.

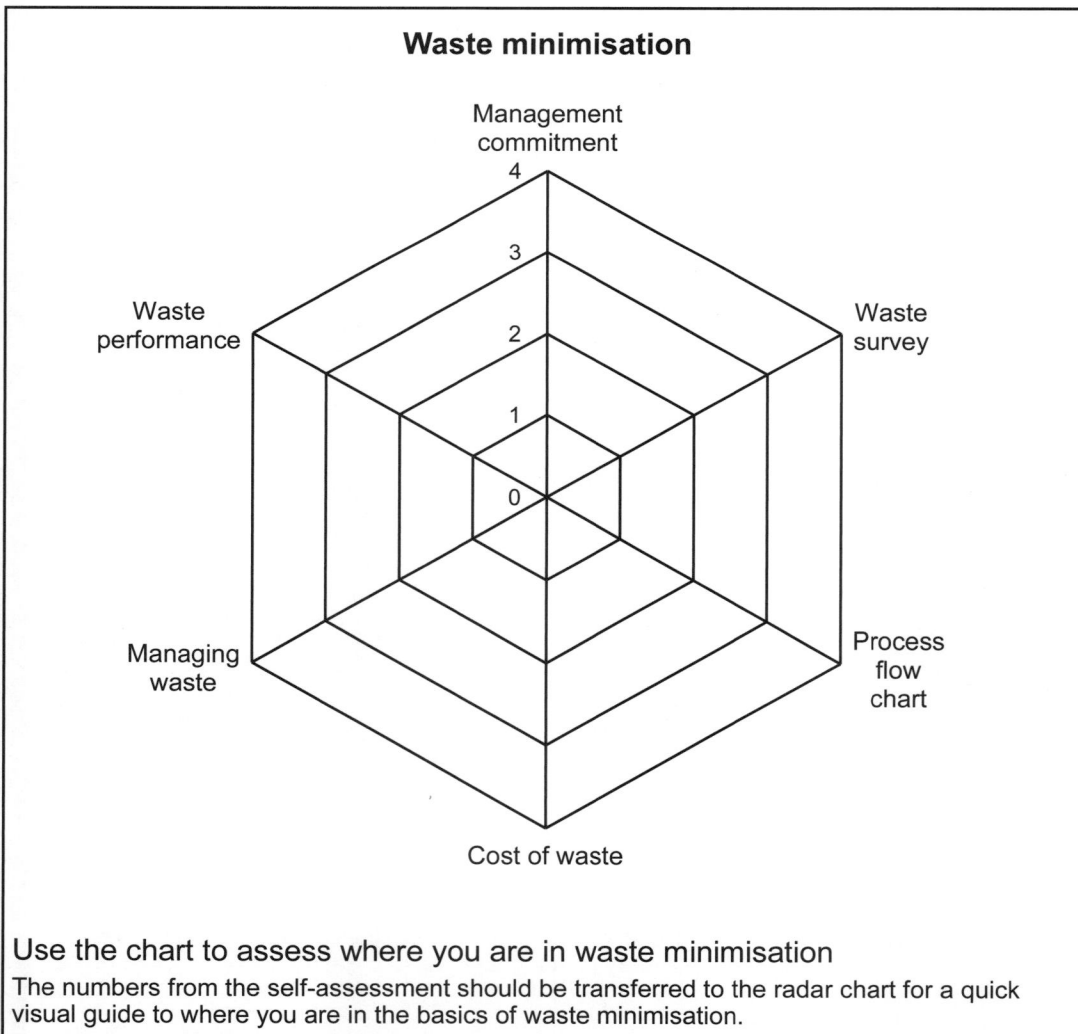

Waste minimisation

(Radar chart with axes: Management commitment, Waste survey, Process flow chart, Cost of waste, Managing waste, Waste performance. Scale 0 to 4.)

Use the chart to assess where you are in waste minimisation

The numbers from the self-assessment should be transferred to the radar chart for a quick visual guide to where you are in the basics of waste minimisation.

	Waste minimisation					
Level	Management commitment	Waste survey	Process flow chart	Cost of waste	Managing waste	Waste performance
4	Waste management is seen as important to improving profits & environmental performance. Action taken on all identifiable concerns.	Regular waste surveys carried out to identify new opportunities. Action taken on all opportunities identified.	Full process flow charting for complete site (including office processes) to enable waste targeting. No code words accepted for waste.	Full cost of waste assessed, targets set & monitored for performance.	Full waste management program in place. Program has proven effective in reducing costs.	Waste performance is visibly better than the industry average. Monitoring & targeting used to further improve performance.
3	Waste management is an explicit & stated business goal. Action taken on easily visible concerns.	Initial waste survey carried out. Action taken on all of the opportunities identified but no further survey carried out.	Good process flow diagrams developed for all processes. Processes have few areas that are not considered.	Full cost of waste assessed but no targets set for performance.	Full waste management program in place. Program effectiveness in reducing costs is not yet proven.	Waste performance is slightly better than the industry average. Monitoring & targeting being introduced to improve performance.
2	Waste management is not an explicit goal. Sporadic action taken when concerns are very visible.	Initial waste survey carried out. Action taken on some of the opportunities identified.	Good process flow diagrams developed for most processes. Processes have some areas that are not considered.	Good knowledge of the cost of waste for most areas.	Partial & largely ineffective waste management program in place.	Waste performance is similar to the industry average. Monitoring & targeting being introduced to improve performance.
1	Waste management is not a goal. Visible & obvious waste is openly tolerated by management. No improvement techniques used.	Initial waste survey carried out. No action taken on opportunities identified.	Outline process flow diagrams developed for some processes. Processes have considerable areas that are not considered.	Vague knowledge of the cost of waste. Knowledge is primarily in the cost of disposal.	No waste management program in place but planned for implementation.	Waste performance is slightly worse than the industry average. No monitoring & targeting used.
0	Waste management not considered by management. Getting the product out the door is the only goal.	No waste survey carried out.	No process flow diagram produced.	No concept of the cost of waste to the company.	No waste management program in place & no plans for action in the future.	Waste performance is visibly worse than the industry average. No monitoring & targeting used. High use of 'code words' for waste.
Score						

Key tips

- Calculate the potential benefits of waste minimisation.
- Find some of the code words used for waste in the company and look at how much these can be reduced.
- Waste minimisation can be used with an EMS to reduce environmental impact and costs.
- Waste minimisation should follow the waste hierarchy to produce the most sustainable results.
- Preventing waste is the most sustainable option.
- Process flow charts are the best tool to map resource use in processes and to identify waste.
- Use process flow charts to identify detailed inputs and outputs to process steps.
- A waste tracking model for each process can be used to put numbers on the waste identified by the process flow chart.
- Use the waste tracking model to locate potential savings by waste minimisation.
- Use cause and effect analysis (see Section 9.4) to identify the root causes of waste.
- Most of the cost of waste is hidden.
- Accountants can be very helpful in locating the hidden costs of waste.
- Calculate the total cost of waste.
- The cost of packaging is not simply the purchase cost.
- The flow of packaging needs to be mapped to gain an understanding of the true cost of packaging.
- Much of the packaging used can be eliminated or reduced by simple measures.
- Re-using packaging improves sustainability but can also improve operations and logistics.
- Re-using packaging requires management skills and effort.
- The value of packaging to be recycled and recovered can be maintained by good segregation of materials.
- Oil and hydraulic fluid stocks can be rationalised based on the viscosity.
- Oil and hydraulic fluid use can be reduced by testing and good housekeeping.
- Solvent use in plastics processing is generally small but solvents are expensive to use and costly to handle.
- Good practice can eliminate many of the uses of solvents.
- Good housekeeping can reduce fugitive solvent loss.
- Carry out a site waste survey (walk-around) to quickly identify areas of high waste.
- Examine the contents of the skips to see what is being wasted.
- Use the Opportunity Checklist to identify some areas for improvement.
- Select a 'waste minimisation champion' and establish a waste minimisation team.
- Identify possible projects for waste minimisation.
- Make the first savings to improve sustainability and reduce costs.

Chapter 10

Use and end-of-life

This was probably the most difficult of all the chapters in this book to write. As one of my old bosses used to say "Sometimes you just sit and think". As I sit in my kitchen (late at night), thinking and writing this book I wonder if I really ought to say the things that I will say on the next few pages.

In many ways, I am resolute in my defence of the plastics processing industry but I am also realistic enough not to ride to 'the defence of the indefensible'. As an industry we have sometimes made the wrong choices, albeit some of these were at the request of our customers but sometimes we should have stood up and said 'No'.

Many years ago, I knew a man in Toronto who had several extruders on his garage (a big garage), in the morning he loaded the hoppers and the extruders produced drinking (sipping) straws. It was beautifully automated and as the straws were cut the handling systems cycled full and empty boxes. In the event of a fault, the line stopped. I was in awe as he sent these straws all over Canada and made huge profits. Now I am frightened that many of these straws ended up in the Great Lakes, in some other watercourse or simply in landfill. What a waste and how times have changed.

For my first 25 years in this industry, I was primarily concerned with production efficiency and getting the product out the door. Now there are other considerations and we are searching for the best methods to deal with these. As plastics processors, just as parents, we lose control of our products as soon as they go out the door. That doesn't mean that we don't care, either as processors or as parents (we do hope that they come back with grand-children) simply that our control is much diminished. In the new circular economy, the concept of product stewardship will become increasingly important and we have the choice of accepting this or having it forced upon us.

The areas of 'use' and 'end-of-life' should, perhaps, be treated separately but it is hard to separate the two when the industry has little control over the product after it has gone out the door. As this book is written for plastics processors, I have combined the two areas to give processors an overview of the issues that we face.

It is not going to get easier and the pressures are mounting. It seems everybody has something to say about plastics, even when they know very little about the products, the benefits or the science. In the UK, we have had 'celebrity chefs' presenting a programme titled 'War on Plastic', the very title showed they had made up their minds before starting. This was not the epitome of unbiased reporting.

Sustainability is much broader than plastic waste (although that is important). We must not lose sight of the overall aim by focusing on a single visible effect.

This chapter tries to examine plastics products at the 'use' and 'end-of-life' phases in a systematic way to identify what is happening and how processors can contribute to sustainability and prepare for what is coming.

10.1 Use and end-of-life timescales for plastics products

Use defines end-of-life

The use life of a plastic product defines how we treat the product at the end-of-life stage and the product use classification that we will use and the relevant use timescale is shown in the table on the lower right.

Very short-life products need, and will, be treated very differently to long-life products at the end-of-life phase.

- Very short-life products and many short-life products have a short time to provide value to the consumer and society. These are largely regarded as 'disposable' products, e.g., packaging. They can enter the general MSW stream (when not sorted correctly or disposed of as general waste), enter the recycling stream (when sorted and treated correctly) or become litter (when treated incorrectly or not at all).

- Medium-life and long-life products have a long time to add significant value to the consumer and society. These are regarded as 'valuable' products and will generally be captured by external recycling schemes, e.g., WEEE or ELV, or by industry operated closed-loop schemes, e.g., VinylPlus®.

The degree of society's control varies with the product use timescale and the industry response also needs to vary. There is no 'one size fits all' solution to this challenge.

Most of the current public concerns with plastics products are with those which have a use life of < 2 years and these are sometimes referred to as 'single-use' plastics. This term can be confusing as it joins two concepts. Does it mean that the product only has one use life before it is discarded, e.g., a coffee stirrer, or does it mean that the product only has a single use function, e.g., a plastic pipe? A coffee stirrer is a 'very-short life product' and a plastic pipe is a 'long-life product' – they will be treated very differently when they reach end-of-life.

Defining a product in terms of the use timescale provides greater clarity and allows us to discuss use and end-of-life for the complete range of plastics products.

How do we define the use timescale?

The 'use' timescale is not simply the time between production and use, i.e., the time in the supply chain. It is the time that the product is required to fulfil the product requirements. Medical products and their packaging may appear to have a very short use life, i.e., they are used and discarded, but the packaging is required to keep the product clean before use. The packaging therefore fulfils a vital use function for all the time between production and use. Similarly, food packaging that fulfils product preservation requirements can have a significant use life between filling or wrapping and the time of consumption. Other products can have a very short use life even if they are in the supply chain for a long time, e.g., a coffee stirrer can be in the supply chain for a long time but the use life is very short.

Very short-life products

Very short-life products have a use timescale of < 1 day but these can be sub-divided into small and medium sized products.

Small very short-life products

These small products are primarily those that are deliberately designed to be disposed of after one use, e.g., cotton buds, straws, coffee stirrers, wet wipes and plastic coffee pods. These disposable products are not valued by society.

These small items will rarely enter the MSW stream for recycling. If they do enter the recycling stream, separation is difficult and they will generally be treated as unrecyclable waste and sent to landfill. If used 'on the go' they can be visibly littered to potentially enter watercourses and the oceans.

These are potentially the most problematic products in sustainability terms.

Medium very short-life products

These are larger products used in disposable goods or packaging applications where

Focusing on the use timescale starts to tell processors what they need to do for a sustainable future.

Use classification	Use timescale
Very short-life	< 1 day
Short-life	< 2 years
Medium-life	2-15 years
Long-life	> 15 years

A classification for the use phase of plastics products
The treatment of plastics products varies with the use life of the product. Very short-life products need, and will, be treated very differently to long-life products at the end-of-life phase. The degree of control varies and the industry response also needs to vary.

packaging functionality is not important for product preservation, e.g., general goods packaging, disposable cups, cutlery and plates and plastic bags. Although larger, these disposable products are still not valued by society.

When disposed of correctly many of these products can successfully enter the MSW stream for recycling (see Section 4.5) but many will be used 'on the go' to enter a mixed waste stream and be landfilled or be littered.

These are also problematic products in terms of sustainability.

Very short-life products are covered in Section 9.2.

Short-life products

Short-life products have a use timescale of < 2 years and are primarily used in packaging applications where the package functionality is important for product preservation, e.g., food and drink packaging, medical packaging, pharmaceuticals packaging and cosmetics packaging. The use phase timescale for this type of product depends on time between the packaging process and the use of the packaged product. These products vary in value to society depending on the application, their functional benefit and the individual's view of the need for the product. They generally only become visible when disposal is needed, e.g., few people notice pharmaceutical packaging until it is empty despite the functionality during use.

This type of product will often be medium sized and they are rarely part of an assembly. The materials used may not be obvious even if the base material is marked on the packaging (especially in the case of materials which use a functional barrier layer). At end-of-life most of these products will enter the domestic MSW stream (see Section 4.5).

This type of product is not as problematic as the very short-life products due to their high utility value but they are under investigation around the world and both consumer and government pressure on the industry is growing.

Short-life products are covered in Sections 9.3-9.5.

Medium-life products

Medium-life products are those with a use life of 2-15 years and are primarily found in household durables, automotive and electrical/electronic applications, e.g., household appliances (both white and brown goods), car parts, computer parts, storage and recreational products (toys and outdoor goods). These products are regarded by society as valuable during the use phase

even if they are largely 'invisible', e.g., even the most ardent environmental campaigner against plastics rarely sees their keyboard or phone as being made of plastics or totally dependent on plastics for their operation.

This type of product will also generally be medium sized, even if it is part of a larger assembly. The material used is often not obvious or not marked on the product and this makes recycling more difficult. At end-of-life many of these products will fall under legislation such as WEEE or ELV and recycling will be controlled.

Medium-life products are covered in Sections 9.6 and 9.7.

Long-life products

Long-life products are those with a use life of > 15 years and these are primarily used in construction or in capital equipment applications, e.g., windows, pipes, thermal insulation, electrical insulation, furnishing fabrics and membranes. This means that they are also regarded by society as valuable during the use phase but, in most cases, they will be incorporated in a larger product and again be 'invisible'.

These are normally large products where the material is either obvious or clearly marked on the product. At end-of-life, handling is generally well controlled by the established recycling infrastructure or via closed-loop recycling schemes, e.g., windows and pipes.

Long-life products are covered in Sections 9.8-9.10.

Why focus on the use timescale?

Focusing on the use timescale allows us to think about where we add value to society and how we are going to deal with the issues of the end-of-life of products.

- **Tip** – Where are your products in the use timescale? What does this tell you about your future?

I have tried to be consistent in setting the 'use life'. Adjust the use timescale for a product if you do not agree, it will generally only move a product by one classification.

Face masks for COVID-19 protection have made an appearance as litter on beaches and in oceans around the world.

A vital product tainted by littering but is it the fault of the face mask?

Possibly we ought to ban the use of face masks?

10.2   Very short-life products

The products under threat

Very short-life products with a use life of < 1 day are under threat around the world. These products are already subject to increasing legislation and regulation everywhere. Many, if not most, of these products do not have an obvious future and it is increasingly hard to defend their continued production in the current climate.

- **Tip** – Processors making these products need to think carefully about their future because these are the most problematic of all plastics products in sustainability terms.

Plastics microbeads were developed in the 1970's and have some excellent medical applications due to their size and very high surface area. They were then used in cosmetics as exfoliants, e.g., face scrubs. These applications released tonnes of engineered microbeads into water systems and the oceans just so that people could exfoliate. As evidence and opinion mounted against the use of microbeads in cosmetics, the plastics industry could not support this egregious application of plastics. The removal of microbeads by cosmetics manufacturers and bans on the use of plastics microbeads in cosmetics were justified and widely supported by the industry.

The message is clear: Products which damage the environment and are unjustified are indefensible. The wider industry will 'throw you under the bus' and move on.

Legislation in Europe

The primary legislation in Europe is EU Directive 2019/904 – "on the reduction of the impact of certain plastic products on the environment" also known as the 'Single Use Plastics Directive'. This becomes law in the EU on 03 July 2021 (although some aspects are delayed until later). The legislation groups actions into several categories:

Restriction on placing on the market

The following plastic products will be banned from the EU market from 03 July 2021:

- Disposable plastic cutlery.
- Disposable plastic plates and bowls.
- Plastic straws.
- Cotton buds with plastic stems.
- Plastic stirrers.
- Food containers made from EPS.

- Sticks to be attached to and to support balloons.
- Products made from oxo-degradable plastic.

Similar legislation is being introduced in the UK but the actual products affected vary with the specific region.

Consumption reduction

Countries must take measures to reduce and monitor the consumption of certain products where there is no alternative, e.g., drinking cups including covers and lids, and containers of prepared food for immediate consumption.

Compulsory marking

Some products must carry a marking on the product or packaging to inform consumers about the waste management options (or what type of waste disposal should be avoided), the presence of plastics in the product and the environmental impact of littering. These are:

- Sanitary items.

"We're doomed"
Private Frazer in 'Dad's Army' (BBC 1968-1977)

Typical 'under threat' products
Many very short-life products are already under threat from legislation or consumer action. The direction of travel is clear and this type of product does not have an obvious future in a circular economy where sustainability is a key issue.

- Wet wipes.
- Tobacco products with filters.
- Drinking cups.

Collection and design of bottles

The directive sets a collection target of 90% recycling for plastic bottles by 2029 (77% by 2025) and bottles should contain at least 25% recycled plastic by 2025 (PET bottles) and 30% by 2030 (all bottles).

Extended producer responsibility

The directive includes the 'polluter pays' principle. Producers will have to cover the costs of waste management clean-up, data-gathering and awareness raising for:

- Food and beverage containers.
- Bottles.
- Cups.
- Packets and wrappers.
- Light-weight carrier bags.
- Tobacco products with filters.
- Wet wipes.
- Balloons.
- **Tip** – The directive also requires extended producer responsibility for plastic fishing gear.

Awareness raising

The directive includes a requirement that countries inform consumers and encourage responsible behaviour to reduce litter from these products, to make consumers aware of reusable alternative products and the impact of inappropriate disposal of plastic waste on the sewerage system.

- **Tip** – The 'Single Use Plastics Directive' will have far-reaching effects for producers of very short-life products.

Industry action in the UK

One of the best industry actions in Europe is the UK Plastics Pact. This is operated by WRAP (www.wrap.org.uk) and has > 120 members including plastics processors, food producers and most of the UK supermarket chains. It is especially interesting because the ≈ 80 plastics processor members are responsible for $\approx 85\%$ of the plastics packaging of products sold in UK supermarkets and $\approx 50\%$ of the total plastics packaging sold in the UK. This is a coalition of the willing.

The Plastics Pact has 4 main targets for 2025:

- Eliminate problematic or unnecessary single-use packaging. The Plastics Pact defines problematic or unnecessary packaging as:

  "Single-use plastic items where consumption could be avoided through elimination, re-use or replacement and items that, post-consumption, commonly do not enter recycling and composting systems, or where they do, are not recycled due to their format, composition or size."[1]

- Make packaging 100% reusable, recyclable or compostable.
- Ensure that 70% of plastics packaging is effectively recycled or composted.
- Ensure that 30% average recycled content is used in all plastic packaging.

The Plastics Pact has a 'Roadmap to 2025'[2] to define the actions needed to achieve the targets and regularly reports progress.

The 'problematic plastics' identified by the UK Plastics Pact are mainly very short-life products and very similar to those covered by the Single Use Plastics Directive (although the Plastics Pact also includes PVC packaging because of the effect on the recycling stream).

The initiative does not stop with very short-life products, it also has a list of ≈ 20 products which are 'under investigation' for potential elimination. These products are to be investigated in terms of their utility value and any unintended consequences of elimination.

- **Tip** – The Plastics Pact provides a template for coordinated action through the supply chain.

Legislation in the USA

Legislation in the USA is very complex, very local and changing rapidly. Some states and cities have imposed bans on some very short-life products, e.g., plastic bags, straws and PS food containers, but these are local or regional actions and there is no coordinated approach to legislation. This patchwork of local bans and charges has driven a counter-legislative response to produce 'bans on bans', e.g., some states have produced state-level legislation preventing city-level bans on very short-life products.

There are some calls for a federal approach to short-life products but there is no indication that individual states are prepared to give up their control in this area.

Very short-life products do not have an obvious future in a sustainable world. Their utility value is low, their resource consumption is high and they are unlikely to fit into the developing circular economy.

Most of the initiatives against very short-life plastics conflate the issue of 'single use' and the 'use timescale'.

They then put food packaging where the packaging is functional for a considerable period (the effective use life) into the same groups as plastic straws and similar items where the effective use life is very short.

We prefer to work in terms of the use timescale to allow functional plastic packaging to succeed (or fail) on the merits of the functionality.

This is echoed in the Plastics Pact which considers the utility value and unintended consequences of elimination, e.g., increased food waste.

These are not the only initiatives and legislative responses.

It is only going to get worse for very short-life plastics.

- 1. 'Eliminating Problem Plastics', Version 3 December 2019, WRAP, www.wrap.org.uk.
- 2. 'A Roadmap to 2025', 2019, WRAP, www.wrap.org.uk.

10.3 Short-life products

The next candidates?

It appears inevitable that many of the very short-life plastics products will soon be faced with increased legislation and outright bans. The focus will then inevitably move to short-life products and some of these products are already on an 'under investigation' list published by WRAP[1]. As noted in Section 9.2, WRAP is very conscious of the utility value of many short-life products and the potential for unintended consequences of their elimination.

Short-life products may initially appear to be very short-life but the use timescale is not simply the time during which the product is 'used' by the consumer. For most products, the use timescale is the time between the packaging process and the use of the packaged product, i.e., the timescale during which the packaging is performing a function. As an example, medical packaging performs a vital function in terms of product preservation and sterility from the time of packing until the time of use and this may be many months.

A sense of perspective?

Food packaging serves a particularly vital function in prolonging the life of foods from farm to consumer and in reducing food waste.

WRAP[2] estimated that in 2018 food waste in the UK was valued at > £19 billion and would be associated with > 25 million tonnes of CO_2e. To put this into context, the total estimated CO_2e emissions of the UK plastics industry are $\approx$ 4 million tonnes of CO_2e (see Section 7.8) of which packaging is estimated to be responsible for $\approx$ 2 million tonnes. Most of the food waste is in the household (> 70%) and effective plastic packaging can extend the life of food products by 3-10 days (depending on the product). If short-life plastic packaging can reduce food waste by only 10% then the use of the packaging is effectively carbon neutral, i.e., it saves more carbon emissions than it uses.

- **Tip** – Whilst many food products may be considered to be carbon neutral, the growing, treatment, transport and processing of the food are not carbon neutral. These are all lost to the system if the food is wasted.

- **Tip** – Food waste in the household is decreasing (down 18% in 2018 since 2007). Is the increasing use of plastics in packaging assisting in this?

It is entirely technically feasible to replace plastics packaging with alternative materials, e.g., glass, tin plate, aluminium and paper-based materials. However, multiple studies of the impact of plastics packaging on life cycle energy consumption and greenhouse gases in Europe[3] and the USA and Canada[4] have shown that replacing plastics packaging with alternative materials would:

- Increase the total mass of the products.

- Increase the energy consumption in the total life-cycle.

- Increase the GHG emissions in the life-cycle.

The benefits of plastic packaging increase further as recycling increases and the material is recovered for future packaging use (see Section 4.6).

- **Tip** – Plastics packaging for short-life applications performs a vital functional role in product preservation. Replacement with alternative materials would lead to increased environmental impacts.

The main applications

Typical short-life products

Many packaging applications are short-life products where the use timescale starts when the packaging begins to be functional. For meat the protection provided to the product starts at packaging and continues until the meat is consumed, i.e., not quite 'farm to fork'.

- If plastics in packaging were replaced with 'alternative materials' then European energy consumption would double[3].

- If plastics in packaging were replaced with 'alternative materials' then European GHG emissions would increase by a factor of 2.7[3].

Food

Food packaging is the largest sector for short-life products and these products will often have a functional life much greater than the < 1 day of very short-life products.

We have carried out an assessment of the carbon impact of a variety of foods and their packaging and the results, in terms of carbon footprint, are shown on the lower right. Even in the worst case, the packaging carbon footprint is < 6% of the contents carbon footprint. This is for milk where PE-HD bottles are currently reaching a recycling rate of > 80% in the UK. For higher value foods, such as steak, the packaging carbon footprint is << 1% of the contents carbon footprint.

- **Tip** – What is being protected by the packaging is more important and has higher environmental impacts than the packaging itself (even if you are vegan).

- **Tip** – The calculations are approximate but even large errors do not change the output significantly.

Non-food

Packaging for non-food items, e.g., cosmetics, cleaners and general product packaging, also provides product protection and ease of distribution but, unlike foods, there is generally little product life extension.

As with food products, the carbon impact of the product is significantly higher than the carbon impact of the packaging. Material substitution[3,4] may appear to offer opportunities to reduce impacts but these are illusory and product losses will be higher through damaged or lost goods, e.g., glass containers can shatter.

Medical

Medical devices are inherently short-life products as a result of patient safety and product requirements. As with food products, the value of the packaging is significant in preventing degradation or contamination of the contents.

- **Tip** – The plastics industry is working to develop recycling for non-infectious waste. This requires new approaches as almost all of the material is classed as 'hazardous waste' and sent directly for incineration.

Films

It is not only rigid packaging that reduces food waste, flexible packaging (film) extends the life of products, e.g., bread and cucumbers, and reduces food waste. Silage wrapping film not only makes transport and storage easier but also promotes fermentation in the silage to break down the product it is made from, e.g., hay or oats, and to make the contents much easier to digest for the animals.

Films are less likely to be recycled due the erroneous belief that films are 'non-recyclable' whereas the real issue is inconsistent collection systems.

Let's start to tell the story

Short-life products provide substantial reductions in food waste and added value retention for non-food products. This is a story that is not often told and the plastics industry needs to start to tell this story instead of allowing products to be demonised through misinformation.

It is undoubtedly true that we need to investigate many of the short-life products to ensure that they deliver value but we must also recognise the functional value of packaging.

- 1. 'Eliminating Problem Plastics', Version 3 December 2019, WRAP, www.wrap.org.uk.
- 2. 'Food surplus and waste in the UK – key facts' WRAP, January 2020, www.wrap.org.uk.
- 3. 'Impact of plastics packaging on life cycle energy consumption and greenhouse gases in Europe', Denkstatt, 2011, www.plasticseurope.org.
- 4. 'Impact of plastics packaging on life cycle energy consumption and greenhouse gases in the United States and Canada', Franklin Associates, 2014, plastics.americanchemistry.com.

In many cases, the brand owners drive change and not the plastics industry.

The industry responds to demands for increased product life and value retention with the best tools and techniques available and the development of new tools and techniques to meet the demands of the brand owners.

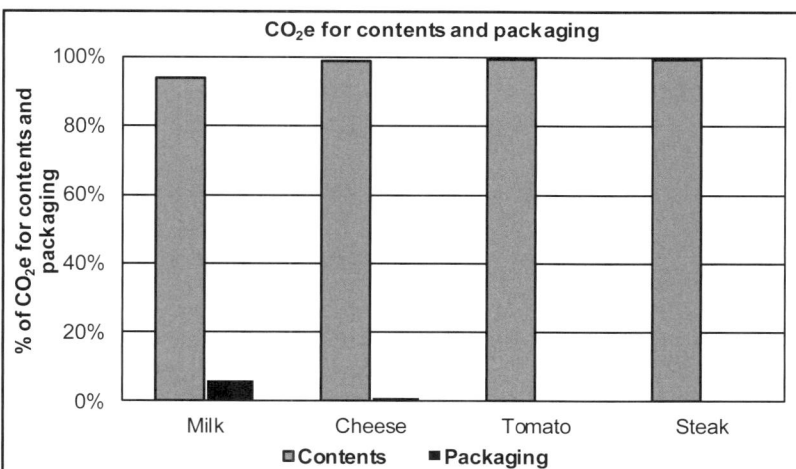

CO$_2$e for contents and packaging

% of CO$_2$e for contents and packaging — Milk, Cheese, Tomato, Steak — ■ Contents ■ Packaging

The contents are more important than the pack

Food packaging serves a vital function in reducing food waste and the carbon footprint of the pack is minor in comparison to the carbon footprint of the contents. For a steak, the carbon footprint of the pack is < 0.5% of the carbon footprint of the complete product.

10.4 Short-life products – preparing for a sustainable future

Start work now

It is easy to show that most of the carbon footprint for short-life products is in the raw materials and manufacturing phases (see Section 7.7 and the diagram on the right) but a carbon footprint is not designed to show the carbon footprint avoided by the product. The value of short-life products in terms of 'avoided' carbon is not calculated and the result is a growing demand for their removal, re-use or replacement with alternative materials even when this will result in increased environmental impacts.

The demand, legislative and otherwise, for a circular economy means that plastics processors making short-life products need to start work now to prepare for a sustainable future (or else).

- **Tip** – The industry needs to address short-life products for economic sustainability as well as environmental sustainability.

Note: Chapter 3 looks at the design aspects of sustainability across the life cycle and gives detailed actions, this section looks at how processors can respond at the company level.

Make it easy to collect

Most short-life products will be used in controlled situations, e.g., home or industry, and at end-of-life will enter the MSW stream (see Section 4.5). If a product is not easy to collect then it will not be collected, if it is not collected then it will not be sorted and it will not be recycled. Processors need to produce products to be easy to collect so that they can enter the recycling stream.

This is the logic of EU Directive 2019/904 (see Section 10.4) which requires that caps and closures remain attached to bottles from 2024. This is to prevent small caps and closures becoming separated from the bottle and not being recovered and has driven a re-design of caps and closures across Europe.

- **Tip** – Examine the product range to ensure that products are easily collected and enter the recycling stream.

Inconsistent collection strategies around the world (and in individual countries) pose a significant barrier to industry making products that are easy to collect. In many areas, polyolefin flexible packaging is labelled 'not widely collected' and 'not widely recycled' due to a lack of consistency and recycling infrastructure – even though > 80% of these products are already 'recycling ready'.

- **Tip** – In Europe, CEFLEX (ceflex.eu) has produced 'Designing for a Circular Economy' guidelines (guidelines.ceflex.eu/resources). These give design best practice for the recycling of polyolefin-based flexible packaging. In the USA, Materials Recovery for the Future (MRFF – www.materialsrecoveryforthefuture.com) is working to establish kerb-side collection and recycling of flexible packaging.

Make it easy to sort

Even if collected, a product will fail to be recycled if it is not easily sorted. Most short-life products are sufficiently large that automatic sorting is possible but the primary material type still needs to be clearly marked and preference should be given to mono-material products made from widely recycled materials such as PET, PE-HD and PP.

- **Tip** – Detailed recommendations on choosing raw materials for recycling are given in Section 3.5.
- **Tip** – Examine the product range to ensure that products are easily sorted in the recycling stream.

Make it easy to recycle

Recycling process streams (see Section 4.6) depend on clean materials with minimal contaminants. This means that processors need to reduce the number of materials to a minimum, to reduce the use of additives, e.g., colourants, coatings and stabilisers, and

Lightweighting is already reaching the limits of what can be done and still retain product functionality, e.g., compare the material used in today's PET bottle with that of 10 years ago. There is a vast difference.

The industry needs new strategies for the future.

If it is not collected, it is not sorted.

If it is not sorted, it is not recycled.

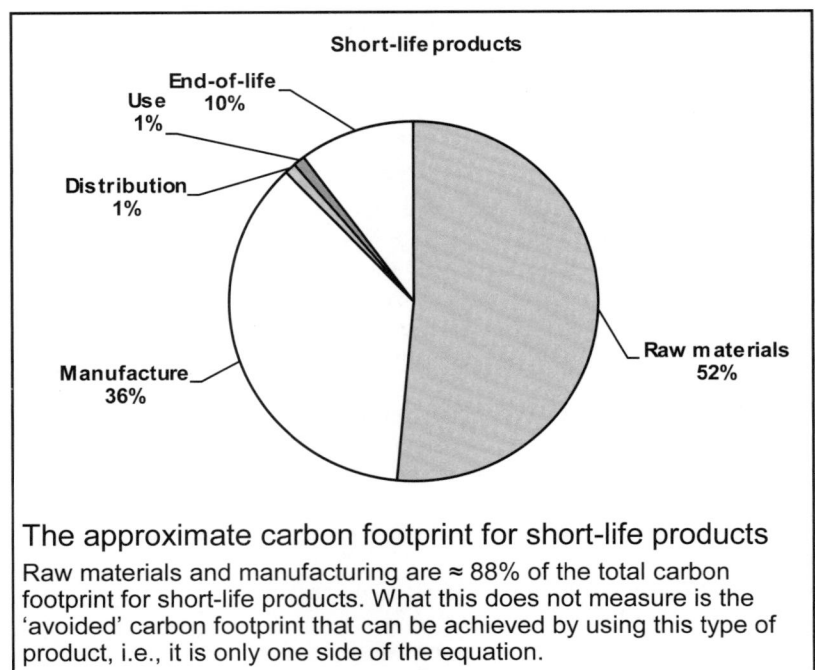

Short-life products

- End-of-life 10%
- Use 1%
- Distribution 1%
- Manufacture 36%
- Raw materials 52%

The approximate carbon footprint for short-life products

Raw materials and manufacturing are ≈ 88% of the total carbon footprint for short-life products. What this does not measure is the 'avoided' carbon footprint that can be achieved by using this type of product, i.e., it is only one side of the equation.

to reduce the use of adhesives and inks that can reduce the value of the recyclate.

A simple example is the use of tinted PET for drinks bottles. This may differentiate the product but the resulting recyclate is worth less than a clear PET recyclate.

- **Tip** – Recommendations on choosing raw materials are given in Section 3.5. Start now to examine the product range to ensure that products are easily recycled.

- **Tip** – Brand owners have the largest influence on pack and product design but the industry must give advice on product design to enable brand owners to make the right choices. They will accept good advice and many are now driving the change.

- **Tip** – Use some of the many examples of good packaging design on the Internet to create a portfolio of good design for discussion with customers. A good resource is the 'Recyclability by Design - Case Studies' from RECOUP (www.recoup.org).

Make it easy to include recycled materials

Products should not only be recyclable but should also be designed to incorporate recycled materials. This is not always easy but companies need to start thinking about how recycled materials are going to be included in products at the design stage.

- **Tip** – It is always easier to include recycled materials at the design stage than to try to include them later.

Get 'recycled content ready'

It is not enough to design products for recycling, plastics processors must also prepare their products and sites to be ready for recycled content. Processors will be familiar with using regrind material (see Section 5.3) but this will normally be in low percentages, e.g., < 5%. There are exceptions such as EBM and ISBM where PE-HD and PET are being recycled in increasingly high volumes.

Sustainability for short-life products means that all sites must be ready to dramatically increase the amount of recycled content in the coming years.

Recycled materials will not always have the same processing properties as virgin materials. Processors need to start selecting their new materials, running the trials and preparing their sites and processes for running recycled materials at much higher concentrations (>70%).

- **Tip** – Start this now, it may be a long process to get it right and to get customer acceptance.

- **Tip** – Many of the short-life applications are 'food contact', the industry needs to

ensure product and material safety at all levels.

Develop closed loop systems

The ideal development for processors is the use of a closed loop process. For PE-HD milk bottles in the UK, the recycling rate is already very high and this is almost a closed loop. For many products that are not as ubiquitous, this will be a difficult proposition but the discussion must start soon.

- **Tip** – Start to talk to the rest of the industry and the waste management companies about the potential for closed loop systems.

Start to think about bio-based and biodegradable materials

Processors need to start to think about using bio-based materials (see Section 4.14) and biodegradable materials (see Section 4.15). Many brand-owners and supermarkets are driving change in the use of these materials.

Environmental issues and their effects can easily develop momentum and society can remove a company's licence to operate.

Getting ahead of the curve is the only survival mechanism.

Recyclability is not the same thing as sustainability.

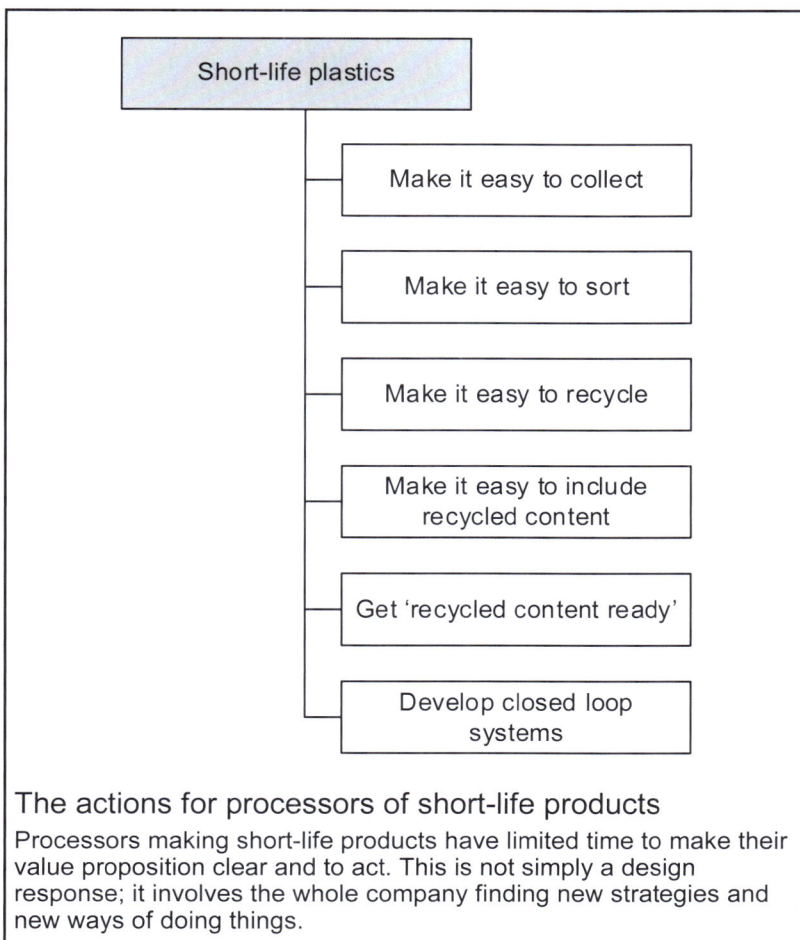

The actions for processors of short-life products

Processors making short-life products have limited time to make their value proposition clear and to act. This is not simply a design response; it involves the whole company finding new strategies and new ways of doing things.

10.5 Short-life products – where are you now?

The last chance saloon

Short-life products provide undoubted functional value and can reduce carbon emissions, particularly when they increase the usable life of foods. This functional benefit is not captured by the current calculation methods. Even using the current calculations, short-life products are not only essential but are also often the best environmental solution. When designed correctly short-life products can be easily captured in the MSW recycling stream and be recycled multiple times to add value and functional benefits over many life-cycles.

Despite this, short-life products are under attack throughout the world because they are not seen as the valuable resource that they are and issues such as littering make them a very visible target.

The industry has a limited window of time to start to move the discussion on short-life products. It needs to ensure that these products are captured in the MSW stream and that their full value is realised.

Completing the chart

This chart is completed and assessed as for the previous charts.

Note: If you do not make short-life products then do not fill this chart.

Getting this right is crucially important.

Materials substitution is not the answer but the industry has to show that product stewardship is important and to ensure that we look after the valuable materials that we have been entrusted with.

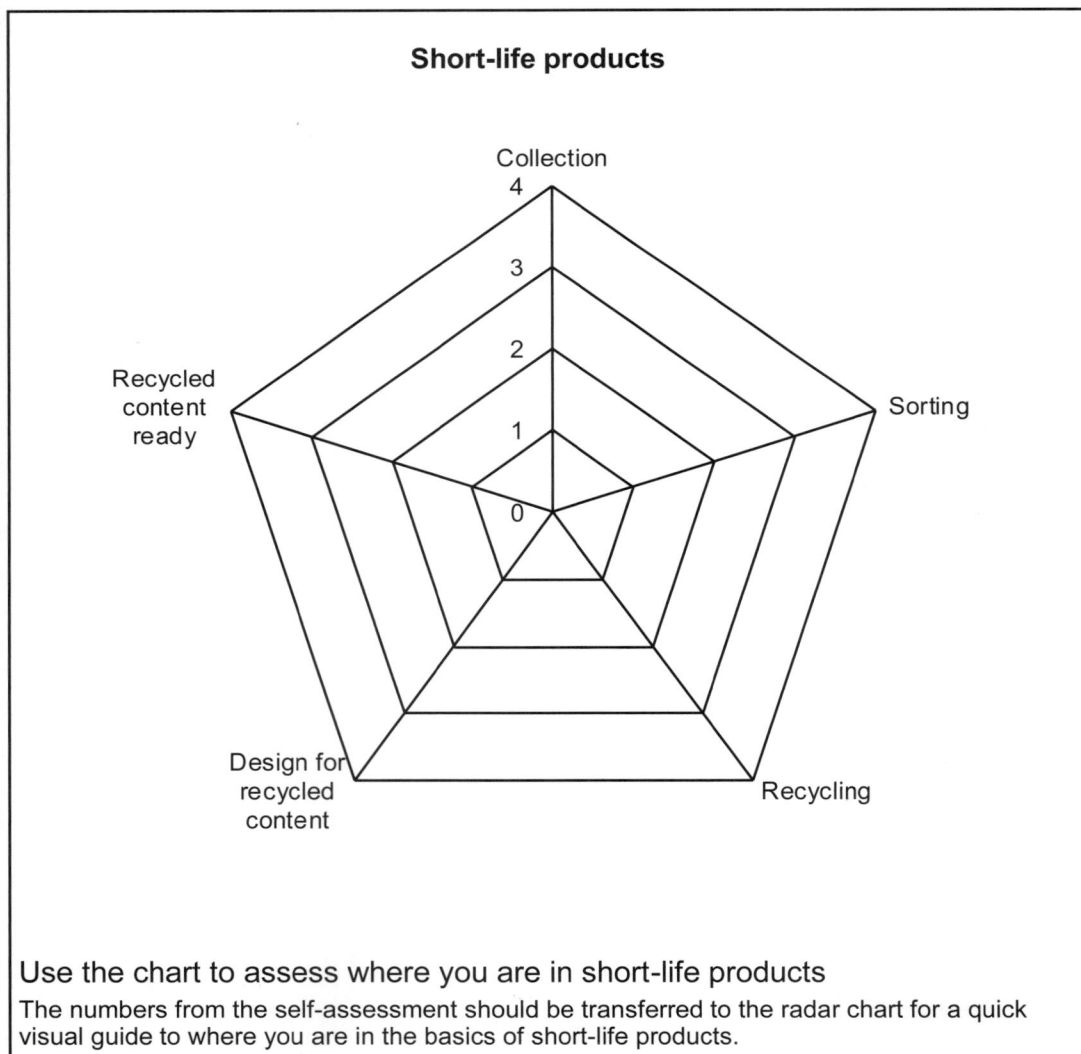

Short-life products

Collection

Sorting

Recycling

Design for recycled content

Recycled content ready

Use the chart to assess where you are in short-life products

The numbers from the self-assessment should be transferred to the radar chart for a quick visual guide to where you are in the basics of short-life products.

The benefits of short-life plastics are clear.

The difficulty lies in promoting the benefits in ways that society can recognise and in doing this quickly enough.

				Short-life products	
Level	Collection	Sorting	Recycling	Design for recycled content	Recycled content ready
4	Full review of all products carried out to make collection in the MSW stream easy. No small items present.	Full review of product range carried out to make MSW sorting easy. Products use mono-materials that are commonly recycled & are clearly marked.	Additives adhesives & inks removed/ replaced to improve recycling potential. Labels meet design requirements for size & removal. Easy to recycle.	All current designs reviewed for recycled content potential & changes made. Design process for new designs includes requirement for recycled content	All new & existing products & processes checked & revised to make suitable for recycled content use. Ready for recycled content.
3	Full review of all products carried out to make collection in the MSW stream easy. All small items trapped/tethered to allow easy collection.	Most products reviewed for ease of sorting in MSW stream. Products use mono-materials that are commonly recycled & are clearly marked.	Additives adhesives & inks reduced to a minimum to improve recycling potential. Labels meet design requirements for size & removal. Easy to recycle.	Some current designs reviewed for recycled content potential & changes made. Design process for new designs includes requirement for recycled content.	Majority of new & existing products & processes checked & revised to make suitable for recycled content use. Well prepared for recycled content.
2	Majority of products reviewed to make collection in the MSW stream easy. Most small items trapped/tethered to allow easy collection.	Limited review of products for ease of sorting in MSW stream. Products use compatible material mixtures to allow for easy recycling & are clearly marked.	Additives adhesives & inks reduced but still considerable. Very limited number of small labels used. Some products difficult to recycle & may be sent to landfill.	Some current designs reviewed for recycled content potential, changes still to be made. Design process for new designs does not include requirement for recycled content.	Limited consideration of recycled content in processes. No consideration of recycled content in products.
1	Some products reviewed to make collection in the MSW stream easy. Some small items trapped/tethered to allow easy collection.	Limited review of product range carried out to make MSW sorting easy. Products use compatible material mixtures to allow for easy recycling but not clearly marked.	Number of additives adhesives & inks not considered. Limited number of labels used. Products difficult to recycle & may be sent to landfill.	Very few current designs reviewed for recycled content potential, changes still to be made. No requirement in design process for new designs to include recycled content.	Limited consideration of recycled content in products. No consideration of recycled content in processes.
0	No consideration of product collection in the MSW stream. Product generates many small items that are not likely to be sorted or recycled.	No attempt made to make products compatible with sorting. Products use multiple & incompatible materials that make sorting difficult & landfill likely.	Large number of additives adhesives & inks used. Large labels that are difficult to remove. Almost impossible to recycle & will be sent to landfill.	No consideration of including recycled content in current or future designs.	No consideration of recycled content use in any products or processes. Not recycled content ready.
Score					

10.6   Medium-life products

Security in value

Typical medium-life products have a use timescale of 2-15 years and are used in most household appliances, cars, transport, computers, telecommunications and toys. Medium-life plastic products are generally components of a larger product, e.g., the plastic parts that go to make up a vacuum cleaner, and these components tend to be larger than short-life products.

The functional demands on medium-life products are much higher than for short-life products and they often require the use of engineering plastics rather than the commodity plastics that are most often used for short-life plastics. Indeed, some medium-life products do not use thermoplastics but use high-performance thermosets to achieve the desired performance (these products are not recyclable due to the permanent cross-linking in the material).

- **Tip** – Not all medium-life products are manufactured from engineering plastics. As my friend and colleague Alan Griffiths once said, "PP is the new mild steel". I often wish I had said that.

The impact is in the use

The approximate carbon footprint of a medium-life product is shown on the right. This shows the carbon footprint for the complete product rather than simply for the plastic components and as for many medium-life products most of the carbon footprint is in the use phase, e.g., a vacuum cleaner will have an electricity use that will be larger than any of the other impacts during the complete lifecycle. This will be similar for many medium-life products but, as for short-life products, the carbon footprint is not designed to identify carbon 'avoided' due to the use of medium-life products, e.g., the increase in fuel efficiency of cars due to weight reductions from using plastics.

- **Tip** – The use of plastics in many medium-life products improves their sustainability but they must be used well and disposed of carefully.

Medium-life products are largely ignored by environmental groups because of their high value proposition. Without medium-life products our current lifestyle would not be possible.

Materials simplification

The variety of materials and compounds used in medium-life products means that action to simplify the number of materials used is a critical issue. Whilst medium-life products represent only $\approx$ 20% of the volume of plastics used, they represent more than $\approx$ 80% of the materials and compounds used. In fact, it appears that almost every application needs a specific plastic type and the materials suppliers continue to generate new compounds and blends at a dizzying rate.

The processing industry needs to start to simplify the materials used to allow the critical mass for collection and recycling to be achieved.

- **Tip** – Even if it involves slightly more material then it may be wiser to opt for a 'standard' material than to choose a 'new' material with slightly better properties.
- **Tip** – Many of the materials used for medium-life products are fibre-reinforced and this can add additional difficulties or prevent recycling. Avoid fibre reinforcement, if possible, to improve recyclability.
- **Tip** – Avoid other material mixtures, e.g., bi-injection mouldings, flocked plastics and inset or outsert mouldings, to reduce materials contamination in recycling.

Material replacement

The substitution of existing materials with bio-based materials (see Section 4.14) to reduce the environmental impact may initially appear attractive but the raw

---

Medium-life products are caught in the middle.

They do not often use commodity plastics that are easily collected and they are not suitable for closed loop collection.

However, they do provide enormous benefits to society.

Materials simplification also:

- Reduces the effort of stocking of large numbers of materials.
- Reduces the potential for redundant stocks of materials.
- Reduces the difficulty involved in distributing large numbers of materials.

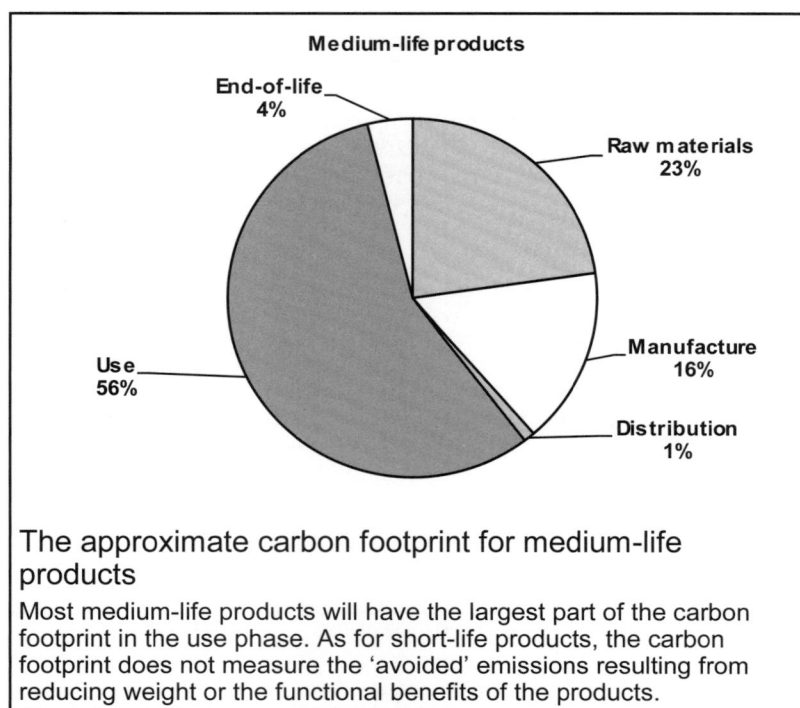

Medium-life products

End-of-life 4%
Raw materials 23%
Manufacture 16%
Distribution 1%
Use 56%

The approximate carbon footprint for medium-life products

Most medium-life products will have the largest part of the carbon footprint in the use phase. As for short-life products, the carbon footprint does not measure the 'avoided' emissions resulting from reducing weight or the functional benefits of the products.

material contribution to the carbon footprint for this type of product is relatively low (≈ 23%). 'Drop-in' bio-based materials, e.g., PA, PUR and PTT, may reduce the raw materials carbon impact but will still need to meet the high functional requirements of the products and material replacement is not going to be an easy task.

- **Tip** – The biggest benefit will be achieved by reducing the use contribution through product design improvements and there is more to aim for in the first place.
- **Tip** – LEGO™ is investing heavily in a search for materials that have a lower environmental impact to replace ABS in their products. Although, in my experience, the ubiquitous brick is more of a long-life product and does not have a use emission.

Re-use potential

The longer life of these products increases their potential for re-use where products can be made multi-purpose and modular to reduce the need to dispose of the complete product due to the failure of one part. The rise of the 3D printer and additive manufacturing may mean that medium-life products can be 'refreshed' at the end-of-life to become long-life products.

- **Tip** – Is anyone ready for distributed manufacturing and what would that do to our industry? I have my 3D printer ready.

Disposal, recycling and recovery

The wide range of plastics used for medium-life products means that most will not be collected in the MSW stream for mechanical recycling and, at end-of-life, will go to EfW plants or landfill (see Sections 4.11 and 4.12). In future, the high value of many of these materials may make mechanical recycling possible but the small volumes and current lack of effective collection systems means that at present the best option is landfill (or 'storage for the future' – see Section 4.12).

- **Tip** – These materials are too valuable for simple EfW, we need to retain their value for the future.
- **Tip** – Due to the functional requirements, it is unlikely that medium-life products will be made from biodegradable plastics (see Section 4.15).

Where the products are used in cars or electrical products, they will be captured by the WEEE and ELV waste streams (see Section 3.14) for treatment. After the removal of potentially salvageable large parts, e.g., engines, the product is treated to remove oils, air conditioning gases, copper in wiring looms and other valuable materials. Large plastic parts that are easily identified and removed are extensively mechanically

recycled, e.g., refrigerator liners (PS), batteries (PP – although this is mainly driven by the recycling of the lead in the battery), fuel tanks (PE-HD), bumpers and other large parts.

The remaining carcase is then shredded and sorted to produce a mix of recyclable metals and Automotive Shredder Residue (ASR). Most of the remaining medium-life products, e.g., interior trim, will be part of the ASR. This can be sorted and recycled to produce new compounds, e.g., ABS, PC or PC/ABS, for a variety of applications (see MBA Polymers – mbapolymers.com) but the majority of the plastics rich ASR will go to EfW or landfill (see Sections 4.11 and 4.12).

WEEE and ELV are not fully closed-loop processes despite the initial impression.

- **Tip** – Processors should maximise the recovery potential by clearly marking the material type on the product (see Section 3.5) and avoid using the 'Other' category (7).
- **Tip** – Consider using an extended marking system such as ISO 11469 which uses a symbol of the type >ABS< for single polymers or >PC+ABS< for blends. This can extend to high filler loadings with symbols such as >PP-MD30< to indicate a PP containing 30% mineral filler or products containing difficult to separate polymers which are shown as >PVC,PUR,ABS<. This is generally mandatory in the car industry.
- **Tip** – Try to use the most common materials for these applications, e.g., ABS, PC or PC/ABS to maximise recycling.

Get 'recycled content ready'

Processors of medium-life products also need to ensure that their products and sites are ready for increased recycled content. This means not simply using internal regrind but developing products and processes to incorporate commercial regrind at higher dose rates. This involves selecting suppliers, selecting materials and running the trials to make sure that processes and products function correctly.

- **Tip** – Start this now; it may be a long process to get it right.

Start to think about bio-based and biodegradable materials

Processors need to start to think about using bio-based materials (see Section 4.14) and biodegradable materials (see Section 4.15). The pressures for medium-life products may be less than for short-life products but it may be best to get ahead of the game.

Plastics are an integral part of all mobile phones, tablets, computers, telecommunication equipment and data storage media.

Can you imagine life without them?

It would not be too much fun.

Plastics have helped reduce the average weight of a car by 200 kg and this saves 500 litres of fuel/100,000 km.

Lightweighting in cars has not yet reached the limit but it is approaching it.

Polystyrene liners allow refrigerators and freezers to be kept clean and hygienic. Plastic foams are used to improve their insulation and reduce energy use.

10.7　Medium-life products – where are you now?

Caught in the middle

Medium-life products typically have the largest environmental impact during the use phase of the product lifecycle, i.e., when they are incorporated into an assembly. In many cases they have huge societal benefits and add greatly to our quality of life.

In no way does this mean that they are exempt from the need to improve sustainability or that their lifecycle impacts cannot be improved. There are significant actions that the industry can take to improve the value of these products to society, to retain their value at the end-of-life and to minimise their environmental impact.

These products are not currently subject to the pressures that are faced by products with a shorter functional life but this can easily change and the industry needs to be prepared for these changes. The high added-value of these products means that it is relatively easy to justify the changes needed to improve their sustainability credentials.

Completing the chart

This chart is completed and assessed as for the previous charts.

Note: If you do not make medium-life products then do not fill this chart.

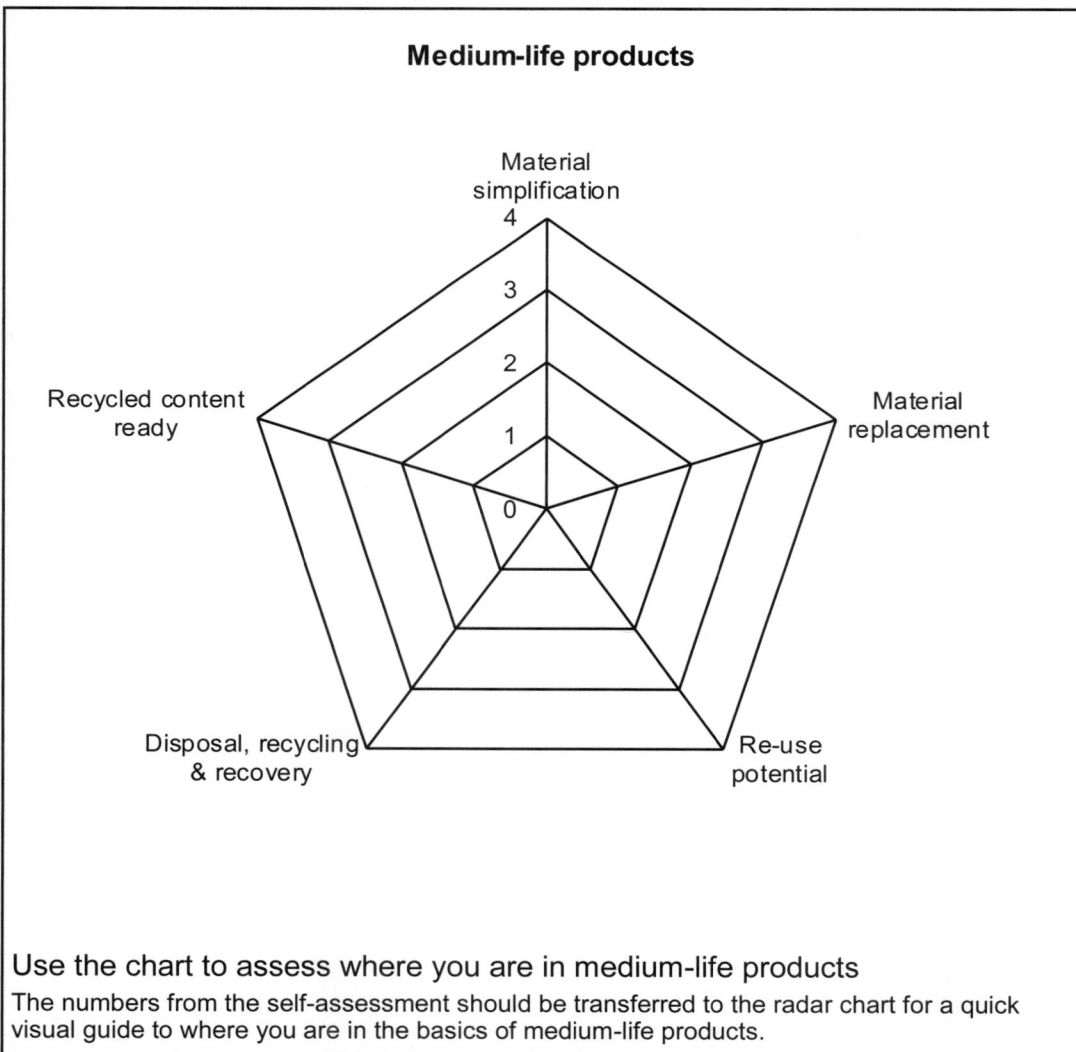

Medium-life products

Radar chart with axes: Material simplification, Material replacement, Re-use potential, Disposal, recycling & recovery, Recycled content ready; scale 0 to 4.

Use the chart to assess where you are in medium-life products

The numbers from the self-assessment should be transferred to the radar chart for a quick visual guide to where you are in the basics of medium-life products.

	Medium-life products				
Level	Material simplification	Material replacement	Re-use potential	Disposal, recycling & recovery	Recycled content ready
4	Standard materials used in all applications. Fibres, additives & material mixtures not used.	Bio-based materials investigated & used wherever possible & with agreement of customer. Recycled materials already used for many products.	Products are modular & allow component replacement in the event of failure or damage.	All products clearly marked with material type, grade & material designation. Extended marking system used to identify material.	All new & existing products & processes checked & revised to make suitable for recycled content use. Ready for recycled content.
3	Very few non-standard materials used. Fibres, additives & material mixtures reduced but potential to reduce further.	Bio-based materials investigated & used in some products as requested by customer. Recycled materials used in most products.	Products are largely modular & allow most components to be replaced in the event of failure or damage.	Extended marking system used on most products & basic marking system used on remaining products.	Majority of new & existing products & processes checked & revised to make suitable for recycled content use. Well prepared for recycled content.
2	Standard materials used in some applications. Fibres, additives & material mixtures reduced to a minimum.	Bio-based materials investigated but not currently used in any products. Recycled materials used in some (limited) products when required by customer.	Products are partially modular & allow limited components to be replaced in the event of failure or damage.	Basic marking system used on all products to aid recycling. Large number of grades used makes recycling difficult.	Limited consideration of recycled content in processes. No consideration of recycled content in products.
1	Significant number of non-standard materials used. Significant number of products use amounts of fibre, additive & multi-material combinations.	Bio-based materials not investigated or used in any products. Recycled materials not often used & only when required by customer.	Some products are suitable for limited re-use & repair in the event of failure or damage.	Basic marking system used on few products and only when required by customer.	Limited consideration of recycled content in products. No consideration of recycled content in processes.
0	Wide range of materials used, many of which are specialist grades. Most products use large amounts of fibre, additive & multi-material combinations.	No consideration or use of bio-based materials to date. No use of recycled material in any product.	Products are not suitable for re-use & repair. Product is obsolete as a result of failure of any part.	No marking used unless mandatory by customer. Wide range of engineering plastics materials used. Mechanical recycling is extremely unlikely.	No consideration of recycled content use in any products or processes. Not recycled content ready.
Score					

10.8   Long-life plastics products

You don't see most of them

Most long-life plastics products are used in the construction sector and these are not, in general, visible products. They are hidden from view and silently carry out their vital work. These products typically have a use timescale of > 15 years and, in some cases, up to and significantly over 50 years when the products are buried, e.g., pipes and membranes. Typical applications are:

- External/internal components – long-life plastics products are used extensively in the building industry and typical products are:
  - Windows and doors – PVC-U windows and doors.
  - Soffits and fascia boards.
  - Gutters and downpipes.

  The primary material used for these products is PVC-U.

- Pipes – long-life plastics products are used extensively in the pipes industry, typical products are:
  - Distribution water pipes (pressure) to the house or inside the house.
  - Drainage pipes (non-pressure) for wastewater, sewerage and stormwater drainage.
  - Gas distribution pipes.
  - Flood control and water management pipes.
  - Fibre cable distribution pipes.

  The primary materials used for these products are PE-HD, PE-MD and PVC-U with some other materials such as PB and PE-X.

- Carpets

  The primary material used for these products is PP.

- Power distribution and insulation.

- Damp-proof membranes for houses and sealing sheets for water reservoirs

- Insulation boards and insulation products.

  The primary materials used for these products are EPS and foamed PUR.

The list is almost endless and modern construction relies on long-life plastic products to save energy, reduce water losses and distribute vital services.

Using or losing energy

The environmental impact as measured by the carbon footprint for a long-life product will depend on whether the product uses/ loses energy during the use phase.

Energy products

A typical carbon footprint for energy using products, e.g., windows and pressure pipes, is shown in the diagram below. In both cases, the largest component of the carbon footprint is in the use phase.

- Windows have a use life of ≈ 25-50 years and windows use/lose energy throughout their lifetime. The energy lost through the window is far greater than the impacts of the other phases. Any action taken by the processor to reduce thermal losses, e.g., reduce the U-value (or increase the R-value for the USA), will be beneficial.

- Pressure pipes have a use life of > 50 years and need an energy input to pump water through the pipe. The impact of this pumping energy is greater than the impact of the other phases. Action taken by the processor to reduce pumping pressure losses will be beneficial.

For energy products, reducing energy use in the use phase is beneficial not only to the environment but, in the case of windows, also to the consumer even if it uses more material or costs more to produce.

- **Tip** – This is also the case for insulation products.

Non-energy products

Many long-life products do not have an energy component associated with the use phase and a typical carbon footprint for non-energy using products ,e.g., guttering and

Plastic gas and water pipes have guaranteed life of 50 years.

PVC-U doors and windows remove the need for chemical treatment of wood and the paint used to preserve the wood.

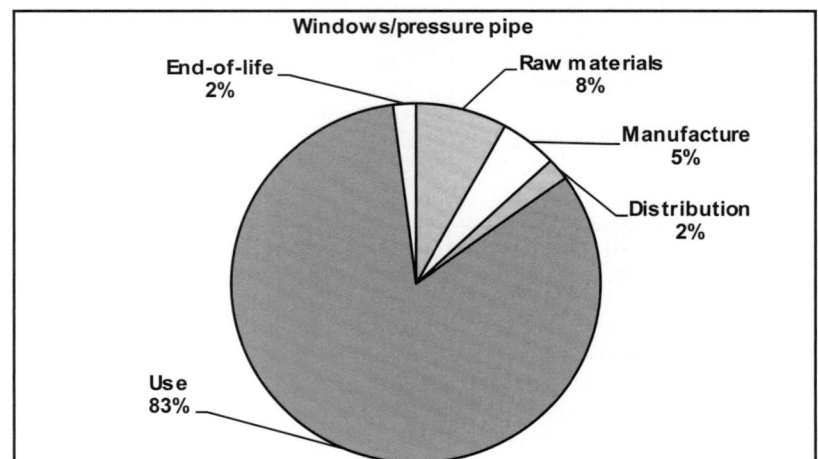

Windows/pressure pipe

End-of-life 2%
Raw materials 8%
Manufacture 5%
Distribution 2%
Use 83%

Carbon footprint for long-life 'energy' product

Long-life 'energy' products where there is an energy use/loss during the use phase will have the largest part of the carbon footprint in the use phase (even larger than for medium-life products due to the extended life cycle). Reducing energy use/loss during use is critical.

non-pressure pipes, is shown on the right. In these cases, the carbon footprint closely resembles that of a short-life product, i.e., most of the carbon footprint is in the raw materials and manufacturing phases (see Section 10.4). For this type of product, any action taken to reduce the amount of material used or the manufacturing impacts is most beneficial in carbon footprint terms but this is not always the most important issue.

- **Tip** – This is also the case for most membrane products and cable products.

This analysis highlights the limitations of carbon footprinting as a method of assessing product environmental impacts, i.e., it ignores the huge functional benefits of using long-life plastics products such as pipes and waterproof membranes, e.g., drainage water removal, flood protection and water ingress/egress prevention.

Are they better?

There have been a multitude of LCAs (see Section 1.5) carried out for long-life plastics products, both energy and non-energy, e.g., windows, pipes and gutters. In most cases, long-life plastics products more than compete with traditional materials and often perform significantly better.

- **Tip** – As with any LCA, one of the largest issues is in the assumptions and in the choice of the boundary conditions.
- **Tip** – LCA studies carried out to compare materials have found that the impact varies as much from plant to plant as it does from material to material.

The lifetime is the key

The use lifetime of these products is the key to their market success, adoption and environmental impact. Long-life plastics products need to be designed for durability and performance over the long term to be effective. These products are often invisible because they are built into the infrastructure or buried and access for replacement or maintenance is difficult or impossible without high disruption or cost. This means that the product longevity must be guaranteed through design and materials to survive and perform over an extended period.

Minimising the carbon impact of a waterproof membrane used to prevent water ingress to a building by reducing the thickness of the membrane might appear to be a logical step but this would be counter-productive if it led to water ingress in the short term through puncturing of a very thin membrane. The cost, and environmental impact, of having to replace a failed membrane would far exceed any savings

made by the thickness reduction.

- **Tip** – The long use phase and the need for validated durability mean that any material substitution or changes must be undertaken with care.

Plastic insulating foams in houses can save up to 150 litres of heating oil per year.

Plastic pipes have a low surface roughness that reduces the pumping energy needed, they are naturally resistant to corrosion, they can provide long-life leakproof joints and they are damage resistant and flexible enough to be laid in arduous conditions.

They contribute to delivering clean drinking water throughout the world and help to deliver UN SDG Goal 6 (see Section A1.6).

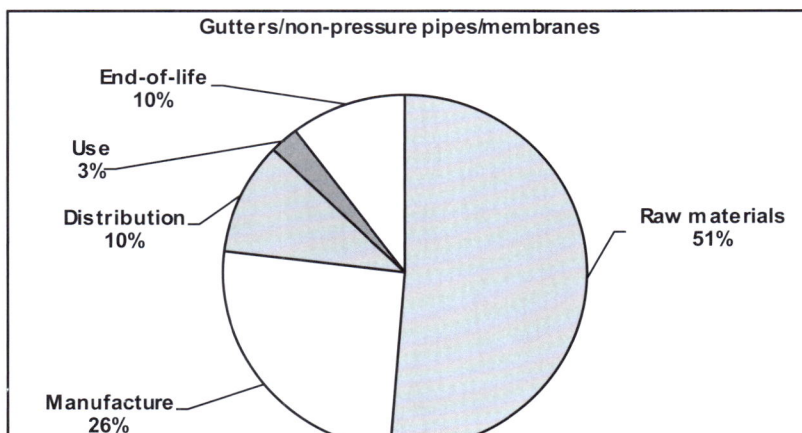

Gutters/non-pressure pipes/membranes

- End-of-life 10%
- Use 3%
- Distribution 10%
- Manufacture 26%
- Raw materials 51%

Carbon footprint for long-life 'non-energy' product

Long-life 'non-energy' products have most of their carbon footprint in the raw materials and manufacturing stages. The carbon footprint does not consider the other functional benefits, e.g., resistance to rusting and flexibility, that drive the uptake of plastics products.

10.9   Long-life products – leading sustainability

Building for the future

The use of long-life products in construction means that the products can effectively use the longevity of plastics to deliver functional benefits for many years. The long life cycle makes these products highly sustainable and they can compete with other materials in terms of either LCA or carbon footprint.

- **Tip** – In many cases, long-life products will be in place for the life of the structure or until it is significantly refurbished.

Reuse

Processors of long-life products internally re-use significant amounts of material (see Section 5.3) from scrap or head waste to feed back into production. This is clean material and the formulation can be precisely defined.

External re-use is an issue because most long-life products are in the construction sector and are built into the structure. This means that they are either produced or cut to size to fit the structure, e.g., windows are individually cut to fit the aperture and pipes are cut and jointed to fit the required layout. This means that the products are not generally suitable for re-use at end-of-life. Where these products are not cut to size, e.g., industrial machinery, then the life cycle is normally so long that reusing the product is not an option, i.e., the product is no longer made.

- **Tip** – Reusing long-life products is possible but difficult because they are custom-made.

Recycling

Most of the materials used for long-life products are commodity thermoplastics that are easily recycled using standard recycling technologies (see Section 4.6) and most should be capable of being marked with the appropriate recycling symbol (see sidebar).

Only a small proportion of the MSW stream (estimated at < 1%) is made up of long-life products, this is due to:

- The nature of the products, i.e., they can be in service for up to 100 years and will not be replaced during that time.
- The lack of collection at the MSW level, i.e., most of these products are not collected in standard recycling schemes because of the low volumes and they are treated as general waste.
- Most of the products are handled by the construction industry and are treated as industrial or construction waste.

- The presence of industry-led collection and recycling schemes to capture these valuable materials, e.g., VinylPlus®.

Industry-led

Most of the materials sectors have either set up, or are in the process of setting up industry-led recycling schemes to trap long-life materials as they are removed from buildings.

VinylPlus® is the scheme for recycling PVC in Europe and builds on work started in 1995 when PVC resin producers in Europe set out the first European Voluntary Charter to improve environmental performance in the industry. In 2010 VinylPlus® set out a new 10-year commitment and in 2011 set a target of recycling 800,000 tonnes of PVC/year by 2020.

The 2020 Progress Report showed a rate of 771,313 tonnes in 2019 (vinylplus.eu/). The group has now committed to recycling 900,000 tonnes/year by 2025.

- **Tip** – VinylPlus® gets things done across the value chain to improve the sustainability of long-life products and is rightly held in high esteem by industry and government.

Similar schemes are being implemented for other materials and products but have not yet reached the scale and success of VinylPlus®.

- **Tip** – Sorting or collecting at source is

PVC

PVC came under much criticism from Greenpeace in the 1990s. This was based on belief that chlorine was 'the Devil's Element'.

If you take chlorine (a toxic gas) and compound it with sodium (a reactive metal) the resulting compound is NaCl. This is that most harmless of all products – common table salt – without which we would die. The properties of a compound are not always related to the properties of the elements that make up the compound!

Typical long-life products

The windows, doors, guttering and drainpipes on this house will be in place for 25-50 years. When they are replaced the industry-led scheme will recycle the material back into new PVC-U products. The invisible pipes and membranes will be there for the life of the house.

always more effective than sorting afterwards and this requires care at all levels of the industry.

- **Tip** – A prerequisite for efficient recycling is the identification of long-life products before they enter the solid waste stream and become part of general waste.

Legacy additives

A potential issue for recycling any long-life product is that of regulatory changes during the life of the product. This is an issue for PVC-U where changes to regulations in the EU require the removal of lead from recycled materials produced many years ago (see Section 4.8). This is an impossible task and threatens the recycling of PVC-U products which contain firmly bound lead-based stabilisers from over 40 years ago.

- **Tip** – Long-life products should always be labelled (preferably permanently) to allow future separation and recycling. They will be there a long time so make sure that future generations know what they are and what they have in them. Let us not repeat the legacy additives debacle and let us be ready for legacy additives issues – even though this may need a 'crystal ball' at some stages.

Chemical recycling

The PVC industry has successfully established a complete recycling ecosystem for long-life PVC products and this encompasses not only mechanical recycling but also chemical recycling processes (see Section 4.10). Other materials industries need to emulate this type of system to close the loop in recycling.

Get 'recycled content ready'

Processors of long-life products need to ensure that their products and sites are ready for increased recycled content. This means developing products and processes to incorporate commercial regrind at high dose rates, selecting suppliers, selecting materials and running the trials to make sure that processes and products function correctly.

- **Tip** – Start this now; it may be a long process to get it right.

Disposal

If a long-life product is not captured by an industry-led recycling scheme at end-of-life then recycling is unlikely, i.e., very few of these products will be collected as part of the MSW stream. In this case, disposal is the remaining option and can be by:

- EfW (see Section 4.11) to recover the embodied energy. PVC, in particular, has been criticized in the past for potential HCl and dioxin emissions during incineration but these are easily scrubbed from exhaust gases in modern incinerators and removing PVC from a modern incinerator has no effect on dioxin emissions, which are minimal.

The industry is taking this further with processes designed to recover HCl from the exhaust gases for re-use in the chemical sector or to recover the salt (NaCl) for use as part of the PVC production process.

- Landfill (see Section 4.12) is a potential destination for long-life products as 'storage for the future'.

Disposal is not the preferred option for these valuable materials.

- **Tip** – Disposal via biodegradability is not an option for long-life products.

- **Tip** – Disposal by littering is not relevant for long-life products because they are not normally found in a state to be littered.

What do we do next?

Producers of long-life products may think that they do not have to worry about sustainability and can safely ignore it. Nothing could be further from the truth, the decisions made today will influence the actions the industry needs to take in the future. You may be gone but remember the acid test "What would my grand-children think of this?".

Window fabricators are a vital part of recycling PVC-U. They collect their offcuts from production and most have a recycling skip for installers to return windows (PVC-U and glass) that have been replaced with new PVC-U windows.

They can then sell this to recyclers and help both the environment and themselves.

The PVC industry, from a standing start, has managed to turn the market around and is now recognised as a leader in the field.

Congratulations to all.

The window industry is ready

The European standard for PVC-U window profiles (EN 12608:2003) defines several types of externally reprocessed and recyclable material (ERM_a, ERM_b, RM_a and RM_b).

- ERM_a and RM_a are materials produced from offcuts during fabrication of a window or from used PVC-U windows that have been recycled in a closed-loop scheme. These may be used for the core of new windows provided they are covered by at least 0.5 mm of virgin or internally re-used material.

- ERM_b and RM_b are similar to ERM_a and RM_a but may contain materials from other PVC-U products. These may not be used for new window production.

Internally re-used material (see Section 5.3) may be used up to 100% in a profile provided it meets the same specification as the virgin material.

This standard was released in 2003 and the industry had been developing the recycling processes since well before then.

10.10   Long-life products – where are you now?

A great advertisement for plastics

Long-life products use the benefits of plastics to great effect. They have a long functional life and deliver huge benefits to society across a wide range of areas and help to deliver the UN SDGs. They are probably the last of the plastics products that will come under pressure for environmental reasons.

This should not lead the industry to be complacent and not attempt to improve its sustainability credentials. It is always possible to improve and the industry needs to prepare now for the future. The long-life products have, potentially, the time to improve but the experience of the PVC-U industry shows that the landscape can change rapidly so that what was acceptable becomes unacceptable and society can remove your licence to operate.

The industry needs to start work now to lay the foundations for a sustainable future where plastics products are seen as fundamental to achieving sustainability in the broadest sense.

Completing the chart

This chart is completed and assessed as for the previous charts.

Note: If you do not make long-life products then do not fill this chart.

Long-life products are the flag-bearers of a sustainable plastics processing industry but only if they can fully embrace the circular economy.

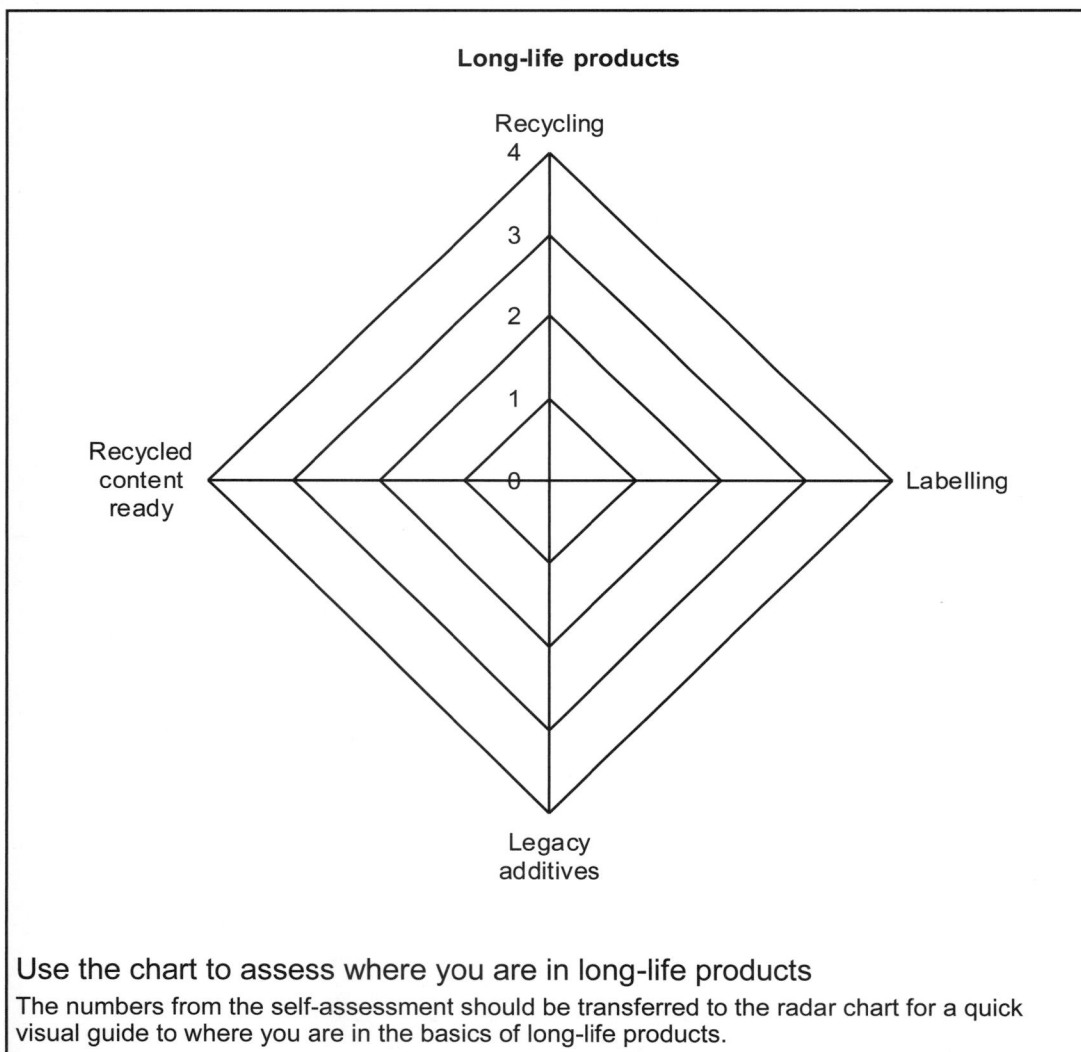

Long-life products

Recycling

Recycled content ready

Labelling

Legacy additives

Use the chart to assess where you are in long-life products
The numbers from the self-assessment should be transferred to the radar chart for a quick visual guide to where you are in the basics of long-life products.

Water contaminated with bacteria and faeces was once the largest cause of human disease and death.

This was not improved by doctors but by the engineers who provided pipes and clean water to cities and people.

Plastics pipes carry on this grand tradition and provide people around the world with clean and safe water.

			Long-life products		
Level	Recycling	Labelling	Legacy additives	Recycled content ready
4	Internal re-use of all available & acceptable material. External recycled material used where possible & acceptable. Company is member of industry-led scheme for recycling.	All products clearly marked with material type, grade & material designation. Extended marking system used to identify material.	Excellent knowledge & consideration of potential for legacy additives. Precautionary principle used in relation to additives. Advice sought from suppliers.	All products & processes checked & revised to make suitable for recycled content use. Ready for recycled content.
3	Internal re-use of all available & acceptable material. External recycled material used where possible & acceptable. Company is not a member of industry-led scheme for recycling.	Extended marking system used on most products & basic marking system used on remaining products.	Good knowledge & consideration of potential for legacy additives. Advice sought from suppliers.	Majority of products & processes checked & revised to make suitable for recycled content use. Well prepared for recycled content.
2	Internal re-use of all available & acceptable material. Very limited use of external recycled material.	Basic marking system used on all products to aid recycling. High number of grades used makes recycling difficult.	Poor knowledge of potential for legacy additives. Advice taken from suppliers only when offered.	Limited consideration of recycled content in processes. No consideration of recycled content in products.
1	Internal re-use of all available & acceptable material. No external recycled material used.	Basic marking system used on few products & only when required by customer.	Minimum current legislative conformance. Legacy additives & future legislative issues only considered in relation to business activities.	Limited consideration of recycled content in products. No consideration of recycled content in processes.
0	No consideration of recycling schemes. Virgin material used for all products. No recycled material used	No marking used unless mandatory by customer. High number of material grades used. Mechanical recycling is extremely unlikely.	Minimum current legislative conformance. No consideration of legacy additives or future legislative issues.	No consideration of recycled content use in any products or processes. Not recycled content ready.
Score				

Key tips

- The use timescale of a product defines what happens at end-of-life.
- The use timescale of a product tells processors what they need to do to improve sustainability.
- The use timescale of a product defines the value that society puts on the product. Even though the same materials can be used, short-life products are valued by society less than medium-life or long-life products.
- Very short-life products are under threat around the world because they are not valued by society and the resulting littering makes them very visible.
- Legislation and societal pressure are rapidly forcing the removal of very short-life products from the market
- The carbon footprint of short-life products is mainly in the raw materials and manufacturing phases.
- Short-life products such as food packaging have a functional life from the moment they begin to protect the food.
- The carbon footprint of food packaging is much less than the carbon footprint of the food it is protecting.
- Short-life products must be easy to collect to make it into the recycling system.
- Short-life products must be easy to sort to allow recycling.
- Short-life products must be easy to recycle to recover the materials.
- Short-life products must be designed to include recycled material.
- The carbon footprint of medium-life products is mainly in the use phase.
- The materials used for medium-life products should be as simple as possible to allow collection and recycling.
- Medium-life products should have a re-use potential to extend their life.
- Medium-life products should always be marked with the material designation to allow sorting and recycling.
- Producers of medium-life products need to get 'recycled content ready' to prepare for the circular economy.
- The carbon footprint of long-life products that use or lose energy is mainly in the use phase
- The carbon footprint of long-life products that do not involve energy is mainly in the raw materials and manufacturing phases.

- The long use life of long-life products makes them very sustainable and they use the properties of plastics to the best effect. These products are valued by society because of their utility and long life.
- Long-life products need industry-led recycling schemes because they will rarely appear in MSW.
- Well-run and industry-led schemes benefit both industry and society. They are far preferable to legislation.
- Producers of long-life products need to get 'recycled content ready' to prepare for the circular economy.

Viewing plastics in terms of the use timescale allows the industry to develop the right strategy for each product.

Whilst processors do not always control the end-of-life phase of their products, they have a duty of care to use their skills to make their products conform to the principles of the circular economy as much as possible.

If they don't do it then governments will.

Chapter 11

Social responsibility

Sustainability encompasses three aspects; financial sustainability, environmental sustainability and social sustainability. Financial sustainability is largely covered in a previous book[1] and the bulk of this book covers environmental sustainability as it applies to plastics processors. However, it would be remiss not to cover the important aspects of social sustainability and this chapter looks at the social responsibility aspect of sustainability.

In the plastics processing industry, sustainability is often viewed simply from the 'product' perspective, i.e., if we get the material and product right then we have solved the sustainability issue. For some plastics processors this may be appropriate but, for most processors, social sustainability is just as important. The UN SDGs (see Section 1.16 and Appendix 1) are not simply concerned with products and production but also with making the world a better place for more people. As an industry we can contribute to the UN SDGs through our products and the many benefits that they bring but we can also contribute by being 'good citizens' and by contributing to the social aspect of sustainability.

Company staff can make an invaluable contribution to efforts to improve social sustainability within a company but it is also incumbent on every company to improve social sustainability as part of the social responsibility that every company is encouraged to have towards the staff, stakeholders and general society.

In the past, the responsibility of any company was seen as being primarily to maximise value for the shareholders, but the drive for social sustainability has broadened the concept of good corporate governance to recognise the interests of other stakeholders. This is central to the concept of social sustainability and social justice and is part of the larger drive to internalise all the impacts of a company's operations.

The rise of world-wide supply chains means that many companies are purchasing products from remote suppliers instead of local suppliers. Whereas in the past it was possible to visit and inspect suppliers because they were 'down the road', today many suppliers are on a different continent in very different countries with very different social norms. This raises potential concerns in areas such as human rights, labour practices and fair operating practices which can damage a company's reputation and licence to operate.

Social responsibility is broader than the Corporate Social Responsibility (CSR) agenda which began in the 1960s for companies. It covers all types of organisations, not simply companies. CSR never really went away but it is now part of the sustainability agenda.

Things just got more complicated and plastics processors need to be in a position to respond to new pressures and demands.

• 1. Kent, R.J. 2018. 'Cost management in plastics processing', Elsevier.

11.1   The social responsibility framework

A multitude of initiatives

The UN SGDs described in Chapter 1 and in detail in Appendix 1 are a strong driver for social responsibility. They are a framework for considering social responsibility in the broad sense and a method of measuring human progress at the world scale. At the company scale there are several initiatives to help companies develop social responsibility and to integrate this into their operations. These have been developed at different times and have slightly different models but are complementary in most aspects, i.e., they all seek to provide guidance on how companies can develop and manage social responsibility.

Organisation for Economic Cooperation and Development (OECD)

The OECD 'Guidelines for Multinational Enterprises' (www.oecd.org/daf/inv/mne) were first adopted in 1976 and updated for the fifth time in 2011. These cover:

- Human rights.
- Employment and industrial relations.
- Environment.
- Combating bribery, bribe solicitation and extortion.
- Consumer interests.
- Science and technology.
- Competition.
- Taxation.

The guidelines are non-binding but they provide a set of principles and standards for social responsibility in a global environment and have been agreed at government level by all OECD member governments.

- **Tip** – The OECD guidelines are obviously aimed at government level and at multi-nationals but the concepts are still applicable to all companies.
- **Tip** – The 2011 update and addition of a human rights clause to the OECD guidelines brought them more into line with ISO 26000 which was published a year earlier (see below).
- **Tip** – ISO 26000 has many more practical guidelines for integration into a company's operations and for implementation than the OECD guidelines.

UN Global Compact (UN GC)

The UN GC (www.unglobalcompact.org) was launched in 2000 with 9 principles and

updated in 2004 with an additional anti-corruption principle. The UN GC is based on ten principles in four sections:

Human Rights

- 1. Businesses should support and respect the protection of internationally proclaimed human rights.
- 2. Businesses should make sure that they are not complicit in human rights abuses.

Labour

- 3. Businesses should uphold the freedom of association and the effective recognition of the right to collective bargaining.
- 4. The elimination of all forms of forced and compulsory labour.
- 5. The effective abolition of child labour.
- 6. The elimination of discrimination in respect of employment and occupation.

Environment

- 7. Businesses should support a precautionary approach to environmental challenges.
- 8. Businesses should undertake initiatives to promote greater environmental responsibility.
- 9: Businesses should encourage the development and diffusion of environmentally friendly technologies.

Anti-Corruption

- 10: Businesses should work against corruption in all its forms, including extortion and bribery.

The UN GC is designed to establish a culture of ethics and integrity so that companies are not only financially successful (profit) but also successfully manage environmental (planet) and social (people) impacts.

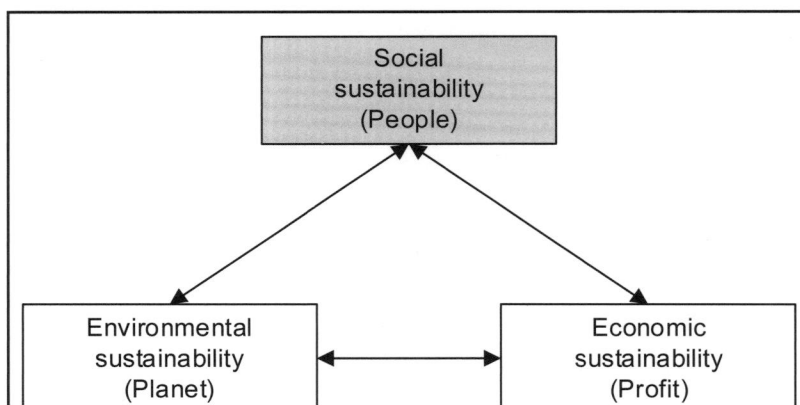

The sustainability aspects

Sustainable companies focus on the three key areas of profit, planet and people. It is only by achieving a balance between these areas that true sustainability can be achieved. Being good at one of these does not make up for being poor at one of the others.

The UN GC promotes responsible business, i.e., social responsibility, and is the largest initiative in the world with over 10,000 companies in 166 countries and over 70,000 public reports submitted (2020). The advantage of the UN GC is that it is strongly focused on business (although it also has a 'cities programme') and takes a very focused approach to social responsibility. The UN GC has two excellent publications that give details of the programme, these are:

- UN GC Guide to Corporate Sustainability (www.unglobalcompact.org/library/1151).
- UN GC Management Model (www.unglobalcompact.org/library/231).

The UN GC is non-binding and provides no assessment, monitoring or regulation of claims. Joining the UN GC requires a simple Letter of Commitment from the CEO and an annual 'Communication on Progress' which can be linked to an annual report (see Chapter 12).

- **Tip** – The UN GC principles are flexible and aspirational rather than prescriptive but they provide a set of fundamental principles.

ISO 26000

ISO 26000, 'Guidance on social responsibility', was published in 2010 and has seven principles of social responsibility (see Section 11.3):

- 1. Accountability.
- 2. Transparency.
- 3. Ethical behaviour.
- 4. Respect for stakeholder interests.
- 5. Respect for the rule of law.
- 6. Respect for international norms of behaviour.
- 7. Respect for human rights.

These seven principles are then applied to seven core subjects:

- 1. Organizational governance.
- 2: Human rights.
- 3. Labour practices.
- 4. The environment.
- 5. Fair operating practices.
- 6. Consumer issues.
- 7. Community involvement and development.

Included in the seven core subjects are 37 issues to be addressed. Each issue is described in general terms and a set of related actions and expectations provided for reference. For some of the issues, examples of practical theoretical actions are provided for reference.

The detail and guidance provided in ISO 26000 is extensive and it provides an excellent template for addressing, identifying and taking action in the social responsibility area of sustainability. The remainder of this chapter will focus on understanding ISO 26000 in the context of plastics processing.

- **Tip** – We strongly support the use of ISO 26000 as the most appropriate and accessible template for managing social responsibility for plastics processors.
- **Tip** – ISO 26000 can easily be linked to all of the other main social responsibility frameworks, e.g., the OECD guidelines, to the UN GC, to the UN SDGs and also to the GRI standards (see Chapter 12).

The business case for most economic and environmental sustainability issues is relatively clear, it is less clear for social responsibility issues and this sometimes makes it harder to motivate management.

This doesn't make it less important, just less easily quantified.

The most fundamental contribution a company can make towards achieving social responsibility is to be financially successful while having a well-defined and high standard of ethics in the treatment of the environment, employees and the wider community.

11.2 Social responsibility – ISO 26000

It is unusual

ISO 26000:2010 was an unusual standard in ISO terms because:

- Social responsibility had not previously been covered by ISO and was a totally new subject area.

- ISO standards have traditionally been related to either technical issues and testing, e.g., 'ISO 489:1999 Plastics – Determination of refractive index', or management systems standards, e.g., 'ISO 14001:2015 'Environmental management systems' (see Section 2.6).

ISO 26000 is neither a technical nor a management system standard.

- ISO standards have traditionally been developed as a result of national standards bodies or trade bodies putting forward recommendations for the development of a standard. In the case of ISO 26000, the recommendation for development came from within ISO itself.

- This was the first time that ISO itself had used a 'balanced stakeholder' approach which sought input from industry, NGOs, consumers, governments, labour organisations and other interested parties.

ISO 26000 is a guidance document for audiences ranging from SMEs to NGOs (and everything in between), It is applicable to the plastics processing industry and provides a template for improving social responsibility.

The structure of the standard

ISO 26000 is not a traditional 'management systems standard' and does not follow the Annex SL structure (see Chapter 2). The broad outline of the standard is:

- Clause 1: Scope.

- Clause 2: Terms and definitions.

- Clause 3 Understanding social responsibility – an introduction to social responsibility.

- Clause 4: Principles of social responsibility – this introduces the seven principles of social responsibility (see Section 11.1).

- Clause 5 Recognizing social responsibility and engaging stakeholders – this covers the two fundamentals of social responsibility, i.e., recognition of social responsibility and identification and engagement with stakeholders.

- Clause 6: Guidance on social responsibility core subjects – the bulk of the standard and covers the seven core subjects and the issues to be addressed for each core subject (see Section 11.4 and subsequent sections).

- Clause 7: Guidance on integrating social responsibility throughout an organization – this provides guidance on implementing social responsibility in a company, e.g., understanding, integration, communications, progress review and improvement (see Section 11.10).

ISO 26000 is available from the many standards organisations and we recommend buying a copy. There are also many text books on ISO 26000, most of these cost less than the actual standard (and are easier to read).

Quick start

The subsequent sections cover the seven core social responsibility subjects and the relevant issues in detail, but what do you need to get started quickly? Some quick start actions are:

- Read the seven main principles (see Section 11.3) and analyse your company's performance in each of the core subjects and issues. It may be helpful to look at each core subject through the product life-cycle to more easily identify issues at each stage (see diagram on opposite page).

- **Tip** – A much longer and detailed 'Issue Matrix' using a risk analysis approach is available at www.learn2improve.nl.

- Create a stakeholder map listing the main stakeholders (see Section 11.10), how the company's activities affect them, what they expect of you, what you are already legally required to do and what you do on a voluntary basis.

- **Tip** – Get the view of the organisation through internal and external stakeholder engagement and don't forget the importance of two-way communication.

- Analyse what you already do, what ISO 26000 guidance recommends and what you can do to get better. Due diligence or 'gap analysis' can be used for this.

- **Tip** – The "Issue Matrix" referred to above can also be used for this.

- Define and refine objectives and targets for core subjects and appropriate issues. This should include consideration of the relevance and significance of the issues, how they will be managed and how they will be reported (see Chapter 12).

- **Tip** – Do not forget to look at the business case for social responsibility.

Every company must create value for the key stakeholders – just satisfying one stakeholder is the road to ruin.

You are probably being 'socially responsible' in some way already.

Why not list what you already do and link these actions to the core subjects?

Providing paid time off for volunteer work could be linked to Issue 1 of Community involvement and development (see Section 11.9).

You might be amazed at what you already do but have never thought about in terms of 'social responsibility'.

The two social responsibility fundamentals:

- Recognise social responsibility.

- Identify and engage with stakeholders.

- Start to integrate social responsibility into the company. This needs to start with the top management.
- **Tip** – Section 7 of ISO 26000 provides more guidance on starting out.

Certification to ISO 26000

ISO states quite clearly "ISO 26000 does not contain requirements and is not a management system standard. It is not intended or appropriate for certification purposes or regulatory or contractual use"[1] and "...ISO 26000 offers guidance and is not appropriate for certification. Any company that claims to be ISO 26000-certified would be misrepresenting the intent and purpose of the standard"[1].

Whilst it would be hard to make a clearer statement, even before ISO 26000 had been released there were companies offering certification and, in some cases, national standards bodies were developing 'son of ISO 26000 management systems' to allow certification and to specify national social responsibility requirements. This continues.

Verification to ISO 26000

Clause 7.5.3 of ISO 26000 does state: "...claims can be verified through internal review and assurance. For enhanced credibility, these claims may be verified by external assurance."

This verification process (see Annex A of ISO 26000) could include reporting to the GRI standards and external auditing of the report (see Section 12.11).

- **Tip** – ISO 26000 is for guidance only. It was deliberately designed not to be certified against.

ISO 26000 is easily linked to the Global Reporting Initiative (GRI) standards for reporting (see Chapter 12).

Map the process and products to find the value chain for the company both in terms of money and in terms of social responsibility.

- 1. 'ISO 26000 and OECD Guidelines', ISO 2019.

	Raw materials	Manufacturing	Use	End-of-life	Logistics
Governance (Clause 6.2)					
Human rights (Clause 6.3)		Working conditions (health and safety). Fair wages. No forced labour.		Working conditions (health and safety). Fair wages. No forced labour.	
Labour practices (Clause 6.4)	Healthy and safe work place.	Healthy and safe work place. Staff development.		Waste export. Healthy and safe work place.	Healthy and safe work place.
Environment (Clause 6.5)	Resource depletion. Granule escape.	Energy efficiency. Water efficiency. Waste minimisation.	Energy efficiency. Single use products.	Waste collection. Recycling options.	Reduce CO_2 emissions. Improve efficiency.
Fair operating practices (Clause 6.6)	Bribery or cartels in pricing.	Fair competition. Property rights (incl. intellectual).	Bribery or cartels in pricing.		
Consumer issues (Clause 6.7)	Sustainable consumption.	Honest marketing. Accurate labelling. Food safety.	Food safety. Single-use products.	Take back programme. Recycling.	Food safety.
Community involvement (Clause 6.8)		Local suppliers. Local community initiatives.		Littering and dumping. Waste export.	

High impact	Low impact	Not applicable

The seven core subjects of ISO 26000 related to the life-cycle of a plastic product

A short example of how the seven core subjects of ISO 26000 can be related to the life-cycle of a plastic product. Not all the issues listed will be relevant for all products and the table should be modified for each of the company's specific products, i.e., issues should be added or deleted depending on the product life-cycle. The table should be coded to show the relative impact of the issue for the company.

11.3　Organisational governance

Sustainability starts at the top

Unlike the other core subjects of ISO 26000, organisational governance does not have specific issues and this is because the subject is at the heart of social responsibility. Organisational governance is 'how the company is run'. This involves owners and top management leading the company by practicing (and being seen to practice) the seven principles of social responsibility. It also means promoting these values to all levels of the company.

- **Tip** – This requires 'leadership' and not simply 'management'. Management is following the rules; leadership is about defining direction and setting the standards.

Many companies will already have mission statements, or similar, that define the company's values. They will also have a variety of policies, e.g., health and safety policies, environmental policies, procurement policies, quality policies and employment policies. Good governance is about integrating the key issues of social responsibility into all of these policies and practices so that it is automatic and simply part of the way the company does business.

Good governance is not simply about being seen to be good, it is also about helping companies to improve the 'triple bottom line' of environmental, social and economic sustainability. This benefits the business, all of its stakeholders, the natural environment and the community.

The seven principles

Good governance affects all the other core subjects and the application of the seven principles in governance is a key to social responsibility. The seven principles provide the framework for social responsibility and are particularly important in governance and management. The seven principles are:

Accountability

Accountability is taking clear responsibility for the company's policies and decisions. It is obviously most relevant for the top management but also applies to everyone in the company who makes decisions. Anybody acting for the business must be accountable.

Transparency

Transparency is being open about policies, decisions and activities with all stakeholders. It is also about reporting these appropriately to the stakeholders (see Section 12.2).

Ethical behaviour

Ethical behaviour is deciding what is the 'right thing to do' on a day-to-day basis. Ask yourself: "Would I be comfortable if my actions were to become public knowledge?" If the answer is "Yes", then you are probably being ethical. Transparency means that you will have to answer the question sooner or later.

- **Tip** – Being ethical in decision making does not stop you from making a bad decision. You might make the wrong decision for all the right reasons. As long as you did not do it deliberately then you have still been ethical (as long as you are accountable and transparent).

Respect for stakeholder interests

Social responsibility involves being aware of stakeholders and their interests (see Section 11.10). It is also about respecting their interests and responding to their concerns, it does not mean that you have to agree with them or be governed by them.

Respect for the rule of law

Respect for the law means complying with all applicable local laws. In some countries these may not be enforced adequately but companies should still respect and comply with them.

The core issues are interdependent and there is no specific 'best' way to address the core issues. This depends on the company but it is recommended that the high impact areas be addressed first whilst at the same time considering how these affect the other issues.

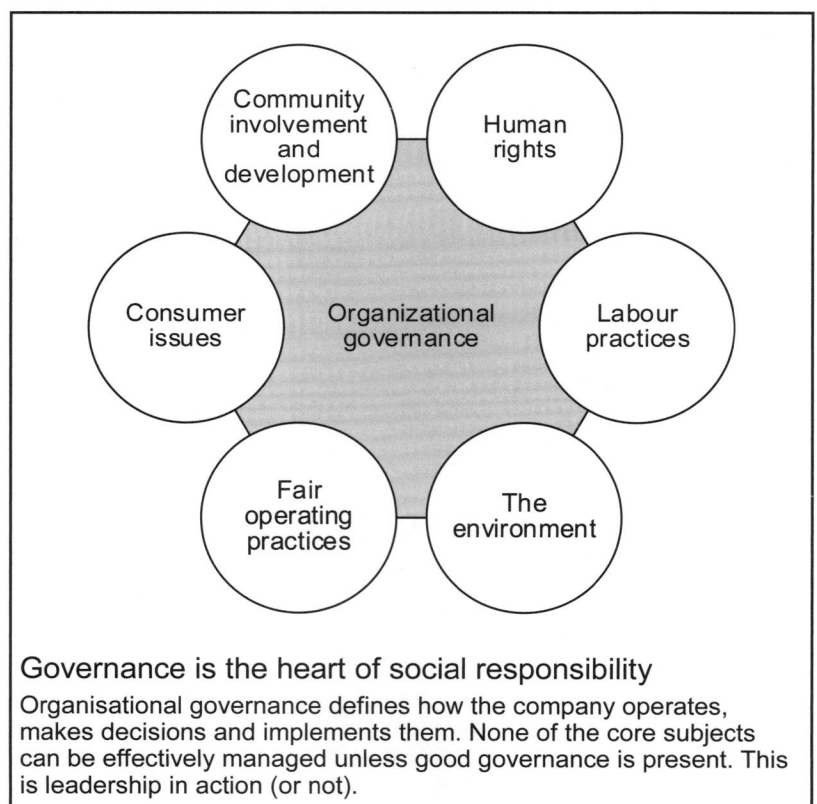

Governance is the heart of social responsibility

Organisational governance defines how the company operates, makes decisions and implements them. None of the core subjects can be effectively managed unless good governance is present. This is leadership in action (or not).

Respect for international norms of behaviour

Companies should respect international norms of behaviour. If a country does not have adequate laws (or inadequate enforcement) this does not mean that companies can ignore the principles of international law or intergovernmental agreements, e.g., UN or ILO.

Respect for human rights

Human rights are a key concept of ISO 26000, companies should have respect for the human rights of all stakeholders and ensure the fair treatment of any vulnerable populations in the stakeholders.

- **Tip** – Employee 'Codes of Conduct', when supported by top management and managed well, can help to embed the seven principles in a company's operations. Just having a 'Code of Conduct' will not. This is still about 'leadership'.

- **Tip** – Extend the 'Code of Conduct' to suppliers and contractors as part of the purchase order or contract (see Section 5.11). Driving social responsibility down the supply chain is essential.

Things to do

- Examine all of the current company policies and practices with a 'social responsibility hat' on and examine how these can be modified to include the seven principles.

- Set some short and long-term goals for integration of the seven principles into the existing company policies and practices.

- Improve accountability by recording decisions and responsibility for completion.

- Ensure that power and responsibility are linked, known and delegated appropriately at all levels of the company to improve accountability and transparency.

- Start to develop incentive systems for participation and performance in social responsibility and encourage staff to contribute to decisions and actions.

- **Tip** – This can also improve staff involvement and contribution in other areas.

- **Tip** – Never forget that there is a balance here. The company needs to survive and prosper now and into the future. It is no use being socially responsible if you go broke.

- **Tip** – If organisational governance is thought to be lacking then ISO 19600: 2014 'Compliance management systems', provides good guidance on planning, operating and evaluating compliance systems.

SMEs – does it matter for you?

Many SMEs do not see social responsibility as a key issue and think that it is only something for large companies. This is understandable but perhaps short-sighted when many SMEs are also part of long supply and value chains. These supply chains often extend to countries where labour practices are a concern and where the SME has no direct contact, unlike large companies. In such cases, the SME and the owner are at risk of reputational damage if they fail to adequately manage social responsibility (see Section 5.10).

Good organisational governance and social responsibility are not difficult when approached using an integrated approach and can add value to an SME's investors and customer base by reassuring them that any risk is minimised and controlled. Implementation and communication may use resources and cost money but it is the 'right thing to do'.

- **Tip** – Owners with strong personal values will find that good organisational governance and social responsibility are easy to implement because they are probably already present in the company.

Reporting

Reporting on organisational governance can be carried out by reporting the GRI disclosures GRI 2-18 to GRI 2-39 (see Section 12.7).

For most people there is an instinct to be fair to our neighbours. Organisational governance is simply formalising this.

Do we really have to say this?

I have worked in, and with, many plastics processing companies.

Some I liked and respected, some not so much, but I have never yet been in one which did not meet the basics of good corporate governance.

Still, everybody can get better and ISO 26000 is all about proving it.

11.4 Human rights

We are all human

We are all human and the respectful treatment of all human beings, irrespective of their race, gender, ethnicity, religion, age or immigration status is fundamental. Human rights are basic rights that have been agreed upon by the United Nations and apply all over the world.

Companies have a duty, within their sphere of influence, to respect human rights and to avoid actively or passively infringing the rights of others.

The issues

The eight issues to be considered under human rights are:

Issue 1: Due diligence

In many countries, human rights are guaranteed by legislation and well protected but there are countries where, despite local legislation, the human rights situation is less than acceptable. Companies need to be careful operating in countries such as these and carry out 'due diligence' on their operations. Due diligence is simply finding out about and identifying human rights impacts that result from the company's activities and taking action to resolve or prevent negative impacts.

- **Tip** – Due diligence applies not only to human rights but also to all of the core subjects. As such it is addressed in greater detail in Clause 7.3.1 of ISO 26000.

- **Tip** – Integrating human rights throughout the company is part of due diligence.

Issue 2: Human rights risk situations

The risks of human rights abuse are not equally spread and they are far higher in some situations than in others, these are 'risk' situations and companies need to be aware of these and take particular care. Risk situations may include (but are not limited to): civil war, natural disasters, mining or other industries that affect natural resources, operating in corrupt environments, operating in countries with poor legislation and poor control or operating in countries with strict but abusive control.

- **Tip** – It is not difficult to identify these countries, simply read the newspaper.

Risk situations make it particularly important to consider a company's potential impacts on human rights, and to plan to have a positive, not negative, impact.

Issue 3: Avoidance of complicity

It is not enough to focus only inside the company, companies should also avoid being complicit in abuse, i.e., staying silent when risks are seen or clearly identified, assisting or benefiting from abuse by others. This involves finding out about human rights in suppliers or business partners to ensure that they are acting responsibly.

- **Tip** – If in doubt, ask for a declaration or 'proof of compliance' with human rights requirements from suppliers (and follow it down their supply chain). Still, check compliance because people have been known to lie.

Issue 4: Resolving grievances

It is inevitable that grievances, from staff or other stakeholders, will arise and companies need to set up mechanisms to resolve these. The mechanisms should be fair and effective but not prejudice access to other legal rights or mechanisms should this be necessary.

Issue 5: Discrimination and vulnerable groups

Discrimination based on race, gender, ethnicity, religion, age or immigration status etc. is prohibited by international law and is a fundamental human right. Groups that have suffered from historic and repeated discrimination are also vulnerable to further discrimination.

Vulnerable groups can include (but are not limited to): women, children, indigenous peoples, and migrant workers and their families. ISO 26000 encourages the identification of vulnerable groups within a company's sphere of influence and the establishment of mechanisms to ensure that the company does not discriminate against them or take unfair advantage of them.

- **Tip** – A positive attitude to personnel diversity in a company can lead to improved performance – this is a chance to be good and also to benefit.

Issue 6: Civil and political rights

This about respecting the fundamental civil and political rights of every individual within a company's sphere of influence. Typical rights that need to be respected are (but are not limited to): the right to freedom of opinion and expression, the right to peaceful assembly and association, the right to own property, the right to seek and impart information, the right to due process and a fair hearing before disciplinary measures.

It is not just you!

The risk of human rights violations is not simply about whether you operate or participate in business in particular countries or not. It is also about whether your raw materials or product suppliers operate in these countries.

'Due diligence' requires an analysis of the risks and you should consult widely and purchase wisely (see Section 5.10).

Providing an internal or external contact or even a 'whistle blower' scheme may not be sufficient in countries with restricted or no freedom of expression.

Issue 7: Economic, social and cultural rights

Alongside the other rights, these are regarded as fundamental rights. Typical rights that need to be respected are (but are not limited to): health, education, food, clothing, social protection and culture. Companies should use due diligence to ensure that its activities do not infringe on these rights, particularly for any vulnerable groups identified in Issue 5.

- **Tip** – As with any other right, this needs to be considered in the local context but without breaching any of the fundamental rights.

Issue 8: Fundamental principles and rights at work

These are the labour issues of human rights. Typical rights that need to be respected are (but are not limited to): the right to freedom of association, the right to collective bargaining, the elimination of forced or compulsory labour, the abolition of child labour and the elimination of discrimination in employment and occupation.

- **Tip** – These rights are also considered in the labour practices core subject.

Reporting

Reporting on human rights can be carried out by reporting the GRI disclosures GRI 406 to GRI 412 (see Section 12.10).

Mapping human rights in the value chain

Mapping human rights to the value chain (as shown on the right) gives a global view of areas of weakness. Improvement action can be taken to remove or reduce exposure in the vulnerable areas.

Note: The areas where a company has reduced influence are shaded.

Human rights and the value chain		
Area	Good practice	Poor practice
Raw materials and procurement	ISO 20400 used for sustainable procurement. Human rights clause in standard contract and checked for compliance.	No supplier engagement on human rights concerns. Purchases made from high-risk areas.
Strategy and planning	Clear corporate guidance on human rights concerns. Implications for all operational areas assessed.	No human rights policies or guidance established. Employees not aware of required conduct.
Investor relations	Investors assured of human rights policies. New investor requirements identified and assessed.	No consideration of human rights requirements of existing investors and no consideration of requirements of new investors.
Sales and marketing	Assessment of effect of any human rights concerns to company, brand, competitive advantage or product success.	No assessment or awareness of human rights concerns.
R&D	R&D strategy considers long-term risks and opportunities of products in context of human rights.	R&D is simply interested in the product.
Production	Procedures and policies in place to prevent forced labour and/or discrimination via due diligence. Fair wages paid.	Potential concerns not considered.
Information	Human rights performance information available and widely distributed.	No information on human rights performance available.
Internal audit	Guidelines for facilitation payments in place. Validated as part of audit.	High potential for illegal funds transfer, i.e., bribery.
Reporting	Reporting via GRI 406 to GRI 412.	Statutory financial reporting only.
Human resources	Procedures and policies in place to manage any concerns or local issues.	No consideration of human rights in HR policies or practices.
Use	N/A.	N/A.
End-of-life	Conditions (human rights) of workers in recycling areas considered and verified.	Potential concerns not considered.
Logistics	Sustainable procurement practices used. Human rights clause in standard contract and checked for compliance.	Potential concerns not considered.

11.5   Labour practices

They are not commodities

Social responsibility does not see labour as a commodity but as human beings and there is, naturally, a cross-over between labour practices and human rights. Labour practices are regulated in most parts of the world and governments have the prime responsibility for good labour practices. This does not mean that companies can shirk their responsibility for implementing good international labour practices even if local law or enforcement is inadequate.

- **Tip** – Responsibility goes beyond locations that the company owns or controls

The issues

The five issues to be considered under labour practices are:

Issue 1: Employment and employment relationships

Employing people should never be a coercive relationship and the employment relationship involves both the rights and obligations of the employee and the rights and obligations of the employer. These are often legal rights and obligations and employers must recognise and fulfil these. A significant gap often exists in the legal rights and obligations between companies and either self-employed workers or staff employed by suppliers or sub-contractors. Companies should be very careful of relationships with suppliers who may use unfair labour practices (see Sections 10.4).

- **Tip** – Consider giving priority to local people or companies in employment and procurement.
- **Tip** – Wages should be paid directly to directly employed staff.
- **Tip** – Discrimination, child labour and forced labour are part of overall human rights and dealt with in Section 10.4.

Issue 2: Conditions of work and social protection

As with employment relationships, conditions of work are often set by local regulations and companies should comply with these. Irrespective of the local requirements, companies should provide good work conditions and social protection for all staff, e.g., living wages from freely chosen work, working hours, and fair contracts etc.

- **Tip** – As always, the activities of key suppliers should also be examined.

- **Tip** – Collective bargaining, discrimination and the elimination of forced or compulsory labour are part of overall human rights and dealt with in Section 10.4.

Issue 3: Social dialogue

Social dialogues are the negotiations, consultations and information exchanges that occur between employers and other stakeholders, e.g., staff, government, trade unions and trade bodies. These dialogues are vital in avoiding disputes, managing change and in gaining the trust of staff and stakeholders. Social dialogue, when carried out responsibly and effectively, can improve staff satisfaction and productivity whether it is through trade unions or elected worker representatives.

- **Tip** – Social dialogue does not have to be only about working conditions, it can also be about sustainability.
- **Tip** – Companies should respect the right of staff to join unions and to engage in collective bargaining as part of social dialogue.

Issue 4: Health and safety at work

A healthy and safe workplace is an essential in any industry and plastics processing is no exception. In most countries, there is strict legislation for health and safety but companies also have a social responsibility to promote and maintain the health and safety of their staff. This means assessing risks, removing or reducing hazards, providing training to reduce the risk and providing safety equipment if the risk cannot be removed.

Good health and safety systems and practices are not simply part of labour practices, they are part of good business practice wherever the company is located.

- **Tip** – Health and safety should also consider mental health.
- **Tip** – Health and safety systems are covered in detail in Section 2.12.

Issue 5: Human development and training in the workplace

Developing the capabilities of staff to encourage personal and professional growth is not simply about the staff, it is about equipping the company to grow now and in the future. Companies need to provide opportunities for staff for personal development, for skills development and for career training.

- **Tip** – One company in the UK provides

In 2020, the corona virus shut-downs around the world brought out a range of responses.

Most companies treated their staff well and attempted to manage the situation. Others treated their staff appallingly.

We will remember your response.

There is often a debate about the cost of training and the riposte is:

If you think the cost of training is high then consider the cost of not training.

£500 for training for every staff member every year. They can spend it on any type of training from flower arranging to rock climbing to work related training. It is rated as one of the best places to work in the UK. Staff turnover is effectively zero, I wonder why?

Wouldn't you want to work there and what would you spend your 'training budget' on?

Reporting

Reporting on labour practices can be carried out by reporting the GRI disclosures GRI 401 to GRI 405 (see Section 12.10).

Mapping labour practices in the value chain

Mapping labour practices to the value chain (as shown on the right) gives a global view of areas of weakness. Improvement action can be taken to remove or reduce exposure in the vulnerable areas.

Note: The areas where a company has reduced influence are shaded.

Labour practices and the value chain		
Area	Good practice	Poor practice
Raw materials and procurement	ISO 20400 used for sustainable procurement. Labour practices clause in standard contract and checked for compliance.	No supplier engagement on labour practice concerns. Purchases made from high-risk areas.
Strategy and planning	Clear and fair corporate policy on labour practices. Implications for all areas assessed.	No labour practices policy or guidance established. Employees not aware of required conduct.
Investor relations	Investors assured of labour practices policy. New investor requirements identified and assessed.	No consideration of labour practices requirements of existing investors and no consideration of new investor requirements.
Sales and marketing	Assessment of effect of any labour practices concerns to company, brand, competitive advantage or product success.	No assessment or awareness of labour practices concerns at any level.
R&D	R&D strategy considers long-term risks and opportunities of products in context of labour practices.	R&D is simply interested in the product.
Production	Health & safety system in place. Structured training for all staff in both H&S and in production skills.	Potential concerns not considered.
Information	Labour practices performance information available and widely distributed.	No information on labour practices performance available.
Internal audit	Guidelines for facilitation payments in place. Validated as part of audit.	High potential for illegal funds transfer, i.e., bribery.
Reporting	Reporting via GRI 401 to GRI 405.	Statutory financial reporting only.
Human resources	Procedures and policies in place to manage any concerns or local issues. All staff contracts include conditions of work. Training plan available for every staff member.	No consideration of labour practices in HR policies or practices.
Use	N/A.	N/A.
End-of-life	Conditions (labour practices) of workers in recycling areas considered and verified.	Potential concerns not considered.
Logistics	Sustainable procurement used. Labour practices clause in standard contract and checked for compliance.	Potential concerns not considered.

11.6 The environment

This is part of social responsibility too!

This is part of ISO 26000 but it is also part of the overall sustainability agenda, i.e., environmental, social and economic sustainability. ISO 26000 has four guiding principles and these are:

- Environmental responsibility – a company should accept responsibility for any environmental impacts caused by its activities and to try to improve performance and those of other companies within the sphere of influence, e.g., suppliers.

- Precautionary approach – where there are threats to the environment then a company should not wait for certainty before taking measures to avoid or limit any damage, i.e., it is better to be cautious rather than do serious damage to the environment.

- Environmental risk management – risk management techniques (see Section 2.18) should be used to assess and reduce environmental risks.

- Polluter pays – companies should internalize the cost of any pollution caused by activities, products or services, i.e., companies should pay for cleaning up any pollution.

ISO 26000 recommends the assessment of the relevance of various approaches:

- Life cycle assessment (see Section 1.5).
- Cleaner production and eco-efficiency (see Chapters 4–8).
- A product-system approach.
- Use of environmentally sound technologies and practices (see Chapters 4–8).
- Sustainable procurement (see Section 5.10).
- Learning and awareness raising.

All of these approaches are also recommended by this book and dealt with in more detail in the sections referenced.

- **Tip** – Environmental issues will vary in relevance with the company operations.

The issues

The four issues to be considered under the environment are:

Issue 1: Prevention of pollution

This includes any conventional type of pollution, e.g., air, water, soil, and waste, as well as other forms of pollution such as noise, odour, light, vibration or electromagnetic radiation. As with an Environmental Management System to ISO 14001, the standard recommends the assessment of 'aspects and impacts' as an improvement methodology (see Section 2.5), .

- **Tip** – Chemicals or substances identified as being of concern, e.g., carcinogens, ozone-depleting substances and other hazardous chemicals, should be avoided and replaced with environmentally friendly alternatives.

- **Tip** – Companies should have accident prevention plans for any identified risks and emergency plans to manage any incidents that occur.

Issue 2: Sustainable resource use

All of the actions and expectations in this issue have been dealt with in more detail in other parts of this book, e.g., the reduction in resource use through actions to such as energy efficiency (see Chapter 6), water efficiency (see Chapter 8), materials efficiency (see Chapters 4 and 8).

- **Tip** – Carrying out the actions listed in other areas of this book should adequately prepare a company to comply with this issue.

Issue 3: Climate change mitigation and adaptation

This issue is in two parts: the first is mitigation of climate change, i.e., reducing the company's contribution to climate change, and the second is adaptation to climate change, i.e., taking action to reduce the company's vulnerability to climate change.

Reducing the amount of energy used in processing is one of the most effective methods of mitigating climate change for a plastics processor and this is covered in detail in Chapter 6. Adapting to climate change is something that the plastics industry does very well and many products, e.g., plastic pipes are already used to help adapt to climate change.

- **Tip** – Recording the carbon footprint (see Section 7.1), reporting this (see Section 12.9) and reducing it (see Chapter 6) are all recommended actions to mitigate climate change.

- **Tip** – Adapting to climate change can include developing technology that will ensure the security of drinking water supply (see Section A1.6) and other actions to increase the resilience of society (see Section 1.16).

The environmental area of social responsibility is broadly stated and in some ways is less specific and useful than other standards for environmental protection. There are no metrics and simply broad statements.

Indeed, this is a criticism of the whole of ISO 26000 – the broad range of 'motherhood and apple pie' statements make it hard to disagree with the sentiments behind the standard but also hard to fulfil the high expectations of the standard.

Still, this is the best standard we have so far and it is the start of a journey towards achieving the social aspect of sustainability.

Issue 4: Protection of the environment, biodiversity and restoration of natural habitats

Ecosystems around the world are adversely affected by human development and yet we depend on these for our quality of life. This issue covers areas such as: valuing and protecting biodiversity, valuing, protecting and restoring ecosystem services, using land and natural resource sustainably and advancing environmentally sound urban and rural development.

- **Tip** – Companies should protect local ecosystems and natural habitats and promote environmentally sound development in their development plans.

Reporting

Reporting on the environment can be carried out by reporting the GRI disclosures to the GRI 300 series (see Section 12.9).

Mapping environmental issues in the value chain

Mapping environmental issues to the value chain (as shown on the right) gives a global view of areas of weakness. Improvement action can be taken to remove or reduce exposure in the vulnerable areas.

Note: The areas where a company has reduced influence are shaded.

Environment and the value chain		
Area	Good practice	Poor practice
Raw materials and procurement	Raw materials assessed for environmental impact and resource depletion. Suppliers assessed for granule escape.	No supplier controls.
Strategy and planning	Environmental impact assessed as part of strategy. Environmental and energy management systems in place.	No assessment of environmental impacts. No management systems in place.
Investor relations	Climate change risks, e.g., stranded assets, assessed and quantified.	No assessment of climate change risks.
Sales and marketing	Products designed to mitigate climate change. Product range assessed for climate change risk.	No consideration of environmental issues in sales and marketing.
R&D	Design for sustainability and recycling carried out for all products.	No consideration of sustainability or recycling at the design stage.
Production	Energy and water use minimised. Waste minimised and recycling maximised.	Poor use of energy and water use in production. Process generates large amounts of waste.
Information	Carbon footprint and environmental impact data available.	No carbon footprint or environmental impact data available.
Internal audit	Energy and water use minimised. Waste minimised and recycling maximised. Systems regularly audited (internally and externally).	No systems = no audit.
Reporting	Reporting via GRI 300 series.	Statutory financial reporting only.
Human resources	Energy and environmental training for all staff.	No training.
Use	Product energy and water use minimised. Resource depletion and environmental pollution through single-use products minimised.	No consideration of energy or water use in use phase. High amount of single-use of products.
End-of-life	Products designed for recycling and easy waste collection/segregation. Recycling options/labels marked on product. Take-back scheme developed.	Products are difficult to recycle. Products not labelled for recycling.
Logistics	Logistics CO_2 impacts minimised by efficient vehicles and routing.	No environmental consideration of logistics. Price is the only factor.

11.7 Fair operating practices

It's not fair!

Like most children, as a child, my daughter always had a sense (sometimes misplaced) of fairness and was often heard to say "It's not fair". We all have a personal sense of 'fairness' and this should be translated into a 'company sense of fairness'. This means ethical treatment of all stakeholders in the company's operations and promoting fairness both up and down the supply and value chain.

Promoting and leading in fair operating practices is not simply showing social responsibility, it is also good business. We would all rather deal with companies that demonstrate fairness and they will retain our business because we will trust them, and trust is the foundation of a good business relationship.

The issues

The five issues to be considered under fair operating practices are:

Issue 1: Anti-corruption

Corruption can cover a wide variety of unfair practices, e.g., bribery (both paying or taking), conflicts of interest, fraud, money laundering and abuse of power. The provision of, and maintenance of, 'whistle-blower' schemes is highly recommended to allow staff to report instances of corruption without being subject to recriminatory action by superiors.

- **Tip** – Employees need clear rules of conduct but even more important is company leadership that makes it clear that corruption in any form is not acceptable and will be dealt with very firmly.

- **Tip** – Never ever, ever try to find out who the whistle-blower was. Jes Staley, the chief executive of Barclays bank attempted to track down a whistle blower. He was fined more than £640,000 by the Financial Conduct Authority, was issued with a formal written reprimand and had his bonus cut by £500,000. He apologised but was this 'leadership from the top'?

- **Tip** – Fortunately the practice of suppliers giving 'Christmas gifts' to customers has decreased in recent years. These traditionally went to purchasing and whilst in most cases were innocent, there was always a risk. There are two possible strategies:
  - Every 'gift' (wine and food) is collected and sent to the local home for the

elderly for their Christmas party. A great way to brighten up their day.
  - Every 'gift' is collected and raffled amongst the staff with the proceeds of the raffle donated to charity.

Either strategy is good and the staff will respect you for being fair.

Issue 2: Responsible political involvement

Politics affects us all and it is legitimate for companies to support public politics and help to develop policy but it is not legitimate to try to use undue influence or to use manipulation or to try to undermine the political process.

- **Tip** – Be transparent about political involvement in all forms and avoid financial contributions that could be interpreted as undue influence or corruption.

Issue 3: Fair competition

Every company would like to be the sole supplier in a market because it would make things much easier, but this is rarely the case and the principle of fair competition deals with issues such as cartels (which are illegal in most countries), bid rigging and dumping products on the market to drive competitors out of business (also illegal in many countries).

Companies must comply with all local legal requirements for competition and never engage in cartels to ensure fair competition.

- **Tip** – If you had to cheat to win then you really didn't win at all.

Issue 4: Promoting social responsibility in the value chain

The concept of social responsibility is new and needs promotion up and down the supply chain to prosper. This can be as simple as treating suppliers fairly by prompt payment (it is good business anyway) and by treating customers fairly by dealing with their concerns promptly (equally good for business). Equally, it can involve using sustainable procurement (see Section 5.10) to help drive social responsibility to suppliers.

Simply being seen as a role model for social responsibility can increase awareness and create a successful company.

- **Tip** – We know who the socially responsible companies in our society are. They are also the successful ones that are envied for their great reputation. Try to be

---

If it doesn't feel fair then it probably isn't.

"He was a cruel man, but fair".

Stig O'Tracy on Dinsdale Piranha after having his head nailed to the floor for an unknown reason.

Monty Python sketch.

I once overheard a main contractor in the construction industry saying "We'll never make a profit on this job; it looks like we'll have to push a few subbies (sub-contractors) under to make any money at all".

Ethical conduct?

I walked off the job, took the loss on the chin and hung up on them the next time they called.

---

like them.

Issue 5: Respect for property rights

Property rights, both physical and intellectual, are the foundation of any successful company or society. They are also a fundamental human right. Without the right to own and control physical and intellectual property then no company would ever prosper; who would invest in a company or strive to make a company successful if it could be arbitrarily taken away from you. This means that companies should respect the property rights of others and pay fair compensation for any property that it acquires or uses.

- **Tip –** Counterfeiting or piracy is an infringement of property rights.

Reporting

Reporting on fair operating practices can be carried out by reporting the GRI disclosures GRI 202 to GRI 207 (see Section 12.8) and GRI 415 (see Section 12.10).

Mapping fair operating practices in the value chain

Mapping fair operating practices to the value chain (as shown on the right) gives a global view of areas of weakness. Improvement action can be taken to remove or reduce exposure in the vulnerable areas.

Note: The areas where a company has reduced influence are shaded.

Fair operating practices and the value chain		
Area	Good practice	Poor practice
Raw materials and procurement	Code of Conduct provided to suppliers. Guidelines for facilitation payments in place. Validated as part of audit.	No controls or policies in place for procurement and fair operating.
Strategy and planning	Anti-corruption and other fair operating practices in Code of Conduct. Whistle-blower process in place and respected. Led from the top and enforced.	No Code of Conduct in place. No whistle-blower process in place. Top management unconcerned about fair operating practices.
Investor relations	Clear communication of fair operating practices to investors.	No investor communication about fair operating practices.
Sales and marketing	Code of Conduct provided to customers and enforced. Fair competition used in all sales and marketing processes. Anti-corruption measures applied in sales.	Only measure of sales and marketing is success in sales. Potential for corrupt practices.
R&D	High awareness of potential for property rights infringement. Property rights respected in new product design.	No consideration of property rights in new product design.
Production	Property rights respected in machinery and process developments.	No consideration of property rights in process design.
Information	Whistle-blower incidents investigated by third party with full access to information.	No information available on fair operating practices.
Internal audit	Whistle-blower incidents investigated by third party with full access to information.	No system = No audit.
Reporting	Reporting via GRI 202 to GRI 207.	Statutory financial reporting only.
Human resources	Code of Conduct provided to all employees and enforced.	No Code of Conduct = no enforcement.
Use	N/A.	N/A.
End-of-life	N/A.	N/A.
Logistics	Code of Conduct provided to logistics suppliers and enforced.	No Code of Conduct = no enforcement.

11.8   Consumer issues

The ultimate user of the product

Most plastics processing companies do not have consumer issues. This is not because our products are not used by consumers but because the plastics industry is an 'enabling' industry and most of our products are sold as business to business (B2B) products where fair operating practices (see Section 11.7) are more important.

Where companies do sell directly to consumers then there are specific actions, e.g., protecting the health and safety of consumers and data protection and privacy, that must be carried out to meet legal requirements, and there are other actions that should be carried out to meet consumer expectations and to grow the business, e.g., provide good customer service and complaint resolution.

- **Tip** – In terms of ISO 26000, a consumer does not have to purchase the product, if somebody makes use of the product then they are a 'consumer'.

- **Tip** – Consumer issues will vary in relevance with the company operations.

The issues

The seven issues to be considered under consumer issues are:

Issue 1: Fair marketing, factual and unbiased information and fair contractual practices

Consumers need to make informed decisions about products and to do this they need clear and accurate information on the product or service. This does not mean providing excessive information on tests, etc. but providing appropriate and practical information that will allow consumers to compare products. Fair marketing also means not targeting marketing or advertising at vulnerable groups, e.g., children.

- **Tip** – 'Greenwashing' (see Section 1.20) is obviously not fair marketing or unbiased information.

In addition to clear information, consumers should also be provided with clear and easy to understand information about the cost, purchase conditions (if any) and warranties. Unfair contract terms such as limiting liability or excessive contract periods should never be used.

- **Tip** – Examine contracts for the 'fine print'; contracts should be written in clear and easily understandable language.

Issue 2: Protecting consumers' health and safety

Any product or service must not put the consumers' health at risk and must be safe to use. Companies should design and test products (including appropriate risk analysis) to achieve this and to avoid any potential risk to consumers. If necessary, user information (see above) should make consumers aware of any potential risks but this is not a substitute for a well-designed product.

- **Tip** – Risk analysis should look at all the potential users and their specific circumstances throughout the whole life cycle and not simply during use, e.g., during storage, transport or disposal. Think broadly, from long experience, consumers do the strangest things.

- **Tip** – Ensure that all materials used are safe and approved.

Issue 3: Sustainable consumption

Sustainable consumption is effectively one of the 17 SDGs (see Section A1.12) and the plastics industry has both much to contribute and much to improve. We have to offer consumers products that are both socially and environmentally beneficial and that operate as efficiently as possible. Sustainable consumption begins at the design phase and getting this right can greatly increase the sustainability of our products.

- **Tip** – Design products to minimise material use in production (see Section 3.4).

- **Tip** – Design products to minimise energy use during the use phase (see Section 3.11).

- **Tip** – Minimise packaging use and provide recycling or disposal systems for it (see Section 9.8).

- **Tip** – Design products for re-use, repair or recycling at end-of-life (see Section 3.12).

Issue 4: Consumer service, support, and complaint and dispute resolution

No matter how good your product, there will always be situations where customer service or support is needed and sometimes this will escalate into a customer complaint or even a dispute. In many countries there are legal requirements for consumer protection and these must be met by every company. However, meeting the legal minimum requirement is never a recipe for success.

> Protecting the health and safety of consumers is a paramount duty for any company.

> Good customer service is not only socially responsible but also simple good business practice, whether you deal in B2B or direct with consumers.

- **Tip** – Customer service and complaint handling is an opportunity to 'delight the customer' and to forge a great relationship with them. Research shows that exceptional customer complaint handling can actually lead to a more satisfied customer than one who never had a complaint.

- **Tip** – ISO 9001, ISO 10001, ISO 10002 and ISO 10003 all provide guidelines for good practice in dealing with complaints handling and dispute resolution.

Issue 5: Consumer data protection and privacy

In most countries, consumer privacy is regulated and controlled by law, e.g., the General Data Protection Regulations in the EU, and every controller or processor of personal data must be registered and conform to these regulations. Companies should only collect necessary personal data (after permission has been sought and explicitly granted), inform the consumer that it is being collected and what it will be used for, protect the data and keep it private and allow consumers to see what information you keep on them when requested.

Issue 6: Access to essential services

The plastics industry does not fall under 'essential services' in the terms of ISO 26000.

Issue 7: Education and awareness

Companies should try to educate consumers on their rights and responsibilities with regard to the product. This means providing the necessary information to allow informed choices.

Reporting

Reporting on consumer issues can be carried out by reporting the GRI disclosures GRI 416 to GRI 419 (see Section 12.10).

Mapping consumer issues in the value chain

Mapping consumer issues to the value chain (as shown on the right) gives a global view of areas of weakness. Improvement action can be taken to remove or reduce exposure in the vulnerable areas.

Note: The areas where a company has reduced influence are shaded.

Consumer issues and the value chain		
Area	Good practice	Poor practice
Raw materials and procurement	Raw materials safety data checked and approved. MSDS sheets available for all materials.	Poor raw materials acceptance, supplier decides with little evidence.
Strategy and planning	Sustainable consumption strategy in place. Minimal consumer sales and strategy is B2B.	No strategy in place.
Investor relations	Investors informed of any business risks from consumer issues.	Investors not informed of business risks from consumer issues.
Sales and marketing	Sales and marketing ensure that all marketing is fair and unbiased. No greenwashing of product advertising. Packaging minimised and disposal considered.	Advertising is not checked against 'fair, factual and unbiased' criteria. 'Greenwashing' used to promote product.
R&D	Honest and accurate labelling considered as part of design. Material use minimised during design. Energy use during use phase minimised. Risk analysis carried out for products during design.	No consideration of labelling, material or energy use during design phase. Risk analysis not used.
Production	Quality management and control used to control product quality and safety.	Product quality and safety poorly controlled.
Information	Controls in place for appropriate use and protection of consumer information.	No controls for personal data. Wide access available.
Internal audit	Internal audit of customer complaints to improve product.	No system = No audit.
Reporting	Reporting via GRI 416 to GRI 419.	No information available on fair operating practices.
Human resources	N/A.	N/A.
Use	Risks during use quantified and minimised by design or by consumer instructions.	Poor or non-existent consumer instructions.
End-of-life	Products designed for recycling and easy waste collection/segregation. Recycling options/labels marked on product. Take-back scheme developed for post-consumer use products.	No consideration of end-of-life phase at any stage.
Logistics	N/A.	N/A.

11.9 Community involvement and engagement

Part of a community

Companies do not operate in isolation but as part of a local community. One of the biggest contributions that a company makes is in providing employment which promotes economic development. Companies need to recognise that their long-term success depends on the success of the community in which they operate and actively contribute to community development. This can range from sports sponsorship to educational initiatives but these programmes must be carried out sensitively to gain the most benefit for both the company and the community.

- **Tip** – Relationships that benefit both the company and the community make the company a valued part of the community.

The issues

The seven issues to be considered under community involvement and development are:

Issue 1: Community involvement

Many companies are already involved in their community and this is to be praised. This issue is concerned with working with the community stakeholders to determine what is best for the community rather than imposing what the company thinks is best. Community involvement can make existing good work more effective.

- **Tip** – Consult with stakeholders to see what they really want you to do.
- **Tip** – This is an opportunity to assess existing initiatives (which may be 'ad hoc') and see if they can be improved.

Issue 2: Education and culture

Promoting and supporting education is not only good for the community but also provides a company with skilled workers for the future. Working with local schools through open days, tours or even scholarships provides excellent community links and supports your industry.

- **Tip** – Take what you have learnt from this book, make a quick presentation on sustainability and what you are doing and show this to local schools.
- **Tip** – There are schemes in the USA and the UK for 'Polymer Ambassadors' and both use teachers or industry people to demonstrate the benefits of plastics to society.
- **Tip** – My university education was paid, in part, by a scholarship from a local

company. It wasn't a lot of money to them but I still feel a strong loyalty to them.

Local culture is very important and companies can increase social cohesion by supporting and valuing local culture and traditions. Do not underestimate the power of culture in defining a company's reputation in the community.

- **Tip** – I once worked for an international company located in a rural area. There was no understanding, or attempt to understand, the local culture. Relationships were rocky!

Issue 3: Employment creation and skills development

Creating local employment is one of the biggest benefits that any company brings to the local community but this can be made even better by encouraging the development of skills in the community. This can happen naturally through staff training in technology and skills but companies can think wider, e.g., training local young people via internships, apprenticeships or a 'training academy'. In addition to being socially responsible, these can act as 'talent spotting' opportunities, i.e., everybody gains skills but great ones get jobs.

- **Tip** – Think about disadvantaged people in the community and do not just look for people 'who look like you'.
- **Tip** – Think local when selecting suppliers and outsourcing activities (see Section 5.12).

Issue 4: Technology development and access

Technology, and access to it, has the potential to help local communities develop and thrive. Companies can help with jobs and skills but can go further by developing technology to help local communities.

- **Tip** – Is it is possible to work with local organizations on technology development, with partners from the local community, employing local people in this work?
- **Tip** – Do you have any technology that helps local issues and how can it be deployed?

Issue 5: Wealth and income creation

Wealth and income creation from employment can be multiplied by preferentially sourcing locally, i.e., using local suppliers and giving them preference. Outsourcing may make short-term financial gains but does not help the resilience of the

Some companies give financial donations to charities. The best model I know of is that of an American tech company which links donations to their employee's commitments: If an employee spends an hour volunteering for a charity then the company donates ≈ $20 directly to the charity. The destination of the support is not decided by the company but by the employees. If they care enough to spend their own time then the company supports this.

Great linkage of employee commitment and company spending.

Other companies help employees donate time to charities. One of my fellow Cub Scout leaders was granted time off by his company to take Cub Scouts on camp. It was not classed as 'holiday' but came from the 'volunteering' budget. He gave back to the community and his company helped him do so. I had to use my holiday allowance.

Guess who was more dedicated to their company?

local community upon which you depend (see Section 5.12).

- **Tip** – Work with local suppliers to ensure that they conform to good labour and fair operating practices but if they do not comply then move on rather than be complicit.
- **Tip** – Ensure that organisational governance principles (see Section 11.3) are followed and treat local suppliers fairly (see Section 11.7).

Issue 6: Health

Every company has a duty of care towards their staff to protect their health and safety (see Section 11.5) and to protect the health and safety of the product's users (see Section 11.8) but this issue is about promoting good health and encouraging healthy lifestyles. Public health is the responsibility of the government but companies can contribute to this by raising awareness about major diseases and their prevention. The plastics industry worked hard to produce products to help defeat the COVID-19 outbreak, e.g., protective equipment and ventilators.

As an industry we can contribute enormously, it is our social responsibility and we are meeting the challenge.

Issue 7: Social investment

Many companies invest in local 'social investments' but these can also be considered as part of a wider programme to improve community sustainability.

- **Tip** – Rather than isolated 'one-off' actions then consider a long-term programme to better contribute to the community.
- **Tip** – Focus on what you are good at: If you make windows then donate windows to local communities and not financial advice.

Reporting

Reporting on consumer issues can be partially carried out by reporting the GRI disclosures GRI 413 and GRI 414 (see Section 12.10).

Mapping community involvement and development issues in the value chain

Mapping community involvement and development issues to the value chain (as shown on the right) gives a global view of areas of weakness. Improvement action can be taken to remove or reduce exposure in the vulnerable areas.

Note: The areas where a company has reduced influence are shaded.

Community involvement and development and the value chain		
Area	Good practice	Poor practice
Raw materials and procurement	Local suppliers chosen where possible. Local suppliers encouraged to meet good labour and fair operating practices as part of contract and performance audited.	No controls or policies in place for local procurement.
Strategy and planning	Community involvement budget and programme established and reviewed with stakeholders. Long-term and local are focal points of community engagement programme.	No community relationship budget.
Investor relations	Investors clearly aware of community involvement.	Ad hoc donations and not declared to investors.
Sales and marketing	Sales and marketing involved in promoting community engagement. Publicity managed by sales and marketing.	No publicity for any initiative.
R&D	R&D encouraged to check for potential technology transfer to local community.	No involvement with community engagement.
Production	Skills training for all staff organised and apprenticeship scheme running.	No staff training.
Information	Clear reporting of community involvement to all staff. Staff aware of programmes and results.	No reporting.
Internal audit	Supplier performance audited. Community engagement programme audited for effectiveness.	No internal audit.
Reporting	Reporting via GRI 413 and GRI 414.	Statutory financial reporting only.
Human resources	Operates and coordinates community engagement programme. Community engagement linked to employee interests. Open days, tours and local scholarships run by HR as part of function.	No involvement, all initiatives run by CEO.
Use	N/A.	N/A.
End-of-life	N/A.	N/A.
Logistics	N/A.	N/A.

11.10 Engaging people

It is a process

Approximately 15% of ISO 26000 (16 pages) is devoted to integrating social responsibility in a company. This covers issues such as practices for integration, communications, credibility and improvement – the working group recognised that this was not a traditional standard. Also, consistent with the other unusual aspects of ISO 26000, after the publication of the standard the drafters of the standard set up a Post Publication Organisation (PPO) which ran for eight years between 2010 and 2018. The PPO worked to support the implementation of ISO 26000 and produced training material, linkages to other initiatives and conferences on ISO 26000.

Nobody on the working group was under any illusion that implementing ISO 26000 was going to be easy and neither should you be.

One of the key issues in implementing social responsibility is how to engage people (both internal and external) in the process, i.e., stakeholder engagement, and an outline of some potential stakeholders is given in the box on the right. Engaging stakeholders is a process that is both challenging and rewarding, you will learn how your company is seen internally and externally (be prepared for some 'home truths') but importantly you will learn where you can improve (if you listen more than you talk).

Communication opens up the potential for knowledge exchange, suggestions and solutions – when the inevitable crisis comes along then you will be better prepared, more respected and have credibility as a company.

Starting out

Before starting out there are some key points:

- This is a long-term process and you need to be prepared for the long-haul, this is not about quick fixes.
- Look at the core subjects in the standard and set some priorities, you will not be able to action all of the core issues at once.
- **Tip** – It sounds like cheating but choose only two subjects to start with and choose those that you think you can make rapid progress with because:
  - You are already doing something.
  - It is easy to implement.
  - There is little resistance in the company, i.e., it is legally required or customers are already asking for it.

- Look at the potential stakeholders and set some priorities. Do not try to start with all the stakeholders at the start.
- **Tip** – It is OK to cheat again whilst you get some experience in managing stakeholders, their expectations and the possible outputs of the process.

The process

For the two priority subjects:

- Analyse what you already do and what ISO 26000 guidance (actions and expectations) recommends for each issue in these subjects. Use due diligence or 'gap analysis' for this.
- Examine the company policies (or lack thereof) and processes for each issue with a social responsibility 'hat' on.
- Check existing company policies and processes for each issue against the seven principles (see Section 11.3) and update/modify them to comply with the principles (if required).
- Create an initial list of stakeholders for these issues, how the company's activities affect them, what they expect of you and what you are legally required to do.
- Identify the key stakeholders for these issues. As a first step, start with only one subject, e.g., labour practices, and start with the internal staff.

> The essence of social responsibility is awareness and consideration.
>
> The acid test is: "Would I be happy if this decision and the reasons behind it were in the public domain?"

Stakeholders

Social responsibility involves identifying 'stakeholders', these are the people or groups who are most affected by the impacts of a company's decisions and actions.

Potential stakeholders are:

- Suppliers.
- Employees.
- Employees families.
- Contractors.
- Shareholders/owners.
- Customers.
- Trade bodies and other organisations.
- Local residents.
- Local government.

Not all stakeholders will be affected by every issue and there may be more stakeholders for your specific company.

Communicating (and more importantly listening) to stakeholders is one of the best ways of telling them about your actions and finding out what you could do better. They will happily tell you if you are socially responsible or not.

Note: Being aware of stakeholder concerns, responding and reacting to these is not the same thing as letting them make decisions for you. Every company still needs to make decisions and implement them.

- Meet with the stakeholders to get their views on the subject, the company's activities and potential improvements, i.e., start to involve them in the process.
- Define the initial objectives, targets and actions for the chosen subject.
- **Tip** – Focus on objectives and targets that can be achieved with the available resources.
- **Tip** – Meetings should be kept to the social responsibility agenda, i.e., do not let this type of discussion wander into general areas.
- Define how the process will be managed and how it will be reported (see Chapter 12).
- Act, measure and report the results.
- Repeat the process for the next subject and start to involve external stakeholders in the established process.

This is actually a simple project management approach to delivering social responsibility but the 'listening' part is unique to the social responsibility area. It follows the Plan-Do-Check-Act (PDCA) process of most of the management systems standards.

- **Tip** – There are several useful tools available at www.learn2improve.nl to help in this process although they tend to go for all of the subjects at once.

Moving forward with social responsibility

At the end of the starting out process there will have been a lot of lessons learned and it is possible to start on the other harder subjects where real behaviour changes are needed.

Company

- Analyse the company's performance in each of the core subjects and related issues with reference to the seven main principles.
- **Tip** – It may be helpful to look at the core subjects through the product life-cycle to more easily identify the relevant issues in each subject (see Section 11.2).
- **Tip** – Do not forget that this covers all areas of the company. Map the subjects to the value chain (see Sections 10.4 to 10.9) to be sure of covering all the areas.
- Start to integrate social responsibility using the existing management systems.
- **Tip** – Do not forget to balance between the needs of the business and the stakeholders in the present and the future.
- **Tip** – ISO 26000 suggests tools for integrating social responsibility into core organizational decisions.

- **Tip** – Consider external reporting and auditing of social responsibility performance.

Culture

- Ensure that the seven principles are followed at all levels of the company and that decisions and actions are always accountable and transparent.

Resources

- Ensure that the resources (time and money) and budget are available to deliver the full programme.
- **Tip** – Never start without a budget.

Staff

- Encourage staff to take part in social responsibility decisions.
- Delegate authority and responsibility to the staff and implement performance incentives for social responsibility.

Communicating with stakeholders

In the starting out phase, the communication needs will be mainly internal but as the process accelerates and moves into other areas then there will be a need to communicate with external stakeholders. It gets more difficult at this stage and it will be necessary to include different stakeholder groups who will not always share the same background or views, it will also be necessary to involve this broader community to get different views.

The secret is to ensure that the communication is two-way, i.e., listening to them is as important as explaining your position to them, and that conversations are both realistic and positive, i.e., it is always better to 'under promise and over deliver' than to 'over promise and under deliver'.

- **Tip** – Engaging with stakeholders is not simply (or ever) a publicity exercise. It must be serious about hearing their views.

ISO recommends that business users say,

- "We recognize ISO 26000 as a reference document that provides guidance for integration/ implementation of social responsibility / socially responsible behaviour."

or

- "We have used/ applied ISO 26000 as a guide/ framework/basis to integrate/implement social responsibility into our values and practices."

ISO 26000 PPO SAG N 15 rev 1

11.11 Social responsibility – where are you now?

It is part of the landscape

Social responsibility was often a neglected part of overall sustainability and the focus was on environmental and economic sustainability. The introduction of the UN SDGs and their importance in the drive to improve overall sustainability has now raised the profile and importance of social responsibility.

The subjects in the social responsibility area are not really contentious, in most cases the issues should have already been covered due to local legal requirements or simply due to good practice. For many plastics processors, this is an easy area to excel in because of their restricted operations but it is also an area where they can have a very positive impact both locally and around the world. Not only that, but improved social responsibility can be used to promote products and even to negotiate improved prices.

Completing the chart

This chart is completed and assessed as for the previous charts.

Socially responsible producers can achieve better prices because of their investments to protect the customers and the supply chain.

Do not forget that we can be the 'good guys' too!

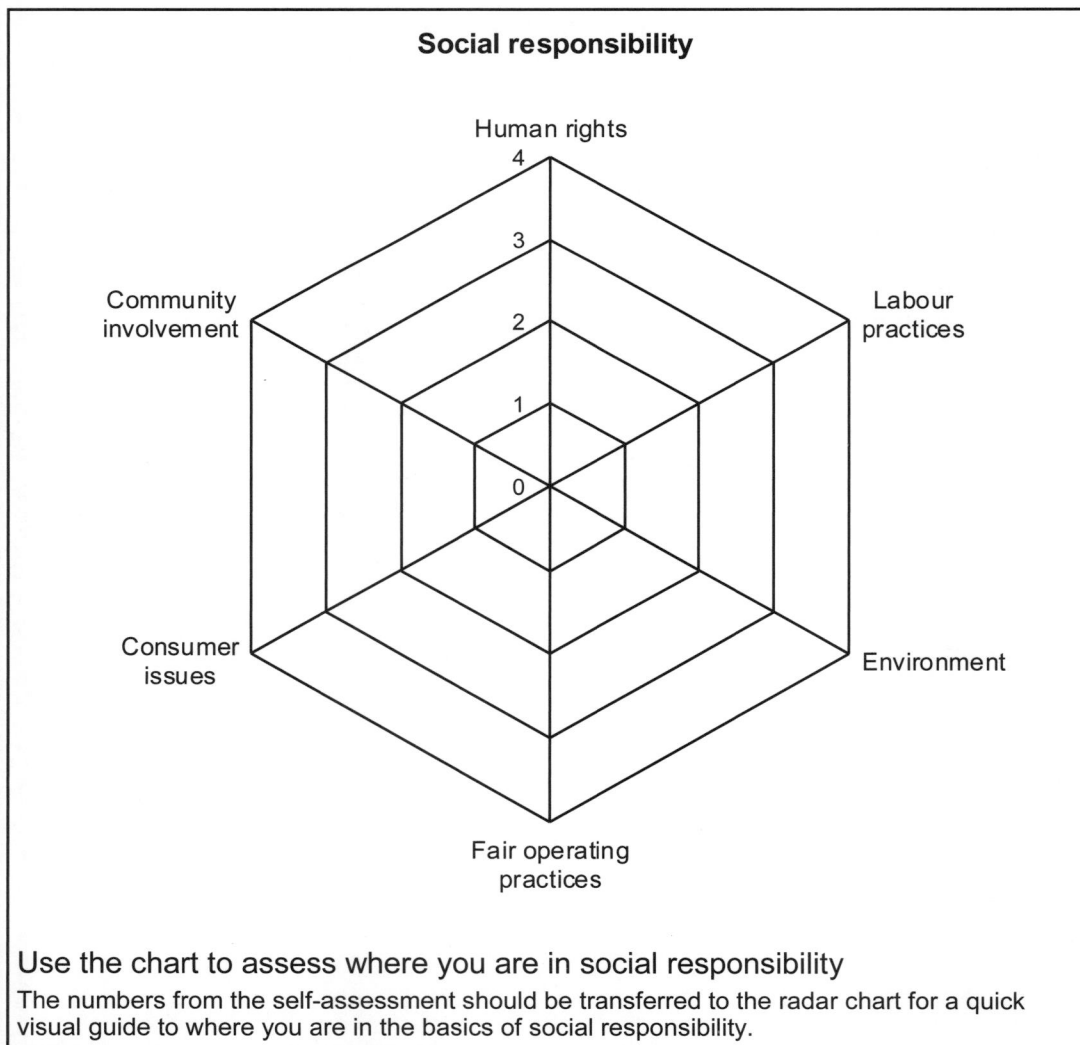

Social responsibility

Human rights
Labour practices
Environment
Fair operating practices
Consumer issues
Community involvement

Use the chart to assess where you are in social responsibility

The numbers from the self-assessment should be transferred to the radar chart for a quick visual guide to where you are in the basics of social responsibility.

Use the social responsibility programme to tell a compelling story.

Prove compliance (or better still, exceed compliance) with local laws and gain recognition for this but never, ever, lie about it.

			Social responsibility			
Level	Human rights	Labour practices	Environment	Fair operating practices	Consumer issues	Community involvement
4	Clear human rights guidance for all staff. Due diligence carried out for all risks. Performance regularly verified & reported.	Exceeds legal requirements in all areas. Excellent staff relations & an exceptional place to work. Training in place for all staff.	Environmental leader. No pollution, good resource use & planning for climate change. Performance verified & reported.	Robust Code of Conduct in place & applied throughout the company. Led from the top with respected 'whistle-blower' programme.	Regarded as an industry model for treatment of consumers. Excellent reputation for ethical treatment of consumers.	Excellent community involvement. Widely seen as an asset to the community. High social investment in community.
3	Clear human rights guidance for all staff. Due diligence carried out for all risks. Performance regularly reported but not verified.	Exceeds legal requirements in most areas. Good staff relations & a good place to work. Training in place for most staff.	Good environmental reputation. No pollution & good use of resources. Performance regularly reported but not verified.	Code of Conduct in place & applied throughout the company. No 'whistle-blower' programme in place.	Good reputation for fair treatment of consumers. Consumers are treated fairly & consistently.	Good community involvement. Seen by some as an asset to the community. Moderate social investment in community.
2	Some guidance on human rights in most areas. Due diligence carried out for some risks. Performance not reported or verified.	Exceeds legal requirements in some areas. Acceptable staff relations & 'simply a job'. Training in place for some staff.	No environmental reputation. Low recorded pollution & average use of resources. Performance not reported or verified.	Code of Conduct in place but inconsistently applied or not applied throughout the company.	No sales direct to consumers, business is B2B only.	Some community involvement but on an ad hoc basis. Little known in the community & relationship is neutral.
1	Human rights not understood or applied in most of the company. Due diligence not carried out. Performance not reported or verified.	Minimum legislative compliance only. High staff turnover. Little training in place.	Minimum legislative compliance only & some recorded incidents of pollution.	Minimum legislative compliance with some potential for breaches of fair operating practices.	Average reputation for treatment of consumers. Few incidents of poor treatment of consumers.	No community involvement & unknown in the community. Community relationship is neutral.
0	No consideration of human rights.	Failure to meet legislative requirements in significant areas, including health & safety.	Poor environmental practices & significant pollution incidents recorded.	Poor legislative compliance & high potential for corruption or other breaches of fair operating practices.	Known for poor treatment of consumers. Poor reputation & only used for low prices. Multiple incidents of poor treatment of consumers.	Negative community involvement & very poor relations with community.
Score						

Key tips

- Social responsibility is an important, but often ignored part of the overall sustainability agenda because most companies are already doing many of the right things.
- There are competing initiatives but ISO 26000 offers a good framework that links to many other initiatives and reporting frameworks.
- Reporting social responsibility is possible via the GRI reporting framework (see Chapter 12) and a narrative report.
- ISO 26000 is an unusual standard because it is not a technical or management system standard. Instead, it is a guidance document.
- It is not possible to be certified to ISO 26000 and ISO makes it clear that any company claiming to be ISO 26000 certified would be misrepresenting the intent and purpose of the standard.
- ISO 26000 makes many references to 'stakeholders', these are the people who are affected by the impacts of a company's decisions or actions.
- A list of the stakeholders in a company is a first step in social responsibility.
- ISO 26000 has seven principles; accountability, transparency, ethical behaviour, respect for stakeholder interests, respect for the rule of law, respect for international norms of behaviour and respect for human rights. The seven principles are the lens through which all of the rest of the standard should be viewed.
- ISI 26000 identifies seven core subjects; organisational governance, human rights, labour practices, environment, fair operating practices, consumer issues and community involvement and development.
- Organisational governance is at the heart of social responsibility and is about applying the seven principles throughout the company so that a company has a framework for social responsibility.
- Social responsibility needs 'leaders' rather than 'managers'.
- Human rights is basically about treating everybody as a human being irrespective of their race, gender, ethnicity, religion, age or any other differentiating factor.
- Companies have a duty not only to protect human rights but also to avoid being complicit in other people or companies infringing human rights, e.g., child labour

or forced labour in suppliers.
- Labour practices are governed by legislation in many parts of the world but companies still have a duty to treat staff fairly whatever the law says or however diligently it is enforced.
- Staff should be fairly employed, allowed to organise, have their health and safety protected and be given the opportunity to develop their skills via training.
- Companies have a social duty to avoid polluting the environment, to minimise the use of resources and to protect the environment. This is covered in many other parts of this book.
- Companies have a social duty to use fair operating practices and to promote these up and down the supply chain.
- Anti-corruption, responsible political involvement and respect for property rights are all part of fair operating practices.
- Consumer issues are not often an issue for plastics processing because of the structure of the industry. However, providing fair marketing, factual information and fair contracts as well as good consumer service is simply good business for any company.
- Many companies are already involved in and contribute to their community. The standard encourages community involvement at a range of levels and can act as a template for good practice.
- Social responsibility is not a about 'quick fixes', it is about a long-term programme to improve the sustainability of a company.
- Communicating with stakeholders is a key to gaining their cooperation and support.
- Being aware of stakeholder concerns and listening to them is not the same thing as letting them make decisions for you.
- Every company needs to make decisions and implement them.

Chapter 12

Reporting

Business has a vital role in driving sustainability and sites or companies that have taken action to improve their sustainability will naturally want to report on their efforts and successes (they must also report on their failures). Sustainability is a process and part of any process is setting goals, measuring performance and reporting the actual performance.

Sustainability reporting can be regarded as part of the larger and older issue of Corporate Social Responsibility (CSR) reporting and a good sustainability report will cover all the resources used by a site, i.e., social, financial, human and natural. In this sense, the sustainability report is also a report on the use of 'natural capital' by the site (see Section 1.4).

Internal sustainability reporting is important but external reporting is rapidly becoming mandatory in many parts of the world. Governments are increasingly legislating on sustainability reporting, most started with energy reporting but this has been extended to environmental reporting of Key Performance Indicators (KPIs). Full sustainability reporting is not yet mandatory around the world but this is predicted to become part of the landscape in the future.

External reporting for sustainability is information disclosure on the significant economic, environmental, social and governance impacts and performance of the business. This is part of the larger movement towards internalising the impacts of business on society.

Financial reporting is mandatory around the world and, although sustainability reporting is largely non-financial, it complements financial reporting by disclosing the impact of a site, or a company, on the natural capital of the world.

As the preceding chapters have shown, sustainability is a multi-dimensional issue and reporting is not simply about a single issue such as energy, it is about the complete system, i.e., get it right in one area and you might get it wrong in another due to non-obvious links. This makes sustainability reporting difficult but no less important. The good thing is that standards for reporting are rapidly becoming common across the world and most voluntary or mandatory reporting schemes are becoming consistent in the type of information that they require, although significant local differences will remain for some years.

The good thing is that sites do not need to report on everything, they can assess the important aspects for their business, i.e., the materiality, and report on those aspects of sustainability that are the most relevant to their activities and society. This allows sites not only to include the numbers but also to include a narrative aspect describing their activities. The type of report is also changing and companies are increasingly using web-based tools to support or replace the more traditional paper-based reports. This gives more flexibility in the style of the report whilst maintaining consistency with the reporting requirements.

We would like to thank GRI for permission to quote extensively from the GRI standards. These are all freely available from www.globalreporting.org.

12.1 Why report sustainability?

It is the right thing to do

The rise of Corporate Social Responsibility reporting in the 1980s has now been overtaken by a rise in Corporate Sustainability Reporting (CSR). Sustainability reporting covers most of the topics previously covered in the Corporate Social Responsibility report but this was largely voluntary whereas Corporate Sustainability Reporting is increasingly becoming mandatory.

Sustainability reporting covers not only the standard environmental issues but also covers the economic and social issues dealt with in Chapter 11. ISO 26000 (see Section 11.2) and the UN SDGs (see Chapter 1) can provide a framework for actions and expectations but the sustainability report provides the narrative and data for target setting and performance assessment. The sustainability report therefore covers all the resources used by a site, i.e., social, financial, human and natural, and reports on the use of 'natural capital'.

- **Tip** – Companies are being subjected to increasing pressure from a variety of stakeholders, both internal and increasingly external, to report on sustainability issues and, in the current climate, sustainability reporting is the 'right thing to do'.

The drivers for reporting

There are many drivers for reporting, both external and internal. Some of these are:

External drivers

- To comply with mandatory reporting requirements, e.g., stock exchange listing requirements or government legislation.
- To comply with voluntary reporting requirements, e.g., Carbon Disclosure Project (CDP).
- To respond to stakeholder demands for information and accountability, e.g., UN SDGs.
- To improve access to external capital.
- To increase operational transparency.
- To publicly communicate sustainability activities, goals and achievements to stakeholders.
- To improve performance in financial markets.
- To increase stability.
- To strengthen the company's licence to operate.

Internal drivers

- To improve company efficiency and competitiveness.
- To manage and report on a wider range of resources than traditional financial resources.
- To demonstrate company commitment to sustainability.
- To track progress against sustainability commitments.
- To set and monitor targets and evaluate performance.

Not all companies will be subject to the same drivers but all companies will be subject to some of the drivers and sustainability reporting is rapidly moving from a 'nice to have' to a 'business essential'.

- **Tip** – In 2009/2010, we helped to implement CDP disclosure for a multi-national company. This was a major project but, after a hard campaign, it was decided that this was 'the right thing to do'. The discussion would be much easier today.

> Reporting is an opportunity to build trust. Do not betray it. Provide feedback loops to show that you are listening as well as telling.

> If you publicly release targets then achieving these is no longer voluntary – it is compulsory.

> "You talkin' to me?"
> **Travis Bickle: Taxi Driver**

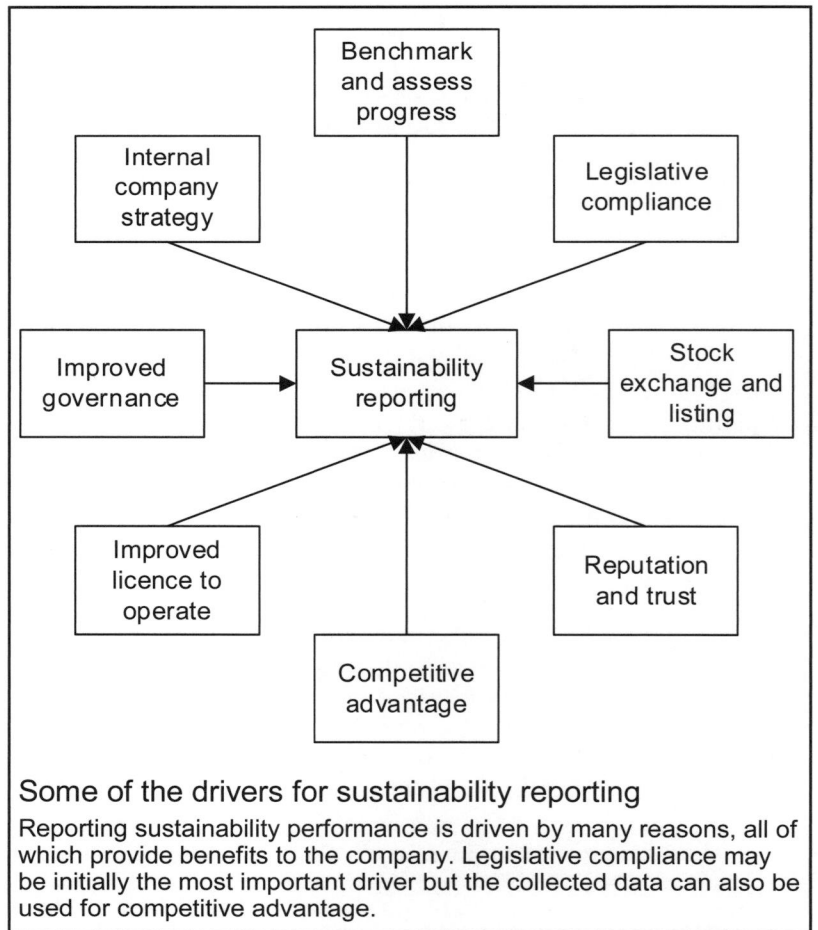

Some of the drivers for sustainability reporting

Reporting sustainability performance is driven by many reasons, all of which provide benefits to the company. Legislative compliance may be initially the most important driver but the collected data can also be used for competitive advantage.

Who are you talking to?

The audience for the sustainability report is also internal and external and reporting will have both internal and external impacts.

National governments around the world, e.g., Australia, Canada, EU, France, Japan, the UK and the USA, have developed guidance, policies and regulations to promote sustainability reporting. These are not necessarily consistent in the data they require or the format in which it is supplied and many are in the process of development, i.e., the requirements are changing. In most cases, national schemes are designed to consolidate data, to report on national progress and to inform government actions rather than to benefit individual companies.

Therefore, we will concentrate on the more consistent and stable internationally recognised reporting formats that are of greater use to individual companies. These formats contain most of the data required for national reporting but are recognised around the world and also provide for reporting on a variety of sustainability issues.

- **Tip** – Do not confuse regulatory reporting with broader sustainability reporting, they generally have different objectives.

- **Tip** – Companies will need to conform to national reporting regulations but this does not stop them producing a wider ranging report for other internal or external stakeholders.

- **Tip** – The data required for national reporting will, in most cases, be the same as the data required for broader external reporting. It may be reported differently but the numbers should be the same.

Reports and numbers drive improvement

Sustainability reports and the information they provide will inevitably produce benefits for the company by providing a focus on long-term strategy, improving governance and planning, increasing awareness of risks and opportunities, better benchmarking and transparency and improving reputation. The simple act of reporting will drive improvement in almost every company.

- **Tip** – Companies do not have to start with a full report. They can start with a high-quality but small report using good data and then grow the report with time as the systems develop.

Other people are already doing it (and benefiting)

The number of companies reporting on sustainability is growing rapidly due to both the pressures to report and the benefits of reporting. Many large companies are already

required to report by legislation or investor pressure (see sidebar) and these numbers are rapidly increasing. Most plastics processors are Small and Medium-Sized Enterprises (SMEs) and the current engagement at this level is low. This will undoubtedly change as the legislation grows to encompass SMEs and pressure is driven down the supply chain.

It is now almost impossible to view a company web-site without seeing a specific section devoted to sustainability but the full benefits will only be achieved when reporting is based on well-defined standards and formats.

- **Tip** – A pressing issue for many sustainability statements is that there is a total absence of data. Many consist of statements that are effectively 'greenwashing' (see Section 1.20). Following good practice guidelines and accurate reporting is essential to maximise the value of sustainability reporting.

- **Tip** – This is not simply for large companies; SMEs have a choice, they can do this voluntarily, gain experience and the benefits or wait until it is forced upon them.

The benefits of sustainability reporting

Sustainability reporting:

- Improves long-term management strategy and business planning.
- Supports improvements in governance and accountability.
- Increases awareness of risks and opportunities.
- Reduces the risk of environmental, social and governance failures.
- Improves reputation and brand loyalty.
- Builds relationships with stakeholders.
- Enables external stakeholders to understand a company's true value, including tangible and intangible assets.
- Enables users to decide how to allocate their own resources through investment, procurement and policy decisions.
- Strengthens the link between financial and non-financial performance.
- Streamlines processes, reduces costs and improves efficiency.
- Benchmarks and assesses sustainability performance.
- Minimises environmental, social and governance impacts.

12.2   Reporting principles

It is like any other reporting

Reporting sustainability is similar to reporting any other activity and the various formats (see Section 12.5) all demand rigour in reporting. Just as financial reports are controlled by legislation and standards to make them trustworthy and reliable; sustainability reporting is also controlled by legislation and standards to make it trustworthy and reliable; after all, it is simply non-financial reporting.

The basic principles

The basic principles for sustainability reporting are:

Transparent

Transparency is essential for any report. The data are important but the methods of getting the data are just as important, flawed or opaque methods of data collection can easily make data worthless. Transparency drives reliability by providing quality control and evidence of a clear data trail.

External assessment (similar to the auditing of financial reports) can improve transparency (see Section 12.11).

Credible

The sustainability report must be placed in overall context to allow impacts to be credibly assessed. This will encourage internal and external acceptance of the report.

Material

Reports should focus on topics that are material to the company in terms of environmental, social and economic impacts (see Section 12.4).

In addition to the standard aspects of materiality, reports should be appropriate to the management efforts of the company, operational performance and the environmental conditions (see below for definitions).

The report should also be relevant and understandable to internal and external interested parties.

- **Tip** – Focusing on materiality will make reporting achievable and practical by preventing reports from becoming too long, containing extraneous information and being expensive to prepare.
- **Tip** – Reporting should always be cost-effective and timely.

Quantitative

KPIs are measurable and are therefore quantitative. This allows them to be used to set targets, assess performance and document sustainable practices.

Quantitative data should be supported by a narrative describing the purpose, the impacts, the data collection process, calculation methods (if any) and any relevant assumptions. Performance should be assessed noting progress and any significant events (including any violations or fines).

Quantitative data should be measured in units appropriate to the KPI.

- **Tip** – Data should serve the needs of the users (internal and external) and the report should contain information that allows management and investors to make decisions.
- **Tip** – When using conversion factors, e.g., vehicle emissions for distance travelled, always report on the source of the conversion factor and use reliable conversion factors.

Comparable

Companies should report using accepted KPIs to allow benchmarking of the company (both internally and externally). Reporting should be over comparable periods, ideally linked to the Annual Report and ideally in the same format and at the same time. Reporting can be absolute or normalised, depending on the user needs (see box on the opposite page).

- **Tip** – Never use 'self-developed' KPIs, they can hide poor performance or mask good performance. Always use standard KPIs. Where these are not absolutely relevant then the narrative can be used to explain the reasons.
- **Tip** – Try to choose KPIs that are responsive and sensitive to changes in the

A KPI does not need to satisfy all of these considerations to be useful to a company (but it is best if it does).

General guidance on reporting:

- Track absolute and normalised metrics.
- Track the metrics at various levels, i.e., process, site and total.
- Track the metrics over time.
- Collect data at the lowest point in the chain and consolidate upwards.
- Automate data collection and processing.
- Get the numbers into the accounts package at the start and get the numbers from accounts.
- Get the right metrics for each level to prompt action. They will be different for each level.

company's performance and that give information on current or future performance.

Consistent

Reports should use consistent methods from period to period to allow performance to be assessed and benchmarked. If there are material changes to the organisation or other factors then these should be noted in the narrative.

- **Tip** – The report should also be consistent with the company's stated environmental, quality and other policies and representative of the company's performance.

Accurate and complete

Reports should be sufficiently accurate and complete to allow users to assess performance. If any significant factors have been excluded then these should be disclosed and justified.

Action oriented and linked

The report should encourage improvement action and be specifically linked to national and international priorities and goals, e.g., the UN SDGs.

Key Performance Indicators

Most reporting standards use some type of KPI to assess past performance and set future targets. However, KPIs are not all the same and within sustainability reporting there are different types of KPIs for different purposes. This is different to standard business KPIs which are normally based on the key business objectives, i.e., overall business performance for high-level KPIs or process effectiveness for low-level KPIs. Sustainability is more closely aligned to environmental performance evaluation which is covered in ISO 14031. Sustainability performance can be evaluated in terms of three types of KPI:

- Management Performance Indicators (MPIs) – these give information about a company's efforts to manage performance, i.e., the effectiveness in implementing sustainability. They measure the effectiveness of system implementation, the conformance of the system in achieving the goals, the financial performance of the sustainability programme and the community response to the sustainability programme. Typical metrics could include: number of achieved sustainability targets, degree of regulatory compliance (number of violations, fines or penalties), savings from improved resource use and number of community engagement programmes.

- Operational Performance Indicators (OPIs) – these give information about the

performance of a company's operations, i.e., the company's effectiveness in converting inputs to outputs. These indicators measure the effectiveness of the operation, materials use, energy use, water use and effluent output, emissions, hazardous waste and recycling rates. These are the traditional type of indicators and typical metrics could include: material use, energy use, water use, effluent output and recycled materials use. OPIs can be absolute or normalised (see below) and are chosen based on the materiality of the topic.

- Environmental Condition Indicators (ECIs) – these give information about the condition of the local, regional or other environment. They are not generally used by individual companies but are used by governments, regulators or researchers to measure areas such as air, water, flora, fauna, biodiversity and human issues.

These reporting principles are also covered in GRI 1 (see Section 12.6).

- 1. ISO 14031, ISO, www.iso.org.

Absolute or normalised?

Any KPI can be absolute or normalised.

Absolute KPIs give an overall total, e.g., kWh of electricity used, m^3 of water used, tonnes of waste to landfill, etc. They give the absolute impact but do not reveal any relationship between the impact and time or other factors, e.g., the kWh of electricity used may have decreased due to decreases in production volume rather than to energy efficiency improvements. In any situation where the activity changes over time then absolute KPIs are less than ideal. Absolute KPIs are primarily used in government reporting schemes (see Section 12.1) where the data is to be consolidated to report on national progress.

Normalised KPIs (sometimes called 'intensity ratios') give a ratio, e.g., kWh/kg of production, m^3 of water used/kg of production, tonnes of waste/tonne of production, etc. Normalised KPIs provide information on performance relative to the activity metric, i.e., the production volume. Normalised KPIs are useful for a specific company or site and provide a comparison over time.

Normalised KPIs can be misleading as a monthly indicator if the activity metric changes rapidly and if there is a base load or fixed element to the impact – see Section 6.5 for details of how the normalised electricity use (kWh/kg) can be extremely misleading as a monthly metric because of the base load.

The moral of the story is that there is real peril in using a single number in complex situations – KPIs must be chosen carefully, interpreted even more carefully and reviewed using an appropriate narrative. The sustainability report should never consist solely of numbers.

12.3 Reporting standards

A multitude of standards

Whilst the need for and the basic principles of reporting are not contentious, the actual task of reporting is not made easy by an increasing number of reporting frameworks, protocols, reporting systems, standards and guidelines. Some are statutory government requirements, some are voluntary declarations, some deal with specific impacts, e.g., GHG emissions, some deal with measurement and others are very broad in their approach, i.e., more a declaration of intent than reporting. It is possible to spend weeks reading these documents and still have no clarity about which is the most suitable for a company.

Therefore, in this section, we will only cover the main schemes and try to provide an overview of the possibilities.

Government reporting

As already noted (see Section 12.1), many governments around the world have mandatory schemes for reporting, although these have, to date, concentrated on energy use and GHG emissions. Some of these schemes[1] are in the process being extended to include other resource metrics, e.g., water, but none currently require reporting on the complete range of sustainability metrics.

The existing schemes are also largely designed for national or sector data consolidation and reporting in absolute terms is primary with the use of intensity ratios being secondary in importance (see Section 12.1).

- **Tip** – Most of the national reporting schemes are useful to governments but less useful to industry and also less useful in terms of overall sustainability reporting.
- **Tip** – Companies must fulfil their local legal reporting obligations but this does not mean that they cannot use the same data to report impacts in a wider sustainability report.

Other reporting

The main protocols and reporting schemes have considerable overlap, interlinking and even duplication. Alphabetically these are:

Carbon Disclosure Project (CDP)

CDP is a voluntary global disclosure and reporting system for cities, states and companies that is used to collect information relating to climate change risks, opportunities, strategies and performance. The CDP is also used by investors to assess the carbon risk of companies and investments. Over 5,000 companies (2020) have voluntarily reported via the CDP on-line system and most of these used the GHGP as a reporting methodology.

Climate Disclosure Standards Board (CDSB)

CDSB is a consortium working to provide a framework for reporting environmental information with the same rigour as financial information and integrating environmental information into mainstream corporate reporting. CDSB works with a multitude of other organisations, e.g., CDP, IIRC (see Section 12.11), WBCSD and WRI.

Global Reporting Initiative (GRI)

GRI was founded in 1997 and produces a global framework of sustainability reporting standards. These cover the complete range of sustainability issues and not simply GHGs. These standards are comprehensive, continuously developed and relevant for a range of organisations. The GRI standards specify the requirements but do not provide a reporting method.

A large benefit of the GRI standards is the collaboration and synergies with other organisations and standards, e.g., UN SDGs, UN GC, CDP, CDSB, ISO (for ISO 26000), IIRC (see Section 12.11) and Earth Charter.

- **Tip** – The original GRI G4 Sustainability Reporting Guidelines were superseded by the GRI Standards in 2018 and current reporting must be to the GRI Standards.

Greenhouse Gas Protocol (GHGP)

GHGP is a partnership between the WRI and the WBCSD to establish a standard accounting framework to measure and manage GHG emissions. The 'Corporate Accounting and Reporting Standard' (2004)[2] provides the accounting platform for most corporate GHG reporting and the 'Product Life Cycle Accounting and Reporting Standard' (2011)[3] can be used to understand a product's full life cycle emissions and to focus efforts on the greatest GHG reduction opportunities.

The standards do not require reporting but many government and voluntary reporting programmes use the standards or are compatible with them, e.g., EU Greenhouse Gas Emissions Allowance Trading Scheme (EU ETS). They are the de-facto standards for GHG monitoring, measurement and reporting.

Report everything and not simply the good news.

Sustainability is a process and not a destination – but you still have to know where you are. One of the first things that the author did as part of the BPF Sustainability Committee was to try to assess where the industry was at the time.

Only by knowing where we are starting from can we assess how we are doing.

Ignore continual improvement – the leaps must be large and transformational. Don't just recycle those things you can make a profit on, recycle everything. Calculate the total returns to justify it – not just the things you could have made a profit on.

Task Force on Climate-related Financial Disclosures (TCFD)

TCFD was established by the Financial Stability Board (FSB) in 2015 to develop recommendations for climate-related disclosures that would promote informed investment, credit, and insurance underwriting and enable stakeholders to understand better the concentrations of carbon-related assets in the financial sector and the financial system's exposures to climate-related risks.

The first recommendations were published in 2017 and other organizations, e.g., CDP, CDSB, have worked to align with the TCFD recommendations.

TCFD aims to provide better information to allow companies to incorporate climate-related risks and opportunities into risk management and strategic planning.

UN Global Compact (UN GC)

UN GC (see Section 11.1) has a transparency and accountability policy known as the 'Communication on Progress' and an annual report is an essential demonstration of a participant's commitment to the UN GC and its ten principles. Reporting for many companies is based on the GRI standards.

World Business Council for Sustainable Development (WBCSD)

WBCSD is a global coalition of over 200 companies and aims to be a catalyst for innovation and sustainable growth. The WBCSD was a co-founder of the GHGP in 1997 and founded 'The Reporting Exchange' in 2017 (www.reportingexchange.com) to allow companies to track trends, share best practices and compare performance for a range of sustainability indicators. WBCSD also works with GRI and the UN GC to provide information on the UN SDGs.

World Resources Institute (WRI)

WRI is an environmental think tank founded in 1982 to provide science-based and practical policy research and analysis on global environmental and resource issues. WRI is focused on seven global challenges; climate, energy, food, forests, water, sustainable cities and the ocean. Research and analysis into these challenges is based on four centres; business, economics, finance and governance. WRI is a co-founder of the GHGP and the CDSB.

Which one to use?

My personal preference for sustainability measurement standards is the GRI standards. These provide for a wide range of metrics across the whole sustainability area and these metrics are also compatible with most of the main reporting systems. They

are also easy to read, understand and apply for most companies.

Access to the standards is free and they are widely used for reporting in the plastics industry (see Section 12.5).

Linking the standards

The variety of standards and reporting requirements around the world is daunting but some clarity is provided by:

- An excellent table produced by CDSB, available at www.cdsb.net/connections. This provides the links between CDSB and many other standards such as CDP, GRI, UN GC and many country specific requirements.

- A guide to linking GRI and CDP ('Linking GRI and CDP') is available from GRI at www.globalreporting.org/standards/resource-download-center.

- A guide to linking the GRI standards and the UN SDGs ('SDG Compass') developed by GRI, WBCSD and UN GC is available at sdgcompass.org.

- GRI also has information on linking and reporting progress towards the UN SDGs in:

  - Analysis of the Goals and Targets.

  - Integrating the SDGs into Corporate Reporting: A Practical Guide.

Both are available at www.globalreporting.org/information.

Get good data

The plethora of reporting schemes highlights the need for good data. Good data can also be sustainable, i.e., it can be re-used and recycled for more than one purpose.

Use the linkage documents to re-use and recycle data.

- 1. Environmental Reporting Guidelines: (including SECR), UK Government (DEFRA and BEIS), 2019, www.gov.uk.

- 2. Corporate Accounting and Reporting Standard, ghgprotocol.org/corporate-standard.

- 3. Product Life Cycle Accounting and Reporting Standard. ghgprotocol.org/product-standard.

12.4 Materiality and reporting

Reporting essentials

The process of mapping materiality was discussed in Section 1.8 as a fundamental for identifying effective sustainability actions and the same concept is used in reporting. Materiality in reporting is a concept that will be familiar to every accountant. The International Accounting Standards Board defines materiality as "Information is material if omitting, misstating or obscuring it could reasonably be expected to influence decisions that the primary users of general-purpose financial statements make on the basis of those financial statements...". A similar concept of materiality is embodied in most sustainability reporting frameworks, i.e., it is not essential to report everything but it is essential to report material matters.

- **Tip** – Identifying materiality is essential in deciding what to report.

- **Tip** – Decrease the workload, decrease the cost, decrease extraneous information and increase the relevance by reporting only on those impacts that are material to the company and stakeholders.

- **Tip** – Not all material topics are of equal importance and the detail in the report should reflect their priority.

- **Tip** – Reporting for governments is based on their concept of materiality and not on yours. There is a large difference.

As noted in Section 11.3, we believe that the GRI standards are the best measurement standards for reporting sustainability and this is validated by the wide use of the GRI standards throughout the world for reporting to key stakeholders.

Materiality determines the shape of the report

Using materiality as a filter for reporting means that the report will focus on material topics. Other topics of interest to the company can be reported but the material factors should be the most prominent. We recommend that the GRI reporting framework is used as a base for materiality and that other topics are added if they are significant to the company, e.g., significant material spills should be disclosed under GRI 306-3 but work on projects such as Operation Clean Sweep (see Section 5.4) can also be reported separately.

Whilst determining materiality may seem a large programme, it is not necessary to carry out the analysis every year. If the situation remains the same then it is possible to simply revalidate the materiality analysis, i.e., only repeat the exercise when materiality has changed and keep reporting consistent through time.

Every company should be prepared to disclose and explain the method used to determine materiality and topic priority. This is explicitly required in GRI 2-46 which requires an explanation of how materiality has been applied and records should be kept of these processes for disclosure as appropriate.

Materiality factors

Determining the relevant materiality factors for a company can be from a range of activities such as:

- Company strategies, policies or management systems, e.g., ISO 9001, ISO 14001 etc.

- **Tip** – Impacts that the company already considers important enough to need management are likely to be material. If you are working on it then it is probably material (or else you are wasting your time).

- Consultation with internal stakeholders, e.g., staff (at all levels).

- Consultation with external stakeholders, e.g., shareholders, local communities or NGOs.

- Customer requirements and issues.

- **Tip** – Customers are a vital source of information on materiality and they have a lot of power too. Just as companies can drive sustainability downstream to suppliers (see Section 5.10), expect to have sustainability driven downstream to you from your customers (or your customer's customers).

- Supplier information on potential material factors, e.g., materials MSDS.

- Industry analysis of other company sustainability initiatives, e.g., competitors and comparable companies.

- Existing or proposed legislation.

- Existing or proposed country or international standards.

- Existing or proposed voluntary agreements (for the sector or wider society).

- Reasonable estimates of environmental, social and economic trends and impacts.

- Areas which have been identified as a concern by experts or by tools such as impact assessment or life cycle assessment.

Materiality looks at the trends and issues. It prioritises the opportunities and threats to a company and selects those that add the most value or have the highest risk for action.

A material aspect will:

- Reflect the company's significant environmental, social and economic impacts

or

- Substantively influence the assessment and decisions of stakeholders about the company.

Considering materiality is not just about reporting, it involves scanning the broader landscape for issues that can affect the company .

A certain sporting goods company probably never thought that the working conditions in their sub-contract manufacturing sites was an issue until people started making documentaries about them.

Are you also exposed?

These activities will give a range of potential materiality factors that must then be prioritised for reporting.

Assessing materiality

The simplest method of prioritising materiality uses a materiality grid (see Section 1.18) and the box on the right is a simplified materiality grid showing the two main factors in determining materiality, i.e., the significance of the impact and the relevance to the stakeholders. The grid uses these two factors to assess materiality. A topic can be material for reporting if it ranks highly on only one of these factors, i.e., it does not have to rank highly in both of them.

In the example, we have differentiated and marked the issues according to whether they relate to the governance, environmental, social and economic area. This is to highlight materiality in various areas.

Note: This grid is an example only, every company needs to do their own analysis.

- **Tip** – The materiality issues covered in the report should reflect the company's governance, environmental, social and economic impacts and be sufficient to allow stakeholders to assess the company.

- **Tip** – Materiality can be assessed in many ways but one of the largest distinctions is between the internal and external stakeholder viewpoints.

Taking materiality further

After the materiality factors have been identified, they should be prioritised not only for reporting but also for action. This can be done by:

- Risk analysis can be carried out (see Section 2.17) as a method of prioritising and ranking material concerns as well as identifying corrective or preventative action to reduce the impact.

- SWOT analysis can be carried out for each material factor to determine the improvement actions to be taken.

- Mapping actions through the product life cycle (see Section 1.7) can be carried out to highlight the areas where action would be most effective and appropriate.

- **Tip** – Ranking and reporting on materiality also implies that there is a parallel set of actions to reduce the impact (see Section 1.7). These can also be ranked as for risk management.

- **Tip** – This process can also aid in reporting compliance with IIRC (see Section 12.11).

- **Tip** – Materiality reporting can be linked to the UN SDGs (see Sections 1.16 and Appendix 1) for a broader perspective on the actions. An easy method of determining the linkages is to use the 'Inventory of

Business Indicators' available at sdgcompass.org/business-indicators/.

Report on what is material and important. Do not report on things that are not relevant – it simply increases the workload, increases the cost and makes the sustainability report less relevant.

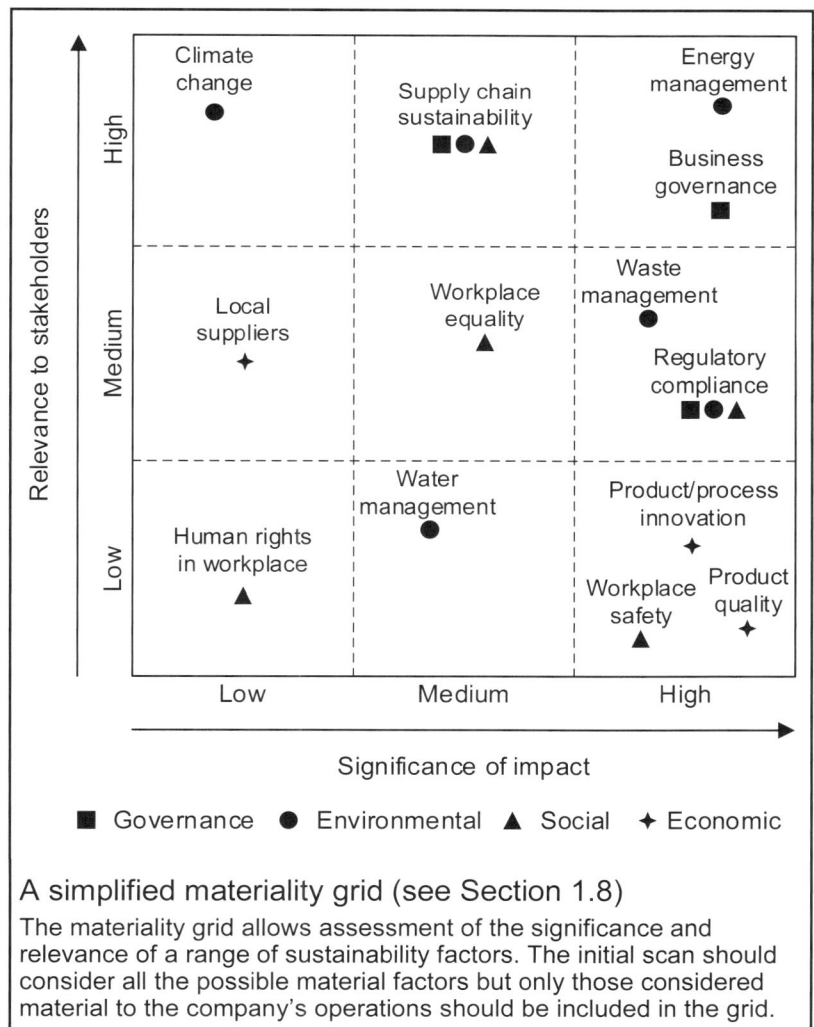

A simplified materiality grid (see Section 1.8)

The materiality grid allows assessment of the significance and relevance of a range of sustainability factors. The initial scan should consider all the possible material factors but only those considered material to the company's operations should be included in the grid.

12.5 Reporting formats

The numbers are not the only thing

The standards specify the numbers to include in a sustainability report, e.g., how to reference the standards and other details about reporting (see Sections 11.6 to 11.10). However, they do not tell you how to format the report or how to deliver it to the stakeholders so that it is concise, understandable and relevant. This is the decision of the reporting company.

Whilst the numbers must be reported consistently there is still flexibility in the narrative and how it is reported.

Reporting is an opportunity

The process of generating the sustainability report, from collecting the data to getting feedback on the report from the stakeholders, is an opportunity to measure and check performance and to set new targets.

- **Tip** – Reporting is as much about communicating internally as it is about communicating externally. Take advantage of the process to tell staff what you are doing to make the world a better place.
- **Tip** – The sustainability report can be seen as a chore or as an opportunity. Focus on the positive and treat it as an opportunity.

Report the non-numbers

Any sustainability report will contain many numbers because the numbers tell stakeholders how the company is doing in a quantitative way. Simply reporting the numbers misses the opportunity to report the non-number data, e.g., the qualitative reporting, on what the company is doing to improve sustainability.

- **Tip** – Use diagrams and infographics to make it visual. Every report must connect with the reader to be useful and the sustainability report is no exception.

Publishing the report

It is possible to simply report the relevant numbers to produce a functional sustainability report but this misses much of the potential of sustainability reporting. This is an opportunity to show what the company is doing and how it is making a difference.

Some ideas for publishing are:

- Print it as an appendix to the Annual Financial Report or as an Annual Non-financial Report to show stakeholders that it matters to the company.
- Put it on the company web site for the world to see.
- Register the report with the GRI database to get international exposure.

Good practice reports

Starting out in reporting is a daunting task and it is always worthwhile looking at other reports for ideas and inspiration. The GRI Sustainability Disclosure Database (database.globalreporting.org/) is an excellent resource which catalogues, and makes freely available, more than 60,000 global company sustainability reports. These reports can be GRI complaint or non-GRI compliant and are clearly marked as such in the database.

- **Tip** – Any company new to reporting is advised to view some of the sustainability reports on the database for ideas.
- **Tip** – Report to GRI and your report will also appear here. Get used to becoming public.

Some of the best reports in the plastics area, in alphabetical order, are:

Alpla

Alpla produce a good print-ready report (sustainability.alpla.com/en) and also a web-based GRI index (sustainability-report18.alpla.com/en/service/gri-content-index) with full web-based crosslinking between the index and the relevant web information. Much of this data is also included in the GRI Sustainability Disclosure Database.

Covestro

Covestro produce a print-ready report covering both the annual report and sustainability with a complete GRI index (report.covestro.com/annual-report-2019/servicepages/downloads/files/covestro_ar19_supplement.pdf). The information is also provided as a web-based report (report.covestro.com/annual-report-2019) with a web-based GRI index (report.covestro.com/annual-report-2019/management-report/further-supplementary-information-on-sustainability/gri-index.html). Prior to 2019, Covestro released separate print-ready reports on sustainability (including a GRI index) and much of this data is also included in the GRI Sustainability Disclosure Database. The reports are independently audited (see Section 12.11).

GRI offers several paid for services in reporting. These are:

- Content index service.
- Materiality disclosures service.
- SDG mapping service.
- Kick-off service.
- Training workshops.

GRI also offers several tools for managing disclosures and improving reporting. These are:

- A content index tool for improved report navigation.
- Certified software and tools for report creation.
- The Sustainability Disclosures Database for report registration or searching.
- A digital reporting tool where all reporting is via a web-based reporting service.

Elix Polymers

Elix is a Spanish producer of ABS resins and produced a print-ready report for 2018 (elix-polymers.com/uploads/2018_Sustainability_Report_ELIX_Polymers.pdf) that shows even a medium-sized company (≈ 250 employees) can produce an excellent report. This covers materiality and has good information on local initiatives and an excellent GRI index that links the GRI disclosures to the UN SDGs and the UN GC principles.

Henkel

Henkel produced a print-ready report (2018) and a print-ready GRI Index and UN Global Compact Index 2018. Both of these are available at (henkel.com/sustainability – go to the bottom of the page) and show how reporting can add value to sustainability efforts. Much of this data is also included in the GRI Sustainability Disclosure Database and many of the reports are independently audited (see Section 12.11).

A particularly useful chart showing the company's contributions towards the UN SDGs is provided at (henkel.com/sustainability/positions/sustainable-development-goals).

- **Tip** – Sustainability is not about one thing and all of these example reports take a broad view of sustainability and reporting.
- **Tip** – Integrated reporting (see Section 12.11) drives a company to report on activities affecting financial, social and natural capital. Start to think about <IR> now and link it to the sustainability report.

GRI standards cover the most common disclosures but there is nothing to stop a company using their sustainability report to disclose other non-financial information or topics that are of interest to the company or the stakeholders.

GRI Disclosure Number	Disclosure title	Section/Page in Annual Report	Section/Page in Sustainability Annex	Explanation and omissions	External Audit	Relevant UN SDG	Relevant UN GC
General Standard Disclosures							
GRI 2 General disclosures							
02-Jan	Name of organisation	Section 2: Company structure	Publication details	–	–	–	–
02-Feb	Activities, brands, products and services	Section 2: Company structure	–	–	–	–	–
02-Mar	Location of headquarters	Section 2: Company structure	–	–	–	–	–
–	–	–	–	–	–	–	–
Economics							
GRI 201 Economic performance							
201-1	Direct economic value generated and distributed	Section 3: Financial statement	–	–	–	–	–
201-2	Financial implications and other risks and opportunities due to climate change	–	Section 1: Sustainability strategy	–	–	–	7
–	–	–	–	–	–	–	–
Environmental							
GRI 300 - Materials							
301-1	Materials used by weight or volume	Section 4: Operations (Pg. 67)	Section 2: Materials use	–	✓	12	7, 8, 9
301-2	Recycled input materials used	–	Section 2: Materials use	–	✓		
–	–	–	–	–	–	–	–
302-1	Energy consumption within the organization	Section 4: Operations (Pg. 68)	Section 3: Energy use	–	✓	7, 9, 12, 13	7, 8, 9
302-3	Energy intensity	Section 4: Operations (Pg. 68)	Section 3: Energy use	–	✓	7, 9, 12, 13	7, 8, 9
–	–	–	–	–	–	–	–

A simple matrix approach to reporting

Using a simple matrix approach allows all the material topics to be reported and indexed concisely to provide a reference to the essential the UN SDGs and the UN GC. This is very similar to the output of the GRI content indexing tool.

12.6   The foundation – GRI 1

The basics of reporting

GRI 1: Foundation (2021) sets out the basics of the GRI reporting format and can be applied to any organisation or company that wants to report its environmental, social and economic impacts to the GRI standards.

It is the starting point in using the standards and is one of the three 'Universal Standards' that apply to every company using the GRI standards.

This standard covers the 'how' of reporting, i.e., the principles of reporting, rather than the 'what' of reporting which is covered in all the other standards. The 'how' of reporting MUST be read first because it sets out:

- The principles for context.
- The principles for quality.
- The requirements for report preparation to the GRI standards.
- How the GRI standards can be used and referenced.
- The specific claims required for companies using the GRI standards.

These are all important to ensure robust and GRI compliant reporting.

- **Tip** – The GRI standards use conventional standards language.
  - A mandatory reporting requirement uses 'shall' and is in bold type.
  - A reporting recommendation or good practice uses 'should' and is in standard type.
  - All 'guidance' is clearly highlighted as such and is in standard type.
- **Tip** – A company is not required to follow the GRI 'guidance' to claim Conformance with GRI standards. However, the guidance notes are very valuable to reporting companies, ignore these at your own peril.
- **Tip** – Where a term in a GRI standard is underlined, e.g., <u>stakeholder</u>, then the definition of the term for reporting purposes is given in the standard or the GRI Standards Glossary (available from GRI).

Reporting principles

Section 1 of GRI 1 covers the basic reporting principles for sustainability reporting (see Section 12.2). These are divided into two categories that define the context of the report and the quality of the report.

Principles for context

The basic requirements for context are:

- Stakeholder inclusiveness.
- Sustainability context.
- Materiality.
- Completeness.

Most of these concepts have already been discussed in previous sections and GRI 1 gives additional valuable guidance.

Principles for quality

The basic requirements for quality are:

- Accuracy.
- Balance.
- Clarity.
- Comparability.
- Reliability.
- Timeliness.

As for the context principles, most of the quality principles have already been discussed in previous sections and GRI 1 gives additional valuable guidance.

Whilst there is no actual required disclosure for the reporting principles GRI 1, this section does have mandatory requirements, e.g., Section 1.1 requires that 'The reporting organization shall identify its stakeholders, and explain how it has responded to their reasonable expectations and interests'.

The guidance for all sections includes 'Tests' which are designed to confirm that the requirements are met. The tests are designed to prove that the company has carried out the requirements, e.g., for Section 1.1, one of the tests is 'The reporting organisation can describe the stakeholders to whom it considers itself accountable'.

- **Tip** – All the tests should be considered and documentation that they have been met retained for potential external examination (see Section 12.11).
- **Tip** – GRI 1 should be consulted for a full discussion of the reporting principles and the valuable guidance.

Using the GRI standards

Section 2 of GRI 1 covers the process of GRI reporting and compliance with all sections is required to claim that a report meets GRI standards. This section covers:

- Applying the Reporting Principles.
- Reporting general disclosures.
- Identifying material topics and their boundaries.
- Reporting on material topics.

All of the GRI standards are free to download at www.globalreporting.org/standards.

They can be downloaded individually or as a complete set.

A company does not have to report to the complete set of GRI standards but can use parts of the standards as applicable BUT must reference them correctly and notify GRI if the GRI standards are used.

- Reporting required disclosures using references.
- Compiling and presenting information.

As for the reporting principles, there are no actual disclosures required (although some aspects of this section will require disclosure under GRI 2) but there are mandatory requirements, i.e., 'shall' is used extensively. The guidance information is extensive but, unlike the reporting principles section, there are no 'tests' to confirm that the requirements are met.

- **Tip** – Whilst there are no 'tests', companies should consider each requirement and retain documentation to show that the requirements have been considered and met for potential external examination (see Section 12.11).

Making GRI related claims

There are three possible methods to report using the GRI standards:

- Use GRI standards to prepare a report 'in accordance with the GRI standards'. This can be done via one of two options:
  - Core option – a report to this option contains the minimum amount of information needed to understand the company. The company can select the disclosures to be made for each material topic.
  - Comprehensive option – a report to this option requires additional information and requires the company to make all the disclosures for each material topic.

It is important to note that the requirements for context and quality do not change. There is also no requirement to move from 'core' to 'comprehensive' and the choice is at the company's discretion.

GRI 1 provides an excellent table listing the criteria for claiming that a report is 'core' or 'comprehensive'.

- Use only selected sections of the GRI standards to report specific impacts. This is termed a 'GRI-referenced' claim and can be used when a company wants to report only specific impacts. This type of report is not 'in accordance with the GRI standards' and there are specific statements and requirements for referencing the GRI standards. The requirements for context and quality do not change.

Notifying GRI

Whichever method is used to report using the GRI standards, the reporting company shall notify GRI of the use of the GRI standards and any claims made by either:

- Sending a copy of the report to GRI.
- Registering the report with GRI.

It is not that daunting

Whilst the use of the GRI standards initially appears daunting, GRI has done an excellent job of making the reporting process clear and as easy as possible. The GRI standards provide a comprehensive, flexible and widely recognised framework for sustainability reporting. It is strongly recommended that all companies use this framework for reporting.

> The GRI standards are a logical grouping of sustainability disclosures that provide a consistent declaration of what a company is doing in sustainability.

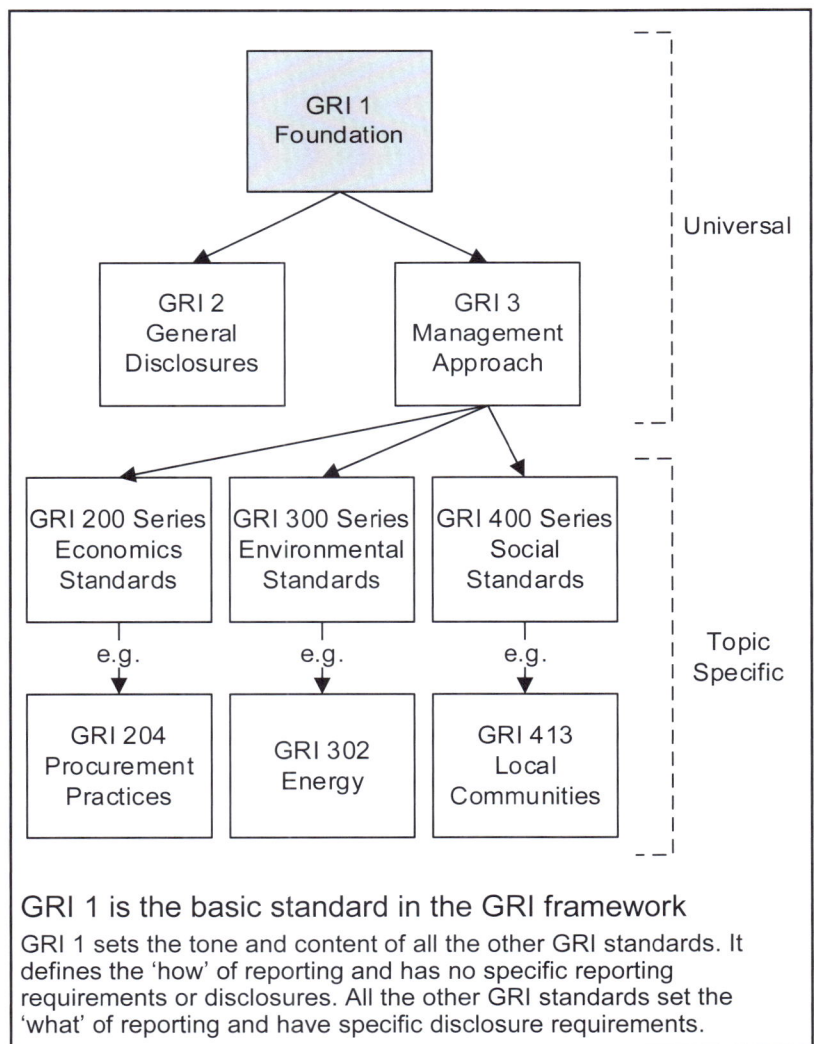

GRI 1 is the basic standard in the GRI framework

GRI 1 sets the tone and content of all the other GRI standards. It defines the 'how' of reporting and has no specific reporting requirements or disclosures. All the other GRI standards set the 'what' of reporting and have specific disclosure requirements.

12.7 The other universal documents – GRI 2 and GRI 3

The start of the disclosures

GRI 1 (see Section 12.6) is the Universal Standard that defines the 'how' of reporting but GRI 2 and GRI 3 are the other two Universal Standards that start to describe the 'what' of reporting, i.e., they are the universal foundations of the factual reporting.

GRI 2 and GRI 3 include reporting requirements, reporting recommendations and, where appropriate guidance on making the required disclosures.

- **Tip** – These standards apply to every company reporting in 'accordance with the GRI standards' whether the core or comprehensive option is chosen.

GRI 2: General Disclosures (2016)

The general disclosures set out the general context of the company. Most of these disclosures should be relatively easy to make and many will already be disclosed in the company's Annual Report or on the company's web site. There is no requirement to repeat these disclosures separately, the Annual Report or the web site can simply be referenced in the GRI content index (see Disclosure 102-55 in GRI 2 and Section 11.5).

Organizational profile

This covers the basics of the organisation, i.e., name, activities, locations, ownership, markets served, scale, employee data, supply chain, external initiatives and membership of associations.

There are 13 disclosures (Disclosure 102-1 to Disclosure 102-13) and all disclosures are required whether the core or comprehensive options are chosen, after all, this is basic company data (and most of it is already available for cross-referencing in the Annual Report).

- **Tip** – All of these disclosures will be routine to most companies.

Strategy

This covers the company's sustainability strategy to give context when reporting on specific GRI topics. It is important that this section deals with the company strategy rather than simply repeating information provided in other sections of the report.

There are only two disclosures in this section (Disclosure 102-14 and Disclosure 102-15) but Disclosure 102-15 is non-core and only required for the comprehensive option.

Ethics and integrity

This covers the company's approach to ethics and integrity with all business partners, i.e., suppliers, customers and anybody else who has a business relationship with the company.

There are only two disclosures in this section (Disclosure 102-16 and Disclosure 102-17) but Disclosure 102-17 is non-core and only required for the comprehensive option.

Governance

The disclosures in this section give an overview of the governance of the company and range from governance structure through to remuneration and incentives for the governing bodies.

Of the 22 disclosures (Disclosure 102-18 to Disclosure 102-39), only one (Disclosure 102-18 Governance structure) is required for the core option and the other 21 are only required for the comprehensive option. Despite the temptation not to report on many of these topics, these are very important disclosures for many stakeholders as they relate directly to the effectiveness of the governance body in driving sustainability and assessing impacts and risks.

Stakeholder engagement

Stakeholder engagement is key to using the GRI standards and, in fact, is the first of the principles for context in GRI 1 (see Section 12.6). This makes stakeholder engagement an important part of the process of using the GRI standards for disclosures.

There are five disclosures in this section (Disclosure 102-40 to Disclosure 102-44) and all are required for both the core and comprehensive options.

- **Tip** – Stakeholder engagement is not simply about the sustainability report, it is about the broader concept of stakeholder engagement and how the company engages with the broader community.

Reporting practice

GRI 1 gives the details of context and quality and this section gives additional details on the content of the report. The definition of the report content, topic boundaries and material topics are essential information in considering if the report is robust and considers the most important topics. This section also considers the topics of reporting claims, the GRI content index and the use of external assurance or audit.

GRI 2 is mostly basic information that should already be available in the Annual Report or on the company web site as information for investors.

Using the 'core' option requires 33 disclosures for GRI 2 and using the 'comprehensive option requires 56 disclosures.

The majority of the additional 23 disclosures are with respect to governance (21 out of 23).

There are 13 disclosures in this section (Disclosure 102-45 to Disclosure 102-56) and all disclosures are required whether the core or comprehensive options are chosen. Again, this is about reporting the basics.

GRI 3: Management Approach (2016).

This standard is about the general management approach for sustainability, it is not about the general management approach of the company.

All disclosures are required whether the core or comprehensive options are chosen.

Unlike most of the GRI disclosures, which are quantitative, these are primarily qualitative and narrative disclosures that describe:

- Why each topic is material – this should be the result of the materiality considerations (see Section 12.4) previously carried out (Disclosure 103–1).

- The boundary for each topic – boundaries for material topics are important because these describe where the impacts actually occur and how the company is involved with the impacts (Disclosure 103–1).

- **Tip** – The topic boundary need not be the same for each topic, in fact, in most cases it probably won't be.

- How the company manages each topic identified (Disclosure 103–2).

- **Tip** – This is an opportunity for the company to describe specific actions, e.g., projects and results, taken to manage each topic. As a primarily narrative section, this is a chance to discuss targets, progress towards the targets and how the company is delivering sustainability.

- The methods that the company uses to evaluate, monitor and modify the management approach – this can cover methods such as auditing, benchmarking, stakeholder surveys and communicating targets and progress (Disclosure 103-3).

- **Tip** – This is an opportunity for the company to describe how they are assessing progress and changes that are being made to improve the processes. Again, as a primarily narrative section, this is a chance to discuss the changes that are being made to improve effectiveness in sustainability management.

GRI 3 is an opportunity to add value to the quantitative information being reported in other disclosures and not simply as a requirement. This is where companies can engage with stakeholders through a narrative discussing their management methods, progress, successes and evaluation methods.

- **Tip** – Do not miss the opportunity that GRI 3 provides to add value to the whole sustainability report. It is possible to produce the narrative as stand-alone information or to cross-reference information in other parts of the sustainability report or the Annual Report.

Each topic in the GRI standards also has a 'Management Approach' clause and reporting to the GRI standards requires companies to report to GRI 3.

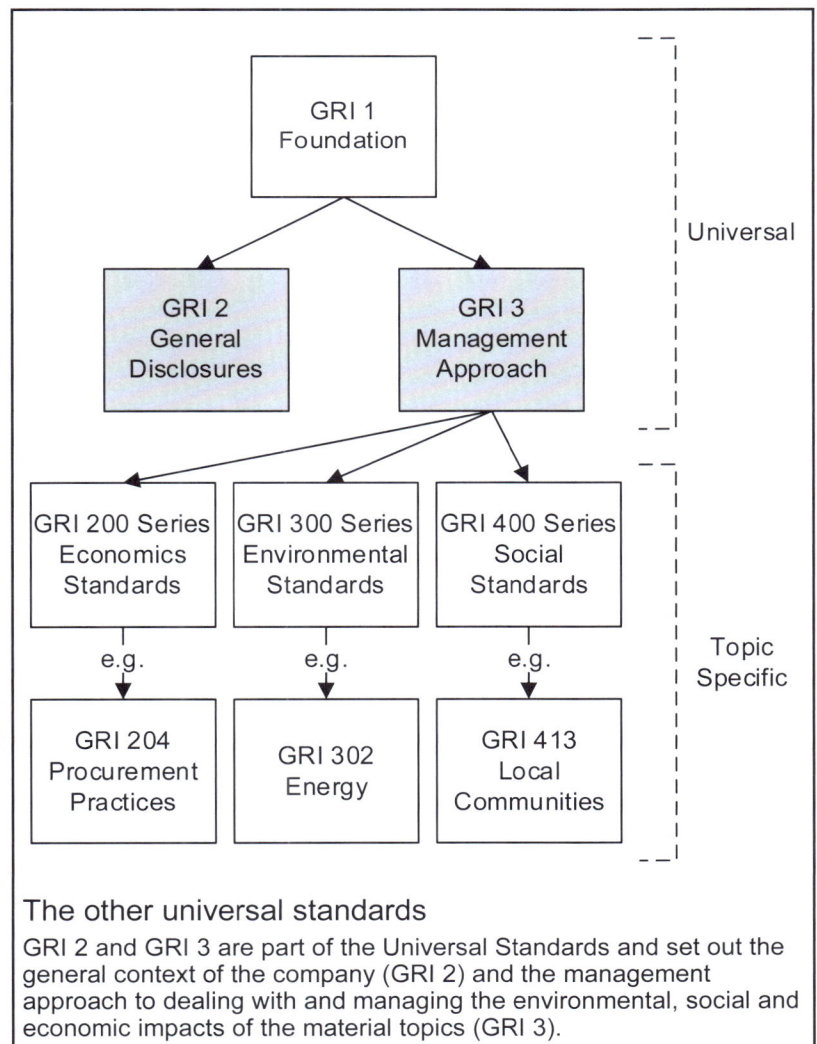

The other universal standards

GRI 2 and GRI 3 are part of the Universal Standards and set out the general context of the company (GRI 2) and the management approach to dealing with and managing the environmental, social and economic impacts of the material topics (GRI 3).

12.8   Economics – GRI 200 series

The economics standards

The 200 series of the GRI standards include the topic-specific standards used to report information on a company's material impacts related to economic topics. These are not the standard economics or financial reporting numbers, these will be given in the standard Annual Report, this is the economic aspect of sustainability. It is how the company's material impacts affect the economics of the range of stakeholders at the local, national and global levels.

There are 6 separate standards with a potential total of 17 disclosures (shown in the box on the right). The requirements for disclosure are:

- Companies reporting via the 'core' option are only required to disclose for material topics, i.e., if a topic is not material then it does not need to be reported. Additionally, they can choose which disclosures to make within a material topic, e.g., if Anti–corruption (GRI 205) is a material topic then they can choose to make Disclosure 205-1 only.

- Companies reporting via the 'comprehensive' option are also only required to disclose for material topics, i.e., if a topic is not material then it does not need to be reported. However, for a material topic, they must make all the disclosures within the topic, e.g., if Anti–corruption (GRI 205) is a material topic then they must make Disclosures 205-1 to 205–3 (inclusive).

- **Tip** – Each topic has a requirement for a management approach disclosure using GRI 3 and this can be used to add a narrative to the quantitative information (see Section 12.7).

The GRI 200 topics

The GRI 200 series covers a range of topic-specific standards relating to the economic prosperity of both the company and the local area. Companies need to survive and prosper economically but they can only do this if they operate in communities which also survive and prosper economically.

GRI 201: Economic Performance (2016)

This topic covers economic value generated and distributed by the company and the risks and opportunities of climate change. It also covers the liabilities resulting from defined benefit retirement plans (if operated by the company).

Additionally, Disclosure 201-4 requires disclosure of any financial assistance received from governments such as tax credits, subsidies or financial incentives.

- **Tip** – Disclosure 201-2 provides an opportunity to assess the risks of climate change on the company and should be used as a framework to assess climate change issues, including the direct and indirect impacts and the methods used to control the risks. Standard risk assessment processes (see Section 2.17) can be used of this.

- **Tip** – Do not forget that for some companies, climate change may be an opportunity.

Exceptionally, if a disclosure cannot be made then it is possible to omit the disclosure within tight requirements and the reason for omission must be explained (see GRI 1 Para 3.2).

The GRI 200 series disclosures

GRI 201: Economic Performance (2016).
- Disclosure 201-1 Direct economic value generated and distributed.
- Disclosure 201-2 Financial implications and other risks and opportunities due to climate change.
- Disclosure 201-3 Defined benefit plan obligations and other retirement plans.
- Disclosure 201-4 Financial assistance received from government.

GRI 202: Market Presence (2016).
- Disclosure 202-1 Ratios of standard entry level wage by gender compared to local minimum wage.
- Disclosure 202-2 Proportion of senior management hired from the local community.

GRI 203: Indirect Economic Impacts (2016).
- Disclosure 203-1 Infrastructure investments and services supported.
- Disclosure 203-2 Significant indirect economic impacts.

GRI 204: Procurement Practices (2016).
- Disclosure 204-1 Proportion of spending on local suppliers.

GRI 205: Anti-corruption (2016).
- Disclosure 205-1 Operations assessed for risks related to corruption.
- Disclosure 205-2 Communication and training about anti-corruption policies and procedures.
- Disclosure 205-3 Confirmed incidents of corruption and actions taken.

GRI 206: Anti-competitive Behaviour (2016).
- Disclosure 206-1 Legal actions for anti-competitive behaviour, anti-trust, and monopoly practices.

GRI 207: Tax (2019).
- Disclosure 207-1 Approach to tax.
- Disclosure 207-2 Tax governance, control, and risk management.
- Disclosure 207-3 Stakeholder engagement and management of concerns related to tax.
- Disclosure 207-4 Country-by-country reporting.

After GRI.

GRI 202: Market Presence (2016)

GRI 202 looks at wages paid in relation to any local minimum wage and the degree of local hiring at the senior management level. Every company has an impact on the local area and these disclosures provide an insight into the market presence in the local area.

GRI 203: Indirect Economic Impacts (2016)

This topic is primarily a narrative disclosure and is an opportunity to disclose activities that support the local economy and community. A company does not simply have economic impacts and GRI 203 looks at the additional impacts such as infrastructure investment and can include a range of community investments such as social facilities, health centres and sports facilities that measure a company's contribution to the local economy. Other indirect impacts on the stakeholders can include improving job markets, supporting jobs in the local area and improving education and skills in the local area.

- **Tip** – The impacts can be financial or non-financial and reporting on the indirect economic impacts is an opportunity to disclose activities that benefit the local community but the evidence must be there and be robust.

GRI 204: Procurement Practices (2016)

An important indirect impact for any company is the procurement practices used and this disclosure covers local procurement that will support local communities. The management approach disclosures can include procurement practices such as lead and payment times for small suppliers to support SMEs (see Section 5.12).

GRI 205: Anti-corruption (2016)

GRI 205 covers issues such as bribery, fraud and extortion. Corruption in any form can pose a substantial risk for any company. GRI 205 requires a risk assessment for corruption as well as disclosure of anti-corruption policies and training. Disclosure 205-3 requires reporting of any incidents of corruption and the corrective actions taken as a result of these.

GRI 206: Anti-competitive Behaviour (2016)

Anti-competitive behaviour is illegal in most parts of the world and GRI 206 covers issues such as price fixing, cartels, and monopoly practices that restrict market competition and fair markets. Disclosure 206-1 requires the reporting of any incidents of anti-competitive behaviour and the corrective actions taken as a result of these.

GRI 207: Tax (2019)

GRI 207 is the most recent of the GRI standards to be released. Stakeholders expect companies to comply with the relevant tax legislation in the countries in which they operate and not to engage in tax evasion or aggressive tax avoidance which can increase risk to the operations.

Many of the disclosures can be taken directly from other company documents.

Look for ways to link the disclosures to the UN SDGs and the UN GC so that you can include the linkages in the report matrix (see Section 12.5).

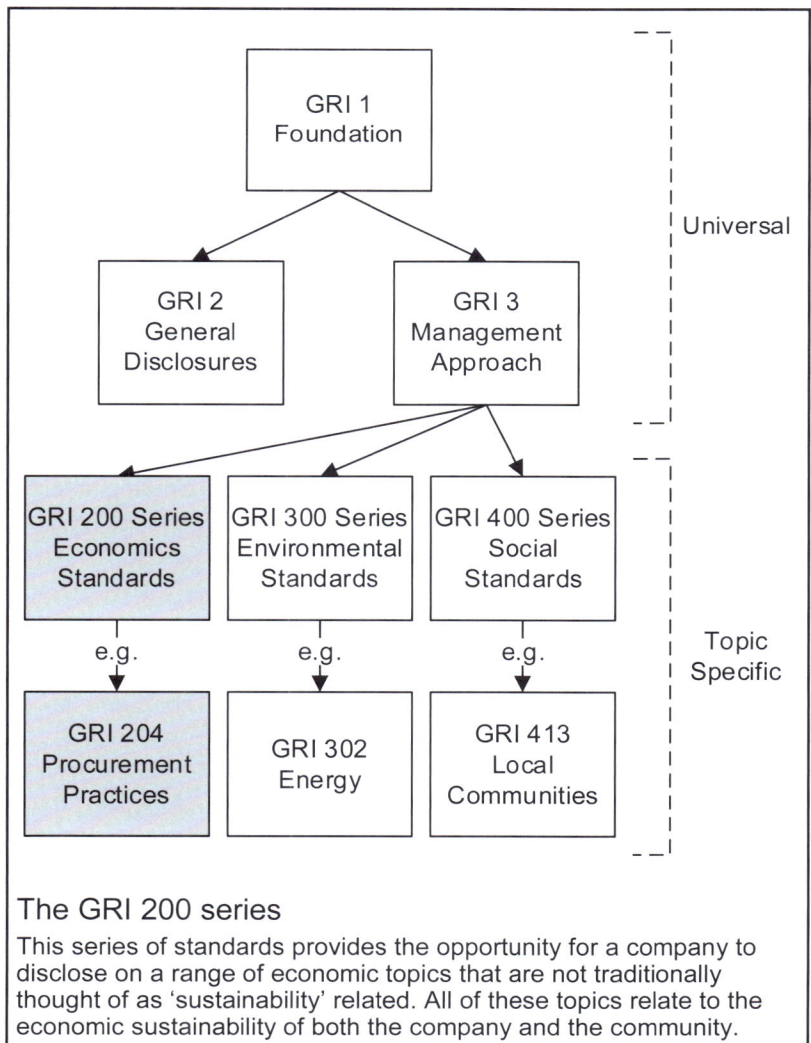

The GRI 200 series

This series of standards provides the opportunity for a company to disclose on a range of economic topics that are not traditionally thought of as 'sustainability' related. All of these topics relate to the economic sustainability of both the company and the community.

12.9 Environmental – GRI 300 series

The environmental standards

The 300 series of the GRI Standards include topic-specific Standards used to report information on an organization's material impacts related to environmental topics. These are, perhaps, the standard disclosures that most people will have thought of for sustainability, i.e., materials, energy, effluents etc., but those who have read the rest of this book will realise that sustainability is a broader topic.

There are 6 separate standards with a potential total of 32 disclosures (shown in the box on the right). The requirements for core and comprehensive disclosures are the same as for the Economics series (see Section 13.8) and, similarly, each topic has a requirement for a management approach disclosure using GRI 3 that can be used to add a narrative to the quantitative information.

The GRI 300 topics

The GRI 300 series covers a range of topic-specific standards that relate to the environmental impacts of a company. Managing the inputs and outputs of a company, particularly those that have environmental impacts is essential for sustainability at both the company and world scales. Companies that reduce their inputs by converting more efficiently and reduce their outputs by good management will not only contribute to sustainability at the environmental level but also be more profitable and sustainable as a company.

GRI 301: Materials

Disclosure 301-1 requires information on the total materials use by the company and sub-divides this into renewable and non-renewable materials.

- **Tip** – The materials used are not simply the plastics materials, it also includes all inputs, e.g., packaging (see Section 9.6).

The amount of recycled materials (which may include packaging) is declared in Disclosure 301-2 and the amount of reclaimed products or packaging materials is declared in Disclosure 301-3.

- **Tip** – The standard defines recycled input materials as those which replace virgin materials. This does not include internal regrind (see Section 5.3).

GRI 302: Energy (2016)

Disclosures 302-1 to 302-4 are all standard disclosures and should present no issues for

The GRI 300 series disclosures

GRI 301: Materials
- Disclosure 301-1 Materials used by weight or volume.
- Disclosure 301-2 Recycled input materials used.
- Disclosure 301-3 Reclaimed products and packaging materials.

GRI 302: Energy (2016).
- Disclosure 302-1 Energy consumption within the organization.
- Disclosure 302-2 Energy consumption outside of the organization.
- Disclosure 302-3 Energy intensity.
- Disclosure 302-4 Reduction of energy consumption.
- Disclosure 302-5 Reductions in energy requirements of products and services.

GRI 303: Water and Effluents (2018).
- Disclosure 303-1 Interactions with water as a shared resource.
- Disclosure 303-2 Management of water discharge-related impacts.
- Disclosure 303-3 Water withdrawal.
- Disclosure 303-4 Water discharge.
- Disclosure 303-5 Water consumption.

GRI 304: Biodiversity (2016).
- Disclosure 304-1 Operational sites owned, leased, managed in, or adjacent to, protected areas and areas of high biodiversity value outside protected areas.
- Disclosure 304-2 Significant impacts on biodiversity.
- Disclosure 304-3 Habitats protected or restored.
- Disclosure 304-4 IUCN Red List species and national conservation list species in areas affected by operations.

GRI 305: Emissions (2016).
- Disclosure 305-1 Direct (Scope 1) GHG emissions.
- Disclosure 305-2 Indirect (Scope 2) GHG emissions.
- Disclosure 305-3 Other indirect (Scope 3) GHG emissions.
- Disclosure 305-4 GHG emissions intensity.
- Disclosure 305-5 Reduction of GHG emissions.
- Disclosure 305-6 Emissions of ozone-depleting substances.
- Disclosure 305-7 Nitrogen oxides (NO_X), sulphur oxides (SO_X), and other significant air emissions.

GRI 306: Effluents and Waste (2016).
- Disclosure 306-1 Water discharge by quality and destination.
- Disclosure 306-2 Waste by type and disposal method.
- Disclosure 306-3 Significant spills.
- Disclosure 306-4 Transport of hazardous waste.
- Disclosure 306-5 Water bodies affected by water discharges.

GRI 307: Environmental Compliance (2016).
- Disclosure 307-1 Non-compliance with environmental laws and regulations.

GRI 308: Supplier Environmental Assessment (2016).
- Disclosure 308-1 New suppliers screened using environmental criteria.
- Disclosure 308-2 Negative environmental impacts in the supply chain and actions taken.

After GRI.

companies who have implemented good energy management. The information and actions should be present from the energy management system and easily accessible.

- **Tip** – Obviously we have an issue with reporting based on kWh/kg (see Section 6.5) when there is a base load.
- **Tip** – Disclosure 302–2 can use the GHG Protocol Corporate Value Chain (Scope 3).
- **Tip** – If conversions are made, the source of the conversion factors should be given.

Disclosure 302-5 offers interesting opportunities for plastics processors as it covers energy use reductions from products. Suppliers of products such as PVC-U windows and light-weight packaging can report on energy saved as a result of their products.

GRI 303: Water and Effluents (2018)

The disclosures in this standard follow the framework for water management developed in Chapter 8 and should present no issues for companies with good water management.

GRI 304: Biodiversity (2016)

The disclosures in this standard give information on a company's impacts on biodiversity and how this is managed.

GRI 305: Emissions (2016)

The disclosures in this standard are based on the GHG Protocol (see Section 7.1) and should present no issues for companies already reporting to the CDP using the GHG Protocol.

GRI 306: Effluents and Waste (2016)

The disclosures in this standard give information on a company's handling and actions in the area of liquid and solid waste and how this is managed and controlled. Disclosure 306-1 relates to water-based effluents (see Section 8.13) and Disclosure 306-2 relates to solid waste (see Section 9.1).

GRI 307: Environmental Compliance (2016)

This standard requires disclosure of any significant fines or penalties for non-compliance.

- **Tip** – Compliance is also covered in GRI 419 and it is possible to combine the management approach disclosures for these if both are material.

GRI 308: Supplier Environmental Assessment (2016)

A company's supply chain will have an environmental impact and management of suppliers is essential. This can, in some cases in fact, be larger than the internal impacts.

This part of the GRI standards covers supplier environmental assessment and is consistent with ISO 26000 in terms of sustainable procurement (see Sections 5.10-5.14).

Tip – Supplier social assessment is covered in GRI 414 and it is possible to combine the management approach disclosures for these if both are material.

Look for ways to link the disclosures to the UN SDGs and the UN GC so that you can include the linkages in the report matrix (see Section 12.5).

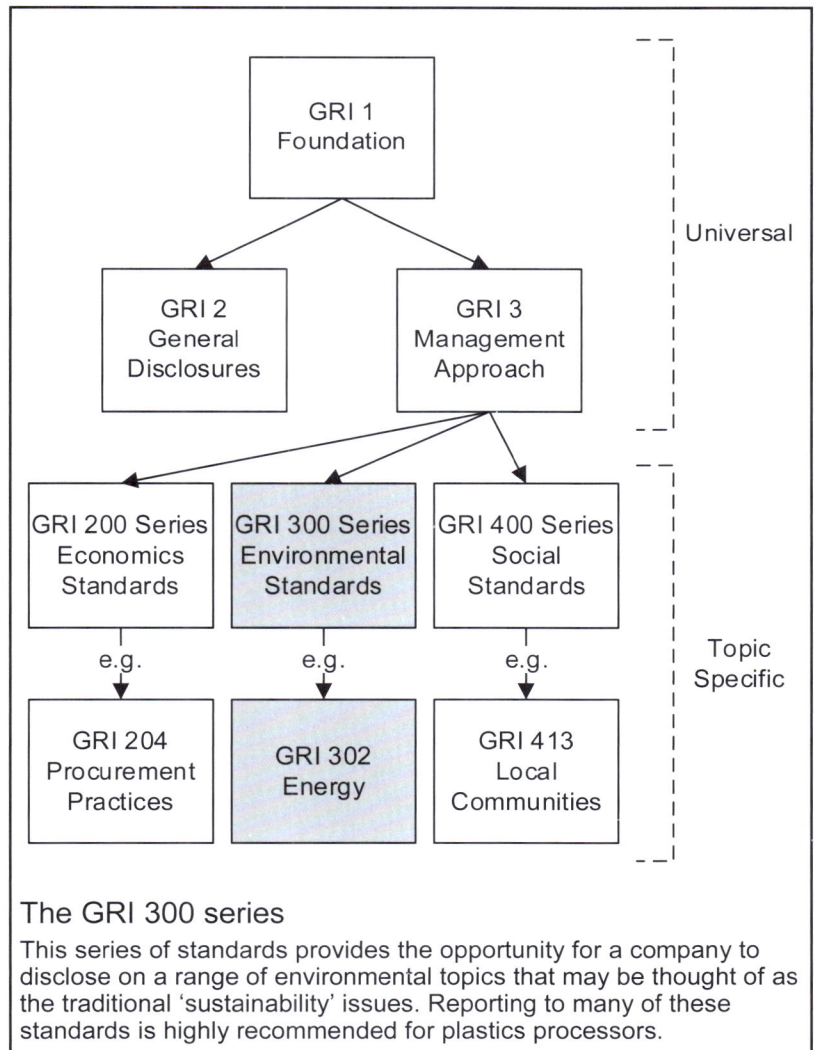

The GRI 300 series

This series of standards provides the opportunity for a company to disclose on a range of environmental topics that may be thought of as the traditional 'sustainability' issues. Reporting to many of these standards is highly recommended for plastics processors.

12.10 Social – GRI 400 series

The social standards

The 400 series of the GRI Standards includes topic-specific Standards used to report information on an organization's material impacts related to social sustainability. The majority of these topics have already been discussed in Chapter 11, i.e., social responsibility.

There are 19 separate standards with a potential total of 40 disclosures (shown in the box on the right and on the opposite page). The requirements for core and comprehensive disclosures are the same as for the Economics series (see Section 12.8) and, similarly, each topic has a requirement for a management approach disclosure using GRI 3 that can be used to add a narrative to the quantitative information.

The GRI 400 topics

There are too many topics and disclosures to cover individually in this section and we will only try to cover the broad intent of the series.

The GRI 400 series topics fit broadly into 4 main categories (although there is a great deal of crossover between categories).

- **Tip** – There are considerable linkages possible between the GRI 400 topics and ISO 26000 (see Chapter 11). ISO 26000 gives good guidance on the actions and expectations for organizations to address each of these topics.

- **Tip** – ISO 26000 is not intended to be used for certification but reporting to the GRI 400 series will, in effect, disclose information relevant to ISO 26000.

- **Tip** – Most of the disclosures in the GRI 400 series can also be easily linked to the UN SDGs and the UN GC. Do not miss the opportunity to add value to reporting by linking to these and including this in the report (see Section 12.5).

Labour practices

GRI 401 to GRI 405 are all concerned with labour practices and this is the largest category with a total of 19 potential disclosures. These disclosures can broadly be linked to Clause 6.4 of ISO 26000 (see Section 11.5).

Ten of these disclosures are in the area of occupational health and safety and a company operating an effective health and safety management system to ISO 45001 (see Section 2.12) should have no difficulty in providing the required information.

The GRI 400 series disclosures

Labour practices

GRI 401: Employment (2016).
- Disclosure 401-1 New employee hires and employee turnover.
- Disclosure 401-2 Benefits provided to full-time employees that are not provided to temporary or part-time employees.
- Disclosure 401-3 Parental leave.

GRI 402: Labour/Management Relations (2016).
- Disclosure 402-1 Minimum notice periods regarding operational changes.

GRI 403: Occupational Health and Safety (2018).
- Disclosure 403-1 OH&S management system.
- Disclosure 403-2 Hazard identification, risk assessment, and incident investigation.
- Disclosure 403-3 Occupational health services.
- Disclosure 403-4 Worker participation, consultation, and communication on OH&S.
- Disclosure 403-5 Worker training on OH&S.
- Disclosure 403-6 Promotion of worker health.
- Disclosure 403-7 Prevention and mitigation of OH&S impacts directly linked by business relationships.
- Disclosure 403-8 Workers covered by an OH&S management system.
- Disclosure 403-9 Work-related injuries.
- Disclosure 403-10 Work-related ill health.

GRI 404: Training and Education (2016).
- Disclosure 404-1 Average hours of training per year per employee.
- Disclosure 404-2 Programs for upgrading employee skills and transition assistance programs.
- Disclosure 404-3 Percentage of employees receiving regular performance and career development reviews.

GRI 405: Diversity and Equal Opportunity (2016).
- Disclosure 405-1 Diversity of governance bodies and employees.
- Disclosure 405-2 Ratio of basic salary and remuneration of women to men.

Human rights

GRI 406: Non-discrimination (2016).
- Disclosure 406-1 Incidents of discrimination and corrective actions.

GRI 407: Freedom of Association and Collective Bargaining (2016).
- Disclosure 407-1 Operations and suppliers in which the right to freedom of association and collective bargaining may be at risk.

GRI 408: Child Labour (2016).
- Disclosure 408-1 Operations and suppliers at significant risk for incidents of child labour.

GRI 409: Forced or Compulsory Labour (2016).
- Disclosure 409-1 Operations and suppliers at significant risk for incidents of forced or compulsory labour.

- **Tip** – The definition of workers in GRI 403 (Occupational health and safety) includes workers directly employed by the company, workers who are not employees but whose work or workplace is controlled by the company and workers who may be affected by the company's products or services. This is a very broad, but appropriate, concept of workers.

Human rights

GRI 406 to GRI 412 are all concerned with human rights and this is the second largest category with a total of 9 potential disclosures. These disclosures can broadly be linked to Clause 6.3 of ISO 26000.

Community involvement and development

GRI 413 to GRI 415 are all concerned with community involvement and development. These disclosures can broadly be linked to Clause 6.8 of ISO 26000.

- **Tip** – Supplier environmental assessment is covered in GRI 308 and it is possible to combine the management approach disclosures for GRI 414 with those of GRI 308 if both are material.

Consumer issues

GRI 416 to GRI 419 are all concerned with consumer issues. These disclosures can broadly be linked to Clause 6.7 of ISO 26000.

- **Tip** – Compliance is also covered in GRI 307 and it is possible to combine the management approach disclosures for GRI 419 with those of GRI 307 if both are material.

Linking to other initiatives

Many of the GRI social disclosures can be linked to the UN SDGs and good guidance on this is given in the 'Inventory of Business Indicators' available at sdgcompass.org/business-indicators/. At the time of writing this linkage was to the now obsolete GRI G4 standards but the linkages remain clear.

The GRI 400 series disclosures (cont.)

GRI 410: Security Practices (2016).

- Disclosure 410-1 Security personnel trained in human rights policies or procedures.

GRI 411: Rights of Indigenous Peoples (2016).

- Disclosure 411-1 Incidents of violations involving rights of indigenous peoples.

GRI 412: Human Rights Assessment (2016).

- Disclosure 412-1 Operations that have been subject to human rights reviews or impact assessments.
- Disclosure 412-2 Employee training on human rights policies or procedures.
- Disclosure 412-3 Significant investment agreements and contracts that include human rights clauses or that underwent human rights screening.

Community involvement and development

GRI 413: Local Communities (2016).

- Disclosure 413-1 Operations with local community engagement, impact assessments, and development programs.
- Disclosure 413-2 Operations with significant actual and potential negative impacts on local communities.

GRI 414: Supplier Social Assessment (2016).

- Disclosure 414-1 New suppliers screened using social criteria.
- Disclosure 414-2 Negative social impacts in the supply chain and actions taken.

GRI 415: Public Policy (2016).

- Disclosure 415-1 Political contributions.

Consumer issues

GRI 416: Customer Health and Safety (2016).

- Disclosure 416-1 Assessment of the health and safety impacts of product and service categories.
- Disclosure 416-2 Incidents of non-compliance concerning the health and safety impacts of products and services.

GRI 417: Marketing and Labelling (2016).

- Disclosure 417-1 Requirements for product and service information and labelling.
- Disclosure 417-2 Incidents of non-compliance concerning product and service information and labelling.
- Disclosure 417-3 Incidents of non-compliance concerning marketing communications.

GRI 418: Customer Privacy (2016).

- Disclosure 418-1 Substantiated complaints concerning breaches of customer privacy and losses of customer data.

GRI 419: Socioeconomic Compliance (2016).

- Disclosure 419-1 Non-compliance with laws and regulations in the social and economic area.

After GRI.

12.11 Accounting for sustainability

Make it automatic

Reporting sustainability should add value and the best way to do this is to make the reporting automatic and to re-use existing data. Most of the data needed is somewhere in the company's systems but the major challenge is to find it, access it and format it for publication and disclosure. Equally, the costs and benefits of reporting will be hidden all over the systems and making them visible will generate savings and be a driver for improvement.

- **Tip** – Concepts such as 'the cost of quality' and 'energy accounting' can be used to capture the real costs and benefits by using accountants to generate the reports.

The reporting process

Reporting is a process and this is shown in the box on the right. Throughout this process, the company strategy is an important factor. Sustainability is rapidly becoming central to the strategy of any business and sustainability reporting also needs to be aligned with the company strategy, decision-making and risk assessment processes. The reporting process is as follows:

Identify legal requirements

Some sustainability reporting is covered by country-specific legislation or Stock Exchange requirements. Companies must, at a minimum, identify and report on these topics to comply with the law.

Engage stakeholders

Seeking the views of both internal and external stakeholders is vital to determining materiality.

- Internal stakeholders can be from any part of the company and sustainability needs wide input to capture all of the views. An initial survey of prospective internal stakeholders can provide both motivation and engagement.

- **Tip** – Don't just talk to management.

- External stakeholders can be difficult to identify but engaging with them gives essential input on what they consider to be material and how a sustainability report can serve their needs as well as the company's.

- **Tip** – It may be worthwhile to set up an on-going external stakeholder engagement forum to understand their perspectives and what is most material to them. At the worst it is great PR.

Determine materiality

Materiality flows from stakeholder engagement (see Section 12.4) and is key in deciding what to report, i.e., the significant material factors are the ones to report on.

Set boundaries

Materiality decides the topics and, in some cases, also determines the topic boundaries. In other cases, the company needs to decide the reporting boundaries, e.g., will data be collected from suppliers? These decisions should be made before reporting so that subsequent reports are consistent.

Identify data sources

The key to reusing data for the report is in locating the actual data. Few companies have fully integrated systems and in most

Sustainability must be integrated with the company strategy. Aligning the reporting with the strategy is the key to driving change and improving sustainability.

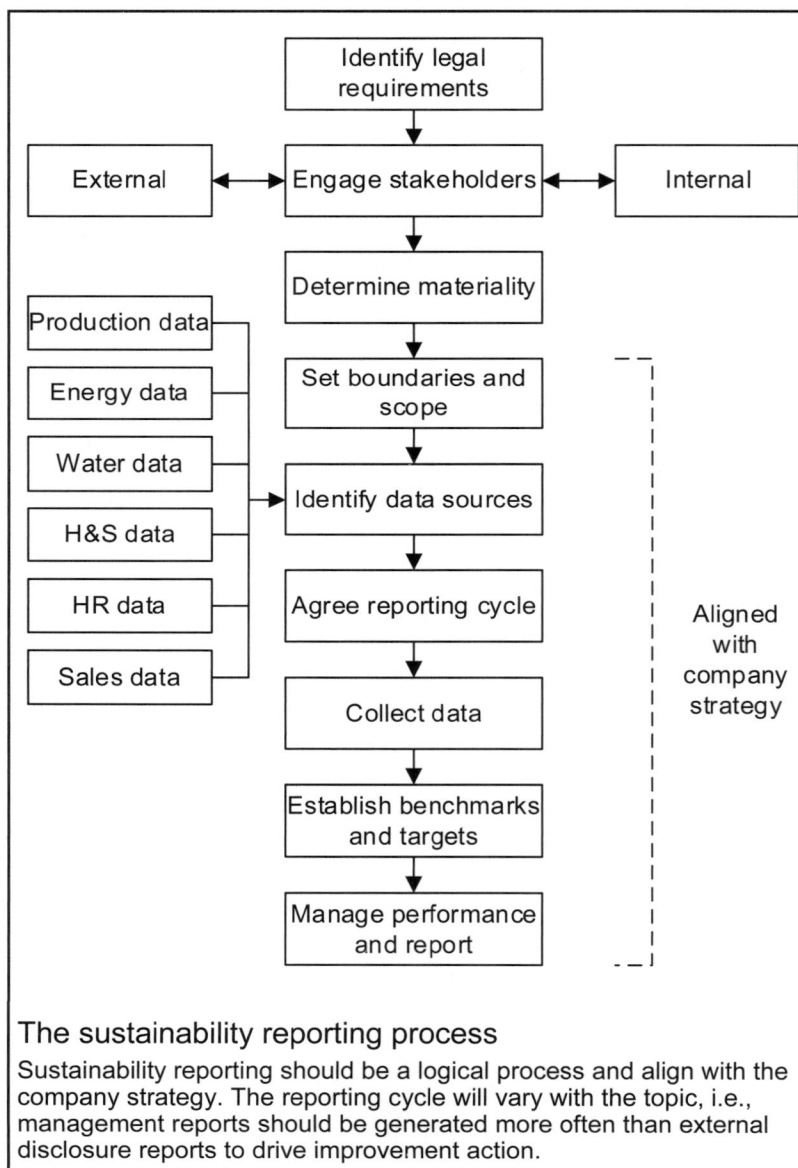

The sustainability reporting process

Sustainability reporting should be a logical process and align with the company strategy. The reporting cycle will vary with the topic, i.e., management reports should be generated more often than external disclosure reports to drive improvement action.

there will be 'islands of data' that can be used to populate the report. The data may come from a variety of sources (see box on the left) and these need to be identified, validated and formatted so that the data can be used for reporting.

Agree reporting cycle

A full sustainability report will probably be on an annual cycle to match the financial reporting cycle (although it may be different). However, individual topics can be reported more often for operational reasons, e.g., reporting on energy use should be on a monthly or weekly cycle to allow for rapid management action. This means deciding the reporting cycle and the data flow for the various elements of the report.

Collect data

Collecting the actual data and information is relatively mechanical having found the source data and agreed the reporting cycle.

Establish benchmarks and targets

Initial data collection will provide data on the current status, and perhaps past performance if historical data is available, but there is a need to assess performance against benchmarks and to set targets for the future. This may be difficult in some areas but targets are needed to drive improvement. Targets needs to be SMART (Specific, Measurable, Achievable, Relevant and Time-limited).

Manage performance and report

The sustainability report is simply part of the overall performance management system. It should drive improvement through a long-term strategy and disclosures in the report allow all stakeholders to see clear progress.

External auditing

Many companies producing sustainability reports have chosen to undertake external auditing (in the same way as financial reports are audited). This adds considerably to the cost of producing a report and in most cases will be the major cost of producing the report. External assurance adds credibility to the report but there is no requirement for external auditing. GRI recommends external auditing but it is not a requirement for a report to be 'in accordance with the GRI standards' (see Section 12.6).

- **Tip** – If sustainability information is a 'risk' then it may need to be included in the Annual Report and audited as part of the financial audit to ensure that there is no material misstatement.

- **Tip** – GRI publishes 'The external assurance of sustainability reporting' (GRI, 2013) and this gives guidance on external assurance.

There no current global agreement on external audit standards and the most commonly used standards are:

- ISAE 3000 (isae3000.com).
- AccountAbility: AA1000AS www.accountability.org/standards/).
- **Tip** – These are guidance for the auditors and not for the reporting company.

External auditors should:

- Be independent.
- Be competent.
- Be professional.
- Assess data quality.
- Provide a written publicly available report.

Integrated reporting <IR>

Integrated reporting is designed to report how a company creates value, both now and in the future, in terms of strategy, business model, governance, performance and prospects (see integratedreporting.org). <IR> was developed by the International Integrated Reporting Council (IIRC) and aligns capital allocation and corporate behaviour with the wider goals of financial stability and sustainable development. This is in contrast to the traditional Annual Report which focuses on historical financial information.

<IR> and sustainability reporting are different but complementary. Sustainability reporting is designed so that a range of stakeholders can understand the company's impacts in the environmental, social and economic areas. <IR> is designed for investors and providers of financial capital to see how a company creates value and how it will continue to do so in the future. <IR> focuses on how a company's activities affect the six <IR> capitals: financial capital, manufactured capital, intellectual capital, human capital, social and relationship capital and natural capital. Considering the company's impact on these allows investors to consider all resources and relationships in the value creation process.

<IR> has no specific KPIs or methods of disclosure and allows companies to decide materiality and what disclosures are to be made. GRI and <IR> work together and GRI supports the <IR> concept.

Sustainability Accounting Standards Board (SASB)

SASB also produces reporting standards that are focused on financially material sustainability issues that are relevant to investors but these are less well used than <IR> and GRI. The SASB standards are focused on specific industry sectors, e.g., 'containers and packaging'. The standards give the disclosure requirements for the most likely material topics in each sector. However, the company decides what is the most appropriate standard, what is financially material and what to report.

The SASB materiality map (materiality.sasb.org/) is a very useful chart showing the most likely material issues for the relevant sectors. This can be used as a starter for identifying materiality in a range of industries.

12.12 Reporting sustainability – where are you now?

Tell the world

Reporting sustainability is an opportunity to tell the world what you are doing in the sustainability area. This can add value to the company by improving relationships with a variety of stakeholders ranging from internal staff to investors and NGOs. Reporting is not an option for many companies, it is covered by legal requirements, but companies simply meeting the minimum legal requirements are missing the benefits of good sustainability reporting – there are many other benefits to high-quality reporting.

However, reporting should never be an exercise in greenwashing, reports should meet agreed principles and standards and cover the topics that are material to the operations of the company. Reporting is not something that is done once, it should become part of the overall reporting schedule for a company, in the same way that

financial data is reported, this means setting up systems to collect the data automatically and effectively.

Completing the chart

This chart is completed and assessed as for the previous charts.

Reporting for many companies means doing the legislative minimum. There is so much more that can be gained from external reporting (disclosure) – do not miss this opportunity to add value to the company.

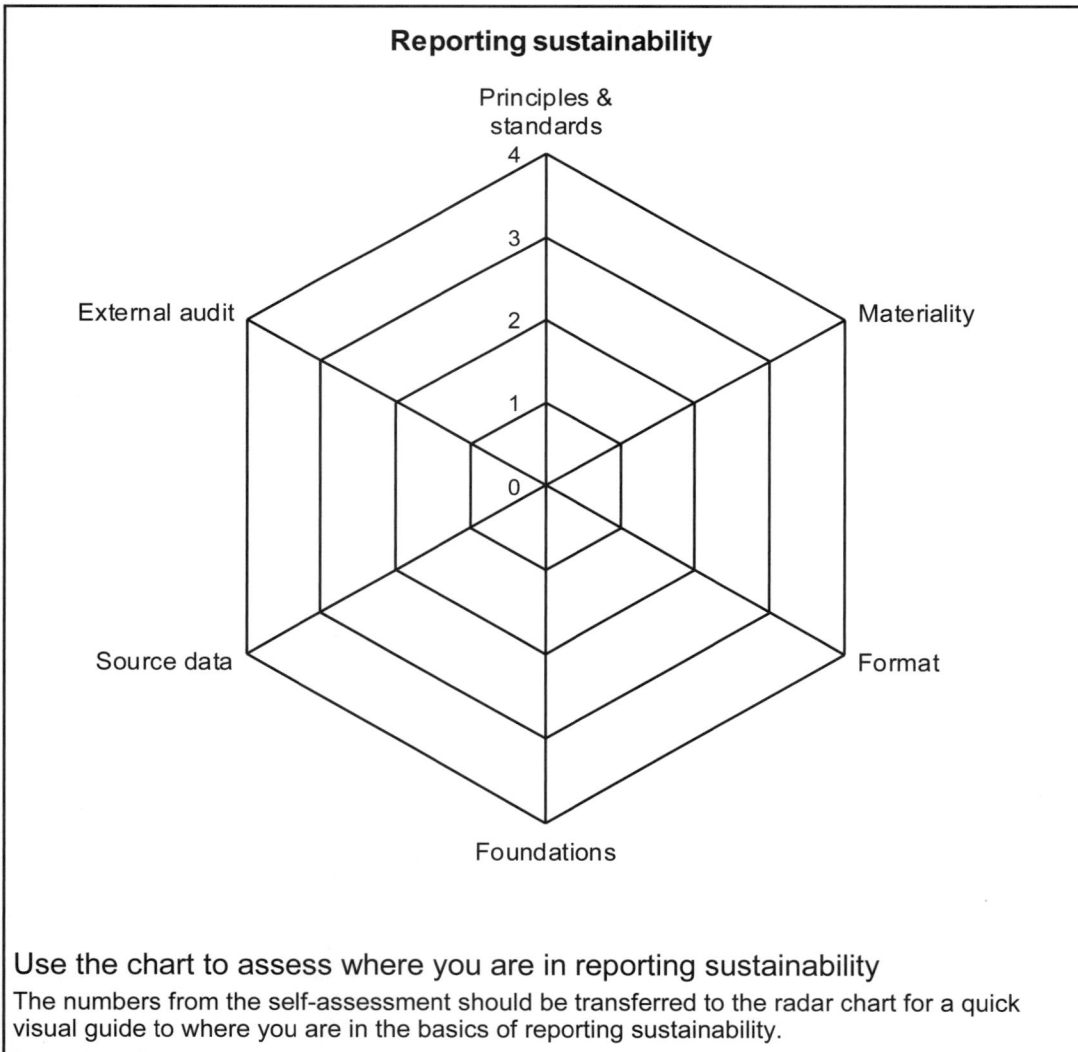

Reporting sustainability

Use the chart to assess where you are in reporting sustainability

The numbers from the self-assessment should be transferred to the radar chart for a quick visual guide to where you are in the basics of reporting sustainability.

Whilst we disagree with single number reporting as an operational performance metric, it can have value when used over the long-term, i.e., greater than 12 months, as this smooths out the effect of production volume changes.

			Reporting sustainability			
Level	Principles & standards	Materiality	Format	Foundations	Source data	External audit
4	Report meets all principles & standards. Covers all material impacts, encourages improvement & acknowledges failures.	Materiality study with all stakeholders. Impacts identified & reported. Risk analysis & SWOT used to create action plan.	Available in all formats. Significant narrative & numerical disclosures. GRI index linked to SDGs & UNGC.	Report is 'in accordance with the GRI standards' & uses 'comprehensive' option.	All source data for material topics identified, validated & formatted for easy access. Data collection for report is automatic.	Full external audit to ISAE 3000 or AA1000AS carried out, reported & publicly available.
3	Report meets most principles & standards. Report covers all material impacts but does not encourage improvement.	Materiality study with all stakeholders. Impacts identified & reported.	Available in some formats. Good narrative & some numerical disclosures. GRI index with no links to SDGs or UNGC.	Report is 'in accordance with the GRI standards' & uses 'core' option.	Source data for material topics are fragmented but most data are validated. Some manual data collection necessary for report production.	Full external audit to ISAE 3000 or AA1000AS carried out & reported but not publicly available.
2	Report meets some principles & standards. Report covers most material impacts & does not encourage improvement.	Materiality study with internal stakeholders only. Impacts identified & reported.	Available in some formats. Mainly narrative reporting & some numerical disclosures. No index of GRI disclosures.	Report uses 'GRI-referenced' claim for some specific topics. Meets GRI context & quality requirements.	Source data for material topics are fragmented but some data are validated. All data must be collected manually for report production.	Internal audit carried & audit report publicly available.
1	Report does not meet all principles & standards. Report covers few material impacts & does not encourage improvement.	Materiality study with internal stakeholders only. No report available or reporting is poor.	Available only in pdf (on web) format. Narrative reporting only with few numerical disclosures.	Report meets legally required disclosures.	Source data for material topics are fragmented & unvalidated. All data must be collected manually for report production.	Internal audit carried out but no report available.
0	Report is vague & aspirational rather than credible & does not focus on material topics. Primarily 'greenwashing' rather than reporting.	No materiality study carried out.	Format does not meet any established standard.	Report does not comply with any recognised disclosure reporting structure.	No validated source data easily available.	No external or internal audit carried out.
Score						

Key tips

- Reporting sustainability is the right thing to do. It assures compliance not only with legislation but also adds value to the company.

- There are many drivers for reporting but reporting is moving from a 'nice to have' to a 'business essential'. Do not be left behind, if you start now then it will be easier later.

- Always remember who you are talking to in sustainability reporting. Different audiences need different information.

- Sustainability reporting is like any other external reporting, it must be transparent, credible, quantitative, comparable, consistent and complete.

- KPIs can reflect management performance, operational performance or environmental conditions. Most sustainability reports will use a mix of the first two types of KPI.

- KPIs can be absolute or normalised. Most government reporting requires absolute KPIs but companies will often get the most benefit from normalised KPIs as these will provide a comparison of activity over time.

- Single number normalised KPIs may be affected by base loads or costs and may need adjustment because of this.

- In most countries there is legislation for some type of reporting. This must be produced but in most cases the schemes are designed to benefit the government and not the company.

- There are also multitude of independent (non-governmental) reporting schemes and there is considerable overlap and duplication in what they require for reporting.

- Most of the external reporting requirements can be linked so that reporting to one meets the requirements of reporting to another.

- The GRI standards are probably the most widely recognised and linked standards for sustainability reporting. The GRI standards can be linked to most of the other schemes, e.g., CDP, GHGP, UN SDGs and UN GC.

- Assessing materiality is one of the first tasks in reporting. Material topics are those topics which are both relevant to the stakeholders and which have a highly significant impact on the business.

- Materiality should be assessed using the viewpoints of both the internal and external stakeholders to get a wide understanding of materiality.

- Reporting should focus on the material topics. Do not report on topics that are not material unless required to do so.

- Most material topics will be covered by a standard GRI disclosure but if there are material topics that are not covered by these then it is possible to report on these to add value to the report.

- Materiality does not have to stop with identification, it is possible to take it further using risk analysis, SWOT analysis and an action plan.

- There are many reporting formats but an index of what the report contains and how it links to GRI standards, UN SDGs and UN GC provides most of the necessary linkages.

- The GRI standards are an excellent format and process for disclosing sustainability information.

- The GRI standards have rigorous requirements for materiality, management approach and general disclosures.

- The GRI standards cover economics, environmental and social issues and offer a flexible approach to disclosures to a well proven and accepted series of standards.

- Reporting should be a process that is both automatic and uses already available information.

- It is possible to have sustainability reports externally audited in much the same way as financial reports are externally audited.

- Sustainability reporting is part of a world-wide process of increasing non-financial information as well as the traditional; financial information.

- <IR> offers a framework for reporting that gives investors an insight into how a company creates value and is highly complementary to sustainability reporting to the GRI standards.

A high-quality sustainability report will give an insight into not only what a company is doing today but also what it did in the past and what it plans for the future.

The financial report is important in terms of investors valuing a company but the sustainability report is rapidly becoming as important in terms of investors valuing the future prospects of a company.

Appendix 1

The UN Sustainable Development Goals

The UN SDGs were covered in outline in Sections 1.16 and 1.17 but they are a comprehensive framework for sustainability action and deserve to be examined in detail.

This Appendix looks at the individual SDGs, the targets for achievement by 2030 and the role that plastic products can play in achieving these targets.

Note: Some of the targets were scheduled for achievement by 2020.

The industry has already contributed to achieving the MDGs (Millennium Development Goals) and there is much that we can do to contribute to the SDGs.

We have placed this discussion of the SDGs in an Appendix not because they are unimportant but to preserve the flow of the book in the early sections.

The bulk of Sections A1.1 to A1.17 was originally developed by the author as part of his work with the Sustainability Committee of the British Plastics Federation. The critical input of the committee and their permission to pirate my own work is acknowledged.

The targets for most of the goals are interlinked and almost all activities will affect one or more of the goals. These are cross-referenced in the form: 'see UN SDG XX' where XX refers to the relevant goal rather than referring to the section number.

Portions of these sections are taken from the UN Sustainable Development Knowledge Platform (sdgs.un.org/goals). Their copyright is acknowledged. A visit to the platform is very worthwhile and highly recommended.

A1.1   Goal 1: No poverty

End poverty in all its forms everywhere

Extreme poverty rates have been cut by more than half since 1990 but 20% of people in developing regions still live on less than $1.90 a day, and there are millions more who make little more than this daily amount, plus many more people who risk slipping back into poverty.

Poverty is more than the lack of income and resources to ensure a sustainable livelihood. Its other manifestations include hunger and malnutrition, limited access to education and other basic services, social discrimination and exclusion as well as the lack of participation in decision-making. Economic growth must be inclusive to provide sustainable jobs and promote equality.

Challenges for plastics

The challenges to the plastics industry in helping to reduce poverty are:

- The investment costs for a plastics processing site are high and there is a need for a substantial local or export market for the products to cover these costs.

- The investment costs mean that plastics processing sites tend to be located in areas where the rule of law is strong enough to prevent appropriation of the substantial assets involved (see UN SDG 16).

- The market scale and transport difficulties in developing countries inevitably lead to sites being located in urban areas and the benefits of industrialisation are not seen in rural areas, even if the benefits of the products are.

- Some of the skills required are specific to the industry and the rise of Industry 4.0 may see a concentration of skills being located centrally and delivered via the Internet to the detriment of local skill development.

- The plastics industry is a mobile industry and recent years have seen production migrate to low-labour cost countries. This has benefited local development but as local costs rise, the industry can migrate once again to a lower cost country.

Opportunities for plastics

The plastics industry has much to contribute in terms of reducing poverty throughout the world.

- Plastics are light-weight materials and transport costs mean that production sites are generally located relatively close to the customer. This gives decentralised development opportunities and increased economic activity in the local area.

- The small size of the typical processing site means that it is a high-volume employer (in the UK, the industry provides more jobs than the automotive and pharmaceutical industries combined). High employee numbers not only provide direct jobs but also reduce poverty in surrounding areas through the need for services and increased economic activity. The plastics industry is responsible for the economic health of entire regions in parts of the world.

- The industry brings employment to all parts of the world (see UN SDG 8). Plastics processing sites are predominantly SMEs (< 250 employees) and tend to be located close to their markets. These small companies provide local employment that is good for developing countries but more importantly they provide a base for skills training that can aid development (see UN SDG 4). Employment opportunities can range from low-skill machine operators through to high-skill maintenance and management positions.

> The use of plastics has dramatically reduced the cost of many household products and made them widely available. Plastic products are low-cost, durable and easily transported.

Goal 1 targets

- By 2030, eradicate extreme poverty for all people everywhere, currently measured as people living on less than $1.25 a day.

- By 2030, reduce at least by half the proportion of men, women and children of all ages living in poverty in all its dimensions according to national definitions.

- Implement nationally appropriate social protection systems and measures for all, including floors, and by 2030 achieve substantial coverage of the poor and the vulnerable.

- By 2030, ensure that all men and women, in particular the poor and the vulnerable, have equal rights to economic resources, as well as access to basic services, ownership and control over land and other forms of property, inheritance, natural resources, appropriate new technology and financial services, including microfinance.

- By 2030, build the resilience of the poor and those in vulnerable situations and reduce their exposure and vulnerability to climate-related extreme events and other economic, social and environmental shocks and disasters.

- Ensure significant mobilization of resources from a variety of sources, including through enhanced development cooperation, in order to provide adequate and predictable means for developing countries, in particular least developed countries, to implement programmes and policies to end poverty in all its dimensions.

- Create sound policy frameworks at the national, regional and international levels, based on pro-poor and gender-sensitive development strategies, to support accelerated investment in poverty eradication actions.

- Plastics products are not only low-cost but, when used correctly, have a high-utility value. They can be used to improve crop yields (see UN SDG 2), increase the availability and distribution of clean water (see UN SDG 6) and therefore increase the productive capacity of areas to reduce poverty.

Action taken to date

The action taken to date has been driven mainly by economic pressures but the industry has already contributed to poverty reduction through actions such as:

- There are plastics processing sites in almost every country in the world. These provide employment and poverty reduction through stable jobs and skills development (see UN SDG 8).

- The ability of plastics to increase the availability and distribution of clean water (see UN SDG 6) has enabled other economic activity to replace the time required to source clean water. This other economic activity, e.g., education and social, has reduced poverty and enabled community development.

- The benefits of improved water supplies and improved sanitation through plastics products has increased the resilience of communities (see UN SDG 6) and reduced the effects of climate change (see UN SDG 13).

- Improved crop yields (see UN SDG 2) have allowed communities to have adequate food and to move beyond subsistence farming to trading and consequent poverty reduction.

- Improved product protection in the transport phase mean that more of the harvested product reaches markets in saleable condition and improves the trading position of small-scale farmers (see UN SDG 2).

- The plastics industry provides high-quality employment opportunities for all genders and gender is not a barrier to progression in the industry (see UN SDG 5).

- Product development for developing countries means that plastics are widely used in the refugee and disaster relief camps where they can provide low-cost shelter and protection from the elements (see UN SDG13).

Future action

The plastics industry will continue to contribute to poverty reduction in a variety of areas:

- The increased use of plastics will continue to improve crop yields and retain product life to market (see UN SDG 2 and UN SDG 8). This will increase the income of small-scale farmers and further reduce rural poverty.

- The development of cheap, easy-to-maintain dwellings (see UN SDG 11) will reduce housing costs and therefore poverty in wide areas.

- Continued economic growth (see UN SDG 8) will increase local markets for plastics products and increase the need for local production to provide employment and reduce poverty.

- The recycling of plastics will involve whole communities in the collection, transport, sorting and recycling processes. This is a new business area that will stimulate economic growth.

- The continued use of plastics in the 'cold chain' will enable medicines to be transported greater distances (see UN SDG 3) and increase economic activity in rural areas.

Summary

The plastics industry has contributed significantly in the past to the goals of UN SDG 1 and will continue to contribute in the future. Plastics products (both short- and long-life products) enable increased economic activity to reduce poverty and improve living conditions.

Plastics are used extensively in refugee camps throughout the world to provide low-cost housing and water distribution.

The plastics industry supports 166,000 jobs in the UK and millions globally.

The Circular Economy is predicted to create over 200,000 gross jobs and reduce unemployment by 54,000 in the UK alone.

Source: Green Alliance – 'Employment and the circular economy.'

Even simple single – use products such as PET water bottles have value after use in some countries. In Northern Ethiopia these are collected and traded in markets as an efficient way of collecting and transporting water.

A1.2   Goal 2: Zero hunger

End hunger, achieve food security and improved nutrition and promote sustainable agriculture.

Correctly done, agriculture, forestry and fisheries can provide nutritious food for all and generate decent incomes, while supporting people-centred rural development and protecting the environment.

Soils, freshwater, oceans, forests and biodiversity are being rapidly degraded. Climate change is putting even more pressure on the resources we depend on, increasing risks associated with disasters such as droughts and floods. Many rural people can no longer make ends meet on their land, forcing them to migrate to cities in search of opportunities.

A change in the global food and agriculture system is needed if we are to nourish today's 795 million hungry and the additional 2 billion people expected by 2050.

The food and agriculture sector offers key solutions for development, and is central for hunger and poverty eradication.

Challenges for the plastics industry

The challenges to the plastics industry in helping to deliver zero hunger are in the appropriate selection, use and, increasingly, disposal of plastics:

- The inappropriate use (and overuse) of plastics in the food system – this not only includes the use of excessive or difficult to recycle plastics (and combinations thereof) in retail food packaging applications but also includes the use of excessive or difficult to recycle plastics earlier in the food system (see UN SDG 12).

- The inappropriate disposal of plastics products used in the food system – this not only includes retail packaging but also includes the plastics used earlier in the food chain. In most cases, this is due to a lack of an appropriate waste management system to cope with re-use or recycling but can also include consumer behaviour aspects such as littering.

- The plastics industry is also being challenged by environmental campaigners for 'plastics-free aisles' in supermarkets and the issues of plastics litter and plastics in the oceans (see UN SDG 14) are areas of rising consumer concern.

The plastics industry delivers immense benefits in achieving this goal but major challenges remain, particularly in product re-use and/or disposal at end-of-life.

Opportunities for plastics

Plastics products have already enabled more food to be produced from the available resources but there is more that the industry can do at all levels of the food system, such as:

- Packaging can be used to increase farm productivity and crop yields through the use of 'poly-tunnels' to extend growing seasons.

- Plastics pipes can be used to increase food systems resilience to drought by reducing

Without plastics packaging, food waste during transport can be as high as 50%. Food waste during storage can also be reduced by more than 20% using well-designed plastics packaging.

Goal 2 targets

- By 2030, end hunger and ensure access by all people, in particular the poor and people in vulnerable situations, including infants, to safe, nutritious and sufficient food all year round.

- By 2030, end all forms of malnutrition, including achieving, by 2025, the internationally agreed targets on stunting and wasting in children under 5 years of age, and address the nutritional needs of adolescent girls, pregnant and lactating women and older persons.

- By 2030, double the agricultural productivity and incomes of small-scale food producers, in particular women, indigenous peoples, family farmers, pastoralists and fishers, including through secure and equal access to land, other productive resources and inputs, knowledge, financial services, markets and opportunities for value addition and non-farm employment.

- By 2030, ensure sustainable food production systems and implement resilient agricultural practices that increase productivity and production, that help maintain ecosystems, that strengthen capacity for adaptation to climate change, extreme weather, drought, flooding and other disasters and that progressively improve land and soil quality.

- By 2020, maintain the genetic diversity of seeds, cultivated plants and farmed and domesticated animals and their related wild species, including through soundly managed and diversified seed and plant banks at the national, regional and international levels, and promote access to and fair and equitable sharing of benefits arising from the utilization of genetic resources and associated traditional knowledge, as internationally agreed.

- Increase investment, including through enhanced international cooperation, in rural infrastructure, agricultural research and extension services, technology development and plant and livestock gene banks in order to enhance agricultural productive capacity in developing countries, in particular least developed countries.

- Correct and prevent trade restrictions and distortions in world agricultural markets, including through the parallel elimination of all forms of agricultural export subsidies and all export measures with equivalent effect, in accordance with the mandate of the Doha Development Round.

- Adopt measures to ensure the proper functioning of food commodity markets and their derivatives and facilitate timely access to market information, including on food reserves, in order to help limit extreme food price volatility.

the cost and effort of transporting water to crops.

- Plastics drainage pipes can reduce the effects of flooding by channelling excess water to drainage systems without excessive soil losses.
- Plastics products can reduce spoilage and wastage in the food system and allow more food to be delivered from the producer to the plate.
- Plastics products are fundamental to the operation of seed and plant banks that protect genetic resources and diversity.
- Plastics products can improve fishing yields and protect stocks through sensitive and appropriate design of fishing nets.

Action taken to date

The plastics industry has had clear benefits for the majority of the world's population through actions such as:

- Plastic pipes deliver a reliable source of water to crops and villages to increase crop yields and decrease repetitive labour tasks (see UN SDG 6 and UN SDG 8).
- Crop yields are significantly improved through the use of 'poly-tunnels' to improve yields and extend growing seasons.
- Plastics products are used to reduce food losses and degradation from vermin and other organisms at the producer level.
- Plastics films are used to protect silage for winter animal feed.
- Plastic fish boxes allow fresh fish to be protected, transported and consumed large distances from the delivery port.
- Correctly designed food packaging increases shelf life and greatly reduces waste between producer and plate. In most cases the environmental case for plastics is poorly made but research[1,2] shows that substitution of plastics with alternative materials can result in increased environmental impact over the life of the product.
- Plastics enable food packaging, such as dispensing or child resistant caps, that are simply not possible with alternative materials.
- Modern harvesting equipment depends on plastics to function and these have increased crop yields and reduced the manual labour required (see UN SDG 1).
- Water-logged and unproductive land can be drained and then cultivated using land drainage piping.

The UN estimates that 800 million people suffer from hunger world-wide but this number would be far increased if not for the contribution of plastics in the food system.

Future action

The actions taken to date have had clear benefits for most of the world's population but extending these benefits to those most at risk of hunger requires further action such as:

- Reducing the amount of packaging used to the minimum necessary for product protection to reduce the amount of plastic entering the waste stream.
- Reducing the diversity of plastics in individual food packages. This does not imply reducing the number or types of plastics used but reducing the number or types of plastics used in a single package. This will avoid commingling of plastics types in the waste stream and make recycling easier.
- Establishing recycling networks to recover valuable plastics materials before they enter the waste stream. This can assist with UN SDG 8 by providing work in developing countries provided these are not used as 'sinks' for plastics waste from developed countries.

Summary

The plastics industry produces products that are vital to relieving hunger and both improving and changing the global food system. The industry produces the 'enabling products' to improve yields, improve harvesting practices, reduce food wastage and deliver more and better-quality food to the 800 million hungry people in the world.

Plastic fishing nets have decreased the cost of nets and increased the yield of fishing. Fish protein is essential for food security in many developing countries and for small island states and in coastal regions, fish provides over 50% of the animal protein intake.

Sustainable fishing practices and the inappropriate disposal of fishing nets remain a significant issue (see UN SDG 16).

Refrigerators and freezers are essential for food storage. Both use plastics for insulation and ease of cleaning.

- 1. 'Impact of plastics packaging on life cycle energy consumption and greenhouse gases in Europe', Denkstatt, 2011, www.plasticseurope.org.
- 2. 'Impact of plastics packaging on life cycle energy consumption and greenhouse gases in the United States and Canada', Franklin Associates, 2014, plastics.americanchemistry.com.

The threat of climate change (see UN SDG 13) and the contribution of the plastics industry are often linked. However, plastics bring enormous benefits to improving food production and reducing hunger.

The question is: Would you rather be definitely hungry today or possibly dead tomorrow?

A1.3 Goal 3: Good health and well-being

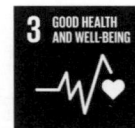

Ensure healthy lives and promote well-being for all at all ages

Ensuring healthy lives and promoting well-being for all at all ages is essential for sustainable development. Significant strides have been made in increasing life expectancy and reducing some of the common killers associated with child and maternal mortality. Major progress has been made on increasing access to clean water and sanitation, reducing malaria, tuberculosis, polio and the spread of HIV/AIDS. However, much more effort is needed to fully eradicate a wide range of diseases and address many different persistent and emerging health issues.

- **Tip** – Can anyone imagine the response to COVID-19 without plastics?

Challenges for the plastics industry

The challenges to the plastics industry in ensuring healthy lives and well-being are:

- Medical plastics are traditionally single use to reduce cross-contamination and the spread of disease. Reducing the amount of waste through effective sterilisation processes whilst still preserving the benefits of plastics is a challenge for the industry.

- The health care industry generates large amounts of potentially contaminated waste that must be disposed of even it is re-used before disposal. Potentially contaminated products cannot be treated as part of the standard waste stream. Medical waste is treated separately to standard waste and incineration is currently the standard method of disposal. Managing contaminated waste to recover the material or embodied energy requires improved processes and development.

- The use of medical plastics in developing countries presents specific issues with the long supply chains for delivery and recycling, if this is chosen as the disposal route.

- Exploring and quantifying the issue of micro-plastics. These are the result of physical or chemical breakdown of plastics products. Micro-plastics potentially enter the ecosystem as a result of wear from plastics in normal use or from inappropriate disposal of plastics and wear in the environment. Whilst plastics are inert 'en masse', the effect of micro-plastics

on the ecosystem needs further research, even though there currently no clear evidence of ill effects.

- Reducing the by-products and hazardous chemicals used in raw materials production or in the processing of plastics.

The overall challenge is to continue to deliver the recognised health benefits of plastics products without compromising the future.

> 600,000 pacemakers are fitted every year and these are all insulated with plastic.

Goal 3 targets

- By 2030, reduce the global maternal mortality ratio to less than 70 per 100,000 live births.

- By 2030, end preventable deaths of new-borns and children under 5 years of age, with all countries aiming to reduce neonatal mortality to at least as low as 12 per 1,000 live births and under-5 mortality to at least as low as 25 per 1,000 live births.

- By 2030, end the epidemics of AIDS, tuberculosis, malaria and neglected tropical diseases and combat hepatitis, water-borne diseases and other communicable diseases.

- By 2030, reduce by one third premature mortality from non-communicable diseases through prevention and treatment and promote mental health and well-being.

- Strengthen the prevention and treatment of substance abuse, including narcotic drug abuse and harmful use of alcohol.

- By 2020, halve the number of global deaths and injuries from road traffic accidents.

- By 2030, ensure universal access to sexual and reproductive health-care services, including for family planning, information and education, and the integration of reproductive health into national strategies and programmes.

- Achieve universal health coverage, including financial risk protection, access to quality essential health-care services and access to safe, effective, quality and affordable essential medicines and vaccines for all.

- By 2030, substantially reduce the number of deaths and illnesses from hazardous chemicals and air, water and soil pollution and contamination.

- Strengthen the implementation of the World Health Organization Framework Convention on Tobacco Control in all countries, as appropriate.

- Support the research and development of vaccines and medicines for the communicable and non-communicable diseases that primarily affect developing countries, provide access to affordable essential medicines and vaccines, in accordance with the Doha Declaration on the TRIPS Agreement and Public Health, which affirms the right of developing countries to use to the full the provisions in the Agreement on Trade Related Aspects of Intellectual Property Rights regarding flexibilities to protect public health, and, in particular, provide access to medicines for all.

- Substantially increase health financing and the recruitment, development, training and retention of the health workforce in developing countries, especially in least developed countries and small island developing states.

- Strengthen the capacity of all countries, in particular developing countries, for early warning, risk reduction and management of national and global health risks.

Opportunities for plastics

The opportunities for the plastics industry in improving health and well-being are:

- Child mortality can be dramatically reduced through the provision of clean water and sanitation provided by plastics pipes (see UN SDG 6). The delivery of clean water is a major factor in improving health and well-being.

- Immunisation programmes are capable of saving up to 1 million lives/year (mainly children) and these depend on plastics not only for immunisation equipment such as disposable syringes but also for the 'cold chain' that is necessary to get the drugs to the point of delivery.

  This has become particularly evident in the national vaccination programmes for COVID-19 which requires billions of disposable syringes for vaccination.

- Epidemics such as COVID-19 and Ebola require barrier treatment that can only be provided by plastics products.

- Epidemics such as COVID-19 and HIV/AIDS require specialist treatments that depends on plastics products.

- Traffic accidents can be reduced by transport applications (see sidebar) to make roads safer for drivers and other road users.

- The effect of traffic accidents can be reduced by the use of plastics in areas such as air bags, seat belts, soft bumpers and safety helmets.

Action taken to date

Despite the challenges, health care is an area where plastics have changed the landscape and enabled modern healthcare since the invention of surgical gloves in 1889. The plastics industry has contributed through:

- Pipes for clean water (see UN SDG 6) to reduce water-borne disease and reduce the labour intensity and dangers of water collection. Plastic drinking water pipes are extensively tested to ensure that they do not affect the quality of the transmitted water.

- Closed storage tanks or covers for water storage to provide water that is not only safe from contamination by water-borne disease, insects and animals but also does not provide breeding grounds for insects, e.g., mosquitoes.

- Barrier plastics such gloves, gowns and face masks and to prevent cross-contamination or infection in all areas of health care and especially in surgery or the treatment of COVID-19.

- Barrier plastics such as condoms for safe sex to prevent HIV/AIDS transmission and for birth control and family planning.

- Blood bags and tubing for the collection, transport and transfusion of blood.

- Medical tubing for drug administration, drainage and surgery.

- Stents and implants for surgery applications and life extension.

- Medicine containers providing controlled dosages through the use of plastics, e.g., inhalers.

- Child resistant caps and closures preventing the inadvertent ingestion of drugs or other substances (liquid or solid).

- Insulation and protection of medical devices and implants such as heart pacemakers for quality-of-life extension.

- Plasters, adhesives and wound dressings for wound treatment and assistance in recovery.

- Packaging and protection of drugs both individually (blister packaging) and in bulk.

Future action

The plastics industry will continue to contribute to the delivery of good health and well-being in the areas of:

- The development and improvement of the recycling or incineration of contaminated plastic waste to recover the material or the embodied energy.

- Improving the sterilisation capabilities of plastics to allow re-use and a reduction in the single-use applications of plastics.

- Reduction in the by-products and hazardous chemicals used in the production of plastics raw materials.

Summary

The past contribution of the plastics industry to improving health and well-being is one of the great achievements of the industry. Without plastics products the current standard of health care and well-being achieved in the developed world would not be possible. Plastics will continue to contribute as new applications are developed and as the standard of health care in the developing world rises to that of the developed world.

Safety helmets for workers, cyclists, motorcyclists and sports are essential for protection and safety. They save hundreds of lives every year throughout the world.

Protective clothing such as high visibility vests and outdoor clothing keep people safe and warm in poor conditions and allow people to enjoy active life styles.

Refrigerators and freezers are essential for storage of medical products. Both use plastics for insulation and ease of cleaning.

Drivers and road users are kept safe by plastics products such as safety belts, air bags, collision/impact foams, road markers and road cones.

Roads are kept drained and water-free by plastic drainage pipes and drainage products.

A1.4 Goal 4: Quality education

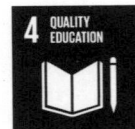

Ensure inclusive and quality education for all and promote lifelong learning

Quality education is the foundation to improving people's lives and sustainable development. Major progress has been made in increasing access to education at all levels and increasing enrolment rates in schools particularly for women and girls. Basic literacy skills have improved tremendously, yet bolder efforts are needed to make even greater strides for achieving universal education goals. For example, the world has achieved equality in primary education between girls and boys, but few countries have achieved that target at all levels of education.

Challenges for the plastics industry

This is a key SDG as it can break the cycle of poverty, unemployment, poor health and provide a route to economic growth and stability. The challenges to the plastics industry in helping to deliver quality education are:

- Education needs to be for all and lifelong to provide continued opportunities in a changing world.
- The plastics industry is a single sector industry and it is difficult, if not impossible, for such an industry sector to change the world. However, all progress is made by a single step at a time.
- The plastics industry provides the tools and basics that enable quality education for all and promote life-long learning.
- The industry needs well educated and trained staff to continue to develop processes and products. Skills shortages are becoming apparent in all areas of the world. Plastics processing is a world-wide industry and the challenge is not simply to provide quality education but also to provide the trained staff that the industry needs.
- The industry needs a pool of trained staff and this means being inclusive and working with everybody in recruitment and training (see UN SDG 5 and UN SDG 10).

Opportunities for plastics

The opportunities for the plastics industry in helping to deliver quality education and plastics enable modern education in a variety of ways:

- Plastic pipes provide clean water and sanitation (see UN SDG 6) but also reduce the daily labour of collecting clean water. This allows women and girls in developing countries more time to engage in education, decent work and innovation. This provides for improved growth and educational opportunities at the most needed level.
- Plastics allow the development of infrastructure and telecommunications (see UN SDG 9) and there are opportunities to use plastics to change the learning landscape. In the past, 1 teacher was needed for every 28-30 students. The availability of affordable and clean energy (see UN SDG 7) and improved infrastructure (see UN SDG 9) means that

Electronic whiteboards have replaced chalk and blackboards and distributed learning will replace conventional teaching.

Goal 4 targets

- By 2030, ensure that all girls and boys complete free, equitable and quality primary and secondary education leading to relevant and Goal 4 effective learning outcomes.
- By 2030, ensure that all girls and boys have access to quality early childhood development, care and pre-primary education so that they are ready for primary education.
- By 2030, ensure equal access for all women and men to affordable and quality technical, vocational and tertiary education, including university.
- By 2030, substantially increase the number of youth and adults who have relevant skills, including technical and vocational skills, for employment, decent jobs and entrepreneurship.
- By 2030, eliminate gender disparities in education and ensure equal access to all levels of education and vocational training for the vulnerable, including persons with disabilities, indigenous peoples and children in vulnerable situations.
- By 2030, ensure that all youth and a substantial proportion of adults, both men and women, achieve literacy and numeracy.
- By 2030, ensure that all learners acquire the knowledge and skills needed to promote sustainable development, including, among others, through education for sustainable development and sustainable lifestyles, human rights, gender equality, promotion of a culture of peace and non-violence, global citizenship and appreciation of cultural diversity and of culture's contribution to sustainable development.
- Build and upgrade education facilities that are child, disability and gender sensitive and provide safe, non-violent, inclusive and effective learning environments for all.
- By 2020, substantially expand globally the number of scholarships available to developing countries, in particular least developed countries, small island developing states and African countries, for enrolment in higher education, including vocational training and information and communications technology, technical, engineering and scientific programmes, in developed countries and other developing countries.
- By 2030, substantially increase the supply of qualified teachers, including through international cooperation for teacher training in developing countries, especially least developed countries and small island developing states.

---

1 teacher can communicate and teach 28–30 million students through Internet technology. Learning about anything is possible through the plastics-enabled Internet. Plastics will inevitably be used to extend the reach of distributed teaching and to provide quality education for all through life-long learning.

- Solar panels using plastics materials can help to deliver power to remote areas and increase educational opportunities (see UN SDG 7).

- Quality education will help countries to improve and move from agrarian to industrialised nations. This will drive an increased demand for plastics products in water and sanitation (see UN SDG 6), industry and innovation (see UN SDG 9) and in sustainable cities and communities (see UN SDG 11). The plastics industry has much to gain and much to contribute from improving education and encouraging development.

- The development of societies from agrarian to industrialised will provide increased demand for plastics products across a range of areas from industrial to consumer products. The plastics industry has much to gain from supporting the UN SDGs provided this is done in a sustainable manner.

Action taken to date

The plastics industry has contributed to quality education through:

- Supporting the basic developmental requirements that enable teaching and learning, e.g., reduced hunger (see UN SDG 2), good health and wellbeing (see UN SDG 3), clean water (see UN SDG 6), and energy (see UN SDG 7). Without these fundamental building blocks, education and advancement are always less important than the daily struggle for simple survival.

- The industry is world-wide and employs significant numbers of people. The industry is primarily concerned with employing people who are willing to work, contribute to and improve the industry. It is largely indifferent to class, caste or gender (see UN SDG 5 and UN SDG 10) in employment. Vocational training is provided as required by companies to meet internal needs.

- The industry is not well served in education terms, i.e., there are few traditional training programmes to prepare for joining the industry. Most sector-specific training is provided on-site and in-house by companies and this helps to raise the general educational levels of the population. The industry also gains from distributed learning which allows key skills to be distributed to a diffuse industry spread around the world.

Future action

The plastics industry will continue to contribute to the delivery of quality education in the areas of:

- Developing the education infrastructure for the learning technologies of the future that can leverage the skills of teachers to reach more people without regard to social status.

- Improving living standards throughout the world to enable learning for all people to become the norm rather than the exception.

- Improving and achieving gender equality at all levels of the industry (see UN SDG 5) to remove the current gender imbalance at technical and management levels.

- Supporting education initiatives at the local level. This is primarily taking place in the developed countries where organisations such as the Worshipful Company of Horners' support teacher training about the plastics industry to 'promote positive perceptions of the plastics industry'.

Summary

The plastics industry has much to contribute and much to gain from achieving quality education throughout the world. Quality education will allow everybody to contribute to achieving the UN SDGs.

Learning is changing and plastics are essential for this transformation.

2020 has shown the effects of distributed learning. It is not yet perfect but the progress has been amazing.

"An investment in education always pays the highest returns."

Ben Franklin

A1.5 Goal 5: Gender equality

Achieve gender equality and empower all women and girls

While the world has achieved progress towards gender equality and women's empowerment under the MDGs (including equal access to primary education between girls and boys), women and girls continue to suffer discrimination and violence in every part of the world.

Gender equality is not only a fundamental human right, but a necessary foundation for a peaceful, prosperous and sustainable world.

Providing women and girls with equal access to education, health care, decent work, and representation in political and economic decision-making processes will fuel sustainable economies and benefit societies and humanity at large.

Challenges for the plastics industry

The challenges to the plastics industry in achieving gender equality are:

- The plastics industry is international in outlook but local in operations. It must work with local conventions and attitudes. Overcoming these to promote gender equality is a task that is probably beyond the plastics industry alone (see UN SDG 17).

- The plastics industry already has low discrimination barriers and women are currently working throughout the plastics industry all over the world at a range of levels. The challenge is to transform and improve this work and the role of women in the industry to benefit not only women but also the industry.

- The plastics industry is engineering based and achieving the right education and training to deliver the necessary skills (see UN SDG 4) is crucial in achieving gender equality. The industry needs to invest in gender-blind training not only to achieve this goal but also to increase the trained labour pool.

- Engineering was not traditionally seen as suitable for women and girls but this has changed dramatically. It now offers opportunities regardless of gender. This outdated view of engineering needs to change throughout the world.

Opportunities for plastics

The opportunities for the plastics industry in improving gender equality are:

- The plastics industry is well suited to flexible working and can provide opportunities for everybody, not only women, to balance work and family commitments and still contribute to the industry.

- The plastics industry's assistance in achieving good health and well-being (see UN SDG 3) will improve family planning and sexual and reproductive health to allow integration of women into the workforce and assist in achieving this goal.

- The plastics industry's assistance in achieving clean water and sanitation (see UN SDG 6), will release women and girls from repetitive water collection to allow them to take part in education (see UN SDG 4) and work (see UN SDG 8). This will improve gender equality in society and in the industry and help to achieve this goal.

- Women are already working in the industry but in many cases, this is in low-skill roles, e.g., machine minding and assembly. Giving women the training and opportunities to advance could transform the sector, increase the labour pool and

> The international aspect of the plastics industry will help transfer best practice on gender equality across the world.

Goal 5 targets

- End all forms of discrimination against all women and girls everywhere.
- Eliminate all forms of violence against all women and girls in the public and private spheres, including trafficking and sexual and other types of exploitation.
- Eliminate all harmful practices, such as child, early and forced marriage and female genital mutilation.
- Recognize and value unpaid care and domestic work through the provision of public services, infrastructure and social protection policies and the promotion of shared responsibility within the household and the family as nationally appropriate.
- Ensure women's full and effective participation and equal opportunities for leadership at all levels of decision making in political, economic and public life.
- Ensure universal access to sexual and reproductive health and reproductive rights as agreed in accordance with the Programme of Action of the International Conference on Population and Development and the Beijing Platform for Action and the outcome documents of their review conferences.
- Undertake reforms to give women equal rights to economic resources, as well as access to ownership and control over land and other forms of property, financial services, inheritance and natural resources, in accordance with national laws.
- Enhance the use of enabling technology, in particular information and communications technology, to promote the empowerment of women.
- Adopt and strengthen sound policies and enforceable legislation for the promotion of gender equality and the empowerment of all women and girls at all levels.

help to achieve decent work and economic growth (see UN SDG 8). This is already being seen in some areas of the world where women are taking an increasingly important role in the industry.

The plastics industry already includes many women in the workforce and offers safe work and good pay.

Action taken to date

The plastics industry has contributed to gender equality through actions such as:

- The plastics industry employment ratio is already very even between the genders and it is thought that women already make up nearly half of the industry employees (see UN SDG 8). The issue is that the majority of the women work in lower-level manual or support roles and that there are not enough women in technical and management roles.

- The plastics industry is not the only manufacturing industry where women are under-represented in technical and management roles. This is changing for most manufacturing but progress needs to be more rapid to achieve this goal.

Future action

The plastics industry is already contributing to gender equality but there is still much to do:

- The industry needs to provide safe work, good pay and equal opportunities for advancement for all staff. This is already happening in many areas but there is still much to be done around the world.

- The industry needs to train, train and train again. This training needs to be gender-blind not only for the benefit of women but also for the benefit of the industry. The industry already suffers from skill shortages in many areas of the world. Skills need improvement at all levels of the industry to allow everyone to achieve their full potential and to allow the industry to meet the challenges of the future.

- The industry does not simply need to train and develop the people already in the industry. It needs to make all areas of manufacturing more attractive to all genders to increase the number of people in the industry. Women are a largely untapped resource for the technical and management skill base and their inclusion will benefit the industry and the people.

Women in Plastics is an initiative that recognises achievement, encourages development and supports diversity and equality across the plastics industry.

A great initiative but read more at: www.womeninplastics.com/

Summary

Gender equality is vital for progress and will be a key component in achieving all of the goals. No industry sector can hope to do more than influence achieving this goal but the plastics industry is pervasive across the world and can contribute more than many other sectors in achieving true gender equality.

By contributing to the sexual and reproductive health of women and assisting with family planning, the plastics industry is already helping to improve gender equality.

A1.6   Goal 6: Clean water and sanitation

Ensure access to water and sanitation for all

Clean, accessible water for all is an essential part of the world we want to live in. There is sufficient fresh water on the planet to achieve this. But, due to bad economics or poor infrastructure, every year millions of people, most of them children, die from diseases associated with inadequate water supply, sanitation and hygiene.

Water scarcity, poor water quality and inadequate sanitation negatively impact food security, livelihood choices and educational opportunities for poor families across the world. Drought afflicts some of the world's poorest countries, worsening hunger and malnutrition.

By 2050, at least one in four people is likely to live in a country affected by chronic or recurring shortages of fresh water.

Challenges for the plastics industry

There is enough water for everybody in the world but it is not always in the right place at the right time and the infrastructure to remedy this is not present in much of the world. Plastics products are key to delivering this goal. The versatility and longevity of plastics products can provide many infrastructure solutions but this is not without challenges, some of these are:

- The initial solutions and systems for clean water and sanitation were enormously successful and without these the modern city would not be possible. These solutions were initially implemented with ceramics or metals but these systems are now being replaced or extended using plastics products. The challenge is transferring the existing solutions to other areas of the world using the appropriate technology.

- Providing sanitation and hygiene to all and ending open defecation needs both major infrastructure and local projects to deliver the goal.

- Infrastructure projects have high capital costs and implementation requires high-level skills and technology that are not always available in the developing world. Developing the required skills needs quality education (see UN SDG 4) and industry to drive innovation and infrastructure (see UN SDG 11).

- Infrastructure projects are often centrally planned and do not always take into account the needs of local people. There is also a history of large infrastructure projects having unintended and deleterious local environmental impacts (see UN SDG 10).

The challenge is to deliver infrastructure projects that deliver the goal at a local level and do not have unintended impacts.

Opportunities for plastics

There are enormous opportunities for the plastics industry to help in delivering this goal:

- The plastics industry has the technology to deliver large or small infrastructure projects that capture, store, treat and distribute clean water to wide areas without affecting water quality.

- The plastics industry has the technology to deliver large or small infrastructure projects to collect, control, remove and treat contaminated water, particularly faecal contamination. These projects can be at the local level, e.g., septic tank technology, or at the city or state level, e.g., centralised municipal treatment systems. Effective treatment of contaminated water allows it to be recycled with low risk of disease transmission.

- Plastics products for clean water delivery, contaminated water removal and grey

> The range of benefits of plastics towards achieving this goal is extensive, proven and quantifiable.
>
> It is almost inconceivable that this goal can be achieved without the plastics industry.

Goal 6 targets

- By 2030, achieve universal and equitable access to safe and affordable drinking water for all.

- By 2030, achieve access to adequate and equitable sanitation and hygiene for all and end open defecation, paying special attention to the needs of women and girls and those in vulnerable situations.

- By 2030, improve water quality by reducing pollution, eliminating dumping and minimizing release of hazardous chemicals and materials, halving the proportion of untreated wastewater and substantially increasing recycling and safe re-use globally.

- By 2030, substantially increase water-use efficiency across all sectors and ensure sustainable withdrawals and supply of freshwater to address water scarcity and substantially reduce the number of people suffering from water scarcity.

- By 2030, implement integrated water resources management at all levels, including through trans-boundary cooperation as appropriate.

- By 2020, protect and restore water-related ecosystems, including mountains, forests, wetlands, rivers, aquifers and lakes.

- By 2030, expand international cooperation and capacity-building support to developing countries in water- and sanitation-related activities and programmes, including water harvesting, desalination, water efficiency, wastewater treatment, recycling and re-use technologies.

- Support and strengthen the participation of local communities in improving water and sanitation management.

---

water recovery have low losses to the environment and are long-life products suitable for all environments.

- The plastic pipe sector has the technology to provide these solutions using local labour (see UN SDG 8) to provide a robust infrastructure (see UN SDG 9) that can deliver sustainable cities and communities (see UN SDG 11).

Action taken to date

The plastics industry, particularly the pipes sector, has already demonstrated the effectiveness of plastics products through actions such as:

- Water capture and retention through the development and use of pond liners and covers to reduce losses from leakage and evaporation from local dams and ponds.

- Water delivery and distribution using plastics pipes and/or retro-fitted pipe liners with reduced leakage, reduced transmission losses and reduced micro-organism retention or growth. This can include large-scale trans-boundary schemes to transfer water from areas of surplus to areas of shortage.

- Water control and re-direction though guttering and down-pipes in housing and industrial developments.

- Flood risk abatement through drainage pipes, flood run-off pipes, flood water retention and storage schemes and flood water barriers/defences.

- Septic tanks and local sewerage treatment to reduce open defecation, the contamination of waterways and the spread of disease. Septic tanks need a good septic drain field and also maintenance and regular treatment to prevent groundwater pollution but can be an effective local solution to an infrastructure deficit.

- Sewerage and grey water piping to collect wastewater locally for connection to a centralised municipal waste treatment system. Plastics pipes are a long-life product and also have minimal losses or leakage and therefore avoid and reduce groundwater contamination.

Future action

The actions taken to date show the effectiveness of plastics products in delivering clean water and sanitation to the developed world. The future actions are:

- To roll-out the appropriate parts of proven plastics technology to the developing world.

- To improve and update the water and sanitation infrastructure (see UN SDG 9) of the developing world to provide clean water and sanitation. This will reduce

disease transmission and improve health in developing countries (see UN SDG 3) as well as helping to achieve this goal.

- To reduce water use in the plastics industry through the development of closed loop water systems at sites using water for cooling or other parts of the process. The plastics industry is already very good at this but water management can always be improved (see Chapter 8).

Summary

Providing clean water and sanitation is an area of real strength for the plastics industry. Plastics products are supremely effective in use and also have a long service life. Disposal is not yet an issue for products that have a service life of over 100 years and is not likely to be one for some time. All of the products used in the sector have a relatively low cost and a high utility value. The successes in the developed world provide a proven template for improving the infrastructure of the developing world.

Plastics pipes used for most applications have proven benefits over those of other materials.

Plastics make it possible to safely store and transport the essential chemicals used for water treatment.

A1.7   Goal 7: Affordable and clean energy

Ensure access to affordable, reliable, sustainable and modern energy for all

Energy is central to nearly every major challenge and opportunity that the world faces today. Be it for jobs, security, climate change, food production or increasing incomes, access to energy for all is essential.

Sustainable energy is opportunity – it transforms lives, economies and the planet.

Former UN Secretary-General Ban Ki-moon is leading a 'Sustainable Energy for All' initiative to ensure universal access to modern energy services, improve efficiency and increase use of renewable sources.

Challenges for the plastics industry

The challenges to the plastics industry in achieving affordable and clean energy are:

- The plastics processing industry is an energy intensive industry, although not the largest, and requires approximately 1–2 kWh/kg to process a product from the raw polymer. The industry needs a high-capacity electricity supply to function and grow.

- The plastics processing industry needs a reliable energy supply to function because the operational pattern is normally 24/7 or 24/5. This can be an issue in some countries where the electricity supply is neither stable nor consistent. Plastics processors in such countries can use gas for tri-generation but this is still a barrier to progress.

- The high power and 24/7 requirements of the plastics processing industry can restrict the use of local renewable sources where there is a lack of substantial local energy storage. This can be a barrier to local industrial development.

- A key issue for the plastics industry is decoupling the link between growth in GDP and energy use. This can be achieved by increased efforts in energy efficiency in the industry (see Chapter 6). The methods to reduce energy use in plastics processing are clear but their implementation to date is relatively poor.

Opportunities for plastics

The opportunities for the plastics industry in achieving affordable and clean energy are:

- Unless affordable and clean energy is achieved it will not be possible for growth to occur, some areas of the world will be restricted to an agrarian economy and progress towards any of the other UN SDGs will be severely limited.

- Decarbonising the energy supply is simply not possible without the use of plastics.
  - Wind energy relies on plastics for the production of wind turbine blades using fibre-reinforced composites. These are the only materials with the right combination of stiffness, density and fatigue resistance to produce economical blades.
  - Solar energy relies primarily on silicon technology but most solar cells are encapsulated in a polymer resin to protect the silicon cell. New technologies using printed polymers offer lower weight, flexible solar cells that will be easily transported and assembled. This technology would revolutionise the provision of low-cost distributed solar for the developing world.
  - Energy storage technology is based on batteries (lithium-ion) and these can suffer from poor performance and fire issues. New technologies using plastics can provide higher energy densities and inherent safety.
  - Hydrogen gas (a decarbonised energy source) distribution will be a new opportunity for plastics pipes.

- Irrespective of the generation method used, the resulting energy must be transported from the generation site to the user and plastics remain the only suitable insulation material for electricity transport.

- Distributed (local) generation may reduce the distribution distance but will still rely on plastics for transport and use. Without

> The energy reaching the earth from 2 minutes of sunlight is enough to satisfy the complete demands of the world for one year.
>
> It is simply not evenly distributed and is not necessarily where it is needed.

Goal 7 targets

- By 2030, ensure universal access to affordable, reliable and modern energy services.
- By 2030, increase substantially the share of renewable energy in the global energy mix.
- By 2030, double the global rate of improvement in energy efficiency.
- By 2030, enhance international cooperation to facilitate access to clean energy research and technology, including renewable energy, energy efficiency and advanced and cleaner fossil-fuel technology, and promote investment in energy infrastructure and clean energy technology.
- By 2030, expand infrastructure and upgrade technology for supplying modern and sustainable energy services for all in developing countries, in particular least developed countries, small island developing states, and land-locked developing countries, in accordance with their respective programmes of support.

plastics it will not be possible to create and transport energy for applications such as schools (see UN SDG 4), local water pumps (see UN SDG 6), local refrigeration for cooling vaccines and medicines (see UN SDG 3), local solar for recharging mobile communication and enabling mobile payments and Internet access (see UN SDG 8 and UN SDG 9).

Action taken to date

The plastics industry has contributed to achieving affordable and clean energy through actions such as:

- The distribution network in every country relies on plastics for insulation and protection. Without modern plastics, there would be no modern world.

- The modern world relies on plastics in almost every energy application from mobile phones through to heavy industry. Without modern plastics, there would be no modern world.

- The development and implementation of industrial-scale wind energy farms relies on plastics for turbine blades.

- The development and implementation of local-scale wind power relies on plastics for cable insulation for both generation and distribution.

- The development and implementation of large and local-scale solar power relies on plastics for both generation (solar panels) and distribution (cable insulation).

- The cables and fibres needed for fast Internet connections are protected by specially designed plastics pipes that not only protect the fibres but also allow for re-cabling or network extension when required.

- Plastic pipes are extensively used for long-life and safe network gas distribution.

Future action

Without plastics there is no possibility of achieving affordable and clean energy or delivering this to people but future action is required in the areas of:

- Improved energy efficiency in the plastics processing industry to reduce the overall energy intensity of the industry.

- An acceleration in the decarbonisation of industry through the use of renewable energy in plastics processing.

- The development of improved, distributed and more resilient power networks to provide local power for schools, medical facilities and water distribution (see UN SDG 9).

- The development of improved battery storage technologies to allow renewables to

meet the high power and continuous demand of the plastics processing industry and of society.

Summary

It is estimated that 1.2 billion people are without reliable power supplies and this dramatically affects their standard of living and development prospects. The plastics industry already enables the use and production of affordable and clean energy in many parts of the world but this can be extended to all parts of the world using the existing techniques and tools and developing new techniques and tools.

Affordable and clean power is a requirement for the transformation from an agrarian society into one where the UN SDGs can be met for the whole world.

Over 80% of the world's population have some access to electricity.
Without plastics this would not be possible but we still have lots of people to connect.

Paradoxically, some of the least developed nations also have the greatest available resources for the development of affordable and clean power, e.g., wind and solar.

A1.8 Goal 8: Decent work and economic growth

Promote inclusive and sustainable economic growth, employment and decent work for all

Roughly half the world's population still lives on the equivalent of < US$2 a day and in too many places, having a job doesn't guarantee the ability to escape from poverty. This slow and uneven progress means we must rethink and revise the economic and social policies aimed at eradicating poverty.

A continued lack of decent work opportunities, insufficient investments and under-consumption lead to an erosion of the basic social contract underlying democratic societies: that all must share in progress. The creation of quality jobs will remain a major challenge for almost all economies well beyond 2020.

Sustainable economic growth will require societies to create the conditions that allow people to have quality jobs that stimulate the economy while not harming the environment. Job opportunities and decent working conditions are also required for the whole working age population.

Challenges for the plastics industry

The challenges to the plastics industry in achieving decent work and economic growth are:

- The plastics processing industry is the development of a third-order society. The initial development is from an agrarian society to a pre-industrial society while the infrastructure necessary for an industrial society is created. This allows the growth of a plastics processing industry which is not possible without the necessary infrastructure. The plastics processing industry can contribute to infrastructure development (see UN SDG 9) but without an effective industrial infrastructure a plastics processing industry cannot exist.

- The plastics processing industry needs peace, justice and strong institutions (see UN SDG 16) to protect the rights and property of investors, employers and employees. The industry is not unique in needing other drivers to promote stable and well-paid work.

- The plastics processing industry is a relatively mobile industry that follows the major customers and industries and these tend to seek out low-wage economies. The industry has developed rapidly in low-wage

economies around the world but the trend is also for these economies to grow and for wages to rise. The presence of the plastics processing industry is thus a driver for decent work, skills training and economic growth.

Opportunities for plastics

The opportunities for the plastics industry in achieving decent work and economic growth are:

- As an industrial society is established, economic growth and employment rise and there is a growth of a middle-class, e.g.,

> Plastics processing companies in many countries train their staff to take on new jobs or responsibilities.
>
> Life-long learning and advancement does not always mean changing employers.

Goal 8 targets

- Sustain per capita economic growth in accordance with national circumstances and, in particular, at least 7 per cent gross domestic product growth per annum in the least developed countries.

- Achieve higher levels of economic productivity through diversification, technological upgrading and innovation, including through a focus on high-value added and labour-intensive sectors.

- Promote development-oriented policies that support productive activities, decent job creation, entrepreneurship, creativity and innovation, and encourage the formalization and growth of micro-, small- and medium-sized enterprises, including through access to financial services.

- Improve progressively, through 2030, global resource efficiency in consumption and production and endeavour to decouple economic growth from environmental degradation, in accordance with the 10-year framework of programmes on sustainable consumption and production, with developed countries taking the lead.

- By 2030, achieve full and productive employment and decent work for all women and men, including for young people and persons with disabilities, and equal pay for work of equal value.

- By 2020, substantially reduce the proportion of youth not in employment, education or training.

- Take immediate and effective measures to eradicate forced labour, end modern slavery and human trafficking and secure the prohibition and elimination of the worst forms of child labour, including recruitment and use of child soldiers, and by 2025 end child labour in all its forms.

- Protect labour rights and promote safe and secure working environments for all workers, including migrant workers, in particular women migrants, and those in precarious employment.

- By 2030, devise and implement policies to promote sustainable tourism that creates jobs and promotes local culture and products.

- Strengthen the capacity of domestic financial institutions to encourage and expand access to banking, insurance and financial services for all.

- Increase Aid for Trade support for developing countries, in particular least developed countries, including through the Enhanced Integrated Framework for Trade-Related Technical Assistance to Least Developed Countries.

- By 2020, develop and implement a global strategy for youth employment and implement the Global Jobs Pact of the International Labour Organization.

China and India. This drives a growth in consumer activity and increased opportunities for the plastics processing industry to supply the plastics products that are necessary for consumers. Initially, these products will be low value items but development increases the value of the products and of the industry.

- The decent work, fair income (see UN SDG 1), security and social protection (see UN SDG 16) that the plastics industry can provide enables our employees to become consumers and to use more of our products.
- The opportunity does not stop with our employees but extends to the industries and to the supply chain that the plastics industry supports.
- Consumption figures show that the use of plastics is a good indicator of the developmental status of a country and that there are many countries where consumption and activity can increase to benefit both the industry and the country.
- As an enabling industry, the plastics processing industry produces products that not only help countries achieve this goal but also to lever progress in many of the other UN SDGs.

Action taken to date

The plastics industry has contributed to achieving decent work and economic growth through actions such as:

- As one of the largest employers in the world, the plastics industry already provides a huge amount of decent work around the world.
- The plastics industry is a highly regulated industry either through country-specific legislation or through the purchasing pressure of our customers. This means that working standards are generally high and consistent throughout the world.
- Many plastics processing companies around the world provide significant staff benefits to employees. These can range from free or highly-subsidised meals for employees to family health protection. The industry values and invests in people because healthy employees are good employees.
- The plastics processing industry trains employees (see UN SDG 4) for job-related tasks and this allows everybody to overcome education gaps and barriers (see UN SDG 5 and UN SDG 10).
- Plastics enable mobile telecommunications and the Internet. This allows micro-payments and the economic development of some to the poorest people on the planet and provides access to banking, insurance and financial services to allow their

integration into an economic society. This provides not only decent work but also economic growth.

Future action

The plastics industry will continue to contribute to the delivery of decent work and economic growth in the areas of:

- Training of staff that is not specifically job-related needs to increase to increase the available labour pool, e.g., literacy education. Such training must be blind to gender, class or caste but based on ability and willingness to contribute.
- The industry needs to work with all suppliers and sub-contractors, as our customers do with us, to raise standards of employment and worker benefits.

Summary

The plastics processing industry is a decentralised industry that employs large numbers of people throughout the world. It can contribute to decent work and economic growth by both facilitating the necessary infrastructure for development and also providing jobs and financial stability in both the initial processing and in the recycling and recovery of the raw materials.

Plastics processing companies in countries as diverse as Bangladesh, Sri Lanka and Brazil regularly provide free or highly subsidised meals to employees.

Plastics processing companies in countries as diverse as Tunisia, Russia and Poland regularly provide assisted transport to reduce the cost and impact of commuting to work.

A1.9 Goal 9: Industry, innovation and infrastructure

Build resilient infrastructure, promote sustainable industrialization and foster innovation

Investments in infrastructure, e.g., transport, irrigation, energy, information and communication technology, are crucial in achieving sustainable development and empowering communities in many countries. Growth in productivity and incomes, and improvements in health and education outcomes require investment in infrastructure.

Inclusive and sustainable industrial development is the primary source of income generation and allows for rapid and sustained increases in living standards for all people, and provides the technological solutions to environmentally sound industrialization.

Technological progress is the foundation of efforts to achieve environmental objectives, such as increased resource and energy-efficiency. Without technology and innovation, industrialization will not happen, and without industrialization, development will not happen.

Challenges for the plastics industry

The challenges to the plastics industry in assisting the delivery of this goal are:

- In some areas, there is a perception that industry is 'the enemy' but industry is the main driver of economic growth, improved sustainability and is an essential in attaining all of the UN SDGs. Industry is a core driver of development to overcome poverty, to release people from low-quality labour and to achieve the other development goals. Overcoming this perception is a major challenge.

- The circular economy concept to improve resource efficiency for materials, water and energy to reduce environmental and societal impacts is a relatively new concept for the industry and needs to be rapidly developed and implemented.

- Infrastructure developments must be developed and implemented to be resilient and sustainable with deleterious environmental effects minimised. In most cases, the plastics products used in infrastructure projects are enabling products. They have much to contribute but they must be managed correctly, particularly at the end-of-life phase.

- All elements of the plastics industry are innovation driven. The challenge is to focus innovation on reducing impacts and in transferring appropriate and accessible innovations and skills to the developing world.

Opportunities for plastics

The plastics industry is already contributing to this goal but the opportunities for further contributions are huge:

- As a provider of enabling technologies, virtually no industry, innovation or infrastructure development is possible without using plastics products. At the most basic level, the use of plastics for cable and wire insulation enables modern electrical and electronics engineering to function.

- The plastics industry is already present in almost every country in the world (I have worked in 46 of them). The development of

> Innovation, research and development in plastics is world-wide and helping to achieve this goal by developing people and skills.

Goal 9 targets

- Develop quality, reliable, sustainable and resilient infrastructure, including regional and trans-border infrastructure, to support economic development and human well-being, with a focus on affordable and equitable access for all.

- Promote inclusive and sustainable industrialization and, by 2030, significantly raise industry's share of employment and gross domestic product, in line with national circumstances, and double its share in least developed countries.

- Increase the access of small-scale industrial and other enterprises, in particular in developing countries, to financial services, including affordable credit, and their integration into value chains and markets.

- By 2030, upgrade infrastructure and retrofit industries to make them sustainable, with increased resource-use efficiency and greater adoption of clean and environmentally sound technologies and industrial processes, with all countries taking action in accordance with their respective capabilities.

- Enhance scientific research, upgrade the technological capabilities of industrial sectors in all countries, in particular developing countries, including, by 2030, encouraging innovation and substantially increasing the number of research and development workers per 1 million people and public and private research and development spending.

- Facilitate sustainable and resilient infrastructure development in developing countries through enhanced financial, technological and technical support to African countries, least developed countries, landlocked developing countries and small island developing states.

- Support domestic technology development, research and innovation in developing countries, including by ensuring a conducive policy environment for, inter alia, industrial diversification and value addition to commodities.

- Significantly increase access to information and communications technology and strive to provide universal and affordable access to the Internet in least developed countries by 2020.

the industry helps to develop other industries and to drive progress towards this goal.

- Food production, processing and protection between producer and plate will allow not only more food to be produced but more food to be treated and arrive with the consumer (see UN SDG 2).

- Plastics products in phones, phone towers and telecommunications will allow developing countries to 'leapfrog' developed nations in communications technology and applications to drive economic growth, e.g., Morocco has gone from difficult to get land-line phones direct to easily obtainable mobiles which are faster than the residual/legacy copper wires.

- The use of plastics products in water and sanitation infrastructure projects (see UN SDG 6) reduces water-gathering labour (see UN SDG 8), reduces disease transmission (see UN SDG 3) and improves crop yields (see UN SDG 2). There is a major opportunity to use plastics to improve water management (both in the collection and disposal).

- The plastics industry will contribute to transport infrastructure development through sustainable transport systems ranging from mass-transit systems to road and materials transport systems.

- Vehicles used in transport systems will inevitably use plastics for lightweight and safe solutions.

Action taken to date

The plastics industry has already contributed strongly to the development of the developed world's industry, innovation and infrastructure:

- Without flexible, durable and low-cost cable insulation, modern society would simply not be possible. Without cable insulation, the current industry, innovation and infrastructure of the developed world would cease to exist and there would be no possibility of achieving this goal.

- The contribution of the plastics industry to areas such as power generation, power distribution (see UN SDG 7), electrical products (including telecommunications, mobile phones and internet technology), clean water and sanitation (see UN SDG 6), medical technology (see UN SDG 3) are such that the current developed world society would not be possible without the plastics industry.

- The plastics industry is a large-scale employer both directly and indirectly. The UK plastics industry is estimated to consist of over 5,200 companies and to

employ more than 166,000 people. This makes it larger than the automotive and pharmaceutical sectors combined. However, as an enabling industry, the plastics sector is often overlooked in terms of the economic benefits it brings to industrial development.

Future action

The future is bright for the plastics industry and it can contribute enormously and effectively to achieving this goal:

- At the industry level, it is estimated that every job in manufacturing will create 2.2 jobs in other sectors. For the UK, this means that the plastics industry creates another 365,000 jobs in other sectors. A small plastics factory can be a valuable addition to a region's economy.

- Recycling is a 'new' industry that will increase investment in industry and require a new infrastructure for waste management and re-processing.

- The industry is actively working towards the circular economy and as an inclusive industry (see UN SDG 5) the industry can, if appropriately directed, use small-scale, easily transferred and sustainable industrialisation to lever the economic benefits of industry to the wider society.

Summary

The plastics industry can help to achieve this goal in many ways. As an industry of itself it can contribute to the development of industry but, perhaps more importantly, the products can be used in a wide range of infrastructure projects to raise the quality of life for millions of people.

Mobile technology allows developing countries to 'leapfrog' developed countries and go straight to mobile communications without having to go through the fixed line technology of the developed world.

In some developing countries the mobile network has enabled internet access, banking, micro-payments, insurance and social development that would otherwise not be possible.

Plastics products are essential for data storage and retention. Magnetic tapes, CDs, DVDs and most storage media rely on the unique properties of plastics. Without plastics, the information age would not be possible.

A plastics processing site Malaysia directly employs 200 people but creates another 440 jobs in the local economy.

A1.10 Goal 10: Reduced inequalities

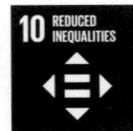

Reduce inequality within and among countries

The international community has made huge strides towards lifting people out of poverty. The most vulnerable nations, i.e., the least developed countries, the landlocked developing countries and the small island developing states, continue to make inroads into poverty reduction. However, inequality still persists and large disparities remain in access to health and education services and other assets.

While income inequality between countries may have been reduced, inequality within countries has risen. There is growing consensus that economic growth is not sufficient to reduce poverty if it is not inclusive and if it does not involve the three dimensions of sustainable development, i.e., environmental, social and economic.

To reduce inequality, policies should be universal in principle, paying attention to the needs of disadvantaged and marginalized populations.

Challenges for the plastics industry

The challenges to the plastics industry in reducing inequalities are:

- The plastics processing industry cannot develop directly from an agrarian economy (see UN SDG 8) but needs a developed and effective infrastructure.

- The plastics processing industry can assist the development of the necessary infrastructure (see UN SDG 9) but it is impossible for the industry to develop before the infrastructure is in place. This requires imported products to develop the infrastructure before the local industry can develop to support local needs and assist in the reduction of inequalities. The challenge is to reduce inequalities during the development phase.

- The plastics industry is global in scope but local in character. This means that the local ethos can dominate even when multi-national companies are involved. The challenge is to ensure that policies to reduce inequalities are applied evenly and fairly around the world.

Opportunities for plastics

The opportunities for the plastics industry in reducing inequalities are:

- Reducing inequalities drives progress at the country level and improves development in areas where plastics can contribute, e.g., poverty reduction (see UN SDG 1), good health (see UN SDG 3) and clean water (see UN SDG 6).

- Reducing inequalities means developing the infrastructure to support growth and this increases the opportunities for plastics to contribute to infrastructure development (see UN SDG 9) although this may mean importing whilst the local industry develops.

- Increasing the income of the bottom 40% of the population will create more consumers for plastics products in all sectors. People living in poverty do not consume plastics products, although they may use them.

- Reducing inequalities allows selection of the best candidate for the job regardless of gender (see UN SDG 5), race, class or caste. This drives progress and can improve performance in any business sector.

> The plastics processing industry enables progress and reduced inequalities.

Goal 10 targets

- By 2030, progressively achieve and sustain income growth of the bottom 40 per cent of the population at a rate higher than the national average.

- By 2030, empower and promote the social, economic and political inclusion of all, irrespective of age, sex, disability, race, ethnicity, origin, religion or economic or other status.

- Ensure equal opportunity and reduce inequalities of outcome, including by eliminating discriminatory laws, policies and practices and promoting appropriate legislation, policies and action in this regard.

- Adopt policies, especially fiscal, wage and social protection policies, and progressively achieve greater equality.

- Improve the regulation and monitoring of global financial markets and institutions and strengthen the implementation of such regulations.

- Ensure enhanced representation and voice for developing countries in decision-making in global international economic and financial institutions in order to deliver more effective, credible, accountable and legitimate institutions.

- Facilitate orderly, safe, regular and responsible migration and mobility of people, including through the implementation of planned and well-managed migration policies.

- Implement the principle of special and differential treatment for developing countries, in particular least developed countries, in accordance with World Trade Organization agreements.

- Encourage official development assistance and financial flows, including foreign direct investment, to states where the need is greatest, in particular least developed countries, African countries, small island developing states and landlocked developing countries, in accordance with their national plans and programmes.

- By 2030, reduce to less than 3 per cent the transaction costs of migrant remittances and eliminate remittance corridors with costs higher than 5 per cent.

Action taken to date

The plastics industry has contributed to reducing inequalities through actions such as:

- The plastics processing industry is already helping the reduce poverty (see UN SDG 1) and hunger (see UN SDG 2) which serves to reduce inequalities.

- Fiscal, wage and social protection is widely supported by the plastics processing industry and its customers. Customers can drive reduced inequalities through procurement policies and the industry can also drive this down the supply chain with procurement policies that include non-discriminatory requirements (see Section 5.10).

- The industry already largely has non-discriminatory policies for hiring, training, development and progression but this is also an area for improvement in consistency across the world. Reducing inequalities cannot be driven from 'head office' but must be a local solution.

- The industry enables infrastructure development in areas such as mobile telephony and Internet connection to provide micro-payments and access to banking, insurance and financial services. This reduces inequalities by integrating the poorest people into an economic society.

Future action

The plastics industry will continue to contribute to reducing inequalities in the areas of:

- Despite the current progress in non-discriminatory policies there is more work to do in gender equality (see UN SDG 5) in the plastics processing industry in order to remove the gender imbalance at higher levels of the industry.

- The industry needs further documented action in the area of open hiring and progression to select the best employees for all levels of work. The industry needs to be seen to be taking positive action to eliminate discriminatory laws, policies and practices in all areas of the business.

- The industry needs to provide documented action in the area of paying fair wages to all employees, especially in low-income countries.

Summary

Reducing inequalities is not something that the plastics processing industry can achieve alone. The contribution of the industry to the other UN SDGs helps to drive progress and the reduction of inequalities but there is more work to be done in this area.

A plastics processor in Malaysia has 4 separate religious worship areas on site to cater for the variety of religions of staff employed at the site.

A1.11 Goal 11: Sustainable cities and communities

Make cities inclusive, safe, resilient and sustainable

Cities are hubs for ideas, commerce, culture, science, productivity, social development and much more. At their best, cities have enabled people to advance socially and economically.

However, many challenges exist in maintaining cities in a way that continues to create jobs and prosperity whilst not straining available land and resources. Common urban challenges include congestion, lack of funds to provide basic services, a shortage of adequate housing and declining infrastructure.

The challenges cities face can be overcome in ways that allow them to continue to thrive and grow, while improving resource use and reducing pollution and poverty. Cities should provide opportunities for all, with access to basic services, energy, housing, transportation and more.

Challenges for the plastics industry

Already over 50% of the world's population live in cities and in the next decades most of the growth will be through urban expansion. This places pressure on existing, often crumbling, infrastructure for clean water and sanitation (see UN SDG 6), on health and well-being (see UN SDG 3) and on infrastructure (see UN SDG 9). Whilst many of the UN goals affect rural people, in the future they will affect the increasing number of urban dwellers even more:

- Delivery of clean water is difficult in rural areas because of 'distance', in urban areas 'volume' is the issue. Poor sanitation can lead to epidemics and health crises in rural areas but in urban slum areas, epidemics will spread faster and affect many more people. The challenge is to use the properties of plastics to provide the volume of clean water and subsequent sanitation requirements necessary for the development of sustainable cities and communities.

- Cities need the healthcare benefits of plastics (see UN SDG 3) to provide the resilience that enables high-quality routine healthcare as well as a rapid response to disease and epidemics.

- A sustainable city needs transport systems that reduce pollution, energy use and travel times. The plastics industry must enable the improvement and delivery of the improved and integrated transport infrastructure needed for cities.

- Waste management is a major issue for cities and the plastics industry. Cities generate huge amounts of waste that is often poorly collected and treated. It is estimated that 2 billion people around the world do not have access to an effective waste management system. This waste is then poorly handled and often finds its way into the wider ecosystem (see UN SDG 14 and UN SDG 15). It is estimated that 82% of the plastics waste in the oceans comes from Asia and that 90% comes from 10 rivers (8 of which are in Asia). The plastics industry must accept the challenge of enabling and facilitating waste management systems to recover valuable resources, either as materials or as energy, and to prevent them from entering the ecosystem. We must recognise the

> Cities and societies depend on the unique properties of plastics to support the current lifestyles.

Goal 11 targets

- By 2030, ensure access for all to adequate, safe and affordable housing and basic services and upgrade slums.
- By 2030, provide access to safe, affordable, accessible and sustainable transport systems for all, improving road safety, notably by expanding public transport, with special attention to the needs of those in vulnerable situations, women, children, persons with disabilities and older persons.
- By 2030, enhance inclusive and sustainable urbanization and capacity for participatory, integrated and sustainable human settlement planning and management in all countries.
- Strengthen efforts to protect and safeguard the world's cultural and natural heritage.
- By 2030, significantly reduce the number of deaths and the number of people affected and substantially decrease the direct economic losses relative to global gross domestic product caused by disasters, including water-related disasters, with a focus on protecting the poor and people in vulnerable situations.
- By 2030, reduce the adverse per capita environmental impact of cities, including by paying special attention to air quality and municipal and other waste management.
- By 2030, provide universal access to safe, inclusive and accessible, green and public spaces, in particular for women and children, older persons and persons with disabilities.
- Support positive economic, social and environmental links between urban, peri-urban and rural areas by strengthening national and regional development planning.
- By 2020, substantially increase the number of cities and human settlements adopting and implementing integrated policies and plans towards inclusion, resource efficiency, mitigation and adaptation to climate change, resilience to disasters, and develop and implement, in line with the Sendai Framework for Disaster Risk Reduction 2015-2030, holistic disaster risk management at all levels.
- Support least developed countries, including through financial and technical assistance, in building sustainable and resilient buildings utilizing local materials.

difference between those areas of the world where littering and the poor disposal of plastics is largely 'personal' and those areas of the world where littering and the poor disposal of plastics is largely 'structural' because of the lack of waste management systems.

- Resilience to natural disasters such as floods can be mitigated by effective control and management of resources. The plastics industry will need to work with infrastructure partners to deliver effective resilience to cities.

Opportunities for plastics

The major opportunities for the plastics industry are:

- Providing essential pipes and infrastructure products to enable clean water, sanitation and grey water recovery (see UN SDG 6) to improve living standards and to reduce the incidence of epidemics.

- Providing healthcare products to combat routine healthcare issues as well as disasters such as epidemics (see UN SDG 3)

- The plastics industry will be necessary to improve and deliver the improved transport infrastructure needed for the modern city (see UN SDG 9).

- Designing, implementing and assisting in waste management systems to reduce the leakage of plastics to the environment (see UN SDG 14 and UN SDG 15).

- Designing and implementing food processing, packaging and transport systems to maximise the utility of food produced at some distance from cities (see UN SDG 2).

- Designing and producing plastics products for buildings to reduce energy use and improve comfort (see UN SDG 7).

- Assisting in the development of connected and SMART cities to reduce the impact of urban living (see UN SDG 9 and UN SDG 7).

Action taken to date

The plastics industry and the infrastructure it enables are essential for the existence of the modern city:

- Without clean water and sanitation, the city as we know it would not be sustainable, or indeed possible. The role of the plastics industry in helping to deliver clean water and sanitation (UN SDG 6) and good health (see UN SDG 3) have been crucial in allowing the development of urban environments.

- Transport infrastructure and vehicles depend on plastics to function. Areas such

as water control, electrical insulation for mass-transit systems and lightweighting of vehicles are only some of the areas where plastics have contributed to the development of effective cities.

- Communications in an urban environment are vital to provide the connectivity necessary for development at both the city and the state level. Without plastics, the essential telecommunications infrastructure would not be possible.

- Cities and buildings use enormous amounts of energy for climate control. Plastics are widely used for insulation to reduce energy use caused by heat gains and losses. Products such as PVC-U windows save energy not only in heating but also in air conditioning use (one of the fastest rising uses of energy in hot countries).

Future action

The plastics industry will continue to be integral to the development of sustainable cities and communities. Most of the opportunities will be fulfilled naturally due to the unique properties and applications of plastics. The biggest challenge in the future is to manage the use, re-use and potential disposal of plastics at the end-of-life (see Chapter 10).

Summary

The high utility and long life of plastics products have enabled the world to create modern cities and to support the world's growing population but the use of plastics in packaging needs to be managed to reduce pollution of the land and sea.

The 'plastics-free life' of the 21st century is bound to be as short and as unpleasant as the one in the Middle Ages.

Bloggers give many tips but they all revolve around packaging. Few realise that:

- Their keyboard is plastic, their computer relies on plastics to function, their blogs are distributed, stored and read over wires insulated with plastics.
- Their healthcare relies on plastics.
- Their clean water and sanitation rely on plastics.

Let's try to avoid throwing the baby out with the bathwater!

Recycling rates for plastics in Europe are already high and increasing rapidly.

PVC doors and windows avoid the use of chemicals such as paint and varnishes as well as the painting maintenance load.

A1.12 Goal 12: Responsible consumption and production

Ensure sustainable consumption and production patterns

Sustainable consumption and production are about promoting resource and energy efficiency, sustainable infrastructure, and providing access to basic services, green and decent jobs and a better quality of life for all. Its implementation helps to achieve overall development plans, reduce future environmental, social and economic costs, strengthen economic competitiveness and reduce poverty.

Sustainable consumption and production aim at 'doing more and better with less', increasing net welfare gains from economic activities by reducing resource use, degradation and pollution along the whole life cycle, while increasing quality of life. It involves different stakeholders, including business, consumers, policy makers, researchers, scientists, retailers, media, and development cooperation agencies, among others.

It also requires a systematic approach and cooperation in the complete supply chain, from producer to final consumer and back to the recycler. It involves engaging consumers through awareness-raising and education on sustainable consumption and lifestyles, providing consumers with adequate information on recyclability through standards and labels and engaging in sustainable public procurement.

Challenges for the plastics industry

The challenges to the plastics industry in achieving sustainable consumption and production are:

- The plastics industry has been demonised in the press for a variety of reasons but primarily because of plastics in the oceans (see UN SDG 14). This SDG is not simply focused on industry but on society as a whole and the plastics industry must improve communications to show that plastics are 'part of the solution' and not 'part of the problem'.

- Ocean waste is not simply a plastics issue, it is also an issue of rapidly rising middle-classes in countries, e.g., India and China, without the concomitant construction of a waste management infrastructure, systems and ethos. This has led to poor waste management and large-scale leakage of post-consumer use plastics into the environment.

- The plastics industry needs to work harder and faster to improve waste management, particularly in developing countries. This will enable increased recycling rates and improved production and consumption. In-house re-use is already very high in the industry but the recycling of 'post-consumer waste' needs improvement. This can be physical recycling where the plastics are cleaned and reprocessed for new plastic products (see Section 4.6) or chemical recycling where the plastic is broken down into the component chemicals and reassembled into new plastics (see Section 4.10) . The less desirable alternative is 'energy from waste' to recover the embodied energy (see Section 4.11), as is the 87% of petrochemicals that are simply burnt with no prior use.

Plastics are not the issue – they can be re-used almost endlessly. The issue is using them irresponsibly and this must change for the industry to survive.

Goal 12 targets

- Implement the 10-year framework of programmes on sustainable consumption and production, all countries taking action, with developed countries taking the lead, taking into account the development and capabilities of developing countries.

- By 2030, achieve the sustainable management and efficient use of natural resources.

- By 2030, halve per capita global food waste at the retail and consumer levels and reduce food losses along production and supply chains, including post-harvest losses.

- By 2020, achieve the environmentally sound management of chemicals and all wastes throughout their life cycle, in accordance with agreed international frameworks, and significantly reduce their release to air, water and soil in order to minimize their adverse impacts on human health and the environment.

- By 2030, substantially reduce waste generation through prevention, reduction, recycling and re-use.

- Encourage companies, especially large and transnational companies, to adopt sustainable practices and to integrate sustainability information into their reporting cycle.

- Promote public procurement practices that are sustainable, in accordance with national policies and priorities.

- By 2030, ensure that people everywhere have the relevant information and awareness for sustainable development and lifestyles in harmony with nature.

- Support developing countries to strengthen their scientific and technological capacity to move towards more sustainable patterns of consumption and production.

- Develop and implement tools to monitor sustainable development impacts for sustainable tourism that creates jobs and promotes local culture and products.

- Rationalize inefficient fossil-fuel subsidies that encourage wasteful consumption by removing market distortions, in accordance with national circumstances, including by restructuring taxation and phasing out those harmful subsidies, where they exist, to reflect their environmental impacts, taking fully into account the specific needs and conditions of developing countries and minimizing the possible adverse impacts on their development in a manner that protects the poor and the affected communities.

- The biggest challenge to the plastics processing industry is to start a 'fact-based' discussion about plastics where the real contribution of plastics to modern society can be recognised.
- The perceived challenge of oil depletion and plastics is mythical. Plastics use only ≈ 4-6% of global petrochemical production where ≈ 45% is used for transport and ≈ 42% for energy and heating. Transport, energy and heating are the largest single-use of petrochemicals whereas almost all plastics can be recycled for multiple uses before finally being burnt for energy or heating.

  The raw materials for plastics can also be sourced from biomass and, whilst these may be currently more expensive, the technology is readily available. Running out of oil is not the real issue for plastics (see Section 1.9), the real issue is the destination of the products at the end-of-life (see Section 1.10).
- Responsible consumption and production are not simply issues for the plastics industry, these are also a societal issue that depends on the consumer.

Opportunities for plastics

The opportunities for the plastics industry in sustainable consumption and production are:

- Improving waste management and increasing recycling rates are major opportunities for the plastics industry. Recycling technology is largely mature but improvement is needed in sorting technology to adequately separate the various plastics. Consumer attitudes to waste also need improvement so that plastics are regarded as a resource and not as a waste.
- Improving chemical recycling methods to recover the chemical constituents for the production of new plastics.
- Improved waste management is particularly needed in developing countries where the rising numbers of middle-classes is driving consumption without a suitable waste management infrastructure for recycling or recovery.
- It is estimated that one third of all food produced is wasted. Improved food packaging can reduce waste. Packaging averages only 1-3% of the total product weight and much less in terms of the carbon footprint of the complete product (see Section 10.3).

Action taken to date

The plastics industry has contributed to sustainable consumption and production through actions such as:

- Plastics products are used extensively for insulation products and windows to reduce the use of heating or air conditioning and save energy. Heating is a single-use of petrochemicals in almost every case.
- The use of plastics in almost any application is more resource efficient than that of other materials.
  - Plastics reduce the use of water for cleaning and processing most alternative materials, e.g., cotton.
  - Plastics reduce the volume and weight of packaging and save transport fuel (a single-use of petrochemicals).
  - Plastics reduce food waste during all phases of the production from farm to fork (see UN SDG 2).
- Improved design for lightweighting and recycling to improve both resource efficiency and recycling rates. This will continue into the future to improve effectiveness and profitability.
- The plastics processing industry is working to reduce energy use during production. This was initially driven by economics but is increasingly being driven by sustainability issues.

Future action

The plastics industry will continue to contribute to the delivery of sustainable consumption and production in the areas of:

- The plastics processing industry has worked to reduce plastics leakage to the environment at the site level. More work is needed to prevent consumer leakage and improve waste management.

  The solution is not to ban plastics but to establish the waste management and recycling infrastructure to recover and recycle these valuable materials.
- Improved design for reduced materials content and recycling.

Summary

Plastics are an integral part of achieving responsible consumption and production.

"Oil in the future will not be burnt away and wasted in energy and transport but reserved for high-value processes and products such as plastics manufacturing...... and the energy trapped within the plastics can either be recycled or recovered and used for heat generation."
Ray Hammond
'The World in 2030'

Plastics pipes used for gas and water have a guaranteed service life of 50 years.

Recycling of plastics at the end of their service life is growing rapidly but must increase faster and go further.

Plastics are widely used and visible because of their long-life and durability, these factors must be used to the benefit of society and achievement of this goal.

A1.13 Goal 13: Climate action

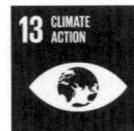

13 CLIMATE ACTION

Take urgent action to combat climate change and its impacts

Climate change is now affecting every country on every continent. It is disrupting national economies and affecting lives, costing people, communities and countries dearly today and even more tomorrow.

People are experiencing the significant impacts of climate change, which include changing weather patterns, rising sea level, and more extreme weather events. The greenhouse gas emissions from human activities are driving climate change and continue to rise. They are now at their highest levels in history. Without action, the world's average surface temperature is projected to rise over the 21st century and the rise is likely to surpass 3°C this century – with some areas of the world expected to warm even more.

The poorest and most vulnerable people in the world are the ones who will be affected the most.

Affordable, scalable solutions are now available to enable countries to leapfrog to cleaner, more resilient economies. The pace of change is quickening as more people are turning to renewable energy and a range of other measures that will reduce emissions and increase adaptation efforts.

But climate change is a global challenge that does not respect national borders. Emissions anywhere affect people everywhere. It is an issue that requires solutions that need to be coordinated at the international level and it requires international cooperation to help developing countries move toward a low-carbon economy.

Challenges for the plastics industry

The challenges to the plastics industry in combating climate change and its impacts are:

- The plastics industry must seek to reduce the overall carbon footprint (a measure of the direct and indirect greenhouse gas emissions) to reduce the industry's impact on the climate. This is an internal action that involves reducing energy use, the major cause of greenhouse gas emissions in the industry, as well as other emission factors such as heating and transport.
- The industry needs to increase the amount of recycled materials used (see UN SDG 12) to use the carbon locked up in most plastics materials as effectively as possible.

- The industry needs to increase the use of bio-based materials (see Section 4.14) to reduce additional greenhouse gas emissions to the atmosphere.
- However, reducing the greenhouse gas emissions of the industry is not enough and the industry needs to contribute to the development of the necessary infrastructure (see UN SDG 9) for disaster risk mitigation.
- As with many other SDGs, the value of plastics in achieving a sustainable world and climate action needs to be based on facts rather than on emotions, plastics products are needed to help combat climate change and this needs recognition.

Opportunities for plastics

The plastics industry has much to offer in combating climate change and in disaster risk reduction. The opportunities for the plastics industry are:

- Reducing greenhouse gas emissions is not only good for climate action but also reduces costs (both actual and compliance costs) and is financially attractive to the industry. Companies that supply customers with carbon footprint data, plans to reduce their carbon footprint and the costs will have a competitive advantage over companies that continue with business as usual.
- The upgrading and improvement of infrastructure to reduce or mitigate climate change risk (see UN SDG 9) will need plastics to be feasible. This includes projects such as:

> The plastics industry is already taking action to reduce greenhouse gas emissions.

Goal 13 targets

- Strengthen resilience and adaptive capacity to climate-related hazards and natural disasters in all countries.
- Integrate climate change measures into national policies, strategies and planning.
- Improve education, awareness-raising and human and institutional capacity on climate change mitigation, adaptation, impact reduction and early warning.
- Implement the commitment undertaken by developed-country parties to the United Nations Framework Convention on Climate Change to a goal of mobilizing jointly $100 billion annually by 2020 from all sources to address the needs of developing countries in the context of meaningful mitigation actions and transparency on implementation and capitalise the Green Climate Fund as soon as possible.
- Promote mechanisms for raising capacity for effective climate change-related planning and management in least developed countries and small island developing states, including focusing on women, youth and local and marginalized communities.

- Improved and efficient low-cost housing for developing countries.
- Improved water collection, e.g., dam liners to reduce water leakage.
- Improved water distribution, e.g., pipes, to provide clean water (see UN SDG 6) and combat the effects of climate change, e.g., desertification.

- Improved flood management and drainage capability using plastic pipes and geo-cellular sustainable drainage systems (SuDS) to store or re-use surface water at source.
- Plastics will be an essential part of impact reduction by enabling the telecommunications networks for monitoring and early warning systems.
- Disaster relief in the event of severe climate change events will use plastics for essential items such as tents, water containers, tarpaulins and packaging for medical supplies. These will all be packaged in a strong reusable boxes for transport and delivery to the point of need.

Action taken to date

The plastics industry has contributed to combating climate change and its impacts through actions such as:

- Decoupling the growth in greenhouse gas emissions from the growth of the industry. This is part of the larger efforts to decouple the growth in greenhouse gas emissions from the growth in GDP. This is the result of improved energy efficiency and cost competitiveness in the industry.
- The products of the plastics industry are mainly lightweight and many plastics processors are close to their customers. This reduces transport emissions (and costs) in product delivery.
- Plastics are essential in reducing vehicle weight and improving transport fuel efficiency. This is not simply for automobiles but also for aeroplanes which use plastics to achieve improved fuel performance.
- Plastics have been responsible for lightweighting of many products to reduce the carbon emissions resulting from product transport.
- Plastics have excellent thermal insulation properties and are widely used for building insulation and windows to reduce heating or air conditioning use and the resulting greenhouse gas emissions.
- The plastics industry is essential for the production, storage and transport of clean energy (see UN SDG 7) to reduce greenhouse gas emissions from energy production.

Future action

The plastics industry will continue to contribute to combating climate change and its impacts in the areas of:

- Increasing the resource efficiency of the industry. This covers areas as diverse as increasing energy efficiency, decreasing water use and improving material efficiency through product lightweighting.
- Decreasing the raw material carbon impact by increasing the use of bio-based materials that do not increase the amount of short carbon in the biosphere (see Section 1.10).
- Developing products for the production, storage and transport of clean and affordable energy (see UN SDG 7).
- Developing products and systems for disaster risk early warning systems.
- Developing products for disaster relief in the event of climate change events.

Summary

Climate action is concerned with both reducing greenhouse gas emissions to reduce climate change and with strengthening resilience in the event of climate change events. The plastics processing industry can contribute to both of these areas and help to achieve this goal.

Plastics products provide the tools necessary for mitigating the effects of climate change events and disasters.

Plastics products in buildings, e.g., insulating foams and windows, reduce heating oil or gas use significantly.

"If plastics in packaging were replaced by traditional materials then world energy consumption would double."
Plastics Europe/ Gesellschaft für umfassende Analysen.

Plastics are part of the solution to mitigating climate change. They are not simply part of the problem.

A1.14 Goal 14: Life below water

Conserve and sustainably use the oceans, seas and marine resources

The world's oceans, their temperature, chemistry, currents and life, drive global systems that make the Earth habitable for humankind.

Our rainwater, drinking water, weather, climate, coastlines, much of our food, and even the oxygen in the air we breathe, are all ultimately provided and regulated by the sea. Throughout history, oceans and seas have been vital conduits for trade and transportation.

Careful management of this essential global resource is a key feature of a sustainable future.

Challenges for the plastics industry

The challenges to the plastics industry in managing and protecting life below water are many and this is one of the most contentious issues facing the plastics industry in assisting with the UN SDGs. The challenges are:

- Ocean plastics are a very visible and public issue and there are many groups discussing this issue and lobbying for action. This consumer pressure has led to pressure for 'plastics-free aisles' in stores, legislation for 'bag taxes' and bans on single-use and short-life plastic products such as straws and coffee cups. The industry is not well placed to defend certain applications and supports many of these initiatives.

However, some of the initiatives are based on emotion rather than science and the industry needs to be robust in providing science-based evidence and mounting a defence of plastics on the basis of their overall utility to mankind.

If this challenge is not met then the future for the entire industry will be clouded by this issue and the benefits of plastics will be lost in the resulting noise.

- Most of the plastics entering the oceans are from countries with a rapidly growing middle-class but with no effective waste management system (see sidebar on this page). This includes most of the developing world and particularly China. The challenge is to reduce leakage of plastics into the environment (see UN SDG 15) and into the seas through effective waste management.

- The packaging industry is often accused of 'over-packaging' products. This needs to be balanced with the need to protect products (often foodstuffs) and the need for hygiene. The challenge is to economically and environmentally justify the use of all plastics in packaging and to reduce their use to the minimum.

- Plastics do not degrade in the same manner as many other materials. The longevity of plastics is now potentially a bad thing – how times change. Plastics do not degrade conventionally but gradually break down into smaller and smaller particles which can then be ingested and

- 82% of the plastics entering the oceans comes from Asia.
- 2% comes from the USA and Europe.
- 16% comes from the rest of the world.

Jambeck et al., 'Plastic waste inputs from land into the ocean'. Science, 2015.

Goal 14 targets

- By 2025, prevent and significantly reduce marine pollution of all kinds, in particular from land-based activities, including marine debris and nutrient pollution.

- By 2020, sustainably manage and protect marine and coastal ecosystems to avoid significant adverse impacts, including by strengthening their resilience, and take action for their restoration in order to achieve healthy and productive oceans.

- Minimize and address the impacts of ocean acidification, including through enhanced scientific cooperation at all levels.

- By 2020, effectively regulate harvesting and end overfishing, illegal, unreported and unregulated fishing and destructive fishing practices and implement science-based management plans, in order to restore fish stocks in the shortest time feasible, at least to levels that can produce maximum sustainable yield as determined by their biological characteristics.

- By 2020, conserve at least 10 per cent of coastal and marine areas, consistent with national and international law and based on the best available scientific information.

- By 2020, prohibit certain forms of fisheries subsidies which contribute to overcapacity and overfishing, eliminate subsidies that contribute to illegal, unreported and unregulated fishing and refrain from introducing new such subsidies, recognizing that appropriate and effective special and differential treatment for developing and least developed countries should be an integral part of the World Trade Organization fisheries subsidies negotiation.

- By 2030, increase the economic benefits to small island developing States and least developed countries from the sustainable use of marine resources, including through sustainable management of fisheries, aquaculture and tourism.

- Increase scientific knowledge, develop research capacity and transfer marine technology, taking into account the Intergovernmental Oceanographic Commission Criteria and Guidelines on the Transfer of Marine Technology, in order to improve ocean health and to enhance the contribution of marine biodiversity to the development of developing countries, in particular small island developing states and least developed countries.

- Provide access for small-scale artisanal fishers to marine resources and markets.

- Enhance the conservation and sustainable use of oceans and their resources by implementing international law as reflected in UNCLOS, which provides the legal framework for the conservation and sustainable use of oceans and their resources.

incorporated into the marine food chain. These particles then make their way up the food chain (fish are a major source of protein in developing countries) to humans. The effects of micro-plastics need clarification. Plastics are inert materials and the major risk is from the additive packages that many plastics contain.

- Whether the loss is to the sea or the land (see UN SDG 15) the challenge is in preventing the loss of valuable raw materials from the processing and use system and into the wider environment. The circular economy provides a model system but the plastics industry must engage more, whether voluntarily or not, in ensuring that plastics are seen as a vital part of the circular economy.

Opportunities for plastics

The opportunities for the plastics industry in managing and protecting life below water are:

- Preventing the loss of plastics materials to the marine environment. These are valuable current and future raw materials and represent potential lost profits (both current and future). This needs a combination of industry and societal action but industry can lead the way through internal industry programmes such as 'Operation Clean Sweep' (see Section 5.4) and external awareness programmes that highlight the real areas of loss to the system and methods of stopping waste entering the oceans.

- Assisting in the removal of existing plastics in the oceans. Plastic in the oceans is not simply a pollutant, it is a potential future raw material. There are a variety of schemes being developed throughout the world to harvest plastics raw materials from the oceans but all of these depend on plastics (for nets, flotation devices or other applications). It should not be forgotten that plastics are not simply the problem, they are also part of the solution.

- Improving recycling methods to allow recovery of plastics materials from the waste stream to prevent leaching into the oceans. This may need improved recycling or materials technology to allow full recovery.

Action taken to date

The plastics industry has contributed to managing and protecting life below water through actions such as:

- The industry is already exceptionally good at post-process waste recycling, i.e., internal re-use (see Section 5.3), because of the recognition that this is valuable raw material.

However, the industry and society are not good at recycling post-consumer waste. The issues here are:

- Post-consumer waste is often dirty and/ or contaminated.
- Post-consumer waste is often widely distributed on land and in the oceans.
- The industry is already widely engaged in programmes such as 'Operation Clean Sweep' to prevent pellet loss from the system.

Future action

The plastics industry will continue to contribute to managing and protecting life below water in the areas of:

- The industry must lead in eliminating single-use and trivial applications of plastics, e.g., coffee stirrers and micro-beads where the use of plastics is debateable or indefensible. Consumer and legislatory pressure will help to remove unsuitable applications and will not be resisted by the industry.

- The industry must take action to determine the exact effect of micro plastic particles on the ecosystem. Plastics are generally inert (see above) but 'absence of evidence is not the same as evidence of absence'.

- The industry must work with recyclers to identify and assist in removing difficult to recycle materials (or combinations thereof) from the supply chain.

- The industry must work with, and develop, projects to assist in the future removal of plastics from the oceans.

Summary

This goal is mainly concerned with preserving and managing marine ecosystems but the current publicity is focused mainly on plastics pollution as a result of poor waste management. This cannot be ignored and places the industry at severe reputational risk. The industry needs to do more in this area to assist in achieving this goal.

- 90% of the plastics entering the oceans comes from 10 river systems.
- 8 are in Asia: the Yangtze; Indus; Yellow; Hai He; Ganges; Pearl; Amur; Mekong
- 2 are in Africa: the Nile and the Niger.

Schmidt et al. 'Export of Plastic Debris by Rivers into the Sea', Environ. Sci. Technol. 2017.

It is our environment too!

Fishy numbers?

"The ocean is expected to contain 1 tonne of plastic for every 3 tonnes of fish by 2025, and by 2050, more plastics than fish (by weight)."

The New Plastics Economy

Ellen Macarthur Foundation

The only problem with this is the source of the numbers:

- The plastics data is based on a study of the San Francisco Bay area that made projections until 2025. These were then extrapolated to the world and again to 2050.

- The fish data is based a phytoplankton study which estimated 899 million tonnes of fish but was later revised to between 2 billion and 10.4 billion tonnes.

www.bbc.co.uk/ news/magazine-35562253

A1.15 Goal 15: Life on land

Sustainably manage forests, combat desertification, halt and reverse land degradation, halt biodiversity loss

Forests cover 30% of the Earth's surface and in addition to providing food security and shelter, forests are key to combating climate change, protecting biodiversity and the homes of indigenous populations. Thirteen million hectares of forests are being lost every year while the persistent degradation of drylands has led to the desertification of 3.6 billion hectares.

Deforestation and desertification, caused by human activities and climate change, pose major challenges to sustainable development and have affected the lives and livelihoods of millions of people in the fight against poverty. Efforts are being made to manage forests and combat desertification.

Challenges for the plastics industry

The challenges to the plastics industry in managing and protecting life on land are very similar to those for protecting life below water and this is also a contentious issue for the plastics industry. The challenges are:

- Whilst ocean plastics are currently a public issue, 80% of the plastic in the oceans originates from land-based sources (the remaining 20% consists of products such as discarded fishing nets). Reducing or eliminating land-based sources will therefore dramatically affect the plastics entering the oceans (see UN SDG 15). This will benefit both life on the land and in the oceans.

- Not all of the plastic that escapes from the processing/use system enters rivers and oceans and the materials that remain on land represent a significant challenge to the plastics industry.

- The challenge of reducing material leakage from the system by improved waste management and reducing casual disposal, i.e., littering, is the crucial issue.

The industry is already working hard to reduce leakage at the process level and countries with efficient waste management systems are starting to achieve excellent recycling rates for many products (although there is still much to be done to reduce leakage through poor behaviours such as littering).

A significant issue is that many of the applications of plastics that help to achieve other UN SDGs will result in products reaching their end-of-life phase in countries which do not have an adequate waste management system and that these materials will not be handled properly at the end of their life. The challenges to the industry are to develop methods to recover the material or the embodied energy and to prevent material leakage into the environment and subsequent material loss.

- Whether the loss is to the sea (see UN SDG 14) or the land the challenge is in preventing the loss of valuable raw materials from the processing and use system and into the wider environment. The circular economy provides a model system but the plastics industry must

Plastics entering the oceans also indicates a failure of waste management on the land. The plastics materials are discarded on the land and then make their way to the oceans.

Goal 15 targets

- By 2020, ensure the conservation, restoration and sustainable use of terrestrial and inland freshwater ecosystems and their services, in particular forests, wetlands, mountains and drylands, in line with obligations under international agreements.

- By 2020, promote the implementation of sustainable management of all types of forests, halt deforestation, restore degraded forests and substantially increase afforestation and reforestation globally.

- By 2030, combat desertification, restore degraded land and soil, including land affected by desertification, drought and floods, and strive to achieve a land degradation-neutral world.

- By 2030, ensure the conservation of mountain ecosystems, including their biodiversity, in order to enhance their capacity to provide benefits that are essential for sustainable development.

- Take urgent and significant action to reduce the degradation of natural habitats, halt the loss of biodiversity and, by 2020, protect and prevent the extinction of threatened species.

- Promote fair and equitable sharing of the benefits arising from the utilization of genetic resources and promote appropriate access to such resources.

- Take urgent action to end poaching and trafficking of protected species of flora and fauna and address both demand and supply of illegal wildlife products.

- By 2020, introduce measures to prevent the introduction and significantly reduce the impact of invasive alien species on land and water ecosystems and control or eradicate the priority species.

- By 2020, integrate ecosystem and biodiversity values into national and local planning, development processes, poverty reduction strategies and accounts.

- Mobilize and increase financial resources from all sources to conserve and sustainably use biodiversity and ecosystems.

- Mobilize significant resources from all sources and at all levels to finance sustainable forest management and provide adequate incentives to developing countries to advance such management, including for conservation and reforestation.

- Enhance global support for efforts to combat poaching and trafficking of protected species, including by increasing the capacity of local communities to pursue sustainable livelihood opportunities.

engage more, whether voluntarily or not, in ensuring that plastics are a vital part of the circular economy.

- A significant challenge is in the use or substitution of bio-based polymers. The challenge here is to enable the efficient and cost-effective substitution of petrochemicals without driving excessive land use for bio-based precursor production or driving up the cost of food production in the developing world.

Opportunities for plastics

The opportunities for the plastics industry in in managing and protecting life on land are:

- Preventing material leakage and recovering either the material or the energy before plastics escape into rivers or oceans. This is better, and easier, than action after the material has been dispersed and distributed by rivers (see UN SDG 14).

- Good water management (see UN SDG 6) can help to prevent habitat degradation, desertification and biodiversity loss. The opportunity is to use plastics wisely to provide clean water for growing populations without habitat destruction.

- The contribution of plastics to sustainable cities (see UN SDG 11) will allow increasing urbanisation to take place with reduced pressure on forests and reduced climate change (see UN SDG 13). Plastics have much to contribute in this area.

- The retention and protection of biodiversity relies on repositories such as seed banks and DNA banks to retain and protect species for the future. These would be impossible to develop, operate or manage without the unique properties of plastics.

- Improving recycling methods to allow recovery of plastics materials from the waste stream. This may need improved recycling or materials technology to allow full recovery for recycling or small-scale incineration technologies to recover the embodied energy without excessive harmful emissions.

- Plastics can be used to substitute many of the threatened hardwoods in construction applications and the plastics can be re-used or recycled at the end-of-life stage.

Action taken to date

The plastics industry has contributed to managing and protecting life on land through actions such as:

- Good, and minimal packaging, improves the yield of food sources and protects the food on the journey to the consumer. This can help to protect forests by reducing the amount of land needed for food production and reducing the need for 'slash and burn' agriculture.

- Technologies such as tree protectors and area protectors to encourage and manage tree and plant growth whilst protecting the trees from animal depredation.

- The plastics industry provides the enabling technologies for seed banks and DNA banks to protect species from extinction and to provide the basis for future reintroduction of species.

Future action

The plastics industry will continue to contribute to managing and protecting life on land in the areas of:

- Leading the way in eliminating single-use and trivial applications of plastics (see UN SDG 14).

- Working with recyclers to identify and assist in removing difficult to recycle materials (or combinations thereof) from the supply chain (see UN SDG 14).

Summary

This goal is concerned with preserving and managing land-based ecosystems to stop desertification, land degradation and biodiversity loss. This can be achieved by improving and refining waste management systems to recover valuable raw materials and/or their embodied energy, preventing materials loss from the production/use system (the circular economy) and assisting in providing the tools for environmental recovery.

The presence of 'plastic bag trees' (or trees covered with plastic bags) indicates a failure of the system to capture waste at the most basic level. This can be from personal littering or from institutional waste management failure.

It is our environment too!
Again.

A1.16 Goal 16: Peace, justice and strong institutions

Promote just, peaceful and inclusive societies

This goal is dedicated to the promotion of peaceful and inclusive societies for sustainable development, the provision of access to justice for all, and building effective, accountable institutions at all levels.

Challenges for the plastics industry

The challenges to the plastics industry in helping to achieve peace, justice and strong institutions are:

- The plastics processing industry is a capital-intensive industry that requires high skill levels and secure supply chains. It is thus unlikely to develop or prosper significantly in countries which do not have peace, justice and strong institutions.

- The plastics processing industry needs a robust and dependable infrastructure to function. Countries which do not have peace, justice and strong institutions are also unlikely to have an infrastructure which can support a viable plastics processing industry.

- The plastics industry is a world industry and companies will choose to locate and develop plastics processing sites in countries which have peace, justice and strong institutions.

These factors limit the industry's ability to contribute to this goal because the industry is not generally present in countries which do not already have peace, justice and strong institutions.

Where countries with existing peace, justice and strong institutions fail to maintain these then the plastics processing industry's response is limited due to the diffuse nature of the industry.

Opportunities for plastics

The opportunities for the plastics industry in helping to achieve peace, justice and strong institutions are:

- The plastics processing industry can help to achieve many of the other UN SDGs, e.g., UN SDG 4, UN SDG 6, UN SDG 8 and UN SDG 9. Achieving these goals increases sustainable development and will improve a country's stability. This will also indirectly help to achieve peace, justice and strong institutions.

- The plastics industry needs to treat all workers with respect (see UN SDG 5 and UN SDG 10) irrespective of seniority, gender, race, class or caste. Reducing inequality will assist in the development of peace, justice and strong institutions.

Action taken to date

The plastics industry has helped to achieve peace, justice and strong institutions through actions such as:

- The industry has already contributed to development by helping to achieve many of the UN MDGs. This development has helped to promote the achievement of peace, justice and strong institutions.

- The industry has a generally good record of labour relations and equitable treatment of workers irrespective of seniority, gender, race, class or caste (see UN SDG 10).

- The plastics processing industry provides good jobs, rewarding work and a structure for development. All of these contribute to the development and retention of peace, justice and strong institutions.

Future action

The plastics industry will continue to help to achieve peace, justice and strong institutions by assisting in achieving the other UN SDGs to promote sustainable development.

> Peace, justice and strong institutions are a necessary precursor for achieving almost all of the UN SDGs.

Goal 16 targets

- Significantly reduce all forms of violence and related death rates everywhere.
- End abuse, exploitation, trafficking and all forms of violence against and torture of children.
- Promote the rule of law at the national and international levels and ensure equal access to justice for all.
- By 2030, significantly reduce illicit financial and arms flows, strengthen the recovery and return of stolen assets and combat all forms of organized crime.
- Substantially reduce corruption and bribery in all their forms.
- Develop effective, accountable and transparent institutions at all levels.
- Ensure responsive, inclusive, participatory and representative decision-making at all levels.
- Broaden and strengthen the participation of developing countries in the institutions of global governance.
- By 2030, provide legal identity for all, including birth registration.
- Ensure public access to information and protect fundamental freedoms, in accordance with national legislation and international agreements.
- Strengthen relevant national institutions, including through international cooperation, for building capacity at all levels, in particular in developing countries, to prevent violence and combat terrorism and crime.
- Promote and enforce non-discriminatory laws and policies for sustainable development.

The industry can help by:

- Improving efforts in social sustainability.
- Reporting on efforts and action taken to improve social sustainability.

Summary

It is difficult for the plastics processing industry to directly and significantly contribute to the achievement of this goal but the contribution of the industry in achieving the UN SDGs will help in this area.

The author has worked in the plastics processing industry in 46 different countries. Peace, justice and strong institutions are needed for a strong plastics processing industry.

A1.17 Goal 17: Partnerships for the goals

Revitalize the global partnership for sustainable development

A successful sustainable development agenda requires partnerships between governments, the private sector and civil society. These inclusive partnerships built upon principles and values, a shared vision, and shared goals that place people and the planet at the centre, are needed at the global, regional, national and local level.

Urgent action is needed to mobilize, redirect and unlock the transformative power of trillions of dollars of private resources to deliver on sustainable development objectives. Long-term investments, including foreign direct investment, are needed in critical sectors, especially in developing countries. These include sustainable energy, infrastructure and transport, as well as information and communications technologies. The public sector will need to set a clear direction.

Review and monitoring frameworks, regulations and incentive structures that enable such investments must be retooled to attract investments and reinforce sustainable development. National oversight mechanisms such as supreme audit institutions and oversight functions by legislatures should be strengthened.

Challenges for the plastics industry

The challenges to the plastics industry in revitalizing the global partnership for sustainable development are:

- Some people see plastics as the opposition and demonise the plastics industry. The reality is that it is an enabling industry. It produces the products that the customers want and as their desires and needs change then so will the industry. The challenge for the industry is to change quickly and responsibly so that it can continue to produce products that produce vast benefits for society whilst at the same time phasing out products that are detrimental to society.

- The plastics industry is not an independent entity. It consists of people who are doing their best to advance themselves, their employer and society. The plastics industry is part of society and it is our world too! The challenge is for the industry to act in a clear and responsible manner and for all of the people involved in the industry to individually act in a responsible manner.

Goal 17 targets

Finance

- Strengthen domestic resource mobilization, including through international support to developing countries, to improve domestic capacity for tax and other revenue collection.
- Developed countries to implement fully their official development assistance commitments, including the commitment by many developed countries to achieve the target of 0.7% of GDP to developing countries and 0.15 to 0.20% of GDP to least developed countries. Providers are encouraged to consider setting a target to provide at least 0.20% of GDP to least developed countries.
- Mobilize additional financial resources for developing countries from multiple sources.
- Assist developing countries in attaining long-term debt sustainability through policies aimed at fostering debt financing, debt relief and debt restructuring, as appropriate, and address the external debt of highly indebted poor countries to reduce debt distress.
- Adopt and implement investment promotion regimes for least developed countries.

Technology

- Enhance regional and international cooperation on and access to science, technology and innovation and enhance knowledge sharing on mutually agreed terms, including through improved coordination, in particular at the United Nations level, and through a global technology facilitation mechanism.
- Promote the development, transfer, dissemination and diffusion of environmentally sound technologies to developing countries on favourable terms, including on concessional and preferential terms.
- Fully operationalize the technology bank and science, technology and innovation capacity-building mechanism for least developed countries by 2017 and enhance the use of enabling technology, in particular information and communications technology.

Capacity building

- Enhance international support for implementing effective and targeted capacity-building in developing countries to support national plans to implement all the sustainable development goals, including through regional cooperation.

Trade

- Promote a universal, rules-based, open, non-discriminatory and equitable multilateral trading system under the World Trade Organization, including through the conclusion of negotiations under its Doha Development Agenda.
- Significantly increase the exports of developing countries, in particular with a view to doubling the least developed countries' share of global exports by 2020.
- Realize implementation of duty-free and quota-free market access for all least developed countries, consistent with World Trade Organization decisions, including by ensuring that preferential rules of origin applicable to imports from least developed countries are transparent and simple, and contribute to facilitating market access.

- The plastics processing industry has, for far too long, been reticent about discussing or publicising the benefits of plastics to society. The challenge in this area is for the industry to promote the benefits that plastics bring to society and to show that plastics make modern society possible.

- The plastics processing industry is traditionally a science-based industry and criticism of the industry is usually emotion-based. These very different approaches have hampered communication between the industry and its critics. The challenge for the industry is to engage with the critics in a reasonable manner, to respect their motivations and to form partnerships to meet the common challenges.

Opportunities for plastics

The opportunities for the plastics industry in revitalizing the global partnership for sustainable development are:

- Achieving the UN SDGs will have a transformative effect on the world and living standards across the world. This will change the face of the plastics industry. Raising living standards across the world will raise the total demand for many of the products of the plastics processing industry. There will obviously be casualties as certain products are phased out due public or legislative pressure but the industry is flexible and adaptive. It will change to produce new products that help to achieve the UN SDGs and to do so sustainably.

- The plastics processing industry is multi-faceted and adept at forming multi-stakeholder initiatives. The opportunity for the industry is to be part of the process of change and to drive change (which it does very well) rather than to be purely reactive and have change forced on it.

- The plastics processing industry is world-wide and can use the technological strengths to assist the developing world in capacity building and in sustainable and appropriate technology development.

Action taken to date

The plastics industry has contributed to revitalizing the global partnership for sustainable development through actions such as:

- The plastics processing industry is already working with organisations and programmes such as 'Operation Clean Sweep' to prevent pellet loss.

- The plastics processing industry has already developed Voluntary Commitments at the European level throughout the polymer industry e.g., VinylPlus, PCEP, SCS etc.

- The plastics processing industry is already working to create a science-based discussion of how to achieve the UN SDGs through collaborative action.

- The plastics processing industry is working to improve energy efficiency, resource efficiency and carbon emissions. Some of this is driven by economics, some is driven by legislation and some is driven by a desire to improve sustainability. The motives don't really matter, the results do.

- The plastics processing industry is already working to transfer technology and skills to developing nations. This may be driven by an initial desire to reduce costs but again, the motives don't matter. Developing the technology base of a country will raise living standards and help to achieve the UN SDGs.

Future action

The plastics industry will continue to contribute to the revitalizing of the global partnership for sustainable development in the areas of:

- Working with partners to improve the environmental credibility of the industry based on science and evidence rather than on emotion.

- Working with external partners to show improved performance through unbiased external judgement.

Summary

A single industry sector such as the plastics processing industry obviously cannot achieve the UN SDGs alone. We must form partnerships to achieve these. The plastics industry is a vital partner for many of the UN SDGs and is willing to assist in achieving the goals.

"No man is an island, entire of itself; every man is a piece of the continent, a part of the main. If a clod be washed away by the sea, Europe is the less, as well as if a promontory were ......"

John Donne

It is our world too!

Postscript

This is may well be the last full book that I will write on plastics processing and suitably it is a 'love-letter' to an industry that I love more than any other. It has not only provided me with a living for nearly 50 years but it has challenged me, changed me, succoured me, provided me with lasting friendships and has been a source of constant inspiration throughout these years.

The industry is an integral part of society and the people in it play a vital role by providing the many products that enable modern life and also provide the essential glue to hold our society together.

This book contains work adapted from my other books and I make no apologies for this. It feels that, in some way or another, I have been writing about 'sustainability' from the very start, whether it be about financial sustainability, energy sustainability or the other aspects of sustainability that will allow the plastics processing industry to prosper into the future and provide the vital products that our society needs.

Hopefully, putting all the bits together in one book makes sense and I hope that I have covered all the main points of sustainability for plastics processors. This is a workbook for plastics processors and the ultimate proof of the value will lie in the factories where I have spent many happy and rewarding days in steel toecaps and hi-vis vest (although not in the early days).

A simple structured approach to sustainability in environmental, social and economic terms will allow the industry to protect the environment, provide good jobs into the future and to make money. This is indeed the fabled 'triple bottom line' that has sometimes proved elusive for the industry but which is well within reach using the strategies, targets, techniques and tools presented here.

If there are significant omissions then please let me know and I will update the text for future editions. I now have another 2 grandchildren who need books in their name so at least two more editions are needed before I can 'call it a day'.

I have acknowledged as many of the sources as possible in the text but it is impossible to acknowledge all those people who contributed practical assistance and ideas (good and bad) during my time in the plastics industry. If I have forgotten anybody then please feel free to contact me and I will rectify this as soon as possible.

I would like express my thanks to my friends and colleagues in the plastics industry who kindly reviewed draft sections of this book. They all provided insights, valuable criticism and encouragement to the solitary author. They are (in alphabetical order to avoid favouritism): Rory Brazier, Jonathan Churchman-Davies, Terry Cooper, Jesús Fernández Torres and Keith Freegard.

I would also like to thank my wife, Vivienne, for editing the final text. She found the areas where I was confused and made it a better book despite the fact that she never really signed up for editing a book on plastics.

Any errors are mine alone. My colleagues, and my wife, were probably right.

"This is the paradox of historical knowledge. Knowledge that does not change behaviour is useless. But knowledge that changes behaviour quickly loses its relevance."

Yuval Noah Harari – Homo Deus – A Brief History of Tomorrow

"No book can ever be finished. While working on it we learn just enough to find it immature the moment we turn away from it".

Karl Popper

Abbreviations and acronyms

A/C – Air Conditioning
AC – Alternating Current
AD – Anaerobic Digestion
AMR – Automatic Meter Reading
ASR – Automotive Shredder Residue
B2B – Business to Business
BPR - Business Process Re-engineering
BSI – British Standards Institution
CDD - Chlorinated dibenzo-*p*-dioxin
CDF – Chlorinated dibenzofuran
CDP – Carbon Disclosure Project
CDSB – Climate Disclosure Standards Board
CiP – Cleaning in Place
CSR – Corporate Social Responsibility
CVA – Cistern Volume Adjuster
C2C – Cradle-to-cradle
DBE&IS – Department for Business, Energy & Industrial Strategy (UK)
DfD – Design for Disassembly
DfA – Design for Assembly
DfM – Design for Manufacturing
DfX – Design for X
DOE – Design of Experiments
EBM – Extrusion Blow Moulding
ECI – Environmental Condition Indicator
EfW – Energy from Waste
EMS – Environmental Management System
EMAS – EU Eco-Management and Audit Scheme
EnMS – Energy Management System
EPR – Extended Producer Responsibility
EPS – Expanded Polystyrene
EU – European Union
FIFO – First In First Out
FLT – Fork Lift Truck
FMEA – Failure Modes and Effects Analysis
FTY – First Time Yield
GHG – Greenhouse Gas
GHGP – Greenhouse Gas Protocol
GRI – Global Reporting Initiative
GWP – Global Warming Potential
HEM – High Efficiency Motor
HGV – Heavy Goods Vehicle
IBC – Intermediate Bulk Container
IEA – International Energy Agency
IIRC – International Integrated Reporting Council
ILO – International Labour Organisation
ISBM – Injection Stretch Blow Moulding

ISO – International Organisation for Standardisation
JIT – Just-in-Time
KPI – Key Performance Indicator
LCA – Life Cycle Assessment
LCI – Life Cycle Inventory Analysis
LCIA – Life Cycle Impact Assessment
MBY – Mass Balance Yield
MEK – Methyl Ethyl Ketone
MPI – Management Performance Indicator
MRF – Materials Recovery Facility
MSDS – Materials Safety Data Sheet (also referred to as Safety Data Sheet or SDS)
MSS – Management Systems Standards
MSW – Municipal Solid Waste
M&T – Monitoring and Targeting
NGO – Non-governmental Organisation
NIR – Near-infrared (NIR)
OCS – Operation Clean Sweep®
OECD – Organisation for Economic Cooperation and Development
OH&SMS – Occupational Health and Safety System
OPI – Operational Performance Indicator
PAF – Prevention, Appraisal and Failure
PDS – Product Design Specification
PC – Polycarbonate
PCF – Product Carbon Footprint
PCM – Post-Consumer Material
PCR – Post-Consumer Regrind
PCW – Post-Consumer Waste
PDS – Product Design Specification
PE – Polyethylene
PEF – Polyethylene Furanoate
PE-HD – Polyethylene – High Density
PET – Polyethylene Terephthalate
PHA – Polyhydroxyalkanoate
PIR – Post-Industrial Regrind
PLA – Polylactic Acid
PM – Permanent Magnet
POP – Persistent Organic Pollutant
PP – Polypropylene
PPE – Personal Protective Equipment
PRN – Packaging Recovery Note
PS – Polystyrene
PU – Polyurethane
PVC-U – Unplasticised Polyvinyl Chloride
QFD – Quality Function Deployment
REACH – Registration, Evaluation, Authorisation and restriction of Chemicals

RoHS – Restriction of Hazardous Substances

RPN – Risk Priority Number

SME – Small and Medium Enterprise

SPC – Statistical process control

SWOT – Strengths, Weaknesses, Opportunities and Threats

T_g – Glass Transition Temperature

T_m – Melting temperature

TCE – Trichloroethylene

TDS – Total Dissolved Solids

TEQ – Toxic Equivalent Quantity

TPM – Total Productive Maintenance

T&D – Transport and Distribution

UN GC – UN Global Compact

UN SDG – United Nations Sustainable Development Goal

UV – Ultraviolet light

VE – Value Engineering

VI – Viscosity Index

VGP – Vulgar General Public (as used by my colleague Iain Brown).

VOC – Volatile Organic Compound

VSD – Variable Speed Drive

WBCSD – World Business Council for Sustainable Development

WIP – Work in Progress

WRI – World Resources Institute

WTE – Waste to Energy

WTT – Well-to-Tank

Tranquillity Leading to Insight

Exploration of Buddhist Meditation Practices

Chulan Sampathge

authorHOUSE®

AuthorHouse™ UK
1663 Liberty Drive
Bloomington, IN 47403 USA
www.authorhouse.co.uk
Phone: UK TFN: 0800 0148641 (Toll Free inside the UK)
        UK Local: (02) 0369 56322 (+44 20 3695 6322 from outside the UK)

Published by AuthorHouse 12/19/2022

ISBN: 978-1-6655-9820-0 (sc)
ISBN: 978-1-6655-9819-4 (e)

SYNOPSIS

Distilled into a concise and comprehensive twenty-two chapters in connection with a range of similar texts, *Tranquillity Leading to Insight* reveals how to extract the pure gold of consciousnesses as taught by the enlightened Buddha. Grounded in profound scholarship and philosophical psychology, the book brings the ancient teaching to the modern world. Delve deep into the mind, which explains how to overcome obstructions, stress, and hindrances and how to remove defilements to find true happiness and peace of mind. Written in readable language, giving examples and similes, the book takes you back to the historical time of the Buddha and the subsequent cultural awakening across the South East of Asia after his enlightenment.

PREFACE

Tranquillity Leading to Insight is intended to be read and studied by those who want to learn more about Buddhism and meditation practices. It contains extracts of the Abhidhamma philosophy of the Theravada School of Buddhism, focusing on the theoretical understanding and practical application of the ultimate truth, with emphasis on its relevance for meditation practice as taught by the Buddha to his analytically gifted disciples. The main theme of the discussion is the transformation of consciousness and how to purify the consciousness in search of liberation and of the knowledge that has been passed down in unbroken continuity from a historical perspective.

I was inspired by the book *Abhidhamma Studies: Researches in Buddhist Psychology*, written by Venerable Nyanaponika Thera, a German monk. The book was published in 1965 by the Buddhist Publication Society (BPS), Kandy, Sri Lanka. This book was the starting point for my further search into the subject of Abhidhamma. Since my youth I have been inclined towards this subject and attempted to learn it by myself by gathering materials from various sources, including publications of the Burmese master Venerable Ledi Saydow, who contributed much of the earlier literature on this subject.

The groundwork for *Tranquillity Leading to Insight* started in 1994, when I was practising insight meditation (vipassana) at the Buddhapadipa Temple in Wimbledon, London. I was practising mostly under the guidance of Venerable Amara Thera, the abbot and meditation master of the temple, who formally introduced me to Abhidhamma. In parallel to meditation practice, Venerable Amara Thera conducted very rich Dhamma discussions, including questions and answers, and offered residential retreats for trainees. He also kindly offered structured classroom lessons on this profound philosophical subject over a period of three and half years. The structure of his teaching method was well balanced, giving equal weight to both theory and practice.

Since 2004 I have been consistently teaching insight meditation at the Buddhapadipa Temple and have followed in the master's footsteps by conducting these meditation classes along with Dhamma discussions. To enhance my own understanding of the Dhamma and for the benefit of other meditation students, I began to read the Visuddhimagga (Path of Purification) and the Sutta Pitaka (discourses on the Nikayas) and drilled deep down into the rich mine of the Buddha's teachings. I had previous experience with practising tranquillity (samatha) meditation, and in my search to comprehend the Buddha's enlightenment, I found that tranquillity and insight meditation converged into one coherent system of practice.

The essence of the Buddha's doctrinal teaching is formulated in the Four Noble Truths. During his long ministry of teaching, the Buddha explained the doctrine to many in a form of dialogues and used eloquent similes to explain the complex dynamic phenomena that are documented in the Thripitaka—three vast collections of teachings known as the Pali Canon. The Fourth Noble Truth, which is the Noble Eightfold Path of liberation embedded in discipline, simplifies the doctrine, translating the abstract formulas of the doctrine into a practical method to unfold the Noble Truths. Abhidhamma literature, the third collection, zooms into the Four Noble Truths and the Noble Eightfold Path. It provides a valuable analytical view of life to glean the truth, giving practical guidelines for one who is searching for the truth and for the liberation from mistakenly grasped false notions of how to purify the mind and to see things clearly as they are.

Tranquillity Leading to Insight emerged unbidden in my enquiry into the philosophical and psychological teachings of the Buddha. I should like to dedicate this book to Venerable Amara Thera as a mark of gratitude for the great kindness and wisdom he has shown to me over the years. His guidance has made my journey through Buddhism thoughtful and rewarding.

In this task also, I am thankful to Suraya Dunsford and Danyal Dunsford for their valuable suggestions, along with their proofreading, editing, and helping me prepare the final manuscript. I would also like thank the Buddhist Publication Society, Kandy Sri Lanka, for giving me kind permission to use their materials for reference.

I would like to express my special thanks and gratitude to the Buddhapadipa Temple, Wimbledon, London, for kindly giving me permission to use their mural painting images for the cover design.

Photography by Alexandra Kovacs, AK.Foto@outlook.com

Chulan Sampathge
United Kingdom
December 2022

INTRODUCTION

The purpose of *Tranquillity Leading to Insight* is to discuss meditative absorption in relation to cognitive processes of consciousness as a gateway to final deliverance and reaching supreme happiness. Meditation absorption, also known as trance, is a state of mind that enables worldly people or trainees in the practice of meditation to acquire deep mental stability that opens a path to insight wisdom. The great majority of compilers and commenters who narrated the ancient Pali texts of Buddhism follow a unifying theme that a correctly performed meditation practice can result in the attainment of sainthood. The texts show that there is a change of lineage, from worldly to noble (i.e. puthujjana to Ariyan), in those who pursue liberation by following the path propounded by the Buddha. The Buddha gave these instructions in many geographical locations, a majority of which were found in the greater Ganges basin of north-eastern India, where since thousands of his followers have successfully become liberated through understanding of the Four Noble Truths.

This area was the centre of several empires, notably those of the Mauryan and Gupta dynasties, between the sixth century BCE and the eighth century CE. The language used by the Buddha is said to have been Magadha, a dialect of Pali. The words and terms used by the Buddha have etymological meaning, and the narrative is reasonably consistent with that of several early Sanskrit equivalents. His basic technical vocabulary has been adapted with stemmatic principles by the Thai, Sinhala, and Burmese languages, cultures that are much influenced by Buddhism. The thematic scheme of the teachings has been translated into English and other European languages directly from Pali by the Pali Text Society of the United Kingdom and by many other scholars. The Pali terms are inserted throughout this work for verification of their meanings.

The writer explores the Buddha's enlightenment and the meditation practices he taught: the way of tranquillity (samatha) leading to insight (vipassana), and its reverse, that of insight leading to tranquillity. The higher philosophical psychology of the mind and the process of mental training towards enlightenment, as taught by the Buddha, has been compiled into a set of teachings by his enlightened disciples and learned monks. Over a period of time, this branch of Buddhism was further developed by scholarly monks who made commentaries and sub-commentaries, now known as the higher teachings of Buddhism, the Abhidhamma. Most of the material used in *Tranquillity Leading to Insight* is about this subject and gives an overview and detailed explanations of the nature of the mind and how it can be trained to liberate a person from suffering through self-discipline.

Meditation practice is essentially for intelligent people with a keen interest in questioning and discerning what is hidden. The Buddha's invitation to those with enquiring minds is to come and see. Meditation towards enlightenment is a technique of investigation into one's own mind. There are two main types of meditation practices in Buddhism: samatha and vipassana *(tranquility and insight)*. The latter was a long-lost technique rediscovered by an ascetic, Siddhattha, which enabled him to achieve the supreme enlightenment, *samma sambodhi*. By this endeavour he came to be the fourth Buddha of our aeon, or Bhadra Kalpa (a timescale of an indefinite but very long period).

According to samatha, there are two types of concentration, known as approaching and access, both leading to the attainment of *jhana* (mental absorption and a state of trance). This concentration is foundational to vipassana meditation. Attaining jhana temporarily purifies the consciousness from all known defilements. Vipassana is a technique used to develop momentary concentration; hence, it helps one to investigate the ever-changing mental phenomena from moment to moment and thereby helps one gain insight into the true nature of life. There are many levels of jhana in samatha, and there are sixteen vipassana knowledges. High levels of concentration are required in samatha practices to stabilise the mind and prepare for vipassana in order to gain knowledge. A student may acquire intuitive wisdom through his or her own direct vision in meditation, but neither by thinking nor reading about it. Realisation of the truth occurs sequentially, stage by stage, while one lets go of defilements, which leads to the purification of the mind. However, there are occasions where defilements of the mind manifest in meditation practice, for example light and illuminations appearing. These experiences might be very pleasant, but clinging to them could hinder progress by causing the person experiencing them to mistakenly believe that his or her mind has now been liberated. Therefore, it is recommended to practise under an experienced master.

Purity of the mind gives rise to attainment of insightful knowledge, and this opens the pathway to deliverance. Insight wisdom coexists with insight meditation, and by virtue of moral purity and concentration, the meditator will realise the Four Noble Truths. The Buddha vividly explained that "such and such is the morality, such and such is the concentration, such and such is wisdom, great are the fruits of wisdom" (Mahaparinibbana Sutta).

The meditation techniques taught by the Buddha are documented in the suttas and commentaries known as the Mahasatipatthana Sutta—the Four Foundations of Mindfulness. In his advice, the Buddha proclaimed that there is only one path to deliverance, with these Four Foundations being the only path. In this declaration the Buddha rejected the other practices that prevailed in contemporary India and introduced the Noble Eightfold Path as the Middle Path for liberation (*ekayano maggo*— Madhyma Pratipada).

The Noble Eightfold Path can be divided into three sections. The latter three factors of the path are in regard to meditative practice. An intense practice of meditation is required to observe the true nature of both mind and matter, and their combinations. These are the complex phenomena of mental and physical events that, as life experiences, present themselves and manifest at any given time. Our usual observation by means of perceptions, concepts, rational thinking, or intellect could be incorrect because of errors in perception. For this reason, in the Four Foundations of Mindfulness, perceptions are not included as a means of understanding. Instead, one must rely on a method of direct observation and direct realisation of truth.

The Buddha understood the dynamic reality of life and condensed it into its two components: elements and processes. The Buddha was the first teacher to succeed in capturing a thought moment. He was able to see a thought's conditions and what it was made of. From this point the Buddha began his journey to solve the problems of the wheel of samsara (cycles of birth and death). For six years he undertook this vigorous research, known as "Ariya Pariyeshana" (the Noble Experiments), then on the night of the full moon of Vesak (the month of May), the Buddha attained all-knowing wisdom, that is enlightenment.

In the Mahasatipattanhana Sutta, the Buddha teaches how to understand experienced reality as it presents itself in introspective meditation. In his analysis, ultimate reality consists of a multiplicity of elementary constituents called Dhammas. These fall into two main categories of Dhamma: conditioned (*sankatha*) and unconditioned (*asankatha*). A detailed study of the Dhammas and their intricate relations are given in the books of Abhidhamma, the higher philosophical and psychological teachings of the Buddha. These are discussed under four main categories of ultimate reality: consciousness, mental factor, matter, and nibbana (respectively *citta*, *cetasika*, *rupa*, and nibbana). The first three classes/categories are called conditioned Dhamma because they are the momentary mental and material phenomena that constitute the process of experience. An ontological view of conditioned Dhamma has the power to lead one to full liberation. This liberation allows humankind escape from the bondage of mistakenly grasped conceptual constructs of the self and the world around them and enables them to see things "as they really are". This refined knowledge and successive stages of purity would progress the mind to states of meditative absorption (jhana). Here the term 'jhana' (*pali; jhāna*) is often translated as "absorption" because, as the process unfolds, one's attention is drawn more naturally, easily, and deeply into the calm inner reaches of the mind. This would be then applied intermittently in reverse order of insight leading to tranquillity in successive stages until full liberation is achieved.

It was through this insight and wisdom born of meditation that the Buddha attained enlightenment. Jhanas are the keys to opening the supramundane path and its fruits (*magga* and *phala*). Jhana purifies the mind temporarily and provides a setting and grounds for intuitive wisdom. Pure observation of both mental and physical phenomena gives rise to right knowledge and right thought. These are the first two factors of the Noble Eightfold Path. The middle three factors are concerned with one's moral foundation, a prerequisite to support one's meditation. By virtue of following this path, the Buddha's disciples also attained enlightenment.

The whole of the Buddha's teaching stems from three main doctrines, namely impermanence, suffering (unsatisfactoriness), and nonself. The result (phala) of walking the Noble Path is the realisation of the doctrine, changing the lineage of a worldly person's or trainee's mind to that of the supramundane stage. Those who reach this stage are emancipated from samsara and attain nibbana if so wished. Nibbana is the highest goal in Buddhist cosmology. It also known as the supreme happiness, which is an absolute and a singularity. An uncompounded element hence also defines parramatta Dhamma, meaning the undifferentiated ultimate reality of unconditional truth.

Mindfulness is a beautiful mental state. It signifies presence of mind and attentiveness to the present moment. The Buddha regarded mindfulness as the supreme function of the mind, which manifests as a guardian; without mindfulness, life is in vain and has no protection. Mindfulness is a gift we are born with which helps us enjoy life in a more passive way. Being mindful has lots of benefits to our physical and mental well-being. Mindfulness is a factor associated with consciousness, the process which relates to our experience of sensations (sight, sound, smell, taste, and touch). To see this process clearly, one must apply mindfulness and awareness to the present moment as it happens. By being mindful and living in the present moment, it is easier to experience the simpler things that we otherwise might not notice, such as the feeling of the feet touching the ground as one walks. The application of mindfulness together with compassion has been recognised by psychologists and academics as a useful therapy to improve awareness and to help us better understand ourselves. Improving momentary awareness will allow a person to see the present moment clearly, thereby allowing him or her to take steps to develop mindfulness in his or her daily life.

Consciousness is a series of moments which occur in succession. All life experiences happen according to this principle of cognition and the five senses. Buddhism recognises mind as the sixth sense. It follows the same principle of cognition, forming feelings, thoughts, emotions, and other mental states. With improved mindfulness and regular practice, one can directly know what is going on inside one's mind from moment to moment without getting entangled. To be disentangled is to observe the changes of the mind without reacting to thoughts or emotions and without identifying with one's behavioural patterns. Psychologists and clinicians have questioned and tested "What is mindfulness?" and "How is it clinically applicable?" They have come to redefine its meaning in an attempt to elucidate its multifaceted nature as both a meditation practice and a fundamental way of treating patients with various disorders. They have explored applications of mindfulness in psychotherapy and counselling and as a discipline; this has been fostered as a tool for self-help for patients and professionals alike. The method and approach have been altered to suit a clinical environment. The core concepts and practices of mindfulness have been studied in regard to common therapeutic factors and therapeutic outcomes. Mindfulness offers a new way of envisioning healthcare and mental health, expanding the field to include well-being within clinical practice. Mindfulness has now become a pivotal therapeutic technique which has been successfully integrated into cognitive behavioural therapy (CBT) and dialectical behaviour therapy (DBT).

Learn to be mindful by living in the now and by clinging to nothing. Then you can free yourself from unnecessary pain by observing the pattern of nagging thoughts and cultivating a non-judgemental attitude. Regular daily practice of this method will come to reveal the impersonal nature of the mental process. Understanding that one is devoid of a permanent self will help to reduce the antagonistic attitude of the ego. This can lead to a change in behaviour without further clinical intervention. The presence or absence of mindfulness is the determinant factor of healthy or unhealthy behaviour respectively. It is fundamental in behavioural psychology to understand the conditions that contribute to human behaviour. Mental factors, although not directly observable and measurable in clinical settings, are the underpinning conceptual framework of the analysis of all types of human behaviour. This includes child psychology and the development of personal traits from a young age. The human life cycle consists of growth in phases. Each phase is characterised by a distinct set of physical, psychological, and behavioural features. The Buddha saw many cycles of life and identified suffering as the common denominator to all stages of life. By responding to this suffering with wisdom, one can successfully mitigate the problems of life.

The devastating effects of many social issues are still hidden. The long-term psychological effects of violence and social disruption take a big toll on the average family in any society. As part of the suffering of victims and their families, the psychological impact on children is going unnoticed or unheard. The public consensus is that something has to be done to change the current notion and paradigm of crime and punishment. Behavioural psychology can provide this wider scope of study and application. Its usefulness can make a profound, positive, and decisive impact on a range of burning social issues, from supporting millions of victims of crime and domestic abuse to helping improve responses in setting industrial health and safety standards to prevent work-related accidents at the earliest opportunity. This must all be done with a view to alleviating concealed suffering and avoiding man-made catastrophes.

Tranquillity Leading to Insight is well grounded in the knowledge of Buddhist higher-philosophical psychology and seeks to help professionals in experimental research to understand the psychological ramifications and neural mechanisms relevant to human behaviour. It addresses the root causes of suffering from a different perspective. A good medical doctor, to truly understand the physiology of the human body, must have an appreciation of the mechanisms of cellular biology to perform his or her treatment process well; similarly, this book provides crucial knowledge that explains the functions of the mind so one might comprehend the foundation and driving force of the mind. Such literature like this has not been published before; therefore, it is presented for the advantage of researchers and clinicians of psychology. The intricate knowledge and analytical structure of the entire edifice of the mind, as explained and narrated in this way, could be useful in clinical practice, allowing clinicians to quickly and accurately interpret symptoms when treating patients so as to appropriately translate emerging test results and findings into evidenced-based treatment processes to benefit the mental health and well-being of many vulnerable groups suffering from episodes of depression, anxiety, worry, and nagging headaches. These theoretical methods of analysis, synthesis, and synchrony can include critical information about the functioning of various mental factors and their relation to mind–body systems (and, by extension, to the unit of mind that composes them). It is important to know what exactly is happening to devise the best course of treatment and functional efficiency of the mind. It would be very difficult to understand a patient's cognitive behaviour without knowing enough about the mental functions that are being tested. *Tranquillity Leading to Insight* also includes higher scientific education to inspire and enrich theoretical rigour so as to enable clinicians to view illnesses and the meaning of life from an intellectually fascinating and philosophical standpoint. Currently there is no literature like this available to explain the depths of the human mind. Professor Mark Williams of Clinical Psychology and Honorary Senior Research Fellow at the University of Oxford has stated that such in-depth knowledge is certainly needed. *Tranquillity Leading to Insight* aims to narrow the knowledge gap of psychology by disseminating a higher Buddhist philosophy. It will contribute to understanding suffering better and to advancement in the field of treatment to end suffering.

CONTENTS

AUTHOR'S NOTE

RELEVANCE OF BUDDHISM AND ABHIDHAMMA STUDIES

The fundamental error of human thought is deeply engraved in ignorance, which is the root cause of suffering. Not knowing this fact can sadly add more to injury and inflict more suffering. If anyone wishes to improve the quality of thought to become illuminated and make a fruitful contribution to the society in which he or she chooses to live, it can be done only through careful reflection to understand the human experience.

In the twenty-first century, we still witness disturbing incidents of social injustice, such as those in the middle of the Covid-19 pandemic in United States of America, the land of democracy, liberty, and free speech. On 1 June 2020, *BBC News* reported that violence erupted in cities across the United States on the sixth night of protests sparked by the death of African American George Floyd while in police custody. Dozens of cities imposed curfews, but many people ignored them, leading to stand-offs and clashes. Riot police faced off with protesters in New York, Chicago, Philadelphia, and Los Angeles, firing tear gas and pepper bullets to try to disperse crowds. Police vehicles were set on fire and shops were looted in several cities. The country experienced the most widespread racial turbulence and civil unrest since the backlash from the assassination of Martin Luther King Jr. in 1968.

The outpouring of anger began on Tuesday, 26 May 2020, after a video showed Mr Floyd being arrested in Minneapolis, Minnesota, and a white police officer continuing to kneel on his neck even after he pleaded that he could not breathe and fell unconscious.

The Floyd case reignited deep-seated anger over police killings of black Americans and racism. For many, the outrage also reflected years of frustration over socio-economic inequality and discrimination, and not only in Minneapolis itself, where George Floyd died.

This incident was given large media coverage and drew attention away from other Covid-19-related deaths, economic hardship, and poverty in poorer countries that are struggling to cope. Despite lockdown measures and social distancing guidelines still in place, the protest spread to the city of London, where thousands of young people gathered and marched in the streets. Their concerns were not so much about the killing of Mr Floyd as they were projected from their own uncertainties and fears of racism and discrimination.

This incident was a stark reminder of the importance and urgency with which we all must address these inequalities. When our society is broken for some, it is broken for all of us. And where we have the power, the influence, the tools, and the opportunity to be a part of fixing this, we must stand together and must act together.

Covid-19 arrived as a health crisis, soon evolving into an economic and a social crisis. While it impacts everyone, we must be open in acknowledging that it didn't affect everyone in the same way. In every instance, black and coloured people in the United States and multicultural cities have been disproportionally impacted, adding to the inequalities and injustices that have already existed for too long. This is a reminder that we still have a long way to go to make our communities and the world more equitable, more inclusive, and better for all. Social unrest is always a breeding ground for opportunists looking to loot, riot, and commit acts of anarchy, but this is in no way a justification for overlooking the real cause, ignoring the moral justification of the righteous cries of peaceful protesters, and allowing the situation to escalate.

This sequence of events highlights a pattern of behaviour which reminds us that history repeats itself. If this weren't true, then we would have learned lessons from past events and what we had done to prevent such calamity. We must admit that we do not know all the answers to rectifying many decades of racism and inequality. But the pattern clearly indicates a deterioration and diminishment of human values.

In Buddhism, we believe that every human being has the potential to be enlightened, and we advocate peace, nonviolence, respect for life, and individual acceptance of social responsibility. The Buddha rejected the social injustice prevalent in India at the time and believed that no one became upper class or lower class by birth but only by one's own actions in choosing one's social position.

A peaceful protest can easily turn into violence depending on how each side of the conflict responds to the other. Leadership style could make a big difference in resolving a dispute. If one chooses to meet aggression with equal or more powerful aggression, then major destruction and further loss of life is inevitable. A crisis, if mishandled, could be a waste or, if wisely responded to, can lead to the reaching of a lasting peaceful settlement.

For the past few years, strengthening diversity, equality, respect, and inclusion have been hot topics. There is evidence of this concept of social engagement in the way communities live and operate. However, it has not yet yielded results and no visible changes have happened. It appears that aggressive responses have resulted in the tearing apart of the social fabric of the USA.

When Russia invaded Ukraine in February 2022, under the political phrase 'Special Operation', many observers expected that Russia's military would make a quick win on their mission: to capture the country's capital, Kyiv, depose its democratically elected government and restore Ukraine to Moscow's control. But nearly six months later, after Russian forces failed to take Kyiv, the war has evolved into one of attrition, grinding on with no end to heavy losses on both sides. Since the war started, staggeringly high rate of casualties was reported in Ukraine almost daily, in addition, sadly millions of Ukrainians fled the country. Since the armed conflict began in the region, fighting has caused nearly three thousand civilian deaths and internally displaced more than seven million people, according to the United Nations. This war, displacement of civilians and loss of lives would have been avoided should Russia chose a diplomatic path to resolve the conflict and mitigated the perceived threat maintaining peaceful relationships between nations. It only added unnecessary suffering and increased uncertainty not only in the warzone but extended to the rest of the world economy by sending shockwaves because it severely interrupted the global energy markets. Ukraine and Russia are major producers of wheat, corn and barley, the conflict has exacerbated a global food crisis too

(The New York Times, August 2022). The Buddhist doctrine of "nothing arises singularly" suggests that finding a solution to a crisis requires deeper analysis of causes and their effects, in particularly, a conflict situation needs broader understanding of synthesis and synchrony of dynamic relations.

Studies shows that majority of soldiers in battlefield suffer from Post-Traumatic Stress Disorder: the unending echo of battle etched in the brain affects 15% of soldiers. It can destroy families; it can leave its sufferers unable to work and addicted to substance misuse. It is difficult to understand what they endure. There were issues of soldier suicide. Statistics from past and present wars tell the sad story of the magnitude of this problem. Unlike in the historical time, much of the suffering in the modern world is complex phenomenon, aggravated by human-caused disasters. All those events and incidents have an element of human intent, negligence, or error involving a failure of a man-made systems. Correction of those errors need ethical, pragmatic and wise interventions.

We also know from our experience in history the futility and the devastating result of war. As an alternative, we can bring about peace through dialogue, which we must engage in using all available talent. We must also ignite human potential to sustainably grow prosperity around the world, not flames of fire. It is the right thing to do, and it is what we must do to prosper.

We know that we should never stop working towards creating better and more equal societies around the world where everyone feels welcome, listened to and seen, and able to unleash their human potential. We will never shift from our purpose, namely that meaningful and sustainable economies be given the power to change the world for better. We should always take seriously our responsibility and our commitment to reach, assess, train, and provide opportunities for people of all communities to acquire skills and knowledge, because we know that work, education, skills, and aspiration are critical parts of community cohesion and prosperity.

These are things that Buddhism can offer to the world at times of difficulty. Buddhism is a light in the darkness. It gives us a choice: we can choose to walk in the darkness and fall into the same pit-hole again and again, or we can use a torch and avoid the pit-holes. The Northern School of Buddhism says that the Buddha met with eighty-four thousand different people and gave them advice to resolve their issues concerning life and liberation. The Theravada School of Buddhism has a collection of Buddha's teachings and has arranged these into three divisions which are the source of their knowledge.

The Abhidhamma, the third great division of early Buddhist teaching, expounds a valuable system of philosophical psychology ascribed to the unimpeded insight of the omniscient Buddha, the way he explained the nature of the mind. His learned disciples organised the master's teaching explaining the entire spectrum of human consciousness, including how it is defiled and how to purify it to see things clearly as one seeks to escape from bondage and move towards liberation. The Abhidhamma systematically maps out, with remarkable rigour and precision, the inner landscape of the mind and points to the escape route, the Noble Eightfold Path.

According to Sigmund Freud, "The inclination to aggression is an original self-subsisting instinctual disposition." The human mind is a complex phenomenon. Indignation is an accepted social norm of all popular cultures; however, generous, caring impulses, while they may exist, are generally weak, fleeting, and unstable in human beings. But dig a little beneath the surface and you will find a ferocious and persistently profoundly selfish person. This insight into the behaviour of an untrained person is, in Buddhism, simply termed as "ignorance".

The Abhidhamma texts and manuals provide priceless and timeless insight into the entire substratum of the human mind. Bhikkhu Bodhi tells us, "However, is that familiarity with the manuals is not sufficient. Illuminating and fruitful lines of thought lie hidden in the original texts, and it is only by unearthing these through deep inquiry and careful reflection that the riches of the Abhidhamma can be extracted and made available, not to Buddhist studies alone, but to all contemporary attempts to understand the nature of human experience. The road is open but not pursued."

Nyanaponika Thera adds, "The inclusion of the path factors in the analysis of wholesome consciousness means the raising of the spiritual eye from the narrow confines and limited purposes of everyday consciousness to the horizon of the ideal. It means that, in the midst of life's dense jungle, amid its labyrinths and blind alleys, the glorious freedom of a Way is open."

We extend our goodwill and invite you to come and see. The Abhidhamma explains the unexplored terrain of the mind. From these studies you may gain knowledge and insight, not only useful for individual liberation but also applicable in the wider context of the development of psychology, criminology, and behavioural science. By improving your critical thinking, analytical, and communication skills, you might be able to make a compelling impact on policymakers and influence decision makers both inside and outside the socioeconomic system to shape up society and individuals' lives.

CHAPTER 1

THE BUDDHIST DOCTRINES

THE BUDDHA WAS BORN IN 563 BCE. The conventional dates for Buddha's life are around 566–486 BCE. He was an Indian prince named Siddhattha Gotama, his family were of noble royalty and lived in a small kingdom near the foothills of the Himalayas just near the border of modern-day Nepal. In his young life as a prince, he was educated at a well-known university named Taxila. Geographically, this location corresponds to the Achaemenid territories in the north-western reaches of the ancient Indus Valley. During this time, he was also taught military warfare in preparation for his future kingship. He was married and had a baby son when he left his family's palace in search of liberation.

The chronology of the Buddha's life proceeded to narrate the five hundred years after his death, and biography did not exist as a literary genre in ancient India until Western explorers and archaeologists discovered Emperor Ashoka's monuments which were built in 268 BCE. Ashoka, also known as Ashoka the Great of the Mauryan dynasty, who ruled almost all the Indian subcontinent from about 268 to 232 BCE, was the grandson of the founder of the Maurya dynasty, Chandragupta Maurya. Ashoka later laid off arms and demilitarised the empire, turned to religion, sent out his son and daughter to neighbouring countries on diplomatic missions, and promoted the spread of Buddhism across ancient Asia.

The term *Buddha* means "one who has been enlightened". Some writers use the term Awakened One, and the Northern School uses the term Shakyamuni. Buddha's followers addressed him as "Blessed One" or

"Exalted One". One of the doctrines unique to Buddhism is that of nonself. In Pali, this is known as *anatta*. The etymological root of *anatta* equates to the negation of atman *(soul)*, the permanent self or ego. Because of this, in his teaching the Buddha did not refer to himself as "I", "me", or "myself"; instead, he used the word *Tathagata*. The meaning of this word is "one who does not resist but incessantly moves *with* the moving world". Herein, the writer will use the terms *Buddha* and *Tathagata* synonymously.

The ultimate realities taught by the Tathagata were fourfold: consciousness, mental factors, matter, and nibbana. There is no equivalent English word for the Pali word *nibbana*; therefore, to maintain the authentic meaning, all writers of Buddhism use the word *nibbana* untranslated. In Sanskrit it is known as nirvana. The root word was derived from the two words *ni* and *vana*. *Ni* is a negation, as in "free from" or "in absence of"; *vana* means "entanglement", a figurative expression of craving. As long as one is entangled by craving, one remains bound in samsara, which is also known as the wheel of life—the cycle of birth, decay, and death. Nirvana could also be interpreted as the end of craving. The Pali word *nibbana* was derived from its root verb *nibbiti*, meaning "to be extinguished", thus signifying the extinguishment of worldly "fires" or defilements. There is an unconditioned absolute element of nibbana relating to complete transcendence from all other mundane realities. The practice of meditative mental absorption facilitates this transcendence into the supramundane world by opening the threefold gateway to nibbana, voidness *(sunnata)*, sign-lessness *(animitta)*, and desire-lessness *(appanihita)*.

One of the masterpieces of the Tathagata's teaching is that of dependent origination *(paticca samuppada)*. Dependent origination simplifies the complex phenomenon of samsara. It was in the context of dependent origination that the Buddha explained the conditions through which beings may realise their place in this "wheel of life"[1]. There are eleven factors connected to one another that make up the wheel. Each individual factor supports the conditions that give rise to the next in sequential order. Out of compassion, the Buddha shows us the path that leads to the cessation of suffering and to the ultimate reality of nibbana.

The heart of the Buddha's teaching is founded on his first sermon. This was given to his former colleagues in Issipathana, a place near the

---

[1]   Damien Keown. (1996). Buddhism a Very Short Introduction. Oxford University Press.

confluence of the River Ganges in north-east India, now known as Saranath. After attaining supreme enlightenment, the Buddha was hesitant to teach because he doubted whether anyone could understand the deep and profound Dhamma. According to the legend, when the chief god Sakka (*a deity, the ruler of heaven according to Buddhist cosmology*) saw that the Buddha was not teaching, he appeared before the Buddha and invited him to teach. The first sermon is known as the Dhamma Chakka Pavattana Sutta (Turning the Wheel of Dhamma). It is formulated upon three main characteristics of all phenomena: suffering, impermanence, and nonself. On the full moon of Asala (the month of July), the Tathagata revealed the Four Noble Truths to the world. He had the profound ability to make clear the underpinning laws and principles of Dhamma and was able to detail all the intricate relationships that constitute a single phenomenon. These principles are not mere elaborations or speculations of contemporary knowledge but are irrefutable original thought as perceived by an enlightened mind. The First Noble Truth says that all phenomena are impermanent and are therefore subject to suffering. *Dukkha* (suffering) is central to Buddhism; it is very broad in its definition and includes the unsatisfactoriness of experience. The Second Noble Truth says that there is a cause for such suffering. The Third Noble Truth says that the suffering can be overcome and ended. The Fourth Noble Truth shows that there is a path to take leading to the cessation of suffering. By virtue of hearing the Buddha's first sermon, his former colleague Konanda attained the first stage of enlightenment and became his first disciple.

The Fourth Noble Truth exemplifies the greatness of the Buddha's teachings, the unsurpassed beauty of understanding that leads to escaping the wheel of samsara. The first factor of the wheel is ignorance. Born out of this ignorance, beings are trapped in samsara and go through suffering. There are three basic subsets of ignorance: greed, hatred, and delusion. These three factors are the absolute causes of suffering and the very reason why beings are bound to samsara. There are eight factors in the path leading to the cessation of suffering; this is known as the Noble Eightfold Path. Anyone who can see this truth will accept the Four Noble Truths and understand Dhamma as taught by the Buddha.

During his forty-five-year-long ministry of teaching, the Buddha made the Dhamma known, recorded in a mammoth collection of eighty-four thousand discourses called the Pali Canon. There is speculation as to whether Buddhism is a religion, philosophy, or psychology; irrespective of these definitions, the Buddha taught Dhamma. Knowledge of the Dhamma may be obtained by reading, listening to, or believing the Buddha's expositions. However, this alone cannot lead to full comprehension or understanding. There are very special circumstances, as was the case for Konanda, but he had equal ascetic practice of meditation close to the Buddha and was ready. The level of understanding one gains through reading or listening is that of "knowing accordingly" or, in Pali, *anubodha*. However, one is required to penetrate experienced phenomena to gain the absolute insight knowledge of Dhamma. This level of knowledge is known as penetrative knowledge or, in Pali, *pativeda*.

In one of his sermons, the Buddha was advising a group of people in the city of Kalama whom were perplexed by many contradictory teachings. He told them not to believe anything just because it was written in a scripture and passed on from generation to generation, not to believe based on mere hearsay or out of respect for the teacher, and not to believe just because something corresponded to a logical argument. He said, "Believe only if you can comprehend it by your own mind." Hence, the only way to correctly understand Dhamma is to practise the Noble Eightfold Path.

The aim of the Noble Eightfold Path is to make the practitioner discover the truth by himself or herself and to see the Dhamma. This requires *vinaya*, meaning the discipline needed to understand the truth. It was true at the time of the Buddha and is also true in our present time that many people are seeking the truth, but the Buddha only shows us the way; he cannot travel it for us. For this reason, the Buddha maintained a tradition of teaching that invites one to come and see the truth. Because of this, the Buddha was known as *ehipassiko*, one who invites others to come and see.

There are two kinds of realities. The mind's normal, untrained level of consciousness experiencing conventional reality is *sammuti*. This holds reference to ordinary conceptual thought (*pannatti*) which is interpreted according to the conventional mode of expression (*vohara*). In this gross manifestation of phenomena, we refer to entities as living beings—people, men, women, and animals. It also includes apparently stable persisting objects such as mountains, rivers, and trees. Our perceptions are based on opinions placed upon these conceptual observations. Most of the time we like to see things the way we want to see them, colouring our perceptions according to our preferences. Our perceptions constitute our picture of the world without us having proper awareness of things as they are. However, skilful observation and greater awareness would prove that those notions possess no ultimate validity and are instead products of our own mental constructs (*parikappana*). With closer observation, these modes of conception can be further reduced to their respective components and processes. Perfect and absolute insight knowledge must be gleaned from this conventional truth. One who discerns the truth in this way and culminates insightful wisdom will grasp Dhamma and abolish ignorance. By practising the Noble Eightfold Path, one can develop the penetrative knowledge required to become emancipated and escape the misery of samsara. There are sixteen insight knowledges to be gained. The first four are in regard to purification. The middle nine are the core insight knowledges or realities of existence. The next two are in regard to the path of knowledge and its fruits. The last one is knowledge arising from reflection on the path travelled.

CHAPTER 2

❖

THE NOBLE EIGHTFOLD PATH

THE NOBLE EIGHTFOLD PATH IS the fourth of the noble truths. It is the path that leads to the cessation of suffering. As the name implies, there are eight connected factors that constitute the path. These are not steps to be technically followed in sequential order but are intricately connected psychological factors that originate in the mind. They could be grouped into three main categories known as *sila* (morality), samadhi (concentration), and prajna[1] (wisdom). These can be understood as components of a path towards purifying one's own actions, developing concentration and wisdom. In the context of Buddhist culture, customarily morality is first encouraged as an entry point for training the mind. However, these three groups do not exist in isolation; they are supporting factors for one another. For example, there is no concentration without morality and there is no wisdom without concentration, hence these integrated factors need to be practised as a whole set. In *Tranquillity Leading to Insight*, the writer has stated this path first because its practical applications will help the reader to realise the first three noble truths without any speculative or subjective view. The usual order for grouping the path factors in the schema of classification is wisdom, morality, and concentration, starting with right view.

---

[1] *Prajna - direct insight into the truth taught by the Buddha, as a faculty required to attain enlightenment. Its Pali equivalent paññā, is a Buddhist term often translated as "wisdom", "intelligence", or "understanding". It is described in Buddhist commentaries as the understanding of the true nature of phenomena.* Buddhaghosa. (1991). The Path of Purification: BPS translated by Bhikkhu Nanamoli.

Right understanding (or view) and right thinking are the first two constituents of wisdom (prajna). Wisdom is the ability to think and act using knowledge, experience, understanding, common sense, and insight. Wisdom is associated with attributes such as unbiased judgement, compassion, experiential self-knowledge, self-transcendence, and nonattachment. Wisdom holds such virtues as ethics and benevolence. Once this factor is fully developed, it manifests in great personality attributes of compassion, loving-kindness, and joy in others' success. Right understanding and right thought will help to correct morality. There are three factors to morality (*sila*), which are right speech, right action, and right livelihood. One makes a living by one's speech and bodily actions, which actions should be regulated so that one does not break the code of morality, which is required to develop concentration. Concentration/meditation (samadhi) is composed of right effort, right mindfulness, and right concentration, and is the most important on the path because nibbana cannot be attained without meditation.

Although *right* is not defined, it implies that there is a wrong path. Choosing between right and wrong depends on the purity of knowledge and insight wisdom. These eight factors are not mere physical exercises but are broad mental faculties of the mind that need to be developed by diligent practice. The subsets of each factor need to be wisely applied in one's practice to purify all mental, verbal, and bodily actions, until perfection of all factors is reached. Perfection of wisdom will emerge from perfection of concentration.

Right View (Sammaditthi)

There are three kinds of exposition to right view. It could be argued that it is not possible to have a complete understanding of the three expositions at the beginning however, it is possible to begin with a right view of them. Right understanding (also *sammaditthi*) develops as one practises the path. This can be compared to climbing a mountain range: at the beginning you cannot see all the mountains, but the higher you climb, the more you can see, until it is possible to see the whole range. Right view can be interpreted as knowing the clear direction of the path, in other words, knowing where you are going as while trekking. It should not be interpreted as an opinion but as something to be seen. Hence, right view is a transparent or penetrative view of one's own mind and body. This journey to right understanding begins with saying, "I am now the product of my past kamma."

Kamma is the result of intentional action, which includes the volitional mental, verbal, and physical actions performed by all beings. Beings inherit their own kamma, it cannot be destroyed by fire or water, nor taken by thieves or robbers. The nature of these three types of kamma, mental, verbal, and physical, is that they have duality. They can be good and bad, so the effects are twofold according to the timing. Good and bad kamma have an effect in both the present and the future. Kamma is manifested as its results (*vipaka*); good kamma produce good results, and bad kamma produce bad results. Some people have a natural tendency to practise *dana*, *sila*, and *bhavana* (generosity, morality, and meditation), habits cultivated through their experience of samsara. For example, most of the people in Thailand, Sri Lanka, and Burma (Myanmar) practise principles of Buddhism. Those principles are closely interwoven into the entire fabric of their cultures, with the value systems of those societies being the value of Buddha Dhamma. The people in those societies live according to their kamma. The permutations of kamma have twelve different combinations when considering the timing and effect:

$$(\text{Mental/verbal/bodily}) \times (\text{good/bad}) \times (\text{ripening in} \\ \text{this existence or in future}) - (3 \times 2 \times 2 = 12)$$

Good and *bad* are relative terms. The Buddha used the terms *kusala* and *akusala*, meaning "moral" and "immoral" respectively. There are ten kinds of moral kamma and ten kinds of immoral kamma, hence the permutations of kamma increase into one hundred and twenty combinations. These calculations show only the broad categories of kamma. They do not include the mental factors or degree of intention associated with them, nor many other factors that would greatly complicate the possible effects of kamma. There are sixty kinds of kamma that yield results in one's future existence; kamma that ripens in future existence becomes incalculable and immeasurable in its results (vipaka). Accumulated kamma is carried forward to future lives with the rebirth of one's consciousness (*pati sandhi vinnana*). Any happiness or unhappiness one may experience directly corresponds to the kusala and akusala kamma.

Buddha explained that beings are wandering in endless samsara because of their kamma, but out of compassion he taught how to reduce accumulating bad kamma. All physical, verbal, and mental actions that are free from the ten kinds of immoral conduct are the way to reduce any accumulation of bad kamma. This is the foundation of morality (*sila*) and wisdom (prajna) on the path. On the other hand, by virtue of performing ten moral acts of kamma, one can advance in the path.

There is no external or supernatural agency that can reward or punish us, so holding on to such beliefs as mercy is wrong view (*micchaditthi*). The Buddha advised us to abandon such beliefs and rely on our own kamma, knowledge, and wisdom. Hence, Buddha declared, "I am the owner of my actions, heir to my actions, born of my actions, [and] related through my actions, and [I] live dependent on my actions. Whatever I do, for good or for evil, to that will I fall heir" (Sabbe satta kammasaka, kammadayada, kamma yoni, kammabandhu, kammapatisarana, yam kamman karissanti, kalyanam va papakam va tassadayada, bhavissanti).

Kamma could be further understood in terms of wholesome and unwholesome, leading to a state which respectively is or is not conducive to progress in life, thus manifesting as either comfort or suffering. Kamma is the inherited properties of all beings who are born according to their kamma, which will always accompany them. All three types of kamma, physical, verbal, and mental, both wholesome and unwholesome, are inherited by all beings throughout many world cycles, including those to come. Such a view is known as the right view.

The seed of all types of vegetation and trees carries all the qualities and features of the vegetation or tree. When planted, it will grow and form into such a plant or tree. It will then bear flowers and fruits according to the conditions of the environment and other circumstances. Similarly, kamma ripens its results in endless future world cycles when right conditions are met. This exemplifies the timing aspect of kamma. Such a view is known as the right view.

Extended knowledge of kamma and its operation forms the next level of exposition to right view. Tathagata explained ten kinds of right understanding in accordance to specific kamma and its modes of operation. An understanding of these modes helps one to avoid unconducive actions and to progress in the right direction of the path. One should understand the benefits of almsgiving, liberality, and giving gifts and the opposing negative results for cruelty. Immensely positive results (vipaka) exist for *sangha-dana* (almsgiving to monks). These beneficial results stem from acknowledging, with right understanding, the monk's past kamma, faith, and virtuous qualities. The Buddha didn't limit giving only to Buddhist monks; he used the terms *samana* and *Brahmana*, meaning good monks of other schools. Similarly, any small gifts given to a worthy recipient will yield good results in the future. On the other hand, any cruelty or violence produces bad results in this life and many lives to come. Further still, good and evil deeds done towards one's father and mother respectively yield positive and negative results in subsequent existences. Also, in accordance with these laws, there exists other worlds into which beings are born apparitionally according to their kamma. These are known as *devaloka* and *apayaloka* (heaven and hell respectively), which exist among the thirty-one planes of existence. Such a view is known as right view. By having this knowledge, people can refrain from engaging in bad conduct and change it to good conduct. Herein we can see that the Tathagata emphasises the importance of right understanding in accordance with morality and its consequences. These are the foundational stones of wisdom.

Prior to the Buddha's enlightenment, some other schools of thought held that kamma is fixed at birth and cannot be changed. People in those societies were segregated into a hierarchical caste system. The Buddha disagreed with this unfair treatment and stated that no one became higher or lower caste by birth but that a person's position within a society was determined by his or her present action. In this line of analysis, kamma can be divided into a timescale of past, present, and future. The determinant factor is the volition of an action in the moment, which can influence the future result. By being aware and wisely choosing between different causes of action, one can change the effect of past kamma. For example, the Buddha advised a young man, Sigala, not to adhere to outdated traditions but instead to engage wisely and productively with contemporary society. He further explained by the matrix of relations that in society one is able to perform his or her duties righteously, constructively, and correctly in a timely manner.[2]

The principal meritorious deed in Buddhism is giving. Food is a particularly common gift found in all popular Buddhist cultures. The worthiness of the recipient is of more concern than the donor, so the donor supports and sustains the life of another. The field of merit is an important concept in these cultures, and the Buddha is regarded as the unsurpassed field of merit. Today Buddhist monks represent this field. It is built on a principle of kamma, namely that the high spiritual quality of the recipient amplifies the kammic fruit generated from the deed. Even a seemingly insignificant deed can result in large kammic benefits if it is directed towards a good field of merit. This phenomenon is explained by the analogy that even a small seed can grow into a large tree bearing good fruits if the field it is sowed in is fertile. For example, the story of Sujatha, a woman from a wealthy family, offering milk rice to the bodhisattva (Buddha-to-be), finally resulted in her attaining nibbana.[3] This general idea of merit making is embraced throughout Pali literature, including in the Abhidhamma. The aim of a good deed is to shorten the time of suffering in samsara rather than to merit being reborn in the heavenly realms or gaining material wealth. The reward for merit is strongly inferred by the frequent appearance of these stories in the canonical discourses which often depict a donor securing a good rebirth in the heavenly realms. The quality of the donor's mind during the act of giving is emphasised. This shows that faith or confidence in one's mind at the very moment the deed is performed will eventually produce great fruit in the future. In Buddhism, faith is the seed that can, in time, yield to attaining final liberation.

The discourses contained in the Sutta Pitaka (*a collection of Buddhist scriptures*) were expounded by the Buddha under diverse circumstances, sometimes to householders and other times to the dedicated seekers of the truth in their day. The Buddha was able to see the mental capacities of his listeners and their ability for comprehension. His primary intent was to reduce their suffering; thus, within the lexical framework of merit and kamma, he would educate them on the value of intent. By adopting a suitable approach for his audience, he could guide the listener effectively

---

[2] Sigalovada Sutta: "The Discourse to Sigala", DN 31.
[3] Khuddaka Nikaya, Buddhavamsa 2.63. Chronicle of Buddhas.

through his complex knowledge and arrive at the penetrating truth. The Buddha used poetic adjectives to help define the features of an act, for example "fruit of giving fruit" for someone faithfully making merit by donating fruits to the Buddha and his monastic order with a wish to attain nibbana. Here the word *fruit* is used figuratively to mean both fruit and the ripening of a kammic result.

In one of the Buddha's previous lives (according to the Jataka stories: series of non-Canonical literature), when he was an ascetic, he, as Sumedha, gave an offering of a Buddha Dipankara's bowl with clothing, perfumes, ghee, oil, and other things, with a wish to be enlightened. The act resulted in his being born after many thousands of aeons and after a great deal of spiritual practice, eventually crystallising his own awakening as Gotama Buddha. The Buddha describes that on the night of his awakening, he firstly remembered his former rebirths (the first knowledge), then by his purified divine eye perceived the rebirth of beings according to their past deeds (the second knowledge), and finally realised the Four Noble Truths of awakening after his mind was liberated from defilement (the third knowledge). The sequence of events narrated in this passage of the Jataka stories mirrors the principle of kammic concatenation. The Buddha continued to give examples of what he had received for what he had given on that day: "Today my bed is the application of mindfulness, my pasture is concentration and meditative absorptions, my food is the factors of awakening; this is the fruit of giving ghee."[4]

Criminology

Some actions produce results in this very lifetime, for example crime is punishable by law; the term *crime* has a legal definition rather than a universal value. Criminal acts are investigated with reference to any history of violent behaviour, existing case law, and the statute of jurisdiction. Thoughts, intentions, evidence of action, and a series of patterns to connect the incident all play their part in the prosecutor's developing an assertion of truth.

Buddhist psychology looks deeper than the surface to understand

---

[4] The Great Discourse to Saccaka, MN 36.

what makes someone commit a crime by relating to past and present conditioning: a grudge, an attitude held towards others, an unresolved grievance, or an opportunity to take revenge or steal—or there could be a premeditated factor hidden in the individual's personality. Whatever the reason may be, the generic mental factors that drive criminal behaviour are greed, hatred, and delusion. These are deeply rooted in the mind and surface as a personality disorder, with an egotistical trait characterised by persistent antisocial behaviour, bullying, and impaired empathy. These factors make up the mental disposition of a mentally unstable person who is likely to commit a crime and whose aggressive behaviour is predictable. Early detection of sinister motives could prevent a far more serious incident. If Buddhist principles are correctly and systematically applied, it is possible to reform a criminal.

The Buddha has taken a completely different approach to the contentious issue of crime and punishment. There is a striking story in which the Buddha met with an armed serial killer, Angulimala, who was roaming the streets hunting to kill and chopping the fingers off from innocent people. As part of his daily routine to help someone, the Buddha decided to meet face to face with this ferocious murderer. Despite the warnings of the villagers, he walked down to where the murderer was hiding. Angulimala saw the Blessed One coming from afar and decided to ambush him as an easy target. He began following him with his sword. However, by some psychic feat, he could not reach the Blessed One, though the latter was walking at a normal pace. Angulimala demanded that the Buddha stop. The Buddha took this opportunity to start a conversation with the murderer and pointed out to him that he was the one who needed to stop his violence. With those words of the Buddha, Angulimala evidently realised his violent nature, abandoned his evil, and surrendered to the Buddha.

Angulimala was his given name. Formerly he had been a brilliant student at the University of Taxila. Some were jealous of his success, accusing him of treachery, and he was brainwashed to become a murderer. His rivals set out a plot for him, hoping that he would be killed by the king's army. The story depicts how an innocent student turned into a criminal. It unravels the rationale of the relational philosophy of psychology, showing how a tendency towards criminal behaviour can lay hidden, latent in the personality, and yet still one who accepts his or her mistakes retains the possibility of reforming that behaviour. If someone is willing to accept his or her mistakes and the consequences of his or her actions, then it is possible for him or her to be reformed. William Godwin, an English political philosopher, even believed that torturous punishment was simply a case of the elites controlling the masses. Godwin and thinkers like him subscribed to John Locke's theory that humans are born with some mental inhibitions but that other inhibitions are mainly a result of one's experiences and influences (nurture over nature) (John Locke, 1652–1675, Oxford University, England). Many even argued that criminals could change and be reintegrated into society through greater practice of reform.

Crime and capital punishment still remain a controversial issue. We have been moving away from a preclassical view advocating criminal behaviour as evil intuition or a demon's dark spell. In terms of punishment, imprisonment was introduced as a deterrent to prevent crime in place of corporal punishment and public humiliation, the original forms of barbaric capital punishment having been replaced by a new penological system. Socrates, a philosopher of Ancient Greece who lived in the city of Athens, in 500 BCE famously turned down his conviction to life imprisonment, instead choosing to sentence himself to compulsory suicide rather than spend the rest of his life in prison enslaved to the elite class.

Many ideologies have contributed to a new way of thinking. Eighteenth-century Italian criminologist Cesare Lombroso is credited as being the father of classical criminology. He introduced the theory that people are rational with free will and judge pleasure vs. pain. He therefore favoured determent and did not advocate torture as a means of punishment. He argued that punishment should be to prevent future crime. This argument led to rethinking the question "What is punishment?" Offenders are sent to prison *as* a punishment, not *for* punishment. Punishment is one of the purposes of sentencing and may serve additional instrumental functions, primarily the reduction of crime. Most criminal court judges, prosecutors, public defenders, and other justice practitioners believe that the prevalence and severity of crime depends mainly on factors that affect most individuals long before they are taken into custody. Most justice practitioners understand that they can rarely do for their clients what parents, teachers, friends, neighbours, clergy, biogenetic inheritance, and economic opportunities have failed to do. As John Locke wrote, "Government has no other end, but the preservation of property." The state has a legal obligation to protect its citizens and property.

The media amplifies crime as happening on every corner, with police raids in the style of those depicted in Hollywood movies. However, according to the 2018 volume of the FBI's annual Uniform Crime Report, the number of violent crimes has decreased over the years. The declining trend in the crime rate is mainly attributed to the evolution of the US criminal justice system, which is consistently improving its efforts to reduce criminal activity. Rates of crime and recidivism have long served as a critical measure for the performance of any criminal justice system. These statistics indicate that it is possible to reform a criminal through a systemic approach to bring about a profound reversal and psychological changes.

The Buddha had only one indicator: suffering. In the scenario of Angulimala, the Buddha successfully used reverse psychology by reflecting the pain and fear Angulimala had inflicted on others towards the perpetrator himself, so that he might feel others' suffering. The right view in Buddhism is decided on the basis of actions and consequences in relation to the way out of suffering.[5]

The phrase *papa kamma* is used to express what is unwholesome, and its opposite, *punna kamma*, is used for what is wholesome—this is in regard to one's destination after death. Actions are thus evaluated according to the results, and often those results manifest in future rebirths. The correct understanding of kamma influenced the attitudes of people and reflected broader social changes within early Indian Buddhism. This has since spread to the countries which adopted Buddhism. The Buddha advocated a culture of nonviolence. In rejecting confrontation, he instead showed the benefits of benevolent and peaceful means to resolving conflicts. This, in the Buddha's teachings, is the way to avoid accumulating bad kamma. Wholesome kamma, on the other hand, makes possible the opportunity to have a good birth. Through the journey of life, one comes to understand the deep philosophical teachings and realises that nibbana is the highest achievement and of paramount importance to Buddhists. The Buddha specifically addressed lay Buddhists by reinforcing their faith through

---

[5] *Angulimala Sutta: About Angulimala (MN 86)*, tr. Thanissaro Bhikkhu, Access to Insight (2013), http://www.accesstoinsight.org/tipitaka/mn/mn.086.than.html.

devotion and donations, showing them the great rewards of giving. This created a system of reciprocal interdependency between lay communities and the monastic order; necessary material support was exchanged for the teaching of Dhamma. In the Abhidhamma point of view, nibbana is not a result of merit, though it could be a contributing factor. A passage in the allegory of Mara, a mythological demon, tells of an unsuccessful attempt to distract the bodhisattva while he was meditating prior to his awakening. The Mara challenged him and forcefully suggested that he should instead seek merit. The bodhisattva replied, "I do not have use of even a little merit." This dialogue ended with the bodhisattva pointing to the earth as the witness for his meritorious deeds done in his past lives.[6] Nibbana, according to the Abhidhamma, is the fruit of a purified consciousness with knowledge attaining to a supramundane level; it is not a state of cause and effect, but an "unshakeable state" of mind free from craving. After one reaches this stage, no newer kamma is produced and all actions become defunct.

The third exposition to right view is the right understanding of the Four Noble Truths. There is no path to liberation from samsara if one cannot understand suffering. Suffering (*dukkha*) must be explored until one gains a penetrative insight into the truth of its arising, its origin, and the path leading its cessation. Suffering arises in the six senses, namely by way of sight, hearing (sound), smell, taste, and touch, and in the mind. One must gain the right understanding that all those senses are suffering, impermanent, and not-self (respectively *dukkha*, *anicca*, and *annatha*). The attachment to or craving of sensory pleasures is the root cause of suffering. Hence the Buddha declared that "detachment of craving is the cessation [of suffering]" (*thanha sokayo jayathi*), which means inversely that attachment is the root cause of suffering. One may gain understanding of this truth about suffering by gradually developing insight through vipassana meditation, then eventually attachment will cease in this very life. The knowledge and understanding of the true path that leads to the cessation of suffering is known as *sammaditthi nana*—right knowledge.

In this vast ocean of samsara, kamma is like a ship that carries us; mindfulness is like a radar system that helps us to navigate carefully

---

6 SN 43.61.

through rocks; and wisdom is like a compass that directs the course of life towards the noble path to the end of suffering.

Right Thinking—Samma Sankappa

Thinking is the formation of thoughts in the mind. Thoughts are mainly reflections, daydreams, perceptions, or memories. Ideas and opinions are forms of thought, and we understand things by means of thought. The mind is the faculty of ideation; by means of interpreting perceptions through a language, it produces rational or irrational thinking, coupled with associated emotions. There are three avenues from which thoughts flow to the mind. They are the sense avenue, the mind avenue, and the free avenue. When an object makes contact with a sense, it produces feeling and can begin to form a thought process. On the other hand, numerous objects appear in the mind from memory and form thoughts. All kinds of mental activities transpire in the mind through these two processes. The mind grasps both sense and mental objects to construct ideas, opinions, and judgements. Most of the time the mental formations of an untrained mind are accompanied by greed, hatred, and delusion by way of discursive thinking, which may even escalate into anger, hatred, and revenge of a violent nature.

There are thoughts of a special nature that arise at the time of death through neither the sense avenue nor the mind avenue; hence these are classified as "free avenue". They appear as symbols (*kamma nimitta*), followed by signs of destiny (*giti nimitta*); these death signs determine one's rebirth through terminal thought processes and rebirth consciousness respectively (*maranasana javana vithi* and *patisandi vinnana*).

The exposition of right thinking is threefold. Right thinking has a wisdom factor, and enough wisdom for right understanding provides the supporting conditions for right thinking to arise in the mind. Right thinking can renounce the five senses as it is free from greed and hatred. The second and third subfactors of right thinking are the thoughts of loving-kindness and compassion towards all beings. When there is right thinking, only wholesome thoughts arise in all three avenues; right thinking is the mental development of sublime thoughts in all three avenues. Thoughts are the result of past and present action and the cause of future action. It is, therefore, important to have right thinking to accumulate good kamma and have a good rebirth.

Right thinking arises in the mind as a result of right understanding. It provides the supporting conditions for morality (*sila*). Since thoughts are the forerunner of all action, and vice versa, right thinking helps to stabilise the mind, freeing it from discursive thinking.

Right Speech—Samma Vacha

Right speech, the first factor of morality, concerns maintaining moral conduct of speech, which, as kamma constitutes, is a volitional verbal action. The word *right* implies that the act should be wholesome and thus not contrary to the progress of acquiring wisdom. The Buddha's teaching of moral discipline consists of training rules. These rules provide guidelines for manner, conduct, and behaviour, thereby regulating one's actions. The rules forbid all immoral actions. In the context of right speech, this involves refraining from four types of speech: false speech, backbiting, use of offensive or abusive language, and frivolous talk. On the other hand, the speaking of words relating to Dhamma and vinaya are the right speech, that is words of wisdom and morality.

(These disciplines are extended to cover right action and right livelihood, listed below.)

Right Action—Samma Kammantha

- Refrain from killing or injuring living beings.
- Refrain from taking property which is not given.
- Refrain from sexual misconduct and taking intoxicants.

These rules are based on the underlying philosophy of accepting responsibility for one's own actions. Any form of killing is unacceptable in Buddhism because that action produces irreversible kamma that has consequences in this life and many lives to come. Being honest and truthful, respecting others' lives, being mindful, and respecting others' property rights are based on the principle of the law of action. For example, in both criminal law and Buddhism, first-degree murder is considered as the most deviant behaviour that is both willful and premeditated, meaning that it was committed after planning.

Right Livelihood—Samma Ajiva

- Refrain from making a livelihood by means of immorality by way of physical or verbal actions.
- Refrain from making a livelihood by merchandising goods such as meat, intoxicants, poisons, and weapons or by slaughtering living animals.

These rules are set for one to be led by pure conduct by avoiding corruption and harm to others. There is a much more extensive set of rules for monks and nuns. Rules are practised as precepts. They are not imposed but are voluntarily, accepted by the practitioner according to the degree of his or her understanding. For example, laypeople practise five training rules. The lineage of the practitioner changes with the number of precepts he or she has undertaken to practise. Novice monks and nuns observe eight precepts; *anagarika* (trainee) monks and nuns observe ten; whereas fully ordained monks observe two hundred and twenty-seven rules.

The purpose of these rules is to prevent any offensive behaviour to others and also to ensure one's own protection. Those who are striving to develop their wisdom in Dhamma should abstain from all kinds of imprudent and unskilful actions. Thoughts are the forerunners of all intentional actions and vice versa; therefore, good actions produce wholesome thoughts and bad actions produce unwholesome thoughts. By preventing bad actions and practising good actions, one can accrue good kamma and peace of mind simultaneously. Such a mind is free from the guilt of wrongdoing. The practice of morality is essential to developing a good mental culture. A mind free from ill will and worry is a basic requirement for meditation and to achieve higher mental development. Observation of these moral precepts helps one not to commit bad kamma or crimes, which would otherwise disturb the mind. They serve as a foundation to stabilise the mind in preparation for practising meditation. The Pali word *bhavana* means "mental culture"; this is the mental training one undertakes to develop a pure consciousness, free from greed, hatred, and delusion. It is a standard cultural practice in Buddhism to take five precepts before meditation. It helps to clear the mind.

Right Effort—Samma Vayama, or Viriya

The second category of mental training is concentration (samadhi). Right effort is the first factor of this. The transformation of a worldly character must precede transformation of thought. However, this transformation requires effort. In seeking enlightenment, one must make a strong determination to suppress evil and develop discipline (*samvara vinaya*). On the full moon of Vesak under the peepul tree, the future Buddha made a firm decision: "May my blood and flesh go dry before I desist from such effort." By this historical effort, he attained the all-knowing wisdom, *samma sambodhi*.[7] His enlightenment was the culmination of the ten virtues that make up Buddhahood training (*Buddha karaka Dhamma/ paramitha*). These ten virtues are generosity, moral habit, renunciation, wisdom, effort, patience, truth, resolution, love, and equanimity. During his life as a bodhisattva, he practised those ten qualities until perfecting them. Such perfection of morality was the result of the virtuous kamma he had accumulated throughout 550 lives as recorded in the Jataka, the prehistorical legend of the Buddha. It could be argued that virtuous kamma itself does not produce the result of enlightenment without an effort towards it. The Buddha set an example of unprecedented effort to liberate his mind.

Note: The sacred fig tree (*Ficus religiosa*) is a species of fig native to the Indian subcontinent and the Indochina region. It belongs to the fig or mulberry family (*Moraceae*). It is also known as the bodhi tree, pippala tree, peepul tree, peepal tree, or ashwattha tree. The sacred fig is considered to have religious significance in three major religions that originated in the Indian subcontinent, Hinduism, Buddhism, and Jainism. Hindu and Jain ascetics consider these trees to be sacred and often meditate under them.

---

[7] AN 2.12, "Unremitting Effort".

It is this type of tree under which Gotama Buddha is believed to have attained enlightenment.

There are four kinds of right effort which could be categorised further as wholesome or unwholesome and born or potential. Unwholesome kamma refers to the ten kinds of evil conduct: threefold bodily actions, fourfold verbal actions, and threefold mental actions. Unwholesome bodily and verbal actions were discussed under right action and right speech. It is by way of right effort in practising right speech, right action, and right livelihood that one acquires productive mental development and gains wisdom. By practising those precepts, one can prevent the arising of unwholesome kamma (akusala). It means making an effort to eradicate the roots of evil. There are threefold mental actions, namely covetousness, ill will, and worry—wrong view. Each of these has subsets, such as envy, jealously, attachment, hatred, and delusion. An untrained mind is deeply rooted in these defilements.

There is always an intention behind all mental, verbal, and bodily actions. For example, false accusation, slander, and defamation of the character of a rival for personal gain would result in violent disputes. The effort to suppress the wrong intentions that trigger defilements must continue until one attains *anupadisessa* nibbana (nibbana without the constituent groups of existence remaining). On the other hand, a second kind of right effort is to suppress the arising of unwholesome kamma. This can be done by making an effort to practise tranquillity and insight meditation. During the practice, trainees must make an effort to identify any wrong intention in the mental continuum and replace it with wholesome kamma.

Wholesome kamma refers to the seven kinds of purification (*visuddhi*). These are (1) purification of virtue, such as observation of the five precepts; (2) purification of consciousness; (3) purification of view; (4) purification of overcoming doubt; (5) purification of what is and what is not the path, by knowledge and vision; (6) purification of the course of practice, by knowledge and vision; and (7) purification of knowledge and vision.

The third kind of right effort is to strive to practise the above-mentioned seven factors of purification. By the power of such purification, there arise seven factors that lead to enlightenment (*bodhipakkhiya Dhammas*), which may arise in this present existence.

The fourth kind of right effort is to maintain the development of arisen, or arising, wholesome kamma until attaining the path (*magga*) and its fruit (*palah*).

These four kinds of right effort must be applied in order to purify the mind. The relevant Dhamma or mental faculty is known as *viriya* (effort). The development of this faculty, striving for perfection and searching for excellence, is fundamental to Buddhism; it is one of the main qualities and examples set by the Buddha.

Right Mindfulness—Samma Sati

Right mindfulness is the second factor of meditation and the most important faculty of the mind. Without mindfulness, the mind may lapse into a continuous state of aimless wandering, with thoughts successively arising in the mind one after the other. The Buddha explained the way thoughts come into being through a manner of causal relations known as *pattana*. Pattana explains the conditioning factors and the relations of thought formation. By practising mindfulness, one can observe this process.

The first condition is **object relation**, which states that thoughts cannot arise in the mind without an object. There are many different thought processes that can occur in relation to any given object, depending on its character and quality. For example, when the eye contact with a flower there arise pleasant thoughts in the mind in relation to the flower at sight.

The second category of thought is known as **past action relations**. When the mind contacts an external object through the door of a sense, there is another, independent thought process that occurs to investigate the object. These thoughts are mainly past memories that determine the nature of the object. They are the result of kamma.

The third type of thought arises in the mind only in the presence of an **associated relation**. Some thoughts arise in the mind only in the presence of other thoughts; in the same way, a lotus flower appears only in the presence of water. When the water dries out, the lotus also dries out.

The fourth category is **suffusing relation**. This describes thoughts that are conditioned by mental states; this relation between the thought and its conditioning state is inseparable. An analogy of this is of a painting existing only on a surface; the quality of the painting depends on the canvas or paper surface.

Altogether there are twenty-four relations describing how the mind and body interact with each other to form thoughts. The mind is in contact with the body and the external world through consciousness. It is a process of a series of thoughts. As a matter of fact, the mind can only be conscious of one object at a time. According to the process of consciousness, the way thoughts arise in the mind can be divided into two main categories. The eyes, ears, nose, tongue, and body comprise the five-sense consciousness avenue. These senses are grouped here by the common principle of their processing. The sixth sense is the mind-consciousness avenue. Thoughts are channelled into the mind via these two avenues. Without mindfulness, this could easily give rise to discursive thinking. Mindfulness is a higher level of the mind that stands above observing the thoughts; a thought cannot be seen by the thought itself. "Mindfulness has been called universal by the Blessed One. Because a mind has mindfulness as its refuge and mindfulness manifests as its protection and there is no exertion nor restraint of the mind without mindfulness."[8] Mindfulness protects the mind from lapsing into agitation or idleness. Mindfulness can be defined as the ability to remember, thus its function is not to forget. As a mental factor, it signifies a presence of mind—attentiveness to the present rather than a faculty of memory regarding the past.[9] When there is mindfulness, it produces an awareness of the object which is in contact with a sense. Therefore, full awareness means the clear comprehension of the object (*sampajana*). The characteristic of sampajana is nonconfusion, and sampajana has the function of making a judgement through objective investigation. It manifests as a guardian. Its foundation is morality (*sila*). The proximate cause is a strong perception of right and wrong.

In "Establishing the Four Foundations of Mindfulness" (Mahasatipattanhana Sutta), the Tathagata explained how to develop mindfulness. As explained in terms of its causal relations, the mind needs an object to be mindful of. The Four Foundations are categorised as body,

---

8 Buddhaghosa, *The Path of Purification: Visuddhimagga*, tr. Bhikkhu Nanamoli (Buddhist Publication Society, 1999).
9 Ibid.

feeling, mental formation (mind—*citta*), and mental objects or Dhamma. These groups are both the subjects and objects of mindfulness. This sutta (discourse) explains in great detail how to establish mindfulness. It was formulated according to the principles of causal relations and natural law (cosmic order). Developing mindfulness in accordance to the sutta is to develop the right mindfulness. By practising *satipatthana* (establishment of mindfulness), one is able to discover the ultimate truth of the mind and body. The nature of the mind is that it is never steady but fleeting. In meditation, the mind is bound to a fixed object by the rope of mindfulness.

To demonstrate the importance of mindfulness, the Tathagata used the parable of a little bird that wanted to fly away from her territory. The bird had a desire to explore the world, so, therefore, ignoring the warning of her mother bird, flew beyond her territory and became vulnerable to hawks. Establishing mindfulness by the Four Foundations marks the territory of the meditator. Contemplating each group helps to keep the wandering mind within the territory of its own body. Moving away from this territory into thoughts of external objects will result in dwelling on those thoughts and losing one's mindfulness. Sleepiness can also result in the mind's falling into a dreamy state. If either of these things happen, then mindfulness is lost, and thoughts will take its place. Fixing the mind on the Four Foundations and contemplating those phenomena arising and passing away belongs to the class of insight meditation (vipassana). Mindfulness can be regarded as a pillar because it is firmly founded, or like a doorkeeper in the way that it guards the senses. When mindfulness is firmly established, the meditator can clearly comprehend an object by seeing its true nature, disassociating it from thoughts that are mere contents of the mind.

Application of Mindfulness

Contemplating the Body (Kayanupassana Satipatthana)

In kayanupassana satipatthana, the mind is firmly fixed on the corporeal realm. Examples of this are focusing on the rising and falling of the abdomen, meditating on the breath (*anapana sati*), and reflecting on the physical composition and particular elements of the body. There are four postures for meditation: standing, walking, sitting, and lying down.

Walking meditation is based on foot movements. It helps train one to stay in the present moment while observing physical movements. Over time, when regularly practised, mindfulness improves.

Contemplating Feelings (Vedananupassana Satipatthana)

In vedananupassana satipatthana, the mind is firmly fixed on the feelings. Some of such feelings in this group are pain, sensations, sadness, happiness, pleasantness, unpleasantness, and equanimity. There are nine different kinds of feelings that arise at various moments in one's life. With right mindfulness, one can contemplate these feelings as they are. They all have the same characteristics of arising, maturing, and passing away. Developing this observational practice for some time, the meditator can overcome restlessness, worry, or anxiety of the mind and establish mindfulness.

Contemplating Thoughts (Mental Formations)
(Cittanupassana Satipatthana)

In cittanupassana satipatthana, the mind is firmly fixed on one's rapidly changing thoughts. To achieve right mindfulness of thought is to experience one's thoughts through pure observation. As previously discussed, thoughts arise in the mind because of various relations and conditions. When the mind is calm, the meditator can see those fleeting objects and their relations clearly, as when looking through the calm water of a lake. One can develop the ability to see the associated conditions of each thought such as greed, hatred, and delusion. By practising mindfulness on thoughts, restlessness of the mind will in time disappear. This practice involves paying bare attention to the mental process as it is presents itself in the moment. To do this without forming opinions, judgements, or analyses, and not dwelling on one's thoughts, is right mindfulness.

There are four kinds of unwholesome states of mind and their opposites of four wholesome states. Unwholesome states are lust, anger, delusion, and distraction. When thoughts born of these states appear, right mindfulness is nonreactive, observing them leave as if playing the role of a doorkeeper (guard). Thoughts are more often coupled with emotions, and these create mental formations and make the mind vacillate between memories to imagine future events. This is a state where the mind relapses into discursive thinking, aimlessly wandering here and there. It is very common and is a state of conceptual proliferation known in Pali as *papancha*. The process by which such formations arise is explained by consciousness (citta). This details what conditions support a continuous stream of thought and how the thoughts come into being. The next step is to investigate the observed phenomena and clearly comprehend them. These are understood by their appearance, characteristics, and proximate conditions. In this way the meditator is able to understand the impersonality, knowing that the thoughts are fleeting phenomena, and let go of the attachment or aversion to the object. Progress in this practice gradually reduces entanglement in mundane issues. Entanglements are caused by mistakenly grasped ideas arising from errors in perception because of ignorance. Wholesome states arise from a great, unsurpassable, concentrated, and liberated mind without entanglements. They are the result of reinstating good kamma by making an effort to generate wholesome thoughts of loving-kindness, compassion, generosity, and ego-lessness.

Contemplating the Dhamma (Dhammanupassana Satipatthana)

In dhammanupassana satipatthana, the mind is firmly fixed on the meditative hindrances, the five aggregates, the five senses, the constituents of enlightenment, and the Four Noble Truths. These are some of the subjects of meditation in this group. Establishing mindfulness on this foundation is mainly achieved by observing the formation and dissolution of the phenomena not included in the foregoing three foundations. The key characteristic to observe is that whatever subjected to rise must followed by fall and cyclic nature of phenomena.

The hindrances of meditation are attachment, aversion, restlessness, worry, sloth, torpor, intoxication, and sceptical doubt. The meditator must clearly understand these phenomena and make an effort to overcome them by practising mindfulness.

The five aggregates are body, feeling, perceptions, mind (formations), and consciousness. The first three aggregates were covered in the first three groups. However, as a foundation of Dhamma, contemplation is deeper.

The meditator gains a deeper awareness of his or her composite nature. By further investigation and practice, the meditator gains a clear comprehension of all six kinds of consciousness and of the limitations of sensory perceptions.

The higher level of Dhamma involves contemplating the constituents of enlightenment and the Four Noble Truths. The mind is firmly bound by the rope of mindfulness amid the rising and vanishing of all phenomena. When the meditator's practice matures, he or she will be able to pay qualified attention to an object without being overcome by its overwhelming impact. At this stage, mindfulness allows the meditator to stay in the present moment while impermanence, suffering, and nonself become clear.

Right Concentration—Samma Samadhi

Right concentration is the eighth and final factor of the Noble Eightfold Path. According to Buddhist doctrine, the word *right* means "what is conducive to deliverance".[10] This definition of *right* is applicable for all the other factors too. Concentration means the unification of the mind.[11] Like mindfulness, concentration also requires an object to concentrate on. We experience an object through a process of consciousness; therefore, right concentration could be defined as centring the consciousness on a single object until the mental factors arising with that consciousness dissolve, to achieve mental one-pointedness.

Concentration is synonymous with mental one-pointedness (*ekaggata*). Its function is to eliminate distraction and to closely contemplate an object. In this context, mindfulness is a supporting factor for concentration. Its proximate cause is blissfulness. "Being blissful, his mind becomes concentrated." In insight meditation, mindfulness and concentration are applied together. Once the object is framed by mindfulness, concentration is like a spotlight shining on that object.

Concentration is threefold: momentary concentration (*khanika samadhi*), access concentration (*upacara samadhi*), and attainment concentration (*appana samadhi*). As the name implies, momentary

---

[10] Nyanaponika Thera, *Abhidhamma Studies: Researches in Buddhist Psychology* (Kandy, Sri Lanka, Buddhist Publication Society, 1965).
[11] Buddhaghosa, *The Path of Purification: Visuddhimagga*.

concentration is concentration on the object only for a moment. This is the technique of insight meditation, contemplating on a variety of momentarily changing objects to gain insight knowledge into the formation and dissolution of phenomena. In Buddha's time, he taught choiceless awareness and momentary awareness of objects to advanced students. Access and attainment concentration are the result of tranquillity (samatha) meditation.

There are forty subjects (*kammatthana*) used to develop concentration; the choice of which to implement depends on the meditator's temperament. The mind becomes very calm as a result of temporary suppression of defilement. These defilements are the cause of distraction and distress. When the mind is free from distraction, concentration gets stronger. By intense concentration, the mind becomes absorbed into the object and reaches a state of jhana. This state of mind is known as *kusala citta ekaggata samadhi*, a wholesome state, where the one-pointedness of a (moral) thought is well fixed to the object. By practising tranquillity meditation (samatha), the meditator can achieve five levels of jhana samadhi. Further, the mind could be developed on an immaterial plane until it reaches its limit of neither perception nor nonperception. However, jhanas alone do not lead to insight wisdom; their use is purely to develop concentration and to prepare for insight knowledge.

CHAPTER 3

MINDFULNESS IN PSYCHOLOGY AND BEHAVIOURAL SCIENCE

THE CULTIVATION OF MINDFULNESS AND the concept of the present moment is deeply rooted in Buddhist meditation practices. As taught by the Buddha, these ideas are established in the Four Foundations of Mindfulness. Anyone is able to benefit from these psychological principles and techniques, irrespective of their beliefs. Mindfulness is not a religious ritual. Its usefulness has recently been identified by psychologists. Mindfulness is now widely applied to develop emotional intelligence and is part of stress-reduction therapies. Mindfulness helps to shift preoccupied thoughts, negative feelings, and depressive emotions. It brings a greater awareness, which allows for clarity of mind and an appreciation of beauty within the present moment. The practice of mindfulness is energising and compels a person to pursue his or her own goals. It can improve performance in all domains of life and everyday experiences. It makes the practitioner aware of his or her inner sentiments, while allowing him or her to remain detached from any subjective judgements. This can help a person to make informed decisions on what matters most, causing him or her investing his or her time and effort into improving well-being rather than chasing illusions.

Mindfulness is a self-help therapy used in cognitive behavioural therapy (CBT) and dialectical behaviour therapy (DBT). Apply in practice can get you to the root cause of your anxiety, depression, or poor relationships.

Understanding your thoughts and feelings can help you correct a personality disorder, which in turn allows you to integrate into society in a more constructive manner. The correct understanding of Buddhist doctrines can empower the meditator to resolve the suffering brought on by deceitful misconceptions. These Buddhist doctrines and principles can successfully be applied to improve therapeutic and professional relationships, which can be critical to the success of the treatment. It also makes for an assertion of truth in a variety of multidisciplinary professions, such as child psychology, behavioural psychology, and criminology.

Mindfulness and the Present Moment in Therapeutic Practice

Mindfulness is related to the present moment. By the present moment, I mean this very moment of time, here and now, neither the past nor the future. It takes training of the mind to stay in the present. The practice of insight meditation is one such method. This method always requires an object to aid in directing the attention to what is happening right now and at the same time cultivating an attitude of unbiased and detached observation of the chosen object. Application of mindfulness is now a popular therapy adapted by psychologists in CBT and DBT. The professional psychological institutions agree fairly well with the outcomes of these treatments.

There are a variety of neutral techniques that can help train the mind to stay present; simple things such as standing, walking, breathing, and observing the rise and fall of the abdomen are some of these. Although in explanation we discuss a moment as static, in practice it is a dynamic phenomenon, changing all the time. This can be explained by projecting a timeline representing the past, the present, and the future. The present moment needs constant renewal and replacement; therefore, we bring awareness to the chosen object, which is both dynamic and neutral. Dwelling in past memories or future imaginings is not useful because these things are not real; although such mental formations are happening, they are something that deceives the mind. The past has gone, and the future has not yet happened. Memories portray things differently from the way in which they actually happened, and future imaginings are often only what a person wants to have happen. These can be compared to a mirage, the deceptive appearance of a distant object, an optical illusion caused by atmospheric conditions, a common experience in scorching deserts. Similarly, some medications such as tranquillisers can distort perceptions by causing delusive mental formations. These mental formations are illusions which bring only anxiety, emotional regret, worry, and depression on a daily basis. Those who cannot purge their emotions or discuss them with someone else suffer in silence. The only solution is stay in the present and learn to let go of one's emotions by cultivating an attitude of bare observation.

Mindfulness allows for an objective observation rather than a subjective judgement. This opens up an opportunity to let go of any attachment or aversion to the traumatic past and uncertain future. Mindfulness is therefore a wise choice, leading to the understanding that holding on to illusions will only make life miserable. The goodness it brings can make someone want to be better and more loving. It creates feelings of elevation. Mindfulness is beneficial for anyone with problems managing their feelings, such as those showing withdrawal symptoms or signs of character flaws, personality disorders, or impulsive behaviours. It is also useful for those who are experiencing more pain or sadness than they can cope with and might think that their unhappiness will never end. These people often keep revisiting past memories with regret or think of fixing the problem in the future rather than taking any corrective action in the present. To dwell in those time domains is to hide from reality, seeking a comfort zone outside the present moment. Time travel is only fictional; fixing on it can make life more difficult and entrap a person in an illusory tunnel vision. Buddhism teaches one that by being kind to oneself there is always a way out of a problem and is a solution to every crisis, no matter how difficult it may appear to be. Buddhist wisdom can help a person choose between a path that leads to self-destruction and one that leads to true happiness; meditators can decide for themselves that there is hope and progress.

The aspect of self-help in mindfulness practice allows clinicians to be detached from their clients yet continue to engage constructively by developing a therapeutic relationship; a clinician needs to create the necessary space and interpersonal distance to yield an accurate diagnosis and better outcomes. The most effective approach is to have a live framework of a caring relationship based in a broader systematic perspective with which to diagnose. The clinician also needs to be mindful, having insight into behavioural patterns that relate to the inner psychological terrain of the client. To achieve this, the clinician must avoid confrontation and engage with empathetic listening skills, trying to understand his or her client's perceptions. This is to explore the client's long-term personal goals; educate him or her on his or her symptoms; identify any emotions that trigger from the client's past; build trust; encourage a confession; help the client to consider the effect of his or her behaviour on realising either success or failure in the future; and work on alliances with other members of the client's social network to build confidence.

Cognitive behaviour is directly related to the consciousness; therefore, any treatment options or solutions to alleviate suffering must be compatible with psychological principles. The Abhidhamma's analysis of suffering and of both the conscious and unconscious mind provides the necessary knowledge for evidence-based psychological treatments for mental disorders. Innovative developments in treatments for clinically vulnerable patients become feasible when the contributory factors are identified along with domains for future research. For example, development of hypnotherapy treatment can be used for a range of disorders from anxiety to panic attacks or insomnia.

Today the prevalence of mental disorders is high, and more people appear to be looking for alternatives to drug therapy. Buddhism has a holistic view of the mind and body. The internal relations between body and mind and their external relation to the outer world are explained in the philosophy of relations and the analytical methods. It is worth pointing out here that the Buddhist attitude to suffering and the way out of it is not based on any inducement or stimulating of the senses but on withdrawing from it. A renunciation of the senses, embedded in self-discipline, can open up a completely different solution to the problems of life. Suffering can occur in multiple realms, physical, cognitive, emotional, and social. Any treatment process needs to take into consideration all aspects of suffering. The patient has to cope with both the illness and the impact of treatment. The treatment may also affect their family, friends, and colleagues and the healthcare teams.

The First Noble Truth in Buddhism is suffering. The correct

understanding of this term would greatly help in patient care, the use of treatments, and the acceptance of any possible changes in lifestyle. Suffering is a fact of life. Not accepting this truth could result in having the wrong attitude towards life, adding more injury by further aggravating any mental suffering. The Buddha provided a complete solution to suffering in his formula of the Four Noble Truths. These are founded on wisdom and compassion, by which we can understand the true nature of life, carefully reflecting on life's experiences and wisely responding to the challenges of life.

Prevention is better than cure. These simple words of wisdom can help to build inner strength. Managing your own emotions in positive ways allows you to engage with society in a constructive manner; the otherwise negative emotions can make you inept at work or break down your relationships and can eventually make you a social reject. You must take steps to understand who you really are, rather than who you think you are, and discover your individual abilities and hidden talents. If you do nothing about the stress caused by uncontrolled emotions, it can make a severe impact on your mental well-being, in turn making you vulnerable to anxiety, panic attacks, and depression, further exacerbating any psychosomatic conditions you may have. Mindfulness meditation helps bring about a self-awareness. This is key to recognising your own emotions. It is important to know how your emotions affect your thoughts and patterns of behaviour. You can then self-evaluate your strengths and weaknesses in order to boost self-confidence.

Recent developments in psychotherapy and emotional intelligence (otherwise known as emotional quotient (EQ)) are based on these ancient teachings of centuries-old meditation techniques. In Buddhist cultures, meditation is a daily routine, not only for monastics but also for laypeople. This practice helps establish social cohesion by building stronger social relationships and empathy with family, friends, and social acquaintances. There is a specific meditation practice of loving-kindness (metta) which goes hand in hand with mindfulness; when your mind is clear, you can spread love and kindness to all beings. Buddhism strongly encourages you to love yourself first before giving love to others, not as a self-centred idea, but to increase confidence and bring positive changes to your character. Self-love can also improve effective communication with others by connecting you to your feelings and giving you the ability to express them. It is not mindfulness per se that matters, but correct understanding of the psychological principles that are intrinsically related to the mind. To facilitate positive changes is what fundamentally matters in Buddhism. The greatest happiness comes from sharing and caring and helping others, not living in self-isolation.

Western Perspective of Consciousness

The English word *conscious* originally derives from the Latin *conscius* (*con*, meaning "together", and *scio*, meaning "to know"), but the meaning of the Latin word differs from that of modern English. It means "knowing with", in other words, "having joint or common knowledge with another". It is not clear how the word *consciousness* is understood in modern Western thought. The origin of the modern concept of consciousness is attributed to the seventeenth-century philosopher John Locke in his "Essay Concerning Human Understanding" published in 1690. Locke defined consciousness as "the perception of what passes in a man's own mind". Locke equates consciousness to perception. His ideas greatly influenced the eighteenth-century view of consciousness, and since then many others have contributed to the development of this concept.

The meaning of the English word *perception* overlaps with *intuitive judgement*, whereby intuition is knowledge without conscious reasoning. Today, with the development of research into the brain and psychology, there are new understandings and definitions concerning the functions of the mind. For example, *cognition* refers to "the mental action or process of acquiring knowledge and understanding through thought, experience, and the senses". Cognition therefore encompasses almost all functions of the mind, such as the ability to make judgements, to evaluate, to reason, to problem-solve, and to make decisions, and includes the use of language and gathering of knowledge. Different professions have adapted its meaning to limit its scope of application. For example, in medicine, consciousness is assessed by observing a patient's arousal and responsiveness in the way that strokes or comas reduce response, or after surgical anaesthesia when a patient is closely monitored during recovery until full consciousness is regained.

Eighteenth-century German philosopher Friedrich Schelling went somewhat deeper than the surface in his investigation into the nature of consciousness and introduced the word *unconsciousness*. Unconsciousness is a process in the mind which is not apparent; it happens beneath any awareness of the mind. By this, a clinician is able to examine the psychological aspects of thoughts, feelings, memories, interests, motives, and possible hidden complexes. This concept was further developed and popularised by the Austrian neurologist and physician Sigmund Freud (1856–1939), who laid the foundation for psychoanalysis. His theory states that dreams and instantaneous behaviours relate directly to these unconscious processes. Freud believed that the unconscious mind is the source of dreams and the repository of forgotten memories that may still be accessible to consciousness at some later time. This is similar to the Buddhist concept of the life continuum (*bhavanga*), which is a function of the unconscious mind. There is another, intermediary state of semiconsciousness, manifesting in sleepwalking, implicit memories, trance, and hypnosis. This reduces peripheral awareness, which enhances the capacity of response to suggestions. Hypnosis is now used in controlled environments for psychological treatment.

The Western study of consciousness has evolved over a period of time, and from these studies a new area of discipline has emerged. Cognitive science has integrated science, philosophy, psychology, neuroscience, anthropology, linguistics, and artificial intelligence with the hope of creating better intelligent devices.[1] It appears that the interest in understanding the mind and consciousness has moved towards industrial use, presupposing that machines can make our lives better than we could achieve ourselves by our own capacity. The search for fulfilment and happiness still seems to be vested in the outside world.

There are similarities to be found in the etymological analysis of *consciousness*, yet there is a vast gulf between Western thought and the Buddhist understanding of the mind. This might be because of the Buddha, who had only one goal, to understand suffering and a way out of suffering. Buddhist thought has evolved from this one source, the enlightened Buddha. The writer strongly believes that Buddhism has much

---

[1] Benjamin Bly, ed., *Cognitive Science* (1999). Elsevier Science Publishing Co Inc.

to contribute to an understanding of the true nature of life and to finding true happiness through tranquillity and insight.

Barriers to True Happiness

The practice of generosity (*dana*), central to Buddhist cultures, is often encouraged. It helps reduce attachment and self-ego and develops great human qualities. One of the causes of suffering is greedy attachment coupled with ego. The happiness that comes from giving is more satisfying than possessiveness. In the present day, media coverage and social media fuels competition by token of one's identity. By an association with a popular personality, it projects a dominant position in society, making the person inseparable from his or her ego. Personality is a mental construct that is largely a product in itself and much influenced by one's society. Through the personality, happiness is derived from social recognition or comparison rather than having any substance to one's character, even conventionally so. Most of the projections are fake or fictional stories, which if questioned or cross-examined are unable to uphold the person's position. Many people are mentally suffering through problems rooted in a wrong identity within their minds.

In 1943, American psychologist Abraham Maslow introduced a paper concerning the hierarchy of human needs. This proposed that human beings were social animals, and their lower psychological needs must be satisfied before achieving the higher needs of self-esteem and self-actualisation. The denial of social recognition would result in someone who was motivated by an egocentric and antagonistic response striving for the survival of his or her ego, yet eventually failing when that ego was socially destroyed, challenged, or stripped off, because there was no safety net to fall into. Buddhism teaches us that ego is a self-deception and its proximate cause is the mental factor of delusion, usually coexisting with greed and hatred. It is the root cause of mental suffering that makes one believe in the existence of a permanent personality. Absolute freedom comes from letting go of the attachment to a false identity. This stems from the factor of wisdom, which results from mindfulness meditation. The simple technique suggested by masters of meditation is that of nonreactive observation of thoughts and emotions. One can come to manage one's core feelings by letting go of the attachment and aversion to emotionally resentful thought patterns. Meditation is a skill that when practised can bring tranquillity and insight, aiding in the management of emotional stress for lasting happiness.

Behavioural Psychology

Physical, verbal, and mental actions are preceded by thinking. What goes on in the mind manifests as the behaviour of a person. The character of an individual is much related to his or her mental disposition. It is the predominant trait that determines the way in which one acts or conducts oneself, especially towards others. There are two possible outcomes of behaviour: how a person reacts or how he or she responds. Reactions are coupled with other aspects of psychology such as emotions, thoughts, and aggressiveness, along with other internal mental processes, often irrational. On the other hand, response is a sign of intelligence associated with knowledge. If the person's disposition is known, then the outcome of his or her behaviour is predictable.

In a controlled environment, cognitive behaviour in relation to an external or internal stimulus can be observed more restrictively, and any action or function can be objectively measured in response to controlled stimuli. Historically, behaviourists considered mental activities as subjective and thus unsuitable for scientific study, as opposed to objective behaviour. The Buddhist philosophy of relations reveals how mental activities are interrelated, are synchronous, and perform the act of cognition. Over time, a person's thinking pattern becomes fixed, settling as a regular tendency or practice. It becomes especially hard to change these habitual tendencies. For example, smoking and drinking become a habit.

Personality is largely a product of the society in which one lives, and the mind gets conditioned by cultural influences, beliefs, and peer groups. The hope is that if we can use behavioural psychology to help us predict how humans will behave, we can build better habits as individuals and develop better relationships as communities rather than seeking the intervention of law enforcement agencies to resolve disputes.

Behavioural psychology is the study of the connection between our minds and our behaviour. The researchers and scientists who study behavioural psychology are trying to understand why we behave the way we do. They are concerned with discovering patterns in our actions and behaviours. A simple action such as giving a helping hand, holding someone's cup of coffee for a moment, or behaving with consideration or generosity tells a lot about that person's character.

There is an interesting article published by James Clear[2]. In the 1960s, a Stanford professor named Walter Mischel began conducting a series of important psychological studies. During his experiments, Mischel and his team tested hundreds of children, most of them around the ages of four and five. The study, known as the Marshmallow Experiment, revealed what is now believed to be one of the most important characteristics for success in health, work, and life. The experiment began by bringing each child into a private room, sitting them down in a chair, and placing a marshmallow on the table in front of them. At this point, the researcher offered a deal to the child. The researcher told the child that he was going to leave the room and that if the child did not eat the marshmallow while he was away, then the child would be rewarded with a second marshmallow. However, if the child decided to eat the first one before the researcher came back, then the child would not get a second marshmallow. The choice given to the child was simple: one treat right now or two treats later. The researcher left the room for fifteen minutes. The observation was entertaining. Some kids jumped up and ate the first marshmallow as soon as the researcher closed the door. Others wiggled and bounced and scooted in their chairs as they tried to restrain themselves, but eventually gave in to the temptation a few minutes later. And finally, a few of the children did manage to wait the entire time.

Published in 1972, the popular study of the Marshmallow Experiment wasn't made famous by the treat. The interesting part came years later. As the years rolled on and the children grew up, the researchers conducted follow-up studies and tracked each child's progress in several areas. What they found was surprising. The children who were willing to delay gratification and had waited to receive the second marshmallow ended up having higher SAT (Scholastic Assessment Test) scores. They were

---

[2] James Clear. https://jamesclear.com/

confident, self-reliant, and resilient, and had lower levels of substance abuse, a lower likelihood of obesity, better responses to stress, better social skills as reported by their parents, and generally better scores in a range of other life measures. Those kids who ate the marshmallow, in their later age as adults, were frustrated, indecisive, disorganised, and underachieving.

The researchers followed each child for more than forty years, and over and over again, the group who had waited patiently for the second marshmallow succeeded in whatever capacity was being measured. In other words, this series of experiments proved that the ability to delay gratification was critical for success in life. The child's ability to delay gratification and display self-control was not a predetermined trait, but rather was impacted by the experiences and environment that surrounded him or her. In fact, the effects of the environment were almost instantaneous. Just a few minutes of reliable or unreliable experiences were enough to push the actions of each child in one direction or another. This is an observable behaviour pattern of every child whether in a controlled or an open environment.

The foregoing study began by measuring the child's ability to resist temptation. The insight we gain from this study is important in our understanding the underlying tendencies of the mind, evident in the choices we make from an early leaning stage. Not every child gives in to temptation. There were other children in the same age group who had discipline in delaying gratification. One thing seems clear: if you want to succeed at something, at some point you will need to find the ability to be disciplined. You will need to act instead of becoming distracted and doing what's easy, looking ahead for a better reward in the promising future. This is true in nearly every field, that success requires you to ignore doing something easier in favour of doing something harder. One who is willing to accept challenges will certainly do well in life.

Human behaviour is more complex than this because there are many other factors that contribute to our choice making. However, there is a relation to one's childhood habits. It therefore follows the premise that you can train your child to become better simply by making a few small improvements. One such suggestion would be to build a more caring parent–child relationship, and as an adult your child can continue to make the necessary effort towards improvement and better habits. In the case of the children in this study, the defining variable for changes in behaviour was in being exposed to a reliable environment, where the researcher promised something and then delivered it. The trust one places in another individual or in society at large makes a significant contribution to behavioural patterns; on the other hand, breach of trust, betrayal, and treachery may result in violent revenge.

Buddhism traces tendencies of behaviour not only to early learning habits or persistent conditions but also to previous lives. The second section of the Pali Canon is a collection of discourses given by the Buddha in relation to a code of conduct and disciplinary rules. One such example is the restraint of the senses, improving behaviour by not allowing sense gratification. There is rule that prohibits monks from helping themselves to food, even if it is left on the table; they should practise self-restraint until the food is offered to them by the donor. Buddhist monks are allowed only two meals a day, breakfast, and lunch with no solid foods in between. In the Thai tradition, young children are given the opportunity to be temporarily ordained so that they can practise and develop these valuable character-building habits during school holidays. The Buddha has shown that beyond temporary gratification, there is a higher degree of mental satisfaction to be reached, which is peaceful and happy. In Buddhism one must regulate behaviour and cultivate virtues to gain social acceptance and recognition and finally to achieve success in life.

Human Behaviour and Health and Safety

In March 2020 there was an outbreak of SARS-CoV-2 virus, leading to the Covid-19 pandemic. It was a global health crisis and took the world by surprise, causing widespread disease and a high number of deaths. Even now it does not seem to be going away anytime soon. There is evidence from recent studies that self-management in relation to this disease improves the community's general health status and reduces hospital admissions for Covid-19 patients. It is critical in a situation where the transmission rate is very high to contain the spread of the disease by modifying behaviour, which is an epidemiological strategy to fight an epidemic. Self-management and modifying one's lifestyle has become an individual responsibility.

Regulating your own behaviour is vital, not only in an outbreak of disease but also in any health crisis. To be successful in self-management requires a multifaceted approach that incorporates self-discipline, mindfulness, and living in the present, while managing your own thoughts and emotions. Covid-19 had a profound effect on everyone's life. It may not be affecting everyone in the same way, but each person must manage his or her own behaviour, adapt, and be flexible to cope with this contagious infection. Social distancing rules introduced by governments compelled people to follow basic mannerisms, respecting the space of others, and keeping up hygiene standards to contain the virus's spread. The ripple effects of this pandemic are not yet known, but the global economic downturn is visible. This unusual situation highlighted many of the issues that have been neglected, matters of concern for our civilisation. It compels us to revaluate our value systems, prioritise what is most important to us, and learn how to live in harmony with nature.

Mindfulness is increasingly becoming an important theme when it comes to personal protection from infectious diseases. On July 23, 2022, the World Health Organisation (WHO) Director-General declared the escalating global monkeypox outbreak a Public Health Emergency of International Concern (PHEIC). Monkeypox[3] is an illness caused by the monkeypox virus. It is a viral zoonotic infection, meaning that it can spread from animals to humans. It can also spread from person to person through close contact with someone who has a monkeypox rash similar to Covid-19. Vast majority of reported cases were in the WHO European Region. Mindful living and taking up control measures recommended by health professionals we can prevent spread of the infection.

Buddhism is in affirmation of the safety of all. It focuses on the human factor involved in risk management. Individuals are responsible for their own safety and that of others in the society. People are expected to behave in a manner that does not cause danger to others. Despite plans, procedures, laws, and systems in place, people can, and do, make mistakes. Buddhism clearly shows the limitations of a human mind that is rooted in latent ignorance, having an inability to see the consequences of an error of judgement. Nature will provide an answer to questions according to the

---

[3] WHO. (2022, July 23). Monlypox: https://www.who.int/multi-media/details/monkeypox--what-you-need-to-know?

way we ask or test it. For example, if we knock on a door, the door panel will return the answer immediately with a sound unique to its surface material (e.g. wood or metal). A person hearing the sound behind would open the door, thinking someone is calling him. The door, the sound, the wood, the metal, the caller, and the receiver are all perceptions. If we ask the wrong question, then we might get the wrong answer. Thus, we can unknowingly take an erroneous assertion as the right answer. As a result of asking the wrong question of nature, we are confronted today with the biggest consequences and challenges of our living memory, caused by Covid-19. A double-edged blade cuts the hand if mishandled, but even so, an experiment in mishandling could be a disaster.

For example, in April 1986, an accident at the Chernobyl Nuclear Power Plant caused a highly destructive steam explosion. Radioactive contamination precipitated onto parts of the former USSR and nearby Western Europe, resulting in the evacuation of large numbers of people. This accident happened during a safety test on a nuclear reactor. Similarly, in March 2011, there was an accident at the Fukushima Daiichi Nuclear Power Plant in Japan, caused by an earthquake and tsunami—the most severe nuclear accident since the Chernobyl disaster. It was classified a Level 7 event according to the International Nuclear and Radiological Event Scale (INES), which was introduced by the International Atomic Energy Agency (IAEA) in 1990 for prompt communication of significant safety information. The aftermath of laboratory and industrial plant accidents and incidents is often characterised by calls for reflection and re-examination of the academic discipline's approach to safety research and policy, by measures for transparency, by constraints, and by the implementation of such constraints through regulations and conventions.

Policies and procedures are established after careful study of decades of experience in human behaviour. Behavioural safety approaches, also known as behaviour modification, consider the number of human factors that contribute to risk. Behaviours and actions influence culture through attitudes and change of perceptions and vice versa. Individual or collective behaviour determines the performance of systems. Having a health and safety framework for hazardous environments will help to identify dangerous situations, adjust performance, and manage safety problems. It can influence change in an individual's behaviour and achieve measurable results. The key element of safety at work or in society at large is changing behaviour and the individual's responsibility for doing so. The level of severity of a man-made disaster is difficult to interpret. It's a question of what we learn from these incidents and how we adjust ourselves to make our lives better. It compels us to define what is advancement of science, what is quality of thought, and what benefits humankind. Mindfulness is an intelligent approach to reflect on the impact of industrial disasters or personal crises to redefine what is most important to us and most sustainable. At the end of the day, we are responsible for our own actions, and the consequences are also ours.

The Buddha has made clear that the knowledge of the universe is infinite, and the human mind is incapable of comprehending it. Instead, humankind should spend their energy improving welfare, finding liberation from ignorance, and attaining peace of mind. Buddhism is like the light that dispels the darkness. Its ethics and values can be useful, not only for correcting individual behaviour and achieving personal liberation, but also, when applied universally, for creating sustainable socioeconomic prosperity.

Behavioural Psychology and Abusive Behaviour

The responses to resolve the issues of domestic violence, and street violence appear to only ignite the flames of hatred. Fuelled by emotion, impulse, and aggression, these unacceptable behavioural sequences of action often end in someone's getting hurt and occasionally end in irreversible traumatic incidents, even unlawful killing. The underlying psychological purpose of the abuser is to gain and maintain total control over his or her victim—and there are many methods and techniques an abuser can use to keep this tight control. Abusers generally seek to maintain this control in domestic situations and in the other aspects of their lives. In domestic violence, it is women who are more often victimised in the relationship, starting with threats and verbal abuse, and ending in physical injuries. As a form of deterrent, the law has recently taken a serious view on domestic violence and unlawful killing; however, many researchers believe that the best way to prevent abuse is to stop people from becoming abusers in the first place.

To achieve this kind of long-term behavioural change, we need strong cultural messages throughout the education system introducing the basic decency of people, strong value systems, and good social ethics.[4] During May 2020 mass street violence erupted in the United States following a death of a man while he was in police custody. The incident captured the public's attention and raised troubling questions regarding the limits of legitimate police authority in a democratic society. It was evident that black and minority groups were disproportionally affected and unfairly treated during the peak of the pandemic. There is rightfully public outrage over it, enforcing change and demanding justice.

Justice seems to be greatly needed in this era of darkness where basic values are appearing to diminish. Buddhism stands like a lamp post amid this darkness. The values of Buddhist principles can be successfully integrated into schools, dominant social institutions, and the institution of marriage to prevent unnecessary suffering. Nonviolence, loving-kindness, ethical values, mindfulness, and awareness of one's own behaviour and the consequences of one's own actions are all well explained in Buddha's teaching. The message is clear: abuse, bullying, and harassment of any kind is no longer acceptable. The world is still waiting for such a signature accomplishment.

---

[4] Joe Biden, "20 Years of Change: Joe Biden on the Violence Against Women Act," *Time* (10 Sept. 2014), https://time.com/3319325/joe-biden-violence-against-women/; Nancy Lemon, Violence Against Women Act (1990); US Department of Justice, Office of Justice Programs, National Institute of Justice, "Research in Brief", May 2020.

SPIRITUAL HEALING (SPIRITUALLY INTEGRATED PSYCHOTHERAPIES)

MINDFULNESS IS A NOVEL AND simple method of treating sufferers of depression. Its clinical applications are sensible as an acute treatment to help patients out of chronic unhappiness, mood swings, intense migraine headaches, mental tension, and various other depression-related illnesses without the administration of conventional medications. The process of mindfulness treatment, responsiveness to it, and its outcomes have been studied through the ancient wisdom of Buddhist meditation practices. It is fundamentally a technique of observing one's own mind with greater awareness. In a controlled environment, uniquely qualified clinicians guide patients in how to carry out their practice, and after few sessions they are able to continue the practice in their own time. It is a way of self-transforming mental states through self-observation, providing one with a sense of greater resilience so that the sufferer can benefit from its therapeutic effect.

The application of mindfulness meditation as a treatment can be traced back to 1955 and the International Meditation Centre in northern Yangon, Myanmar (formerly Burma). It was established by a civil servant in the government of Myanmar, Mr Sayagyi U Ba Khin. At that time, a certain wealthy gentleman of Indian descent was suffering from debilitating migraine headaches and was treated with doses of highly addictive morphine. He was an extraordinary industrialist and a business tycoon, but his wealth and fame gave him no peace of mind. After travelling around the world to find a doctor, he found no cure for his illness. One of his good friends recommended that he see Mr U Ba Khin, who was renowned for his integrity and effectiveness in teaching meditation. He was also a lay teacher of vipassana (insight meditation), a technique of self-introspection that had been handed down from ancient times by the community of Buddhist monks in Myanmar. After attending a ten-day meditation course under the guidance of Mr U Ba Khin, by his perseverance and diligent practice, this gentleman found a benefit he had never dreamed possible and cured his migraines. Thereupon, and for the rest of his life, he had deep sense of gratitude for his teacher. In 1963 came a turning point when the newly installed communist military regime in Myanmar launched a programme of nationalisation. Overnight this gentleman lost the industries he had established and much of his fortune as well. His name also appeared on a list of capitalists targeted for execution. He accepted this situation smilingly and urged his former employees to keep working hard for the good of their country. This gentleman is none other than Mr S. N. Goenka, who established an insight meditation centre in India which later spread across Western Europe, known as Goenka Meditation Centre. It is a non-sectarian school of mindfulness that offers mediation as a universal remedy for mental well-being.

Clinical interventions based in mindfulness have their limitations. Mindfulness-based cognitive therapy (MBCT) has been developed by Oxford Neurosciences at the University of Oxford with the aim of reducing relapse and recurrence for those who are vulnerable to episodes of depression. There is a great need to reduce the risk of relapse and recurrence in those who have been depressed because the amount of triggering required for each subsequent episode becomes lower each time depression recurs. Collaborative research conducted by Zindel Segal (Toronto), Mark Williams (Oxford), and John Teasdale (Cambridge) investigated how meditation may help people stay well after recovery from depression.[1] Their work is based on the observation that once a person has recovered from an episode of depression, a relatively small amount of negative mood can trigger a large number of negative thoughts (e.g. *I am a failure*; *I am weak*; *I am worthless*), together with bodily suffering of weakness, fatigue,

---

[1] Mark Williams et al., (2007). *The Mindful Way through Depression*. The Guilford Press.

or unexplained pain. Both the negative thoughts and the fatigue often seem out of proportion to the situation. Professor Mark Williams says that patients who believed they had recovered may find themselves feeling back at square one. They end up inside a rumination loop that constantly asks, "What has gone wrong?"; "Why is this happening to me?"; "Where will it all end?" Such rumination feels to the person as if it ought to aid him or her in finding an answer, but it only succeeds in prolonging and deepening the mood disturbance.

The American Psychiatric Association (APA) describes depression (major depressive disorder) as a common and serious medical illness that negatively affects how one feels, the way one thinks, and how one acts. Fortunately, it is also treatable. Depression causes feelings of sadness and/ or a loss of interest in activities one once enjoyed. It can lead to a variety of emotional and physical problems and can decrease one's ability to function at work and at home.

The APA lists a variety of symptoms that vary from mild to severe and can include the following:

- Feeling sad or having a depressed mood
- Experiencing a loss of interest or pleasure in activities once enjoyed
- Experiencing changes in appetite and weight loss or weight gain unrelated to diet
- Having trouble sleeping, or sleeping too much
- Experiencing a loss of energy or increased fatigue
- Seeing an increase in purposeless physical activity (e.g. inability to sit still, pacing, hand-wringing) or slowed movements or speech (these actions must be severe enough to be observable by others)
- Feeling worthless or guilty
- Having difficulty thinking, concentrating, or making decisions
- Having thoughts of death or suicide

Many who suffer from depression also suffer from eating disorders (such as anorexia) and insomnia. They become vulnerable to a deficiency in immunity. Symptoms must last at least two weeks and must represent a change in one's previous level of functioning for a diagnosis of depression. There are other medical conditions that can cause depression (e.g.

thyroid problems, a brain tumour, or a vitamin deficiency), which can mimic symptoms of depression, so it is important to rule out general medical causes. There are also other life situations that can change one's circumstances. Death of a family member, loss of a job, relationship breakdowns, separation, and divorce are common challenges that make people sad. It is normal to experience sadness in such real-life scenarios as an emotion that expresses the inner feelings of a person who is going through such a trauma.

In March 2020, the Covid-19 situation had a severe effect on mental health sufferers. People's emotions became heightened with uncertainty. The familiar social landscape disappeared, which caused anxiety. Fear comes with an impulse to fight, run away, or freeze with paralysis. Emotions can swiftly override one's slower rational and logical thinking. This leads to displaying irrational behaviour, although the reasons underlying such behaviour are unique to each person. Grief and bereavement are not the same as being clinically depressed. In major depression, feelings of worthlessness and self-loathing are common. A person suffering from waves of painful feeling, bipolar depression, or frequent mood swings needs clinical intervention for correction. Professor Williams's team observed in their study that during an episode of depression, a negative mood occurred alongside negative thinking and bodily sensations of sluggishness and fatigue. When the episode was past and the mood had returned to normal, the negative thinking and body sensations tended to disappear as well. However, during an episode the association had been learned between various symptoms. This means that when a negative mood occurs again (for any reason), it will tend to trigger all the other symptoms in proportion to the strength of association (this is called "differential activation")[2]. When this happens, the old habits of negative thinking will start up again. Negative thinking gets into the same rut, and a full-blown episode of depression may be the result.

Neuroscientists view their patients suffering from depression as having a chemical imbalance condition in their brains, and the patients are consequently treated with drug therapies to rebalance their chemical composition. However, the side effects of chemical-based pharmaceuticals

---

[2] John Sommers-Flanagan, Rita Sommers-Flanagan. (2016). *Clinical Interviewing.* John Wiley & Sons, Inc.

have led to the search for an alternative treatment. Applications of mindfulness as a therapy and its clinical applications are first found in *Seeking the Heart of Wisdom: The Path of Insight Meditation* by Joseph Goldstein and Jack Kornfield (Boston, Shambala, 1987). Both these writers have similar academic backgrounds and an exposure to Buddhist meditation practices. They were the pioneering contributors in introducing mindfulness meditation practices for mental health patients in North America. Mindfulness-based cognitive therapy (MBCT) helps participants to see more clearly the patterns of the mind and learn how to recognise when their moods are beginning to go down. It helps to break the link between negative mood and the negative thinking that it would normally have triggered. Participants develop the capacity to allow distressing moods, thoughts, and sensations to come and go without having to battle with them. They find that they can stay in touch with the present moment without having to ruminate about the past or worry about the future.

Buddhism explains in detail, with a comprehensive overview, suffering, the root cause of suffering, and the cessation of suffering. During my time as a mindfulness meditation teacher, I have met many young adults suffering from psychological problems, plagued with anxiety, depression, suicidal tendencies, and psychosis. I have successfully helped them to return to a normal, active life without relapse and to be successful in their professional careers. Among these are university students who were overambitious and unable to meet their life goals, people who therefore were stuck struggling to reach unattainable heights. Some of my colleagues have extensive experience working with drug addicts and alcoholics and visiting secure prisons to reform convicted criminals as defined by the Mental Health Act. The most important and effective ingredient in these types of therapies is to convince the sufferer, and have him or her self-realise, that there is hope. This allows the sufferer to rediscover his or her own potential to fight back, regain life, and become self-reliant. This is done by opening the person's eyes to a new vision, a positive and sustainable view of life that is the Middle Way as taught by the enlightened Buddha. When hope is gone, a person feels an emptiness and void in his or her life. In terms of the frustration of not finding the satisfaction and esteem he or she once enjoyed, the person will begin to feel anxiety-ridden, insecure, unsafe, lonely, and/or isolated. He or she may begin to fall into a bottomless pit without having anything to hold onto. The person may lose confidence to face the society wherein he or she once had a place and instead hide away to avoid social contact. An early detection of these symptoms would greatly help for correction by spiritually integrated psychotherapies.

Currently the transfer of spirituality into applicable treatment processes is underdeveloped. Treatment outcomes of such processes are not easily measurable. People's beliefs and moral backgrounds are very diverse, and early healthcare professionals often ignore the role of belief, faith, and moral values in therapeutic healing. Recently, though, in cosmopolitan cities where multiple cultures and numerous faiths are recognised, people are open to new ways of thinking and are considering valuable spiritual ideas to alleviate suffering. Buddhism is a spiritual tradition of teaching that has greatly influenced the West because of its openness, friendliness, and empathic approach to universal suffering. Regardless of belief, ethnicity, social differences, and geographical differences, it is open to all. The application of Buddhist values such as compassion and kindness towards another being's suffering and proactively doing something to help is the primary focus in Buddhism. This central teaching successfully addresses suffering and end of suffering.

In Buddhism, questions about the use of spiritually integrated approaches for healing are clearly recorded in a collection of discourses wherein the meta-analysis of morality and its relation to suffering is well examined. This explains the functions of psychological factors in the higher Buddhist philosophical psychology towards ultimate liberation of the mind. The process of spiritual healing is a sequence of mental activities that changes attitudes towards suffering and opens up positive thinking. This secures the confidence of the one who practises Buddhism and takes multiple corrective actions by letting go of wrong views to arrive at a stable mindfulness that can penetrate through the suffering and allow the practitioner to see its cause and its cessation. Suffering is the key element of spiritual awakening. This has been proven true in the history of Buddhism as well as in modern times. The world is awakening to this new knowledge, which was a hidden treasure in the East for centuries and now has arrived in the West. The great spiritual teacher the Buddha awoke to the truth of suffering and laid down a clear path for its cessation. Understanding suffering correctly is important as an early signal to diagnose the cause before intervention with any treatment. Misunderstanding would only lead to an incorrect or insufficient treatment process or to developing treatment procedures inappropriate for the cause because of the diversity of clients and other varying clinical issues.

There are two types of suffering: *"suffering that leads to more suffering and suffering that leads to end of suffering"*[3]. According to one of the main Buddhist doctrines, nothing arises singularly; there are multiple causes. Without understanding, neither structure nor stability could be reached.

I have applied a three-pronged approach to help students who are vulnerable to episodes of depression: tranquillity leading to insight meditation, psychotherapy based on Buddhist principles, and gradual withdrawal of medication recommended by their doctors. Clinicians currently only recommend one method of mindfulness because they lack the knowledge of Buddhist philosophical psychology or there are no effective therapies based on further research. The philosophy of science helps all scientists and researchers for their successful discoveries. If anyone wants to become a good and successful doctor, to reach his or her full potential, the person needs to understand the basic structure of a cell and the biological functions of different cells. Similarly, understanding of the mind, its structure, consciousness, functions of various mental factors, and the mind's relation to the body is important when treating psychological illnesses. Without a profound value and substance based on truth, psychotherapy might not be effective. Although *Tranquillity Leading to Insight* does not follow a strict academic approach as a research paper might, the following chapters give insight into the human mind, its structure, and the entire edifice of the mind as understood by Buddhism. It fills that gap of knowledge in the field of psychology with a different view, strong enough to make a paradigm shift.

Enlightenment according to Buddhism is a psychological process that completely purifies the mind from all defilements, hindrances, and negative mental factors that contribute to suffering. The knowledge of Buddhist higher-philosophical psychology would help professionals in experimental research to understand the psychological ramifications and neural mechanisms relevant to human behaviour by following the root

---

[3] Ajhan Chah Quotes. (2014). Amaravati Publications. UK.

causes of suffering by way of a different perspective. Intricate knowledge and analysis of the entire edifice of the mind could be useful in clinical practice to enable a clinician to interpret symptoms quickly and accurately and appropriately translate emerging results and findings into evidenced-based treatment processes. One can modify the process to benefit the mental health and well-being of many vulnerable groups suffering from episodes of depression, anxiety, worry, and headaches. Research could be focused on issues such as preventing depression and realising one's human potential across the lifespan.

The novel term used in researched-based methods of increasing human potential is *flow psychology*. This recognises the flow of positive energy when a person is performing some activity and accomplishing it to the best of his or her ability. Flow state is the mental state or condition in the present moment where the mind is fully immersed in a feeling of energised focus on the task at hand. The agent is happy, engaged, and confident with his or her participation in the process of that activity. In essence, flow is characterised by the complete absorption in what one does. It has a resulting transformation of one's sense of time. The concept of flow state has been well recognised by professionals in occupational therapy to improve the performance of a workforce.

The flow state shares many characteristics with a hyperactive mind. When the focus is intense, it can lead to negative behaviour when side-tracked; this is to the detriment of the overall assignment. In some cases, hyper-activeness can burn a person out, perhaps causing him or her to appear unfocused or to start several projects but complete few, thus underperforming and defeating the objectives. Overwork can result in poor performance, inability to make critical decisions, negligence, inability to manage time, absenteeism, and high turnover of workforce. Employers are now responsible for providing a healthy work-life balance for their workforce, improving productivity, and providing better employee relations. Positive psychology is the study of "positive subjective experience, positive individual traits, and positive institutions—promises to improve the quality of life". Positive psychology focuses on both the individual's well-being and societal well-being.

Positive psychology began as a new domain of psychology in 1998 when Martin Seligman chose it as the theme for his term as president of the American Psychological Association. It was a reaction against past practices which tended to focus on "mental illness", meanwhile emphasising maladaptive behaviour and negative thinking. The concept was further developed towards achieving life goals and the fulfilment of one's life, which encourages an emphasis on happiness, well-being, and positivity. Study in this area of psychology aims to understand the causes of work-related stress, anxiety, worry, apathy, breakdown of relationship, and boredom, all of which block the flow state. This is often in relation to one's occupation or profession with a hope of changing behavioural patterns towards better productivity and improving one's quality of life through one's happiness, confidence, creativity, peak performance, and control of the job.

Creativity is arguably one of the most important tools in the twenty-first century, and similar to flow states, it involves a state of consciousness rather than a discrete set of skills. There are massive opportunities to promote creativity in education, in the workplace, and beyond to reach the full human potential. The University of Southern California Flow and Creativity Flow Research Centre defines creativity as a recombinatory process where novel information is combined with older ideas to produce something new and useful. It is subdivided into four dimensions: product (what is produced), process (the creative methodology), personality (the creative personality traits), and press (the environmental conditions). *Flow state* is a modern term given to mindfulness, a concept which existed many thousands of years ago as rediscovered by the Buddha. Decades of neuroscientific studies show the power of mindfulness to relieve stress, awaken our positive capacities, and bring balance to our lives. Mindfulness gives us the power to meet any situation wisely, to be fully present, and to operate with compassion. When we can look more clearly at ourselves and our life situation, new possibilities naturally open up, transforming difficulties and allowing us to emerge stronger and wiser. Today's competitive employment market is looking for positive, creative, high-performing individuals with stable personality traits who have the ability to compete in the face of challenge and to resolve issues with a variety of uncertainties.

Buddhist higher-philosophical psychology provides us with invaluable discoveries of the functions of the mind. These theoretical methods of analysis, synthesis, and synchrony may include critical information about the functioning of various mental factors and their relationships with the mind and body. It is important to know what exactly is happening to devise the best course of treatment. It would be very difficult to understand a patient's cognitive behaviour without knowing enough about the mental functions that are being tested. The Buddhist method of mindfulness, the four noble truths and the theory of purification of toxic elements from the mind can also include higher scientific education to inspire and enrich theoretical rigour. By this, one can view illnesses and the meaning of life from an intellectually fascinating and philosophical standpoint. Currently there is no literature like this available to explain the depths of the human mind. It could narrow the knowledge gap of psychology by disseminating the higher Buddhist philosophy and contribute to a better understanding of suffering. Professor Mark Williams has also stated that such in-depth knowledge is certainly needed.

The effectiveness of intervention and the applied course of treatment depends on empirically evaluating what is described in the literature. In addition, because the majority of studies have focused primarily on outcome questions, many descriptive and process questions about the use of spiritually integrated approaches remain unclear. Not every phenomenon can be explained by reasoning alone. Miracles happen too. A miracle is a phenomenon that we are unable to explain. It does not necessarily fall on reason to reject it because, after all, it could be the truth we have been searching for. Most of the pure scientific findings were accidental discoveries. When reasoning is combined with empirical evidence, there comes a new method of understanding. What is needed now is more evidence-based information about the best practices for professionals to incorporate spirituality into their clinical practices.

Quoted below are a few comments from Professor Allison Harvey of the University of California at Berkeley's Department of Psychology's Golden Bear Sleep and Mood Research Clinic:

> *Although an evidence-based treatment for most types of mental illness has been developed, there is substantial room for improvement. The effect sizes of most available treatments are small to moderate, gains may not persist, and there are too*

many people who derive little or no benefit. Even under optimal conditions, treatment failure is alarmingly common.

Traditionally the development of psychological treatments has involved consensus between groups of skilled clinician researchers and many medication treatments have been discovered by serendipity. Hence, there have been calls for "increased attention to science" in the treatment development process[4].

The following chapters of *Tranquillity Leading to Insight* investigate the structure of the mind from an analytical point of view, consider the synthesis and synchrony of mental factors, and provide a broader definition of consciousness. They also explain historical developments, show how tranquillity and insight meditation practices complement each other, investigate the theory behind enlightenment and the discovery of truth, and explain the psychological functions in bringing suffering to an end. The nature of the mind is explained in terms of a rich philosophical psychology known as Dhamma. Investigating this on an intellectual level contributes to the advancement of this field of psychology, which is benefitted by the discovery of more refined and effective clinical responses and therapies. This will open new avenues for further research and be a vehicle to convey Buddhist psychology further to cognitive behavioural science.

Buddhism has lot to offer to the world. It can provide more evidence-based information about centuries-old practices. Professionals can incorporate, among other things, Buddhist spirituality into their clinical practices. The Buddhist practices of generosity, loving-kindness, and counselling based on wisdom, compassion, and moderation are very effective in addressing intersecting religious, spiritual, and mental health concerns. This gives an alternative treatment process for illnesses. These

---

[4] Allison G. Harvey and Nicole B. Gumport, "Evidence-Based Psychological Treatments for Mental Disorders: Modifiable Barriers to Access and Possible Solutions", doi: 10.1016/j.brat.2015.02.004; Golden Bear Sleep and Mood Research Center, https://www.ocf.berkeley.edu/~ahsleep/gbsmrc_mock/.

Suárez, L. M., Bennett, S. M., Goldstein, C. R., & Barlow, D. H. (2009). Understanding anxiety disorders from a "triple vulnerability" framework. In M. M. Antony & M. B. Stein (Eds.), Oxford handbook of anxiety and related disorders (pp. 153–172). Oxford University Press.

traditional practices have a profound effect on the process of spiritual healing, which is quantifiable and measurable.

In a paper published on 15 October 2020, "Millennial Health", the Blue Cross Blue Shield Association of the United States reported a downward trend in the health of the population of young adults between twenty-two and thirty-eight years of age in the year 2019. Generation Y are the demographic cohort following Generation X and preceding Generation Z; population studies often use those who were born between 1981 and 1996 (or sometimes 1980 and 2000) as the method of grouping people for statistical analysis. Unfortunately, there is a high level of youth unemployment in this age group because of the Great Recession of 2008 and the March 2020 Covid-19 recession. Additionally, Covid-19 had other consequences that severely impacted this generation. Ninety-two per cent of millennials said Covid-19 had a negative impact on their mental health. According to this report, there was a 34 per cent increase in alcohol consumption, a 20 per cent increase in smoking, a 17 per cent increase in vaping, and a 16 per cent increase in nonmedical drug use. These significant percentage increases in unhealthy habits suggest that when a young and active mind loses a purposive engagement and an opportunity to actively participate in the mainstream economy, to support itself and to be independent, it creates a vacuum in the young person's life that needs to be filled in one way or another. This generation is generally known for its elevated use of technological devices such as mobile phones and laptops, of entertainment such as computer games, and of social media for socialising. They depend on technology for information, for constant amusement, and for excitement and are thus hooked on those devices, games, and apps. They consume a great deal of online data for entertainment, yet they are unable to fill the psychological void they experience. They are very distanced from traditional values and religious faith, let alone any knowledge of history. It is novelty that has taken hold in this modern era, searching Google for information rather than engaging in active study to gain any depth of knowledge. This age group is vulnerable to social engineering that they could be easily manipulated to accept new social and political tends.

Blue Cross Blue Shield[5] stressed that the Covid-19 pandemic is likely to further accelerate the rise of behavioural health conditions among millennials. Certain lifestyle behaviours have risen within this group since the outbreak began, and these behaviours can lead to the development of behavioural health conditions or worsen existing ones. Today the world is facing a health crisis bigger than ever before in history, and it is our responsibility to respond with compassion for these vulnerable young people. They need to feel and think inclusively, giving them a sense of belonging to the society in which we live. If we do not address this wisely and identify appropriate solutions for these declining health issues and despairing habits, a looming situation could soon go spiralling out of control and these young people will miss the whole opportunity to do good in what time lies ahead of them.

Spiritual Healing

Could this unfortunate situation be reversed, and instead could the vulnerable people of this population be helped with ancient spiritual healing?

Recently there is an acceptance of and renewed interest in spiritually integrated psychotherapies with an emphasis on the clinical process and treatment outcome research. Together with occultist and mystical practices, there are many traditional values that have been lost or deliberately rejected in the past few decades because young generations think that there is no rationale behind spirituality. They are overtly addicted to fashionable technological solutions rather than taking simple remedies for their psychological well-being. For example, many religions and cultures have a concept of "holy water", which is water that has been blessed by a member of the clergy or a religious monk. The use of holy water as a sacrament for protection against evil is common among Lutherans, Anglicans, Roman Catholics, and Eastern Christians. There is a famous sanctuary among pilgrims at the Massabielle Grotto in Lourdes, France, popularly known as the Lourdes Water Fountain, which flows from a spring in the churchyard.

---

[5] Blue Cross, Blue Shield, Federation of United States health insurance companies, (1926).https://www.bcbs.com/the-health-of-america/reports/millennial-health-trends-behavioral-health.

Sick pilgrims are reputed to be miraculously healed by Lourdes water. On 6 September 2019, the *Telegraph* reported that a British man who believes that Lourdes holy water cured his cancer was to have his claims tested by a board of doctors in an upcoming *Songs of Praise* episode. Kazik Stepan feared his life was over after he was diagnosed with an inoperable tumour on his spinal cord when he was just eighteen years old. However, he claims to have experienced a miracle after he bathed in the sacred water at Lourdes, which enabled him to walk for the first time in months because of its reported healing properties.

Similarly, in popular Buddhist cultures, the sprinkling and drinking of holy water is a common sight in temples, where devotees feel themselves blessed. This simple treatment has a profound psychological effect of being recognised, valued, and made worthy through the concept of purification. Buddhist monks visit sick patients at hospitals and care homes with holy water, and these patients sometimes regain their health. Additionally, protection is granted by way of a holy wristband which represents entry onto the Noble Eightfold Path towards the end of suffering. These common practices have proven results. Buddhist spiritual chanting is another method of suffusing love and kindness known as *pirith* (*pariththa*), literally meaning "the protection". Chanting conveys the spiritual qualities of the Buddha. These chants are the recitation of the Lord Buddha, and they protect the listener from evil influences. Playing these chants every day in one's home, one's workplace, or one's place of business, or while one is travelling, brings benefits. It is the constant listing of the Buddha's qualities that can change the mental disposition of the listener with the vibration of change. Its rhythm has a positive effect on the mind of the chanter and allows for the release of negative energy and the cleansing of the mind, causing one to feel good and invigorated. Buddhist monks have a custom of morning and evening chants. It helps shift one's mindset if one is feeling low, or is just a daily practice to promote healthy thinking. Focusing on the sound, contemplating on great qualities of the Buddha develops deep concentration and sets the ground for meditation practice interconnecting the mind and body. Sensations of the body and mind can be experienced directly by a disciplined attention to the physical sensations that form the life of the body, as that which continuously interconnects and conditions the life of the mind. It is this observation-based, self-exploratory journey to the common root of mind and body that dissolves mental impurity, resulting in a balanced mind full of love and compassion.

Beliefs help us make sense of our perceptions by way of mental representations of the external world. This is what makes an opinion or judgement of what has been perceived to be true. Strong beliefs can lead to acceptance of new perceptions, and this sudden change could alter past tenets, allowing one to move forward and take corrective action. Having spiritual beliefs might also lead to enjoying a longer, healthier life.

A large body of research finds that religious people live longer, are less prone to depression, are less likely to abuse alcohol and other drugs, and even are less likely to need a doctor. On the other hand, circumscribed and delusional beliefs can create injury or pose a danger to life. These involve anomalous perceptual experiences created by a deficiency or a failing of the neurological or psychological function of the person's perceptual system. Misinterpretation of these experiences, due to biased reasoning related to cultural background, upbringing, or an impairment, leads to mental construction and imagining. The Buddha advised us not to accept anything for its face value but to investigate further before acceptance. Therefore, the position of religion is open to research, so that one can integrate issues involved in the philosophy and psychology of belief and examine the scope for a mutually beneficial interaction.

Recorded in the Pali Canon SN 2.1 (222–238), the Buddha convened to a city beset by an epidemic. The occasion for this discourse, in brief, according to the commentary, was as follows: The city of Vesali was afflicted by a famine, causing death, especially to the poor. Because of the presence of decaying corpses, evil spirits began to haunt the city; this was followed by a pestilence. Plagued by these three fears of famine, nonhuman beings, and pestilence, the citizens sought the help of the Buddha, who was then living at Rajagaha in India. Followed by a large number of monks, including the venerable Ananda, his attendant disciple, the Buddha came to the city of Vesali. With the arrival of the Master, there were torrential rains which swept away the putrefying corpses. The atmosphere became purified; the city was cleaned. Thereupon the Buddha delivered this Jewel Discourse (Ratana Sutta) to the venerable Ananda and gave him instructions as to how he should tour the city with the Licchavi citizens, reciting the discourse as a mark of protection to the people of

Vesali. The venerable Ananda followed the instructions and sprinkled the sanctified water from the Buddha's own alms bowl. As a consequence, the evil spirits were exorcised, and the pestilence subsided. Thereafter the venerable Ananda returned with the citizens of Vesali to the public hall, where the Buddha and his disciples had assembled, awaiting his arrival. There the Buddha recited the same Jewel Discourse to the gathering.[6]

Spirituality is defined as conditions that positively affect the human mind. It is a mental function that brings a profound relationship or emotional communication with the source that triggers belief. It is often coupled with religions, interposed, and interpreted as divine experience, which makes a devotional connection deserving veneration as the source from which it originated. These mystical experiences make people believe that there is hidden meaning in life, a spiritual symbolism or allegorical significance that transcends human understanding, inspiring unity with a superior being or divine power. This can, however, refer to any kind of ecstasy or altered state of consciousness which is given a religious or spiritual meaning.

There is archaeological evidence that ancient Egyptians and Mayan civilisations were endowed with certain mystical powers unknown to science[7]. This magical and divine power was transformed into aesthetic values by means of organised religions, architecture, and structural engineering. Beautiful sanctuaries, temples, shrine rooms, and altars were built in honour of and to glorify such powers. Some of these can be found at places of worship even today. The Ancient and Mystical Order Rosae Crucis (AMORC) also known as The Rosicrucian Order is part of an ancient primordial tradition that "... seeks to establish a fundamental origin of religious belief in all authentic religious teachings, adhering to the principle that universal truths are a cross-cultural phenomenon and transcendent of their respective traditions, mythologies, and religious beliefs" (Faivre, Voss. 1995).

The psychological effect of vibrations by chanting, praying, and singing hymns, make a profound impact on the mind, allowing

---

[6] *Ratana Sutta: The Jewel Discourse* (SN 2.1), tr. Piyadassi Thera, Access to Insight (30 Nov. 2013), http://www.accesstoinsight.org/tipitaka/kn/snp/snp.2.01.piya.html; https://www.youtube.com/watch?v=vedEwuh92RE.

[7] The Rosicrucian Order. AMORC. https://www.rosicrucian.org/.

it to transcend ordinary perception and move to ecstatic and mystical experiences, superseding the joy and happiness one may gain from other conventions. For example; Gregorian chanting is the central tradition of Western plainchant (plainsong), a sacred a cappella song sung in Latin (and occasionally Greek) of the Roman Catholic Church. Gregorian chanting developed mainly in Western and Central Europe during the ninth and tenth centuries, with later additions and redactions. Although popular legend credits Pope Gregory I with inventing Gregorian chant, scholars believe that it arose from a later Carolingian synthesis of Roman chant and Gallican chant.

The effect of sound on our state of mind as a healing power is more profound with religious chanting and hymns. It has a soothing effect like honey and lemon, and can produce some of the strongest emotional reactions in humans, whether happiness, sadness, fear, or nostalgia. Some melodies and tones can bring tears to the eyes of the singer and the listener alike. They can open up deep-seated emotions and bring them to the surface. Tears soothe our emotions—a clear indication of consciousness producing material phenomena to alter our mental state. Sound is a series of vibrations that travel into the ear and are transmitted to the brain via the cranial nerves, conveying sensory impulses from the organs of hearing and balance in the inner ear to the brain. The brain then tells the listener that he or she is hearing a sound and what that sound is. If you view depression or mental pain as being similar to a physical wound, then you'll accept that when it is cleaned and treated with the correct medication, it will heal naturally. The mind has an ability to heal when treated with natural means. Science has yet to recognise the full liberating power of sound.

These religious melodies follow the natural order and have been learned by monks through observing and listening to natural phenomena such as rain, wind, birdsong, ocean waves, or the echoing low rumble of a distant thunderstorm. The effect of the healing power of sounds was not understood by science until recently. Researchers at Brighton and Sussex Medical School (BSMS) found that playing natural sounds affected the bodily systems that control the fight-or-flight response and the rest and digest autonomic nervous systems, with associated effects on the resting activity of the brain.

They noted: "When listening to natural sounds, the brain connectivity reflected an outward-directed focus of attention; when listening to artificial sounds, the brain connectivity reflected an inward-directed focus of attention, similar to states observed in anxiety, posttraumatic stress disorder and depression." Their research highlights that our thinking, our attitudes, and even the people we surround ourselves with can have a major effect on our state of mind and overall well-being.

Buddhism and ancient Indian religions agree that all physical forms, including our bodies, are made up of four great elements. The universe is in constant vibration and flux, and science agrees that 60 –70 per cent of the human body contains water and that both mind and body are in a process of vibration. When in harmony with nature, or more precisely if one can attune to nature, the body and mind are brought into balance, stimulating the brain and the production of hormones to achieve mental stability. The Indian yoga system is a natural remedy that can reduce the stress of urban life, revitalising a person and improving his or her well-being.

Historically, religions have given humankind hope and comfort in times of difficulty, especially amid the psychological consequences of being disowned and socially ostracised, including being lonely and having unsupportive or absent family members. Religious institutions have successfully mitigated this very important psychological need of humankind for centuries. Buddhist higher-philosophical psychology goes farther and explores the root causes of suffering and the complete cessation of suffering with deeper analytical investigations into the healing power of the mind. Buddhist meditative traditions have much to contribute to the further development of cognitive therapy, combining it with spiritual healing so a person may face life's challenges with greater resilience.

These stories and prayers of healing may sound superstitious in this twenty-first century. However, these practices have given comfort to people in times of difficulty, have freed them from limiting beliefs, and are still alive in Buddhist cultures. Buddhism as a nontheistic religion encourages the idea that humankind's salvation is dependent on one's own effort. Traditionally, faith was fundamental to all religions and its therapeutic factor was devotion. Recently the word *faith* has been replaced with *confidence* by many writers of Buddhism. The Pali original word for faith is *saddha*, literally meaning "reaching out" to get something pure, what one seeks or needs for one's spiritual welfare.

The *Encyclopaedia Britannica* explains that the Theravada School of Buddhism, which claims to adhere most closely to the teachings of the historical Buddha, does not rely upon the supernatural authority or the word of the Buddha literally. Rather, it claims that all his teachings can be experientially verified. *Saddha* indicates one's provisional acceptance of the Buddha's teachings, the placing of trust and confidence in Dhamma as one enters onto the Eightfold Path (the system of spiritual progress). That trust in the Buddha and his teachings is later confirmed by direct experience and the growth of right understanding. The initiative to find answers for whatever the issue may be has to come from the sufferer himself or herself.

Faith is the first mental factor in the group of "beautiful mindset". Its psychological function is to clarify what has been perceived. The Pali commentator gives a simile of water cleaning a gem, which causes muddy water to become clear. The function of faith is to set forth, as one might set forth to cross a flood. It functions as a cleaning agent that removes gross impurities of the mind, namely greed, hatred, and delusion. Faith is the reason for firm resolution and the conviction to learn good Dhamma as firmly established in morality and to live according to the teachings. On the other hand, faith alone without wisdom could become blind faith. Questioning of its validity and engaging in metaphysical arguments will not cure the illness.

At the time of the Buddha, there were many wanderers who provoked him to engage in such useless conversation. The Buddha gave the following simile to them: a man shot by a poisoned arrow has his family and friends take him to a physician for treatment, but he insists that he will not allow them to treat him unless they answer his questions of who shot him, whether he is tall or short, what his clan is, and what type of arrow and poison the person who shot him used. Then the Buddha asked of his disciples, "Monks, what do you think will happen to him?" The monks wisely answered, "Surely, Lord, he would die." If you are ill, you must seek help from a wise doctor who can cure you, whether he or she is scientifically qualified or has a proven track record of traditional healing. It is a question of facts that make you want to believe in spiritual healing.

Within the human mind there is so much undiscovered potential, hidden and preserved. One needs to make an effort to unleash that energy with wisdom to realise one's dreams and goals. Potential is the difference between where one is now and the place one could reach. Potential will often lead to success for the individual who seeks it. If success is about achievement, then it must be a realistic and attainable goal, otherwise unattainable heights could demotivate the person to pursue it. For some it's about finding peace. Faith in Buddha's teaching would allow one to see beyond one's self-constructed walls and see things as they are that Intrinsic nature of all physical and mental phenomena.

Mindfulness and the Noble Eightfold Path as we know it has been proven by experience, with evidence of having healed difficult emotions and encouraging self-compassion that can extend to the wider world. You can manage your emotional state at any time. Meditation as a healing process allows you to bypass your limited thoughts and beliefs by learning not to attach to that which has gone. It allows you to turn your failures into your story of success. Mindfulness mediation can be the key to making a positive change in your life for spiritual awakening. Taking this one small step can change your life for the better and turn vulnerability into nonaggressive strength. It can open you up to discover the deeper, more profound scientific laws that govern your thoughts, feelings, judgements, and sensations and the relation between your body and your mind. Through direct experience, the nature of how you grow or regress, and how you produce suffering or free yourself from suffering, is understood. Life becomes characterised by an increased awareness, nondelusion, self-control, and peace. You feel liberated; rebuild your confidence; rebound with your loved ones, family, and friends; reconnect to the world; and find inner peace to rebuild your life. Buddhism encourages you to be with friends and family who are greatly supportive and who impact you by giving good advice and showing you the right direction, to be with those who inspire you and encourage you to be a self-reliant and productive person.

CHAPTER 5

HUMAN DEVELOPMENT

Human development is a concept defined as the process of enlarging people's freedoms and opportunities and improving their well-being. There are many approaches to defining this concept. The traditional approach spans the major stages of the human life cycle, including prenatal development, infancy, early childhood, middle childhood (puberty), adolescence, adulthood, middle age, and late or senior age. Another approach refers to the biological and psychological development of the human being throughout his or her life. With the advancement of knowledge in psychology, there are five theoretical approaches closely looking at the psychoanalytic, cognitive, behavioural cognitive, social cognitive, ethological, and ecological aspects of life. In this context the concepts of and approaches to human development form complex interconnections in our understanding of the evolution of the mind. Development is an important multidimensional element of the growth process. The scientific study of human psychological development is sometimes known as developmental psychology. This identifies events in our childhood, education, and cultural background in terms of emotions and mental forces that have a great influence on our adult lives. These are defining as mature personality traits and form our ability to face life's challenges with resilience or lack thereof. There is clinical evidence to suggest that difficult childhood experiences have adverse effects in adulthood leading to psychological illnesses.

In the twenty-first century, democratic societies claim to be providing people with political freedom and opportunities for their development. In developed societies there are ample opportunities for personal development; higher education, wealth creation, and one's social status and social class are highly valued, along with a focus on climbing the social ladder. Education is incomplete if there is no ethical or moral dimension; in such a case, it leads only to unfair competition and rivalry, leaving behind underprivileged groups to show the tears in the social fabric. In recent times the gap between rich and poor has widened around the globe.

Research shows that excessive concentrations of wealth and power at the top have negative consequences for nations across the globe and for metropolitan areas, in big cities like Chicago, New York, London and there are visible adverse effects can be seen in Asian cities like Colombo and Bangkok.

"If the income share of the top 20 percent increases by 1 percentage point, GDP growth is actually 0.08 percentage point lower in the following five years, suggesting that the benefits do not trickle-down. In contrast, an increase in the income share of the bottom 20 percent (the poor) is associated with higher GDP growth."[1]

According to recent research by the International Monetary Fund (IMF), such commercial siloes of wealth hurt not only those at the bottom income levels but also the overall national economy. Trickle-down economics, or "trickle-down theory," states that tax breaks and benefits for corporations and the wealthy will trickle down to everyone else[2]. It argues for income and other tax breaks or other financial benefits to large businesses, investors, and entrepreneurs to stimulate economic growth. The argument hinges on two assumptions: All members of society benefit from growth, and growth is most likely to come from those with the resources and skills to increase productive output. However, lack of fiscal prudence, capital flight and loopholes in tax laws allows wealthy tycoons

---

[1] IMF. Income Inequality. (2022). https://www.imf.org/en/Topics/Inequality.
Philip G. Alston (ed.), Nikki R. Reisch (ed.). Tax, Inequality, and Human Rights. (2019). Oxford University Press.
[2] Metropolitan Planning Council. Chicago. (2022). https://www.metroplanning. org/index.html. https://www.metroplanning.org/news/7415/Why-trickle-down-economics-is-still-not-the-answer

and corrupted politicians to hide their money in tax havens[3]. Critics argue that the added benefits the wealthy receive adds to the growing income inequality in the country and no benefit is pass down to many. The Buddhist view of income redistribution is to promote generosity, charity and encourage wealthy people to support monastic communities.

Culture and social norms are major contributory factors that influence people's behaviour at every stage of the life cycle. Such behaviour reflects the people's level of development and is regulated by the law of the land. The ethics of a society is one of the measurable indicators in psychodynamics which goes beyond the law and shapes the human personality. In simple terms, ethics is making decisions based on principles; in society it means using core values to guide people's choices. In civilised societies, civic leaders guide people by setting themselves as examples with integrity, honour, and a noble character, thereby becoming the role models of society. Take for example Jiddu Krishnamurti, an Oxford-educated philosopher of Indian descent who advocated nonviolent means to finding inner peace by using social philosophy for spiritual development.

People can learn from social interaction and use it to shape up their behaviour, character, and personality. Buddhism considers human development as one of the important components of social change. Its main focus is on developing the mind through mental training towards higher values in order to achieve the optimal functioning of mental factors. Love, compassion, caring for others, and the opportunity to engage in worthy and productive activities gives a person a sense of belonging to the community in which he or she lives. These humane qualities are the binding force that bring a community together. Aggression and violence are signs of ignorance that divide people. Buddhism recognises that a person's social position can be improved by education and training. For example, Thailand is one of the countries in South East Asia that is rich with cultural values. Their ethical dimension is much influenced by Buddhist principles, contrasting with Western materialistic values.

Complications can arise due to migration or, in the modern world, immigration issues. Historically, there were mass exoduses of people travelling by foot and carrying their belongings on their shoulders. Such

---

[3] Ronen Palan et al. (2010), Tax Havens: How Globalization Really Works (Cornell Studies in Money). Ithaca, N.Y. Cornell University Press.

exoduses became visible again in the twentieth century because of wars, political instability, and work opportunities. After World War II there was a mass immigration of people from Europe to the United States and regional resettlements in Israel and Palestine. These political settlements exhibited the use of violent aggressive means for decades, long before peaceful political dialogue began. Human migration brings a change to the population and provides an important network for the diffusion of ideas and attitudes that entails a culture change. It can be considered as a human adjustment to economic, environmental, and social problems. The uncertainty associated with the decision to change one's place of residence and adapt to a new environment could make a person displaced and disoriented, thus affecting his or her natural psychological development process. Human nature seeks the best, but if a person is unable to form a constructive relationship with the new environment, then he or she fails to engage productively and to make a valid contribution to the society.

Migration is viewed as a form of individual or group adaptation to perceived changes in the environment. The mixing of the earth's population results in changes to demographics, a dynamic cultural phenomenon directly affecting human development. The effect of this change depends on the stage of one's life when this displacement occurs, as characterised by a person's behavioural pattern. For example, if a family from Africa were to migrate to the United States, different members of the family would be affected by this change in the different ways. Successful adaptation to a new country could take generations, indicating symptoms of development disability and underdeveloped qualities and having major psychological ramifications. Bullying and harassment among children is common, so all schools have measures in place to prevent these behaviours. But it can still result in difficult schooling by undermining a victim's health and safety to produce feelings of isolation, despair, and even fear. In extreme circumstances it can lead to absenteeism or even leaving school altogether.

Central to the human development approach is the concept of capabilities. Capabilities of what people can do and what they can become is exactly what one has to pursue in a life of value. It can be extremely difficult for one to be successful and reach his or her full potential, especially in a different environment from one's native country. However, given the right conditions, success can be achieved. There have been good examples of success in immigrant families in the United States, Australia, Great Britain, and other parts of Europe in the past few decades. This is an indication of fairness and equal opportunity in those lands. Some people have natural talent and are gifted with developed mental factors. This is an aptitude or skill which has economic value and is in demand; one can make a living out of it if used wisely. An example of talent is the ability to sing well; given the right opportunity, a gifted person can become a star or an icon in the society and inspire generations. Commercial success alone is not the fulfilment of life, however. For example, Michael Jackson was an extraordinary singer and a stage performer. He is regarded as one of the most significant cultural figures of the twentieth century, but when middle-aged, Michael had difficulties with various allegations and police investigations into his private life, and suffered from sleep disorders. He died from an overdose of sleeping medication prescribed by his own physician. Talent is something that one is born with; it is one's natural ability to do something without really thinking about it. Skill, on the other hand, is something that one acquires after putting in a lot of hard work; unlike talent, it is not innate, but learned.

As Napoleon Bonaparte said, "Ability is of little account without opportunity." Displacement or having to live in unsettled violent lands could mean losing opportunities. Having nothing, one might turn to aggression or violent means to seek his or her survival. Young people can be easily manipulated to follow destructive ideologies when living under pressurised circumstances. This is a feature of war-torn countries, where many young people are caught up in political divisions, being radicalised simply because they see no other alternative under the circumstances. They suffer traumatic psychological stress, perhaps all their lives, with an increased risk of premature death, which could otherwise be preventable. There is evidence in developed societies such as Britain that young people from the working class at age as young as ten had been radicalised by far-right movements. Opportunity must improve the welfare of people without using violent means and should be aligned with human development according to what professionals agree to be beneficial. In this context, human development is a wise choice that requires mature guidance based on moral principles and values. Buddhism provides opportunity for human growth and development at every stage of the life cycle, thus allowing people to live life with hope and dignity. It can also be a healing power in difficult times. Buddhism is all-embracing and adaptable to the ethos of developed societies. It can enrich the values of human development at a young age to stabilise one's life for the benefit of the individual, adding value to improve attitudes and the aspirations of society at large.

A breakdown in the human development cycle at any stage of life could lead to mental illnesses, and if these are not identified or treated at the earliest stage, they could become a severe mental condition that requires specialised treatment. People with severe mental disorders on average tend to die earlier than the general population, a reduction of between ten and twenty-five years' life expectancy. In addition to suicide, people with severe mental disorders have a higher prevalence of many chronic diseases and are at a higher risk of premature death associated with these diseases. The excess mortality among this group largely relates to cardiovascular, respiratory, and metabolic diseases. Metabolic disease is a collective term referring to diabetes, hypertension, and weight gain. The prevalence of diabetes in people with schizophrenia is two to three times higher than that of the general population[4]. This is in part due to lifestyle and health risk factors, but it is also partly due to unmonitored antipsychotic treatment, which can lead to weight gain. Significant weight gain is one of the main reasons patients do not want to take prescription medication. Weight gain in this population also poses a significant risk of lipid abnormalities and cardiovascular complications. People with severe mental disorders also have higher rates of infectious diseases such as HIV and hepatitis. Studies have indicated that people with severe mental disorders are often at a socioeconomic disadvantage and have a greater prevalence of risky behaviours such as intravenous substance abuse and risky sexual practices. Patients with schizophrenia have been found to be at a higher risk for tuberculosis than the general population on account of factors such as a history of substance abuse, poor nutrition, homelessness, or previous time spent in an institution or prison. Behaviours leading to poor self-care, such as tobacco use and lack of exercise, are associated with depression, schizophrenia, and bipolar disorder and can lead to chronic illnesses such as coronary heart disease and type 2 diabetes. Patients with schizophrenia are more likely to smoke. The prevalence of smoking among

---

[4] National Library of Medicine, The Lancet Psychiatry, Volume 2, Issue 12, 2015.

them is about three times higher than that of the general population. Once chronic illness has developed, the severe mental disorders associated with poor self-care can lead to worsened health outcomes and higher mortality rates[5]. Symptoms of the mental disorders themselves can cause barriers to seeking care, as well as difficulty with following medical advice.[6] Mental health problems affect around one in four people in any given year. Good mental health means being generally able to think, feel and react in the ways that you need and want to live your life. But if you go through a period of poor mental health you might find the ways you're frequently thinking, feeling, or reacting become difficult, or even impossible, to cope with. This can feel just as bad as a physical illness, or even worse.

Human Development Index (HDI)

The United Nations Development Programme defines human development as "the process of enlarging people's choices". It says that choices allow people to "lead a long and healthy life, to be educated, to enjoy a decent standard of living" and to enjoy "political freedom, other guaranteed human rights, and various ingredients of self-respect". The Human Development Index (HDI) is a tool developed by the United Nations to emphasise that people and their capabilities should be the ultimate criteria for assessing the development of a country, not economic growth alone. The HDI can also be used to question national policy choices, asking how two countries with the same level of gross national income (GNI) per capita can end up with different human development outcomes. It helps governments to prioritise their policies. The HDI is a summary measure of average achievements in key dimensions of human development: a long and healthy life, being knowledgeable, and having a decent standard of living. It is the geometric mean of normalised indices for each of the three dimensions. The health dimension is assessed by life expectancy at birth. The educational dimension is measured by the mean years of schooling for

---

[5] The Lancet Psychiatry, Volume 2, Issue 12, (2015), Loneliness 'increases risk of premature death', Association for Psychological Science, https://www.nicswell.co.uk/health-news/.

[6] World Health Organization. www.who.int. National Library of Medicine.

adults aged twenty-five years and expected years of schooling for children of school-entering age. The standard of living dimension is measured by gross national income per capita.

Although there is a close correlation between economics and psychology, this aspect of human development has not been considered as a parameter for the measurement of human development. Neither economic nor political ideas recognise human feelings in their economic equations and bear no relation to human behaviour. The gap between economics and psychology has been bridged by religions. Choosing one's religion is a birthright recognised by the United Nations. All popular religions have a moral order of psychological principles based on a developed ethical system and spiritual values. This provides opportunities for their followers to participate in meaningful living in a communal sense. Religious clergy and leaders are endowed with higher virtues, and these grant prerogative religious rights. These are bestowed by them onto recipients to exercise that right exclusively for the recipients' spiritual development, giving them comfort and strength when faced with the challenges of life. From the seventh century to the twenty-first century, evangelistic religions systematically supported their converts with dignity and successfully contributed to the spiritual dimension of human development. During the tenth century in Anglo-Saxon England, religion was institutionalised and spread, centralised by a cathedral church system. Education was closely connected to these religious institutions and is the foundation of human development in Christendom. Breaking the religious order or code of practice was considered to be a sin or transgression, causing repugnance among church members or family. Repentance, kindness, and forgiveness use reverse psychology to correct transgressions in those religious orders. Knowledge gained through education without an ethical or moral dimension is incomplete for fully fledged human development.

Humankind's beliefs and faith has been challenged by economic, scientific, and technological developments. Overuse of and overdependency on technology and communications has isolated people and put them out of touch with the real issues of life. Today's online technology (*fifth-generation computer technology—5G)* is a social engineering tool capable of monitoring the user's real-life behavioural patterns, infiltrating the privacy of his or her innermost psychology, manipulating the truth by design, and instigating social change in the interests of a few elite classes. The elite seek to maximise the profits of these corporations that manufacture and market these technologies, stealing the user's opportunity for personal growth. Younger generations are addicted to the use of these tools. They are more isolated than ever before from the real world, suffering from anxiety and depression as a result of unfulfilled desires or simply not having an opportunity to constructively engage with society. They have become more distanced from religions and moral values, which can damage the fabric of their cultures. Addictive behaviours similar to any other substance misuse distort the mind, making one unable to make the right choices. In March 2020, the continued resurgence of the Covid-19 pandemic added more to this isolation, increasing demand for psychological treatments and exceeding people's capacity to cope. Researchers pooled the results of previous studies to estimate that loneliness can increase the risk of premature death by around 30 per cent.[7]

The HDI simplifies and captures only part of what human development entails. It does not reflect inequality, poverty, human insecurity, empowerment, or psychological needs. Religious teachings provide valuable principles for ethical behaviour and guidelines for bringing up children in a more refined and disciplined environment, and they serve as a comfort zone for the many challenges of life. Economic and political theory or technology cannot replace humankind's spiritual development. A breakdown in the development process at the early learning to middle-childhood stage would result in weaker personality traits and susceptibility to mental instability in adulthood. Such individuals tend to reject valuable social norms and become antisocial.

Positive Psychology

Positive psychology is a new concept developed by psychologists. It is concerned with the happiness and welfare of people. Positive psychology began as a new domain of psychology in 1998, when Martin Seligman

---

[7] National Health Service, United Kingdom, 2015.Loneliness 'increases risk of premature death', Association for Psychological Science, https://www.nicswell.co.uk/health-news/.

chose it as his theme for his term as president of the American Psychological Association. He did this with a view to moving forward from the old paradigm of treating psychiatric patients with a wider application of psychological principles to improve the mental health of the population. This move can be considered as the first attempt to relate economics and psychology, appreciating humankind's thinking and behaviour as an inseparable aspect of how they choose to meet their ends. It builds further towards other psychological needs to be fulfilled as the self-actualisation of a rational, economic human being. It aims to measure qualitative factors rather than the amount of money in one's bank account; more precisely, it could reflect human flourishing, prosperity, and blessedness. "The good life" is about what holds the greatest value in life—the factors that most contribute to a well-lived and fulfilled life. These values were already recognised and proposed by Socrates and Aristotle as the highest human good. Their concept echoes the religious teaching of "Blessed are those who pure in heart" for whom the gates of haven are widely opened. A person with the heavy burden of work and responsibility or who is subjugated to such conditions by imposed deadlines would be unable to see the world outside his or her work. Becoming a workaholic, such an individual can soon be burnt out by his or her workload or become sick and unproductive.

Economists now agree that recreation has an economic value; it is underwritten into insurance policies. A happy worker is committed to his or her work and is more productive than a clock-watcher, who waits for the bell to ring to escape the workplace. Positive psychologists have suggested a number of ways in which individual happiness may be fostered. Social ties with a spouse, family, friends, and wider networks through work, community associations, or social organisations are of particular importance, while physical exercise and the practice of meditation may also contribute to happiness. Building a good relationship with stakeholders, including employees, is a common feature of Japanese organisational models. For example, Nissan and Toyota motorcar companies have their own company philosophies built on developing soft skills to improve productivity, efficiency, and well-being of employees at all levels, recognising the concept of family values. Happiness may rise with increased income, though it may plateau or even fall when no further gains in income develop, indicating that to rely solely on commercial success may hinder true happiness. The value of spending quality time with one's spouse and children cannot be measured in numbers or money value.

Quality of life (QOL) is the general well-being of individuals and societies, outlining negative and positive features of life. It consists of the expectations of an individual or society for a good life. These expectations are guided by the values, goals, and sociocultural context amid which an individual lives. QOL serves as a reference point against which an individual or society can measure the different domains of a personal life. For example, a family can derive happiness and comfort by attending Sunday service at the local church or visiting a Buddhist temple and listening to a Dhamma talk. The extent to which one's own life coincides with a desired standard level, or put differently, the degree to which these domains give satisfaction and as such contribute to one's spiritual well-being is what determines the benefits.

Quality of life includes everything from physical health to family, education, employment, wealth, safety, security, freedom, religious beliefs, and the environment. QOL has a wide range of contexts, including in fields of diplomatic relations by improving welfare through cultural exchange programmes. For example, Thailand and Sri Lanka are notably proactive in promoting Buddhism overseas through their diplomatic missions. In this modern world, one of the important requirements is spirituality, along with elder healthcare and employment. Most healthcare workers in the United States and the United Kingdom are immigrants seeking employment opportunities to the detriment of their own societies. Health-related QOL (HRQOL) is an evaluation of QOL and its relationship to health. Facilitating healthcare for the elderly has become increasingly expensive and unaffordable for many. Quality of life should not be confused with the concept of the standard of living, which is primarily based on income.

Standard indicators of the quality of life include not only wealth and employment but also the built environment, physical and mental health, education, recreation and leisure time, and social belonging. Flower gardens, parks, and leisure centres provide emotional well-being, while religious institutions give people assurance for their spiritual needs. According to the World Health Organization (WHO), quality of life is defined as "the individual's perception of their position in life in the context of the culture and value systems in which they live and in relation to their goals". It suggests the human evaluation of quality of life is the avoidance of pain and replacing it with comfort at all stages of the life cycle. Buddhism clearly shows in its First Noble Truth that the factor of suffering is included in all stages of the life cycle, irrespective of one's culture or beliefs. One should investigate the cause of suffering so as to avoid adding more suffering, if not to follow the path to eliminate it.

Cultural values are useful only if they contain valuable ethical and moral values. Striking the balance between income and QOL has become difficult in this hyperinflationary economic system with unfair employment conditions, exacerbating factors that contribute to the cause of suffering, including poverty and hardship. Industrial actions and labour disputes are the result of unfair treatment, and if these are not wisely resolved, at the extreme, it could move political movements to the far right or far left. Harsh economic conditions are grounds for exploitation of labour and human trafficking. These activities call into question ethical and moral values. These practices are visible in history during periods of economic decline caused by wrong political ideologies; civil wars forcibly displace people and leave them in despair, affecting their human development cycle.

Buddhist higher-philosophical psychology recognises not only one lifespan, but many cycles of life. It provides basic knowledge and offers practical solutions for the correction of behaviour. It addresses broader psychological and key issues for human development at every stage of life regardless of gender, social standing, or religious belief. It provides the framework and foundation for a specific purpose of life science. Mental development according to Buddhist higher-philosophical psychology is the development of consciousness and its associated mental factors to a higher and sustainable level. Its ultimate goal is the full liberation of the mind. Mindful living helps well-meaning people not to violate the law and establish good social conduct, even unintentionally. The March 2020 pandemic has been a test of our patience and our responsibility to prevent the spread of the virus, thereby protecting each other. Mindfulness is supreme; it encompasses human development to regulate behaviour that has effect, not only at the individual level, but also on a broader global level to help each other.

The Spiritual Importance of End-of-Life Care

People are traditionally focused on the physical aspects of life, namely their economic needs, comfort, luxury, and medical care, while often overlooking their spiritual and psychosocial needs, particularly at the end of the life cycle. The end of life is often where spiritual matters come to the fore and people may wish to re-examine and reiterate their beliefs in order to die peacefully. Most religions involve submission to a divine entity. Buddhism, as an exception, encourages people to accept responsibility for life and provides guidance on how to live purposefully, as well as provides rituals which comfort and influence a person and his or her family at the end of life. Discussion of spirituality is an individual or family responsibility more than it is the responsibility of the healthcare system, although exploring the main faiths and religions of the United Kingdom might allow healthcare professionals to pause and reflect on delivering effective and culturally competent care or allow family and friends to intervene at the end of life for their patients. Historically, people died at home within the comfort of the surroundings of their communities, where their own religious beliefs and traditions were conducted with ease. Since traditional family arrangements have largely changed because of immigration, urbanisation, and other social economic changes, even within a person's own family, there is no prior arrangement to take care of this aspect of life.

A significant number of deaths in the UK occur in hospitals. Those working in hospitals and other community settings do not have a strong understanding of a multifaith approach at the end of life to support the provision of personalised and holistic care in a very demographically diverse society like London. Covid-19-related deaths have highlighted the need for understanding the spiritual dimension of life as many Covid-19 patients died alone without having any family members at their bedside. Buddhism encourages a method of meditation to contemplate death when young and healthy, not to discourage life expectancy, but to give one a more realistic view of the impermanent nature of life. Buddhism says that the time, place, and way in which a person may die is unpredictable, so we had better be prepared to accept death unexpectedly. Understanding life better would certainly help us to live life purposefully and to its full potential without wasting our golden opportunity of being live.

Awareness of the truth of life helps people to incorporate religious and spiritual needs alongside physical, psychological, and social considerations while young and able. This requires people to develop an understanding of their own sociocultural and religious traditions observed by family. This allows people to self-determine by recognising their own idiosyncratic priorities through reconnecting to their source of empowerment—their own religious and cultural platform. To all the main religions in the world, including Buddhism, the end-of-life experience is important and represents the passing from the physical world to beyond. The endeavour to provide a peaceful death is one that is ubiquitously important to all religious communities. Most religions uphold the belief that life is sacred, and therefore there is an assumption that the preference and belief of those with faith would be to prolong life where possible; however, this is not always the case. Such an understanding may enhance the person when he or she reaches the end-of-life experience and allow him or her to avoid fear, anxiety, and worry.

The family is frequently the person's advocate and voice, and as the person is dying, if family is not around, he or she can die peacefully. At the end of life, many people do not wish to be separated from the communities in which they have lived, and those close to them are likely to require local support to cope with their loss. It has long been acknowledged in Buddhism that suffering is not always only physical and can encompass the psychological, social, and spiritual dimensions, which also need to be taken care of. Buddhist temples and monks understand the truth of life and are experienced in providing funeral services. This community-centred focus recognises the need for monks, caretakers, and community leaders to work alongside professional undertakers to ensure that end-of-life care services meet the needs of local communities according to their customs and practices. Buddhist temples understand the importance of faith in comforting affected families and recognise that providing appropriate religious care is important to ensuring that spiritual end-of-life care needs are met.

The Buddha's teaching is primarily based on the law of kamma or karma. Buddhists and Hindus share a similar concept of life and death, believing that death is not the end and that the cycle of birth continues. Kamma guides the Buddhist in how to live an ethical and moral life.

Buddhist temples provide opportunities to acquire good kamma at all stages of the life cycle. They are interpreted as blessings for life. For example, human development is not limited to having a good education and developing a professional career. It is made complete by living a moral, ethical life and performing one's duties, particularly looking after one's elderly parents and re-educating them on Buddhist doctrinal teachings reinforcing traditional and family values.

If one decides to enter monastic life for further training and purification of the mind, this is regarded as a higher blessing to secure a favourable rebirth towards final liberation of the mind. Many Buddhists wish to be reborn in the heavenly realms or in the human realm where Buddhist teachings are prevalent. Most of these values are culturally woven into the social fabric of the society. The law of kamma is a moral one of cause and effect that encourages to live according to moral codes and determines life cycles and rebirth. Kamma follows a hierarchy, based on the accumulation of spiritual merit of good and bad deeds throughout multiple lifetimes. This doctrine strongly influences a Buddhist's attitude to life, particularly as he or she views most life events as the result of kamma. In this regard, Buddhists believe that suffering is born as a result of past and present kamma. For example, smoking can harm pregnant mothers and their unborn children, causing cancer. Suffering and illness is not seen as unique to the individual; rather it is viewed as the result of prior actions in this, or a previous, life. Such suffering may be caused by physical, emotional, and spiritual pain. Some may wish to endure this pain and refuse treatments to avoid reducing consciousness to focus their minds on being present and dying mindfully, which is considered as a good death. Buddhists have successfully mitigated the psychological effect of death by education based on wisdom to show a positive side to it, namely that death is not necessarily a bad event but is a truth of life.

Before the introduction of commercial flights that contributed to large-scale tourism as we know it today, trade was one of the principal means by which people of different religions and cultures met each other. It is also stated in the Christian Bible depicting in stories where Arab, Persian, and Indian traders have travelled, meeting with Middle Eastern religions. Similarly, although Buddhism is not traditionally a religion that actively seeks to 'convert' others, it nonetheless spread across South East Asia and became a widely followed religion in many countries in the Middle Ages, due largely to the voyages of Buddhist traders across Central Asia. There is evidence to suggest that Buddhism has spread along the ancient Silk Road from China as far as Afghanistan, Pakistan, India, Indonesia, Malaysia, Nepal, Thailand, Viet Nam connecting these countries with common values. Buddhist monks travelled on trading ships too, to go on pilgrimage, thus carrying their religious practices far afield sharing cultural festivals whilst promoting economic cooperation. (Silk Roads Highway of Culture and Commerce, UNESCO). In the Far East, Buddhism is much entangled with ancestor worship binding a family strongly together. Family is the unit of society and the comfort and support until end-of-life cycle. Family ties are extended across cross cultures through common beliefs to bring different nations together promoting goodwill.

CHAPTER 6

FLOW PSYCHOLOGY

Introduction

Tranquillity Leading to Insight provides vital insight into the complexities of human behaviour—the product of every neural process. It is well grounded in the higher-philosophical psychology of Buddhism, which delves deeply into the human mind, explaining consciousness and the mental factors that contribute to the peak performance of one who is in flow. One who is in flow has an intuitive knowing but is unable to explain it. For example, a meditator's mind is no different from that of a warrior on the battlefield, but his task is to conquer himself. Confucius wrote, "The greatest warrior is one who conquers himself." During meditation the mind can drift into unknown terrains. These unfamiliar experiences can be frightening, uncertain, and frustrating. It therefore requires courage, determination, and wisdom to succeed in emptying the mind of confusing thoughts and distorted perceptions. As in combat, one has to make split-second decisions between life and death with no margin of error. This comes down to an individual's skill, or put more passively, to being fully aware of the present moment. With knowledge of the psychological processes as bodily and vocal intimations, one can explain this peak performance. In positive psychology this is called the flow state, also known colloquially as "being in the zone"; both are names given to what has been known in Buddhism for thousands of years as mindfulness.

The scientific community has begun to explore this field of expertise more recently. However, there is much to explore when it comes to the neurocognitive underpinnings of flow. Such research can help to better understand the cognitive processes of this elusive state. One element that has begun to gain a growing amount of attention is the peak performance found in flow states, whether it be in sports, business, or other professional endeavours. Flow is described as a state of optimal performance denoted by smooth and accurate endeavour. There occurs an acute absorption in the task to the point of time dissociation, whereby one is free from the distractions that can cause disturbances to the mind. The flow accelerates one's ability to learn new physical skills and use existing skills more productively, thereby amplifying performance and allowing a rising to the challenge in tentative, uneven, and uncertain situations.

The following narration crosses ancient wisdom based on meditative practices with modern flow psychology. This will facilitate a deeper understanding of the relation between Buddhism, psychology, and neuroscience. *Tranquillity Leading to Insight* is primarily framed by the Buddhist path of liberation towards the end of suffering. It also contains the psychological principles that contribute to human development, progression, and the eradication of psychological ramifications to find life satisfaction.

Neuroscience

Neuroscience, also known as neural science, is the study of how the nervous system develops through its structure, its relations, and the functioning of the brain. Neuroscientists focus on the brain and its impact on behaviour by its cognitive functions. Some advances in neuroscience help to prove psychological theory. On the other hand, neuroscience also provides breakthroughs that challenge classical ways of thinking, consequently influencing the way we work and live. New developments complement each of these disciplines. Psychology provides vital insight into the complexity of human behaviour as the product of all those neural processes. Neuroscientists believe that the human brain possesses inexhaustible resources and that we therefore do not use the full potential of our brains, for which there is much scope in improving human capabilities to achieve peak performance. Currently there are studies being carried out in occupational therapy to understand how to best engage employees in eliciting peak performance. A person with the right skill level is intrinsically motivated when the challenges of critical tasks are higher. With a paper published in 1997, Mihaly Csikszentmihalyi introduced flow theory. He is a Hungarian-American psychologist and the former head of the department of psychology at the University of Chicago in the United States. The flow experience relates to the skill set perceived to be possessed by the individual relative to the perceived challenges of the activity. Challenges can be considered as "opportunities for action", thus flow is produced by any situation that requires skill. Csikszentmihalyi's theory was tested in everyday working environments. It proved that high-skilled knowledge workers in flow sate are motivated when the challenge is high, and the outcome is inversely proportionate to the challenge and skill. For instance, if the task is very challenging yet the person possesses no skill level, he or she will experience worry and anxiety, yet if the skill level is very high and the challenge is low, he or she experiences apathy and boredom. Skilled persons, such as brain surgeons, enjoy a difficult task when engaged with it. When they see a breakthrough, they become fully immersed in accomplishing the task. For instance, at first the task might appear boring or anxiety-provoking, but if the action opportunities become clearer or the skill level improves, the task becomes more engaging and finally enjoyable. The optimal level is achieved at the equilibrium point of skill and challenge. The levels of challenge and skill are variable. Research shows that there is an inverse correlation between them. That is to say, the relationship between the two variables changes in opposing directions; as one increases, the other will fall.

Many people are unaware of this fact and suffer psychologically from worry, anxiety, and pressures when faced with challenges. They feel helpless to cope with the perceived challenges within and outside themselves. As a result, they feel miserable during much of the day and night. In this state, their emotions can constantly bounce between extremes of highs and lows. People often describe mood swings as a "roller coaster" of feelings, from happiness and contentment to anger, irritability, and even depression. This phenomenon is a common concept used to describe rapidly and intensely fluctuating emotions. Instead of wasting the day watching television or playing computer games, one can transform boredom by taking a different approach to one's daily routine. Most people's symptoms are mild enough that they do not seek mental health treatment. Because the emotional highs feel nice, they do not realise that there is anything wrong or want to seek help. But mood swings can affect daily life and cause problems with personal and work relationships. There is a risk of developing bipolar disorder[1], so it's important to get help before reaching the latter effects of higher magnitude. The key to changing the habitual pattern of emotional disorders is to engage in tasks that require a high degree of skill and commitment.

Meditation practice is similar, and it is compatible with the flow state described in the literature. When one acquires the required skill level, one enjoys the transcendental experience of unity with the activity. This dissolves one's sense of self-identity, causing the person to become focused and delve deeper into what cannot be expressed in words. One endeavours to continue on this path until one sees its results. The Pali commentators give a simile of a ripe mango fruit: its taste cannot be expressed in words; unless you eat it, you cannot know it. Meditators on intense retreats often report ecstatic flow experiences in their critical tasks, raising the bar above the average level of concentration on the task at hand. Insight meditation is a method of critical investigation into one's own mind to understand the cognitive processes involved in behaviour patterns that block the flow in order to eradicate defilements from the mind. This allows one to carry on under difficult conditions, irrespective of how challenging the task may be, until one sees its beneficial results.

Research shows that flow experiences can have far-reaching implications in supporting an individual's growth, by contributing both to personal well-being and full functioning in everyday life. Flow psychology is a novel concept of coach, teach and train others in the science of peak performance or reaching optimal performance and flow states by helping clients identify their obstacles, psychological barriers and offering them viable path forward. This can be easily achieved by simply incorporating mindfulness meditation practice into one's daily routine to balance emotional states

---

[1] Beverly Merz, "Six Common Depression Types", Harvard Health Publishing, Harvard Medical School (13. Oct. 2020), https://www.health.harvard.edu/mind-and-mood/six-common-depression-types, accessed 8 July 2022.

building confidence of can do attitude. For example, the X Factor is a television music competition franchise created by British producer Simon Cowell and his company Syco Entertainment. It originated in the United Kingdom around 2004 and has been adapted in various countries around the globe in different names. The "X Factor" of the title refers to the undefinable "something" that makes for star quality. One of the important premises of the show is the quality of the singing talent. Talents are selected in open auditions, where the coaches then train raw talent to that of world-class performance. Four coaches, themselves popular performing artists, train the talents in their group and occasionally perform with them. One of the popular show; America Got Talent gives young people opportunity, a platform to build confidence, express hidden talent and reach to their peak performance through competition. Flow in this sense is limited, it is applied for worldly pleasure, entertainment, fame and not for higher spiritual development or its healing power. Certain frequencies of music have healing quality, an effect of a therapy and can break the emotional psychological barriers to open up mental blocks that influences how you perceive others' actions and prevents you from clearly communicating your feelings in a productive way.

The application of mindfulness focuses attention on the present moment. It has proven successful in psychological studies that now recognise meditation's therapeutic value in treating patients suffering from various psychological problems. Meditation practice is a skill recognised by both psychologists and neuroscientists that can successfully alter cycles of depression, stabilise mood swings, and allow one to enjoy life in the moment and in the sun again. There are a lot of other benefits from meditation practice, from coping with work-related stress to improving the general well-being of employees. Meditation helps to calm one's nerves while improving brain function to help one perform well under pressure, manage workloads, and engage in work practices with interest. It can expand one's social cognition, bringing neuroscience and psychology to everyday work routines, increasing efficiency, productivity, and creativity in a competitive business environment.

Recent theoretical and empirical studies in neuroscience have recognised the concept of flow state and flow experience as a brain function that has the capability of effortlessly processing information and has the advantage of being more efficient. It is supported by skill-based knowledge which contributes to success and winning, such as in athletic performance. Flow states have been shown to help people reach peak performance. Meditation practice helps in attaining this elusive state by stabilising the mind and protecting it from low mental states of apathy, worry, and anxiety. It increases energy, arousal, confidence, control, and flow. The ability to reach high performance can be further enhanced by the induction of brain simulation. This is a method used to train high-calibre individuals for a purposive course of action, such as fighter jet pilots or rescue operatives who possess a skill set to meet the perceived dangers of a given situation. They are focused and think fast, risking their lives to protect others, but they enjoy engaging in risky situations, an essential prerequisite when entering into a well-defined precarious task.

A centuries-old method known to the Eastern philosophies, meditation practice makes the mind flexible and increases the capacity to think fast and to assess and judge a critical situation accurately, thus enabling one to reach peak performance. For example, martial arts is a technique of harnessing mental and physical energy for one's optimal performance by paying acute attention to movements. Lee Jun-fan, commonly known as Bruce Lee, was a twentieth-century American actor of Chinese descent, along with being a director, martial artist, martial arts instructor, and philosopher. He combined martial arts and cha-cha dancing, creating a hybrid martial arts philosophy that drew from different combat disciplines to introduce something new to the world. Flexibility of the mind was the key feature of Lee's success. Lee extended the limits of the kinaesthetic possibilities of an average human by constant practice and perfecting of his art of fighting. His action sequences flowed seamlessly, effortlessly, and perfectly from one to another. At his peak, his best level, high-speed cameras even could not capture his movements. He brought Chinese martial arts to the world stage with the grace of a ballet dancer. Bruce Lee said, "Empty your mind, be formless, shapeless—like water. Now you put water into a cup, it becomes the cup, you put water into a bottle, it becomes the bottle, you put it in a teapot, it becomes the teapot. Now water can flow, or it can crash. Be water, my friend." He died prematurely at the age of thirty-two. According to the autopsy report, an aspirin/meprobamate fixed-dose combination (FDC) for treating short-term pain, tension, and anxiety-induced headache was found in his brain, indicating limitations to the mundane flow state.

A qualitative investigation into the flow experiences of elite figure skaters was conducted in order to gain greater insight into the nature of flow in sport[2]. Sixteen former US National Champion Figure Skaters, who held their titles between 1985–1990 were selected for interviews. Researchers have interviewed champions on an optimal skating experience, and then questioned extensively about factors associated with achieving optimal, or flow states, during performance. The task of performing at high speed in synchrony with a partner requires more physical endurance and controlling ego oriented emotions for smooth and accurate movements. Factors perceived as most important for getting into flow included a positive mental attitude, positive pre-competitive and competitive affect, maintaining appropriate focus, physical readiness, and for some pairs-dance skaters, unity with partner. Those factors which were perceived to prevent or disrupt flow included physical problems or mistakes, an inability to maintain focus, a negative mental attitude, and lack of audience response. The skaters placed very high value on flow-like states, and their descriptions of what was occurring during optimal skating experiences paralleled many of the characteristics of flow described by Csikszentmihalyi (1975, 1990). Drawing on the experience of elite athletes may enhance understanding of flow states as they occur in sport. The bottom line of sport activity is feelings it provides. Winning or record–setting is important only to the extent that they make athletes and spectators feel elated. For professional athlete winning being not the only thing that matter, participation and representation are equally important at national competitions.

Buddhism views all phenomena as ultimately impermanent and all people as subject to suffering. The highest human good is achieved by focusing internally on one's own mind. The process of higher cognitive functions such as love, compassion, and empathy direct the mind to search for excellence in more altruistic values, which alleviates suffering for all beings in spiritual development. Researchers have identified these high virtues as an implicit process of the flow state. There is danger in mundane flow state, namely that one can overwhelmingly be absorbed into

---

[2] Susan A. Jackson, (1992) Journal of Applied Sport Psychology, Volume 4 – Issue 2. *Athletes in flow: A qualitative investigation of flow states in elite figure skaters.*

its blissfulness and become overconfident, craving for pleasure of feeling and going into extreme overdrive. There are risks involved in adventure, including not succeeding at all. Misadventure could make you depressed, resulting in a fatal accident, crippling injury, or some other manner of death. Buddhism encourages the Middle Path so one can learn from one's mistakes or make improvement on one's successes by developing equanimity towards success and failure.

The Buddhist higher-philosophical psychology provides insight into the modern concept of flow state by addressing how mindfulness can balance the associated mental factors in synthesis and synchrony on the Middle Path to make the mind flexible and direct its function towards accomplishing purposive goals beyond the mundane to a supramundane state. According to Buddhism, the biggest challenge in life is suffering. Buddhist teaching offers a way to overcome misery by crossing the floods of life. The Noble Eightfold Path is the bridge that allows one to cross to the other shore and guides one to master the skills of peak performance smoothly and safely every step of the way reaching peace of mind.

CHAPTER 7

———— ✿ ————

UNDERSTANDING THE FOUR NOBLE TRUTHS

THE WORD *TRUE* IS DEFINED by the *Oxford Dictionary* as "in accordance with facts, accurate and genuine".[1] When we are describing something according to its facts, if those facts are accurately derived and are genuine, only then can that thing be said to be true. However, our descriptions are conventional, dependent on our perceptions and level of understanding. According to our understanding of facts, we can speak of man, woman, river, mountain, trees, etc., and other apparently stable objects such as planets, stars, and galaxies. This level of perception is conventional truth (*pariyaya sacca*), true according to its conventions and projections of our mind. However, by closer investigation of conventional truth one may discover that these facts can be further reduced into components and elementary units (*paramatha*). In this line of analysis, the Buddha didn't reduce things to molecular, atomic, or particle level. These are scientific concepts. Nothing exists or manifests in nature in these forms; they are perceptions of a scientific mind. Such a mind attempts through perception to define, measure, and experiment to prove a hypothesis by way of evidence.

The theory of having two truths, conventional truth, and ultimate truth, is first found in history in the sixth century BCE with the emergence of Siddhattha Gautama's enlightenment. It is stated in the Pitaputrasamagama

---

[1] "True", *Lexico*, https://www.lexico.com/definition/true, accessed 4 July 2022.

Sutta that Siddhattha became the Buddha, the "Awakened One", because he fully understood the meaning of the two truths, and then came to make a unique contribution to Indian philosophy.

Nagarjuna was the founder of the Madhyamika School of Buddhism, a branch of Mahayana Buddhism. In the Mūlamadhyamakakārikā, Nagarjuna attributes the two truths to the Buddha as follows: "The Dharma taught by the Buddha is precisely based on the two truths: a truth of mundane conventions and a truth of the ultimate." The Madhyamika School of philosophers claim that the theory of the two truths is the heart of the Buddha's philosophy. According to them, the two truths serve as a mirror to the core message of the Buddha's teachings. Therefrom arose the massive philosophical literature this theory subsequently inspired. However, there is no clear evidence in the Pali Suttas (Nikayas) to explicitly ascribe the two truths to the Buddha's teaching. On the other hand, it is not entirely disconnected from the Buddha's teaching either. For example, in *The Numerical Discourses of the Buddha* (AN 60: Seeking the End of the World, and AN 118: Six Routes of Escape), the Buddha states that the end of world is in the "signless liberation of the mind"—referring to absolute or ultimate truth.

The Sarvāstivāda[2] was one of the early Buddhist schools established around the reign of Ashoka. It was particularly known as an Abhidhamma tradition, with a unique set of seven Abhidhamma works. The rise of the Sarvāstivāda school of Buddhism as a distinct group dates back to the 2nd to 1st centuries BCE. Sarvastivadin claimed that conventional truth is evident when an object is physically destroyed, or its defining properties are stripped away. The relating idea or concept ceases to arise, thereby conceptually excluding its conventional identity. A pot and water are designated as conventionally existent and therefore conventionally real. The concept of a "pot" ceases to exist when the object is physically destroyed. The concept of "water" no longer arises when we conceptually exclude from it its shape, colour, etc. For example, when water is boiled, it transforms into vapour, and when frozen, it transforms into ice or snow. On the other hand, if we add oranges to water, it is perceived as orange juice. Pot and water are therefore not the foundational substance or entity. They are rather composite conditional entities. By composite entity, we

---

[2] Oxford Bibliographies, Oxford University Press (2022).

mean an entity or existence which is not fundamental, primary, or simple. Rather, it is a conceptually constructed composition of various properties and is thus reducible both physically and logically.

The Yogacara School, in response to the question "What causes mental representations?", replies that "subliminal impressions" arise from foundational consciousness, and it is these that cause mental representations. These impressions are only internal phenomena acting as intentional objects, therefore excluding contact between one's senses and the external object. By this, the Yogacara deny even the conventional reality of all physical objects, arguing that all conventional realities are our mental representations, mental creations, cognitions, etc.

Central to the Yogacara philosophy is the theory of two truths. This echoes the assertion that anything conventionally real is only an idea, a representation, an image, or a creation of the mind; there is no corresponding object that exists outside the mind, only objects of cognition. The whole universe is a mental universe. All physical objects are fiction; they are unreal even by the conventional standards, similar to a dream, a mirage, a magical illusion. What we perceive are only products of our minds without any real external existence. The clash of contradictory arguments eventually led to the devolution of these schools.[3]

The final irreducible components of existence, according to Buddhism, are called Dhamma. These constitute the ultimate reality as detectable by the mind. Dhammas are not static objects but are true constituents of the dynamic reality that is our experience. Ultimate truth (*nippariyaya sacca*) is the penetrative knowledge that discerns ultimate reality from conventional truth. According to the Abhidhamma philosophy, there are four ultimate realities: consciousness, mental factors, matter, and nibbana. They are called ultimate because there is no further reduction of their constitution. The first three realities (consciousness, mental factors, and matter) are conditioned and appear also in the five aggregates—body, feelings, perceptions, mental formations, and consciousness. These first three realities are analytical descriptions of our existence, though they are relational in nature. It was argued whether a conditioned Dhamma could

---

[3] Metaphysics Research Lab. *Stanford Encyclopedia of Philosophy* (Stanford University, Stanford, California). *The Theory of Two Truths in India*. https://plato.stanford.edu/entries/twotruths-india/

also be ultimate, but the Buddhist philosophy of relations shows that "nothing arises singly". Any further reduction or deduction of Dhamma would be undetected by the mind and would diminish the purpose; therefore, the Buddha limited his teaching to humankind's liberation from suffering. Buddhists teaching emphasis on the meaning and purpose of dhamma to convey the truth.

When explaining Dhamma, the Buddha used both an analytical method and a synthesis method. The synthesis method explains things with the aid of the philosophy of relations, that is, how conditioned phenomena are interrelated. Whereas the analytical method deduces the experienced phenomena into defined components. The relational method groups these components according to the conditions of their arising. By these two methods, Dhamma can be described as the irreducible components of momentary experience, arising, and passing away in rapid succession. These components are the fundamental units that fabricate our conventional thought. Components are relational; therefore, our perceptions are also conditioned accordingly. The Buddha classified Dhammas according to their character, their mode of occurrence, and their interrelationships. The exception is nibbana, an element that is outside the five aggregates and hence unconditioned.

One's discovery of ultimate truth is the result of philosophical enquiry into observed phenomena. This branch of philosophy is known as phenomenology. It investigates both the world outside and the world inside. The Buddha advised his disciples to first look inwards, then outwards. "Therein, monks, the person who gains both internal tranquillity of mind and the higher wisdom of insight into things should only establish himself in these wholesome states and make further effort for the destruction of the taints" (AN 72, p. 103).

Mental development (*satipatthana bhavana*) stems from the phenomenological investigation into what happens and how it happens. Only a mind trained in introspective meditation can find the answers to these questions. The Buddha succeeded in capturing a moment of thought and was thus able to see what it is composed of and how it had come to be. His description of the unit of mind is like a physicist's model of an atom in particle theory. However, the Buddha didn't teach a "postulation" of Dhamma or create a "model" of the mind, nor experiment on the mind by empirical methods. He used his own experience, sharpening his logical and intellectual reasoning by way of insight meditation of direct observation. For example, by the application of rational and intellectual thought, he discovered the Middle Path in order to reject extremes. This conclusion was made by directly observing the world outside and the world inside to discover the truth.

According to the Buddha, the ultimate reality of the mind and body is composed of units of mind (*sampayutta Dhamma*) and units of matter (*rupa kalapa*) respectively. Both units of mind and units of body are inseparable, arising together at any moment of consciousness and passing away in rapid succession. Further analysis shows that units of mind are composed of cognitive and mental factors. Mental factors condition one's cognition (*citta*). These mental factors are inseparable and arise together at any moment of consciousness and pass away in rapid succession. Units of matter are made of bundles of material elements (*rupa dhatu*); these material elements were further analysed into primary and secondary elements. Primary elements are solidity, liquidity, heat, and motion—the four basic properties of earth, water, fire, and wind respectively (*maha bhuta rupa*—the four great elements). For example, when fire is burning, within the flame all four elements are present (heat, carbon, gas, and water vapour). Secondary elements are a direct cause of primary elements; this causation can be either a direct cause (*nipphanna rupa*) or an indirect cause (*anipphanna rupa*). Examples of secondary elements are hydrocarbon, sugar, protein cells, calcium, and bacteria, which are found in life forms made up of primary elements. All secondary elements have functionality, made possible by the faculty of life. The units of matter, corporeal elements, that constitute our physical body are denoted by the term *rupa*. These are not static particles but are particles that have as part of their nature the ability to change and deform.

The Buddha's teaching is based on his direct experience of meditation. In meditation we experience an object through consciousness. Consciousness is a series (*santati*) of units of mind and matter. For example, when there is a visible object, light, both one's intention and one's eye are in contact with that object, and from this a seeing consciousness arises. In this process of seeing, the eye serves as a channel to convey visible images to the mind. As it is stimulated by the light beams, the eye channel sends an orderly series of mental signals to be received by the brain. A series of mental processes then follows. As with seeing, the principles of this process of consciousness are the same for the four other senses (hearing, smelling, tasting, and touching).

The units of mind and matter are interdependent (*nissaya paccaya*) but are not a mixture of both. For example, oil and water can stay together but at the same time are separate from one another. Each unit has a unique function in the process of consciousness, hence there is a relationship of disassociation (*vippayuttha paccaya*). This disassociation condition may be further divided into four types: coexistence, basic pre-existence, basic objective pre-existence, and post-existence, where consciousness is related to producing matter (physical material qualities) by its conditioning force and vice versa during various functions of the life process. These mental and physical phenomena are causally related by a coexistent disassociation. For example, when a matchstick is struck, light arises. Arising with it are sound, heat, smell, smoke, etc. Such oxidizing chemical properties were coexisted in the match head. Similarly, at the time of rebirth, consciousness produces a physical heart base and, during life, various hormones (biological chemicals) that are derived from the four primary great elements and secondary elements. When we speak of a mental relation, with contact (mental) there arise synchronous forms of feelings, perceptions, mental formations, and consciousness related to that instance. In his analysis, the Buddha found that there is no self or soul that binds these units together and there is no perceiver, only perception. These units can only be separated by insight knowledge and seen by insight wisdom.

The lifespan of these mental units is very brief compared to the speed of light. In the blink of an eye, millions of units of the mind will appear and disappear in great succession. When this process of image construction is completed, the mind then perceives the object. Units of mind have a time duration known as thought moments (*citta khana*). The duration of a thought moment is generally seventeen times faster than the duration of units of matter. Because of this time difference, the observation of a body is relatively stable. The time it takes to notice change is longer than for mental movements. These units are formed according to their causes and conditions, and they cease their effect when the conditions cease. This is the ultimate truth of conditioned phenomena. This is true both according to factual descriptions and in correspondence with experience. The Buddha established this truth with evidence in relation to all life forms.

By vigorously practising the Four Foundations of Mindfulness, one will clearly be able to see Dhamma within one's own body and mind. One who gains the knowledge of ultimate truth is able to see things as they really are (*yatha buta nana dassana*) and therefore is able to understand the composite nature of body and mind.

The Buddha reduced a being into five groups of five aggregates (*pancha skandha*), thus abandoning the notion of being whole. The five aggregates are body, feeling, perception, mental formation, and consciousness (*rupa, vedana, sanna, sankhara,* and *vinnana*). The Four Foundations of Mindfulness are directly related to these five groups or aggregates (skandhas). Perception is sometimes called noting or knowing: its function is to recognise what has been previously perceived. Although the word *sati* (mindfulness) has the same meaning "to remember", its significance is presence of mind and attentiveness to the object. Mindfulness manifests as the state of confronting an objective field. Its proximate cause is strong perception (*thirasanna*) or the Four Foundations of Mindfulness. Using these definitions, we can clearly see that perception is an interpretation process of knowing all other four groups and that mindfulness is the key function that focuses on that process and is not same as perception. Being a mindful, one can see the limitation of perception, gradually awaken to the truth, and gain wisdom through meditation.

The Four Noble Truths are founded on the three characteristics of existence. They are: Anicca (impermanence) - This means instability, or a lack of permanence, transient that everything is changing all the time. This character is true for physical body and mind that they are arising and ceasing. The cells of the body are replicating, grow old, decay, and die. Mind is further analysed into four border groupings of feeling, perceptions, mental formation, and consciousness. These groups or aggregates too rapidly change and there is no unchanging essence in them. At death, body and mind disintegrate. Dukkha (dissatisfaction) - This means that everything leads to suffering because of change which is beyond control of casual self. Anatta (no soul) - This doctrine of impersonality says that a being is composed of temporary groupings and these groups are further analysed into elements and processes in the books of Abhidhamma proving that there is no soul and is the idea that people can and do change in life giving freedom to choose one's destination by own effort.

The Abhidhamma exemplifies the five aggregates (*pancha skandha*) by way of objects, processes, elements, and knowing the object in time. This has a twofold element, namely *nama* (referring to units of mind) and *rupa* (referring to units of matter). The word *nama* is used to represent masses of the mental element (*nama skandha*): feeling, noting, formation, and consciousness. The compounded state or aggregate of a thought unit at any given time is made up of cognition and any of seven other factors. These are contact, feeling, noting, volition, concentration, psychic life, and attention. They move in a pattern of arising (*uppada*), reaching their peak (*thiti*) and then being marked by its cessation (*banga*). The transitory nature of these units is beyond the control of any doer or owner. The mind can be conscious of only one object at a time. Perception, however, appears to include many objects at the same time because of the timing difference between consciousness and perception.

The untrained minds of worldly people are obstructed from apprehending ultimate truth by four kinds of "densities" (*ghana*):

density of continuity	santiti ghana
density of the whole	samuha ghana
density of function	kiriya ghana
density of object	arammana ghana

Because of the obstruction to one's senses caused by densities, the perception of reality is veiled.

Density of continuity prevents one from understanding the series of mental impressions: sound, sight, odour, taste, and touch. For example, sound is a vibration transmitted through a medium, through the ear, then a series of mental impressions are created in the mind to form the audio perception. The principle is the same for all the other four senses which are receptive to their corresponding external stimuli. However, human senses are limited to certain frequencies of vibration. Animals, on the other hand, are sensitive to a wider range of frequencies.

Density of wholeness prevents the understanding of the composite nature of things, for example the five aggregates. More simplistically, a loaf of bread is a collection of flour, water, and yeast that makes it bloom, triggering fermentation. Compounded nature of organic materials is apparent if left out for few days that it will get decomposed by growth of bacteria and fungus in yeast. Every composite material is subjected to disintegrate, decay, and deformed over a period of time which is a universal truth. Mindful observation of this truth reduces attachments to material possession.

Density of function prevents the understanding of function by many movements or actions. For example, walking is the movement of feet. Another example: if a light beam is rotated, it appears as a circle of light.

Density of object prevents one from understanding the elementary nature of an object. For example, a vase is made of clay, water, motion, and heat. The potter has the skill to combine these elements together. General physics explains that white light consists of five colours and can be split if sent through a prism.

The untrained mind cannot see the momentariness of what is perceived. If the true nature of the object is concealed, one will unwisely respond to those mental impressions. The mind thus falls into an illusion or an error of judgement. There are three aspects of the mind which carry out this function. These are awareness of mental images; perceptions' and opinion formation. These illusions or errors of perception caused by distortions are known as *vipallasa*. The mind can be trapped by three related kinds of errors. These are errors in consciousness, errors in perception, and errors in view or opinion (*citta vipallasa*, *sanna vipallasa*, and *ditthi vipallasa*). Each stage of the process of perception constructs an idea to finally grasp the object encountered by the sense. Formation of an idea happens so fast that it gives false notions of the object perceived.

There are four kinds of false notions in perceiving or noting mundane things. These are notions of permanence (*nicca sanna*), for example how the gross manifestation of an object appears to be permanent. The notion of pleasure (*sukka sanna*) is the way one associates pleasure and feeling good with consumption. The notion of beauty (*subha sanna*) is where one's perception places an opinion of beauty upon the object in view which is actually devoid of it. And finally, the notion of an ego self (*attha sanna*) is that by which one perceives a self or person. A combination of false notions gives rise to many illusory perceptions regarding the self, for example perceiving *I like it,* or *I do not like it*; *Such-and-such happened to me*; *I do such-and-such.* To fall into these errors and illusions and to grasp things with false notions is ignorance (*avijja*). These false notions have a compounded effect when associated with unwholesome mental factors. The Buddha shows that beings are trapped in samsara and suffer because of ignorance. The mind encounters an infinite variety of objects and effects; when one is in contact with those objects, there are a variety of illusory concepts and ideas which form in the mind because of ignorance.

The Buddha's findings of perception do not suggest that the world is an illusion. Illusory perceptions are the nature of an untrained mind. Truth is the foundation of all things. Even to admit ignorance would make one wise (according to Socrates, the Greek philosopher from Athens). "Ignorance (*avijja*) is the mental factor of delusion, which obscures perception of the true nature of things, just as a cataract obscures the perception of visible objects."[4] The acceptance of an illusion as real and good is the error. The story of Angulimala shows a good example of a deluded mind. Angulimala was an innocent student when he was brainwashed and became a serial killer who murdered thousands of innocent people. Fortunately, the mind can be trained to overcome ignorance. The basic threefold training of the mind is done by the practice of generosity, morality, and meditation (*dana, sila,* and *bhavana).* The next level of threefold training is in morality, concentration, and wisdom (*sila,* samadhi, and prajna). When insight wisdom develops, the meditator is able to glean what is real from the illusion. Insight meditation is enquiring into the ultimate truth by way of four standard questions: What are its characteristics? What are its functions? How does it manifest? What is its proximate cause? When ultimate truth is examined in this manner, insight wisdom emerges.

The units of mind and the units of matter have a number of factors associated with them. These factors are not static but are functions (*kiriya matta*) which are constantly changing and disintegrating. Having said this, for the purpose of explanation we will discuss them as a static moment. These functions of ultimate truth are "reals". The definition of a real is "something that does exist". This also presupposes the meaning of "alive", one who is living, opposed to one who does not really exist or is dead, for

---

4 Bhikkhu Bodhi. (1999). *A Comprehensive Manual of Abhidhamma* (Kandy, Sri Lanka, Buddhist Publication Society).

example a scarecrow (a scarecrow is a decoy or mannequin, often in the shape of a human. Humanoid scarecrows are usually dressed in old clothes and placed in open fields to discourage birds from disturbing and feeding on recently cast seed and growing crops. Scarecrows are used across the world by farmers and are a notable symbol of farms and the countryside in popular cultures) which is physically present but is not a real person.

A unit of mind is a compounded state of consciousness which arises with the presence of an object. The object could be real or nonreal and further analysed into different types. The five sense objects, colour, sound, odour, taste, and touch are conditional reals; any other perception or image to arise in the mind is a mental construction. For example, the size and shape of an object are projections of the mind, ideas of mental formation (*sankhara*). Objects in the mind's avenue could be both real and nonreal. By virtue of a pure mind, the meditator can gain penetrative knowledge (*pativedha nana*) of reality and therefore overcome the errors and illusions caused by distortions (*vipallasa*). Purity of view is the distinction of identifying between real and nonreal, free from defilements.

Buddhist scholars have primarily distinguished observed phenomena into these two broad categories, real and nonreal. Further analysis into the complexity of phenomena could be exhaustive. Therefore, for simplicity's sake, the subcategories are listed as follows:

Analysis of Reals and Nonreals

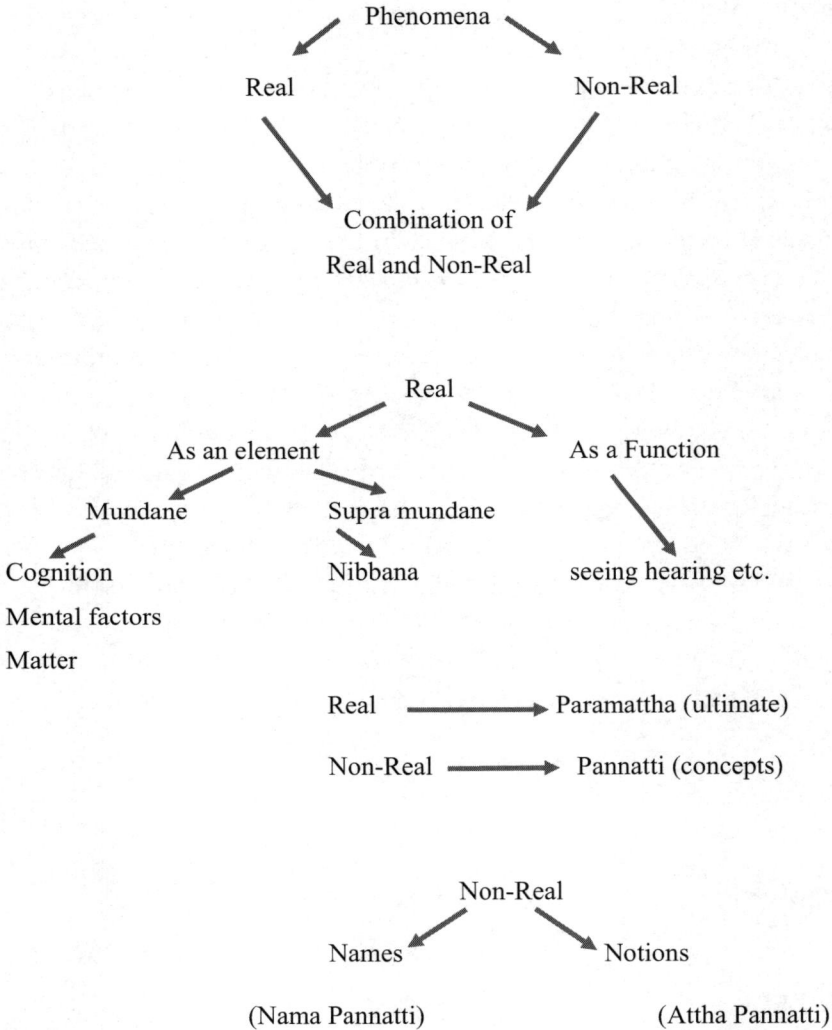

Phenomena

Real → → Non-Real

Combination of
Real and Non-Real

Real

As an element ← → As a Function

Mundane ← → Supra mundane

Cognition   Nibbana   seeing hearing etc.

Mental factors

Matter

Real ⟶ Paramattha (ultimate)

Non-Real ⟶ Pannatti (concepts)

Non-Real

Names ← → Notions

(Nama Pannatti)   (Attha Pannatti)

Nonreal phenomena are explained through conventional usage of language, etymology, and semantics. Therefore, names are given according to the character attributes of the object, to assert a meaning. These names and notions have subjective elements integrated within the meanings because a language has an etymological history and cultural background.

Nama Pannatti

Real

Non-Real

Combination of Real and Non-Real

Real $\longrightarrow$ Vijjamana Pannatti

Non-Real $\longrightarrow$ Avijjamana Pannatti

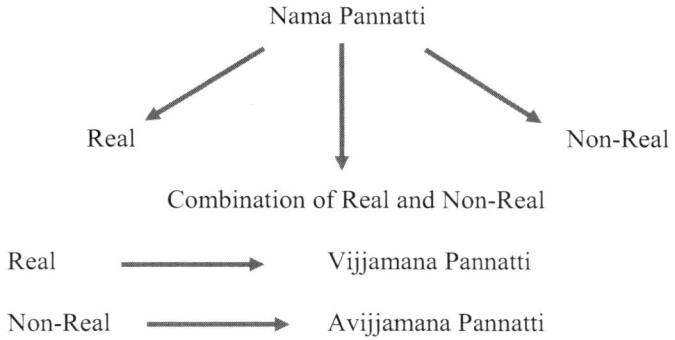

1. Combination of Real with Non-Real Vijjamanena Avijjamana Pannatti

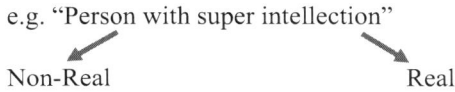

e.g. "Person with super intellection"

Non-Real   Real

2. Combination of Non-Real with Real

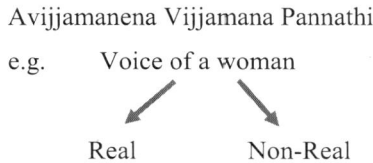

Avijjamanena Vijjamana Pannathi

e.g.   Voice of a woman

Real   Non-Real

3. Combination of Real with Real

Vijjamanena Vijjamana Pannatti

e.g.   Eye cognition (seeing, hearing, etc.)

Nonreal phenomena are ideas or concepts; attributes; or a description of things to which names have been assigned. With any word or sign, one uses to denote or identify an object, one instantly thinks of it as possessing such characteristics. Mass media and advertising can cause fake phenomena to appear real. Computer graphics, virtual reality and special effects can create magical illusions that appear to be real and happening before our eyes.

Following is a list of Pali terms for those nonreal notions:

Santiti pannatti—the notion of apparent continuity as a distinct thing, e.g. river or mountain. The true relations are only the physical elements (water, earth, wind).

Samuha pannatti—the notion that arises from a number of parts, differentiated by mode of construction, e.g. house, chariot, or ship.

Satta pannatti—the notion of being. In reality, the body and mind are the combination of five components or five masses of the aggregate.

Disa pannatti—the notion of direction or relation, e.g. east, west, above, below.

Kala pannatti—the notion of time, e.g. morning, evening, and night. These changes are the result of planetary movements.

Akasa pannatti—the notion of space, e.g. pits, caves, sky. Space is due to noncontact between atoms or materials.

Nimitta pannatti—the notion of meditation symbols, e.g. visualised images, conceptualised images.

Discerning ultimate truth is in the correcting of these false notions and illusions of conventional identity. The following metaphor from the Majjhima Nikaya explains the wisdom that discerns such truth: "A watchman on a lofty tower sees a charioteer urging his horses along the large plain. The driver thinks he is moving rapidly, and the horses, in the prime of their life, seem to scorn the earth, from which they think themselves separated. But the great intelligence of the watchman from above the two horses, the chariot, and the driver sees that they seem to just crawl along the ground, and as the mane of the horses waving in the wind is a part of the horses themselves, so are the horses as much a part of the earth."

Liberation from these illusions is crucial to the attainment of insight. The illusions are corrected in stages as a trainee proceeds through his or her developing practice of meditation. The continuity of a thought process born of illusory perceptions can only be broken by special states of trance or mental absorption. As explained in the philosophy of relations, a thought unit always exists with a thought object. Those thought objects could be real, nonreal, or a combination of both. A series of these units makes up the thought process. The Tathagata uses metaphors such as a stream of water (*nadi*), torrential rain (*sato*), and a flame (*vayo*) to explain thought processes of differing intensities. A combination of tranquillity meditation (jhana) and insight meditation can eliminate a deranged mind, superseding all kinds of mundane thought objects and discursive thought. Those who wish to attain nibbana take the element of nibbana as a thought object. Unlike the mundane elements, a supramundane element is free from cognition and mental factors; by this method, supramundane absorption (trance) will finally transcend even sublime thoughts. This state of mind is called the "cessation of thought" (*Nirodha samapatti*). Cessation of thought is also the cessation of suffering. Attainment of this supreme state is only possible for the highest saints who have mastered all the trances.

"The attainment of cessation is a meditative attainment, in which the stream of consciousness and mental factors are completely cut off temporarily. It can be obtained only by nonreturners and Arahants who have mastery over all the material and immaterial jhanas. Further, it can be obtained only within the sensuous plane or the fine material plane of existence. It cannot be obtained within the immaterial plane. For those who are born in the immaterial plane, it is not possible to attain the four of fine material jhanas, which are the prerequisites for entering cessation."[5]

The state of cessation of thought (*Nirodha samapatti*) is the way to completely overcoming suffering. It is found in the nibbana element. By seeing the impermanence, unsatisfactoriness, and insubstantiality of the ultimate truth of mundane things, the Buddha came to discover noble truth (*ariya sacca*). The liberation of the mind is not necessarily a quality of ultimate truth. As an ascetic, Siddhattha was not satisfied with the answers

---

[5] Bhikkhu Bodhi. (1999). *A Comprehensive Manual of Abhidhamma*. BPS.

he found in the ultimate truth. He found no liberation in conditioned Dhamma, which forced him to seek for a more satisfying answer somewhat outside the given reality.

The early Indian schools held the position that the unique aspects of a unit of mind or of matter are ultimately existent and necessarily produced. After all, only those that are ultimately existent can perform a causal function. A causally efficient and unique element implies constant change; the renewal and the perishing of the antecedent identities is momentary. The proponents of those early schools were unable to see beyond this momentariness because of their strong belief in a permanently transmigrating self (atman) that experiences reality.

The rationale for one's identifying the indescribable nature of reality and insubstantiality of all things is that a conception of self and phenomena presupposes a subject–object duality. Here the self is the subject, and phenomena are the object of experience. It is impossible to sustain self and things as existing substantially without also believing the two to be substantially different. This paradox is a situation from which an individual's reason cannot escape—how the mind can perceive a mental object by way of the mind itself. The correct knowledge of nonself and all the Dhammas negates the interlinked conception that self and object are substantial and dualistic. An understanding of the emptiness of the phenomenal self and personal self is itself a realisation of nonduality. Such a realisation, according to Vasubandhu[6] is like awakening from a deep slumber of ignorance.

The terms of the ultimate need to be understood within the framework of dependent nature. The ultimate, according to Indian philosophy, is "world-transcendent knowledge"—there is nothing that surpasses it. It is the perfect nature, stainless and unchangeable, and therefore it is known as the "ultimate". The Buddha (to-be) was searching for something beyond

---

[6] Vasubandhu (fourth to fifth century CE) was an influential Buddhist monk and scholar from Gandhara. He was a philosopher who wrote commentaries on the Abhidhamma, from the perspectives of the Sarvastivada and Sautrāntika schools. After his conversion to Mahayana Buddhism, along with his half-brother, Asanga, he was also one of the main founders of the Yogacara School.

the world of duality, of birth, of origination, of what is made and what is formed, and he succeeded in attaining that state of mind.[7]

Recorded in Anguttara Nikaya 120, the Buddha describes six unsurpassed things, referring to the six senses of attachment and the hearing of true Dhamma, which would lead to disenchantment, dispassion, cessation, peace, direct knowledge, enlightenment, and nibbana. When one sees emptiness in all six senses, including worldly gains, wrong views, service to them, and recollects the Buddha's teaching correctly and properly without speculation, one transcends the mundane perceptions of the supramundane world of the unoriginated. This is the actual experience of nibbana, constituting everlasting, never fluctuating happiness. "Nibbana is certainly in the sense of the term *lokuttara* a metaphysical or transcendent entity."[8] Here, *lokuttara* is born of the root word *loka-uttara*, meaning "beyond world".

From this line of enquiry, we have now arrived at the discovery of three types of truths. They are conventional truth, ultimate truth, and noble truth. "Nibbana is the Dhamma that is not evanescent or subject to condition[9]." Hence, nibbana is outside the five aggregates, an experience without an experiencer.

Noble truth is the highest wisdom that holds the right view of Dhamma. It is the highest understanding of intrinsic nature, of phenomena, and of their respective constituents. Further, it understands the constituents in respect to characteristics, functions, manifestation, and proximate cause as a means of deliverance allowing one to detached. Noble truths are so called for many reasons. Noble truth is not a concept; it corresponds to facts, and it must be realised by an individual through practice; therefore, it is above intellectual and theoretical truth. The realisation of the Four Noble Truths through practice is the transformation of a worldly person into a noble person (Ariyan).

---

[7] Jay L. Garfield and Jan Westerhoff, eds., *Madhyamaka and Yogacara: Allies or Rivals?* (Oxford University Press).

[8] Nyanaponika Thera, *Abhidhamma Studies: Researches in Buddhist Psychology* (Kandy, Sri Lanka, Buddhist Publication Society, 1965).

[9] Bhikkhu Bodhi. (1999). *A Comprehensive Manual of Abhidhamma.* BPS.

The First Noble Truth (Dukkha Ariya Sacca)

The Four Noble Truths are regarding life—they are about you, a living being. Further, they address the truth of how life comes into being. They investigate in depth how life manifests its true character, its function, and its proximate cause. The Buddha limited his teaching to life's true nature, *dukkha*.

The Simile of Simsapa Leaves

The Buddha limited his teachings to life's true nature, that is suffering and the way out of suffering. A passage narrated in the Samyutta Nikaya (56.31) is the discourse titled "The Simsapa Leaves", which was delivered by the Buddha to his monks while dwelling in a grove. This discourse explains that the knowledge of the universe is infinite and that the ordinary untrained mind is incapable of conceiving it; therefore, it is not profitable to pursue knowledge for the sake of speculation. The Buddha used the analogy of a handful of simsapa leaves, comparing the leaves in his hand with the number of leaves overhead in the simsapa grove. This illustrated the particular nature of his teachings, the Four Noble Truths, and only what is profitable and beneficial for the many; what he did not teach was unrelated to humankind's liberation from samsara. The Buddha asked the monks which were more numerous, the few leaves in his hand or those overhead in the grove. The monks correctly replied that those over in the grove were far greater in number than the leaves in the Buddha's hand. Similarly, the knowledge that the Buddha had discovered was far more than what he revealed, but he only taught what was needed to know in terms of how to end suffering; the rest was unconnected to this overall goal and therefore could only lead to confusion and prolonged suffering.

The Pali word *dukkha* is composed of the root word *du* and of *kha*. Its usage was as *dukkam*. The suffix *-am* gives the meaning of *-ness*, making it a noun. *Du* means "unsatisfactory" or "disgusting". *Kha* means "empty, unsustainable, or unsubstantial". In usage, this word means "unsatisfactoriness, emptiness, or unsubstantialness". These meanings need to be interpreted in an absolute context, according to the word's manifestation and true character.

Primarily, life consists of body and mind. The true character of a body is foul. The body (*kaya*) feels hunger and thirst and is vulnerable to many diseases and illnesses. The body is subjected to decay and finally ends in death. The body must be protected against wind, rain, cold, heat, and draught and from various harmful animals and insects—not to mention muggers, hitmen, and rapists. This body must be continuously fed, cleaned, clothed, and sheltered. Enormous amounts of time and effort must be spent on these matters to keep the body and mind healthy from various illnesses and infections, or more effort is needed in medication and healthcare. Many people have underlying health conditions and fear death. The true manifestations of these events are painful and full of suffering. If no care is taken in terms of simple daily hygiene such as washing the body, then the body's true manifestation will be disgusting. Beyond this, there are various anxieties, worries, and mental sufferings. There is no economic theory or model that has successfully resolved these life difficulties. The Buddha pointed out that life is uncertain, and yet death is certain.

The main cause of all suffering is birth. Therefore birth (*jati*) is termed as suffering; it forms the basis and is the source of all future suffering, from the point of embryo development. Various kinds of human suffering may be classified into two main groups.

Classifications of Suffering

1. Pariyaya dukkha   indirect suffering hidden in the body and the mind
2. Nippariyaya dukkha   direct suffering; intrinsic suffering

Indirect Suffering—Pariyaya Dukkha

At the beginning there is no suffering in some events, yet these are the basis for future suffering, for example association with bad people (*papamittha*), gambling, smoking, and drinking. The cause of suffering is not obvious; it can only be found by deeper investigation. This type of suffering is classified as *pariyaya dukkha*.

Direct Suffering—Nipariyaya Dukkha

This is suffering that is the result of a direct cause, for example intrinsic suffering. The cause of the suffering is directly known, for example an accident, respiratory illness, or flu.

The foregoing two groups could be divided into two subgroups, as follows:

1. Paticchanna dukkha   concealed suffering
2. Apaticchanna dukkha   exposed suffering

Concealed Suffering—Paticchanna Dukkha

The cause of this suffering is not openly evident. It is something painful experienced by an individual, for example toothache, a type of suffering which others would know only if or when it happened to them. This type of suffering is also known as un-evident suffering. This also includes mental or physical suffering. If the conditions are chronic, then it must be diagnosed by a physician for medical treatment, for example cancer.

Concealed Suffering—Apaticchanna Dukkha

This is physical suffering arising from assault or an accident. The cause of the suffering is apparent or exposed. This type of suffering and mental distress is a direct result of an event and is thus classified as exposed suffering, for example cuts, bruises, a broken arm, amputation, wounds, hernia, and ulcers and such pains as exposure to fright or shock, the break-up of a family or personal relationship, exposure to human aggression, domestic violence, long-term illness, or military combat (the Medical Distress Explanatory Model by Maurice Eisenbruch, MD, gives a list of clinical sufferings).[10]

Professor Eisenbruch describes in his paper the background and development of a Mental Distress Explanatory Model Questionnaire

---

[10] Maurice Eisenbruch. (1990). "Medical Distress Explanatory Model", The *Journal of Nervous and Mental Disease.*

designed to explore how people from different cultures explain mental distress.

People explain their distress in a multitude of ways, often blaming social circumstances, relationship problems, witchcraft or other socio-anthropological beliefs and customs of ethnic group to which they are belong. Patient's explanatory models of illness should be elicited using a mini-ethnographic approach that explored their concerns: 'Why me?' 'Why now? Reactionary approach by the patient to suffering adds more to the injury. The clinician can gather a better understanding of the subjective experience of illness, and so promote collaboration and improve clinical outcomes and patient satisfaction if they have a broader knowledge of multiple levels of suffering and illness narratives.[11]

The foregoing two subgroups could be further subdivided into three groups, as follows:

1. Dukkha dukkhata—intrinsic suffering
2. Viparinama dukkhata—suffering due to change or sudden break-up
3. Sankhara dukkhata—suffering due to formation

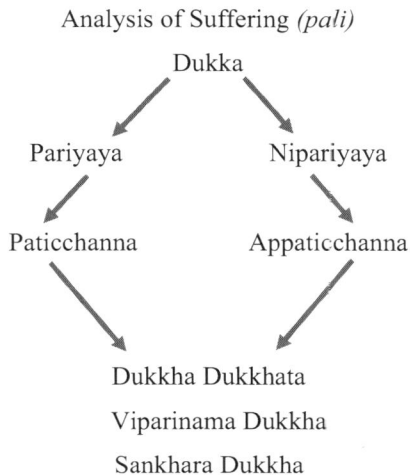

Analysis of Suffering *(pali)*

Dukka

Pariyaya   Nipariyaya

Paticchanna   Appaticchanna

Dukkha Dukkhata

Viparinama Dukkha

Sankhara Dukkha

---

[11] Kleinman, A. (1988). Rethinking Psychiatry: from Cultural Category to Personal Experience. New York: Free Press. Published online by Cambridge University Press: 02 January 2018.

Intrinsic Suffering—Dukkha Dukkhata

All types of bodily pains, aches, and discomforts and unsatisfactoriness are dukkha dukkhata. Also, all types of mental pains are classified as dukkha dukkhata. The term itself means that the very existence of bodily life is suffering. The body and mind undergo constant pains. Sometimes these are visible and sometimes not. Pain is an unendurable or unbearable experience. Its intrinsic nature is oppression and burning.

Suffering Due to Change—Viparinama Dukkhata

Suffering due to change is viparinama dukkha. It could be a sudden, unexpected change like an accident. All cyclic phenomena are impermanent ("Anicca vatha Sankhara"). They are transitory and in a constant state of breaking or disintegrating. All animate and inanimate matter is subjected to this law of change, and nothing is permanent. All kinds of worldly happiness and pleasures that depend on something are not permanent. When that enjoyment vanishes, it is replaced by frustration, grief, and lamentation. For example, loss of wealth, loss of income, or the loss of family member is a cause of great distress.

Suffering Due to Formation—Sankhara Dukkhata

Suffering due to either mental or physical formation is sankhara dukkha. All stages of physical formation are suffering. Birth is suffering. Ageing is suffering. Death is suffering. All kinds of mental formations give rise to suffering because they require constant maintenance of one's mental state. For example, pleasant feelings of happiness that arise in the mind are conditioned by pleasant mental factors. To maintain the happiness requires constant conditioning. For example, listening to music or watching movies keeps the mind entertained. These objects provide the constant conditions necessary to maintain a pleasant mental state. This is true for neutral or equanimous feelings also. Equanimity is a mental factor that requires maintenance. The principle of repair and maintenance is true for the body also. The body must be looked after. The minimum requisites are food, clothes, medicine, and shelter. Apart from that, physical conditions such as temperature, pressure, and humidity must be maintained for a healthy and comfortable life.

The noble truth of suffering is classified under two main groups: *pariyaya dukkha* and *nippariyaya dukkha*. In one way or another, these two states appear in the following forms during one's lifetime:

- birth
- ageing
- death
- sorrow
- lamentation
- physical pain
- grief
- despair
- association without love
- separation from loved ones
- not getting what one wants
- five aggregates of clinging.

This suffering is real. There is no other truth of suffering; it must be experienced and understood to realise *This is the noble truth of suffering*.

In his first level of deduction, the Buddha discovered that life consists of five temporary groupings: body, feeling, perceptions, mental formations, and consciousness. Penetrative truth into each group shows that they are all impermanent, subject to suffering, and without a self.

> The corporeal in which grasping arises, the feeling in which grasping arises, the noting in which grasping arises, the formation in which grasping arises, the cognition in which grasping arises is called the noble truth of suffering. (Rupapadanakkhandhe, Vedanupadanakkhantha, Sannapadanakkhandho, sankharapadankkhandho, Vinnanupadakkhandho idamvaccati bhikkuhave Dukkam Ariya saccam.)

In short, attachment through the five aggregates create suffering. In the Fire Sermon, as narrated in the Samyutta Nikaya 35.28, the third discourse given after his enlightenment, the Buddha pointed to a group of fire-worshipping ascetics who formerly practised a sacred fire ritual showing that the five aggregates burn like fire. He had advised them to look inwards, referring to the six senses, for the fire that burns and blazes with passion, aversion, delusion, and suffering. Enjoyment, happiness, and pleasure, which are derived from the entertainment of one's senses, depend on sense objects. When these objects disappear, happiness also disappears, then turning into disappointment, grief, and sadness. The senses cannot thus be satisfied. It is like attempting to fill a bucket with a hole in it. It will never become full. In meditation one can see this aspect of suffering. By being fully established on the Four Foundations of Mindfulness, the meditator firmly fixes his or her attention on the ever-changing phenomena arising and passing away in rapid succession. To have insight knowledge, which arises as a result of this practice, is to know the true nature of the five aggregates.

The Buddha once explained mindfulness while speaking of his disciple Nanda, his half-brother. The Buddha had once reproached Nanda for going about with well-pressed robes and eyes made up. He advised him to go to a forest and meditate. Eventually Nanda attained enlightenment. The Buddha described Nanda's proficiency in mindfulness and clear comprehension.

> O monks, one may rightly say that he is of good family, that he is strong, handsome, and very passionate. How else could he live a perfectly pure, holy life except by cultivating and having mindfulness and very clear comprehension? This now, O monks, is Nanda's mindfulness and clear comprehension with full awareness giving rise to feelings in him. With full awareness they continue; with full awareness they perfectly cease. With full awareness arise perceptions in him. With full awareness they continue; with full awareness they cease. With full awareness thoughts arise in him. With full awareness they continue; with full awareness they absolutely cease. This, O monks, is Nanda's mindfulness.[12]

---

[12] Nyanaponika Thera. (2005). *Anguttara Nikaya: The Heart of Buddhist Meditation.* (Buddhist Publication Society).

The True Nature of Suffering

There are four characteristics to the truth of suffering. The main characteristic is oppression or pain—*pilanattho* (various kinds of pain). The pain is supplemented in three ways at three stages of arising, maturing, and passing away.

The other three are as follows:

1. *Sankhattho—repair and renewal*
The five aggregates are oppressed by the burden of having to continuously repair and renew during the forming and sustaining of life.

2. *Santapattho—frequent burning*
The five aggregates are oppressed by way of frequent burning. For example, all kinds of dukkha have the character of burning. And there is burning due to defilements such as hatred, jealously, envy, ill will, and conceit; each has the characteristic of burning.

3. *Viparinamattho—constant change*
The five aggregates are oppressed by way of constant change and finally by break-up and decay, for example ageing and related pains.

These characteristics are true of all beings wandering in samsara, including Brahmans, devas, petas, ghosts, asuras (invisible beings), and animals. They all suffer, but to different degrees. All forms are subjected to constant change (*viparinama*). One's plane of existence and the degree of suffering are intricately connected to the individual's previous kamma. At the end, death is also suffering. Death is the result of a displacement or change in the body, disintegrating five aggregates. The Buddha explained, "Corporeality is a murderer, so too are vedana, sanna, sankhara, and vinnana" (feelings, perceptions, mental formations, and consciousness). A new cycle of suffering begins soon after death. In samsara, beings go through an endless cycle of birth and death, so the suffering is endless. The true nature of life is dukkha (unsatisfactoriness, emptiness, or unsubstantialness), which manifests in different ways under different circumstances, but any form of life is dukkha. The term *dukkha* covers a whole range of suffering, from unsatisfactoriness to intense burning. This is the noble truth of suffering. There is no other truth of suffering. If there were another truth, the Tathagata would have revealed this to us.

In meditation retreats, meditators are trained to mindfully observe dukkha in daily life. Many meditators experience direct suffering or discover indirect suffering. The arising and passing away of suffering become clear. Body is suffering. Feelings are suffering. Mental formations are suffering. Consciousness is suffering. Sitting is suffering; walking is suffering; eye, ear, nose, and tongue are suffering. The whole inherited nature of the five skandhas (aggregates) is suffering and impermanence. All phenomena to arise in the body and mind are beyond control. There is no self that binds the aggregates. They are processes without any "self-control", as self after all is an illusion. But the meditator understands that suffering is real.

All phenomena are impermanence, all phenomena are suffering[13].

(Sabba sankhara anicca, sabba sankhara dukkha.)

The Second Noble Truth—Dukkha Smudaya Ariya Sacca

The First Noble Truth (dukkha) is the true nature of life. It is the true manifestation and the true character of life. At the time of his enlightenment, through further investigation of its function and proximate cause, the Buddha discerned the cause of suffering.

The Second Noble Truth is the cause of suffering and its function. The Buddha identified craving as the cause of ever-arising suffering and its continuation.

There is craving that causes an attachment to that particular existence. What is it? It is the desire in sense objects, the desire

---

[13] Nyanatiloka Thera. (2004), Buddhist Dictionary: A Manual of Buddhist Terms and Doctrines. BPS.

that is combined with a false view of externalism and the desire that is combined with a false view of nihilism[14].

(Jayam tanha ponobhavika nandiraga sahagata tatratatrabhi nandini seyyathidam? Kama tanha Bavatanha Vibhava tanha idam vuccati bhikkhave dukkha, samudayam ariyasaccam.)

The Buddha identified three kinds of cravings which bind beings to the wheel of samsara. Craving is by and large the root cause of suffering. The three kinds of craving are:

- craving for sense pleasure (*kama tanha*)
- craving for eternal life (*bhava tanha*)
- craving for nihilism (*vibhava tanha*).

Craving is always accompanied by a false view of the object of desire. It exists because an untrained mind is trapped by illusions and false notions: *citta vipallasa*, *sanna vipallasa*, and *ditthi vipallasa*. Therefore, the roots of craving are entangled with ignorance, causing a compounded error.

Extended Analysis of Suffering

Three subsets of ignorance are greed (*lobha*), hatred (*dosa*), and delusion (*moha*). Ignorance is the result of past unwholesome kamma and the basis of present unwholesome kamma. At the time of supreme enlightenment, the Buddha discovered the law of causal relations and dependant origination. "On ignorance, formations arise" (*avijja paccaya sankhara*); formations are gross manifestations of phenomena appearing to be real.

According to the Abhidhamma's classification of unwholesome consciousness, "Greed and hatred are mutually exclusive; they cannot arise within the same consciousness (citta). The third unwholesome root, delusion, is present in every state of unwholesome consciousness. Thus, in

---

[14] Mahathera Ledi Sayadaw, Aggamahapandita, D.Litt. (1903). The Catusacca-Dipani, The Manual of the Four Noble Truths, translated into English by Sayadaw U Nyana, Patamagyaw of Masoeyein Monastery Mandalay, Edited by The English Editorial Board.

those cittas rooted in greed and in those rooted in hatred, delusion is also found as an underlying root. Nevertheless, there are types of consciousness in which delusion arises without the accompaniment of greed or hatred, which is sheer delusion."[15]

Beings are born in ignorance. They continue to accumulate unwholesome kamma because of ignorance. The principal factor of ignorance is greed. Its function is in causing sensuous desires of various kinds. "The term *sensuous* (*kama*) is the product of the sense object (*vatthu kama*) and the desires in that sense object (*kilesa kama*). They follow a sequence of mental functions in a distinct and discrete order.[16]" Through greed (*lobha*) grows attachment (*tanha*), and attachment leads to grasping (*upadana*). Attachment is the cause of repeated wanting or revisiting (becoming) of those objects. Grasping of the sense object as good and enjoyable is ignorance[17].

The next principle of dependent origination is "on grasping, becoming arises" (*upadana paccaya bhava*). This principle is true for the attachment to worldly things in the present and the future.

The second principle regards the function of craving, which prompts mental, verbal, and bodily action to fulfil that craving. Behind every action there is an intention. Any intentional actions produce kamma; hence, craving is the inexhaustible driving force behind the constant cycle of becoming.

The following metaphor expounded by the Blessed One best explains the phenomena of becoming:

> Now this wheel of cyclical rebirths, with its hub made of ignorance and of the craving for becoming, with its spokes consisting of formations of merit and the rest, with its rim of ageing and death. Joined to the chariot of triple becoming, piercing it with the axle made of the origin of cankers, that has been revolving throughout a time without beginning. All the wheel's spokes (*ara*) were destroyed (*hata*) by him at the place of enlightenment, as he stood firm with the hand of faith and

---

[15] Bhikku Bodhi. (1999). *A Comprehensive Manual of Abhidhamma*. BPS.
[16] Nyanatiloka Thera. (2004), Buddhist Dictionary: A Manual of Buddhist Terms and Doctrines. BPS.
[17] Ibid

the axe of knowledge that destroys kamma. Because the spokes are thus destroyed, he is accomplished (Arahant) also.[18]

Consciousness revolves around the sensory plane, where one craves for sensory pleasure by way of the six senses. The following shows these six types of craving:

Sense organ	Type of craving	Pali phrase
Eye (sight)	Pleasant visible objects	Rupa tanha
Ear (sound)	Pleasant sounds	Saddha tanha
Nose (odour)	Pleasant odours	Gandha tanha
Tongue (taste)	Pleasant tastes	Rasa tanha
Body (touch)	Pleasant bodily impressions	Potthabha tanha
Mind (mental)	Pleasant mental impressions and fantasies	Dhamma tanha

Pleasing the senses and enjoying sensory pleasures gives a false notion of happiness. Worldly beings indulge in sensual pleasures thinking these provide true happiness, but in fact craving is the root cause and it is thus suffering. Craving has the following four natures:

1. *Seeking Satisfaction (Ayuhanattho)*

The function of craving is to drive the senses to gain satisfaction. But these senses cannot be satisfied. They are like a burning fire. Buddha compared craving to oil in a lamp that needs to be constantly refilled in order to keep the flame burning. Worldly beings strive all their lives to please those senses. They are strongly attached to pleasurable objects and cannot envision life without them.

2. *Seeking Delights (Nidanattho)*

Craving has the nature of seeking after more and more delightful objects to replace what has been consumed, for example craving for fashionable clothing, jewellery, or delicious foods and drinks. All kinds of entertainment activities exist. The untrained mind thus has no contentment.

---

[18] Buddhaghosa, (2010,4th edition). *Path of Purification: Visuddhimagga*, p. 193.

Those who are deluded by the idea of nihilism think the purpose of life is to consume as much as possible. They live selfishly in the dark pit of delusion.

3. *Seeking Union with Objects (Samyogattho)*

Craving has the nature of establishing attachment between one's senses and pleasurable objects. Therefore, it seeks continuous union with those objects. It has a function of possessiveness, obsessively revisiting those objects again and again. Inability to depart from pleasurable objects binds beings to samsara.

4. *Seeking Obstacles (Palibodhattho)*

Attachment to sensory pleasure is a hindrance and an obstacle to nibbana. It disturbs the mind and distracts it from attaining calmness; hence, concentration cannot be achieved. Training in meditation is not possible for those who are greedily attached to worldly fires, which obstruct the mind from final deliverance.[19]

The foregoing four natures of craving are common to all types of craving. They are also the four modes of function, the way craving operates. "The grasping of the object, senses, physical or mental, subjectively or objectively as being agreeable and pleasant is the proximate cause of craving."[20]

> In this world, should there be an object that has the nature of being agreeable and pleasant, craving has arisen and takes a foothold. That object may refer to any of the thought bases or sense receptors, to stimuli such as colour and sound etc., and to the various thoughts. (Yam loke piya rupam satarupam tanha uppajjamana uppajjati ettha nivisamana nivasati.)
>
> —Lord Buddha

When the senses encounter an object, either tangible or intangible, and the mind grasps that object, therefrom arises all thought formations.

---

[19] Mahathera Ledi Sayadaw. (1903). *The Catusacca-Dipani: The Manual of the Four Noble Truths*, tr. Sayadaw U Nyana (Mandalay, Myanmar, Patamagyaw of Masoeyein Monastery).

[20] Nyanaponika Thera. (1965). "The Psychology and Philosophy of Buddhism", *Abhidhamma Studies*. BPS.

From these conditions, the consciousness that arises in the sensuous plane gets fuelled by lust, ill will, and ego. The false notion of ego is a powerful motivation to crave and strive for life after life. Other common manifestations appear as *I want this; I enjoy this; These senses are me; This object is mine.* One may strive to lose one's ego, but there are wrong views of a non-existent self. There is the danger that one could use a wrong means to satisfy this never-ending craving, thus accumulating bad kamma until one sinks to the bottom of the ocean of samsara. When attractive external objects are removed, the mind may become monotonous and bored, which can also result in unsatisfactoriness, disappointment, and depression. Unsatisfactoriness and disappointment could lead to quarrelling with others and end up in hatred, thereby achieving what one had set out to avoid. In this way beings are only adding more and more suffering.

Bhava tanha means craving for attainment of jhana and being reborn in a higher plane such as the heavens. When the effect of good kamma comes to an end, the person is reborn in the sensory plane with no escape from suffering.

Vibahava tanha refers to the wrong belief that the material body is annihilated and comes to an end at death. Therefore, this forms the view that one should enjoy as much pleasure as possible while one still has life. Such ignorance materialises in the person's grasping for more and more pleasant objects.

Different levels of desire and ignorance are known as ties, bonds, fetters, defilements, and cankers depending on the complexity of entanglement that forms the attachment to the senses.

By practising insight meditation, one can discover the root cause of suffering. Throughout a meditator's development, various kinds of suffering and craving become apparent to him or her. When insight wisdom matures, one begins to detach from the objects that cause unnecessary suffering. As a result of this detachment, right understanding arises in the mind. One is then able to see things as they are. All compounded things are impermanent and unsatisfactory. Sensory impressions are impermanent, though they may give the false notion of permanence. Attachment to those impressions is the cause of suffering. Impermanent and compounded things cannot give permanent happiness, for they are subject to decay and death.

The Third Noble Truth—Dukkha Nirodha Ariya Sacca

Everything in this universe is symmetrical. For the characteristics of each object, there exists an equal and opposite object. The same is true for every quantified or qualified experience of life, for example good and bad, happiness and sadness. Whatever arises must cease, and to note its appearance is to note its cessation. By further investigation of this truth at the time of enlightenment, the Buddha discovered the Third Noble Truth.

The Second Noble Truth is the cause of suffering; the third is the cessation of suffering. Due to attachment, all beings suffer in samsara. However, the nature of suffering and all other associated phenomena is to arise, mature, and pass away. These are the characteristics of conditioned Dhamma. According to the principles of nature, that which is conditioned should also signify that which is unconditioned and unchanging. The cause of suffering has been identified as attachment or craving; therefore, when those causes or roots (*hetu mula*) are removed, suffering ends and one attains an unconditioned state of peace. When there is peace, there is happiness.

The expulsion of craving is called the cessation of suffering[21].

(Tanhaya pahanam ayam vuccati dukkha Nirodha.)

The Pali word *Nirodha* is a compound word made up of *ni* (absence) and *rodha* (cyclical births of samsara). It could be interpreted as escape from samsara. This is a state of mind to be won by an individual by his or her own effort. It is the "complete annihilation, the abandoning and forsaking of every form of desire."[22]

Cessation is the opposite of arising. The process of cessation can logically be seen as the reverse application of dependent origination.

---

[21] Mahathera Ledi Sayadaw, Aggamahapandita, D.Litt. (1903). The Catusacca-Dipani, The Manual of the Four Noble Truths, translated into English by Sayadaw U Nyana, Patamagyaw of Masoeyein Monastery Mandalay, Edited by The English Editorial Board.
[22] Ibid.

"Dependence on ignorance gives rise to formation." Therefore, "expulsion of ignorance is the cessation of formation.[23]"

If the roots of suffering are not completely destroyed, then there is no escape from suffering.

> Just as a tree that is not cut at its roots grows again, so does craving that is not destroyed at its unconscious existence, thus causing recurrent suffering[24].
>
> (Yathapi mule anupaddave dalhe chinnopi rukkho punareva ruhati Evampi tanhanusaye anuhate nibbattati dukkhamidam punappunam.)

Believe in a permanent self and personality thereof is the main obstacle to understanding the above-noted truth of dependent origination, that "I am the enjoyer or suffer." The roots of suffering are veiled from the self. "The roots of suffering can be clearly seen by unveiling the delusion of self."

Having destroyed all the great fires of defilement, the Buddha declared the Third Noble Truth.

> When there is a complete detachment from, an absolute cessation from, a forsaking of, a deliverance from, a nonattracting to craving, that, monks, is the noble truth of the cessation of suffering.
>
> (Yo tassayeva tanhaya asesa viraga inrodho cago patinissaggo muti analaya idam vuccati bhikkhave dukkha nirodham ariya saccam.)

The state of escaping samsara (Nirodha) is also known as attainment because it represents the element of nibbana. There are three kinds of attainment:

- attainment of fruition (*palah*)
- attainment of cessation (Nirodha)
- attainment of nibbana.

---

[23] Dhammapada Verses 338 to 343, Sukarapotika Vatthu, https://www.tipitaka.net/tipitaka/dhp/verseload.php?verse=338

[24] Ibidi

Attainment is a state of mind free from defilements. As the trainee progresses on the path (*magga*), by virtue of his or her purity of mind and insight wisdom, he or she may attain such fruition. Attainment of fruition is the threshold into the supramundane sphere of consciousness. Entering this sphere also involves a change of lineage. One may go through four stages of sainthood before final attainment of nibbana. These stages are stream enterer, once-returner, nonreturner, and Arahant (Sotapanna, Sakudagami, Anagami, and Arahant). The attainment of Nirodha is only possible for the highest saints, who have mastery over the attainment of jhana. It is also possible for an Arahant who takes the pure vipassana path without attaining samatha jhana. At death, an Arahant attains *anupasisesa* nibbana (nibbana without residue remaining).

The temporary suspension of consciousness and mental factors is the attainment of Nirodha—cessation, entering the bliss of nibbana. The duration of cessation must be predetermined. One emerges from it by bringing objects of the life continuum to the active mind, such as feeling and other perceptions. Saints, those who have attained the bliss of nibbana, have testified to this truth.

The cessation of suffering was presented as the Third Noble Truth. Understanding of cessation is beneficial to understanding the path, and to avoid the danger of following a wrong path and attaining wrong jhanas. It is for this reason that the Four Noble Truths were explained at the beginning of this chapter.

Characteristics of the Third Noble Truth

1. Escape from kammic retribution (*nisaranatto*)

Escaping samsara (Nirodha), as previously explained, has the characteristics of escaping from the cyclical birth of samsara, of being freed from wrong views about the meaning of life, and of being freed from sceptical doubt about the Buddha Dhamma. The attainer of Nirodha escapes from three kinds of cyclic births (*vatta*):

- attachment (*kilesa vatta*)
- unwholesome kamma (*kamma vatta*)
- unhappiness (*vipaka vatta*).

These attributes of attainment are characterised as the same as those of a stream enterer who enters the supramundane sphere.

2. Grounded in resting (*vivekattho*)

Being free from all known defilements, the attainer of Nirodha gains the character of retirement from all unwholesome actions. Such an individual lives in seclusion, without danger of being reborn in lower worlds.

3. Unoriginated (*asankhatattho*)

The attainment of cessation (nirodha-samapatti) is the highest meditational state possible in Buddhism. It has the nature of not being renewed of mental states, for it is the extinction of all feeling and perception, continuing for as long as seven days. It is seen as the actual realization of Nibbana in this life. This is not the case for mundane mental states of happiness such as jhanas; Nirodha has the character of *sa-upadisesa* (nibbana that is eternal). The attainer is no longer subjected to the laws of kamma or vipaka. His or her actions are free from cumulative effect; therefore, his or her actions are known as functional (*kiriya citta*). There is no need for such an individual to perform wholesome kamma because there is no corresponding wholesome result.

4. Deathlessness (*amatattho*)

Nirodha which has the element of nibbana has the characteristic of deathlessness. Sa-upadisesa nibbana means the extinction of defilements with the residue of groups of existence (skandhas—aggregates), such as the body and the functioning of mental groups. Although the five aggregates (skandhas) remain, the mind reaches the state of deathlessness. At the death of skandha, the Arahant attains *sa-upadisesa* nibbana[25].

---

[25] Ven. Taungpulu Tawya Sayadaw, (1979). Vipassana-Insight Meditation - Maha Satipatthana. Department of Religious Affairs, Rangoon Burma.

Summary

The four inherent characteristics of the truth of cessation (*Nirodha sacca*) are:

- nissaranattho—having the characteristic of escape or liberation from suffering
- pavivekattho—having the characteristic of being free from disturbance
- amatattho—a state where there is no more death or dissolution
- asankhatattho—having the characteristic of being unoriginated (nibbana).

Thus, any Dhamma that has the above four characteristics is called Nirodha sacca. Nibbana itself has the above four characteristics, so it is also Nirodha sacca.[26]

---

[26] Mahathera Ledi Sayadaw, Aggamahapandita, D.Litt. (1999). *The Catusacca-Dipani*. Sayagyi U Ba Khin Memorial Trust.

CHAPTER 8

EASTERN GATEWAY TO FREEDOM

Buddhists believe in previous lives and that there is life after death. Rebirth was confirmed by the Buddha; in his enlightenment he has seen his own past lives. The process of rebirth is explained by the continuum of consciousness. Beings die here and are born elsewhere; the next life's form and place correspond to the nature of one's last consciousness, which is determined by kamma. Endless cycles of birth and death are termed as samsara. Beings continue to wander in samsara because of their ignorance. There are three main generic roots (subsets) of ignorance, namely greed, hatred, and delusion. These roots are also known as defilements (*kelesa*). Intentional (volitional) actions performed in ignorance produce a corresponding kamma and its result (*vipaka*). If defilements remain in the mind, the cycle continues. Narrated in a short discourse in the Anguttara Nikaya, the Buddha declared the root causes of kamma that produce a series of unwholesome results.

"O monks, greed, hatred, delusion, is the producer of karmic concatenation."[1]

---

[1] Discourse No. 203, "The Concatenation of Kamma", Anguttara Nikaya, p. 264. Numerical Discourses of the Buddha. (1999). BPS.

The following diagram explains the cyclical nature of continuous becoming and the prolonged existence of samsara:

Analysis of Kamma formation

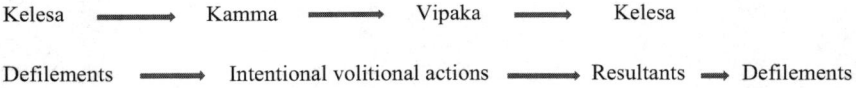

Kelesa ⟶ Kamma ⟶ Vipaka ⟶ Kelesa

Defilements ⟶ Intentional volitional actions ⟶ Resultants ⟶ Defilements

Volitional action (kamma) and its result (vipaka) are intricately related, and they condition the consciousness. The effect of this conditioning passes from one state of mind to the next in succession. Beings are born into this world with defilements, and these inherited defilements cause unwholesome actions and the production of new kamma. As a result, the consciousness is burdened with more defilements. Breaking of this cycle was unknown before the Buddha's enlightenment. The relation of this chain of actions is well explained by the philosophy of relations.

Kamma conditions both mind and matter and produces a result; this is to say that one's volitional actions (kamma) and the results coexist in relation to each other. Therefore, consciousness and resultant consciousness are related to the volition or intention it creates.

Kamma Condition (Kamma Paccaya) (Relation No. 13)

Prenatal kamma (i.e. kamma from a previous birth) is the generating cause of the five sense organs, consciousness, and the other mental and corporeal phenomena in a later birth. Kamma is also a cause of kammic volition. During the active process of consciousness, kamma is produced at the javana[2] stage. The Pali term for the intentional volitional acceleration of consciousness is 'javana', when an action is performed, for example when, in cricket, the bowler runs to gain the momentum to propel the ball towards the wicket defended by the batsman. The bowler's intention is fixed on getting the ball through the wicket. Similarly, javana performs a specific function to complete the function of consciousness. These phenomena are,

---

[2] Bhikku Bodhi. (1999). *A Comprehensive Manual of Abhidhamma*. BPS.

however, in no way kammic results; kamma remains in the life continuum until it meets the right conditions to produce the necessary result.

Kamma Result Condition (Vipaka Paccaya) (Relation No. 14)

For the connascent *(is an adjective; born, produced, or growing simultaneously)*[3] mental and corporeal phenomena, there are five kinds of sense consciousness, which are caused and conditioned by the kammic result. For example, at conception the forming of a new life is the resultant kamma. Transmission of past conditioning to the new life is explained in Chapter 17: Process-Freed Consciousness. The philosophy of relations is explained in Chapter 9: Tranquillity Leading to Insight (Samatha and Vipassana).

There are two types of kamma, *kusala kamma* and *akusala kamma*, which produce their respective results, *kusala vipaka* and *akusala vipaka*. *Kusala* denotes skilful actions that lead to happiness. *Akusala* means unskilful actions that increase suffering. Therefore, one must understand how kamma has a direct correlation to one's happiness and suffering. Happiness is a result of kamma, produced by wholesome roots such as generosity, amity, and wisdom; suffering, on the other hand, is a result of kamma produced by greed, hatred, and delusion. Reactive emotions and not being mindful towards akusala vipaka will result in increasing one's defilements, and so, the cycle will ever increase and suffering continues. Insight wisdom is the ability to see that one's own actions (kamma) result in happiness or suffering, and the resulting ability to choose to cut off the root causes. This is knowledge of cause and effect. The wise person sees that virtue is a firm foundation for liberating the mind and therefore chooses to guard the mind and prevent doing evil; instead, such a person does good and enters on the path of purification of the mind. The core teachings in Buddhism are about breaking the cycle of suffering and redirecting the mind towards one's own welfare and that of others.

Both kamma and its results are not permanent; its respective causes and effects must come to an end. Because of past conditioning and present kamma, the relation of kamma to its results is not linear but forms a complex web. The Buddha advised us not to worry about the past but

---

[3] Collins English Dictionary. (2022). Collins Learning, HarperCollins Publishers.

to be skilful in the present. One must come to see the transient nature of all phenomena, the unsatisfactoriness of all phenomena, and the fact that these phenomena are beyond the control of any permanent self-ego within the mundane world perpetuated through inadvertence. Liberating the mind to the supramundane world is considered to be enlightenment. It is also significant to note that ownership of kamma, while rooted in ignorance, results in the forming of a false perception of oneself.

Below is a metaphorical expression from the Pali Canon (Majjhima Nikaya) that explains the cosmological view and transient nature of a being trapped in samsara:

> The whole cosmos (earth, heavens, hells, and astronomical bodies such as quasars, pulsars, and nebulas. The timescale of these ultrahigh-energy bodies is measured in light years) always trend towards renovation or rather in destruction; it is always in a course of changing by a series of cycles, of which the beginning and end alike are unknowable and unknown. People and gods form no exception to this universal law of composition and dissolution. The unity of forces that constitutes a sentient being must sooner or later be dissolved, and it is only through ignorance and delusion that such a person indulges in the daydream of an existing separate self.

The Buddha in his enlightenment observed things as they are and discovered the transcendental dependent origination of phenomena. He saw how he could destroy the cycle of samsara by addressing the craving that is the central hub of consciousness. The following verses of the Dhammapada, 153–4, explain in poetic form the Buddha's liberation, known as "The Lion's Roar":

> Through many a birth in samsara
> have I wandered in vain,
> seeking the builder of this house of life.
> Repeated birth is indeed suffering.
> O house builder, you have been seen;
> you shall not build the house again.
> Your rafters have been broken up;
> your ridgepole is demolished too.

My mind has now attained the unformed nibbana
and reached the end of every sort of craving.

(This verse was uttered in joy by the Buddha immediately after his awakening. It is not found in any other place in the canon. Venerable Ananda heard these words from the mouth of the Buddha when the two were in discussion.)

At this time of his enlightenment, with his developed mindfulness, clear comprehension, and investigation of Dhamma, the Buddha was able to gain insight into dependent origination, in both forward and reverse order. The Buddha's understanding singled out craving as the central axis of repetitive births and deaths which propel the continuity of samsara. Appearance and disappearance of phenomena become clearer in his attentive mindfulness. The Buddha summarised his experience of the appearance and disappearance of phenomena into the Four Noble Truths. There are many planes and spheres of reality in which consciousness spins. In the human realm, these are the six senses of craving; these are to crave for sensory pleasure, for existence, and for nonexistence, which thus causes beings to wander in samsara in many different forms of life. The human form is said to be the most suitable for liberation because in this form it is possible to experience a wide range of consciousness in the Middle Way. According to the Buddhist cosmological model (adapted from ancient Vedic religion), there are thirty-one worlds, including heavens, hells, and Brahma worlds. Beings born to those worlds may not necessarily be able to understand Dhamma. Correct knowledge and understanding of suffering of samsara and an escape route from it was unknown to the world until the Buddha's enlightenment. People were imprisoned in an interlocking cycle of kamma and vipaka. Out of compassion to help these beings trapped in endless cycles of suffering, the Buddha showed us the gateway to freedom, that is nibbana.

CHAPTER 9

TRANQUILLITY LEADING TO INSIGHT (SAMATHA AND VIPASSANA)

THE NUMERICAL DISCOURSES OF THE Buddha is a selected anthology of discourses from the Anguttara Nikaya (AN) of the Pali Canon. There are several passages in this collection in which the Buddha describes the nature of the mind.

> No other thing do I know, O' monks, that changes as quickly as the mind. It is not easy to give even a simile for how quickly the mind changes. (AN 4.2, "The Mind")

Through tranquillity and insight, the mind can be trained and directed towards full liberation. The meditation methods taught by the Buddha, namely tranquillity and insight, are the most widespread practices in the Theravada tradition. The following discourse gives assurance with regards to how tranquillity and insight can complement the development of the mind.

> Two things, O monks, partake of supreme knowledge: tranquillity and insight. If tranquillity is developed, the mind becomes developed. If the mind is developed, all lust is abandoned. If insight is developed, wisdom becomes developed.

If wisdom is developed, all ignorance is abandoned. A mind defiled by lust is not freed; wisdom defiled by ignorance cannot develop. Thus, monks, through the fading away of lust there is liberation of mind; and through the fading away of ignorance there is liberation by wisdom. (AN 2.3, p. 10; 14, p. 42)

Transformation of one's lineage from a worldly to an Ariyan is possible by virtue of practising the path of tranquillity and insight. To be in a trance (*jhana*) is to attain the state of tranquillity by way of mental absorption. By this the mind is temporarily purified, forming a firm basis for developing insight knowledge. However, trance alone does not change one's lineage. It can be seen as a vehicle to bring the mind to a settled, stable, and tranquil state to allow for further development of insight. At the time of the Buddha in contemporary India, tranquillity meditation leading to trance was a common practice among ascetics. As a prince, Siddhattha left his palace in renunciation of his position so that he might became an ascetic. He learned these methods of meditation and trance under very prominent masters. However, he soon left those schools because he had realised that trance alone was not full liberation. Defilements are hindrances to liberation. They remain dormant in the mind during suppression, resurfacing when there are sufficient conditions to provoke them. Similarly, grass dies under a heavy brick when left for some time in a corner of the garden, but roots of weeds remain under the soil and spring up during the rainy season when the brick is removed.

There are five hindrances that disturb a worldly mind. An untrained mind is prone to the adverse effects of these hindrances, which are taught in pairs. The first is attachment to sensual desires and aversion to unpleasant sensations. Second is anger and ill will. Third is sloth and intoxication. Fourth is restlessness and worry. Fifth is scepticism and doubt. The literal meaning of these terms explains the way hindrances appear in the mind and how they obstruct the progress of mental development. For example, anger is like fire that burns everything around it. It produces a consciousness associated by its function to the unwholesome mental factor. For example, consciousness could be associated with greed, where an object becomes desirable to the new consciousness, and from this a craving for its consumption arises. The function of these factors is to swiftly shift the attention from an object of meditation to a new, more exciting object, therefore entertaining the senses by grasping the new object. Soon one's attention may randomly shift to a different range of objects to further satisfy those senses, thus making the original meditation practice less interesting. To stay with one meditative object requires much discipline and resolution. The Pali term for meditation is *bhavana*, meaning "mental training".

It is advisable to practise both methods of Buddhist meditation under an experienced master who has a good understanding of the Buddha's teachings. There are forty subjects of meditation in samatha (tranquillity), and these can be complicated, so only an experienced master can choose a suitable subject according to the student's temperament. During the time of the Buddha, he himself advised his disciples and gave them meditation subjects accordingly. Here the exposition given by the Buddha should be understood according to what is suitable to the meditator's temperament, as such mindfulness and contemplation is recommended on foulness for one of greedy temperament. The Four Divine Abodes and concentration on colour discs are suitable for one who hates concentration. Mindfulness of breathing and recollection of Buddha's qualities are suitable for one of a deluded or speculative temperament. The meditation manual of Theravada teaching *The Path of Purification*[1] illustrates meditation subjects and methods used at the time of the Buddha and his recommendations.

A skilful meditator may use these meditation subjects and subsequent mental states as conditions to develop insight. For example, contemplation on the foulness of life forms is a condition for direct knowledge. Three consecutive divine abodes, and meditation on loving-kindness, compassion, and sympathetic joy, are conditions for the fourth divine abode: equanimity. Each lower immaterial state is a condition for each higher one. The base consisting of neither perception nor nonperception is a condition for the attainment of cessation. All are conditions for living in a blissful state of mind, for insight, and for the fortunate prize of becoming. In this way tranquillity and insight meditation could be practised simultaneously.

---

[1] Buddhaghosa (1991) The Path of Purification: BPS translated by Bhikkhu Nanamoli *(Oxford)*.

Child Psychology

The following discourse explains the Buddha's advice to his own son, Rahula, who was only seven years old when this discourse was delivered to admonish him. Hence, the discourse has an undertone of fatherly advice to a son. It highlights that from a young age it is important to cultivate the correct and appropriate behaviour conducive to enlightenment. Rahula was referred to the Buddha by his chief disciple Venerable Sariputta, who complained that Rahula was too stubborn a novice to train. This discourse stands as a charter of discipline for children; its lays the foundation of child psychology in Buddhism, to change the course of a child's life for growth and rapid development of good habits. It speaks of what influences a child's development and what cognitive changes occur in childhood, so that we might understand children's true potential. It includes monitoring and mentoring adolescents to become responsible for their own progress through guided training. Understanding of the unique personality trait of his son was the most important thing the Buddha demonstrated in this discourse. Self-esteem is a key factor to success in life. Rahula was too proud of his royal background and of his father's being the Buddha, and because of this he refused to take instructions from the chief disciple. The Buddha understood Rahula's weaknesses and immediately developed a positive parent–child relationship to make him feel special. This provided a framework of support and assurance for the boy to develop a healthy, respectful self-esteem, which is extremely important for the success of children. The Buddha made his point clear, talking it through step by step, to map out the plan for correction and solve the issue; this was firm but fair advice of a father to a son.

Rahulavada Sutta, Majjhima Nikaya 61

> Then the Blessed One, arising from his seclusion in the late afternoon, went to where the venerable Rahula was staying at the Mango Stone. Venerable Rahula saw him coming from afar and, on seeing him, set out a seat and water for washing his feet. The Blessed One sat down on the seat set out and, having sat down, washed his feet. Venerable Rahula, bowing down to the Blessed One, sat to one side.

Then the Blessed One, having left a little bit of water in the water dipper, said to Venerable Rahula, "Rahula, do you see this little bit of leftover water remaining in the water dipper?"

"Yes, sir."

"That's how little contemplation there is in anyone who feels no shame at telling a deliberate lie."

Having tossed away the little bit of leftover water, the Blessed One said to Venerable Rahula, "Rahula, do you see how this little bit of leftover water is tossed away?"

"Yes, sir."

"Rahula, whatever there is of contemplation in anyone who feels no shame at telling a deliberate lie is tossed away just like that."

Having turned the water dipper upside down, the Blessed One said to Venerable Rahula, "Rahula, do you see how this water dipper is turned upside down?"

"Yes, sir."

"Rahula, whatever there is of contemplation in anyone who feels no shame at telling a deliberate lie is turned upside down just like that."

Having turned the water dipper right side up, the Blessed One said to Venerable Rahula, "Rahula, do you see how empty and hollow this water dipper is?"

"Yes, sir."

"Rahula, whatever there is of contemplation in anyone who feels no shame at telling a deliberate lie is empty and hollow just like that.

"Rahula, it's like a royal elephant: immense, pedigreed, accustomed to battles, its tusks like chariot poles. Having gone into battle, it uses its forefeet and hind feet, its forequarters and hindquarters, its head and ears and tusks and tail, but keeps protecting its trunk. The elephant trainer notices that and thinks, *This royal elephant has not given up its life to the king.* But when the royal elephant, having gone into battle, uses its forefeet and hind feet, its forequarters and hindquarters, its head and ears and tusks and tail and its trunk, the trainer notices that and thinks, *This royal elephant has given up its life to the king. There is nothing it will not do.*

"In the same way, Rahula, when anyone feels no shame in telling a deliberate lie, there is no evil, I tell you, he will not do. Thus, Rahula, you should train yourself, 'I will not tell a deliberate lie even in jest' (white lies).

"What do you think, Rahula: What is a mirror for?"

"For reflection, sir."

"In the same way, Rahula, bodily actions, verbal actions, and mental actions are to be done with repeated reflection.

"Whenever you want to do a bodily action, you should reflect on it: 'This bodily action I want to do—would it lead to self-affliction, to the affliction of others, or to both? Would it be an unskilful bodily action, with painful consequences, painful results?' If, on reflection, you know that it would lead to self-affliction, to the affliction of others, or to both; it would be an unskilful bodily action with painful consequences, painful results, then any bodily action of that sort is absolutely unfit for you to do. But if on reflection you know that it would not cause affliction, it would be a skilful bodily action with pleasant consequences, pleasant results, then any bodily action of that sort is fit for you to do.

"While you are doing a bodily action, you should reflect on it: 'This bodily action I am doing—is it leading to self-affliction, to the affliction of others, or to both? Is it an unskilful bodily action, with painful consequences, painful results?' If, on reflection, you know that it is leading to self-affliction, to the affliction of others, or to both, you should give it up. But if on reflection you know that it is not, you may continue with it.

"Having done a bodily action, you should reflect on it: 'This bodily action I have done—did it lead to self-affliction, to the affliction of others, or to both? Was it an unskilful bodily action, with painful consequences, painful results?' If, on reflection, you know that it led to self-affliction, to the affliction of others, or to both; it was an unskilful bodily action with painful consequences, painful results, then you should confess it, reveal it, and lay it open to the Teacher or to a knowledgeable companion in the holy life. Having confessed it, you should exercise restraint in the future. But if on reflection you know that it did not lead to affliction, it was a skilful bodily action with pleasant consequences, pleasant results, then you should stay mentally refreshed and joyful, training day and night in skilful mental qualities.

"Whenever you want to do a verbal action, you should reflect on it: 'This verbal action I want to do—would it lead to self-affliction, to the affliction of others, or to both? Would it be an unskilful verbal action, with painful consequences, painful results?' If, on reflection, you know that it would lead to self-affliction, to the affliction of others, or to both; it would be an unskilful verbal action with painful consequences, painful results, then any verbal action of that sort is absolutely unfit for you to do. But if on reflection you know that it would not cause affliction, it would be a skilful verbal action with pleasant consequences, pleasant results, then any verbal action of that sort is fit for you to do.

"While you are doing a verbal action, you should reflect on it: 'This verbal action I am doing—is it leading to self-affliction, to the affliction of others, or to both? Is it an unskilful verbal action, with painful consequences, painful results?' If, on reflection, you know that it is leading to self-affliction, to the affliction of others, or to both, you should give it up. But if on reflection you know that it is not, you may continue with it.

"Having done a verbal action, you should reflect on it: 'This verbal action I have done—did it lead to self-affliction, to the affliction of others, or to both? Was it an unskilful verbal action, with painful consequences, painful results?' If, on reflection, you know that it led to self-affliction, to the affliction of others, or to both; it was an unskilful verbal action with painful consequences, painful results, then you should confess it, reveal it, and lay it open to the Teacher or to a knowledgeable companion in the holy life. Having confessed it, you should exercise restraint in the future. But if on reflection you know that it did not lead to affliction, it was a skilful verbal action with pleasant consequences, pleasant results, then you should stay mentally refreshed and joyful, training day and night in skilful mental qualities.

"Whenever you want to do a mental action, you should reflect on it: 'This mental action I want to do—would it lead to self-affliction, to the affliction of others, or to both? Would it be an unskilful mental action, with painful consequences, painful results?' If, on reflection, you know that it would lead to self-affliction, to the affliction of others, or to both; it would be an unskilful mental action with painful consequences, painful results, then any mental action of that sort is absolutely unfit for you to do. But if on reflection you know that it would not cause affliction, it would be a skilful mental action with pleasant consequences, pleasant results, then any mental action of that sort is fit for you to do.

"While you are doing a mental action, you should reflect on it: 'This mental action I am doing—is it leading to self-affliction, to the affliction of others, or to both? Is it an unskilful mental action, with painful consequences, painful results?' If, on reflection, you know that it is leading to self-affliction, to the affliction of others, or to both, you should give it up. But if on reflection you know that it is not, you may continue with it.

"Having done a mental action, you should reflect on it: 'This mental action I have done—did it lead to self-affliction, to the affliction of others, or to both? Was it an unskilful mental action, with painful consequences, painful results?' If, on reflection, you know that it led to self-affliction, to the affliction of others, or to both; it was an unskilful mental action with painful consequences, painful results, then you should feel distressed, ashamed, and disgusted with it. Feeling distressed, ashamed, and disgusted with it, you should exercise restraint in the future. But if on reflection you know that it did not lead to affliction, it was a skilful mental action with pleasant consequences, pleasant results, then you should stay mentally refreshed and joyful, training day and night in skilful mental qualities.

"Rahula, all those Brahmans and contemplatives in the course of the past who purified their bodily actions, verbal actions, and mental actions did it through repeated reflection on their bodily actions, verbal actions, and mental actions in just this way.

"All those Brahmans and contemplatives in the course of the future who will purify their bodily actions, verbal actions, and mental actions will do it through repeated reflection on their bodily actions, verbal actions, and mental actions in just this way.

"All those Brahmans and contemplatives at present who purify their bodily actions, verbal actions, and mental actions do it through repeated reflection on their bodily actions, verbal actions, and mental actions in just this way.

"Thus, Rahula, you should train yourself: 'I will purify my bodily actions through repeated reflection. I will purify my verbal actions through repeated reflection. I will purify my mental actions through repeated reflection.' That's how you should train yourself."

That is what the Blessed One said. Gratified, Venerable Rahula delighted in the Blessed One's words.[2]

The method of assuming tranquillity first is also known as *samatha yana*, meaning "tranquillity vehicle". This can take trainees to the pinnacle of the most refined consciousness.

Objects of samatha meditation are static, such as a coloured disk, or they could be concepts such as earth, water, fire, and air. The mind can train by using different classes of objects that correspond to the different levels of trance. Starting from material and moving to fine material, to immaterial, and on to neither perception nor nonperception and so on. An object such as infinite space could be the highest level one may take as an object on this route where there is no perception of gross matter.

There are many levels of concentration: approaching, access, full absorption, and momentary concentration. "A one-pointed mind can be deliberately cultivated as a factor of meditative absorption and can be developed up to the degree of complete absorption of mind." These attainments are considered mundane because they depend upon the chosen object. The effect of this meditation benefits the practitioner once impermanent nature is understood. It then has a liberating quality, and one would have then developed "right concentration".

On the other hand, objects for vipassana are dynamic. These objects and methods are explained clearly in the Mahasatipatthana Sutta, that is the Four Foundations of Mindfulness. Momentary concentration is recommended for this practice. To gain insight wisdom, awareness is crucial when observing changing phenomena within the body and mind. Correct practice of vipassana will lead to supramundane consciousness and final deliverance of the mind, thus uprooting all known defilements permanently. The final object towards enlightenment is the element

---

[2] *Ambalatthika-Rahulovada Sutta: Instructions to Rahula at Mango Stone (MN 61)*, tr. Thanissaro Bhikkhu, Access to Insight (30 Nov. 2013), http://www.accesstoinsight. org/tipitaka/mn/mn.061.than.html, accessed 8 July 2022.

of nibbana itself. At this stage the meditator can see the possibility of an unconditioned world by inferential knowledge. Achievement of the highest jhana transcends perception and confirms the limit of the mundane world. The meditator wishing to attain nibbana must change his or her meditation method to vipassana. This would enable him or her to see things clearly, as they are, without hindrances obstructing the mind. The path to the liberation of one's mind, as taught by the Buddha, is generally known as Dhamma, which is based on his profound intuition of the nature of mind and body.

CHAPTER 10

DHAMMA

Definitions

THE WORD DHAMMA IS DERIVED from the Sanskrit root word *dhar*, meaning "to hold, bare, or support". From this the word *Dharma* was coined and its Pali equivalent, *Dhamma*. The term Dhamma is very broad; it could include anything in the universe, for example the way the earth supports all living beings by sustenance and by its very substantiality, or the way the waters of the world are the home to many fish and other creatures. Within the structure of the earth, there are other elements which contribute to supporting and sustaining life. The study of earth and life forms tells us that the bodies of life contain the same elements of earth, water, wind, and fire—all this falls under the defining term of *Dhamma*.

Dhatu is that which bears its own characteristics; it is translated as "element". Each Dhamma is the "bearer" of a single quality, characteristic, or essence (*sabhava*). *Sabhava* is also translated as "intrinsic nature". The definition of *intrinsic* is "something that is natural or inherent", as opposed to something that has to be learned. For the purpose of Buddhist philosophy, these definitions are used and interpreted to explain life experiences and for the purpose of liberating the mind from wrongly grasped ideas. Hence, anything with an intrinsic nature that supports and sustains life is considered Dhamma. For example, nutriment from food supports and sustains life and is regarded as Dhamma. Further observation of nutrients would reveal that they are derivatives of the four great elements of earth, water, fire, and wind.

According to the Theravada School of Buddhism, the Buddha and his teachings of truth are the embodiment and body of Dhamma (*dhammakaya*). The Buddha understood the true nature of life and reached the full potential of perfection. He also taught others to purify their minds. Therefore, out of respect and veneration, his teaching is now known as Dhamma. Dhamma is channelled down by methods of teaching. It was the master's wish to help people find liberation from suffering.

Life is broadly divided into body and mind. There are many other categories of discretely identifiable units within the body and the mind that constitute the make-up of a living being. For example, as the mind is clearly different from the body, there are ultimates in matter (materiality) and ultimates in mind (mentality). Each of these ultimates has its individual essence (sabhava). These ultimate units of mind and matter are also called Dhamma, for their intrinsic nature, as the fundamentals of existence, must be in sustaining and supporting life.

The following example might make it clearer how Buddhists have adapted definitions and semantics such as these to their purposes:

Lokuttara is broken down into two roots, *loka* and *uttara*. For *loka*, the common-use definition is "world"; however, in Buddhism it is used to refer to the five aggregates, namely body, feelings, perceptions, mental formations, and consciousness. *Uttara* means "above", "beyond", or more commonly in Buddhism, "transcendent". Hence, *lokuttara* refers to the supramundane consciousness that transcends the world of the five aggregates, that is body and mind.

In Buddhist terms, Dhamma is twofold, divided into conditioned and unconditioned Dhamma (referring to a state of mind and body). Conditioned Dhamma is explained in terms of dependent arising and relational methods. It is irreducible in terms of what is detectable to the mind; unlike in science, there are no instruments or apparatuses used to detect or measure it. Dhammas within the body and mind are observed by mindfulness and understood by insight wisdom. "The Dhammas taught by the Buddha are not reducible by further retrogression to any substantial bearers of qualities." For example, even in a single moment of consciousness there is a multiplicity of mental factors and a vast net of relations.

The list of Dhammas which are relevant for meditation practice is given in the book the Dhammasangani, which concerns the psychological aspects of the mind. The sub-commentary to the Dhammasangani (the Mula Tika) says, "There is no other thing than the quality borne by it."[1]

Dhammas themselves, as the Atthasalini expressly says, "are born by their conditions" (p. 40). Any further reduction of the Dhamma would result in its being undetectable by the faculty of mindfulness.

The Buddha limited his teaching to the suffering of samsara and humankind's liberation to achieve nibbana. He has clearly shown us the limitations and what is relevant to end suffering. Directing the mind towards metaphysical things or speculations does not lead us to liberation but only makes things more complicated and confusing. For example, is there self-existence or nonexistence after death? Such questions only make the mind more confused for an untrained and worldly person. It is similar to being asked whether fire goes north or south after it is extinguished; the parameters of the question are nonsensical. The Middle Path in Buddhism is a method to develop these Dhammas so that one's consciousness is purified from latent defilements. For example, one-pointedness of the mind can be cultivated as a factor of meditative absorption (*jhana anga*). This can be developed up to the degree of complete absorption of the mind. One may acquire the aspect of the path "right concentration" if, by one-pointedness, his or her focus is on the quality of liberation. A one-pointed mind can be developed for the purpose of insight (vipassana), only up to access concentration. Or one-pointedness may appear as calmness (samatha).

Buddhism applies the literal meaning of Dhamma widely to explain many life experiences and the effects of those experiences for practical use and to rigorously test its implications. Dhammas are like useful tools. They need to be made efficient and effective by repetitive use and development. This is done in combination with other supportive conditions. For example, with the aid of another tool, one sharpens a knife, maintaining its efficiency for the task of cutting. The following passage from *Abhidhamma Studies* explains the limitation of a single factor and how it could be used for a specific purpose:

---

[1] Nyanaponika Thera, "Dhamma: Reference", *Abhidhamma Studies: Researches in Buddhist Psychology* (Kandy, Sri Lanka, Buddhist Publication Society, 1965), pp. 40–49.

> Even such an elementary factor as "perception" (*sanna*) is not unequivocal. According to the Atthasalini and Mula Tika, its reliability and steadfastness depend on the presence or absence of knowledge and on the higher or lower degree of concentration. (p. 40)

In order to retain its usefulness, Dhamma must be maintained by continuous practice. Dhamma is conditioned by repetition, the relation of habitual recurrence (*asevana paccaya*). For example, concentration can be intensified by the force of repetitive practice. There are other qualitative mental factors that need to be developed and balanced to perform a task, such as tranquillity, lightness, and pliancy.

If any further irreducible residue of identity or quality of Dhamma is found, then that could be kamma, in a form undetectable to the mind until it produces its result. For example, there is a belief among Buddhists that kamma retained in the subconscious (*bhavanga citta*) appears in dreams. This belief is compatible with the kamma (or sign of kamma) that appears at the moment of death. The Dhamma needs to be deciphered from its appearance and manifestation.

The doctrine of anatta (nonself) explains that there is no permanent substance to any of the conditioned or unconditioned Dhammas. In the simile of an onion, after peeling away, one by one, the many layers, at the end there is no final irreducible substance to be found in it.

"All Dhammas are not self" (Sabbe Dhamma anatta). This statement affirms both nonself and the insubstantial nature of all phenomena, including unconditioned Dhamma-nibbana, which is devoid of a self.

"All formations are impermanence" (Sabbe sankhara anicca). This statement is valid for the five aggregates that constitute a being. If everything is impermanent, there cannot be a permanent substance to it. The Buddha understood that the reason people cannot escape from suffering is their wrong understanding and misinterpretation of Dhamma.

The central axiom of the Buddha's Dhamma is the five aggregates, which need to be realised as they are presented and manifested in meditation as its gross manifestation. These are the body, feelings, perceptions, mental formations, and consciousness. These groups are termed aggregates because there are many types of things from each group bundled together. For the untrained mind, these groups manifest, mature, and project an apparently sustained reality, causing either attachment or aversion. The five aggregates combined together can manifest as a whole and conceal the truth. For example, with fire there arises light, heat, smoke, smell, etc. These qualities coexist with the fire element. This reality needs to be seen by insight, whereby with contact comes feeling, perception, mental formation, consciousness, attachment, volition, and grasping. This process becomes clearer when mindfulness and concentration increase.

Different combinations of mental factors perform different functions, and these serve different purposes when developed through the practice of specific meditation methods. For example, the faculty of faith can be developed into the power of faith, which would enable one to enter into "path factors" until it produces the purposed results.

The factors of Dhammas vary in relation to their functions, conditions, and ethical values. They are called ultimate because any further reduction or deduction would be to stray from the path of liberation. Nibbana is the exception, which is an unconditioned Dhamma, and therefore is not subjected to any variable, including time.

The Present Moment

We discuss and analyse a single moment of consciousness as it manifests to us, that is, in the present moment, although reality is actually illusive, constantly "on the move" from the past to the future, both of which are unreal. It is an additional difficulty to identify consciousness; rather, it would be wiser to consider consciousness as a series of phenomena which happen in consecutive order, producing an experience that is receptive to the mind. The true nature, the real facts, of that experience can only be known by insight wisdom. Therefore, it is only for the purpose of crossing the ocean of samsara that one engages in the analysis of Dhamma and the study of its relations, characteristics, and true experience. It is a guide serving as light to the darkness and giving us the right direction. The Dhamma taught by the Buddha is the ferry to help cross this ocean of suffering and come safely to the shore of nibbana.

The present moment is the moment that one becomes aware of an object by way of consciousness. There are three kinds of present moments:

the momentary present; the time it takes to become aware of an object (for example, the time required for an object to become visible after an abrupt change from light to dark in a room); and the time when opportunity arises from the object by association to its past kamma. The Buddha clarified the grammatical confusion of having three time periods (past, present, and future) in relation to the momentary nature of the present moment, because an object under observation might already be a past object.

"Monks, there are three unconfounded appellations, expressions, and designations, unconfounded before, they are now unconfounded, and cannot be confounded; they are not rejected by wise ascetics and Brahmins. Which are these three? For such corporeality, feeling, perception mental formation, and consciousness that is past, gone, and changed, 'It has been' is here the right statement, the usage, the designation. 'It is' does not apply to it; the statement 'It will be' does not apply to it."[2]

The doctrine of impermanence means that all conditioned Dhammas are changing every moment and there can be no two similar situations at all. For example, a rotating wheel touches the ground only one point at a time, and the next moment it touches another point. Similar movements can be explained by a mathematical time series. By this continuum of steady incrementation, the distance travelled can be infinitesimally slow or an exponential growth of speed depending on how fast the wheel turns, as long as it keeps going.

Mindfulness with Clear Comprehension

> In going forward and in going back, a bhikkhu applies clear comprehension.
>
> —Anguttara Nikaya

The second stage of mindfulness is mindfulness and clear comprehension (*sati-sampajjana*). *Sati* is mindfulness; *sampajjana* is clear comprehension, thorough or complete understanding. Derived from these words, *sati-sampajjana* means "one who sees correctly; one who knows correctly, entirely, and equally (or evenly)". Sampajjana is a state of being, and to

---

2 Samyutta Nikaya 22.62; Nyanaponika Thera, (1965). *Abhidhamma Studies*, p. 117.

be is to have sampajjana; hence, it is the act of attainment. From these definitions we can ascertain that mindfulness needs to be established in order to gain a clear comprehension of the object. Through the attainment of that state, one can gain clear comprehension.

The word *sampajjana* can be further analysed by looking at its two root words: *sam* and *pajana*.

Sam has three distinct definitions:

1. To see rightly or correctly

    This term refers to the object of meditation. It means to see the object clearly and precisely, that is not confused by what the body is, and the mind is. It means to see the body as body and the mind as mind.

2. To see entirely

    It means to know entirely by all aspects of both physical and mental phenomena. It also means to know the characteristics, functions, and manifestation of a given object.

3. To see equally or evenly

    This means to apply all five faculties of meditation evenly and to balance them out to develop wisdom.

Five faculties are confidence, effort, mindfulness, concentration, and wisdom.

Clear Comprehension

Clear comprehension of what is of benefit.
Clear comprehension of what is suitable.
Clear comprehension of the meditator's domain.
Clear comprehension of nondelusion.

Benefit: Clear comprehension helps us to understand what is conducive to wisdom and to nibbana. It keeps one from distraction and helps to guide one away from increasing bad kamma.

Suitable: By clear comprehension, we can adopt our practices appropriately, at a suitable time and place and with a suitable object of meditation, that is the Four Foundations of Mindfulness.

Domain: To understand the domain of the meditator, for example a temple, a meditation centre, or a secluded place in one's own territory, without harming anyone or being harmed by anyone. It also means to know the limits and thus the domain of mindfulness.

Nondelusion: To "see things clearly, as they are". It also means to meditate for the purpose of realisation of nibbana.[3]

The nature of the mind is that it has the following characteristics:

- It travels far and wide.
- It travels alone, meaning that there is only one mind.
- It has no fixed form.
- It has no colour,
- It is changing all the time.
- It stays in the body; that is, the body is the house for the mind and has the heart as its base.

These characteristics of the mind are to be observed by mindfulness and comprehended by clear comprehension. Only then might one develop insight by practising vipassana meditation. Further study of the mind, as taught by the Buddha, has been undertaken and compiled by learned monks under the heading of Abhidhamma, which is included in the Pali Canon.

---

[3] Sayadow U Silananda. (2002). *The Four Foundations of Mindfulness.*(Wisdom Publications.

CHAPTER 11

ABHIDHAMMA PHILOSOPHY

THE HIGHER PHILOSOPHICAL PSYCHOLOGY OF Buddhism is an impressive analysis and systematisation of the entire realm of consciousness. The discovery of this subject matter is attributed to the Buddha himself. After he passed away, his enlightened disciples arranged the master's teachings in a manner useful to future students who seek the truth but are not necessarily fortunate enough to learn directly from a living Buddha. The gathering of the First Buddhist Council of enlightened monks convened in 400 BCE. The protocols and the instant written records are now included in the Vinay Pitaka, the second collection of teachings in the Pali Canon. The Theravada scriptures are also known as the Pali canon. They contain teachings of the Buddha on how to reach enlightenment as well as teachings to help guide Buddhists in their everyday life. The Buddha's learned disciples have undertaken further exploration of the subject and have made commentaries to support, clarify, classify, and analyse his philosophical teachings, hoping to make it easier for a beginner who wishes to investigate the Buddha's deeper philosophical teachings. The subject matter is accessible and receptive to those minds who wish to explore an analytical path, and those who endeavour to find true liberation by following the Noble Path. On the other hand, the teaching opens an avenue for academic research in philosophical psychology to help one understand the unexplored terrain of the human mind.

Abhidhamma was conceived by the Buddha during his time. It is traced to immediately after his Great Enlightenment and records the knowledge, wisdom, light, and vision that arose in him at the time of enlightenment, reflecting the mind and the domain of the Buddha. The Dhamma is synonymous with the Truth. The prefix *abi* makes it "higher" or "greater". The term is loosely translated as "the higher truth" or "the higher teaching". There are seven books, commentaries, and sub-commentaries of the Abhidhamma. The seven books of the Abhidhamma demonstrate the penetrative insight into the Buddha's own mind and his ability to conceive the insight of others. Further study will reveal that the Abhidhamma excludes the conventional terminology of perceptions, dealing instead with abstract elements that constitute a being in its ultimate reality (parramatta). On attaining parramatta Dhamma, one reaches a point of changing lineage that would qualify the individual to realise the Ariya Magga (Middle Path) and its fruit. Expansion of this Pali term gives the more profound meaning of ultimate truth, the truth beyond conventions. The Abhidhamma adds two more factors to the Noble Path towards the cessation of suffering, namely right knowledge, and right liberation. Practice of the Middle Way, balancing the potential powers of single and a combination of mental factors, developing and directing them towards final liberation of the mind, is explained in those books of Abhidhamma. Attaining sight and understanding of Dhamma by means of the recommended meditation practices will result in a change of lineage from an ordinary worldly person to a noble person (Ariyan). Abhidhamma classifies ultimate truth into four categories: consciousness (*citta*), mental factors (*citasika*), matter (*rupa*), and nibbana. The full power of liberation comes from putting theory into practice. One who sees Dhamma can, by insight, see the Buddha. Abhidhamma is a useful toolbox for the skilful meditator who practises the Four Foundations of Mindfulness.

The Buddha's penetrative vision and great intuitive wisdom is termed *lokavidu*; this term is translated as "one who has seen the worlds". It implies both the physical world and the conventional world, and the world of the mind. The Buddha always instructed his disciples to look inwards to the inner world and the five aggregates of clinging rather than focusing on the outer world. The generic term used to express his teaching is Dhamma. The method of understanding "the Dhamma within" is given in the great discourse "The Four Foundations of Mindfulness". The Four Foundations of Mindfulness is a fully developed method of meditation practice. It is practically applied in Theravada Buddhism, which is a school of Buddhism based on the Abhidhamma. It is a systematic method to develop the mind through mindfulness and the investigation of Dhamma, allowing a gradual awakening to the truth. This method has been used to gain insight into the manifestations, characteristics, proximate cause, and absolute cause of a given phenomenon, within the boundary of the Four Foundations.

Seven Books of Abhidhamma

Abhidhamma Pitaka consists of seven treatises:

1. Dhammasangani—Classification of Dhammas (Enumeration of Phenomena)
2. Vibhanga—Book of Analysis (Divisions) (Treatises on Modes of Conditionality)
3. Dhatukata—Discussion with Reference to Elements
4. Puggalapannatti—Designation of Individuals
5. Kathavatthu—Points of Controversy
6. Yamaka—The Book of Pairs
7. Patthana—The Book of Causal Relations (The Book of Conditional Relations)

These seven books of Abhidhamma contribute to the massive volume of Buddhist literature making up the third section of the Pali Canon. The Abhidhamma is held in the highest esteem, regarded as the crown jewel of the Buddhist scriptures. Historically, in the tenth century CE, kings in ancient Sri Lanka had the whole Abhidhamma Pitaka inscribed on gold plates and embossed with gems. It contains advanced analytical studies of Dhamma taught by the Buddha to his gifted disciples. The subject matter of these books is written in philosophical terms, superseding conventional usage of language to explain the teaching. Arrangement and treatment of these teachings is expressed in abstract and ultimate terms, positioned in a manner closer to a mathematical model rather than a style of storytelling. It gives the learner an ontological view of phenomena that constitutes an individual's existence of being. It serves the purpose of the master's teaching, to help liberate the mind from wrongly grasped concepts and ideas and to allow one to find supreme happiness, nibbana.

To fully appreciate and comprehend the Abhidhamma, it is recommended to practise meditation as taught in Theravada Buddhism, namely tranquillity and the Four Foundations of Mindfulness. This would develop calm and insight towards final liberation. Over time, masters of meditation have summarised, developed, structured the Abhidhamma and provided more concise applications of it to meditation practices. One of its texts, Dhammasangani, enumerates the full range of consciousness one may experience and the obstacles one may encounter during practice. One can overcome these obstacles by applying the insight wisdom one gains from the practice. A matrix of categories given in the Dhammasangani is the blueprint for the entire edifice that decodes the complexity of the mind. A coherent system with consistent study and practice is useful for steadfast progress. The subject and object of the practice are the body and the mind. Branching farther out from this, the subject becomes feeling, mental formations, and mental objects; these are explored in detail in the Dhammasangani and the Patthana, and in the Abhidhammattha Sangaha, which give clear and comprehensive explanations of Dhamma. Abhidhamma texts serve as a valuable guide for the practice of insight meditation. The limitations of tranquillity meditation are also well-documented in these books. One such limitation is the impersonal nature of perceptions as explained in jhana; while some aspects, such as the intrinsic nature of Dhamma, are for the advanced meditator. This approach of contemplating Dhamma is suitable for an analytical mind.

The Abhidhamma takes a very different approach and applies a different learning style to the teachings of the Buddha. The Buddha freely employed the familiar language required to make the doctrine intelligible to his listeners. He would adapt his approach depending on the capacities of his audience to comprehend and penetrate complex phenomena. He used similes and metaphors to inspire his audience, adjusting the presentation of the teaching so that it would awaken a positive response. Many discourses are recorded in the first collection of the Pali Canon. The Abhidhamma, on the other hand, takes no account of the listener's personal circumstances and ability; it is instead like the scientific study of complex structures, literally the psychological structures of the mind. It could be compared to software architecture that describe a complex system and operating principles. Software architecture is primarily concerned with what the components and relationships are, and software design focuses on how the components interact. Software architecture and design are two separate parts of one process that depend on each other for success. Similarly, the Abhidhamma explains the operating principles of consciousness, mental factors, classify into groups, describe their interrelationship, and take the meditator to his or her goal through its pragmatic applications.

The Buddha often used contemporary beliefs, ideas, and concepts to convey his teachings, for example the allegory of Mara, also a popular theme in Indian mythology. This allegory could be interpreted using the full set of Dhamma enumerated in the Dhammasangani. It depicts the Devil, a tempter, as the demonic celestial king who denies enlightenment and tempts Prince Siddhattha by trying to seduce him. Vividly illustrated in this allegory is the mind of the Buddha as he was about to be enlightened. It shows the temptations and challenges posed by Mara and his three young and beautiful daughters representing lust. The Buddha pointed to the earth as a witness to rescue him and pointed to the ten great perfections (*paramies*, meaning "most excellent"), which are the prerequisite virtues needed to prepare mental faculties for the purpose of self-enlightenment. The virtues he cultivated during training as a bodhisattva produced their resultant kamma when he determined himself to resist both the temptation set before him and the ignorance that remained in latency. According to the principles of kamma (*kamma niyama*), effects of actions remain in one's life continuum (*bhavanga citta*), awaiting the right conditions to produce results. The light, vision, and knowledge of wisdom arose in the Buddha, dispelling the darkness of ignorance. We may interpret that the Buddha defeated Mara by the power of Parami Dhamma. His mental faculties became factors of power, capable of counteracting their opposites, which would otherwise weaken his wisdom, such as temptation towards greed and lust. Defilements remain in the mind in many forms and obstruct the path of enlightenment. In the bhavanga citta, defilements remain in a dormant state and have a latent potential to produce results in a manner obstructive to progress in wisdom. On the other hand, good actions done in the past such as Parami Dhamma also remain in the bhavanga citta, awaiting the right conditions to ripen great fruits (MN 36.31, p. i247).

It is stated in the scriptures that at the last stage of his effort, the bodhisattva turned to practising mindfulness of breathing (*anapana sati*). He had long practised this technique and by it attained first jhana when he was twelve years old. During his period as an ascetic, he had practised tranquillity meditation under two grand masters, who taught him concentration leading to jhana. Those practices helped to develop other mental factors, for example the seven universal and six occasional jhana factors, which are factors of concentration. The differentiation of the levels of jhana is related to the presence or absence of such mental factors. Mastering this method allows one to balance the functions of mental factors and achieve the fine-tuning of not only a single factor but also the function of all the path factors. We can see this clear turning point of direction at the last stages of his practice. Realising the limit of higher jhana, he had directed his mind towards perceiving the object of meditation as it really is—and that is the correct objective of the vipassana (insight) method of meditation. This was a long-lost method rediscovered by his own effort that enabled him to become enlightened.

Vipassana Bhumi

Vipassana bhumi is the term given to the fertile soil from which insight may sprout. These fertile conditions are the establishment of mindfulness, which can be achieved by following the Buddha's teaching of the Four Foundations of Mindfulness. *Bhumi* literally means soil. This grounding is provided in the human realm of existence in an individual's own five groups of existence. An individual's existence is made up of material body, feelings, perception, mental formation, and consciousness. These five aggregates of initial analysis are further expanded into subgroups and analysed. These aggregates need to be understood as they are. Vipassana bhumi describes their true nature in terms of ultimate realities beyond conventional means and conceptual thought. They are grouped for the practical purpose of contemplation, where step by step the meditator discerns the truth of each group until final liberation is attained. These phenomena need to be comprehended by way of insight wisdom. One might then come to see the Dhamma or entities according to their intrinsic nature. Most of these Dhammas are listed in the scheme of classification in the Dhammasangani.

In summary, vipassana bhumi includes the following groups; these are not a set of meditation objects but are the realised truth of Dhammas:

- Five aggregates
- Four great elements
- Consciousness and the process of cognition
- Twelve-sense base, eighteen elements, twenty-two faculties, fifty-two mental factors
- Four Noble Truths; Noble Eightfold Path factors
- Dependent origination
- Insight knowledge

  o Seven stages of purification
  o Sixteen points of insight knowledge.

These groups are subsets of the Abhidhamma's four ultimate truths. They can be taken as the basis for the teaching and study of the vipassana method (*pariyathi*). Both teachers and students can rely on these truths as valid guidance when put into practice. Some teachers choose pure vipassana, known as the "dry vision" of insight meditation that excludes the jhana path. Others take both samatha and vipassana, that is tranquillity leading to insight. The object of insight meditation is the body and mind; these two are contained like oil and water—together but separate. The only method of truly separating these two from each other, as described by the vipassana bhumi, is by insight wisdom. This knowledge is stated as the right knowledge. This method is key, for liberation is reached only in this way and not by any other way. Different disciples attain liberation by realising different groups of insight, thus it is not compulsory to realise all of them. For example, realisation of the four great elements (earth, water, fire, and wind) within the body would be enough for a meditator to let go of craving. The Buddha taught the complete set of Dhamma; he held no knowledge back from students but gave all knowledge needed for the path leading to the end of suffering.

The Buddhist analysis of reality is based on experience, that which corresponds to the facts of life and is in conformity to the natural laws such as in biology, physics, and analytical chemistry. Unlike science, dhamma enumerated in Abhidhamma is to detach from worldly views of nature, find liberation and peace of mind. The Buddha transcended conventional reality and understood ultimate reality. By paying attention to his experience with a method to discern truth, he showed us the limitation of conventional truth. His teaching of Dhammas corresponds to the timeless facts of life; its validity can be seen here and now (*sandhitika, akalika*). Principles of Dhamma are equally applicable to everyone irrespective of geography, social status, ethnic differences, or whether one believes in Dhamma or not. These principles are universal.

CHAPTER 12

DISTORTIONS AND DENSITIES

Is THE WORLD AN ILLUSION? If it is, how could we overcome its deceptive appearance? How could we overcome the beliefs deeply ingrained in our minds? This chapter investigates the Buddhist concept of illusion. It addresses what is real, what is unreal, and what is perceptible to the mind.

The illusion of free will is one of the many illusions and magic tricks that Gustav Kuhn, a magician turned psychology researcher at Goldsmiths, University London, describes in his new book *Experiencing the Impossible: The Science of Magic*. The book explores what magic tricks and illusions can teach us about our brains. Kuhn takes the reader into the psychological underpinnings of tricks, from optical illusions that reveal gaps in perception, to failures of memory that make people think they've seen a ball vanish, when in fact there was no ball to see in the first place.

Philosopher Daniel Dennett is perhaps the leading proponent of the disappearing trick. In his book inaccurately and rather pretentiously titled *Consciousness Explained*, as well as in his talks titled "The Magic of Consciousness", Dennett shows that many of our perceptions and beliefs are illusory in the sense that they do not correspond to consensus facts. There are many thinkers and writers who have attempted to explain how we perceive and embrace reality. The subject is still in its early stages of exploration and is rather limited, though it has application for purposes of amusement and entertainment such as in computer games, movies, and magic shows. Perhaps Shin Lim, the winner of *America's Got Talent* in 2018, is the master of all disappearing tricks on the world stage. This concept of disappearance is also tested in stealth technology, to remain invisible, an improvement of camouflage technics used in combat operations. Use of a decoy to distract or confuse the enemy is a common practice in civil and military activities.

In terms of our beliefs, especially in the modern world, our paradigm is largely a generation of conditioning, by what we are made to believe is true. In today's world, truth is a commodity; it has a commercial value. Through the psychology of marketing, by technology and mass media, people's perceptions and beliefs can be easily manipulated and are often abused to take advantage of this. For example, volatile market indexes and electronic money is projected in graphical charts on computer screens showing us that some markets have collapsed one day and rebounded on the next day.

In the Samyutta Nikaya (4.54), the Buddha describes the notion of emptiness (*sunnata*) as when the aggregates are devoid of all relation to a self. He later describes in poetic similes the nature of the aggregates: form is like a ball of foam, feelings are like a bubble, perceptions are like a mirage, formations are like a banana tree (made of layers upon layers without a defined core), and consciousness is like a magician's trick, an illusion. Speaking again on form (the material image), the Buddha said: "Therefore, O monks, whichever is form, past, present, or future, inner or outer, coarse or subtle, lowly or exalted, far or near, should be observed with correct wisdom as it truly is. … This is not mine, this I am not, this is not myself."

From this piece of teaching, we can learn how to make an assertion from observed phenomena. Illusion is not necessarily in the object. Making a wrong perception from its projection and reaching an incorrect conclusion is the delusion. If one is attached to the object, one may not be able to see it as it really is. Various distortions trick the mind, like a magician blinds his audience to the illusion. Buddhism advises us to avoid blind faith and to be detached from any emotional or subjective response. Instead, we are to be mindful in our observation of both the inner world and outer world.

Recorded in the Anguttara Nikaya 61, Distortions of Perceptions, *The Numerical Discourses of the Buddha* explains four types of distortion:

> Monks, there are these four distortions of perception, four distortions of thought, and four distortions of views:

1. To have seen no change in what is changing.
2. To have seen pleasure in what is suffering.
3. Accepting self where there is no self.
4. To have seen foul or unwholesome as wholesome and good. (p. 91)

This is because an untrained mind has limitations and misconceptions of what is real and unreal. In such a mind, consciousness can at any given moment arise with such mental factors as delusion, greed, and wrong view. These mental factors obstruct the clear vision of the object by "densities" or "crowding". For example, an object appears as being in the collection of either solid or soft, like a stone or a pillow respectively; these projections misconstrue the mind. Further investigation would reveal the mental factor causing the error and how to correct it.

There are four kind of densities: continuity, whole, function, and object. The appearance of the object projects an error in perception, an error in thought, or an error in view.

Continuity

When we see a sequence of events in relation to an object, we assume it to be continuously happening rather than as increments of individual events. This fallacy is often the result of the short space of time between two events. For example, the old-fashioned film projector was made using a collection of celluloid pictures. When run at a speed of twenty-four frames per second, it projects a moving image on the screen. Careful observation would reveal that the moving image is produced by single momentary pictures. Digital technology has developed this process to create electronically stimulated high resolution moving images on compact high density (HD) display screens.

Whole

Objects can appear uniform throughout; that is to say one might assume a quality of the whole from one's perception of the part. This often manifests as an expectation that we do not necessarily recognise.

Function

This is similar to the density of continuity, except rather than a sequence, it concerns a group or set of actions. We fail to notice or perceive the incremental and intricate movements that make up an action being done. For example, in the throwing of a ball we do not recognise the complex movements of the body through which such an action is achieved because attention is on the moving ball.

Object

This refers to seeing the object as a whole and not the parts of which it is comprised. Examples of this include dot matrix printing, digital cameras, and crystal visual display units based on digital technology that has the mechanism of formation by the collection of single units into a gross manifestation of images by increasing its density called pixels. Also, one might see the painter's picture and not the brushstrokes.

By practice of mindfulness and clear comprehension and by investigation of the Four Foundations of Mindfulness, a meditator is able to gain insight into the true nature of the body and bodily movements, feeling, mental formations, mental movements, and consciousness. When meditating on Dhamma, one can overcome the mental obstructions of these densities.

There are three kinds of errors (Pali term: *vipallasa*) mentioned. The term is also translated as "hallucination". The three errors are distortion of perception (*sanna vipallasa*), distortion of thought (*citta vipallasa*), and distortion of views (*ditthi vipallasa*).

Distortion of Perception

Retained in the memory of one's perception might be the image of a man; by the similarity in shape, one might mistake a scarecrow, a snowman, or even something more obscure as being a real man in the distance. Here one misperceives the information received by the visual senses.

Distortion of Thought

This concerns the processing of the information perceived. For example, misperceiving a scarecrow as a man, we might start to think whether the man is a friend or a threat, what his intentions are, etc., hence elaborating into further error from the initial misperception.

Distortion of View

This is often defined by a habitual pattern of thought which, if distorted, will distort the view. In the past there was the view that animal sacrifice would bring good fortune and secure heavenly life. In the example of the scarecrow, we might see that "man" each time we go for a walk and no longer consider it to be anything else and ignore such warning signs in fields or at a cliff's edge. Another example to illustrate this error of view: there could be an urban metropolitan neighbourhood projecting splendid houses, flashy cars, and young adults wearing designer made elegant outfits, which might persuade one to form a view of a more luxurious lifestyle than one might see in suburban areas, but in actual fact it could be a very unhealthy town crowded with criminal gangsters in which not suitable to live and raise children.

These three distortions are cyclical, influencing and affirming each other in turn. Some ideas, beliefs, and observations become erroneous by taking false things as true, and by this method we come to delude ourselves, forming various false syllogisms. Taking a wrong turn can mislead someone to his or her own distraction. Three such views are materialistic; eternal; and nihilistic. These errors can be removed by investigation or by searching into the causes and conditions of things.

The doctrinal teaching of transcendental dependent origination demonstrates the formula of kamma formation, its continuity, its function, and the cycle of birth and death. In this chain of activities, attachment (*thanha*) is the easiest to break. One who sees the three fallacies of phenomena, or at least one of them, might be able to overcome the errors of perceptions, break the attachment, and become liberated. It is said if one of the factors of the chain is broken, then the entire cycle collapses.

The synthesis method of mental formation lists twenty-four relations, explaining how mental factors and consciousness are conditioned. The relation of object (no. 2) and the relation of dominance (no. 3) (discussed in Chapter 12) show that consciousness cannot arise without an object. An object could also be an extrasensory perception, a strong attachment, or an aversion to the object, which is the cause of suffering now or in time to come when craving arises due to unfulfillment of emotional satisfaction. From these analyses we can make the assertion that at any given time our perceptions are conditioned, and the mind is distorted. By contemplating Dhamma and the mindfulness of Dhamma, the meditator is able to discover that there is no permanent substance to our perceptions; they are conditioned, fleeting phenomena, arising, changing, and passing away, although they can become concealed by errors and permeated by illusory perception. From the wisdom gained by the observation of Dhamma, one would be able to correct the error of perception and of the permanent self or soul within.

Mindfulness, clear comprehension, and investigation are the factors of enlightenment. Gradual progression on this path would endow the meditator with thirty-seven knowledges of insight, to reach the final stage of liberation known as Arahant. The Arahant is one who has broken the illusion of perceptions, errors of thought, and wrong views by destroying greed, hatred, and delusion completely. The Buddha emphasises gradual progression and the gradual development of wisdom to understand the profound Dhamma step by step.

"Just as the ocean has a gradual shelf, a gradual slope, a gradual inclination, with a sudden drop-off after a long stretch, in the same way this doctrine and discipline has a gradual training and a gradual performance, a gradual progression, with a penetration to gnosis only expected after a long stretch[1]."

Vipallasa, Density, and Ghana

Density is the degree of compactness of material substances bound together. It causes the crowding of the mind, like clouds moving to fill the sky and

---

[1] Uposatha Sutta: Uposatha (discourse delivered by the Buddha to a community of monks on observance day), Ud 5.5. by Thanissaro Bhikkhu. Access to Insight (BCBS Edition), 3 September 2012. http://www.accesstoinsight.org/tipitaka/kn/ud/ud.5.05.than.html.

distort the vision. The Pali term *ghana* denotes the degree of crowding that distorts the thought process in various ways and how thereupon erroneous perceptions are formed.

The following tables shows types of errors:

	Types of density	Perception	Error
1	Santhiti ghana	Continuity	Permanence
2	Samuha ghana	Whole	Wholesome
3	Kiriya ghana	Function	Passion
4	Arammana ghana	Object	Self

The problems caused by errors are as follows:

- discursive thinking, restlessness, and worry, which can lead to anxiety
- attachment, which can lead to aversion
- ego anger, which can lead to ill will
- personality views, which can lead to the error of superiority, inferiority, or equal status.

To overcome these errors, meditators are advised to pay wise attention— *uniso manasikara*.

The correct attention is to hold to mindfulness while investigating the true nature of cause and effect. If the appearance of the object projects an error in perception, it will be followed by error in thought and error in view. By practice of mindfulness, clear comprehension, and investigating the Four Foundations of Mindfulness, a meditator becomes able to gain insight into the true nature of body and mind, breaking through illusory perceptions of bodily and mental movements. The purpose of insight meditation is to apply momentary concentration to the object. By observing rapidly passing events momentarily with mindfulness, one becomes able to break the error of continuity to see through clearly and identify single units separately that contribute to form a collection of things or a mental event and to let go of the attachment or aversion towards that object without clouding the mind, forming ideas and views. Careful observation would reveal that a moving image is produced by collection of single units. For example, one might be unable to see the single compartments of a high-speed train passing by because of the magnitude of its velocity. This might become clearer if one would observe the outer world with the same level of mindfulness. Then one would be able to discover that white light is a combination of five colours; if sent through a prism, it separates into the spectrum of colours. Similarly, a rotating fan might look like a disk, but if it slows down, one is able to see that it is actually an electromechanical machine made with separate blades.

The table below gives a few examples of how to correct wrong views corresponding to types of density:

	View	Correction	Example
1	Permanent/fixed	Momentary	Thoughts; raindrops
2	Whole/solid state	Collection of many	Sand heap; sunlight; body; chariot
3	Function	Many actions	Walking; dancing; music
4	Self and ego	Compounded	Mental objects: "I am"

The mental factor that causes the error is delusion, and there are types of consciousness involved in sheer delusion. When coupled with other unwholesome factors, this produces hallucinations.

This perception of error takes place because the untrained mind has limitations and misconceives what is real and unreal. Consciousness arises with the mental factors of delusion, greed, and wrong view on a given occasion. These mental factors obstruct the clear vision of the object by way of densities or crowding. For example, an object appears as solid or as soft, like a stone or a pillow respectively, and these projections misconstrue the mind. Further investigation would reveal the mental factor causing the error and how to correct it.

When meditating on Dhamma, one can overcome the mental obstructions listed below:

    a.   taints

    b.   floods

    c.   bonds

d. knots
e. clinging
f. hindrances
g. latent dispositions
h. fetters
i. defilements.[2]

These obstructions manifest in many ways; they are produced by fourteen types of unwholesome mental factors and their combinations, as follows:

1. delusion
2. shamelessness
3. fearlessness or wrongdoing
4. restlessness
5. greed
6. wrong view
7. conceit
8. hatred
9. envy
10. avarice
11. worry
12. sloth
13. torpor
14. doubt.[3]

These fourteen unwholesome mental factors are the basis of the five hindrances and obstacles to insight meditation.

---

[2] Bhikkhu Bodhi, Chapter 8: The Compendium of the Unwholesome, *A Comprehensive Manual of Abhidhamma* (Kandy, Sri Lanka, Buddhist Publication Society), p. 265; Table 7.1, "The Compendium of Categories", ibid. p. 270.
[3] Section 4, Guide 4, ibid. p. 83.

CHAPTER 13

CONSCIOUSNESS

CONSCIOUSNESS IS A CYCLICAL MENTAL process. It is the principle element of experience, and it cognises an object and brings it to awareness. At any given cycle of consciousness, the awareness of the object arises with accompanying mental factors that are operating or occurring at the same time capable of colour the state of mind. One moment of consciousness is denoted by *Cita* and conditioned by those associated mental factors *(Citasika)* and vanishes together at the end of one cycle. Cognitive process consists of a sequence or several cycles of citta's. The quality of consciousness is defined by intention and the type of mental factors associated with it. For example, "seen consciousness" arises when there is light, a visible object, intention, and mental contact present. Subjective attachment or aversion to the object is determined by the mental factors. The process then follows several stages in rapid succession until one fully knows the object. It continues similar to a stream of water and is like a river that flows. This process has to be observed by mindfulness and comprehended by insight wisdom.

Overview of Classification of Consciousness

The chapter "The Compendium of Consciousness" in the Abhidhamma classifies consciousness according to the realm in which it spins. These realms are broadly divided into two classes: mundane and supramundane consciousness. The sense-sphere realm sustains a mundane consciousness; it is related to the six senses of gratification, where suffering or pleasure is experienced. The consciousness of an untrained worldly being moves around in the sense sphere like a bee buzzing around a flower. It could be of either a wholesome or an unwholesome state. Consciousness arises together with mental concomitants (mental factors- *Citasika*). The qualitative factor, or its ethical value of wholesome or unwholesome, is determined by the mental concomitants that accompany the consciousness. These are classed by the factor of intent; in this method of classification, they are regarded as roots. Like roots of a tree feeding into the trunk, branches, and leaves, the mental intent conditions the consciousness and branches out into every action it produces. Good consciousness is classed as beautiful consciousness; it accompanies beautiful mental concomitants such as nongreedy. The quality of consciousness can be deliberately refined and made beautiful by the practice of jhana, using the samatha method. Mental factors apprehend the quality of an object. Depending on the quality of the meditation object chosen, consciousness is classified into fine material and immaterial spheres. When defilements are systematically eradicated, consciousness becomes sublime and lofty, lifted to the irreversible path of supramundane consciousness. In this method of classification there are 89 types of consciousness; adding the supramundane path and fruit jhanas, the figure becomes 121 in total (89 – 8 + 40 = 121).

Jhana Consciousness

In each sphere, jhanas or trances are differentiated according to the combination of mental factors associated with them. In the list of mental factors, there are six which are particularly related to jhana (the occasional six). An initial application of thought towards the chosen object causes the thought to be replaced by sustained application. During this sustained application of thought, one progressively experiences cessation of thought, keen interest (zest), joy, and happiness, until these things come to pass, leaving the mind in a blissful and tranquil state. The mind finally reaches equanimity and settles in a highly tended one-pointedness. There are five levels of jhana consciousness. The method to reach such a state of mind and its resultant fruit is unique to the sphere and plane of consciousness. It is not only trainees, but also fully enlightened ones, who practise jhanas. There are four stages of enlightenment and four corresponding results; hence a total of forty different types of jhana consciousness are examined in the Abhidhamma, corresponding to each stage of enlightenment. State of jhana attained by enlightened ones are class as supramundane jhana to underline the difference of mundane jhanas.

On the other hand, consciousness can be classified according to its nature: unwholesome, wholesome, resultant, or functional. The last two are not rooted in any mental concomitants capable of producing kamma; therefore, they are classed as kammically indeterminate. Since enlightened ones have already surpassed unwholesome and wholesome states of mind, their consciousness is classed as functional (Abayakatha), and no new kamma is produced by their actions.

The Background of the Schema of Classification

The Buddha in his perfect enlightenment (samma sambodhi) was able to capture a single thought moment of a single object and identify the mental factors accompanied with that consciousness.

The schema of classification in the Dhammasangani enumerates fifty-six wholesome mental factors. Nine supplementary factors are given in the sub-commentary the Atthasalini that contribute to enlightened consciousness. The Abhidhammattha Sangaha complied by Acariaya Anuruddha describe fifty two mental factors. The type of consciousness is characterised by the grouping of the mental factors. To study, understand, and practise these Dhammas is important for the spiritual progress of trainees, cultivating both the analytical and synthetical faculties of the mind. The five basic spiritual faculties are confidence, effort, mindfulness, concentration, and wisdom. By the power of these faculties one can develop one's psychological faculties. To burn up defilements, hindrances, and fetters temporarily, it is recommended that one attain the jhanas in the order in which the Abhidhamma classifies consciousness. During these states of absorption (jhana), one is advised to gain wisdom and understanding through insight (vipassana).[1] Craving, false view, and conceit may arise from attaining jhana and the development of tranquillity. This analytical review of mental states is critical to overcoming these impacts of meditative experience. Trainees are able to temporarily burn up defilements, but those advanced in the Noble Path of realisation can permanently burn up subtle defilements by the power factor of their faculties. To avoid falling onto the wrong path, trainees are advised to investigate their meditative experiences by way of analytical retrospection (*paccavekkhana*). Analytical reviewing of one's meditative experiences is an ongoing practice in all schools of Buddhism. For example, trainees are interviewed and discuss their experiences with the master of the school.[2] A difficult analytical enquiry is well illustrated in "Question of King Milinda", which is presented as a dialogue between Bactrian King Milinda and a learned monk, Nagasena:

> "A difficult feat indeed was accomplished, O great king, by the Exalted One."
>
> "Which is that difficult feat, O venerable Nagasena?"
>
> "The Exalted One, O King, has accomplished a difficult task when he analysed a mental process having a single object, as consisting of consciousness with its concomitants, as follows: 'This is sense impression, this is feeling, perception, volition, consciousness.'"
>
> "Give an illustration of it, venerable sir."
>
> "Suppose, O King, a man has gone to the sea by boat and takes with the hollow of his hand a little seawater and tastes it. Will this man know, *This is water from the river Ganges*; *This is water from such other rivers as Jamuna, Aciravati, etc.?*"
>
> "He can hardly do that."
>
> "But a still more difficult task, O King, was accomplished by the Exalted One when he analysed a mental process having a single object, as consisting of consciousness with its concomitants."[3]

---

[1] Nyanaponika Thera, *Abhidhamma Studies: Researches in Buddhist Psychology* (Kandy, Sri Lanka, Buddhist Publication Society, 1965), pp. 11, 30.

[2] Ibid. pp. 10–15.

[3] Ibid. p. 7.

A Comprehensive Manual of Abhidhamma (the Abhidhammattha Sangaha Table 2.1) gives fifty-two mental factors classified under eight groupings of which fourteen are unwholesome factors and twenty-five are beautiful factors. It classifies jhana factors under the occasional group. Closer study of these schemas of analysis reveals the greatness of the Buddha's mind and the profound intuition he gained through direct and penetrative introspective meditation. Before the Buddha's enlightenment, achievement of jhanas was common, but those meditative absorptions were misinterpreted as union with a higher being. A false view of permanency was thus formed by other schools which held onto an externalist[4] view. With the aid of the Abhidhamma, an analytical review of meditative attainment gives a correct and objective knowledge of experience. It informs us that this higher, purified consciousness is also transitory and impermanent and is a conditioned reality caused by developed mental factors. The Buddha instructed his chief disciple, Venerable Sariputta, to analyse his meditative experiences. He proved by the impersonal nature of jhana that they are in fact mundane achievements.

At the time of the Buddha, only a few disciples were analytically gifted. They were able to repeatedly reach the jhanas and then turn their meditative experience towards the development of insight. Right mindfulness and clear comprehension (*sampajjana*) are often emphasised by learned masters to help a trainee avoid falling into false views of his or her meditation practice.

For those who take the analytical path to understanding ultimate truth (paramattha), the Abhidhamma can be a useful aid, by its classification of consciousness and mental factors and other explanations. The path factors of the Noble Eightfold Path are broad aspects of mental faculties. There are five important faculties that need to be developed in order because each path factor is a supporting condition for the next. Once the five factors have been

---

[4] "Externalism is a group of positions in the philosophy of mind which argues that the conscious mind is not only the result of what is going on inside the nervous system, but also what occurs or exists outside the subject. It is contrasted with internalism, which holds that the mind emerges from neural activity alone" (Joe Lau and Deutsch, Max, "Externalism about Mental Content", *The Stanford Encyclopedia of Philosophy* (Fall 2019), Edward N. Zalta, ed., https://plato.stanford.edu/archives/fall2019/entries/content-externalism/).

established, the faculty of wisdom is fully developed, and correct insight knowledge can be attained. It is very useful for trainees to understand the nature of consciousness, how it is defiled by misunderstanding of Dhamma, and how to purify it by wisdom. Through successive attainment of vipassana knowledge, one may be able to completely let go of worldly attachments. When these faculties become power factors, the trainee is able to cut off the remaining defilements in the consciousness to make it purer, leading to a noble (Ariyan) state of mind.

Classification, Method of Treatment, and Cognitive Process in the Analytical Method

Consciousness (*citta*) is the first category of ultimate reality. The Pali word *citta* is rooted in the word *citi*, meaning "to cognise" or "to know". According to the Abhidhamma, consciousness is a process that takes place in the mind, bringing awareness to any object it contacts via the six senses. The Pali commentators classify consciousness in several ways according to the plane of its activity. Further, they have analysed consciousness according to its function.

The first type of consciousness (*kamavacara citta*) is that which arises in the sense sphere. This type of consciousness is related to sight, sound, smell, taste, touch, and mental objects that arise in the mind such as thoughts, feeling, emotions, and ideas.

The second type of consciousness (*rupavacara citta*) is experienced in the fine material sphere.

The third type of consciousness (*arupavacara citta*) is experienced in the immaterial sphere.

The second and third are related to the consciousness that arises in samatha meditation when taking various meditative objects (*kasinas*) for the practice of jhanas.

The fourth type of consciousness (*lokuttara citta*) is experienced in the supramundane (transcendental) sphere.

Other than its relevant plane, consciousness can be classified according to whether it is wholesome (*kusala*) or unwholesome (*akusala*). Wholesome and unwholesome consciousness has either wholesome or unwholesome roots respectively, dependent on the intention. Kamma is formed during these types of consciousness. There are two further dividing branches of consciousness, resultant and functional. According to these classifications, there are a total of 89 types of consciousness. When taking into consideration the jhanas, the number increases to 121 types. This includes consciousness without roots.

The following table gives a summary of roots that condition consciousness:

Wholesome (kusala)	Unwholesome (akusala)
Nongreed	Greed
Nonhatred	Hatred
Nondelusion	Delusion

These roots are the most prominent causes and conditions that determine the type of consciousness. The meditator can experience its manifestations and continuity. Each moment of consciousness goes through three stages of arising, maturing, and passing way. These roots (*mula/hetu*) are the mental factors that arise together and vanish together with the respective type of consciousness. The chapter "The Compendium of Mental Factors" in the Abhidhamma gives a default analysis of how consciousness is classified according to its accompanying mental factor. Other than the few "rootless" exceptions, consciousness cannot arise without mental factors. There are, however, universal mental factors common to every consciousness. The proximate cause of consciousness is body (*rupa*) and mind (*nama*) because it cannot arise in complete absence of mental factors or material phenomena in the sense sphere.

Both wholesome and unwholesome consciousness produce a threefold kamma pertaining to physical, verbal, and mental actions. These intentional volitional actions are termed as kamma. They produce a result immediately or ripen according to timing and when the right conditions are met. Consciousness which arises as a result of kamma is known as *vipaka citta*; this can be divided into kusala vipaka and akusala vipaka. Since these vipaka cittas have no roots in and of themselves, they are classed as rootless and purely mental. There is an exception to this, namely a resultant consciousness accompanied by beautiful mental factors, which hence is not rootless.

During this process, the mental factor that is produced gets further conditioned and could trigger a chain of actions and results. For example, thoughts and emotions in those who have vented anger or experienced a cycle of depression are a result of akusala vipaka. In short, good intentions accompanied with wholesome roots produce good results, and bad intentions accompanied with unwholesome roots produce bad results.

Functional consciousness has no roots, neither wholesome nor unwholesome, and hence is rootless. This consciousness only has the function (*kiriya*) of performing a task. This is to take notice of the object with which it is in contact and bring the mind to its attention at the sense and mind door. Pali commentators use the concept of a door to signify the entry point. There is the special function of a smile producing consciousness, which accompanies joy, arising only in the mind of the Buddha and Arahants (those who have completely eradicated roots and no longer produce new kamma). Resultant consciousness and functional consciousness are kammically indeterminate, known by the special term in Pali *Abayakata*.[5]

Outline of Classification of Consciousness[6]

Classification A	Classification B	Classification C
Sense sphere	Unwholesome	Beautiful
Fine material sphere	Wholesome	Evil
Immaterial sphere	Resultant	Rootless
Supramundane sphere	Functional	

[5] Bhikkhu Bodhi, *A Comprehensive Manual of Abhidhamma* (Kandy, Sri Lanka, Buddhist Publication Society, 1999), pp. 44–50.
[6] Bhikkhu Bodhi *The Abhidhammattha Sangaha of Acariya Anuruddha* (Kandy, Sri Lanka, Buddhist Publication Society, 1999), pp. 45–75.

There is a third way to classify consciousness which divides all types of consciousness into beautiful consciousness, evil consciousness, and rootless consciousness. There are 121 different types of consciousness that fall into these three categories, based on the 25 beautiful and 14 unwholesome mental factors. The word *evil* is a synonym for *unwholesome*, used to contrast some acts that are profoundly immoral in order to discourage them.

Twelve evil consciousnesses refers to ten specified unwholesome mental factors associated with consciousness. Rootless consciousness is the same as previously described; there are eighteen types. The ninety-one remaining consciousnesses are termed "beautiful" for their intrinsic nature. Accompanied by beautiful mental factors (*sobhanacetasikas*), they encompass wholesome, resultant, and functional cittas, which possess beautiful mental factors. This third type of classification supersedes the other two in Buddhist analysis because the beautiful factors are more relevant for mental development. The Noble Eightfold Path is a way to develop those beautiful mental factors and direct them towards liberation of the mind. Unwholesome or evil consciousness is the reason for suffering and a cause of the problems in the world. Developing a beautiful consciousness would purify the mind of all types of bondages, knots, defilements, and hindrances and finally to bring suffering to an end. It should be understood that the beautiful (*sobhana*) has a wider range of factors than the wholesome (kusala).[7]

Consciousness as a Meditation Object

Each moment of consciousness arises and vanishes together with mental factors. A trainee is advised to investigate this process in the present moment as it arises, not before or after. The time domain for consciousness is the present, not the past or future. Mental factors are the second category of ultimate reality. They are mental phenomena of specific characteristics that occur in immediate conjunction with consciousness. Mental factors perform specific tasks in the act of cognition. The type of mental factor gives the consciousness its character and determines its class.

Consciousness requires an object, and it is a process with many stages.

---

[7] Ibid. p. 45.

For example, in the sense sphere, when an object meets a sense it is processed in the sequential order of contact (*passa*), feeling (*vedana*), perception (*sanna*), volition (*cetana*), and consciousness (citta). This process has no time gap other than for the purpose of explanation when we are looking at a static moment and breaking up the process into components. The list of all mental factors arising with consciousness is given in the enumeration of Dhammasangani. The Abhidhamma's explanation gives seven universal factors that are common to all types of consciousness and six occasional factors that are more relevant for the study of jhana consciousness.

This chapter only addresses in detail consciousness and associated mental factors in the domain of beautiful factors and the corresponding wholesome consciousness as in the Dhammasangani. One who wishes to further studies can refer to the Abhidhamma for a complete reference of the other classes of consciousness.

The words *moral* and *immoral* are synonyms for *wholesome* and *unwholesome* respectively. *Moral* (*kusala*) and *immoral* (*akusala*) define actions (kamma) which correspond to wholesome and unwholesome consciousness. To achieve mental development (*bhavana*) is to apply the skill of wholesome consciousness to reach its full potential of liberation in conjunction with the beautiful mental factors and mastery of using their combinations such as loving kindness and compassion.

Process of Consciousness

The occurrence of consciousness in the cognitive process (*cittavithi*) follows a discrete order in accordance with natural law. There are other types of consciousness outside the cognitive process (*vithimutta*) that occur at the time of death, namely death-proximate consciousness, and rebirth-linking consciousness. When there is no active cognitive process taking place, the mind is said to be in a deep dreamless sleep state called *bhavanga*, that is life continuum. A trainee may experience a momentary active cognition process arising and passing away; in times between these occasions of active cognition, one may experience bhavanga. If there is no strong mindfulness, the trainee may fall to sleep.

In addition to having an object, the cognitive process requires a base corresponding to its sense. These bases are named after the sense identity. No cittas occur without a base; consciousness depends on the bases. Similarly, the entry point of the object or point of contact is named according to the conventional means of a door. There are six bases and six doors (for example, eye door) corresponding to the senses and the object presented at the door, dividing consciousness into six types. The table below shows the types of bases and corresponding doors:

Sense	Base	Door	Object	Consciousness
Sight	Eye base	Eye door	Visible	Eye consciousness
Hearing	Ear base	Ear door	Audio	Ear consciousness
Smell	Nose base	Nose door	Odour	Nose consciousness
Taste	Tongue base	Tongue door	Taste	Tongue consciousness
Touch	Body base	Body door	Feel	Body consciousness
Mind–heart	Mind base	Mind door	Mental	Mind consciousness

The sixfold presentation of objects is classed according to how clearly the object appears to the mind. For example, luminosity of light and amplitude of sound is a measurement of the strength of the sensory impression. At the five sense doors, it is classed according to the impact it makes, as very great, great, slight, or very slight. For example, thunder makes a very great sound and makes light. At the mind door it is either clear or obscure.[8] Mental impression could either strong or weak.

Any type of consciousness given in the above table arises in a sense; the object is cognised at the sense door or at the mind door, then follows a discrete cognitive event leading one to the other in a regular and uniform order. This order is one of the five universal orders of mental process related to the consciousness; this order of consciousness is called *cittaniyama* (the law of consciousness), the fixed order of consciousness. There are seventeen stages in one cognition process before an object becomes aware to the

---

[8] Nyanaponika Thera, Guide 4, *Abhidhamma Studies*. p. 157. (2008).BPS.

mind.[9] For a cognitive process to occur, all the essential conditions must be present. If any one of the conditions fails, then the process will fail.

Consider the examples that follow.

For the eye-door process, the following conditions must be met:

a. eye—sensitivity
b. visible—object
c. light
d. attention (*manasikara*).

For the mind door process, the following conditions must be met:

a. the heart bases
b. mental object (Dhamma-rammana)
c. the bhavanga
d. attention.

Seventeen Stages of Mental Processes at the Sense Door

Seventeen stages of mental processes at a sense door are the Fixed Order of Consciousness.[10] When a material object is presented to the mind through one of the five sense doors, a thought process occurs. This thought process consists of a series of separate thought moments, one following another in a particular, uniform order. This order is known as the cittaniyāma, the fixed order of consciousness. For the complete perception of a physical object, through one of the sense doors, precisely seventeen thought moments must occur. As such, the time duration of a fundamental unit of matter is fixed at seventeen thought moments. The process includes seven javana stages where just after determining the object mental factors activate swiftly by its force, performing powerful volitional act producing kamma. After the expiration of that time-limit one fundamental unit of matter perishes, giving birth to another unit. The first moment is regarded as the genesis (uppāda), the last is dissolution (bhanga) and the intervening moments as development (ṭhiti).

---

[9] Bhikkhu Bodhi, *A Comprehensive Manual of Abhidhamma*. (1999). BPS.
[10] Dr. Rewata Dhamma. (2004). Process of Consciousness and Matter. Triple Gem Publications.

When an object enters one of the consciousnesses, through any of doors, one moment of the life-continuum elapses. (This "life continuum" is known as past-bhavanga) Subsequently, the corresponding thought-process runs uninterruptedly for sixteen thought-moments. The object, thus presented, is regarded as "very great" (past-bhavanga) in its intensity.

1. past bhavanga
2. vibrational bhavanga
3. arrested bhavanga
4. five door adverting (one at a time, e.g. eye door)
5. sense consciousness
6. receiving
7. investigating
8. determining
9. javana
10. javana
11. javana
12. javana
13. javana
14. javana
15. javana
16. registration
17. registration.

The table above gives seventeen stages of consciousness, that is, how one becomes aware of an object when consciousness comes into contact with it at a sense door. Kamma forms during the *javana* stage of the process and registers in the bhavanga.[11]

The order of the Noble Path is a formulation according to natural order or natural laws. It is not an invention of the Buddha. The Buddha understood how the mind gets defiled and how to purify it. The Supreme Buddha was endowed with knowledge, power, and confidence that cannot be matched by any other being. He understood the cosmic order by direct knowledge, not from another teacher.

---

[11] Bhikkhu Bodhi, "The Compendium of Cognitive Processes", *A Comprehensive Manual of Abhidhamma*, p. 155.(1999).

The Middle Path propounded by the Buddha is also known as the Noble Eightfold Path, which can be grouped into three sections: morality, concentration, and wisdom. Path factors are arranged in accordance with the great wisdom of the Buddha, particularly his knowledge and wisdom of the order of consciousness (*citta niyama*), the order of Dhamma (*Dhamma niyama*), and the order of the law of action (*kamma niyama*). The order of Dhamma is given in the Dhammasangani. It is traditional at funeral services for Thai monks to chant the order of Dhamma by heart and chant the full set of the philosophy of relations, to remind people that the deceased person had only been a manifestation of these relations. Rebirth follows the laws of kamma; it takes place according to the kusala and akusala kamma performed during one's life.

> By oneself the evil is done, by oneself one suffers; by oneself evil is left undone, by oneself one is purified. Purity and impurity belong to oneself; no one can purify another.

> —Dhammapada, verse 165

Taking place at any single moment are a multitude of processes with a highly dynamic nature operational in the mind. They are not only influential to the immediately successive moment or just to the distant future but are also connected to the multiplicity of past states of consciousness. There are conditions to a process where a beginning or end can neither be found nor ever be known. Instead of looking to the past or to the future, the Noble Eightfold Path provides a practical method to investigate a single moment of consciousness in the present. Meditation practice is not to dwell in memories or to dwell in future imaginations, because both are not real but are mental constructions. The truth can only be found here and now, in the present moment.

To achieve the best results, purification of consciousness should be understood in relation to the factors of morality and wisdom. In the group of concentration, there are three other factors: right effort, right mindfulness, and right concentration. By knowing this, one might develop tranquil, calm concentration (samatha yana) and the attainment of the higher jhanas and then see their limitations. However, if the trainee wishes to attain the supramundane path (*magga*) and fruit (*phala*), then he or she has to change the method of practice to the higher development of insight (vipassana yana). *Yana* in this context denotes a vehicle that takes the trainee to the destination. Samatha practice only temporarily suppresses defilements. The roots of ignorance lay dormant, and their potency to produce bad kamma remains, until sufficient conditions are met to provoke them. The function of the moral factors is to stay within the boundary of the path. Wisdom factors embedded in the path guide one towards the destination. These parameters help one break the cycle of negative kamma so that no new bad kamma is produced. Traditionally, trainees always intentionally adopt moral precepts and are mindful to keep them. It is a method of self-restraint and discipline equivalent to renunciation. These precepts are the foundation for the development of meditative faculties. As the trainee advances in his or her practice, the path factors themselves are the guidelines towards the destination.

The purpose of classifying consciousness and giving a detailed analysis of specific types is to help give an opportunity for self-evaluating one's progress and to assure the trainee to stay on the path towards realisation of the supramundane path and attainment of the noble fruits. These Dhammas are like road signs on a map that gives directions to whom follows it.

Beautiful Consciousness

According to Abhidhamma our external behaviour of body and speech flows primarily from the motivating forces operating within our minds; and our minds, in their turn, follow principally from the activity of their concomitant mental factors depending on the intentions. This process of consciousness is explained in seventeen stages. It therefore becomes extremely important for those who wish to transform their minds to have some understanding the types of beautiful consciousness outlined in Buddhist literature. It is with this recognition that extensive explanations of the mental factors are presented in the Dhammasangani and grouped 91 different types of consciousness conducive to purification of consciousness. Following is the classification of beautiful consciousness according to the sphere it operates.

Sense sphere—24
Fine material sphere—15
Immaterial sphere—12
Supramundane—40

As the name implies, beautiful consciousnesses are those types of consciousness which are accompanied by beautiful (sobhana) mental factors. They are distributed into four classes, as shown above. There are twenty-four in sense-sphere consciousness; these arise in relation to the sense sphere. They can be subdivided equally into three types, each with eight forms: wholesome; resultant; and functional.

The first eight have either two or three wholesome roots which support them and are identified with principles of dichotomy, widely practised in Buddhist cultures, that give participants joy, happiness, knowledge, and profound mental stability and equanimity. They are also known as "meritorious" (punna kamma) because they prevent the arising of defilements, such as impulses or a desire for sensory pleasure. Both moral shame (hiri) and fear (othappa) protect a worldly, a trainee, and a disciple (during the first three stages of ariya). Say that someone joyfully performs a generous deed, understanding that this is a wholesome deed, spontaneously without prompting. Her consciousness is rooted in nongreed, nonhatred, and nondelusion, which manifest as generosity, loving-kindness, and knowledge. Joy is the mental factor corresponding to feeling. There are other beautiful factors accompanied by this consciousness, such as equanimity, wisdom, and their variables, and these produce good results (kusala vipaka). In Buddhist culture, acts of generosity are promoted as a standard method of good practice in accordance with the laws of action such as offering to monks (sangha dana). Since there are eight wholesome consciousnesses, they produce eight corresponding resultant consciousnesses. In this classification, those eight resultant consciousnesses are different from rootless consciousness, classified earlier, because they are "with roots".[12] Both the rootless, wholesome results and the rooted results are produced by the same eight wholesome cittas, but two sets differ in their qualities and functions.

Sixteen types of consciousness in this sense sphere do not arise in

---

[12] Nyanaponika Thera, *Abhidhamma Studies. (1965)*, p. 45. BPS.

the Buddha and Arahants. They have succeeded in breaking all fetters, transcending the cycle of kamma and vipaka; there is no result or rebirth for them because they are no longer in the sense-sphere cycle. The good actions they perform are termed "functional consciousness". One accomplishes some function without residue or further conditioning of mental factors.

Tranquillity and Jhana Consciousness

Jhana consciousness arises as a result of the practice of tranquillity meditation. This is also classed in the group of beautiful consciousness. The Pali term *samatha* means "tranquillity" and is synonymous with peace, settled, and quietness. In Buddhism, happiness is equated to peace of mind. When one's mind becomes calm, the practice of samatha helps one to concentrate. This concentration can be further developed to one-pointedness of the mind (*citta ekaggata*). In the Noble Eightfold Path, the final factor is right concentration (*samma samadhi*). *Right* implies that it is conducive to liberation of the mind and final deliverance to attain nibbana, which is the supreme happiness. Supramundane jhana consciousness attained by the Buddha and Arahants could be regarded as the highest level of concentration (samadhi). It is the gold of consciousness, without blemish or defilement.

The Method of Practice

The Pali word *jhana* has been derived from two roots with the meanings "to contemplate" and "to burn up". Thus, states of jhana contemplate the object and burn up the adverse states opposed to concentration. The adverse states are five hindrances (*nivarana*): sensual desire; ill will; sloth and intoxication; restlessness and worry; and doubt.[13]

A high level of concentration is achieved through close contemplation of meditation devices. There are forty subjects to choose from according to the temperament of the worldly or trainee. The most common method is to begin with the use of meditation objects called "kasinas". For Example, the earth kasina is a small disk of about thirty centimetres in diameter

---

[13] Ibid. p. 58.

made of clay. Other kasinas include a piece of coloured cloth, a flower, a candlelight (fire kasina), or even a clear glass of water (to develop the water kasina). The meditator fixes his or her attention on the object and allows concentration to develop. This physical object will appear in the mind as a mental image, and it develops concentration in stages. Three stages are mentioned in the commentaries; these stages of mental development are preliminary development, access development, and absorption development. These three stages of samatha practice correspond to three levels of concentration: preliminary concentration, access concentration, and absorption concentration respectively. The original object (kasina device) used is the preliminary sign. When concentrating on it at the second stage, the meditator finds a learning sign arising. In the third stage, when concentration intensifies, the counterpart sign arises in the mind.

The consciousness that arises in this practice of samatha meditation is called jhana consciousness. At the preliminary stage, when the five hindrances are successfully suppressed, the counterpart sign emerges. The purity gained at this stage enables one to endure the concentration of the cognitive process further, up to the stage of access concentration through the learning sign and counterpart sign. During this stage absorption develops. The mind gets absorbed into the object is taken to a trance— tranquil, collected, and of a one-pointed mind.

Jhana consciousness is accompanied with five main mental factors:

a. initial application (*vitakka*)
b. sustained application (*vicara*)
c. interest or zest (*piti*)
d. happiness/feeling (*sukka*)
e. mental one-pointedness (*citta ekaggata*).

These five mental factors are the determinants of the levels of jhana. At the first stage of jhana, all five factors are present. As the practice advances, mental factors are dropped in successive order. At the fifth level of jhana, happiness is replaced with equanimity and the mind stays firmly fixed with mental one-pointedness. Jhana consciousness is classified into two classes according to the objects chosen for its practice. They are fine material sphere consciousness and immaterial sphere consciousness.

Fine Material Sphere Consciousness—15

Ruparvacara Citta

The Pali term *vacara* means "frequenting", "moving about in", or pertaining". In this case it "pertains" to the fine material plane of existence. An example is a bee buzzing around flowers seeking pleasure in the sense sphere (*rupa bhumi*).

The fine material realm is the realm in which gross matter is absent and only a subtle residue of matter remains.[14] It is reasonable to say that its material form is rather refined, for example like the petals of a flower or a silk material, which are fine in texture compared to rough sand or stone.

Fifteen cittas fall into this category:

- wholesome fine material consciousness—5
- resultant fine material consciousness—5
- functional consciousness (experienced by Buddha and Arahants)—5.

Each jhana citta is distinct from those in the cognitive process of absorption and are named first to fifth jhana. The jhanas are enumerated in the order given for two reasons:

i. Because when one meditates for the attainment of jhana, one achieves them in this order.
ii. Because the Buddha has taught them in this order.

The order is as follows;

The first jhana is of wholesome consciousness together with initial application, sustained application, zest, happiness, and one-pointedness.

The second jhana is of wholesome consciousness together with sustained application, zest, happiness, and one-pointedness.

The third jhana is of wholesome consciousness together with zest, happiness, and one-pointedness.

---

[14] Ibid. p. 54

The fourth jhana is of wholesome consciousness together with happiness and one-pointedness.

The fifth jhana is of wholesome consciousness together with one-pointedness.

The successive jhanas are in the order of a systematic dissolution of factors.

Parallel to Abhidhamma, there is another system called the Suttanta method compatible with discourses. This enumerates four jhanas of the fine material sphere. The difference is the result of a different combination of mental factors accompanied in the second jhana. It could be seen that the development of jhana is a skill and that different schools and individual meditators could achieve and manifest it in different ways.

Immaterial Sphere Consciousness—12

Arupavacara Citta

"This sphere of consciousness comprises of the citta pertaining to the immaterial plane of existence (*arupa bhumi*), the four realms in which matter has been totally transcended and only consciousness and mental factors remain."[15]

According to the above definition, the immaterial plane is a state of existence without matter. The object for developing immaterial jhana (*arupa jhana*) is the disappearance of the counterpart sign of the object (kasina) used, until it becomes only the space which is "without contact" with matter. Contemplating on space, expanding on it further, it becomes "infinite space". With intense concentration, it will result in being absorbed into the concept of infinite space (*akasapannatti*). The corresponding base which serves for this consciousness is also infinite space. Jhanas are classified in this sphere according to their bases. There are four levels of immaterial jhanas enumerated in Abhidhamma.

1. wholesome consciousness pertaining to the base of infinite space
2. wholesome consciousness pertaining to the base of infinite consciousness

---

[15] Bhikku Bodhi, *A Comprehensive Manual of Abhidhamma*, p. 62.

3. wholesome consciousness pertaining to the base of nothingness
4. wholesome consciousness pertaining to the base of neither perception nor nonperception.[16]

These types of consciousness are all mentioned in relation to a base because they are all transcendental experiences where the object of contemplation is immaterial. Reaching these levels starts from the success of the fifth fine material jhana, where there are no more jhana factors to transcend.

This type of consciousness arises in accordance with the principle of the cognitive process, meaning that no consciousness can arise without a base. In the immaterial sphere, the fine line between an object and its base could be very subtle. Attainment of these fine jhanas is open to those who have mastered the fifth jhana of fine material consciousness. The starting point for the first immaterial jhana is the fifth fine material jhana: "the fifth jhana of wholesome consciousness together with equanimity and one-pointedness".

> At this fifth jhana level (fine material), matter has been totally transcended to a one-pointedness. There is no object to closely contemplate either. Emerging from this, the meditator then directs attention to the concept of space (infinite space) as a meditative object, a transcended object (kasina). ... To progress from the fifth *rupajjahana* to the first *arupajjahana* and from one *arupajjhana* to the next, there are no more jhana factors to be transcended. Instead, the meditator progresses by transcending each subtler object successively. ... This state of mind is totally pure and free from hindrances, it is designated as sublime, lofty or exalted consciousness.[17]

In the sphere of immaterial consciousness, there are not many objects to choose from, unlike in lesser classes of consciousness. Each immaterial jhana (*arupajjhana*) apprehends different aspects of "infinite space", with the meditator experiencing this particular type of consciousness and how each object is contemplated when its perception is fading away.

---

[16] Ibid. pp. 60–1.
[17] Ibid. p. 64.

It is not clear how consciousness itself can be an object for contemplation, or how another type of consciousness could emerge from it. In absorption, the cognitive process is without perception of a material object; only mental factor, equanimity, and one-pointedness are present. Infinite space is the object that appears for the first immaterial jhana (arupajjhana); infinite consciousness is the object for the second. The third and fourth arupajjhanas transcend infinite consciousness to nothingness, and the concept of nothingness transcends to consciousness of neither perception nor nonperception.

Each immaterial sphere of consciousness produces four corresponding immaterial sphere resultant consciousnesses (*vipaka cittas*). The resultant cittas arise through rebirth in the immaterial realms. This is also the case for resultant cittas in the fine material sphere; five such resultant cittas arise through rebirth in the fine material realm. Four immaterial sphere functional consciousnesses arise with the Buddha and Arahants. They attain jhanas in the immaterial sphere because for them there is no rebirth; hence there is no resultant consciousness. There are total of twelve immaterial sphere consciousnesses. It is worth reiterating the fact that the supramundane jhana takes nibbana as its object, which is an unconditional and transcendental experience. Therefore, mundane, and supramundane jhanas are not comparable. Twelve Immaterial jhana conscious are classified as follows. There is not any other type of consciousness other than jhana in this Immaterial sphere, functional consciousness is experienced by saints.

> Immaterial sphere wholesome consciousness—4
> Immaterial sphere resultant consciousness—4
> Immaterial sphere functional consciousness—4
> Total—12

Beyond the four fine material jhanas lie four higher attainments in the scale of concentration, referred to in the suttas as the "peaceful immaterial liberations transcending material form". In the commentaries they are also called the immaterial jhanas, and while this expression is not found in the suttas it seems appropriate in so far as these states correspond to jhanic levels of consciousness and continue the same process of mental unification initiated by the original four jhanas, now sometimes called the fine-material jhanas. The immaterial jhanas are designated, not by numerical names like their predecessors, but by the names of their objective spheres: the base of boundless space, the base of boundless consciousness, the base of nothingness, and the base of neither-perception-nor-non-perception. They receive the designation "immaterial" or "formless" (arupa) because they are achieved by surmounting all perceptions of material form, including the subtle form of the counterpart sign which served as the object of the previous jhanas, and because they are the subjective correlates of the immaterial planes of existence.

Like the fine-material jhanas follow a fixed sequence and must be attained in the order in which they are presented.[18] That is, the meditator who wishes to achieve the immaterial jhanas must begin with the base of boundless space and then proceed step by step up to the base of neither-perception-nor-non-perception. However, an important difference separates the modes of progress in the two cases. In the case of the fine-material jhanas, the ascent from one jhana to another involves a surmounting of jhana factors. To rise from the first jhana to the second the meditator must eliminate applied thought and sustained thought, to rise from the second to the third he must overcome rapture, and to rise from the third to the fourth he must replace pleasant with neutral feeling. Thus, progress involves a reduction and refinement of the jhana factors, from the initial five to the culmination in one-pointedness and neutral feeling.

There are four meditative states that pertain to the immaterial sphere, which come to be called the immaterial jhanas. Practical examination will bring out the dynamic character of the process by which the jhanas are successively achieved. The attainment of the higher jhanas of the fine-material sphere, we will see, involves the successive elimination of the grosser factors and the bringing to prominence of the subtler ones, the attainment of the formless jhanas the replacement of grosser objects with successively more refined objects. From this study it will become clear that the jhanas link together in a graded sequence of development in which the lower serves as basis for the higher and the higher intensifies and purifies mental states already present in the lower.

Once the fourth jhana is reached the jhana factors remain constant, and in higher ascent to the immaterial attainments there is no further

---

[18] Henepola Gunaratana. (1988). The Jhanas In Theravada Buddhist Meditation. Buddhist Publication Society.

elimination of jhana factors. For this reason, the formless jhanas, when classified from the perspective of their factorial constitution as is done in the Abhidhamma, are considered modes of the fourth jhana. They are all two-factored jhanas, constituted by one-pointedness and equanimous feeling.

For trainees, jhanas attained by samatha meditation involve strengthening the faculty of concentration (samadhi). Then the meditation becomes a useful tool to develop path factors. Different schools hold different views on attainment of jhanas. They differ in terms of the contribution of jhanas to attainment of the path and its fruit and in terms of which factors determine the jhana level of the path and its fruit. According to the law of action, intentional volitional actions produce results unless other conditions intervene. Therefore, it is reasonable to ascertain that there is a correlation between concentration development through jhana and the attainment of the supramundane path for trainees who chose tranquillity leading to insight.

Mind-Door Process

The term *mind-door process* refers to a cognitive process that takes place at the mind door through the mental continuum of the mind faculty. When a mental object comes into contact with the mind door, the consciousness that gains access to the object is called the mind door. Unlike the five-door consciousness, which takes place through a sense avenue, the mind-door process has no receiving consciousness.

Different commentators give different opinions on this mind-door process. There is no clear explanation of the stages between "vibration bhavanga" and how the object is received through the mind door. The explanation given in the sub-commentary Vibhavini Tika states that the *bhavanga citta* immediately precedes the mind-door consciousness. Some commentators equate bhavanga to the mind door.[19]

The cognitive process of the mind-door process follows the same stages as the jhana cittas.[20] For example, there are seven javana moments that proceed from determining the object, followed by two registrations, then

---

[19] Bhikku Bodhi, *A Comprehensive Manual of Abhidhamma*. p. 224. (1999). BPS.
[20] Ibid. 17 stages of consciousness (see Table 4.1, p. 155).

falling into bhavanga state. Javana literally means "running after" the object; it denotes the rapid mental moments that occur during one single mind moment. It is said that kamma forms during a javana moment. Similarly, "in the case of a worldly or trainees, there can arise a total of twelve absorption javanas: the sublime wholesome citta of the fifth jhana; the four immaterial jhanas; the four path cittas at the level of the fifth jhana; and the lower three fruition cittas at the level of the fifth jhana." The number of javanas occurring at this jhana level are determined by the equation $5 + 4 + 3 = 12$.[21]

In absorption, unlike the normal cognition process, the number of javanas is fewer than seven in terms of their respective classes of consciousness. The above equation suggests that the consciousness of a worldly or trainee can change to that of path and fruition during jhana practice. The following passage from *A Comprehensive Manual of Abhidhamma* indicates that jhana attainment can result in change of lineage.

> In an individual with average faculties, those preliminary javana occur four times, each one exercising a different preliminary function. The first is called preparation (*parikamma*) because it prepares the mental continuum for the attainment to follow. The next is called access (*upacara*) because it arises in proximity to the attainment. The third moment is called conformity (*anuloma*) because it arises in conformity with both the preceding moments and subsequent absorption. The fourth moment is called change of lineage (*gotrabhu*). In the case of jhana attainment it receives this name because it overcomes the sense sphere lineage and evolves the lineage of sublime consciousness. In the case of the first path attainment, this moment is called change of lineage because it marks the transition from the lineage of worldly to the lineage of the Noble Ones (Ariya). The expression continues to be used figuratively from the moment of transition to higher paths and fruits, though sometimes it is designated by a different name, *vodana*, meaning "cleansing".[22]

[21] Ibid. p. 170.
[22] Bhikkhu Bodhi, "Guide 14: Absorption", *The Abhidhammattha Sangaha of Acariya Anuruddha* (Kandy, Sri Lanka, Buddhist Publication Society, 1999), p. 168.

Note: This passage may be valid only for those trainees who are in the Ariya Magga or those who take nibbana as their meditative object. It is not clear in the texts what the object of meditation is. In sublime jhana consciousness, the object could be the preceding consciousness where meditator would realise characteristics of all phenomena by insight not clinging to jhana attainment.

This explanation is plausible because mundane jhanas take objects such as kasina, unlike the supramundane path jhana, which takes nibbana, unconditioned reality, as its object.[23]

An individual trainee with especially keen faculties can progress in absorption without the preliminary stage, so that only three preliminary sense-sphere javanas occur prior to absorption. Progression in the path starts from stream enterer, to once-returner, to nonreturner, and to Arahant, attaining both fine material sphere and immaterial sphere jhanas. They are termed as the path and fruit of supramundane jhana cittas, so called because the trainee has now superseded the mundane world.[24] For those on the Noble Path, the number of jhana totals 40: (5 jhanas × 4 stages × 2 = (path +fruit) = 40. Because of the additional 40 cittas, the Abhidhamma enumerates 121 cittas (89 − 8 + 40 = 121).

Throughout *A Comprehensive Manual of Abhidhamma*, the Dhammasangani, and other commentaries in conformity with the Sutta Pitaka (discourses), all writers hold the position that the path and its fruition are attained through the practice of vipassana.

"If the meditator aims to reach the path and fruit, then he conveys his mind towards the path and fruit through the development of insight (vipassana). ... All meditators reach the supramundane paths and fruits through the development of wisdom (prajna), insight into the three characteristics of impermanence, suffering, and nonself."[25]

From these statements we can ascertain that the fourth jhana in the immaterial sphere is the limitation of samatha meditation practice. Immaterial sphere jhanas would give rise to the realisation of emptiness of consciousness, that is, nothingness = void = emptiness.

Voidness is one of the gates to nibbana. A trainee who realises the

---

[23] Bhikku Bodhi, *A Comprehensive Manual of Abhidhamma*, p. 73.
[24] Ibid. p. 168.
[25] Ibid. p. 72.

truth of emptiness of phenomena at this point may change his or her meditation towards development of insight, taking voidness as an object, an element of nibbana. Those who develop insight without basic jhana are called practitioners of bare insight (*sukkhavipassaka*). They are also called dry visioned meditators. However, there is no difference in attainment of nibbana for either.

The most important factor in the two different practices of meditation and subsequent achievement of jhanas is the object chosen. For example, mundane jhanas take kasinas and the conceptual images produced by them. Supramundane jhanas take nibbana, unconditioned reality, as their object. On the other hand, mundane jhanas only suppress the defilements. Supramundane jhanas eradicate defilements gradually as the trainee advances. In this way the consciousness is completely purified by balancing wisdom and concentration, conveying the mind towards liberation. In conclusion, samatha meditation takes static objects and concepts. The immaterial sphere jhanas take place through the mind-door process as the object of contemplation is infinite space. Superior jhanas can burn up the remaining defilements of trainees who have already reached stream entry.[26]

Supramundane Consciousness—8

Lokuttara Citta, there are eight in mumber.

The Pali word *lokuttara* has derived from two root words: *loka* and *uttara*.

Loka refers to the world that is a sphere, realm, and plane of existence. This includes the extra celestial world, the physical planetary world, the material body, and the mental body, with more specific reference to mental formation (samsara). When asked, the Buddha pointed out that the five aggregates of clinging are the world.

Uttara means "beyond or transcending the mundane perceptions of the world". The range of consciousness discussed earlier is considered mundane because it binds beings to the world of suffering. All those types of consciousness are conditioned by mental factors. They are called *lokiya* (mundane). Supramundane is "that which transcends the world

---

[26] Ibid. p. 73.

of conditioned things. It is the unconditioned elements, nibbana, and the types of consciousness that directly accomplish the realisation of nibbana, called *lokuttara citta*, supramundane consciousness. These types of consciousness break the cycle of accumulating kamma and thus liberate the mind from the five aggregates of clinging and escape samsara, the cycle of birth and death. Nibbana is the complete cessation of suffering caused by attachment and its supporting conditions."[27]

There are eight supramundane consciousnesses in relation to the four stages of enlightenment. These are classified according to the name of the relating stage of each. These are four wholesome consciousnesses and their four fruits. Instead of using the word *result*, they are termed *fruits* because the former consciousness that produced the fruit is different from the other types of mundane consciousness. However, for classification purposes we use the term *resultant consciousness*. Each path consciousness issues its corresponding fruition consciousness immediately following the path. There is no supramundane functional consciousness (*kiriya citta*). *Functional* (*kiriya*) is one of the four ways of classifying consciousness according to its nature. This type of consciousness is neither kamma nor the result of kamma. It has no wholesome or unwholesome roots. Simply put, its function is to carry out an activity and not produce any kammic result. It performs a specific function in the cognitive process. The principle is same for the Buddha and Arahant. For example:

a. five-door consciousness accompanied by equanimity;
b. mind-door consciousness;
c. a smile producing consciousness for the Arahant and the Buddha, accompanied by joy.

The poetic term *fruit* is used because supramundane consciousness is a meditative attainment by a noble disciple; it results in fruition attainment, a result of the supramundane path, and is not involved in performing any function. For example, higher saints can attain complete cessation—Nirodha, that is cessation of consciousness.

---

[27] Ibid. p. 31.

Supramundane wholesome consciousness—4

Supramundane means beyond the mundane world of consciousness that does not depend on sensory perceptions. It has to be comprehended by insight wisdom.

The Buddha says that just as in the great ocean there is but one taste, the taste of salt, so in his doctrine and discipline there is but one taste, the taste of freedom. The taste of freedom that pervades the Buddha's teaching is the taste of spiritual freedom, which from the Buddhist perspective means freedom from suffering. In the process leading to deliverance from suffering, meditation is the means of generating the inner awakening required for liberation. There are four stages of liberation for those who enter the noble path. Reaching these stages of freedom is marked by change of consciousness from mundane to supramundane and its corresponding resultant, therefore, there are two supramundane consciousness of each stage totalling eight in number.

1. Path consciousness of stream entry
2. Path consciousness of once-returning
3. Path consciousness of non-returning
4. Path consciousness of Arahantship

Supramundane resultant consciousness—4

1. Fruition consciousness of stream entry
2. Fruition consciousness of once-returning
3. Fruition consciousness of non-returning
4. Fruition consciousness of Arahantship

All supramundane consciousness takes nibbana, unconditioned reality, as its object. It is stated in the Abhidhamma that any element of nibbana (nibbana datu) as an object is a factor of wisdom. It should be understood with insight, not as a concept or object in the sense of mundane consciousness. "Each path consciousness arises only once, and only for one mind moment: it is never repeated in the mental continuum of the meditator who attains it."[28]

---

[28] Ibid. p. 66.

Fruition consciousness can be repeated in many mind moments with continuous practice. For example, supramundane absorptions are known as fruition attainment (*phala samapatti*). These states of mind are attained by those who progress in the path.

Practice of both samatha and vipassana is progressive. Vipassana strengthens the faculty of wisdom (prajna). Samatha bhavanga (mental absorption) strengthens the concentration. Vipassana bhavana (insight) choses as an object of meditation the ever-changing phenomena of mind and matter, characterised by impermanence, suffering, and nonself. This line of meditation practice increases insight wisdom, issuing forth in the supramundane paths and its fruits.[29]

The spirit of Buddhist development is inner self-control over the psychological life, meditation, search of freedom and forbearance. There is the vision of invisible solidarity among all living beings in universal life, of all minds in the eternal peace. This has to be practised and promoted in inner and outer lives. Human development in spiritual life is achieved by meditation practices that aim to purify the mind from defilements and hindrances. The abandonment of the hindrances makes the beginning of freedom. It would be like someone has been freed from debt, rid of diseases, out of jail, a free man, and secure. With the hindrances abandon, there is no limit to the possibilities for spiritual growth. Just as gold free from five impurities will be pliant and supple, radiant, and firm, and can be wrought well. It is the precondition, not only for the attainment of jhana, but for achievement of all other higher knowledges and sublime qualities of mind. There is evidence in the scriptures that Buddha and few disciples had psychic powers and special knowledges. Some who wanted to develop psychic powers and failed, for this reason, the Buddha had discouraged such attempts and encouraged them to develop insight for full liberation.

---

[29] Robert E. Buswell Jr. et al. *The Princeton Dictionary of Buddhism*. (2013). Princeton University Press.

CHAPTER 14

THE BUDDHIST PHILOSOPHY OF RELATIONS—PATTHANA

The Method of Conditional Relations

The Twenty-Four Conditions

THE LAW OF PERPETUAL CHANGE is fundamental to Buddhism, and from this inherent character of all conditioned phenomena, the Buddha formulated one of the doctrinal teachings, namely that all formations are impermanent (*sabbe sankhara annica*). This principle of law is applied indiscriminately to all life forms, states of mind, and matter. The Buddhist philosophy of relations is presented in the seventh book of the Abhidhamma, *the Patthana*. Using the basis of twenty-four relations, it expresses how the complex and changing phenomena of mind and matter are related and considers various permutations of these dynamic relations; their interrelatedness forms a web of phenomena. In the mental continuum of a being these phenomena arise, sustain, and pass. The basis for these formations is consciousness, which is conditioned by various mental factors including greed, hatred, delusion, and wrong view, which in turn cause errors of perception, of mental formations, and of views. If consciousness is conditioned by delusion, all perceptions thereupon are erroneous.

Insight meditation is a method to disperse erroneous views, erroneous mental formations, and erroneous perceptions. The gradual development of insight knowledge and insight wisdom would result in a trainee's being capable of reaching the correct path of liberation.

The method of relation examines three aspects of consciousness and determines the functionality of consciousness in relation to the associated mental factors. The pattern of twenty-four modes of conditionality is applied to explain the synthesis and synchrony of relations, how consciousness is formed, and how it performs a specific function. For example, by the synthesis method of relation, a coexistent relation has fifteen different ways it could influence other relations, such as kamma, jhana, and *magga* (path). On the other hand, synchrony explains how all phenomena are related to causes, are conditioned by causes, and arise from interrelated circumstances immediate or in the distance.

There are three factors of this relation to be understood: conditioning state, conditioned state, and conditioning force:

Conditioning State

Conditioning state is a relation of cause and effect. It the cause that by its intrinsic quality has conditioned a state; this is its conditioning force. For example, the hotness of chilli is an inherited force in its conditioning state, and the chilli cannot exist without it. When added to food, the food is conditioned by the chilli. Chilli has an intrinsic quality of hotness. Other examples are salt and paper, which condition food.[1]

Conditioned State

Dhamma arises only by the conditions on which it is dependent. These conditions are all *cittas*, *cetasiakas*, and *rupas (forms)*. All types of consciousness, mental factors, and forms (material forms of the body) are conditioned phenomena.

---

[1] Bhikkhu Bodhi, "The Compendium of Conditionality", *A Comprehensive Manual of Abhidhamma* (Kandy, Sri Lanka, Buddhist Publication Society), p. 292.

Conditioning Force

This is the particular way the conditioning states function as conditions for the conditioned states. Force is an intrinsic quality belonging to the conditioning state. Examples include spices and especial vegetable ingredients used to flavour food such as cumin and cloves. Likewise, all mental factors have their inherited force and operate when the right conditions are met. (Similar to when a bacteriophage infects a bacterium, it inserts its DNA into the bacterial cell so that it might be replicated. The restriction enzyme prevents replication of the phage DNA by cutting it into many pieces.)

Condition

Condition as understood in Buddhism is refereeing to body and mind in particular with reference to feelings, emotions, and mental formations. They are conditioned phenomena interconnected and interactive in constant change. The term is then extended to the outer world. In the first (passive) sense, saṅkhāra refers to "conditioned things" or "dispositions, mental imprint". All aggregates in the world – physical or mental concomitants, and all phenomena, state early Buddhist texts, are conditioned things.

Causes and conditions are co-related. An effect cannot happen without any cause and conditions. The cause of an effect vanishes, then the effect emerges. The cause cannot exist in the effect. The transition of one state, conditioning state to another, conditioned state happens by conditioning force as described in above explanatory notes.

Condition is a state which helps in the arising or persistence of other states. This means that a condition, when operative (active), will cause other states connected to it to arise to maintain and continue its existence. For example, fever is a condition for hypothermia. All conditioned phenomena are included in the category of conditioning states. Conditioning forces operate between conditioning states and the conditioned states.

Conditioning state ⟶ Conditioning force ⟶ Conditioned state

The Philosophy of Relations, Book 7 of the Abhidhamma

This book describes twenty-four relations of conditioning states—Paccaya Dhamma. It is a detailed examination of causal conditioning, the law of cause and effect. The twenty-four types of conditional relations (*paccaya*) are analysed in relation to the classifications in the analysis of the Dhammasangani. Only a very limited application of this gigantic Great Treatise is shown here.

Mental phenomena are conditioned by conditioning states, which in turn condition the mind and the body by their conditioning force. The force has the intrinsic nature of itself, and it conditions the consciousness and its associated mental factors in a cyclical pattern. This is a dynamic reality that must be understood by insight wisdom. According to the analysis of relations, at any given moment both body and mind are the result of past and present conditions. They are not static moments but a dynamic reality of change. Note that conditioning factors may remain dormant until sufficient conditions are met to produce results.

Table 8.3 of *A Comprehensive Manual of Abhidhamma*, p. 308, gives a complete set of relations between conditioning and conditioned states, the latter conditioned by the conditioning force of the former. For example, the three roots of wholesome and unwholesome conditioning states, *greed hatred and delusion* which are the same factors in the analytical method of consciousness:

Unwholesome (greed; hatred; delusion); wholesome (nongreed; nonhatred; nondelusion).

This chapter is focused on errors of perception by way of relations, which influences the consciousness either to defile or to purify.

The Twenty-Four Relations (Paccaya Dhamma)

For the table below, the English term is followed by the Pali, the latter in brackets.

1. Root (*hetu*)	13. Kamma (*kamma*)
2. Object (*arammana*)	14. Result (*vipaka*)

3. Predominance (*abhipati*)	15. Nutriment (*ahara*)
4. Proximity (*anantara*)	16. Faculty (*indriya*)
5. Contiguity (*samanantara*)	17. Jhana (*jhana*)
6. Connascence (*sahajata*)	18. Path (*magga*)
7. Mutuality (*annamanna*)	19. Association (*sampayutta*)
8. Support (*nissaya*)	20. Dissociation (*vipayutta*)
9. Decisive (*upanissaya*)	21. Presence (*atthi*)
10. Prenascence (*purejata*)	22. Absence (*natthi*)
11. Postnascence (*pacchaiata*)	23. Disappearance (*vigata*)
12. Repetition (*asevana*)	24. Nondisappearance (*avigata*)

Method of Analysis

These twenty-four relations are grouped into six combinations of mind and matter.

Conditioning method	Number of combinations
Mind for mind	6
Mind for mind and matter	5
Mind for matter	1
Matter for mind	1
Concepts, mind, and matter for mind	2
Mind and matter for mind and matter	9

Summary of Relations

- Mind alone can condition in three ways (twelve total combinations).
- Matter alone can condition one way.
- Concept, mind, and matter together can condition one way.
- Mind and matter together can condition one way.
- The total is six ways of combination (twenty-four relations in total).

The following table shows the combinations of the relation under each group:

Method of relation	Type of relation
Mind for mind	4. Proximity
	5. Contiguity
	12. Repetition
	19. Association
	22. Absence
	23. Disappearance
Mind for mind and matter	1. Root
	13. Kamma
	14. Result
	17. Jhana
	18. Path
Mind for matter	11. Post-nascence
Matter for mind	10. Pre-nascence

Concepts, mind, and matter for mind	2. Object
	9. Decisive
Mind and matter for mind and matter	3. Predominance
	6. Connascence
	7. Mutuality
	8. Support
	15. Nutriment
	16. Faculty
	20. Dissociation
	21. Presence
	24. Nondisappearance

These are the permutations of relations in which body and mind are conditioned.

Types of Conditioning States within a Relation

- Prenascence—the conditioning state has already arisen
- Connascence—the conditioning states arise simultaneously
- Post-nascence—the conditioning state arises after supporting and strengthening what has already arisen, e.g. rainwater that falls later promotes the existing growth of the paddy.
- Contiguity—the conditioning state is caused to arise immediately after it has ceased, e.g. consciousness and mental factors that arise after the previous citta, like ocean waves. They pass the qualities of the preceding wave to the successive wave and condition it.

Formation of Errors of Perceptions

The function of perception is to recognise what has been previously perceived. Perception manifests when interpreting an object by way of its features, and its proximate cause is the object as it appears. Proceeding from perception is volition, the process of kamma formation. These stages of the cognition process happen at such a speed that millions of mental processes occur in a blink of an eye. Additionally, consciousness, mental factors, and material phenomena are conditioned by at least one of the twenty-four types of conditioning states and their corresponding conditioning forces as listed above. For example, kamma condition (no. 13), whose conditioning state is the volitional (*cetana*) kamma condition, covers all eighty-one types of mundane consciousness and other mental factors associated with those volitions. Similarly, object condition (no. 3) is a condition where the object of consciousness causes other conditioned states to arise. These objects include all visible forms, sound, smell, taste, and touch. Tangible objects are identified with the four primary elements of earth, water, fire, and air. If they pass through the sense door with erroneous perception, the process continues in the mind door, forming thoughts which crystalise in various views (distortions and densities). These conditioning states with multiple aspects make all perceptions receptive by way of the conditioned senses. They can project a distortion of permanence, pleasure, happiness, and self, whereas everything is changing, suffering, and without a self. Correction of these errors is only possible through the Four Foundations of Mindfulness.

Mindfulness is an exclusively wholesome mental factor. The presence of mindfulness supported by right effort and right concentration enables the meditator to comprehend effects of the conditioning forces of unwholesome factors. Once these are identified, the meditator can then cut off and replace them with other, wholesome factors to nourish a healthy mind and discern the truth. Persistent and systematic practice of the Four Foundations of Mindfulness would result in gaining the seven factors of enlightenment. Liberation of the mind would illuminate with beautiful mental factors. Balanced practice of these seven factors of enlightenment is necessary to enter the path factors of the Noble Eightfold

Path. Development of insight parallels the Sixfold Way of Purification. The six mundane stages and the seventh supramundane stage in the purification of consciousness. It is the path of purification recommended in Theravada Buddhism. It serves as the foundation for the standard manual for meditation practice, the Visuddhimagga.[2]

---

[2] Buddhaghosa, (2010, 4th edition). *Path of Purification. Visuddhimagga.* PBS.

CHAPTER 15

TWENTY-FOUR RELATIONS OF CONDITIONALITY

THE SEVENTH AND LAST OF the analytical works in the Abhidhamma is the Buddhist philosophy of relations. Its Pali term is *patthana*, meaning "preeminent or principle cause". It shows the conditionality and dependent nature of all phenomena within a living being that corresponds to the facts of life.

The twenty-four relations explain the synthesis process of consciousness, how the mind and body are conditioned in various permutations. These permutations were described in Chapter 14. Consciousness is a process that arises and vanishes together with its associated mental factors; all consciousness therefore has three phases, arising, reaching maturity, and ceasing. Chapter 13 explains the various classes of consciousness. During the process of consciousness, its conditions as shown in the relation (out of any of the twenty-four) produce a conditioned state at the end of its cycle. This could be towards consciousness itself or a material phenomenon associated with it. These relations accord with the intrinsic nature of consciousness and its inherited force and are applicable to all classes of consciousness, associated mental factors, and material phenomena. Each dynamic relation should be understood in connection to the three aspects— the conditioning state, the conditioning force, and the conditioned state— and to how these are intricately interrelated. The method of relations exemplifies how every occurrence within the body and mind is both determined and determining. Thus, the twenty-four relations explain the nature of perpetual change; since the beginning of time all phenomena has been derived by these relations, constituting every psychic and physical element of being. The Buddha formulated this philosophy from what is known as the Fivefold Law of Cosmic Order, which he acquired by means of wisdom and direct knowledge. According to this philosophy of relations, all events in the universe are interrelated in a variety of ways. The twenty-four relations are the method that examines and expresses these complex steps of the conditioning process into various modes, similar to the role of an oncologist, who uses diagnostic tests to determine a cancer's stage to help him or her plan a course of treatment.

Specific terms are used to help describe these relations, as follows:

- *prenascence*—a conditioning state (usually a material state) which has already arisen and reached the stage of presence.
- *connascence*—meaning to arise together, simultaneously
- *postnascence*—a conditioned state that existed before (had risen prior to) the conditioning state
- *nascence*—being born or coming into existence for the first time
- *contiguity*—meaning sequential
- *proximity*—meaning immediately after
- *predominance*—where the object or condition dominates the mental state
- *sufficing*—meaning supporting
- *reciprocity*—meaning mutual or coexisting
- *basic*—a standing ground, a base, or a foundation supporting the entire realm of consciousness.

The Relation by Way of Roots (Relation No. 1)

There are three unwholesome and three wholesome mental factors. The unwholesome roots are greed, hatred, and delusion; the wholesome beautiful roots are nongreed, nonhatred, and wisdom. These factors condition all types of mundane consciousness. The conditioned states are the mental states associated with each root or a combination of these roots. These are founded on kammic results, including connascence material, which occurs at the moment of rebirth linking, as well as such material qualities during the course of life determined by consciousness.

Just as the roots of a tree firmly fix the tree to ground and sustain it by way of feeding nutrients, supporting, and assisting the growth of its essence, so the ethically variable roots feed the consciousness with the force of those associated mental factors and firmly fix it to define the mental disposition and type of actions reflecting ethically significant qualities and making us who we are.

The Relation of Object (Relation No. 2)

All classes of consciousness, together with their associated mental factors, arise in relation to an object. The Abhidhamma recognises six classes of objects that correspond to the six senses. Tangible objects are material phenomena that manifest as visible forms, sound, smell, touch, and taste. These are identified by a combination of primary elements: earth, water, fire, and wind. For example, the colour of a red rose, its intrinsic beauty of soft petals, and its mesmerising aroma are chemical compositions of earth materials. We cannot see and smell a rose without light and wind respectively.

Concepts are also objects. Mental objects can be anything that comes through any of the avenues: sense avenue, mind avenue, or avenue-free (in the case of the terminal thought process and life continuum, the object is received differently).

An object can inherit qualities of attractiveness or repulsiveness. This conditions the arising consciousness and mental factors. The quality of these objects by their effect can range from decisive to very feeble. The mind and mental factors take hold of the object similar to the way in which a magnet attracts iron: the magnet's apparent desire for metal increases the nearer it gets to it. While the object remains present to any of the six sense avenues, consciousness and an associated mental factor arise; hence, as the contributory factor, they also cease together at the end when consciousness *(Citta)* cease. One moment of *Citta* is treated as a one cycle. The Pali commentators illustrate this function by using the simile of the sound of a violin, produced while the bow strikes its strings but ceasing immediately when playing stops without contact. With the contact arise formations. Neither mundane nor supramundane consciousness can arise without an object. The Four Foundations of Mindfulness take a range of objects as tools for insight meditation practice. Nibbana becomes the object of consciousness, occurring as the mental process of noble individuals.

The Relation of Dominance (Relation No. 3)

The relation of dominance is subdivided into two kinds: object predominance and connascence predominance.

Object predominance takes objects that are most agreeable, lovable, pleasing, highly esteemed, and regarded. In this sense it is identical to Relation No. 9, but it differs by its conditioning force. An example of its conditioning force is the attraction to luxury items, including expensive brand names, cars, perfumes, gold wristwatches and jewellery. These items can dominate an untrained mind, causing a strong response in an arising consciousness and the related mental factors that are conditioned with excitement (the wow factor).

Conascence predominance is where the conditioning state dominates the conditioned state by way of a form of persuasion. This causes one to reach out to the object that is arousing desire, energy, consciousness, and curiosity. Under this relation, one does not shrink from challenges but takes extreme delight in them. For example, someone who pursues psychic powers through the attainment of jhana would endure the difficulties until fulfilling his or her dominant intentions. This psychological factor is the cause of specific behaviours that convey power and dominance.

On the other hand, undertaking a difficult yet virtuous task, such as translating the Pali Canon, is also connascence predominance. In this example, the translator endeavours to achieve his or her goal and works untiringly to do so. This kind of regulated desire in spiritual practice gives faith and confidence to pursue a religious path until the person sees its results. The Buddha had this in abundance.

The Relation of Proximity (Relation No. 4)

The proximity relation supports the view that there is no gap between two consecutive moments of consciousness. All successive classes of consciousness and mental factors are related to the preceding consciousness and mental factors. This relation links the immediately preceding instant to the immediately succeeding instant. The former is the conditioning state, and the latter is the conditioned state—similar to an ocean wave passing its force to the next, propagating its continuity. A commentator articulated this process in the following expression: "Though the proceeding thought ceases, the conscious faculty of it does not extinguish until it has caused the succeeding thought to arise." This principle is true during the course of life and at the moment of death. The one exception is when an Arahant dies: there is no volition or defilements remaining, as these things are completely ceased, ultimately quiescent.

In the event of death, the passing of the current life continuum (bhavanga consciousness) is related to the next life continuum by immediate rebirth without a gap or intervening consciousness. The chain continues in the cycle of samsara.

The Relation of Contiguity (Relation No. 5)

Contiguity relation is the same as Relation No. 4. Sequential occurrence draws the relation between the ceased mental state and the immediately following state. However, because of anomalies to the rule, *contiguity relation* is a term applicable to the proximity of consciousness in a state of continual flux. By this it is differentiated from proximity relation. There are occasions when consciousness can be suspended for a fixed period of time. For example, through mastery of jhana, a higher saint could attain complete cessation (Nirodha), where one is able to suspend consciousness and emerge from it afterwards. In the special case of a nonpercipient being *(one that cannot perceive)*, there is absolutely no cognitive process whatsoever; the person's mental faculties are not annihilated but frozen.[1] For these exceptions the suspension period could be incalculable and measured only in aeons of time.

---

[1] Bhikkhu Bodhi, *A Comprehensive Manual of Abhidhamma* (Kandy, Sri Lanka, Buddhist Publication Society), p. 183.

The Relation of Connascence (Coexistence) (Relation No. 6)

Connascence in this context is synonymous with coexistence, that is phenomena manifesting simultaneously. A phenomenon arises with its inherent effect, causing its effects to arise at the same time. For example, together with the rising sun comes light and heat. Similarly, when a candle is lit, it produces light, heat, colour, odour, and smoke. In these examples the sun is like the consciousness and light and heat are peripheral mental concomitances. Connascence can also be illustrated as an arriving king being accompanied by a retinue of ministers and guards.

The relation of connascence is further extended to the four primary elements (earth, water, fire, and wind); the presence of one of these is a condition for the simultaneous arising of the other three.

This same principle applies to the arising of the four mental groups (feeling, perception, mental formations, and consciousness). At the moment of conception in the mother's womb, kamma produces a heart base (physical base of mind). This and other corporeality groups are related by way of connascence; the commentator extends this relation to five types, including the four great essentials and their derived materials, for example the complex process of photosynthesis, whereby green plants produce carbohydrates, sugar, protein, and minerals using the sun's energy. Similarly, during the course of life, consciousness produces material phenomena by way of the coexistent relation of mind and matter. For example, a healthy mind produces various hormones and enzymes, which are derived matter. Anaemia has symptoms including the craving and chewing of substances that have no nutritional value, which is a sign of iron deficiency, but less commonly nutritional deficiencies are the cause of emotional problems, such as stress or obsessive-compulsive disorder, or a developmental disorder. Symptoms suggest that lack of natural production of nutriments causing deficiency, breaking of body and mind relation. These illnesses can be corrected by cognitive behavioural therapy.

The Relation of Mutuality (Reciprocity) (Relation No. 7)

The mutuality relation is similar to Relation No. 6, but it differs in the particular way that it transmits its coexisting effects. The relation of mutuality is like a mirror that reflects the light back to its source; the effects of the conditioning state not only transfer its force but also reciprocally receive the force via the conditioned state. For example, in the event of a collision, the two bodies get mutually damaged by the reciprocal transmission of forces.

The Pali commentator gives the simile of a tripod that stands upright with three legs, each leg assisting the other two legs, evenly distributing the forces so that the tripod stays upright. This principle of mutuality relation is applicable to all states of consciousness, including its associated mental factors at rebirth and during the course of life. The commentator further elaborates on the relation by quoting the Dhammapada as expounded by the Buddha: "The mind is predominant," and consciousness is not able to arise without mental factors, as the two are reciprocally related.

The Relation of Support (Dependence) (Relation No. 8)

In the context of relations, the term *support* is synonymous with *dependence*, like a child is dependent on the mother and the mother supports the child. This relation has three subcategories of support: connascence support, basic prenascence support, and basic objective prenascence support.

The classification of relations in this category is the same as Relation No. 6 (simultaneous arising), yet it is further subdivided to help analyse how its intrinsic properties support the intended relation.

The term *basic* implies that it is a standing ground, a base or foundation for the entire realm of consciousness. The principle is that no consciousness can arise without a base. It should be understood that the heart base at the moment of rebirth is not physical but is a corresponding mental phenomenon. In the case of rebirth linking, there is no pre-existent heart base at that moment. All phenomena are directly or indirectly attributed to a particular cause except nibbana where there is no rebirth.

Connascence support serves as the foundation to a conditioned state. There is a corresponding conditioning state on which both conditioning and conditioned sates depend. It is similar to an engineering foundation that connects the building to the ground and supports the rest of the structure. Houses, temples, monasteries, mansions, cities, and skyscrapers, and various animals, trees, and so forth, all stand on the pre-existent earth with all its resources, such as clean air. Thus, it should be understood that all things in this universe are causally related to one another by way of dependence or common foundation. It requires a sound moral foundation for meditation practice to be stable. The relation between morality and concentration is connascnce support.

Basic prenascence support explains how consciousness and the associated mental factors are dependent on the pre-existent six bases (eye, ear, nose, tongue, body, and heart). These six sense bases are causally related and dependent on the seven elements of cognition (see Chapter 13). The commentator gives the following example: the violin produces sound only when the player slides the bow across its strings; otherwise, it is silent. In the same way, the five senses waken only the five kinds of sense objects that enter the five sense avenues and touch the five sense bases, and not otherwise. The mind avenue process is also similar, but the objects on which it depends are mental objects. The last consciousness of one's life, death consciousness, takes those death signs as it object.

Basic objective prenascence support is a special relation that arises in the mind door. It is a cognitive process supported by the heart base. Contemplating on Dhamma, the meditator reflects on the wrong view of mental states which depend upon matters that cloud the mind with errors of perception. The cause of this speculative thought includes projection of ownership, of mine, myself, or my soul; craving; and conceit. Errors of crowding arise as a result of errors of perception. The meditator gains *citta-maya-prajna*, wisdom or knowledge gained on reflection and thinking *(the power by which the universe becomes manifest; the illusion or appearance of the phenomenal world)*. This involves coming to understand that all life experiences, whether pleasurable or painful, are causally related to a sense base and ultimately to the heart base. To understand this is to immediately realise the truth of meditative contemplation. By wisdom of impermanence, suffering, and nonself, the meditator realises that experience of the six-sense avenue is dependent on the six sense bases; these are the grounding for, and object of, each of the mind-door cognitions.

The Relation of Decisive Support (Sufficing) (Relation No. 9)

The relation of decisive support has three subcategories: object decisive support, proximity decisive support, and natural decisive support.

Object decisive support is similar to Relation No. 3, where the conditioning object is exceptionally affirmed or highly important and has priority over other objects. The object is decisive; it is firmly fixed on the consciousness and influences other supportive mental factors to arise. This conditions mental states and causes consciousness to arise with a strong dependence on the object. It is sufficient for someone to be seriously dedicated, for instance to study a particular subject or chosen profession, such as developing an unsurpassed mastery of skill in painting, or someone in medical science training to become a heart surgeon.

Proximity decisive support is similar to Relation No. 4 and Relation No. 5 but differs from them in terms of force; it conditions the conditioned state. Proximity is the force that causes succeeding mental states to arise immediately after the preceding state has ceased. Proximity decisive support is the force which causes succeeding states to arise immediately by a strong volition propelled by desire for an object. For example, conviction and strong confidence support the mental states that attain tranquillity through the stages of jhana one after another in sequential order without gaps.

Natural decisive support is very broad. It encompasses the past, present, and future; all classes of consciousness; mental factors; and material phenomena, including the concept and nibbana. This relation has the principle element of kamma; this ethical force has an ability to produce consequences in the present or in future conditioned states. It can follow a hereditary pattern, passing from one generation to another, for example a tradition or social culture that honours killing because of the perpetrators' beliefs. Mental factor hatred is strongly presence as a decisive support of common family view of revenge.

Those with an earnest desire to do good, to be industrious, and to be cultured may plough their fields and sow seeds with a hope of reaping the harvest before winter, or otherwise work diligently to save for later in life. In the same way, many people engage in skilful trades and do many good deeds to serve society and reap the benefits in the future. These are future natural decisive support conditions. Others may receive benefits in the present for the good things they have done in the past; these are present natural decisive support conditions. Those with an ardent wish to attain nibbana in lives to come acquire merits and perfections in this life. Therefore, nibbana is a powerful natural decisive support relation to cultivate in order to do good, not to do evil, and to purify the mind in this present life. The Buddha elaborated the philosophy of relations, showing us the consequences of moral and immoral acts, how they are causally related, and how the dominant object acts as a main basis for subjects.

The Relation of Prenascence (Preexistence) (Relation No. 10)

The word *nascent* is an adjective meaning "starting to grow or develop". With the prefix *pre*, it therefore means "before growth". It has a connotation of pre-existence, like the sun, which first arose in the world, giving light and heat to the people and plants that appeared later. Although Buddhism does not look for a First Cause, the Buddha freely used these comparisons to explain relative phenomena, so as do commentators.

The relation of prenascence has three subdivisions: base prenascence, objective prenascence, and base objective prenascence. (*pre - nascence* in this context is before birth. *These terms need to be understood similar to prenatal and postnatal explained in gynaecology*).

Base prenascence and *base objective prenascence* are the same as Relation No. 8 (Support/Dependence), except that this relates to matter for mind, where physical matter provides the supporting condition for consciousness to arise. A base is a support for the occurrence of consciousness. The principle of base prenascence follows the premise that in the material planes of existence, both consciousness and its associated mental factors arise in dependence of a condition, the base (*vatthu*). The first five sense bases coincide with the first five doors (eye door, ear door, etc.). It is the sensitivity of the eye that serves as the base, not the door, which is only a concept of a channel through which consciousness and mental factors access the object. Each of these physical bases pre-exist as a base prenascence relation for conditioning states, which take its function as material support. At the moment of rebirth there is no pre-existent heart base. On such an occasion the heart base arises simultaneously as a coexistent (connascent) mutual condition. From such a time, the heart base serves as the prenascence base for the beginning of new life, the first bhavanga thereafter including all mind elements and types of consciousness such as attaining higher jhanas during the course of life.

Objective prenascence is a unique relation of matter to mind where matter conditions the mind. All five types of material objects correspond to the five physical senses. Thoughts generated in the five-door cognitive process are thus causally related to their corresponding bases by way of objective prenascence. Objective prenascence has a connotation of pre-existence includes all eighteen types of material qualities that constitute the material form of a being. The commentator explains the process with his famous comparison of the sound of the violin arising when the instrument is played with a bow. The sound pre-exists in both the violin strings and the violin bow, but only a finely tuned instrument and talented player can produce the quality of music that would otherwise lie dormant in the object's potentiality. Thereupon, the quality of thoughts which take part in the five-door process owe their occurrence to the presentation of the five objects and their quality at the five doors. The supporting conditions for the formation of thoughts is therefore the five sense bases, and the quality of the thoughts depends on the quality of the corresponding base.

The cognition process at the five sense bases reveals an important psychological principle. Although there are five senses, the mind can be conscious of only one sense at a time. The presence of intention or attention is a necessary condition, and subsequently the object is presented to the mind door. Presentation of the object at a sense door is only possible when both the object and the base are in their static or equilibrium stage. Compounded distortion could arise due to overload, stress, burnout, lack of sleep, substance misuse, or a preoccupied mind. The cognitive process explained in the analytical method has seventeen stages, which then cause it to fall back into the stream of the life continuum (bhavanga).[2] On account of this process, the life continuum vibrates for two moments and then ceases. This moment, a static instant in the life continuum, gives rise to a new cycle of consciousness, but consciousness cannot arise without the pre-existence of the object or base. These static instances come to define

---

[2] Ibid. p. 155.

the Buddhist concept of the present moment; this is the moment a new object is presented to the sense door at its base, occurring as it happens and recognised as soon as consciousness arises. The past has ceased, and the future has not yet risen; the present is, but only if the object is still existing. In the Abhidhamma, only objective prenascence is classed as the present object, far, near, or concealed from sight, so long as the present object is presented to the mind. In an untrained mind, mental proliferations continue, roaming with circular reasoning, dwelling in the past or future, daydreaming. Only by being mindful and being in the present moment is one able to let go of discursive thinking, either temporarily by tranquillity meditation or permanently by insight wisdom.

The Relation of Postnascence (Post-existence) (Relation No. 11)

Postnascence *(Post-nascence)* is a unique relation of mind conditioning matter, whereby the conditioning state assists the conditioned state which had arisen before the former. The conditioning state supports, strengthens, and helps the conditioned state to develop afterwards. Initially, at conception, all corporeal material qualities are born of kamma, consciousness, and temperature, followed by nutriments. The Pali commentator explains this phenomenon with another example, one of rainwater that falls every season, supporting the crops and vegetation grown in the previous season by way of a post-existence relation. In the same way, consciousness and mental factors condition the pre-arisen material body.

The commentator further elaborates the relation through the prenatal development period. At the time of conception, the rebirth-linking consciousness produces two groups of material qualities born of kamma and temperature. The process then begins its subsequent life continuum, an unconscious state which supports the growth and development of the embryo by feeding nutritive essence from the mother's food. The Buddhist description of prenatal development is compatible with modern explanations in obstetrics and gynaecology of embryo development up until childbirth. The commentator illustrates the process from the time of conception onwards. The groups produced by four origins (kamma, consciousness, temperature, and nutriment) spring up incessantly, like a flame of a burning lamp, to become a full-grown baby. Buddhism holds the view that complications are the result of kamma, which can influence and alter DNA (deoxy ribonucleic acid) structure, immune response, and the general health of the unborn child. For this reason, Buddhists take the natural decisive support relation seriously, believing that the cause for a healthy six-sense base is good kamma (deeds) done in previous lives. In the Metta Sutta, the Buddha's words of loving-kindness include, "Those living near and far away, those born, and to-be-born—may all beings be at ease!" The Buddha encouraged pregnant mothers to give love and kindness to their unborn children so that the children would grow healthy and be born with ease (MN 86).

The Relation of Repetition (Habitual Recurrence) (Relation No. 12)

The word *repetition* is rooted in the Latin word *repetitio*, meaning "to do or say again". Repetition can be very effective in improving proficiency, improving confidence, and giving strength. By repetitive practice one can gain mastery of a skill, for example elocution in speech and articulation in written language or playing a sophisticated musical instrument such as the piano. The conditioning state of this relation can be mundane, wholesome, unwholesome, and functional; there is a total of forty-seven types of consciousness by apperceptions. These are the mental processes by which a person makes sense of an idea by assimilating it into the body of ideas he or she already possesses. Apperceptions are causally related to conditioned states by repetition, which increases their strength and causes the mental process to accelerate through six javana stages, whereby kamma forms and conditions the conditioned states. In this process every succeeding apperception becomes more vigorous by habitual practice. There are seven javana stages in a keen mental process. The recurrence ceases at the seventh stage.

The application of repetition relation varies both in moral and immoral actions, as it forms kamma it can causally relate to previous and future lives. Through an investigation into truth and right aspiration by repetitive practice, one can cultivate factors of enlightenment by an earnest application of mindfulness. Attainment of self-enlightenment is an example of great human endeavour by the relation of repetition. In this world there are many examples of great human achievements benefiting humankind in the fields of arts, science, literature, and philosophy, all of these carried out through much zeal and effort. The pursuit of achieving distinction is performed over a long period with strenuous labour until the task is accomplished. For example, André Rieu, a twenty-first-century Dutch violinist and conductor, has talented orchestra players who have a mastery of their profession; this is achieved by a signal communicated intuitively through their minds, with constant attentiveness; being sensitive to the beat, the pitch, and the duration of the sound; and performing in synchrony with perfect tuning, timing, harmony, and balance.

The Relation of Kamma (Relation No. 13)

The word *kamma* has been adapted by Buddhism from its Sanskrit equivalent *karma*. Before the Buddha's enlightenment, the meaning of this word *karma* was vaguely defined with speculative views. According to Buddhism, kamma is threefold in that it is formed by the intentional (volitional) physical, verbal, and mental actions performed by all beings. The Buddha made the term specific with reference to how an action is performed and its function. The factor of volition (*cetana*) is the function that forms kamma at the javana stage. Kamma and volition are synonymous. There are seven stages of javana in every vigorous cognitive process. The number of javanas can vary with different circumstances.

There are two types of kamma relations: connascent kamma relation and asynchronous kamma relation.

The *relation of connascent kamma* explains how volitions are the conditioning states of all eighty-nine types of consciousnesses, including supramundane path consciousness. Kamma coexists with volition. It determines the dominant mental factors, simultaneously producing material phenomena and conditioning them by its force of volition during the course of life. The conditioned states are the eighty-nine types of consciousness, along with their mental factors and the material phenomena associated with their respective volition. This process is dynamic and similar to a dynamo, which produces electricity by means of rotating coils of copper wire in a magnetic field. The potential to produce electricity coexists with magnetism, just as the potential of kamma coexists with volition. At death, the force of volition transmits the last consciousness and link with rebirth and simultaneously produces material phenomena just as an arrow shot by a skilful archer reaches straight to its target.

The *relation of asynchronous kamma* is when there is a time gap between the conditioning states and the conditioned state. Like an "encoded asynchronous transmission", whereby data is sent intermittently, not at regular intervals, these signals are then recovered at predefined future time periods. In asynchronous kamma relations, volitions that produce the kamma of the conditioning states are actions done in the past, presented as wholesome or unwholesome consciousness. Their effects take place at a different point of time in the future. That is to say that kamma is asynchronous. The conditioning force is the asynchronous kamma, and the conditioned states are the resultant consciousness, associated mental factors, and material phenomena (born at rebirth linking and during the course of life). Asynchronous kamma relations signify a peculiar distinction; kamma does not cease when the volition ceases but will latently follow the sequences of mind until it obtains a favourable opportunity to take effect, immediately seizing that opportunity. If it does not obtain a favourable opportunity, it remains in the same latent mode in the life continuum indefinitely, passing on to the next life through rebirth-linking consciousness and then on to many future existences. The supramundane consciousness also produces its effect in the distant future. A commentator cited that unexhausted kamma can always mature and is only exhausted in the Brahma realm. This relation of kamma also coexists with the Noble Path *(Ariya)* until up to full liberation is accomplished.

The Relation of Result (Effect of Kamma) (Relation No. 14)

The law of kamma dictates that every action has a result. The Pali term for this is *vipaka*. Both wholesome and unwholesome consciousness constitutes kamma, produced by the volition of the action. The states of consciousness and mental factors which arise through the ripening of kamma are called resultants (vipaka); these results are purely mental states.

The analytical method enumerates thirty-six resultant consciousnesses and material qualities born as a result of past kamma. At rebirth these are produced by a resultant consciousness from the past life. This is the basis of the relation of result.[3] In this complex set of relations, the conditioning states are also a result of past kamma. The resultant consciousness and its associated mental factors are passive and inert, lacking the ability to make any decisive impact. The commentator illustrates this quality as thus: "Just as mangoes are very soft and delicate when they are ripe, so also the resultant states are very tranquil, since they are inactive and have no stimulus." On the other hand, in the case of *akusala vipaka*, the quality of the resultant consciousness can be depressive and unpleasurable by experiencing changes in mood or cognition.

The end of each cycle of consciousness falls into the life continuum (bhavanga). The resultant bhavanga consciousness arises and passes away in constant succession. The objects of these resultant consciousnesses are too dim and obscure to trigger an active cognition; it is like a deep sleep without any strength to move on to an active mode. However, there is evidence to suggest that the subconscious mind is revealed as dreams during sleep, appearing as kamma, as symbols of kamma, and as future destiny. These can sometimes cause traumatic nightmares, sleepwalking, and implicit memory. These episodes lack any order or predictability. It's a syndrome of behavioural disorder that originates during deep sleep and results in performing complex behaviours while asleep.

Vipaka should be understood in parallel to the life cycle and the wider context of samsara. Kamma awaits a favourable opportunity to produce its vipaka. We travel hand in hand with our kamma throughout our lives as long as we are trapped in the cycle of life and death. Kamma follows the owner like a shadow that never leaves. The change from infancy to youth and from maturity to old age is the result of kamma. If no opportunity is made to produce results during the course of life, the kamma or symbol of kamma is represented at the moment of death. The dying person is helpless to do anything against this. For these reasons, Buddhism encourages a positive engagement towards doing good things during one's lifetime. The severe effects of past kamma can be reduced by turning towards a moral

---

[3] Ibid. p. 69.

life, much like the strong effect of salt can be diluted by adding more clean water. Rebirth is a result of kamma transmuted into a new form of life.

The Relation of Nutriment (Relation No. 15)

The relation of nutriment is a connection for mind and matter between mind and matter. It is a condition where the conditioning states relate to the conditioned states by way of producing them, nourishing them, and supporting their growth and development by maintaining the biological time of cells. Most physiological processes are influenced by a complex timing system in brain cells. Likewise, mental units also go through a similar process and have a psychic lifespan that needs to be maintained by constantly replacing them. Nutriment is derived from the essence of edible food; metabolism and other internal formations of various biochemical syntheses have properties specified by their associated genes, and these function well under particular internal conditions, for example the synthesis of enzymes. The matrix of the relations that maintain life could be infinitely complex when considering the internal and external conditions. This can be simplified into a twofold conception: material nutriment and immaterial (mental) nutriment.

Material nutriment is the nutritive and dietetic value of food, which is a conditioning state for the sustenance of the physical body. Through the metabolism, new matter is produced, supporting the functioning body, and causing it to exist. This nourishment causes the conditioned state, namely biological cells, feeding them to grow and keeping them renewed with an improved immune response. The internal nutriment contained in the material groups are all born of four causes (kamma, consciousness, temperature, and nutriment). These are already found within the body but also serve as the supporting condition for growth. For example, some vital minerals, vitamins, and glucose, which are essential for growth and puberty, are stored in the bones and liver.

Immaterial (mental) nutriment is included as a cognitive process. There are three stages to this process: mental contact, volition, and the completed process of consciousness. These are considered as mental nutriments because they firmly establish and support the mind, nourishing it with the essence extracted from the desirable object. The mind becomes furnished with the courage to perform actions and becomes predominant in understanding, along with having the ability to take care of life. It is customary for Buddhist monks to recite the Buddha's qualities on a daily basis to give themselves mental vigour. They live long and healthy lives.

A commentator illustrates the life process with a simile of birds flying through the air from tree to tree, from wood to wood, in search of fruits, ascertaining where to nest and sustaining their whole lives; so are beings with the six classes in the sense sphere of consciousness searching for objects for enjoyment and pleasure or suffering and pain. By contact they derive feelings and cravings to perform various deeds, migrating from life to life and wandering in samsara.

The Relation of Faculty (Relation No. 16)

The Pali term for faculty is *indriya*, meaning "lordship over others", "to control". In the context of a relation, the conditioning state relates to the conditioned state by the exercising of control over it. The Pali commentator narrates the explanation with a simile of cabinet ministers, each of whom has the freedom to govern his or her assigned region independently and refrains from attempting to govern or interfere with the other regions of the country.

The relation of faculty has three subdivisions: prenascence faculty, material life faculty, and connascence faculty.

The *prenascence faculty* concerns the five senses and their respective bases (eye, ear, nose, tongue, and body). Consciousness arises along with its mental factors by each of these senses independently. A healthy sensitive organ controls the efficiency of the consciousness that takes it as a base and conditions it. For example, a good ear produces acute hearing without the assistance of a hearing aid. These acute faculties are for a valuable purpose; one should acquire knowledge, social skills, and professional skills in one's respective domains for personal growth, the development of emotional intelligence, and own success during the course of life.

As for the *material life faculty*, the Abhidhamma enumerates twenty-eight types of material phenomena, further classified into eleven general classes. Seven of these are called "concretely produced matter" (such as visible forms) since they possess intrinsic natures and are thus suitable for contemplation and comprehension by insight. The other four classes are more abstract and are classed as "nonconcretely produced matter", such as lightness and malleability, which are secondary qualities of matter. The life faculty is the sixth category of material phenomena. The material life faculty is born of kamma and controls the other nine material phenomena in the same group. It exerts this control by maintaining its vitality, while itself being determined by productive kamma. In the case of material phenomena, these are determined by water and nutriment, just as water lilies run dry and wither during times of drought.

The Pali term for material phenomena is *Rupa*[4]. The material form (kāya) under the physical aspect is an aggregate of a multiplicity of elements which finally can be reduced to the four "great elements", viz. earth, water, fire, and air. The groups of rupa produced by kamma must consist of at least nine rupas: the eight inseparable rupas includes four great elements, colour, smell, taste, nutritive essence, and life faculty (jivitindriya), and such a group is called a "nonad".

Sensitive material phenomena such as eye-sense, ear-sense, smelling-sense, tasting-sense, body-sense, heart-base, femininity, and masculinity are other kinds of rupa (material qualities) produced by kamma and these arise together with the eight inseparable rupas, temperature, life faculty and tangible form, thus, they arise in groups of ten rupas; decad a group or set of ten from the simplest to the most complex forms. These ten inseparable generic qualities of matter are related and form a faculty, for example, eye-faculty, ear-faculty etc. to perform a specific task such as seeing, hearing etc. This should be understood as material qualities presence in a biological cell and its function. All rupas of such a decad are produced by kamma. Thus, one speaks of eye-decad, ear-decad, nose-decad, tongue-decad, body-decad, heart-base-decad, femininity-decad and masculinity-decad. As to the body-decad, this arises and falls away at any place of the body where there can be sensitivity, suitable for wise contemplation and should be understood by insight wisdom.

Therein, life and the (eight) inseparable material phenomena together

---

[4] Bhikku Bodhi. (1999). *A Comprehensive Manual of Abhidhamma*, p. 234.

with the eye are called the eye decad. Similarly, (by joining the former nine) together describe ten material forms.

Kamma produces groups of rupa from the arising moment of the rebirth-consciousness (patisandhi-citta). In the case of human beings, kamma produces at that moment the three decads of body-sense, sex (femininity or masculinity) and heart-base, and it produces these decads throughout our life. The eye-decad and the decads of ear, nose and tongue are not produced at the first moment of life but later on.

In terms of the *connascence faculty*, *nascence* means "being born" or "coming into existence for the first time". The noun *nascence* is commonly used to talk about the development of something new or the emergence of great potential as a result of exertion. Connascence implies the coexistence of an ability or an inherent ability to develop by itself. As a faculty it exercises control over its associated state within its respective domain. By practice this can develop the state's intensity to its peak. There are fifteen qualifying mental faculties that coexist; these control consciousness and their associated mental factors:

1. the life faculty (psychic life)
2. the mind faculty (all eighty-nine consciousnesses)
3. pleasant feelings
4. painful feelings
5. joy
6. grief (displeasure)
7. equanimity
8. faith (confidence)
9. energy
10. mindfulness
11. concentration
12. wisdom
13. knowledge and comprehension of the unknown
14. ability to comprehend the final knowledge
15. the one who has known the unknown.

The compiler of this treatise has excluded femininity and masculinity faculties from this group of relations, as these do not have the function of a condition as defined in the philosophy of relations. They are, however, still classed as faculties since they control the body and its sexual structure, features, character, personality, and outward disposition of the respective domains of femininity and masculinity. These features are determined by kamma at the moment of conception. Similarly, the heart base is not classed as a relation of faculty; though it provides the basis for the mind door's cognition process, it does not control the consciousness.

The Relation of Jhana (Relation No. 17)

Jhana is the Pali term for meditative absorption, a condition where the conditioning state causes the consciousness to closely contemplate an object by observation without wavering. The conditioning states in this relation are the seven jhana factors: the initial application of thought; sustained application of thought; zest; one-pointedness of mind; joy; pleasurable interest; displeasure (grief); and equanimity.

Joy, displeasure, and equanimity are feelings that can be regrouped to make one factor, therefore totalling five jhana factors. These five are considered jhana factors because they enable the mind to closely contemplate the object even outside the meditative framework in the same way a scientist observes a microorganism through a microscope, or an astronomer observes a distant star through a telescope. A commentator illustrates the process in the following manner: "as an archer, from a distance, is able to send an arrow into the bull's eye of a target by holding the arrow firmly in his hand, making it steady, directing it towards the mark, keeping the target in view, aiming, and sending the arrow without wavering right into the bullseye." Similarly, a meditator wishing to attain jhana directs his or her mind towards the object and thrusts his or her mind into it by means of the five mental factors. By closely contemplating the object with diligence, with a steadfast mind, that person can refine his or her mind and bodily and vocal intimations. Jhana factors are wholesome and do not arise in a mind contaminated with unwholesome factors; therefore, keeping the five precepts is important for one who pursues tranquillity and insight through jhanas.

The five jhana factors are:

1.   Initial application (vitakka)
2.   Sustained application (vicara)
3.   Joy (píti)
4.   Happiness (sukha)
5.   One-pointedness (ekaggata)

Practical use of jhana factors are as follows:

1.   Bringing the mind to the object (arousing, applying)
2.   Keeping the mind with the object (sustaining, stretching)
3.   Finding, having interest in the object (joy)
4.   Being happy and content with the object (happiness)
5.   Unifying the mind with the object (fixing).

The initial application of thought directs the mental factors towards the object and fixes the mind firmly upon it. Sustained application of thought causes one to review the object over and over again, and the mind becomes attached to it. Zest has the character of creating interest in observation and makes the mind happy and content. One-pointedness makes the mind concentrated, steadfast, and fixed on the object. The three feelings (joy, displeasure, and equanimity) make the object feel desirable, undesirable, or neutral depending on the chosen object. Successively dropping these factors differentiates between the levels of jhana and makes the mind sublime and refined.

The conditioned states are the consciousness and the jhana factors associated with it. This includes connascent material phenomena that are produced by close contemplation of the object. Although material phenomena make no direct contribution to concentration, they coexist with the attainment of jhana. For example, malleability of a refined material has a quality that generally distinguishes its appearance and intuitive intimations as distinct from others. This is characterised by connascent material qualities produced by the accomplishment of jhana factors. If the mind gets corrupted, one might lose all the gained powers, such as psychic powers, and other qualities.

The Relation of Path (Relation No. 18)

The path relation is where a conditioning state conditions the conditioned state by directing it to a particular destination. The conditioning states are the twelve path factors. This includes the Noble Eightfold Path factors as taught by the Buddha in the Four Noble Truths, which are the means by which one reaches blissful destinations and the final destination of nibbana. The other four are wrong path factors, which are the means to reaching the realm of misfortune and woeful destinations. The corresponding mental factors of the blissful path are the beautiful factors found in the good consciousness. The first mental factor is the right view, a wisdom factor. This is followed by right intention, right effort, right mindfulness, and right concentration. These mental factors correspond to their respective path factors within the group of jhana factors. The remaining three, right speech, right action, and right livelihood, are separated because there are no distinct mental factors assigned directly to them, but instead they are abstinences found in the supramundane and other wholesome consciousness.

Wrong view is exclusively an unwholesome factor. It contributes to all four wrong path factors that lead to misery. The jhana factors of initial application, energy, and one-pointedness are ethical variables which can also be found within wrong speech, wrong action, and wrong livelihood that is to say one applies these for wrong reason. There are no specific factors to the wrong path; reaching it is a result of accumulated unwholesome factors, wrong intentions, wrong conduct, and defilements which direct one to bad destinations. There is no wrong mindfulness, since mindfulness is an exclusively beautiful and wholesome mental factor; it is absent in the unwholesome consciousness. When mindfulness is presence then it dispels defilements and unwholesome mental factors.

Practice of jhana makes the mind malleable, pure, free from defilements, steadfast, firm, and ecstatic towards the object. A commentator illustrates that the attainment of an "ecstatic mind" is similar attainment of a mind that is sinking into the *kasina* object, like a fish in deep water. The sound of silence as the tranquillity gained from jhana sets the conditions for insight. Volition that causes kamma is the way into samsara, whereas wisdom that springs from meditation is the way out of samsara. The way out of samsara is upon the straight and narrow path to prosperity, leading one to develop, to flourish, and to prosper, reaching the higher planes until the final escape to the transcendental state of nibbana. These elastic properties of the mind need to be developed under the guidance of an experienced master.

The conditioned states in the relation of path are all classes of consciousness, their associated mental factors, and connascent material phenomena; rootless consciousness is the exception. The path in this context is the direction in which someone is walking. The insight one may gain from the relation of path is volition of one's own actions in directing life; one comes to understand that unwholesome volitions take one to a destination of suffering and that wholesome volitions take one to a destination of happiness. Path factors are guided by wisdom to avoid suffering and by worldly happiness to awaken the truth of impermanence, suffering, and nonself. It finally directs one out of life's dense wilderness, where one can choose to escape to supreme happiness and arrive at the end of suffering. Its fruit is nibbana. The path is a meditative attainment whereupon, with emancipation to the supramundane sphere of consciousness, the attainer exhibits marked changes in behaviour such as submissiveness, which is overtly noticeable. When such a sage appears in the world, it makes a difference to one's thinking, as observed by the venerable Sariputta when he first met venerable Asanji, one of the pioneer disciples of the Buddha. On sight of venerable Asanji, venerable Sariputta then Upatissa noticed remarkable personality traits and spoke to enquire about his master who trained him.

The Relation of Association (Relation No. 19)

The relation of association is not a specific relation in and of itself. It is a common feature of some of the other relations and is a principle of consciousness. The method of association is the basis of enquiry. It seeks to determine the types of consciousness by each mental factor associated with it. For example, consciousness is classed as wholesome if beautiful mental factors are associated with it, when both consciousness and the associated mental factors have the same object and base. The second method of enquiry is the method of combination; this seeks to determine the type of consciousness by the mental factors that are combined with it.

These principles and methods are applicable to relations where consciousness is a conditioning state that causes other conditioned states to arise. By having the same class of consciousness, mental factors, object, and base, the mental states both before and after are inseparable in association like a close family. Neither the consciousness nor the associated mental factors are independent; the class or group they belong to is characterised by their arising and ceasing together. All classes of consciousness and their mental factors are mutually related to each other by way of association. There are four characteristics of coalescence marked by simultaneous arising, synchronous cessation, mono base, and mono object. For example, eye consciousness and sense-sphere consciousness arise in this way, only by associated with their mental factors.

The Relation of Dissociation (Relation No. 20)

Dissociation is a complex relation of mind and matter for mind and matter. The relation explains how matter born of consciousness stays together with the mind and yet can be separated, exhibiting intrinsic qualities different from those of the mind. This is similar to dissociation in chemistry and biochemistry, which is the general process of molecules separating or splitting into smaller particles such as atoms, ions, or radicals, usually in a reversible manner. For instance, when an acid dissolves in water, a covalent bond between an electronegative atom and a hydrogen atom is broken by heterolytic fission, which gives a proton (H+) and a negative ion. Dissociation is the opposite of association or recombination.[5]

Dissociation is like a mixture of water and oil, two substances which can coexist together but remain separate. Similarly, the mind can condition matter by assisting in its presence and vice versa, yet the two remain separate. This relation is clear at the moment of rebirth when the physical heart base arises simultaneously and disassociates with mental aggregates. At the same time, these mental aggregates condition other kamma-born materials. This relation continues to occur during the course of life as

---

[5] P. Atkins, J. de Paula, et.al., (2017). *Physical Chemistry*. Oxford University Press.

consciousness produces matter and conditions it, related by post-nascence dissociation. Pre-existent matter conditions the mind and stays separate by a prenascence dissociation relation. The philosophy of dissociation shows how seemingly opposite or contradictory things can actually be complementary, interconnected, and interdependent; it further shows that they can even give rise to one another, coming into being at various stages of the life process.

Within the physical body, matter (as described in the Abhidhamma) goes through a variety of biochemical syntheses throughout its life. Though not visible, such synthesis is notably understood by medical science in the pathology of molecular behaviour. Tumours and the overgrowth of cells are evidence of disorderly patterns of cell biology causing distress. Take for instance cancer, a word that fills people with dread—a powerful, devastating illness that has destroyed the lives of many, along with the lives of their families and loved ones. Cancer is a diseases for which scientists and doctors are actively and fervently working towards finding a therapeutic solution, beyond studying the immune response to cancer cells. There is evidence that certain behavioural patterns contribute to cancer, for example intake of synthetic substances through organic crops, processed organic products, livestock products that are not agreeable with the body, and smoking, the last of which can cause lung cancer and respiratory diseases.

Cancer is an abnormal condition in which biological cells in the body divide without control and can invade nearby tissue. Cancer can also spread to other parts of the body through the blood and lymph systems. There are several types of cancer that can even affect children from the age of four. Cells in the body are not permanent, and old cells naturally die to be replaced by new ones. In a healthy body, each and every cell is renewed every seven to ten years. Cancer develops when the body's normal control function stops working. Old cells do not die but instead grow out of control, forming new abnormal cells which are not able to carry out functions as a healthy cell would. This abnormal condition is due to changes to a particular section of DNA *(deoxyribonucleic acid)* called genes. These changes are also called genetic changes. A gene is a basic physical and functional unit of heredity; it is made up of DNA, a larger molecule tightly coiled up to fit in the nucleus of a cell and received from both parents at the moment of conception. Your genes carry all the information that make you. For example, having blond hair and blue eyes is the magical work of genes. Sometimes people inherit certain faulty genes, or mutations in genes are caused by unhealthy habits such as intaking unagreeable chemical substances. When the cell naturally divides, it picks up these faults or mistakes—mutations. Sometimes mutations in important genes cause a cell to no longer understand instructions. The cell can start to multiply out of control, ceases repairing itself properly, and failing to die when it should. This can lead to cancer.

Buddhists believe kamma can alter the genetic code at the moment of rebirth. There is no scientific explanation for how a child born to healthy parents can have faulty genes or how a healthy child can be born to parents with faulty genes. Matter, as described in the Abhidhamma, encompasses all material elements of the human body including its properties. The Abhidhamma does not investigate these material phenomena in any further depth beyond the relation of disassociation, suggesting that material qualities produced by the mind are due to the quality of the consciousness.[6]

Disassociation is similar to separation; its extreme is rejection. Where the relationship between body and mind is broken and showing symptoms of alexithymia, there arises a subclinical cognitive-affective impairment affecting the ability to interpret one's own emotional experiences. In psychological terms, dissociation refers to the alteration of an individual's identity, with a loss of integrating with others, affecting memory, identity, and the perception of the environment and time.[7]

Dissociated phenomena, in their capacity as dissociated conditions, combine according to circumstances with seventeen further conditions. Understanding this relation can prevent both physical and mental distress.

---

[6] Cancer Research UK, https://www.cancerresearchuk.org/about-cancer/what-is-cancer/genes-dna-and-cancer; MedicinePlus, https://ghr.nlm.nih.gov/primer/basics/gene.

[7] American Psychological Association, 2014; Bernstein & Putnam, 1986. Reality versus fantasy: Trauma and Dissociation. Lynn et al. (2014).

The Assessment of Alexithymia in Medical Settings: Implications for Understanding and Treating Health Problems. Contributors: Mark A. Lumley, Lynn C. Neely, Amanda J. Burger. Published in final edited form as: J Pers Assess. 2007. The National Center for Biotechnology Information

These may not be evident at first and can surface surprisingly as concealed suffering. Buddhism helps one to understand what life actually is and how to live a healthy life.[8]

In this relation the two components, mind, and matter, can condition each other while remaining separate, like a mixture of water and oil. There are four subsidiary groups of dissociation: connascent dissociation, base prenascent dissociation, base objective prenascent dissociation, and post-nascent dissociation.

Mind ⟶ Matter   Matter ⟶ Mind

In *connascent dissociation*, the mind is causally related to physical matter by way of coexistent dissociation. The potential of producing matter coexists with the mind and vice versa. For example, the four aggregates of the mind (feelings, perceptions, mental formations, and consciousness) are coexistent with the physical body during the course of life. The conditioning state is either mind or matter, and the conditioned states would be the other, matter or mind respectively. At the moment of rebirth, the heart base and the four mental aggregates arise simultaneously and exhibit physical and mental qualities. They can be distinguished and separately identified; hence it is a relation of disassociation. During the course of life, these mental aggregates condition other conscious-born matter and can be dissociated from it.

Pre- and post-nascent dissociation relates to the order of either matter as a conditioning state for the mind or the mind as a conditioning state for matter; the respective conditioned state is dissociated from each other. These relations are identical with previously explained pre- and post-nascent support conditions (No. 10). The mind has a predominant state of influencing matter, which is always subordinate to the mind. In the order of formation, the mind is supreme, followed by a base and the physical formation of matter, which is an order of cosmic law.

---

[8] TOP 10 MOST PROMISING EXPERIMENTAL CANCER TREATMENTS. (2022), https://www.bestmedicaldegrees.com/experimental-cancer-treatments/.

The Relation of Presence (Relation No. 21)

The relation of presence has multiple subsidiaries and encompasses of a wide variety of other conditions, a range of pre- and post-nascent relations (nos. 6, 7, 8, 10, and 11). Since this is a relation of mind and matter for mind and matter, it has multiple combinations that support both material qualities and mental qualities according to circumstance. In this relation, the conditioning state influences the conditioned state to arise in each phase (arising, persisting, and passing away), persisting in being present and coexisting by temporarily overlapping. It temporarily coexists to support the conditioned state in carrying out its function, assisting by means of fulfilling the necessary conditions to complete the process. An example is in the handing over of responsibility from a leaving officer who remains present and assists until the successor is established in the office.

The Relation of Absence (Relation No. 22)

The relation of absence is a condition which allows conditioned states to arise immediately next to the conditioning states by ceasing all its effects without further influence. This relation is similar to Relation No. 4, where the effect of the conditioning consciousness and its concomitants completely ceases, giving the opportunity for the successive condition to arise immediately next to it. It is like a ceasefire agreement that allows the conditioned state to be peaceful, independent of the influencing turbulence of the former withdrawal of forces. Absence is a relation of mind for mind; the conditioning force is the force of repetitive javana completely ceasing its effect and coming to a standstill. For example, in tranquillity meditation jhanas arise one after the other without the force of the preceding stage.

The Relation of Disappearance (Relation No. 23)

The relation of disappearance is identical to Relation No. 22. The conditioning state simply disappears, giving an opportunity for the next mental states to arise in this mind-to-mind combination. For example, the five sense cognitions are independent, so the eye cognition disappears when the ear cognition arises; it is a psychological principle that only one type of consciousness can arise at a time even within the same class of consciousness. The relations of No. 22 and No. 23 are identical to those of proximity and contiguity conditions (No. 4 and No. 5 respectively).

The Relation of Nondisappearance (Relation No. 24)

The relation of nondisappearance *(not disappearance)* is identical to Relation No. 21. The conditioning state in some way supports the conditioned state to arise by keeping its presence, overlapping its function in all three phases of arising, persisting, and passing away. Not necessarily coexisting but temporarily overlapping for a period, the relation of nondisappearance is mostly applicable to conditioning material phenomena. This phenomena of nondisappeareance give a false notion of permanence.

The Synthesis of Relations

All relations are arranged according to the methodology of a modular system which identifies them under four master conditions: relation of object, decisive support, kamma, and presence. In all cases the connascent material phenomena should be understood as twofold. When life comes into being, to its present life, it forms the dependent nature of all material and mental phenomena, which serves as the basis of analysis and synthesis. At rebirth linking, these material phenomena are born of kamma. According to Buddhism, the presence of rebirth-linking consciousness is a necessary condition with the fertility to trigger conception. During the course of life, such matter *(rupa)* came to life should be understood as being born of consciousness. The Pali commentators have extensively analysed the subsumption of all conditional relations under these master conditions. The Buddhist analysis and synthesis of matter is compatible with pure science, chemistry, physics, biology, biochemistry, etc., explaining the properties of physical matter but excluding any explanation of the presence of the life faculty. There is no comparable study or psychology equivalent to the Buddhist understanding of the mind, but only speculative views and theories formulated on limited observations. The relations expounded in the

Great Treatise (Patthana) are total of twenty-four, omitting nibbana and concepts since they are not variable conditions and are outside the concept of time. Nibbana is permanent and not produced by cause; it is classed as unconditioned (Asankatha Dhamma). Concepts formed on assumptions may become outdated over a period of time unless they are founded on unchanging principles. All other phenomena are subject to impermanence and are liable to dissolution; as shown in the philosophy of relations, they are related to impermanent causes and conditions. The things that beings are dearly attached to and cling to are not permanent and therefore cause distress, bringing affliction and suffering. All things that are impermanent and that dissolve at every moment are devoid of essence, a core substance. Through his philosophy of relations, the Buddha negated the contemporary belief of existence as a permanent self or soul that transmigrates from life to life or as a Supreme Being in control, rejecting the theological position of a supermercy over others. He instead firmly established the Four Noble Truths and the path to end suffering. His exposition of relations laid the foundation for the Buddhist doctrine of nonself. Buddhist philosophy of relations, to elaborate the impermanence as applied to the Law of Perpetual Change, has outset dissolved all things into a continuous succession of events that happenings one after the other in the ultimate sense.

The Synchrony of Relations

Synchrony is defined as the way in which two or more things happen, develop, move, etc., at the same time or speed, for example: "Brain regions which work in synchrony in normal people were slow to respond in autistic subjects."[9]

Autism, or autism spectrum disorder (ASD), refers to a broad range of conditions characterised by challenges with social skills, repetitive behaviours, speech, and nonverbal communication. It has the characteristic of Asperger syndrome, where some activities are asynchronised and there is a time lag between responses. The deficiency emphasises a failure in rapid neuronal synchronisation in the brain's deep neural circuit; hence the focus of study on this subject is the brain.

---

[9] "Synchrony", *Cambridge English Dictionary*, https://dictionary.cambridge.org/us/dictionary/english/synchrony, accessed 6 July 2022.

The Buddhist view of the brain is that it consists of a bundle of material elements (*rupa kalapa*) which are a part of the body. It does not make specific reference to neurons or functions of brain cells. The Abhidhamma explains the relation between these material elements and the mind. For example, a person with acute senses can verbally communicate fluently because of the lightness, malleability, and adaptability of material groups that make up the plasticity of speech. The sound is produced by the elements of body and mind working harmoniously. If the elements are heavy, coarse, not agile, and unadaptable, it would become difficult for someone to maintain bodily postures or perform vocal responses harmoniously; the outcome would reflect as deficiencies in vocal and bodily intimations. These limitations can be circumvented by mindfulness practice and turning to do good things, eventually improving behavioural responses in the social environment.

The Buddha understood the cosmic laws by being attuned to nature and taught us that to find true happiness, one must live in harmony with nature. The truth is everywhere and all around us. Those who discover the natural order are able to provide meaningful solutions to the challenges of life, which otherwise make things more difficult. The most talented mathematicians, scientists, artists, and musicians, from the Middle Ages to the present day, have been studying and applying laws of nature. They have discovered that all events occurring in nature are interrelated. To produce something extraordinary requires a holistic approach, to understand human nature and redefine the human potential with appropriate values and a life purpose. This requires dedication, discipline, time, and practice.

The Buddhist explanation of the synchrony of relations focuses on the psychological aspects of life, the mind, and its mental movements. The concurrence of causal relations that are commonly related is called the synchrony of relations; the Pali term is *paccaya ghatana*. The relation of all conditioned phenomena is explained under three key doctrines of Buddhist philosophy:

1. All phenomena are related to causes.
2. All phenomena are conditioned by causes.
3. All phenomena arise from the conjuncture of dependent origination.

The synchrony of relations sums up all 121 classes of consciousness, the 52 different types of mental factors, and the 28 kinds of material qualities enumerated in the Abhidhamma. These coexist, arising together and passing away together, or are conditioned by one another in common concurrence. These relations show that nothing arises singularly and that nothing created out of nothing; instead, everything is in perpetual flux. One of the laws of physics is that most natural phenomena are reversible. The doctrine of dependent origination intuitively shows the forward order of arising; its reverse order shows cessation.

All one hundred and twenty-one classes of consciousness and their associated mental factors arise in relation to six internal sense bases (*ayatana*) (eye, ear, nose, tongue, body, and mind). Through these sense avenues, the consciousness cognises the corresponding elements of the object. For example, eye consciousness and its associated mental factors arise together to form a relation with the visual element which is its common object. There is not a single class of consciousness which arises without a causal relation to an object. Out of the twenty-four, fifteen relations are common to all mental states. There are eight relations common to only some mental states. Tables 8.3 and 8.4 of the chapter "The Compendium of Conditionality" in *A Comprehensive Manual of Abhidhamma* (pp. 309, 323) give a detailed distribution of synthesis between the twenty-four conditions. It is important to understand that the relation of moral and immoral deeds is a causal relation brought about by asynchronous kamma, which binds a person to cycles of suffering. This relation operates by volition, causing a chain of mental activities from its roots in ignorance, craving, and grasping. This principle of grasping is common to the relations of all six senses. Buddhism views all phenomena, including wholesome (*kusala*) kamma, as ultimately ending up in frustration when their good effects come to an end. Therefore, to perform moral deeds is to seek ultimate happiness, nibbana.

The path of liberation is open to one who practises the jhana route; synchronising the mental factors of the nine stages of jhana prepares the ground for insight wisdom. Effort (energy) is synchronised with dominance, where the faculty and path relations are in combination. Mental one-pointedness may arise as a result of development through the jhanas. Asynchronous kusala kamma form a decisive support relation to attain tranquillity and a calm state of mind from the higher jhanas. By changing direction at the right moment, one can develop the factors of enlightenment by mindfulness. Through investigation of the object of meditation, one can realise the three characteristics of phenomena. On the other hand, *akusala* kamma could obstruct the path of progress and disrupt one's following of that path.

Combinations of unwholesome mental factors operate in the twelve classes of immoral consciousness. These make the mind dull and present an obstruction to enlightenment. For example, delusion, shamelessness, fearlessness of wrongdoing, and restlessness are universal unwholesome factors that cause errors in perception and give rise to perplexity. This causes one to deviate from the function of the faculty relation and the path relation of liberation. There is no relation whatsoever to mindfulness among the classes of unwholesome consciousness.

There are twenty-five beautiful mental factors in different combinations that are related in synchrony to the faculties. The former increases the power of the latter. The majority of mental factors are beautiful factors that operate in combination to contribute to ninety-one "radiant" classes of good consciousness. Mental factors are the conditioning states that condition the consciousness in the relation of root, relation of dominance, relation of jhana, and relation of path. For example, jhana consciousness has seven universal and six occasional groups of mental factors, out of which five, in combination, represent jhana factors. Mental one-pointedness can as a single factor sustain higher jhanas, sharpening the faculty relation. Mindfulness and wisdom are faculties in and of themselves and become power factors to help lead one out of samsara to the path of liberation with the right view. The synchrony of relations should be directed towards attaining higher jhanas and the path. To do this, one must develop wisdom in moral consciousness by repetition relation (habitual recurrence) because path factors are supreme and transcendental, as opposed to sense-sphere consciousness.

The Abhidhamma enumerates twenty-eight types of material qualities and recognises the life faculty of material phenomena. This is causally related to the four great essentials of primary matter and the material quality of nutriment, which are secondary matter and causally related to themselves. These relations are connascence, mutuality, dependence, presence, and nondisappearance, as found in the material body. The material quality of life is causally related to the coexistent material qualities produced by kamma, similar to a parent–child relationship. Sensitive material qualities correspond to the five senses and are related to the five-sense cognitive consciousness by prenascence and dissociation, as matter and mind are identified separately. Similar to a householder being related to the house in which he or she dwells, the heart base is causally related to itself by connascence support at the time of rebirth and to prenascence and dissociation during the course of life. Matter is directly related to the mind by a prenascence relation, whereas mind is directly related to matter by post-nascence relation. This shows the interdependency of body and mind in a holistic way. The object of vipassana meditation is body and mind. Contemplating these with mindfulness is the method to understanding these relations by insight.

Health in life and the sustenance of life is important by the support of nutriments and improving the material quality of wieldiness. These physical and mental relations are shown in nine ways under the permutations of mind and matter for mind and matter. The Buddha rejected extremism and introduced the Middle Path according to the principles of these relations, showing that wrong practices only weaken the body and mind.

Analysis, synthesis, and synchrony shown here are only a limited version of the Great Treatise of Relations. The writer has attempted to include the preeminent conditions, causes, and effects given in the texts. In many conditions there are direct and indirect effects, but the main effect will never fail if the right conditions are met. Indirect effects are incidental. For example, a properly ploughed and sown field in the right season, if treated with the right fertiliser, will definitely produce a good harvest; the farmer reaping the benefit is the direct effect of his or her labour. The straw left after harvesting in the dry season could be used to re-thatch his or her roof, improving the quality of his or her household as an indirect effect.

The Buddha showed us what is right and wrong in order to reduce suffering and prevent the worst-case scenario of failure. The philosophy of relations (patthana) should be understood as a means of self-directing one's life towards arriving at a beautiful and beneficial destination by choosing the right path through life's dense wilderness.

CHAPTER 16

BEAUTIFUL MENTAL FACTORS

Mental factors (*cetasika*) are the second category of ultimate reality. The chapter "The Compendium of Mental Factors" in the Abhidhamma catalogues fifty-two of them. They arise together with consciousness and assist one in carrying out a specific task depending on the factor's function. Consciousness as explained in the chapter 13 is a brief cyclic moment. Mental factors arise in immediate conjunction with consciousness and vanish together when consciousness ceases at the end of that brief moment; these are classified into seven classes. Both consciousness and mental factors share the same object and base. For example, in the case of eye consciousness, visible object such as a lamp and eye base are common for both. In each cycle the mental factors are conditioned according to their own conditioning forces and a cluster of other factors associated with that cycle. After the supporting conditions change, the mental factors subside to the life continuum. The process happens according to the principles of consciousness. This conditioning continues every time that consciousness takes place. In this chapter the writer discusses only the class of beautiful mental factors and their important contribution to mental development.

The commentator of the Abhidhammattha Sangaha lists twenty-five beautiful mental factors in the compendium. These are extracted from the Dhammasangani, which is the source of classification of wholesome consciousness; it lists sixty-five wholesome factors. The order of classification in these two books is not exactly the same, the factors will here be referred to in the order given in the compendium of the Abhidhammattha Sangaha (p. 79). The *F* that precedes the order number stands for "Factor".

The table below lists the twenty-five beautiful mental factors that arise in wholesome consciousness. There are nineteen universal factors found in every wholesome consciousness, further classified into seven and twelve factors for the purpose of highlighting the distinctive features of each. The specific function of the latter six pairs of qualitative factors differs, not only on philosophical grounds, but also on account of their practical importance in refining the gold of consciousness on the path of purification.

Twenty-five beautiful mental factors[1]

Beautiful universal (7)	Beautiful universal (12)
F28. Confidence (Saddhā)	F35. Tranquillity of mental factors (Kāya-passaddhi)
F29. Mindfulness (Sati)	F36. Tranquillity of consciousness (Citta-passaddhi)
F30. Moral shame (Hiri)	F37. Lightness of mental factors (Kāya-lahutā)
F31. Moral fear (Ottappa)	F38. Lightness of consciousness (Citta-lahutā)
F32. Nongreed (Alobha)	F39. Pliancy of mental factors (Kāya-mudutā)
F33. Nonhatred (Adosa)	F40. Pliancy of consciousness (Citta-mudutā)

[1] Bhikkhu Bodhi, Table 2.1, "The Compendium of Mental Factors", *A Comprehensive Manual of Abhidhamma* (Kandy, Sri Lanka, Buddhist Publication Society), p. 79. Column headings denote different type of classes.

F34. Equanimity—neutral mind (Tatramajjhattatā)	F41. Workableness of mental factors (Kāya-kammaññatā)
	F42. Workableness of consciousness (Citta-kammaññatā)
	F43. Proficiency of mental factors (Kāya-pāguññatā)
	F44. Proficiency of consciousness (Citta-pāguññatā)
	F45. Rectitude of mental factors (Kāyujjukatā)
	F46. Rectitude of consciousness (Citta-ujjukatā)

Abstinences (3)	Illimitable (2)	Nondelusion (1)
F47. Right speech (Sammā vācā)	F50. Compassion (Karuṇā)	F52. Wisdom (Paññindriya)
F48. Right action (Sammā kammanta)	F51. Sympathetic joy (Muditā)	
F49. Right livelihood (Sammā ājīva)		

Good Consciousness

Good consciousness is a new term coined to denote consciousness that is accompanied by beautiful mental factors (*sobhana*). This includes *kusala*, *kusala vipaka*, and *kiriya citta* (of an Arahant). Good consciousness is always accompanied by faculties of mindfulness and wisdom, of which concentration and energy are variables. The factor of mindfulness is considered wholly and exclusively as a wholesome factor. It was explained in earlier chapters that equanimity is a factor belonging to wisdom. A trainee who possesses good consciousness has all aspects of beautiful factors F28–F52. It must be stressed that understanding the qualities of the factor of mindfulness and the factor of concentration is imperative before any further refinement of consciousness.[2]

Mindfulness

Mindfulness, according to the Abhidhamma, is an exclusively wholesome mental factor in the list of beautiful mental factors. It arises only in a good consciousness. "In a mind devoid of right faith (*asaddhiya-citta*) there is no mindfulness (*sati*)."[3]

The definition given in the Visuddhimagga outlines "presence of mind" and "recollection" as relating to mindfulness. Its function is awareness of the object of mindfulness. For the awareness to be qualified as mindfulness it should be free from defilements or unwholesome aggregates. For example, mindfulness when contemplating the qualities of the Buddha and of Dhamma (Buddha-anu sati and Dhamma-anu sati respectively), one should practise the Four Foundations of Mindfulness without greed, hatred, and delusion. Unwholesome thought processes occurring in the mind are not considered as mindfulness. Adherents of wrong views also remember their unwholesome actions; therefore, memory is not included in the list of Dhamma, nor is it equated to mindfulness. For example, a killer would clearly remember the murder weapon whether he or she used a gun, used a knife, or strangled the victim when committing a murder. These past experiences are retained in the mind in the form of ethically variable kamma. Memory is also closely related to perception. It has the function of recognising or identifying an object; to remember an object, there should be an experience associated with it. Both these examples of memory surface in the active mode of the mind when there are appropriate conditions. For example, a memory being triggered by an experience which

---

[2] Nyanaponika Thera, (1965), *Abhidhamma Studies: Researches in Buddhist Psychology* (Kandy, Sri Lanka, Buddhist Publication Society, p. 65.
[3] Atthasalini, p. 249; Thera, *Abhidhamma Studies*, p. 68.

has relation to the past, such as visiting an old friend, might bring with it all other childhood memories associated with that friend and place. Memory has at least three aspects: the incident, the time it happened, and the place where it occurred. A memory could be associated with a person, unfair treatment suffered, or an object. All these aspects are perceptions. Similarly, imagining a future incident is also a perception; it is no different from memories other than the time domain of occurrence.

Memories are associated with a mixture of mental factors and emotions. Ultimate happiness depends on eliminating one's negative behaviours and mental states, things such as anger, hatred, and greed. These kinds of emotions seem to be a natural part of our psychological makeup. Many people think that these are natural, and that aggression is necessary to compete with others. The seventeenth-century philosopher Thomas Hobbes saw "the human race as being violent, competitive, in continual conflict and concerned only with self-interest". Before the Second World War, in the earlier part of the twentieth century, a Spanish-born philosopher, George Santayana, wrote that generous and caring impulses, while they may exist, are generally weak, fleeting, and unstable in human nature. Dig a little beneath the surface and you will find a ferocious, persistent, profoundly selfish man. According to Sigmund Freud, "The inclination to aggression is an original, self-subsisting, instinctual disposition."

Memories of past kamma are retained in the life continuum and manifest when an opportunity is made to exact revenge; they remain dormant like a sleeping lion. This past conditioning forms one's mental disposition. Mindfulness, however, is related to the present moment. All objects of the Four Foundations are phenomena happening in the mind and body; they are to be understood with direct knowledge, without the application of perceptions at the moment of their occurrence. The causal application is "knowing" or "noting". One should pay wise attention to the object in order to discover the true nature of the object as it is, that is by mindfulness, investigation, clear comprehension, and wisdom. Perception can be deceptive, projecting a different view and obscuring the true nature. Mindfulness requires one to be detached from the object or, on the other hand, not to make an aversion towards it. To be mindful is to pay bare attention without forming an opinion, judgement, or analysis, to be aware of the intrinsic nature of the object, for example, seeing as seeing or hearing as hearing.[4]

There is a story in the Thripitaka (Pali Canon) that illustrates the power of mindfulness. A wanderer named Bahia was searching for enlightenment. After a suggestion from his friends, he went to see the Buddha and persuaded the Buddha to teach him. Bahia had not much time left to live because of his past kamma. The Buddha, seeing his urgency, advised him to find freedom beyond the self that deludes the mind and gave him these subjects to contemplate: "In the seen, there is just the seen; in the heard, there is just the heard; in the sensed, there is just the sensed; in the thought, there is just the thought." Understanding these words was just enough to dispel Bahia's delusion of having a separate sense of self. Immediately after realising this truth, his illusion of a self which deluded him vanished and he was awakened, finding the freedom he had been seeking. Though Bahia was the first to be enlightened in shortest possible time, according to the story, sadly he died in an accident because of his past kamma (Bahia Sutta Udana 1.10).

Concentration

A higher level of concentration is required to observe the subtle movements of the mind. Though insight meditation uses momentary concentration, it is intended for momentary vigour without waning. It is recommended that a trainee have experience in any of the eight attainments of concentration or make a reasonable effort to pass through the threshold by practising tranquillity. These attainments (*samapatti*) are the eight absorptions of the fine material and immaterial spheres and the four jhanas (the realm of the infinity of space; realm of the infinity of consciousness; realm of nothingness; and realm of neither consciousness nor unconsciousness).

Helpers of the Right View

The right view is the first factor of the Noble Eightfold Path. Its significance is to understand the liberation of the mind by acquiring wisdom. The

---

[4] Nyanaponika Thera. (1965). Appendix, *Abhidhamma Studies*, p. 69.

Anguttara Nikaya (Numerical Discourse of the Buddha No. 97) describes five aspects that help one to acquire wisdom, including tranquillity and insight. The other three helpers are virtue, wide learning, and discussion of Dhamma. Trainees are advised to both study and practise Dhamma. Qualities of learning are given in Discourse No. 98, as "one who is diligent, ardent, and resolute in the path of progress. He ponders, examines, and mentally investigates the Dhamma, which is a factor essential leading to enlightenment."

Potential Defilements (Kilesa)

> Those potential defilements, though in a latent state, may become active at any moment when conditions are favourable.

Potential defilement can come into awareness when other conditions for their arising are also present. The precursive conditions are like fertile soil. This soil (*bhumi*) exists in the individual's own groups of existence, the five aggregates. It lies dormant in the *bhavanga citta*, not in the outside world of tempting objects, but deeply seated in the unconscious mind. One of the fundamental Buddhist doctrines is "that man is not bound by the external world, but only by his own craving. Not only is the actuality, but also the potentiality of bondage centred in the individual." It follows the sequence of contact, feeling, perception, desire, grasping, and consumption in synchrony. The process is the desire to consume or make judgement, it produces kamma, craving, and condition—the related aggregates.

The Dhammasangani classifies factors of mindfulness (*sati*) and mental clarity (*sampajjana*) separately in the group of helpers. It is always important to maintain these two so that one might identify potential defilements. By this, one can avoid unwholesome mental factors arising, with constant effort to supress and finally eradicate them from the cognitive process. Six pairs of mental factors, together with guardian factors, improve the quality of consciousness to a more refined state. Owing to the distinctive features of these factors, they remove defilements and make the mind more palpable.

The Six Pairs of Beautiful Mental Factors—Yugalakani, F35–F46

There are twelve beautiful mental factors that arise together, though only in the good consciousness. Their functions are recognised in both the sub-commentaries of the Mula Tika and the Atthasalini. Explained in these are the operative capabilities of the factors to overcome hindrances. They are described in sets of six pairs. Each pair has distinctive features with its own intrinsic nature. These features provide a unique contribution to curing the pair's opposite and make the mind progress by a transformation of consciousness. In a single pair, it is the same factor that contributes its quality to consciousness and the accompanying mental factors to improve its function.

"Among the sobhana cetasikas (beautiful mental factors) which accompany each sobhana citta, there are twelve cetasikas which are classified as six pairs. Of each pair one cetasika has a quality pertaining to the accompanying cetasikas and one quality pertaining to citta."[5] That is, these special mental factors have dual representation, reflecting on both the body of mental factors and the consciousness which they represent.

Note: The number order given here is with reference to the table presented earlier in this chapter.

The number order of the Dhammasangani is slightly different, which takes into consideration its priorities. The enumeration in the Dhammasangani is as follows: F39–F50 are exactly the same in both texts, differing only in the numbers assigned; F51–F52 list mindfulness and mental clarity (sampajjana) in the group of helpers; F37–F38 are moral shame and moral dread, in the group of "the guardians of the world", listed before and thus prioritising moral values before mindfulness. Together these factors allow consciousness to perform a task smoothly and beautifully to produce excellent results, like harmonics. The Dhammapada best expresses the work of these six pairs: "Irrigators guide water; fletchers straighten arrows; carpenters bend wood; wise men shape themselves."[6] The purpose of these beautiful factors is to protect the mind from evil and

---

5 Nina van Gorkom | (1999), Chapter 31 - Six Pairs of Beautiful Cetasikas. Wisdom Library.

6 Dhammapada No. 80, Khuddaka Nikaya.

develop more sublime qualities in the process of cleansing that increase happiness towards a healthy mind.

I shall now explain the usefulness of the six pairs and revert the numbers back to those used in *A Comprehensive Manual of Abhidhamma*. Each set provides the same function for both mental factors and consciousness equally and respectively. These factors could be compared to a set of surgical instrument. Only a skilful surgent can handle scalpels, forceps, scissors, retractors, and clamps extensively for the success of the surgery. Similarly, skilful meditator knows how to apply these pairwise mental factors together with mindfulness and concentration in balanced manner for the successful purification of the mind.

Tranquillity, F35–F36

Tranquillity is the calm, quiet, and cool condition of consciousness and mental factors. Consciousness and the associated mental factors are then nonreactive. They could otherwise be over-, hyper-, or underactive. The single mental factor of tranquillity can influence the other mental factors to become calm. It is therefore referred to as the keynote of the mind. Tranquillity smooths out the opposites of agitation and restlessness. These are qualities that can cause anxiety and worry. The result of tranquillity's influence appears as tranquil happiness overcoming the hindrances of anxiety and worry. Energy combined with tranquillity "will be a quiet strength" displaying itself in a well-balanced, measured, and therefore effective way, "without boisterousness or uncontrolled exuberance that reacts in unproportioned manner."[7]

"Tranquillity is the inner peace bestowed by any moral act or thought. That is the peace of an unruffled consciousness."[8]

Tranquillity allows "for an unwavering and cool, reliable, and dispassionate judgement". By practice of concentration, tranquillity "prepares entry into the mental absorption (jhana consciousness) and mental one-pointedness". The Atthasalini states that the stability of mind brought about by tranquillity makes good grounds for development of

---

[7] Nyanaponika Thera, (1965). *Abhidhamma Studies*, p. 82. BPS.
[8] Ibid.

insight (vipassana). These two states taken together have the characteristic of pacifying the suffering of both mental factors and consciousness; they have the function of nullifying the suffering of both.

Agility, F37–F38

The agility of a good consciousness makes the mind move swiftly. It releases the capacity to act quickly in grasping a wholesome object— "to siege at once an occasion to do a good deed or to grasp quickly the implication of well-meaning and the consequence of a thought or a situation". An example of agility is identifying intention and being able to change it to wholesome from unwholesome thought. Agility also increases the quality of mindfulness by serving as the basis for presence of mind. When the mind moves away from the object, agility will quickly bring it back, meaning it is a ready wit.

Agility is opposite of the hindrances sloth and torpor, which make the mind heavy, sluggish, dull, and apathetic. For example, sloth and torpor make responses emotionally weak.[9]

Agility, according to the subcomentory Mula Tika, operates by swiftly emerging from the subconscious life continuum (*bhavanga vutthana*). It can bring about good, hidden abilities that result from past kamma.

Pliancy, F39–F40

Pliancy is the quality of mind, which is soft, susceptible, resilient, and adaptable. With these qualities it can perform various functions by having the ability to form its character for the better. Pliancy makes functions of the mind more efficient in performing various tasks and recovers easily to its original state. The opposite of pliancy is rigidity and resistance to change or improvement. Excessive rigidity causes mental insanity. "If this factor refers to the condition of mind in general, it is called pliancy of consciousness (*citta-muduta*)—F40."[10]

"Pliancy of the concomitants (kaya-muduta), F39, is the adaptability

---

[9] Ibid. p. 83.
[10] Ibid.

of the respective functions to their various tasks," for example the ability to attain jhana in sequential order; attainment of Nirodha (cessation); and change from samatha to vipassana at the right time. Pliancy reflects high impressionability in the process of perceiving and improves sensitiveness and the moral emotion of cognitive faculties. It also relates to a sensitivity towards moral values. It improves the intellectual capacity of the mind to learn new knowledge and benefit from new experiences. "It allows one to get rid of inveterate habits and prejudices pertaining to thought, emotion, or behaviour." By pliancy one can bring about change as a result of increased faith or forgiveness—non-hate reflecting as loving-kindness (metta). With increased sensitivity and susceptibility, it increases the capacity of the mind to develop and acquire intuitive wisdom (knowledge or insight without conscious reasoning).

The canonical term as an antonym to pliancy is *obstruction* or *stoppage* (*khila*). This makes the mind dull and increases the tension produced by anxiety. The commentary gives a lot of opposites, such as defilements which adhere to dogmas, and deluded and wrong views. These defilements make the mind hard, tense, or inflexible, like cement that hardens when it dries. "Pliancy appears as open-mindedness." A softer mind responds to appeals to do good charitable works, to broaden one's humanity and move towards selfless generosity. There is a story in the prehistorical legend of the Buddha, when he was practising Parami Dhamma, which recalls how he gave his life to feed a starving lioness and her cubs.

Workableness, F41–F42

Workableness is the middle ground between firmness and softness. To achieve the required results, one must seek the right proportions; like well-polished talent, these perform best. Workableness is "the greatest efficiency of the mental functions, to suit best the formative and transformative work of spiritual development (bhavana)".

According to the Mula Tika, "Workableness signifies that specific or suitable degree of pliancy or softness which makes the gold, that is, the workable mind. While the mind is in the flames of passion, it is too soft to be workable, as molten gold is; if on the other hand the mind is too rigid, then it is like untempered gold."

It is stated in the discourses that when the ascetic Siddhattha was practising self-mortification, he met with a music teacher who was instructing his students on how to tune a musical instrument. When the strings are too tight or too loose, the instrument does not produce the right tone. The right tone is produced somewhere in the middle. This thought made Siddhattha give up asceticism and come to the Middle Way. Since then, any beginner or trainee is advised to practise and develop the mind in the Middle Way. This advice is valid today for anyone who wishes to transform or reform his or her life through Buddhist practices. Effective transformation of consciousness has four stages towards the final goal of liberating the mind. Successive progress needs continued work on meditation practice to develop insight wisdom, until one perfects the complete eradication of hindrances.

The Atthasalini says that the opposites of workableness are "all those remaining hindrances that render consciousness and its concomitants unwieldy".[11]

Sensual desire makes the mind too soft and also makes it "shapeless", effacing its characteristic contours by diluting and dissolving. For example, clay becomes unworkable if too much water is added to it; extreme hate hardens, contracts, imprisons, and alienates the mind, making it unworkable.[12]

The following description of Nirodha is a perfect example of purity and workableness:

Anagami (nonreturners) and Arahants can reach complete cessation and emerge from it as they wish. Attainment of extinction, also called *sanna vedayitha Nirodha*, is the extinction of feeling and perception, and the temporary suspension of all consciousness and mental activity. It immediately follows the semiconscious state called "sphere of neither perception nor nonperception". The absolutely necessary precondition to its attainment is said to be perfect mastery of all the eight absorptions (jhanas) and the previous attainment of a nonreturner or Arahant. According to Visuddhimagga, entering this state takes place by means of mental tranquillity (samatha) and insight (vipassana).

One must pass through all the eight absorptions, one after the other,

---

[11] Ibid. p. 85.
[12] Ibid.

up to the sphere of neither perception nor nonperception, and then one should bring this state to an end in attaining complete cessation *(Nirodha)*.

The disciple passes through the absorption merely by means of tranquillity, that is concentration. He or she will attain the sphere of neither perception nor nonperception, and then come to a standstill. Further, proceeding from absorption to absorption can bring the sphere of neither perception nor nonperception to an end. Such a one, capable and skilful in trance, reaches the state of extinction.

At this state, perceptions, feelings, and verbal and mental functions have been suspended and come to a standstill, but life is not exhausted, the vital heat is not extinguished, and the faculties are not destroyed."[13]

Proficiency, F43–F44

Proficiency, according to the Dhammasangani, is "fitness and competence of mind and mental factors". According to the Atthasalini, it is "opposed to sickness caused by such defilement as lack of faith or confidence", that is sceptical doubt.[14]

Inefficacy is the result of feeble mental and moral constitution caused by a lack of self-confidence. Proficiency can be improved by repeated practice, for example, practice of good deeds such as giving (generosity), observing the five precepts, giving Dhamma talks, and attending meditation retreats. These are considered as the best practices in Buddhism to improve *kusala kamma* and *kusala vipaka*. A culture of good practice improves results, until it becomes a spontaneous response to perform moral and wholesome acts perfectly.

"Inner certainty, assurance, and efficiency in the doing of a good deed are expended by the factor of proficiency." Practice of this mental factor improves confidence to carry on and results in the issuing of supramundane fruits. The relation of repetition (No. 12) explains the process of apperception, gaining greater proficiency, energy, and force by habitual recurrence. The factor proficiency becomes more vigorous on account of each cycle of repetition. For example, Thai Buddhist chanting is made very vibrant, clear, and accurate by habitual practice.

---

[13] "Nirodha samapatti", Nyanatiloka Thera, *Buddhist Dictionary*.
[14] Nyanaponika Thera, *Abhidhamma Studies*, p. 85.

Uprightness, F45–F46

Uprightness means honesty or justness. As a mental factor, it scrutinises the intention to be truthful and become established in ethical value, aligning with what is kammically wholesome. Bhikkhu Bodhi uses the term *rectitude*,[15] meaning "the moral correctness of consciousness". The Abhidhamma provides a deeper analysis with introspective scrutiny to examine hidden motives and the factors associated with morality. For example, someone can offer charitable donations in order to win fame and popularity. The act is rooted in non-greed, but the ulterior motive is not morally correct; it lacks wisdom and is thus classified as being "without right knowledge".

"Uprightness serves to emphasize that ethical value is determined only by an unambiguous intention (*cetena*)." Uprightness provides the necessary conditions to improve the quality of the consciousness in being honest in conduct.[16]

Interrelation between Six Pairs

There are mutual relations between the six pairs. These are in accordance with the principles of relations (*patthana*). Tranquillity and agility balance each other by moderating their influencing conditioning forces. Pliancy sets a limit to workableness, adjusting how much flexibility is required for efficiency and best practice. Uprightness prevents agility and pliancy from becoming unscrupulous; the latter two take care that uprightness does not become too rigid. Proficiency gives agility the required speed and synchrony to coordinate mental movements smoothly. On the other hand, agility would set limits to proficiency so as not to become inflexible by overspecialising on its task.

The Suttanta method is given various references in the sutta (discourse) as a balancing quality to pliancy. It manifests as the "gentleness" of the corresponding act. In this method of description, the dual nature

---

[15] Bhikkhu Bodhi, *The Abhidhammattha Sangha of Acariya Anuruddha* (Kandy, Sri Lanka, Buddhist Publication Society, 1999).

[16] Nyanaponika Thera, *Abhidhamma Studies*, p. 86.

of "straightness and gentleness" is emphasised in place of uprightness. The term is mentioned in the Metta Sutta: "Let him be capable and upright, truly upright." One who is firm and strong in character wishes for liberation not only for himself or herself, but also for other beings and their welfare without ill will. Uprightness is firmly established in a sincere wish for others.

These six pairs of qualitative factors are the functions of the process of consciousness, refining it towards perfection. The Buddha gave the analogy of refining gold, equating the pure mind to pure gold. By emphasising the five defilements which prevent the progress of insight meditation, he thus taught how to purify the mind from mundane to supramundane consciousness. Balancing the six pairs has a great practical use, as "the harmonising of the spiritual faculties". It is therefore complementary to the development of the factors of enlightenment. Right from the beginning of his teachings, the balancing of one's practice and the Middle Path is stressed in the Buddha's instructions.

> With consciousness thus purified and cleansed, without blemish and stain, pliant and workable, steady, and unshakable, he turns his mind to the extinction of passions.

> —Lord Buddha (Majjhima Nikaya 51)

The Buddha revealed the Four Noble Truths and the path of purification to end suffering. It is the systematic culture of the mind to develop beautiful mental factors. Being mindful and eliminating defilements refines the mind as one would refine gold, gradually freeing it from blemish and alien dross. By bringing the mind to its true purity, a person can liberate himself or herself from samsara.

The six pairs of qualitative factors are mentioned here to highlight the important role they play in the path of progress; mindfulness alone cannot purify the mind. Beautiful factors do not engage with an unwholesome consciousness; therefore, if agitation (*uddhacca*) is present in every unwholesome consciousness, beautiful factors would scarcely perform their contributory functions. The most prominent feature in the Buddha's teaching is the Middle Path. It emphasises the cultivating and development of beautiful mental factors by making appropriate effort (*viriya*), like tuning a musical instrument until it produces the right tone. Having a moral foundation, that is exercising restraint by way of the precepts, and adhering to a developed social ethical conduct serves as protection. This is to guard a wholesome character from other, unwholesome mental factors, for only then will good consciousness arise. This group includes moral shame (*hiri*) and moral dread (*ottappa*) (F30 and F31 respectively). In the mode of "power", these serve the function of protection. They are regarded as guardians of the world because they protect the wholesome character of the good consciousness against shamelessness and unscrupulousness. "They are, as it were, the brakes of our mind vehicle and the restraining forces against their opposites."

Mindfulness and clear comprehension in this group are considered "the helpers" (*upakaraka*) (F51–F52). These two factors are mainly applied in the establishment of mindfulness (satipatthana). "They are called helpers because they remove obstacles and enhance spiritual development."[17]

Beautiful mental factors are listed in the Dhammasangani, the Abhidhammattha Sangaha, and the Visuddhimagga, which are extracts and compilations of the Buddha's teaching. The books of the Abhidhamma are Dhamma arranged and explained in the context of both calm (samatha) and insight (vipassana) meditation, otherwise known as the Buddhist mental culture. These two types of meditation practices are complementary. When a meditator enters a period of insight by investigation, he or she spiritually progresses by contemplating impermanence, unsatisfactoriness, and the impersonal nature of all phenomena. The alternative (reverse) method, insight leading to tranquillity, is continued until sanctity (the Ariya Magga) is reached.

Moral Correctness

The six pairs of mental factors, helper, and guardian factors could be considered as balancing factors. They direct the mind towards the purpose of moral correctness and its development. For example, to perform a task in a manner that is agreeable, pleasant, and gentle is productive, efficient, and effective for gaining wisdom from that performance. When training

---

[17] Ibid. p. 92.

a child in Buddhist culture, these mental qualities are taught at an early learning stage. The Buddhist moral values and precepts are set as guidelines to help personal growth, developing the stability of beautiful mental factors to lay a firm foundation from a young age.

Since rejecting the extreme of asceticism, the Buddha recognised the Middle Way. Buddhism promotes living in a community, serving for its welfare, and building good relationships through interaction between people. In the modern world, this sentiment has now extended to international relations, promoting peace and welfare for many. During the outbreak of the deadly Covid-19 pandemic in March 2020, world leaders pleaded for their citizens to accept responsibility, issuing orders to stay at home and enforce self-discipline to combat the contagious disease. The Queen of England requested that her citizens pray or meditate during the lockdown period and respond to this threat wisely. The world has witnessed the remarkable value of cooperation, professionalism, self-discipline, and commitment in people who care for others. Respect for life, compassion, and loving-kindness are the underlying philosophical tenets of Buddhism.

Modern psychometric tests are designed to gauge a person's mental agility and see how that person performs with decision-making under pressure, interacting with others, etc. A mind trained by tranquillity and insight can confidently face these challenges of changing worldly conditions and respond intelligently.

Buddhists believe in rebirth after death; therefore, the process of developing mental factors can continue in the next life. If reasonably developed, one can continue on the same noble path of practising generosity, morality, and mental development, which lead to the practice of morality, mindfulness, concentration, and wisdom, until final liberation is reached. The Buddhist training scheme within the framework of the precepts involves practical methods, applying theory into practice.

PROCESS-FREED CONSCIOUSNESS

Introduction

> There arose in me, bhikkhus, light, vision, knowledge, insight, and wisdom concerning things unknown before.
>
> —The Buddha
> (Dhammacakkappavattana, Sutta 56, Samyuttha Nikaya)

DURING HIS LONG MISSION OF teachings, the Buddha made known to us, from an enlightened perspective, such matters as how life is prolonged in samsara. Prior to the Buddha's enlightenment, knowledge of this matter was incomplete and was open to speculation by various other schools of thought. One of the most striking and puzzling phenomena of life is death and what happens after death. There was a school of thought at the time of the Buddha that denied the law of kamma and rejected the idea of rebirth. Their view was nihilistic, believing that there is no retributory result of action and accepting freethinking as a norm of life. This was the case not only at the time of the Buddha; such views still prevail in the present age.

In the chapter "The Compendium of Process-Freed", the author of the Abhidhammattha Sangaha, Acariya Anuruddha, explains this special function of the consciousness in relation to kamma and planes of existence. The term *process-freed* is differentiated from the normal course of consciousness as a cognitive process that makes us aware. There is no apparent awareness of this function. Process-freed could be considered as a function of the unconscious mind. There are three types of function to this special consciousness, having a specific object and relation to that object:

a. rebirth—linking
b. life continuum (*bhavanga*)
c. death consciousness.

These three types of mental functions are formed as process-freed. They are also termed as *door-freed* because they do not arise through the sense doors or mind door, neither do they follow the principles of cognitive process as experienced during the normal course of life. They are sometimes known as *avenue-freed*. For example, at the moment of death, both sense avenue and mind avenue are closed; whereas the normal active cognitive process always has an object presented at any of the six doors. There can be no specific reference to a door without receiving any new object as a general rule and this type of consciousness is an exception to the rule. It is an object determined only by the last cognitive process of the six-sense door; mind door in the case of death or the preceding existence in the case of rebirth. The object related to the last cognitive process could be a sense object from any of the six senses, a mental object, or a concept of the past or present. Concepts are timeless. An object related to the past is kamma or a sign of kamma.[1]

This bridging *citta* (consciousness) retains the same object from the rebirth moment to the moment of death during a life period. An analysis of this is given below:

- Rebirth-linking consciousness—at birth
- Bhavanga citta—during life
- Death consciousness—at the time of death
- Identical with the object of the last cognitive process in the immediately preceding existence, that is death-proximate consciousness.

At the time of death, an event can appear as a memory; the object presented is the last object in contact at the time of death. Devout Buddhists

---

[1] Bhikkhu Bodhi, (1999). *A Comprehensive Manual of Abhidhamma* (Kandy, Sri Lanka, Buddhist Publication Society), pp. 138–9.

believe it is good kamma to receive a blessing from a Buddhist monk at the time of death. They invite a monk to perform chanting at the deathbed so that the last consciousness becomes purer, and a good destination is expected. This last sign is known as kamma, a sign of kamma, or a sign of destiny.

The determinant factor of process-freed consciousness is kamma. Its casual usage is as an action or deed performed physically, verbally, and mentally during the course of life. These actions have a conditioning effect on the cognitive process throughout one's life. They have a cumulative effect, not in the sense of a total sum, but as a dynamic conditioning factor and conditioning force in continuity. In Buddhist terms, kamma means volitional actions, wholesome or unwholesome, where consciousness is accompanied by an ethical variable, a universal factor of volition. These functions can be understood from experiential and analytical knowledge and eventually comprehended by insight wisdom.

The Buddha stated, "It is volition, monks, that I call kamma. Having willed, one performs an action through body, speech or mind."[2]

"All that we are is the result of what we have thought; it is founded on our thoughts and made up of our thoughts. If a man speaks or acts with an evil thought, suffering follows him, as the wheel follows the hoof of the beast that draws the wagons. All that we are the result of what we have thought; it is founded on our thoughts and made up our thoughts. If a man speaks or acts with a good thought, happiness follows him like a shadow, it never leaves him."[3]

According to the "law of action", every action produces a result, and the one who performs the act inherits its result (*vipaka*). From the Abhidhamma point of view, the resultant consciousness is conditioned by the intention of the action. Some actions produce results immediately; others, after some time. Some actions take a long time to produce a result, even beyond this current lifetime.

Some matter is directly produced by kamma (*kammaja* or *kamma samutthana rupa*). For example, the formation of life at birth requires the presence of kamma. After the consciousness produces matter, it is maintained by food, water, and physical conditions. This process of

---

[2] Anguttara Nikaya 6.63, iii 415. (BPS 1999)
[3] The Dhammapada (No. 1), the Buddhist Society of London.

physical development is explained in the relation of post-existence (No. 11). For example, periodic rainfall supports the growth of vegetation and plants by way of post-existence, as they were planted before the rain. These events happen according to natural laws which are called the law of seed (*bija-niyama*) and law of heat (*utu-niyama*); these govern organic life. This theory of law is also compatible with modern biomedical science.

As the Buddha said, speaking to his closest attendant, the venerable Ananda: "Kamma is the field, consciousness is the seed, and craving is the moisture, for the consciousness of being obstructed by ignorance and fettered by craving is to be established in a new realm of existence either low, middle, or superior."[4]

The Buddha further explained the effect of kamma in the formula of dependent origination. From the beginning of birth to the end of death, he explained the role of kamma in one's life. Kamma also influences the formation of materiality.

> "It is a fundamental idea of Buddhist philosophy that there cannot be existence of matter without a karmic consciousness desiring life in a material world.
>
> "If, Ananda, there was no kamma maturing in the sensuous sphere, could sensuous existence (*kamma-bhava*) appear?"
>
> "Surely not, Lord."
>
> As determined by past kamma, the seed of consciousness falls into an appropriate realm, sends down roots, and is nurtured by its store of karmic accumulations, unfolding according to its hidden potentials.[5]

Beings continue to wander in this samsara because of their ignorance, and not many are aware of this. Given below are extracts relating to dependent origination:

> Dependent on ignorance arise karmic formations.
> Dependent on karmic formations arises consciousness.
> Dependent on consciousness arises mind and matter.[6]

---

4 AN (Anguttara Nikaya) 3.76, i 223, p. 188.
5 Anguttara Nikaya 3.76.
6 Bikkhu Bodhi, *A Comprehensive Manual of Abhidhamma*, p. 294.

The Buddha discovered the operating law of action. The Pali term for this law is *kamma niyama*. Its basic principle states that intentional (volitional) actions produce results.[7]

Kamma Ripening

Kammic results occur when the right conditions are met for the ripening of the kamma to produce the results. Just as an apple seed when planted grows into an apple tree and produces apple fruits, kamma has its own distinct operative functions. It ripens in a particular order, at a specific time and place, according to its qualitative factor, an absolute ethical value which determines its quality. For example, the flavour of an apple depends on the climate, soil, and general environment in which it grows, like English Cox apple is different to New Zealand Gala apple.

Kamma is generally classified into two broad branches of wholesome and unwholesome kamma, according to ethical criteria. These branches are further subdivided into *kusala* and *akusala* kamma according to presence or absence of the wisdom factor. The Abhidhamma classifies a fourfold division of kamma according to its function. This is expanded to another twelve types, totalling up to sixteen distinct types of kamma. The formula of dependent origination recognises ignorance as the factor responsible for the arising of kamma. The reason for ignorant behaviour is wrong view, held individually or collectively.

Those who follow such wrong views degenerate in this existence; even after death a bad destination is expected. On the other hand, there are those who follow the right view and acquire kusala kamma by performing meritorious actions such as being generous, being moral, and developing mental culture. They keep the five precepts and lead happy lives. For them, a good destination is expected after death.

The principles of kamma are closely interwoven into Buddhist cultures. For example, Thai Buddhists have adopted a system of culture that is integral to Buddhist temples. They perform meritorious deeds such as make offerings, and they support and uphold Buddhist values; it is their way of life. Buddhists take the Buddha, Dhamma, and

---

[7] Ibid. pp. 120–2.

Sangha as their refuge. Sangha represents the excellent qualities of the Buddha; the monastics promote the development of good human qualities (*manussa guna*) that enable positive changes individually and as a society at large. From the Buddha's time, through many centuries to the modern day, this promotion of meritorious activities has existed. It is based on mutual interdependency and goodwill between monks and lay supporters, known as "Buddha Sasana", that is Field of Merit. With a view to alleviating others suffering in this life and in future lives to come, some monks and laypeople have devoted their entire lives to the benefit of others, to gaining understanding and seeking the opportunity to gain excellent kamma.

The Buddha uttered the Dhammapada verse (No.1) explaining the relation between the mind and action:

> All mental phenomena have mind as their forerunner.
> They have mind as their chief.
> They are mind-made.
> If one speaks or acts with an evil mind,
> Suffering will follow him,
> Just as the wheel follows the footprint of the ox that draws the cart.

> (Manopubbangama Dhamma.
> Manosettha manomaya.
> Manasa ce padutthena.
> Bhasati va karati va,
> Tato nam dukkhananveti,
> Cakkamva vahato pdam.)

Morally, one might take a nihilistic view which fundamentally denies the validity of ethics and the retributive consequences of action. Three such views are mentioned often in the Sutta Pitaka (pp. 207–8):[8]

1. Nihilism says that both body and mind annihilate at death, thus rejecting the moral and ethical value of action;
2. That given inefficacy of action, deeds are not capable of producing results, and so reject the moral distinction of action;

---

[8] Ibid. pp. 207–8.

3. And the acausality view, which holds the view that there is no condition for defilement and purification of beings; they are defiled and purified by chance, by fate, or out of necessity.

"Wrong view (miccaditthi) becomes a full course of action when it assumes the form of one of the morally nihilistic views which deny the validity of ethics and the retributive consequences of action."[9] Those who follow such views only prolong samsara and add more suffering to their lives. For example, breaking the five moral precepts is one way to acquire akusala kamma.

At any single moment of time, consciousness is a conditioned phenomenon accompanied by several mental factors. These each perform a specific function related to other functions. These relations apply not only in the present but also to the multiple past states of consciousness which are its conditions. For example, concerning *kamma paccaya*, the relationship of kamma, the effect of the volition is asynchronous with the time of its occurrence. Asynchronous volition is volition that differs in its point of time from its effects. The effects do not cease though the volition ceases; instead, they latently follow until favourable conditions are met. If not, until favourable opportunity is found, volition remains in the same latent mode for many hundreds of existences (see *bhavanga citta – below this chapter 17*).[10]

While residing at the Jetavana monastery in Shravasti, the Buddha taught a large retinue of monks about the consequence of action. This story relates to Thera Chakkhupala, who was blind. The story portrays the laws of action, the principle being how the consequence of an action could follow a person like the cart's wheel follows the hoof of the drawing ox.

Chakkhupala was a physician in one of his past lives. He had made one of his patients deliberately blind in revenge for having been cheated. As a result of this evil action, Chakkhupala was blind for many of his following lives, including his last life as an Arahant.

Process-freed consciousness is mainly related to the transitional moment of one life to another. It has been argued whether this is a new

---

[9] Ibid. p. 207.
[10] Mahathera Ledi Sayadaw, *The Buddhist Philosophy of Relations: The Patthanuddesa Dipani*, tr. Sayado U Nyana (1965).

life or the same old life reborn. The Buddhist analogy of this phenomenon compares life to a candle that, with enough wax, continues to burn. If a new candle is lit from the flame of an old one before it vanishes, then the flame will continue in the new candle. We cannot say that the flame is the old one or that it is something new. It is neither old nor new, but only a continuation of the flame—if the wick is strong enough to hold it. Similarly, at death, the preceding consciousness takes birth in the next life immediately after death. The explanation of process-freed consciousness is not complete without explaining where the next life takes birth. There is a time and a place for those who dying here and being born there, each determined according to kamma ripening. The Abhidhamma examines a range of planes of existence and various realms within each of those planes. There are thirty-one different forms that make up the external universe. This knowledge of Indian cosmology existed before the Buddha's enlightenment. However, the Buddha neither accepted nor denied its validity, but used the contemporary belief system as a vehicle to convey his teachings.

The outer world is an appearance of the mind that corresponds to a series of internal consciousnesses and helps us to understand the complexity of those planes and realms. This helps to explain how the external world can influence the inner consciousness to continue and to change and facilitate this happening. The relationship between these two worlds is made by the birth of a new life through the kamma ripening from a previous life. For example, there is a corresponding realm of existence for each type of consciousness. By kamma ripening in the fine material plane, a fine material rebirth consciousness is generated, and fine material existence becomes manifested. This principle is the same for all other realms of existence. Kamma is the force, but more precisely, it is the javana process of consciousness that transits a being dying in one realm to be born in another. Death in Buddhist culture is not necessarily a bad thing as long as a being is born in a better realm where less suffering is expected.

"Consciousness and the world are mutually dependent and inextricably connected, to such an extent that the hierarchical structure of realms of existence exactly produce and correspond to the hierarchical structure of consciousness."[11]

---

[11] Bhikku Bodhi, *A Comprehensive Manual of Abhidhamma*, p. 118

The hierarchical structure of both worlds is defined by the purity of mind. For example, the Brahma world is said to be one of the most exquisitely refined and blissful realms, where meditators of higher attainment are born. When asked, the Buddha referred to the four cardinal qualities of the mind as the dwelling places of Braham, namely loving-kindness, compassion, sympathetic joy, and equanimity. These four great qualities are known as "Brahma Vihara", literally meaning "dwelling places of Brahma". In this statement, the Buddha was referring to a state of mind that could be attained here and now in this life. The Buddha's teaching always directed to inward observation rather than to the outward world. Without the inner development of sublime qualities, no result in the outer world can be gained, as mere speculations and projections of cosmology do not lead to the liberation of the mind. The following Dhammapada shows that the Buddha always advised his disciples to be thoughtful: "Do not live thoughtlessly, nor follow false aims, and do not add to the rounds of rebirth."[12]

Misunderstanding the law of kamma and misinterpretation of its results prolongs samsara because in this one is travelling via the wrong route. "The reason why a living being is reborn into a particular realm is because he has generated, in a previous life, the kamma or volitional force of consciousness that leads to rebirth in that realm."

Four planes of existence are mentioned and examined in the Abhidhamma. They are analysed according to the same classification of consciousness:

    i. the woeful plane
   ii. the sensuous blissful plane
  iii. the fine material sphere plane
  iv. the immaterial sphere plane.

These planes are further analysed into different realms. The broad overview of these existences includes the divisions of hell and heaven, where beings are reborn according to their unwholesome and wholesome kamma respectively. For example, evil deeds result in being born in hell, where pain and intense suffering is expected; wholesome deeds result in being born in pleasurable heavenly realms. In the Buddhist point of

---

[12] Dhammapada 167.

view, the realms of hell and the celestial hierarchy is a measure of pain and pleasure according to the severity of kamma ripening, irrespective of whether anyone is willing to accept the validity of extra-celestial worlds.

The fine material sphere plane and the immaterial sphere plane are the realms of rebirth for those who, at the time of death, possess fine material and immaterial meditative attainment. Each of these planes is named according to jhana consciousness and the resultant citta attained by those individuals during their meditation practices. These planes are the kammic result produced by wholesome jhana consciousness. They perform three functions of rebirth linking, *bhavanga*, and death as process-freed consciousness. The resultant citta of these specific jhana attainments does not occur in the normal cognitive process but only through process-freed consciousness. Nonreturners who attain higher jhanas are reborn into pure abodes and never return to the lower realms but attain nibbana there. There are five such pure abodes determined by the predominant spiritual faculty. For example, those who have the predominant faculty of faith are reborn in the Aviha realm. These abodes are named according to five spiritual faculties. Following is a list of realms and the respective qualities dominant in them:

- Aviha realm—the faculty of faith is dominant
- Atappa realm—the faculty of energy is dominant
- Sudassa realm—the faculty of mindfulness is dominant
- Suddassi realm—the faculty of concentration is dominant
- Akanitta realm—the faculty of wisdom is dominant.

Jhana is not a fixed law that holds to all nonreturners born into those realms, which are open only to nonreturners who have attained the fifth jhana. Dependent on kamma, if any defilements remain, according to the ripening kamma, beings are born into a realm of existence. The same phenomena are explained according to the number of beautiful mental factors present at the time of rebirth, namely nongreed, nonhatred, and nondelusion. These are known as triple-, double-, or single-rooted. Those who have practised jhana are known as *jhana-labi* (gifted with jhana). An extended analysis can be found in the Abhidhamma's fourth book, "Designation of Individuals", which gives an ethical classification of individuals.

Rebirth Linking (Patisandhi Vinna)

Rebirth-linking consciousness is a function of the mind that is active at the time of death. It links two consecutive lives in different planes of existence or in the same plane, determined by the kamma of the previous javana process. It is rooted in the generic mental factors of latent ignorance and latent craving, which propel rebirth around the long samsara, which is well grounded in ignorance. Rebirth-linking consciousness occurs only once in any lifetime, at the time of conception. It takes the heart as its base in the sense-sphere plane, for example birth in the human world. In the case of spontaneous birth, there is no base, for example rebirth linking to an immaterial plane. It apprehends the object presented at the last cognitive process of previous existence. Such processes may appear as a sign of kamma or a sign of destiny. According to the relation of dependence (No. 8), rebirth consciousness arises dependent upon the heart base that coexists with it, as there is no pre-existent physical base at that moment. It is the same for the life continuum, which takes the same heart base in reverse order, creating consecutive previous births according to the relation of basic pre-existent dependence. This phenomenon is different from the concept of *ex nihilo* (out of nothing) because in Buddhism, nothing cannot arise on its own. In this scenario, pre-existent dependence relations track down past human forms; the originating kamma that led to a birth in an immaterial plane would then be exhausted. In earth years, this timescale can be very lengthy and hard to comprehend, but compared to the lifespan of the universe and other celestial worlds, it would be a small number. Relative measures of time and space are not applicable to this phenomenon.

Rebirth linking is first mentioned to make it easier to understand and examine the beginning of one life period. Birth and death, and death and rebirth follow the same sequence. The nature of one's next life could alter radically, for better or worse, at the point of death. For example, in the case of the present life, which is now about to end, it takes as its object death-proximate kamma. The deciding factor is the quality of kamma present in the death-proximate consciousness. It is a purely mental process that does not manifest outwardly. The object of the bhavanga changes only at the time of rebirth-linking consciousness. On a very rare occasion for a rare individual, this object could be the same as death consciousness, maintaining the same personality. This phenomenon is known as reincarnation and is not widely accepted by the Theravada School.

Life Continuum (Bhavanga)

The Pali word *bhavanga* is derived from the root words *bhava* and *anga*. *Bhava* means "existence" (in any of the planes), and *anga* means "factor". It is translated as life continuum because the bhavanga flows like a stream without remaining static for any two consecutive moments. It carries factors unique to an individual life from conception to death, maintaining and preserving the individual's character, conditioned by the kamma of a previous existence. It is the resultant consciousness of previous rebirth-linking consciousness at the time of birth. It has the function of preserving the continuity of an individual's existence. As a factor of existence, it retains all conditioned mental factors of that being and serves as the mind door.

There are different opinions given by Pali commentators regarding the precise definition of the mind door. Whether the mind door is adverting to consciousness, bhavanga, or a combination of both, it could be understood as a state of deep sleep. For example, in the mother's womb, from the time of conception to the time of birth, the baby has no active consciousness and is in a state of bhavanga. Given below are three explanations of bhavanga.

Since time immemorial, all impressions and experiences have been stored up, or better said, functioning. But these are concealed from full consciousness, where they occasionally emerge as subconsciousness (paranormal psychic phenomena). It should be noted that the state of bhavanga citta is a kamma-resultant consciousness (*vipaka*). The main function of bhavanga is death consciousness and rebirth linking, where its object changes from one life period to another by means of kamma.[13]

The factor of existence is the indispensable condition of existence. It preserves the continuity of the individual's existence. Bhavanga citta arises and passes away at every moment of life when there is no active cognitive process taking place.

When an object comes into contact through a sense door or the mind

---

[13] Nyanatiloka Thera, *Buddhist Dictionary*, p. 37.

door during the normal course of life, there occurs a mind moment called *bhavanga calana*. This means "vibration", like an eardrum vibrates as a sound wave comes into contact with the ear. The cognitive process goes through several steps, firstly past bhavanga, then vibrational bhavanga, and on to arrested bhavanga, which cuts off the flow of bhavanga. Then follows another citta adverting to the object. Then comes the rest of the fourteen cognitive steps, before falling back to stream of bhavanga. Since every cognitive process of worldly beings and trainees produces kamma, bhavanga gets conditioned in every cycle of cognition, thus maintaining and sustaining existence. Throughout the course of life, this occurs countless times between active cognition processes, renewing its factors of existence each time. This is similar to a dynamo producing an electrical current.

Essentially the nature of bhavanga for a given lifetime is determined by the last full consciousness process which immediately precedes life. This last process is strongly influenced and conditioned by the kamma performed by the being during his or her life.

Death Consciousness

The Advent of Death

> Death is formally defined as the cutting off of the life faculty (*jivitindriya*) included within the limits of a single existence.[14]

Death can be explained in terms of kamma. When a new life is born into a realm, the lifespan is fixed by the kamma of the previous life. For example, in the human realm an average lifespan is about seventy to eighty years. An individual, however, may exist many aeons within the fine material sphere of existence. A lifespan is determined by one's productive kamma; this generates the rebirth-linking consciousness at the moment of conception. When death occurs, if the productive kamma is still not exhausted, the kammic force can generate another rebirth on the same plane. Premature deaths and child deaths are examples of residual kamma

---

[14] Bhikku Bodhi, *A Comprehensive Manual of Abhidhamma*, p. 220.

generating a rebirth. Kammic force can also prolong the lifespan if there are favourable conditions. When both the lifespan and productive kammic force simultaneously come to an end, death comes by the expiration of both. Powerful destructive kamma can cut the force of productive kamma.

Commentators give the following analogy to illustrate death:

> "An oil lamp, for example, may be extinguished due to the exhaustion of the wick, the exhaustion of the oil, the simultaneous of both, and some extraneous cause, like a gust of wind."[15]

Death (Cuti)

Death consciousness is the last citta to occur in an individual's existence; this is the consciousness which marks the exit from a life. Like rebirth linking and bhavanga, it is in the class of process-freed. It is a passive flow of consciousness outside an active cognition process; its function is the passing away from a preceding life to the next life. This death consciousness takes the same object of bhavanga that the presently ending life had. At the same moment, javana of the death-proximate consciousness generates rebirth-linking consciousness that binds the dying person to an incessant round of renewed life. According to the Theravada School of Buddhism, there is no gap between death and rebirth consciousness.

Death-Proximate Cognitive Process

Signs arise at the time of death in the cognitive process of the dying individual. These vary according to circumstance and by the power of kamma, which can enter through any of the six (sense) doors. They symbolise some action performed during the life or previous lives depending on the severity of that action:

  i. A kamma that is to produce rebirth linking in the next existence.
  ii. A sign of kamma that appears in a particular form, for example as a visual image usually apprehended previously at the time of

---

[15] Ibid.

performing the kamma or something that was used in performing that kamma, such as a knife or gun used to kill someone in the case of a murderer.

iii. A sign of destiny (a symbol of the state) that is to be obtained and experienced in the immediately following existence, for example a heavenly chariot. Transition to an immaterial plane of existence is usually a sign of destiny in the case of a meditator.[16]

These signs appear in the mind of the dying person's javana process and are one of the last objects that are processed by the death-proximate consciousness, so they are weaker than normal. The sign is also the same object to be apprehended by the rebirth-linking consciousness and bhavanga in the next existence.

A sign of kamma and a sign of destiny depends on the purity or corruption of the mind. It is determined by the weight of kamma and is in conformity with the state which one is to be reborn. This is a renewal process of past kamma and appears to the mind door as if it were being done at that very moment; therefore, it is very powerful and beyond the control of the dying person. With the ceasing of the death consciousness, the life faculty is cut off. This stage is known as "clinical death". Certain physical functions may continue, such as body temperature, but these soon end when life has passed, and one is pronounced dead. The corpse is then subject to decay and dissolution.

The Death Consciousness (Terminal Consciousness) (Cuti-Citta)

This is the last consciousness of any life term. The process of this last consciousness is not straightforward. Its function is the passing away from the present life. At this moment of time, both the sense avenue and mind avenue are cut off. The death-proximate consciousness subsides, lacking strength, as all productive kamma has exhausted itself. At the end of death-proximate cognition, death consciousness arises, performing its function of passing away from the present life. The object of death consciousness is rebirth-linking and bhavanga consciousness of the present life that is now about to end.

---

[16] Ibid. pp. 221–2.

Death consciousness should be understood in two parts. First is the death-proximate cognitive process and its object, which is one of kamma, a sign of kamma, or a sign of destiny. This is triggered by one of the six senses at the time of death. The second stage is the function of passing away and taking the object at its birth. There is one exception to the rule: when the object of death consciousness in the presently ending life is not the object of the death-proximate consciousness of that life but instead is that of a previous life. For clarity, death consciousness is termed terminal consciousness.[17]

The three types of process-freed consciousness that activate at the time of rebirth and death are purely determined by kamma. The last opportunity for a person to acquire a good deed is death-proximate (*asanna*) kamma. For this reason, it is customary in Buddhist cultures to remind the dying person of his or her good deeds. It is considered that the dying person is fortunate to receive a blessing from a Buddhist monk because this has a direct influence on generating rebirth in a favourable realm.

Wholesome kamma produces rebirth in fortunate realms, whereas unwholesome kamma will result in a birth in a woeful plane, which ripens more unwholesome kamma, producing painful results. The order of kamma is a fixed law. The four main types are weighty, proximate, habitual, and performed (by a special act/vicarious act); they ripen in this order in accordance with *kamma niyama*.

The functions of death consciousness, rebirth-linking consciousness, and bhavanga citta are not transparent. It could be compared to a train going through a tunnel where the old train becomes a new train with a new engine going to a new destination. The process goes through three stages with unbroken continuity. According to Theravada Buddhism and the Abhidhamma, there is no gap between consciousnesses. At the time of a being's death, there arises death-proximate consciousness taking its object kamma, a sign of kamma, or a sign of destination. It is shorter and weaker than normal consciousness. The mode of consciousness is a similar process to normal consciousness, which is followed by two registrations of the bhavanga citta. Its function is to pass its way to the next life. The life faculty cuts off at the end of the terminal consciousness. The Pali term *cuti* means "vanished", "leaving the body without life

---

[17] Ibid. p. 225.

function". The heart stops and the last breath passes without a following in-breath. In apparent immediacy, the consciousness relinks with a new life in any of the four planes of consciousness. This irreversible process follows the law of Dhamma (Dhamma Niyama) in discrete stages. It is arguably neither a new life nor the old life that is reborn. It is a continuation of consciousness in different time zones and on different planes of existence. It cannot be predicted exactly when or where the rebirth will take place.

The bhavanga citta is not separate but coexistent with the active consciousness. It flows like a river beneath the active mind, an undercurrent. However, bhavanga citta gets interrupted every time an object comes into contact with it. The silent subconscious mind begins to vibrate and shuts off. An active process takes place in several stages, and at the end of that cycle the experience gets registered in the bhavanga. A total number of seventeen stages are explained in this process of active consciousness. Bhavanga citta should be understood as a state of deep sleep, a dreamless state of mind. It could liken to being under anaesthesia or fainting without any use of substance; it is a gap period between two active consciousnesses. The Buddha has stated that he could not find a beginning or end to this flow, only advising his students to be mindful of the present moment. All past experiences, memories, and kamma are flowing in this subconscious mind.

Time and time again the Buddha pointed out that suffering or happiness is a direct or indirect result of one's own actions. The longer one stays in samsara, the longer one's suffering. There are dangers in clinging to the five aggregates of pleasure because these bind an individual to the mundane world. The Noble Eightfold Path propounded by the Buddha is to help transcend the range of clinging, with the right view to end suffering. An explanation of the process-freed consciousness gives a comprehensive and detailed analysis of the cyclical rebirth of samsara and the role of kamma. The law of kamma dictates the range of planes and realms into which beings are born. This accords to kamma ripening, which occurs when the right conditions are met. Everyone, including the Buddha and Arahants, is subjected to the law of kamma, although not all have rebirth-linking consciousness.

Magga Puja Day

Magga Puja is an extraordinary event happened at the time of living Buddha. One February in India, the Buddha addressed a spontaneous gathering of 1,250 enlightened monks who had visited him without a prior appointment to see their master. On this day the Buddha showed them the importance of good conduct as related to purity of the mind. Summarising his teaching, the Buddha gave them the Monk's Code of Conduct. This is based on the principle of nonviolence and on promoting goodwill for the benefit of many. The Monk's Code of Practice is known as Patimokka Samvara Sila, or Vinaya, meaning discipline. It is the upholding of restraint and excellent moral conduct. When practised, this takes a worldly or a trainee to a state of peace and the ultimate happiness of nibbana.

For laypeople, the Buddha gave five precepts as training rules. Purity of mind reflects in good conduct in such a manner that one's actions cause no harm to anyone. The summary of all the Buddha's teachings is to do no evil but to do good and purify the mind. These three summary teachings were given in accordance with the law of kamma. These three main tenets of Buddhism are interrelated, the underlying principle being purity of the mind. Owing to his compassion, the Buddha showed us the way to reduce suffering by cutting off the entanglement born of attachment, craving, aversion, and clinging. He explained the consequence of both wholesome and unwholesome kamma and finally showed us how to break the cycle of birth and death. The Supreme Buddha found the way to deliverance, and his role is nothing other than as a teacher to others.

CHAPTER 18

❊

ANALYTICAL AND SYNTHESIS METHODS OF CONSCIOUSNESS

THROUGH STUDY AND ANALYSIS, ONE comes to understand consciousness, mental factors, functions, and their broad classifications. Meditation practice is a method of tracing those good qualities, harnessing the inherent power of the mind and shaping it for the purpose of liberation. Buddhist philosophical and psychological analysis identifies the mental faculties as corresponding to the Noble Eightfold Path, which provides guidelines to development. By correct meditation practice, one can progress to reach the factors of enlightenment. This chapter discusses specific applications of tranquillity that lead to insight, explaining these in terms of relations between path factors and psychological faculties.

The synthesis method considers the significance of a single factor that can only be understood in terms of its contribution to form a connected system, each factor playing a part in achieving the collective goal. For example, an orchestra is not made from one talented player or one recital, but a large repertoire, a number of trained players, and many types of instruments, all playing in harmony.

There are seven factors of enlightenment: mindfulness; investigation of mental states; energy; zest; tranquillity; concentration; and equanimity. These factors need to be developed as a whole and complete system to have valuable properties, over and above those of the system's single factors.

All of these are beautiful mental factors. Investigation is the designation of wisdom. Zest, tranquillity, and equanimity are specific functions of the path. The rest are faculties of themselves. These three functions of the path continue the work of purifying the consciousness begun by mental absorption (*jhana*). Energy and zest make the mind energised against slow movement and make investigation more vigorous; equanimity is mental neutrality towards changing phenomena, and tranquillity supports its function by neutralising emotions. Mindfulness is the supreme factor which balances all other functions. Path factors that correspond to each faculty guide the development towards a specific destination, namely enlightenment.

The Five Faculties

- Faculty of wisdom
- Faculty of energy
- Faculty of mindfulness
- Faculty of concentration
- Faculty of faith

The Dhammasangani analyses and lists all the phenomena that exist in a good consciousness at any given time. The method is purely descriptive, analytically focusing on a single unit of consciousness, combined with the mental factors associated with it and its stand-alone function.

On the other hand, the Patthana, the book of the philosophy of relations, deals with the conditionality between other phenomena. It explores the influence of a single type of consciousness on its immediate conjunction and on future events. The function of the five faculties aligns according to the single potentiality of each in relation to other faculties. This approach considers how each aspect performs given the interactions between them all, for a positive and beneficial outcome or for the attainment of Arahantship. It is important to understand both the analytical and synthesis method of Dhamma to stay on the path.

Following are the relations between path and faculty:

Path factors	Faculty (spiritual) (indriya)
Right view	Faculty of wisdom
Right thought	Factors of absorption and *vitakka*—thought conception are in the Faculty of concentration
Right speech	Faculty of mind
Right action	Faculty of mind
Right livelihood	Faculty of mind
Right effort	Faculty of energy (*viriya*)
Right mindfulness	Faculty of mindfulness (*sati*)
Right concentration	Faculty of concentration (samadhi), mental one-pointedness

The other faculties to be developed initially are faith (*saddha*), the mind faculty of ideation (*mano-indriya*), and the formation of ideas or concepts. The life faculty (*jivita-indriya*) facilitates life and supports growth and development of all other faculties.

Faculty of Faith (Saddha)

To have faith in Buddhism is to place complete trust and confidence in the triple gem. This gem equates to a gem that is cleansed in water. The triple aspects are the Buddha, Dhamma, and the order of enlightened monks. These are taken as refuge, and a trainee makes a confirmed commitment to undertake the practice of Dhamma. By this the trainee has the capacity to develop faith. It is the confidence that allows one to make an effort, reach out to search for what is needed, and make a firm resolution and have conviction. This psychological faculty of the mind allows the trainee to obtain knowledge and put it into practice to progress and also to overcome doubt. Faith is explained separately because it is the entry point to the path factors.

Pali terms for five main faculties of meditation are given below:

1. Faculty of faith—saddha indriya
2. Faculty of energy—viriya indriya
3. Faculty of mindfulness—sati indriya
4. Faculty of concentration—samadhi indriya
5. Faculty of wisdom—panya indriya.

The development of the first five faculties in the order given above is important for progress because they are in conformity with the order of Dhamma (Dhamma Niyama). For example, mindfulness cannot be established without an effort, and there will be no effort without confidence (faith). Each is a supporting condition for the one below it. They can be understood as psychological factors or positive mental energies that flow in a specific order. Each can be improved by the correct method and practices recommended by masters of Buddhist meditation. Each faculty exercises control over its corresponding hindrances. Through practice of jhana, they become power factors, until they become factors of deliverance (*niyya atthana*), which lead out of samsara.

"Being tranquil in mind, he finds concentration."[1] Right concentration includes the practice of mental absorptions. A purer consciousness brings the tranquillity and stability needed to reflect on the dynamic reality of the object under observation. With such mindfulness, one can see the object clearly without errors of perception, as it is presented; at such a moment the meditator must change to momentary concentration. Intermittent practice of insight and calm cultivates other mental factors towards a perfect balance, conducive to deliverance. By practising jhanas, the meditator can develop perfect calmness and tranquillity of mind to prepare for entry into higher jhanas. Tranquillity is the seed present in every wholesome consciousness that can grow to full stature in the factors of enlightenment.

"Tranquillity (*passadhi*) as a factor of enlightenment (*sambojanga*), when perfected, belongs to a supramundane consciousness."[2]

Factors of Absorption

A clear account of the factors of absorption can be found in the Abhidhamma's classification of jhana consciousness and enumeration of Dhammasangani. The Buddha taught the venerable Sariputta psychological analysis of meditative absorptions. Details of the work he undertook are

---

[1] Digha Nikaya 9.
[2] Nyanaponika Thera, *Abhidhamma Studies: Researches in Buddhist Psychology* (Kandy, Sri Lanka, Buddhist Publication Society, 1965).

documented in the Anupada Sutta.[3] The analysis given in the Atthasalini also supports the same matter in the same order. It was enunciated by the Buddha to his analytically and philosophically gifted disciples. Venerable Sariputta was said to be the first Abhidhamma practitioner under the instruction of the Buddha who undertook the analytical enquiry. In the same discourse, the Buddha gave the name for the order of analysis: Anupada, Dhamma, vipassana, "insight into the things taken one after the other". The venerable Sariputta stated the factors of absorption in the correct order, conforming with the Mahasattipatthana Sutta (Four Foundations of Mindfulness). The Buddha praised the venerable Sariputta for his skill and analytical knowledge of Dhamma. This was expounded in his "Dhammanu Sati"—Mindfulness of Dhamma.

"The things occurring in the first absorption, namely thought conception, deliberation, interest, rapture, happiness, mental one-pointedness, and sense impression, are the order of factors."

From this passage we can propose that insight knowledge and wisdom can be applied in the context of samatha meditation to understand mental states and direct them towards developing and strengthening mental qualities. These are the first five factors and the main characteristics of jhanic consciousness. The qualitative factors and their resultant consciousness differ according to the stage of attainment of the trainee. There are forty-two types of consciousness that are explained in the Abhidhamma.

The following tables provide a summary of jhanic consciousness: Table of Mental Absorption, p. 28, and pp. 64, 70, and 70 of *A Comprehensive Manual of Abhidhamma*.[4]

Buddha gave subjects of meditation, for example the fundamental expositions which are recorded in the books of Mahavagga of the Samyutta Nikaya, to individual monks. He explained the doctrine or these maxims from time to time, but otherwise left them to practise meditation according to their own strengths and abilities.

The Pali term for mental states pertaining to consciousness is rendered as *citta*. This word denotes a specific aspect of the mind for meditation practice. It refers to a function that purifies the mind in the classes of the fine material

---

[3] Majjhima Nikaya 3, ibid. p. 66.
[4] Bhikku Bodhi. (1999), *A Comprehensive Manual of Abhidhamma*. BPS.

and immaterial spheres. Hence those mental states are called wholesome consciousness, *kusala citta*, meaning they are able to cut off hindrances.

The word *kusa* comes from the name of a sharp double-edged blade of grass found in India and tropical countries which can cut through like a surgical knife with precision (botanical name *Eragrotis cynosuroides*). It has come to denote the presence of wisdom in consciousness, capable of cutting through ignorance.

Dhammasangani enumerates fifty-six mental functions which are present at any moment when a wholesome consciousness arises. All forms of wholesome consciousness are accompanied by joy and associated with knowledge (of doing good). In the sensuous sphere (*kama vachara*), consciousness arises by any of the six senses. As the trainee advances in the practice, consciousness may move to different spheres, namely the fine material sphere (*rupa vachara*) or immaterial sphere (*arupa vachara*), and so on. All of these and the associated mental states are exclusively wholesome.

The Dhammasangani further classifies the fifty-six mental functions into thirteen different categories. The Atthasalini gives an additional nine wholesome functions, thus totalling sixty-five functions. These are called the "List of Dhammas". The qualitative or kammic value of wholesome consciousness is determined by way of its wholesome roots. There is a difference between "association with knowledge" and "dissociation from knowledge"—when the factor of nondelusion is present or absent respectively. Wholesome consciousness can arise spontaneously or by being prompted by a mentor. It depends on one's past experiences, one's association with good company, and one's society, and arises for example by discussing Dhamma, being interviewed, or listening to Dhamma talks.

The Five Helpers of Right View

No. 97 of the Numerical Discourses of the Buddha outlines five things that help right view.

> Right view, O monks, if it is helped by five things, has liberation of mind as its fruit and is rewarded by the fruit of liberation of mind; it has liberation by wisdom as its fruit and is rewarded by the fruit of liberation by wisdom.

What are those five things?

Here, monks, right view is helped by virtue, by wide learning, by discussion (of what has been learned), by tranquillity, and by insight.

Similarly, Discourse No. 98 describes five bases of liberation corresponding to factors of absorption and tranquillity: "When he [the meditator] gains such experience, gladness arises. When he is gladdened rapture arises; for one uplifted by rapture the body (mental body) becomes calm; one calm in body feels happy; for one who is happy the mind becomes concentrated."

The Synthesis Method of Consciousness

Paccaya Dhamma

Function and functionality of the consciousness and how it relates to other mental states is explained in the philosophy of relations. There are twenty-four relations explained in this treatise, one of the largest and most profound teachings of the Buddha. This is only a very basic outline of the method and treatment of these relations, given here to help the reader understand some complex mental phenomena during the meditation practice. This involves the process of mental absorption, insight into mental formations, and Dhamma. One can then come to realise the doctrinal teaching of the Buddha without drifting away from the Noble Path.[5]

These relations are explained in the context of consciousness and mental factors. There are various combinations of body and mind in relation to the process of consciousness. There are two methods of explanation, the Abhidhamma method and the Suttanta method. Suttanta means the methods described in discourse.

Consciousness is classified into different classes, planes, and spheres. These are explained in relation to the element of the process and hence how each type differs from another.

Citta regards the state of mind as a dynamic activity. The nature of

---

[5] Mahathera Ledi Sayadaw, *The Buddhist Philosophy of Relations: The Patthanuddesa Dipani*, tr. Sayado U Nyana (1965).

this activity is the most important to investigate in meditation practice. Each cycle of consciousness has at least three stages of arising, maturing, and passing away. At the end of this cycle, it may change to another type of consciousness depending on the intention. Consciousness arises together with a constellation of associated mental factors and vanishes together with them. Each consciousness differs, depending on the associated factors. From observation of this process, a trainee may be able to gain insight. With this, one can correct the wrong views of permanency and self-ego, which are projected by a continuity of activity and delusion.

The twenty-four types of relations make clear how the consciousness is related to other elements within it *(chapter 14)*. In this explanation the mind is referred to as *citta*, translated as "consciousness". There is one exception, where the mind is related to food (nutriment), namely the relation of food, which is a material element. The Abhidhamma examines, classifies, and explains consciousness in detail. Under the "citta" compendium, 121 types of consciousness are explained. On the other hand, the Book of Relations identifies causal relations of consciousness. This explores how consciousness is determined in relation to other happenings. This is implicated by means of three phases of the cycle, which are producing (rising), supporting (maturing) or maintaining, and disappearing. Consciousness is a function, and its relations help to identify its functionality. It thus makes a related impact on the mind according to a specific function.

The twenty-four types of relations in this compendium of the Abhidhamma explain in detail how a given consciousness (state of mind) is determined. This method is a causal classification of consciousness according to the various influences of the mental factors associated with it, for example the relation of root (*hetu paccaya*), where consciousness is rooted in greed, hatred, delusion, and their respective opposites, namely generosity, amity, and wisdom. The state of mind is influenced by those roots and exhibits specific qualities which manifest in behaviour.

"Suppose a woman with a wish to attain nibbana offers food to a monk in generosity. As long as she performs this act of generosity, her words and thoughts regarding the offering are rooted in generosity, amity, and wisdom."

Consciousness is a dynamic process. It has specific characteristics which go through three stages of arising, persisting, and ceasing. This sequence of occurrence follows the principles of cause and effect in momentary succession. One moment of consciousness is determined, for example, by way of its roots. It also determines the next successive moment. Similar to a wave in the ocean passing its influence on to the next wave, each of the three stages produces, supports, and maintains its successor. There is a clear correlation between what follows, but significantly different classes of new consciousness can be born. These appear in relation to their own function and purpose, making the next cycle very diverse because of changing conditions. Being mindful, a meditator can be aware of the changes and continuation of this process.

For an insight meditator who takes samatha–vipassana as a vehicle, any type of life process of consciousness is an impersonal continuum devoid of ego or personality views. The Buddha had instructed his disciples in the Mahasatipattanhana Sutta to be skilful in the observation of Dhammas.

Attention should proceed in two phases, as follows:

Phase 1

a) Contemplation of phenomena (*nama-rupa*), as appearing in individual practice within trainees.
b) Phenomena appearing in others.
c) A combination of both.

This synthesis method is recommended for understanding "nonself" or insubstantiality of phenomena.

Phase 2

a) Arising of phenomena (Samudaya Dhamma).
b) Passing away of phenomena (Vaya Dhamma).
c) Combination of both (Samudaya–Vaya Dhamma).

This analytical method is used to break up wrong identification of personality view. Both the synthesis and analytical method would help one to understand and realise the doctrine of "nonself".

Classification of Dhamma (phenomena) is presented in the

Abhidhamma method and Suttanta method. There are slight differences in the two methods. The Abhidhamma method appears to be more compatible with discourses (Sutta Pitaka), and many teachers prefer to use these explanations. Here the writer has used Abhidhamma and Visuddhimagga as a source of knowledge, perceiving that they are complementary to each other. It is stated in the Atthasalini that the Buddha had taught Venerable Sariputta Abhidhamma in a heavenly realm. The factors of absorption correspond to one of the twenty-four modes of conditionality, namely how the consciousness (citta) is conditioned by way of its relation to absorptions (*jhana paccaya*, No. 17). Fine and immaterial classes of consciousness coexist with their mental factors, therefore exhibiting unique, distinguished characteristics. Like an archer aiming an arrow at the bullseye, the meditator directs his or her mind towards the object, firmly fixes his or her attention, and closely contemplates it. All these body and mind moments are coordinated by way of relations. The factor of sustained application of thought (*vicara*) has the characteristic of reviewing the object again and again and to become able to observe its salient properties. It also explains the first five factors of absorption in the same order capable of lifting the attainer to heavenly realms. Functions F53 and F54 in the Dhammasangani are calm (samatha) and insight (vipassana). These two functions are grouped together under the heading "The Pairwise Combination (Yuganaddha)". It means tranquillity leading to insight and vice versa.

From these analyses it can be ascertained that mental absorptions, if correctly practised, serve as a basis to develop insight. This would improve the quality of mental factors to perform a specific function in continued succession until perfection is achieved.

Tranquillity gained from jhana is the seed present in every wholesome consciousness that, when nourished by insight knowledge, can grow to full stature in the factor of enlightenment. Tranquillity which manifests as calmness brings the stability to continue further development of consciousness. In Samyutta Nikaya 46.51, the Buddha expressed his recognition of tranquillity: "Monks, there is tranquillity of the mental factors (*kaya-pasaddhi*) and tranquillity of consciousness (*citta passaddhi*)."[6]

---

[6] Bhikkhu Bodhi, *A Comprehensive Manual of Abhidhamma* (Kandy, Sri Lanka, Buddhist Publication Society), p. 88.

Its place in the list of Dhamma is in the group of "the six pairs" (yugalaka). "Tranquillity (*passadhi-sambojjhanga*), when perfected, belongs to a supramundane consciousness (*lokuttara-citta*)."[7]

Tranquillity is a state where the hindrances of agitation and restlessness are absent and therefore one is calm, a composed condition given to the six pairs. Then the other ten qualitative factors, for example straightness and gentleness, complement the refinement process of mental factors and the accompanying consciousness. These wholesome factors provide steadfast progress on the path. Their function is to be the harmonising of spiritual faculties. "With consciousness thus purified and cleansed, without blemish and stain, pliant and workable, steady, and unshakable, he [the meditator] turns his mind to the extinction of passion."[8]

The Buddha has given a simile for this process of purification, likening it to the process for refining gold, illustrating the consistency of mind necessary for the purpose of spiritual development.[9]

"Monks, there are five defilements of gold, owing to which gold is not pliant, not workable, impure, [and] brittle and cannot be well wrought. Likewise, monks, there are these five defilements of the mind, owing to which the mind is not pliant, not workable, impure, [and] brittle and cannot concentrate well upon the extinction of passion. Which are those five? Sensual desire, ill will, rigidity, and sloth. Agitation, worry, scepticism, these are the defilements of the mind, owing to which suffering continues."[10]

"Each wholesome thought, but especially the systematic culture of the mind (*bhavana*), is as it were, a process of elimination and refinement by which the gold of consciousness is gradually freed from blemishes and dross, and so brought to its true purity, as stated by the Buddha in the following words."[11]

"Monks this consciousness is pure (or: luminous, *pabhassaram*) (i.e. bhavanga), but is defiled by intrusive (or alien, *agantukehi*) defilements and it is (now) free from intrusive defilements" (Anguttara Nikaya-Nipata).

---

[7] Nyanaponika Thera, *Abhidhamma Studies: Researches in Buddhist Psychology* (Kandy, Sri Lanka, Buddhist Publication Society, 1965), p. 82.

[8] Majjhima Nikaya 51, ibid. p. 89.

[9] Ibid. pp. 89–90.

[10] Samyutta Nikaya 46.33, ibid. p. 90.

[11] Ibid.

CHAPTER 19

MENTAL FACULTIES

The List of Faculties (Indriya)

THERE ARE EIGHT FACULTIES STATED in the Dhammasangani. Their common function consists of exercising a dominating, governing, or controlling influence over the other mental factors associated with them:

- F11—faculty of faith
- F12—faculty of energy
- F13—faculty of mindfulness
- F14—faculty of concentration
- F15—faculty of wisdom
- F16—faculty of mind
- F17—faculty of joy
- F18—faculty of vitality (life).

The Pali word for faculty, *indriya*, is derived from the root *indro* (Sanskrit: *Indra*), meaning "lord", for example, "Faith exercises lordship under the sign of devotion" (the Attasalini).

Meditation practice is the cultivation of the first five faculties. When they are in balance, meditation continues. These factors are also known as the five spiritual faculties. They are called *indriya* because "they master their opposite", that is they keep the five hindrances under control. For example, faith (or confidence) brings keen interest; it is an eagerness to learn that makes for the consistency of faith. Faith also controls indolence; it is from joy that faith derives a good part of its conquering power. Mindfulness controls heedlessness, concentration controls agitation, and wisdom controls ignorance.[1]

The faculty of mind refers to the sixth sense, it is identical with consciousness (mano), where cognition and grasping occurs. Ideas and concepts are formed in this faculty of mind which are also influenced by roots. According to the Abhidhamma's classification and the philosophy of relations, consciousness arises with mental factors and then vanishes with them. Each cycle of consciousness is conditioned by the accompanying mental factors, passing that influence of conditioning onto the subsequent cycle. The indriya quality of consciousness is capable of control over direct purposeful bodily movements and vocal actions, which are coordinated according to the factors associated with consciousness. This quality of the mind can further develop to its peak on one of the four roads to power (*iddhipada*), for example by repetitive practice (*asevana paccaya*, No. 12). When properly trained, the mind has prominence as a "predominant factor" (*adhipati*) over the object (*adhipati paccaya*, No. 3). Purification is the condition of all agreeable objects of good consciousness which have dominant relation. This can be developed by close contemplation of the object and can be improved by knowledge. The function of a faculty is to exercise control over its opposite hindrance that would otherwise retard the functionality. For example, the faculty of mindfulness (F13) has control over distraction; concentration (F14) controls agitation; and wisdom (F15) controls ignorance. Faculties can further develop until they have complete sovereignty over body and mind.

Jhana is the strong absorption in an object. By a practice that uses various *kasina* (meditation objects), the function of the *jhananga* (factors of absorption) intensifies concentration to a level that allows the mind to be absorbed into the object chosen, particularly the factor of mental one-pointedness (*citta ekaggata*), which firmly establishes the object in the mind without any wavering. A high level of concentration can suppress and counteract all the five types of hindrances. Mental one-pointedness is the main force of intensification and final absorption. This faculty

---

[1] Bhikkhu Bodhi, (1999) "The Faculties", *A Comprehensive Manual of Abhidhamma* (Kandy, Sri Lanka, Buddhist Publication Society), p. 65.

can temporarily purify the consciousness. Establishing and maintaining agility of mind to quickly reach concentration is the most decisive avenue for any further development of consciousness. The systematic practising of jhana can improve the function of the six pairs to assist in the removal of hindrances. This develops progressively with noticeable changes in a trainee's personality.

Practice of jhana requires the careful development of concentration and clearly defined objectives. Trainees should be wise enough or should practise under an experienced insight meditation master to understand the limitation of mental absorptions. The Buddha advised and discouraged any further developments of concentration; for example, he didn't justify the acquisition of psychic powers and showed the implication of using such powers for worldly gains. Any further intensification of consciousness should be directed at the realisation of the Noble Eightfold Path and the avoidance of intellectual knowledge. There are three principal categories of the Buddha's teaching:

1. Study of the Dhamma - *Pariyatti (sutta reading)*
2. Practice of Dhamma - *Patipatti*
3. Realisation of Dhamma - *Pativedha.*

To realise Dhamma is to gain right understanding, pertaining to the true characteristics of all phenomena. Trainees need to be guided towards the growth of insight. Sharpening mental one-pointedness only leads to the attainment of full absorption (appana samadhi). Instead of this skill of absorption, concentration should be used to develop the seven powers through developing spiritual faculties. These faculties and their corresponding powers are listed in the Dhammasangani. The grouping of the list is more than a formal principle of arrangement. It gives the function of a factor within a given state of consciousness and within the group of factors. The groups register the common purpose of the various single factors or functions. For example, F11–F18 are faculties and F24–F30 are powers.[2]

There is a danger in developing jhana without realising nibbana,

---

[2] Nyanaponika Thera, (1965). *Abhidhamma Studies: Researches in Buddhist Psychology* (Kandy, Sri Lanka, Buddhist Publication Society), pp. 31, 58.

the final objective of the path. Development of faculties and their corresponding powers should be synchronous with the path factors. In the list of Dhamma, power factors are given after the path factors. These factors can be regarded as a toolkit. Only an experienced and skilful technician would know their functions and how to use them; in the same way, a skilful carpenter would use his or her tools for a specific purpose to shape the wood. Skilful meditation absorption must shift to vipassana to develop insight. Concentration is considered as right concentration when applied to develop insight wisdom. Within the consciousness of a human being there lies a danger of sinking to the downward path, to the animal or hell realms. On the other hand, if properly nurtured to the full growth of insight, one can reach the noble (Ariyan) state of mind. The intensity of consciousness can be improved and increased by spiritual powers known as the Four Roads to Power (*iddhipada*). "Here the intensity of consciousness is increased to such a degree that magical powers (*iddhividha*) may be obtained, giving a far-reaching control over mind and matter."[3]

There are stories in Buddhism that show the acquisition of supernormal powers by certain monks. The story of Devadatta tells us about the abuse of such powers for worldly gains, which led him to burn in hell. The Buddha limited his teaching to suffering and its end. Henceforth, trainees should endeavour to direct their energies to attaining nibbana. There is a fine line between the right and wrong path. Such a decision of one's own choice should not allow for the chance of success or for trial and error at this stage of development. Although the kammic value of all these factors is wholesome, the presence or absence of knowledge depends on direct personal participation and the accumulation of knowledge in the faculty of mind (F16). If anyone lacks knowledge, then it is advisable for him or her to follow the advice and guidance of the enlightened Buddha. It is mainly the fifth factor of mental absorption, mental one-pointedness, which needs special cultivation to direct it towards gaining insight wisdom. For insight meditation, objects are dynamic aspects of body and mind. Momentary concentration, application of mindfulness, and clear comprehension are required to comprehend its true nature; these allow one to overcome the errors that conceal the truth of changing phenomena.[4] This group is

---

[3] Iddhipada Sutta.
[4] Bhikkhu Bodhi, *A Comprehensive Manual of Abhidhamma*, pp. 51–2.

known as "the helpers" *(upakaraka)*. There are two other factors, fear, and shame of wrongdoing. These are termed as "the guardians of the world" *(lokapala)* because they protect the mind from swinging too far from meditation practice and keep it on track. Self- discipline is very important for success in this path of enlightenment.

The Path Factors

The scheme of classification in the Dhammasangani lists only five of the factors of the Noble Eightfold Path. The other three, namely right speech, right action, and right livelihood, are moral training factors. It is assumed that one who seeks a higher development of consciousness already has a firm moral foundation. This is also consistent with the Visuddhimagga (Path of Purification). The great commentator Archariya Buddhaghosa clearly stated in his treatise that "this teaching is for a man who firmly stands on his virtues."

Four path factors are different aspects of the corresponding spiritual faculties. The fifth factor, right thought, is an exception; it is identical with thought conception *(vitakka)*, the factor of absorption, can be included in the faculty of concentration.

> Right understanding ≏ faculty of wisdom
> Right effort ≏ faculty of energy
> Right mindfulness ≏ faculty of mindfulness
> Right concentration ≏ faculty of concentration
> Right thought ≏ thought conception (vitakka)

Path factors are directions towards the spiritual faculties, which are identified by their corresponding powers. The word *right* in this context is a value attribution to the factor; it represents that the faculty is directed towards liberation. The term implies that knowledge and wisdom are present in the consciousness, to guide it to purification from defilements. There is a range of obstacles manifested in different forms, such as defilements, hindrances, fetters, bonds, and knots; each would slow down progress. These are used as mental objects for mindfulness in vipassana meditation, to see clearly without falling to the errors of perceptions. The spiritual faculties are directed in such a way as to completely abolish the effects of those obstacles. By the path factors, they make use of their respective power factors through wholesome consciousness. Without direction from path factors, powers on their own could become corrupted and lose the goal they are working towards.

Mindfulness is called supreme because not only does it have control over its own function but also it coordinates and synchronises control over other faculties. In the same way, the driver of a chariot driven by four horses has supreme control over the horses' speed and direction. To maintain the harmonious performance, the driver has a method of controlling the horses through a harness and ensures that all horses run at the same speed and in the same direction. If any of the horses were to run faster than the others, then the chariot would be tipped over. The horses need to be trained not to compete with the other horses in the team. Similarly, mindfulness watches over the other four faculties to ensure that not one of them overdevelops or suppresses the growth of other faculties. These faculties need to be perfectly balanced. The five spiritual faculties tend to suppress others. For example, strong faith could impair wisdom, and too much energy could weaken concentration and vice versa. Therefore, balanced development of all faculties is important for healthy continuation of meditation.

Power Factors

The five factors of mental absorption (jhananga), discussed previously, serve to intensify the activity of consciousness, particularly mental one-pointedness, as this is the main force of intensification that gives strength to the mind. The texts of the Mula Tika and Skandha Vibanga emphasise the contribution of mental absorption (jhana) to strengthening the mind; they are mentioned here as *bala dayaka*, strength givers.[5] The way this works can be compared to a diamond cutter, which is able to cut through any hard material; the mind trained by jhana has the power to cut through hindrances and penetrate to a more refined state of mind. It serves as a useful tool for insight meditation. Another example is the flame of a welding torch: its intensified strength can cut hard iron rods with

---

[5] Nyanaponika Thera, (1965). *Abhidhamma Studies*, p. 57. BPS.

precision. Only a well-trained and highly skilled technician can handle these tools. Similarly, a trainee needs to undertake the proper mental training to practise insight meditation in order to develop faculties in a skilful, balanced manner for purpose.

In the list of Dhammas given in the Dhammasangani, these spiritual faculties are immediately preceded by the intensifying factors of absorption. This is consistent with the description of Paccaya Dhamma (i.e. jhana paccaya) given in the philosophy of relations; it is the supporting conditions that intensify the consciousness. Factors of absorption, joyful interest (*piti*), and pleasant soothing (*sukha*), when intensified, build faith (confidence). Firm thought conceptions (vitakka) and discursive thinking (*vicara*) allow one to generate a keenness in investigating the object, which gives rise to profound wisdom. A high degree of stimulated interest (piti) also intensifies the faculties of energy, mindfulness, and concentration to their optimum performance. It was mentioned[6] that the intensifying function of mental one-pointedness may influence faith and, if not balanced by mindfulness, may become exclusively devotional. On the other hand, when mindfulness and concentration are well established and progressing well, they sustain and increase "interest"; hence they support each other. "It is from joy (piti) that faith derives a good part of its conquering power; and it is keen and enthusiastic interest that makes for the consistency of faith."[7]

Faith, mindfulness, and wisdom are exclusively wholesome. These are rendered as beautiful mental factors. However, the ethical value of energy and concentration is either wholesome or unwholesome. For example, the hunter aiming at a moving target has concentration and energy, but his or her intention is to kill. Wrong intention can be corrected by wisdom and should be directed by mindfulness. Knowledge requires that correction be drawn from the faculty of the mind that can decide on ethically viable factors.

Through this process of intensification, faculties become spiritual powers, "unshakeable" or, more passively, firmly established. Here power is meant in the sense that it is an ability to overcome its opposite by performing an act. For example, faith or confidence can be increased by undertaking a training course such as attending a meditation retreat. One can further develop these spiritual faculties until firm confidence is established.

---

6 Ibid. p. 65.
7 Ibid.

There are additionally two more powers that are included in the list: moral shame (*hiri*) and moral dread (*ottappa*). These two powers are known as "guardians of the world". Their function is to protect a trainee from falling into the hands of temptation by breaking moral values, committing unethical conduct, or abusing power. These seven spiritual powers are mentioned after the path factors. They serve a meaningful purpose to keep the spiritual faculties focused on the path factors. "Spiritual powers increase the factors of agility and pliancy of the mind and its capacity to effect deliberate inner changes, whether positive, negative, or adaptive. These last features are the basis for any mental and spiritual progress. It is mainly owing to the operation of these five spiritual faculties and corresponding powers that noticeable transformation of character may take place. It even appears as if quite a new personality has emerged."[8]

"It should be understood that the five spiritual faculties and five spiritual powers are simply two different aspects of the same qualities. They function as control and firm desire for liberation to eradicate defilements respectively."[9]

In the following simile, the Buddha explains how the nature of the five faculties is basically the same but how they differ in function: "Suppose there is a river that flows eastward, and in the midst of it is an island. In this case, the stream can be regarded as one when seeing its flow on the eastern and western side of the island; it can be regarded as two when the island's northern and southern sides are considered. The identity of the spiritual faculties and powers are to be understood in the same way."[10]

The power aspect of the faculties is to strive untiringly and not to give up in the face of temptations and challenges, but instead to intensify the function of faculties until they reach their full potential of "unshakeable powers".

Path Factors Continued

> Having been firmly established in morality, one may practise meditation (vipassana bhavana) to cultivate the mind.
>
> —The Buddha (Visuddhimagga)

---

[8] Ibid. p. 66.
[9] Ibid. pp. 73, 74–7.
[10] Ibid. p. 73.

The Noble Eightfold Path is also known as the path of purification. Practice of insight (vipassana) is intended to completely eradicate the defilements rooted in consciousness. The main three types of defilement are greed, hatred, and delusion, generally known as ignorance. Work needs to be done in order to remove defilements from the field of consciousness. There are other defilements which can also take root in the consciousness. The chapter "The Compendium of Mental Factors" in the Abhidhamma lists fourteen unwholesome factors, the main five hindrances, and three subsets of ignorance. These are the focus of the meditation practice. As the trainee progresses in his or her spiritual practice, gradually his or her mind becomes clean and clear. Spiritual faculties together with their corresponding powers can be successfully cultivated and should be directed by path factors. For example, mindfulness is said to be the chief of all because without mindfulness there is no mental development. The next stage of mindfulness is clear comprehension (*sati-sampajjana*). This is the factor of investigation; it is the ability to differentiate phenomena as mind and body clearly without confusion (*nama-rupa*). The Buddha perfected all factors leading to enlightenment by himself. "The Buddha is defined as the embodiment of perfect mindfulness (sati) and perfect clarity of consciousness."[11]

The Buddha attained enlightenment through a long practice of perfecting Parami Dhamma, meaning "Buddhahood training factors". During his search for final liberation, he practised and perfected samatha (tranquillity) meditation under the teachers Alara Kalama and Uddaka Rama Putta, and mastered mental absorptions up to "neither perception nor nonperception". The Buddha's mind was sharp and unshakeable; his spiritual faculties were at their peak. It was said that he had a strong determination not to give up or accept defeat. Although determination is not on the list of Dhamma, the factor of exertion (F55) and its supplementary factor decisiveness (F58) both serve as determination. On the full-moon night of his enlightenment, he revised all his meditation practices, both introspectively and retrospectively, to identify their strengths and deficiencies. At the final stage he realised that those meditation attainments were still mundane and not full liberation, and hence the solution had to be found somewhat outside the mundane sphere. This required his additional effort and force to overcome the temptation to give up.

---

[11] Ibid. p. 64.

Generally, the function of the five powers is to exert a force against their respective opposites. Spiritual faculties under this great force, when not in harmony, tend to dominate and suppress their counterparts. Power factors do not have a directive of their own. It is a danger to allow power to grow without its overall purpose, though it can serve its own purpose and enjoyment could become corrupt. "The goal towards which the respective faculty was originally working and moving will lose its importance, and so its directive influence on that faculty and the entire personality will diminish."[12]

For example, strong faith in and of itself without wisdom can become blind faith. On the other hand, keen intellect without faith and mindfulness could lead to a superiority complex and give rise to the intellectual corruption of ideation and imagination. There is a danger in allowing one faculty and its corresponding power to dominate. For example, the faculty of mind tends to follow speculative views and beliefs. In its extremity, this may lead to grasping hallucinations as real, and one may suffer from psychosis if not careful. Those who are unable to vent their strong emotions may purge violently, taking revenge to satisfy their pride or suffer from suicidal tendency.

The Noble Eightfold Path was discovered by the Buddha after strenuous effort. After his enlightenment, the Buddha formulated the theory that leads to liberation and gave practical advice on how to tread the path towards enlightenment. The Buddha attained enlightenment by changing from samatha to vipassana (insight) meditation. Path factors facilitate a direction towards insight knowledge and wisdom. A trainee who treads this path of purification goes through four stages of change. These are stream enterer, once-returner, nonreturner, and final sainthood as an Arahant. The path and its factors are made known to trainees; this was one of the purposes of the Self-Enlightened Buddha (Samma Sum Buddha) to help others to end their suffering.

The function of spiritual faculties and their corresponding power factors gain strength, stage by stage, to become "unshakeable" (*akampiya*) at their peak. The Theravada School of Buddhism advocates and recommends the Four Noble Truths and the Noble Eightfold Path as the safest path to liberation. It is known as the Middle Path for the same reason. Progress

---

[12] Ibid. p. 76

when following the recommended path can be seen here and now; its results take place as attainment of the stages of sanctity (the Ariya Magga). "Only then, when certain fetters (*samyojana*) and hindrances (*nivarana*) have been completely abolished, do those faculties and their spiritual qualities too become really 'unshakeable', that is, they can no more be lost. For example, faith becomes 'unshakeable' to whom the fetters or hindrances of scepticism (*vicikichha*) are radically destroyed on reaching the stage of stream entry (*sotapatti*)."[13]

Path factors are mentioned in a defined order of hierarchy. The preceding one serves as a supporting condition and a directive to the next factor. A matrix of their intricate relations is given in the philosophy of relations. For example, Magga Paccaya (the eighteenth relation), as well as the eight path factors, explains wrong views, wrong aspiration, wrong endeavour, and wrong concentration. This shows all classes of consciousness and mental concomitants as conditioned by causes (*hetu*) and advises how to direct the mind towards the specific goal. According to this treatise, path factors are to be understood as a means of reaching the final goal, nibbana, whereas constituents of the wrong path lead to realms of misfortune. It further explains the function of jhana and indriya in making the mind straight and steadfast, while also intensifying its quality to sink deep into the meditative object (kasina) like a "fish in deep water". The function of the path factor is to break the cycle of craving and direct the mind towards liberation. The knowledge and vision gained from the practice of insight (*bhavanamaya prajna*) must be applied in successive order. Power factors continue to support this endeavour as the trainee steadily progresses on the path. One who is able to transcend all phenomena (nama-rupa) will see the dependent origination of Dhamma (*hetu-pala*) and succeed in the path. Good actions done in the past and the present will result in good fruits (*pala*). The Buddha's teaching emphasises that one's present actions can change the course of life.

> With path factors we enter the sphere of definite and unmistakable values and value attributions; their directive and purposive energy is consequently greater than that of the spiritual faculties. These features of path factors find expression

---

[13] Ibid. p. 74.

in the commentarial explanation of them (As/P/154) as "Factors of Deliverance" (*niyyan atthena*, literally meaning "leading out" (i.e. from samsara) and as "conditions" (*hetu-atthena*), that is compulsory conditions or requirements for the attaining of sainthood (Arahant). For example, if the factor "concentration" (mental one-pointedness), being in itself neutral (outside the sphere of values), receives the value attribution "right" (*samma*), it then becomes a path factor, that is, a factor of deliverance, because, from the highest standpoint of Buddhist doctrine, only that which is conducive to deliverance is called "right".[14]

---

[14] "The Path Factors", ibid. p. 77.

CHAPTER 20

MEDITATIVE EXPERIENCES

Numerical Discourses of the Buddha

Suttas from the Anguttara Nikaya (Gradual Collection)

ON MANY OCCASIONS THE BUDDHA and his disciples discussed matters directly relevant to improving one's practice. These few suttas inserted in this chapter are in support of meditation practice. They are, as it were, supporting materials for the Suttantha method (discourses given in the second collection of the Pali Canon). These instructions given in discourses are align with the Noble Eightfold Path of *sila*, samadhi, and prajna (morality, concentration, and wisdom). The explanations given in these suttas do not largely differ from those in the Abhidhamma'. Buddhist scholastic enquiry focuses on appearance, analysis, and deeper investigation into the causes and conditions—a method to reach a penetrative understanding of the manifestation of phenomena. Suttas, on the other hand, contain direct instructions from the Buddha as heard and understood by his disciples. These scattered discourses were arranged in an orderly manner by learned disciples after the Buddha had passed away.

Ten Disciplines of the Buddha (No. 202)

The Buddha has given the following ten disciplines that are the framework for purity of mind and clarity of his dispensation of Dhamma. They are the eight factors of the Noble Eightfold Path plus two additional factors given in the Dhammasangani. The meanings and applications of these terms were explained in the foregoing chapters. Their usefulness will be immediately evident at a glance to the earnest reader in this simple format and ultimately converge in the unfolding of insight into the realities of existence.

- right view
- right intention
- right speech
- right action
- right livelihood
- right effort
- right mindfulness
- right concentration
- right knowledge
- right liberation

  - five precepts
  - meditation
  - wisdom.

The Five Helpers of Right View (No. 97)

Right view, O monks, if it is helped by five things, has liberation of mind as its fruit and is rewarded by the fruit of liberation of mind; it has liberation by wisdom as its fruit and is rewarded by the fruit of liberation by wisdom.

What are those five things?

Here, monks, right view is helped by virtue, by wide learning, by discussion (of what was learned), by tranquillity and by insight.

Four Modes of Progress (No. 81)

1. The mode of progress that is painful, with sluggish direct knowledge.
2. The mode of progress that is painful, with quick direct knowledge.
3. The mode of progress that is pleasant, with sluggish direct knowledge.
4. The mode of progress that is pleasant, with quick direct knowledge.

The progress of the trainee depends on the direct knowledge acquired by meditation practice. It may take longer for one who has not developed the faculties but will be quicker for those who have keen faculties.

Ways to Arahantship (Enlightenment) (No. 83)

This discourse was given by the venerable Ananda. He was the Buddha's personal attendant who had the privilege of hearing most of the discourses and discussions.

There are one of four ways a monk or nun attains final knowledge of liberation:

1. Developing insight preceded by tranquillity. While the monk or nun thus develops insight preceded by tranquillity, the path arises in him or her. He or she now pursues, develops, and cultivates that, and while he or she is doing so, the fetters are abandoned, and the underlying tendencies eliminated.
2. Developing tranquillity preceded by insight.
3. Developing tranquillity and insight joined in pairs.
4. After agitation caused by higher states of mind (*jhanas*), the monk's or nun's mind becomes internally steadied, composed, unified, and concentrated, then the path arises in him or her. He or she then pursues, develops, and cultivates that path, and while he or she is doing so, the fetters are abandoned, and the underlying tendencies eliminated.

This is the method of entering the supramundane path and its fruits by tranquillity leading to insight. The object of the supramundane path is nibbana. The meditator takes the nibbana element as the object of meditation. At this stage all erroneous perceptions are discarded.

A Thoroughbred's Meditation (No. 207)

Sutta No. 207 is a discourse recorded in the Anguttara Nikaya, organised in a chapter as sets of eleven, corresponding to the eleven stages of the jhana's path of attainments. This is a dialogue between the Buddha and Venerable Sandha on how a thoroughbred of a person should meditate. The Buddha makes a distinct contrast using the analogy of training a wild colt and a thoroughbred horse. Unlike a colt waiting to get fed, the grown horse is well-disciplined and awaits its trainer's instructions for the day, knowing what it can receive in return. The analogy explains two different levels of meditation. With this, the Buddha disagrees with the venerable Sandha, stating that he was not developing insight and instead was depending on mundane objects for meditation. In return Sandha asks how a senior should progress from mundane to supramundane consciousness. In this discourse the Buddha responds to Venerable Sandha's question:

"But how, Lord, does a good thoroughbred of a man meditate, if he does not meditate dependent on anything else? And yet he meditates."

The Buddha goes to show the limitations of mundane jhanas and the disappearance of perceptions in the same order given in the Abhidhamma. One who wishes to enter the supramundane path takes nibbana as the object of meditation and should move on from using *kasinas* as meditation objects.

The order of disappearance of the phenomenological world is given below the order of disappearance of phenomena when meditating using static meditation objects such as kasinas in order to develop concentration. Usually these objects are earth material, water, fire, or the element of air. When the concentration develops, there arises in the mind of the meditator the counterpart sign of the object. Further development of concentration is achieved by disappearance of the perception of those objects by letting go of the counterpart sign, as explained in the immaterial jhanas, as the perceptions are fading away in the following order.

- Perception of earth
- Perception of water

- Perception of fire
- Perception of air
- Perception of the base of the infinity of space
- Perception of the base of the infinity of consciousness
- The base of nothingness
- The base of neither perception nor non perception
- The perception of this world
- The perception of another world
- The perception of whatever is seen, heard, sensed, and cognised

The order given in the sutta is the way that is first of tranquillity, then, when the mind is settled, the meditator would be able to see the disappearance of phenomena together with its perception. If the meditator depends on anything of the mundane world (kasina, e.g. earth, water, fire, and air), then there is no escape for him or her from this or other worlds.

Space kasina and consciousness kasina are objective supports for the first and second formless meditation respectively, the base of infinity of space and the base of the infinity of consciousness. Jhana samadhi is the attained state of mental absorption based on these objects.

The Buddha's instructions to Sandha clearly show the limitation of formless meditation and when exactly to change the practice of tranquillity without depending on kasinas for the object of insight, which is the element of nibbana. Disappearance of perception happens in relation to the following bases:

- Earth as the base for earth kasina (sample)
- Water
- Fire
- Air
- Infinity of space
- Infinity of consciousness
- Nothingness
- Neither perception nor nonperception
- This world—five senses
- Other world—extrasensory perception

- Sixth sense
- No dependence

Taking the earth as an object, one does not meditate with the perception of the four jhanas. This is because of the absence of any attainment and, thus, neither the pleasure of attainment. "And yet he meditates." All mundane kasinas and sense objects depends on a base, meditation on them is entry level practices. Seniors should move away from dependence. One meditates as fruition attainment, which has nibbana as its object. "Not in dependence on anything else, and yet he meditates." The fully trained meditator surpasses the mundane world objects and directs his or her mind to the supramundane world by directing his or her mind to the element of nibbana. The Buddha explains the thoroughbred of a person in this analogy as a well-trained meditator, one who does not meditate dependent on (*nissaya*) earth or dependent on anything. *Nissaya* means "a relation of dependency". To meditate without dependency is to meditate according to the supramundane meditative attainment. Mundane consciousness depends on a pre-existing base. The supramundane consciousness, on the other hand, is not conditional and does not depend on anything else. The fruit of meditation in this instance is the supramundane resultant consciousness. At this stage, the perception of earth has disappeared in relation to earth, etc. Consciousness in accordance with the path has the function of eradicating defilements; the resultant consciousness has the function of enabling the trainee to experience the degree of liberation made possible by the corresponding path consciousness.

The sutta highlights an important concept of Buddhism, which is to correctly understand the elements (*dhatu*) of the meditation subject. There are six elements to be understood in relation to the subject and object of meditation. They are earth, water, fire, air, space, and consciousness. The first four are primary elements, and matter is a secondary element that is derived from the primary elements. Space is a derived element of noncontacting matter. Consciousness is the element of mind, taken singly (*vinnana-dhatu*) or as an aggregate (*vinnana-khandha*). The difference in experience (*citta*) is the gross manifestation of phenomena; aggregates, feelings, perceptions, and volitional mental formations arise in coexistence with consciousness. These are the five aggregates, formed from four primary elements and four mental aggregates, and they can be separated into matter (*rupa*) and mentality (*nama*) by insight. The meditator understands that beyond these two, there is neither substantial being nor self. Matter has completely transcended when, at the fourth immaterial jhana consciousness, the mind is pure and there are no subjective mental concomitants associated with it. The sutta elaborates the analytical insight of the six elements that lead to Arahantship.

Nibbana as an Object

The nibbana element is an insight wisdom, not a tangible or mental object as such. The generic term used is *Dhamma*. The chapter "The Compendium of Mental Factors" in the Abhidhamma explains that the supramundane consciousness neither recognises compassion nor appreciates joy because these always take the concept of "living being" as their object. Path and fruition citta take nibbana as their object. Nonhatred and mental neutrality are found in the supramundane citta.[1]

---

[1] Bhikkhu Bodhi, (1999). *A Comprehensive Manual of Abhidhamma* (Sri Lanka, Buddhist Publication Society), p. 102.

CHAPTER 21

PURITY OF THE MIND
THAT SEES THE TRUTH

V IPASSANA IS TO SEE THINGS (Dhamma) clearly, as they are, without forming opinions or judgements, without analysing, and without attachment or aversion to them. Samatha provides the necessary tranquillity and intensity to stabilise the mind, to support and strengthen the wholesome mental factors. By his or her own effort and penetrative vision, a trainee is able to discern the truth for himself or herself. The Buddha's role was to show the way; whoever sees the Dhamma within is able to see the Buddha's teachings.

The fundamental principles of Buddhist philosophy are "Nothing arises from a single cause" and "Nothing exists (or moves) by its own power"; these two axiomatic verses are quoted in the Atthasalini.[1] Detailed explanations with direct observation can prove that all phenomena are dynamic in nature, that they are dependent on a supporting condition, that they are impermanent, and that they are subject to suffering and insubstantiality (*anatta*). In Samyutta Nikaya 12, titled "Connected Discourses on Causation", the Buddha explains and elaborates the formula of dependent origination. For example, an unwholesome state of consciousness is dependent on ignorance, and from it the resultant kamma formation takes place. SN 12 also analyses dependent origination

---

[1] Atthasalini, pp. 59–61; Nyanaponika Thera, *Abhidhamma Studies: Researches in Buddhist Psychology* (Kandy, Sri Lanka, Buddhist Publication Society, 1965), p. 22.

as the basis of the different forms of consciousness. In the commentaries, dependent origination is defined as the simultaneous arising of effects that are dependent on the conditions.[2] The Vibhanga, the Book of Analysis, and the Treatise on the Modes of Conditionality apply dependent origination to a single moment with varying types of consciousness.

The synthesis of phenomena given in the book of Patthana elucidates these conditions with reference to twenty-four types of external and internal relations or modes of conditionality. This shows that there are always multiple conditions operating in synchrony which give rise to multiple effects. The Patthana highlights the most important conditions: the preimminent and their direct effects. These two dynamic philosophies of Dhamma are then combined to explain the complexity of conditionality in all phenomena of existence. This remarkable approach of combining these twenty-four modes of conditionality (Paccaya Dhamma) is applicable to "dependent origination within a single moment of consciousness", which links it to a momentary dependent origination. The connection between several distinct moments would be an "external relation".[3]

There is an ingenuity of skill in this method of teaching, ascribed to the Buddha himself, namely, to first apply these two different methods separately and then afterwards combine them to explain Dhamma. In a given single moment of time, however infinitesimally brief the moment of consciousness, there is actually an intricate network of relations. The Buddha was able to magnify a single thought moment, analyse it, and then see the associated relations. Recorded in the Patthana, the "Book of Conditional Relations", is the matrix of the different combinations of relations with detailed analysis such as applied to all phenomena. "This book is a testimony to the Buddha's unimpeded knowledge of omniscience."[4]

Not seeing these ever-changing phenomena, grasping things that are impermanent as permanent and when there is no self to take it as self, is the great delusion (*abhimano*—i.e. great conceit). Applying the analytical and synthetical method with the Four Foundations of Mindfulness can provide the full and correct understanding of all phenomena as impermanent

---

[2] Bhikkhu Bodhi. (1999). *A Comprehensive Manual of Abhidhamma* (Kandy, Sri Lanka, Buddhist Publication Society), p. 295.

[3] Nyanaponika Thera. (1965). *Abhidhamma Studies*, p. 24.

[4] Bhikkhu Bodhi. (1999). *A Comprehensive Manual of Abhidhamma.*

(*annica*), suffering (*dukkha*), impersonal (*anatta*), and unsubstantial/void (*sunnata*). These principles are same for the material body, the material world around us, its objects, and sense perceptions. The perceptive limit of the mundane world is "neither perception nor nonperception"; this can be reached by mental absorptions, which transcend all worldly perceptions. By taking a material body and mental formations as objects of vipassana, a trainee can see the suffering and unsatisfactoriness of the mundane world. By gaining comprehensive understanding of all Dhamma by treading the Noble Path, a trainee can reach the gates to the supramundane world beyond the mundane world, namely the *signless gate* by contemplation on impermanence, the *desireless gate* by contemplation on suffering, and the *voidness gate* by contemplation of nonself. By entering through one gate, one would simultaneously enter the other two gates. One who understands one of these truths understands the two other truths because these three truths are interconnected. One who frees himself or herself through these gates is no longer attached to worldly passions, because he or she has acquired wisdom and won the path to nibbana. The trainee would then have broken the cycle of *kelesa*, kamma, and *vipaka*. Kelesa (defilements) are completely uprooted and have no cause to produce kamma (volition) and hence no vipaka (result). Any further action would be "defunct", no longer operating according to the same laws. This principle is very well-expressed in the reverse order of dependent origination, which is a remarkable formula for how to end suffering.

Insight knowledge and wisdom coexist with insight meditation. There are sixteen stages of insight knowledge which correspond to the seven stages of purity, gained through the practice of samatha–vipassana. On this path, the unwholesome roots weaken, allowing the wholesome roots to flourish, then the lineage of the trainee changes in accordance with the level of purity. There are three generic wholesome roots, namely nongreed (*alobha*), nonhate (*adosa*), and nondelusion (*amoha*). Wholesome states of consciousness are "associated with knowledge" (*nama sampayutta*). Insight knowledge dispels the effects of unwholesome factors just as light dispels its opposite, darkness.

The Noble Eightfold Path, which is the Fourth Noble Truth propounded by the Buddha, is for those who are proactively searching for liberation. This statement is supported by the fact that on many occasions the Buddha refused to engage in conversations if the question put forward to him had no bearing on liberation or the final goal of nibbana. For example, the Buddha refused to answer metaphysical questions of "First Cause" or "an unchanging thing" (recently the definition of metaphysics has changed; see note below). The practice of path factors is for those who understand the Four Noble Truths and seek to bring their suffering to an end. Once this goal of liberation and right understanding is set, further development of wisdom has a purpose. At this stage of progress, a trainee must change the method of meditation and direct it to insight meditation; he or she can then use momentary concentration to observe the ever-changing phenomena, such as those given in the Mahasatipatthana Sutta. Only then can all the eight path factors be harmoniously directed towards acquiring insight knowledge.

> *Note:* The word *metaphysical* is derived from the Greek *meta ta physika* ("after the things of nature"), referring to an idea, doctrine, or posited reality outside human sense perception. In modern philosophical terminology, metaphysics refers to the study of what cannot be reached through objective studies of material reality.
>
> Just as physics deals with the laws that govern the physical world (such as those of gravity or the properties of waves), metaphysics describes what is beyond physics—the nature and origin of reality itself, the immortal soul, and the existence of a Supreme Being. Opinions about these metaphysical topics vary widely since what's being discussed can't be observed or measured or even truly known to exist. So, most metaphysical questions are still as far from a final answer as they were when Plato and Aristotle were asking them.
>
> Metaphysics is the branch of philosophy that examines the fundamental nature of reality. Originally meaning the study of being, First Causes, or unchanging things, it now has a much wider scope.[5]

Desire for liberation comes from the trainee's own understanding of the emptiness (voidness) of the mundane world. Turning away from sense

---

[5] *Encyclopaedia of Philosophy.* Stanford University Press, Metaphysics, https://plato.stanford.edu/entries/metaphysics/.

pleasure, he or she strives for awakening and makes effort towards that goal. It is often advised by masters not to meditate with expectations but instead to see the disadvantage of what is subject to decay and death. From the willingness of a trainee must come the aim of acquiring insight in search of nibbana. Sixteen insight knowledges will arise in a trainee as the result of practising vipassana. Nevertheless, anyone seeking new knowledge can benefit from the Noble Eightfold Path and the timeless truth of the Buddha. This knowledge can aid in resolving issues concerning the mundane world, even if nibbana is not one's goal and one's liberation is determined for a future date.

Sixteen Insight Knowledges

Insight knowledge is a subset of the seven stages of purification of the mind. These are first found in the Pali Canon in the Rathavinita Sutta of the Majjhima Nikaya *(Middle collection)*, where they form the subject of a discussion between the monks Venerable Sariputta and Punna Mantaniputta. Each stage is a "factor of exertion for purity".

Punna Mantaniputta was an eloquent teacher but was filled with pride for his wealthy background and for his own philosophy of life. One day he went to see the Buddha with his students. He had the intention of challenging the Buddha to a debate and sought to win. Having seen his pride, the Buddha persuaded him of the futility of debating, explaining that instead of engaging in debate, people should seek liberation through dialogue. At a later stage, both Punna Mantaniputta and his students became the Buddha's disciples. Venerable Sariputta, who was considered to have understood the Dhamma, second only to the Buddha, one day questioned his friend Punna on the prominent features of the Buddha's teaching. Punna's striking replies highlight the stages of enlightenment and still, to this day, are the essential guidelines to all meditators in the Theravada School of Buddhism. These teachings are well grounded in the latter scholarly work of Venerable Buddhaghosa in his great book *The Path of Purification*. Venerable Punna explained the path of purification by using the colourful simile of a relay of chariots. He made clear to Venerable Sariputta that to advance in the Buddha's path, one must follow the seven stages in sequence, as a king would organise a relay of chariots to get from one place to his destination. One who sought to undertake this journey would mount the first chariot at point A and energetically drive to point B, during this journey fully understanding the nature of both the horses and the chariot. On arriving at B, he would change to the next chariot C, and so on, travelling in sequential order until he arrived at the gate of his destination. Each stage must be energetically experienced and understood until arriving at the final goal of nibbana.

The Relay of Chariots

The path of practice leading to the attainment of nibbana unfolds in seven stages, known as the seven stages of purification. The seven stages are given below as stated by Venerable Punna. The seventh stage is the knowledge and vision of the supramundane path which comes from perfection of the other six previous stages of purity:

1. Purification of virtue
2. Purification of mind
3. Purification of views
4. Purification by overcoming doubt
5. Purification by knowledge and vision of what is path and what is not path
6. Purification by knowledge and vision of the way
7. Purification by knowledge and vision

Along this path of purification, the meditator gains insight into the reality of body and mind corresponding to his or her level of purification. The sixteen stages of insight are stated to review the meditation practice. This ensures that the practice is progressing in the right direction and is not meant for any expectation. The early stages are known as "imperfection of insight" (vipassana upakkilesa); this stage is common to all meditators. Unexpected and pleasant experiences are actually hindrances to progress, for example the meditator could be misled to believe that he or she has reached the supramundane path and fruit. The meditator must understand and should be able to distinguish between what is the path and what is not the path in conformity to impermanence, suffering, and nonself.

The fourth insight knowledge is the knowledge of rise and fall. This constitutes purification by knowledge and vision of the way. All experiences pertaining to body and mind (*nama-rupa*) are going through three stages: arising, maturing, and falling. When insight becomes keen and gains more clarity, the meditator's mindfulness shifts to the falling stage. He or she then contemplates on cessation and destruction, the fall and break-up of phenomena. This knowledge is the knowledge of dissolution; whoever understands this principle of insight will understand the rest of the knowledges. The following dialogue between Venerable Assaji and Venerable Sariputta records one of the foremost disciples of the Buddha testifying to the truth:

Venerable Assaji gave this Dhamma exposition to Venerable Sariputta (then Upatissa, his name before his ordination as a layperson).

> Whatever phenomena arise from cause: their cause and their cessation. Such is the teaching of the Tathagata, the Great Contemplative.
>
> When Sariputta heard this Dhamma exposition, there arose the dustless, stainless Dhamma eye:
>
> Whatever is subject to origination is all subject to cessation.[6]

The knowledge of dissolution results in the realisation of a fearful nature of all existence, not only in the present but also in the past and future existences. This is the knowledge of the fearful nature of existence, that all existences are subject to dissolution. Preceding knowledge combines with this to form a new realisation, namely that all life experiences are without substance and are unsustainable, thus attached to them can be nothing but danger. With this realisation, the meditator will then understand that there is no security in the conditioned phenomenal world and that only the unconditioned world is free from rising and falling.

Key experiences of insight knowledge are not given in the earlier Buddhist literature but can be found in later commentaries on various suttas, where the Buddha gave advice to his monks. Abhidhammattha

---

[6] Thanissaro Bhikkhu, *Upatissa-Pasine: Upatissa's (Sariputta's) Question* (1966). Access to Insight (BCBS Edition), 30 November 2013. http://www.accesstoinsight. org/tipitaka/vin/mv/mv.01.23.01-10.than.html.

Sangaha gives nine insight knowledges (p. 353); this scheme does not include the changing of lineage or entering onto the Noble Path. Other systems give a review of the path and state the fruit of the path. It is generally accepted that there are sixteen knowledges of insight. Some teachers give seventeen. Some commentators expand the sixteen into thirty-seven stages of knowledge. These differences are due to the method of how insight is treated. Some teachers break a stage into different sublevels, as is the case in many ancient commentaries of Dhamma.

Given below are extracts of three schemes. There is no major difference between these schemes; they all stem from the doctrine of the Four Noble Truths and are therefore generally not rejected by wise scholars or meditation masters:

Scheme I—the Sixteen Stages of Vipassana Knowledge

This is the work of the late Venerable Phra Dhamma Theerarach Mahamuni, one of the most renowned vipassana teachers of his time. The original booklet was produced in 1961 by the Division of Vipassana Dhura at Mahadhatu Monastery, Bangkok, and was translated by Helen and Vorasak Jandamit. A revised English version was reprinted in 1988. It is also presented, with commentary, in the book *Insight Meditation: Practical Steps to Ultimate Truth*, by Ajahn Sobin S. Namto (Sopako Bodhi Bhikkhu). In a few places the text has been edited and augmented by the Vipassana Dhura staff.

- Knowledge to distinguish mental and physical states (*namarupa pariccheda nana*)
- Knowledge of the cause-and-effect relationship between mental and physical states (*paccaya pariggaha nana*)
- Knowledge of the mental and physical processes as impermanent, unsatisfactory, and nonself (*sammasana nana*)
- Knowledge of arising and passing away (*udayabbaya nana*)
- Knowledge of the dissolution of formations (*bhanga nana*)
- Knowledge of the fearful nature of mental and physical states (*bhaya nana*)

- Knowledge of mental and physical states as unsatisfactory (*adinava nana*)
- Knowledge of disenchantment (*nibbida nana*)
- Knowledge of the desire to abandon the worldly state (*muncitukamayata nana*)
- Knowledge which investigates the path to deliverance and instils a decision to practise further (*patisankha nana*)
- Knowledge which regards mental and physical states with equanimity (*sankharupekha nana*)
- Knowledge which conforms to the Four Noble Truths (*anuloma nana*)
- Knowledge of deliverance from the worldly condition (*gotrabhu nana*)
- Knowledge by which defilements are abandoned and are overcome by destruction (*magga nana*)
- Knowledge which realises the fruit of the path and has nibbana as its object (*phala nana*)
- Knowledge which reviews the defilements destroyed (*paccavekkhana nana*).

Scheme II—the Progress of Insight (Visuddhiñana-Katha)

The following is according to Venerable Mahasi Sayadaw, translated from the Pali with notes by Nyanaponika Thera, 1994:

- Analytical knowledge of body and mind (*nama-rupa-pariccheda-ñana*)
- Knowledge by discerning conditionality (*paccaya-pariggaha-ñana*)
- Knowledge by comprehension (*sammasana-ñana*)
- Knowledge of arising and passing away (*udayabbaya-ñana*) in its weak stage
- Knowledge of dissolution (*bhanga-ñana*)
- Awareness of fearfulness (*bhayatupatthana-ñana*)
- Knowledge of misery (*adinava-ñana*)
- Knowledge of disgust (*nibbida-ñana*)
- Knowledge of desire for deliverance (*muncitu-kamyata-ñana*)
- Knowledge of reobservation (*patisankhanupassana-ñana*)

- Knowledge of equanimity about formations (*sankhar'upekkha-ñana*)
- Insight leading to emergence (*vutthanagamini-vipassana-ñana*)
- Knowledge of adaptation (*anuloma-ñana*)
- Maturity knowledge (*gotrabhu-ñana*)
- Path knowledge (*magga-ñana*)
- Fruition knowledge (*phala-ñana*)
- Knowledge of reviewing (*paccavekkhana-ñana*).

Scheme III—Purification of the Way

The following is from Bhikkhu Bodhi, A Comprehensive Manual of Abhidhamma, (Abhidhammattha Sangha, 1999, p. 352):

Purification of the mind in the last stages removes the remaining defilements as the meditator begins to see things as they are, letting go of the craving. Insight knowledge and wisdom are synonymous; they correspond to the purity of mind that discerns truth in retrospect, in reflection, or by direct examination at each stage.

The writer will explain all schemes by the following sixteen stages. These stages of knowledge arise in sequential order when the trainee progresses in insight meditation:

1. Analytical Knowledge of Body and Mind

   The realisation of body and mind as separate aspects of the life experience acquired through analytical observation. To know body as body and mind as mind; the ability to see the difference with clarity. The analogy given in Relation No. 20 explains that body and mind stay together like water and oil but are separate aspects of life and must be understood with insight.

2. Knowledge by Discerning Conditionality

   The meditator understands that consciousness and body are conditioned phenomena, understanding by cause and effect that they arise from a cause and produce an effect or product. Conditionality of both mind and body are analysed and explained in the philosophy of relations and analysis of consciousness, which say that at any given time the consciousness is conditioned by mental factors in a variety of ways. All the manifold corporeal phenomena are primarily derived from earth, water, fire, and air elements.

3.   Knowledge by Comprehension

This stage is the understanding of suffering, impermanence, and nonself in all body and mind experiences—the true nature. Several of the individual stages of insight knowledge can be seen to take their inspiration from the early discourses. Thus, a passage in the Samyutta Nikaya speaks of developing "internal comprehension" (SN 2.107). This comprehension stands for contemplating the dependent arising of *dukkha* through craving. The doctrine of transcendental dependent origination spells out the cause of suffering and how craving can bind one to samsara. The meditator can gain the insight into the cause of suffering that is due to defilements.

4.   Knowledge of Arising and Passing Away

At this stage, the Ten Corruptions of Insight should be observed without reaction, such as the fact that illuminations arise because of defilements remaining in the mind. Any misunderstanding of this stage must be discussed with an experienced master. When this comes to pass, insight matures and develops.

5.   Knowledge of Dissolution

The meditator focuses on the falling stage of Dhamma, the dissolution of phenomena. All life experiences are cyclic, rising, maturing, and then falling. For example, someone falling from a tree is struck many times by many branches before landing on the ground. This traumatic experience is compounded by the various injuries suffered, which could be life-threatening. Insight into many dreadful life experiences could be similar; they all break up because of the compounded nature of all life phenomena, physical and mental.

6.   Knowledge of the Fearful Nature of Body and Mind

When the meditator contemplates the dissolution of formation, he or she recognises the fearful nature of all forms of life.

7. Knowledge of Danger

Following the realisation of the fearfulness of all formation, the meditator sees the danger of attachment to formation.

8. Knowledge of Disgust

By seeing the fearfulness and danger, the meditator at this stage takes no delight in any realm of existence and becomes rather disappointed with the mundane world.

9. Knowledge of Desire for Deliverance

In the course of contemplation, the meditator wants to escape from all formations.

10. Knowledge of Reflective Contemplation

This is a review stage of all the realms of existence in relation to the three characteristics of all phenomena. One reaches the confirmed insight of impermanence, suffering, and nonself of all phenomena.

11. Knowledge of Equanimity about Formations

Seeing the dissolution of all formation, the meditator realises that there is nothing to be claimed as "I" or "mine"; he or she therefore abandons attachment and becomes equanimous towards all formations.

12. Knowledge of Conformity

The insight gained at this level conforms with the natural law and the functions of truth. These are seen in accordance with both the preceding twelve stages of insight knowledge and the path attainment to follow. The meditator has discarded wrong views of life and is now endowed with right view. The cognitive process during the knowledge of conformity follows three stages: change of lineage in the cognitive process (*citta*) of the supramundane; supramundane cognitive process of absorption; and subsidence into the life continuum. Change of lineage is also possible for those who practise tranquillity through higher jhana and then change to vipassana.

13. Knowledge of Deliverance from the Worldly Condition

The transformation brought about by this deepening dissatisfaction of suffering (dukkha) develops maturity. The characteristic of nonself becomes increasingly evident (*paisakhañaa, sakharupekkhañaa*, and *anulomañaa*).

This insight becomes a full and direct experience with the breakthrough into stream entry, wherein any sense of selfhood completely disappears.

14. Path Knowledge

At this point, the series of insight knowledge reaches its completion. The mind momentarily withdraws from externals; the trainee leaves the stage of being a 'worldly'. Immediately following are the experiences of the path and its fruitive moments. This is equivalent to gaining liberating insight into the Four Noble Truths by the realisation of the third truth, realisation of nibbana.

15. Fruition Knowledge

On emerging from the experience of the supramundane, the mind naturally looks back on the extraordinary experience that has just happened and reviews what has taken place. At this stage, one is absolutely free from the entire mundane sphere and experiences cessation of formation. It is the noble fruition of nibbana. Since there are four noble stages of attainments, there are four corresponding fruitions of psychological moments that transcend the lower states in the process of evolution, taking nibbana as its definite object.

16. Reviewing Knowledge

This is the last stage of knowledge. It reviews on reflection each of the four supramundane path attainments: the path; its fruits; defilements destroyed; and defilements remaining (if not fully liberated in the case of evolution). During the process of retrospection, the Arahant reflects on the entire journey he or she has travelled, on the stages of attainments and final nibbana that he or she has realised, and on all the defilements that have been destroyed. He or she sees that everything that needed to be done has been done. Thus, comes the knowledge of emancipation when he or she has been emancipated. The final stage marks the culmination of purity of knowledge of insight wisdom. It is the fruit of systematic development of the knowledge of the Four Noble Truths, which have been expounded as the means of self-enlightenment, the only way to the true destiny of humankind.

Moreover, in reviewing the experiences of the ten insight knowledges leading to stream entry; the "change of lineage" from a worldly to a noble person; the "path" as well as the "fruit" of stream entry, one can expand these to additional stages. Hence another four knowledges can be designated (gotrabhu-ñana, magga-ñana, phala-ñana, and paccavekkhana-ñana), resulting in an overall account of the sixteen knowledges.

There are three doors to nibbana dependent on the inclination of the trainee. One who contemplates suffering discards desire through craving, becoming emancipated through the door of desire-lessness. One who contemplates nonself through the faculty of wisdom emancipates through the door of voidness. One who contemplates impermanence emancipates through the door of sign-lessness because he or she abandons permanence ("the sign of perversion" is the deceptive appearance of permanence).

Human development involves studies of the human condition with its core being the capability to transform life from difficult condition that cause suffering.

Human development is defined as the process of enlarging people's freedoms and opportunities and improving their well-being. Human development is about the real freedom ordinary people have to decide for themselves how to live happily. Our fragile happiness depends on things happening a certain way. But there is something else: a happiness not dependent on conditions. The Buddha taught the way to find this perfect happiness. The Buddha's path to happiness requires practising mindfulness until it becomes part of your daily life. Mindfulness is a way of training yourself to become aware of things as they really are to gain knowledge and wisdom for self-realisation of truth and become a fully evolved human being to end suffering.

CHAPTER 22

NIBBANA

Nibbana is the fourth ultimate reality (parramatta Dhamma/ultimate truth) in the Abhidhamma. It is also the fourth type of classification of consciousness belonging to the supramundane consciousness, meaning beyond the mundane world of consciousness. Recorded in the Sutta Nipatha are the Buddha's words explaining what was previously unknown to the world: "Monks, there is an unborn, unoriginated, unmade, and unformed. Where there is not such a state, as the unborn, unoriginated, unmade, and unformed, there would be no such escape for that which is born, originated, made, and formed. Since, monks, there is the state of the unborn, unoriginated, unmade, and unformed, there is an escape for the born, originated, made, and formed."

Samsara is an endless ocean of suffering where beings are perpetuated by desire, seeking pleasure, and not finding satisfaction, which can go on and on through unfathomable cycles of time, to which a beginning and end is unknown. It is ever changing from moment to moment. Like busy bees, beings are trapped in a honeycomb, in a chain of lives, disappearing and reappearing in different forms and in different realms of existence. The Buddha's enlightenment made the transition possible for a worldly being who has been wandering throughout countless rounds of births and deaths, from the mundane consciousness where no beginning could be found to the supramundane consciousness. Nibbana is inner peace, a state of the unborn, unoriginated, unmade, and unformed, which is a passionless, peaceful extinction of worldly fires, a never-changing, never-fluctuating, everlasting, blissful state of supreme happiness.

The great commentator of the Abhidhammattha Sangaha, Acariya Anuruddha, opens his treatise with an in-depth analysis of consciousness and its classifications of gradual evolution, from an untrained worldly mind, through systematic training on the path of purification, to the mind's ultimate liberation. This explains the transcendental reality of the supramundane consciousness as taught by the Buddha. It invalidates the theory of a transmigration of soul and refutes the contemporary belief and wrong view of a permanent self or ego.

"The Compendium of Consciousness" and with "The Compendium of Mental Factors", compiled by Acariya Anuruddha, enumerate 121 types of consciousness under four different headings. This order of analysis follows the principles of the law of consciousness (citta-niyama) and is in accordance with the law of Dhamma (Dhamma Niyama). These universal laws were discovered by the Buddha, who consolidated them into a practical application for the purpose of purifying beings. Out of compassion and in consideration of the suffering in the world, he made his discovery known to others. On the full-moon day of the month of July (Asala), the Buddha set the wheel of Dhamma in motion. By listening to the Buddha's first sermon, a former colleague named Kondanna understood the doctrine of impermanence and attained the first stage of enlightenment. He was one of the royal astrologers who predicted that the newly born prince would one day become the Buddha. When Prince Siddhattha left the royal palace, Kondanna also left his household and became an ascetic, following the prince, who had also become an ascetic. He later came to be known as "Anna Kondanna", meaning "Kondanna, the one who understood". The Buddha personally trained Kondanna and his four other former colleagues who came to practise asceticism with him; soon all the five ascetics were fully enlightened. A new school of thought was established based on the Middle Path discovered by the Buddha, who abandoned the extremes.

The crux of nibbana is the knowledge of the supramundane consciousness that transcends the mundane world of the five aggregates of clinging. When asked what the world is, the Buddha pointed out that the five aggregates of clinging are the world.

Nibbana as an Object

The supramundane consciousness takes nibbana as its object. The three main contemplations for emancipation are nonself, impermanence, and suffering. Within these broad categories, the meditator may choose a suitable practical object, such as the Four Foundations of Mindfulness or the four great elements (earth, water, fire, and air). These objects support the attainment of *jhanas* and Nirodha. The application of mindfulness and wisdom would show all these objects to have the same characteristics. The word *Nirodha* can be equated to complete cessation of feelings, perception mental formations and consciousness comes to standstill while Nibbana is a noun the unconditional state of mind describes as the state of peace and happiness that a person achieves after giving up all personal desires (Oxford Advanced Learner's Dictionary).

Unconditioned

In the Theravada tradition, nibbana is regarded as an uncompounded or unconditioned (*asankhata*) Dhamma (phenomenon, event, or instant), "transmundane" and beyond our normal dualistic conceptions. In the Vibhanga, one of the Theravada Abhidhamma texts, nibbana or the *asankhata-dhatu* (unconditioned element) is defined thus:

> What is the unconditioned element?
>> It is the cessation of passion, the cessation of hatred, and
> the cessation of delusion.

"The Compendium of Mental Factors" explains that the supramundane consciousness does not take compassion as an object, nor is appreciative joy to be found; the concept of a living being is not the object of the supramundane consciousness. Path and fruition citta take nibbana as their object. Non-hatred and mental neutrality are found in the supramundane citta, and these are qualities of tranquillity. Compassion and appreciative joy (*karuna* and *mudita*), together with loving-kindness and equanimity (*metta* and *upekka*), are wisdom factors reflected in an enlightened mind.[1]

---

[1] Bhikkhu Bodhi, (1999) *A Comprehensive Manual of Abhidhamma* (Kandy, Sri Lanka, Buddhist Publication Society), p. 102.

Supramundane Consciousness: The State of Being Unborn, Unoriginated, Unmade, and Unformed

A mundane consciousness takes static objects, sense impressions, or mental impressions as an object. Both consciousness and objects in this sphere are indicating facts of the compounded nature; they depend on various conditions and circumstances and fail when those conditions of support fail. All mundane phenomena have the character of impermanence because the conditions supporting them are unsustainable and transient. For example, a woman joyfully enjoying an ice cream in the hot sun derives happiness and enjoyment from the flavour and the coolness, but her satisfaction might vanish and disappear immediately if someone were to take her ice cream away from her or if it were dropped on the sand.

> Everything that has the nature of arising has the nature of ceasing.
>
> —Venerable Assaji
> (one of the first five enlightened disciples)

This truth concerns all mundane worldly things. It needs to be understood and realised by insight wisdom. In the chapter "The Compendium of Conditionality", Acariya Anuruddha proceeds to explain conditionality in detail, as taught by the Buddha.[2] All types of consciousness that dependently arise, and their connected mental states, are subject to the law of impermanence. An abstract of this formula of dependent arising is quoted thus: "When this exists, that comes to be. With the arising of this, that arises." The formula is also valid in its reverse order: "When this ceases to exist, that also ceases."

The Buddha identified that craving is the centre of worldly consciousness and that it is the main cause of suffering and its transitory nature. There are three types of craving mentioned: for sense pleasure; for eternal life; and for annihilation. These are all born of ignorance and cause prolonging of samsara. All varieties of craving are ultimately rooted in greed, and greed can coexist with delusion. Clinging is dependent on craving; these two conditions operate hand in hand. An untrained being clings not

---

[2] Ibid. p. 292.

only to sense objects but also to wrong views. In all cases this produces unwholesome kamma, which continues its existence. Beings who are deluded or engrossed in greed continue to come into being, being reborn in other realms according to their kamma at the time of death. This chain of activities can only be broken by supramundane path consciousness which takes nibbana as its object. That is to say, the supramundane consciousness eradicates the root cause of greed and wrong views.

In the twelvefold formula, ignorance (*avijja*) is stated as the first of the dependent arisings. Delusion prevents beings from seeing the true nature of existence and accumulates unwholesome kamma.[3] In the Sammaditthi Sutta (MN 9.1, pp. 54–5), the venerable Sariputta explained that the cause of ignorance is the taints, and vice versa which is an interlock. The sutta further explains that ignorance continues in transference from life to life, establishing the round of rebirth in samsara. The Buddha had the light, vision, knowledge, insight, and wisdom to break the cycle of samsara and escape to nibbana. The reverse application of the twelvefold formula of dependent arising is the method of deliverance. Cessation of ignorance ceases kamma formations and breaks the cycle of *kelesa*, kamma, and *vipaka*; it has a knock-on effect on the rest of the factors of the formula and brings the cycle to an end.

Acariya Anuruddha explains how the twenty-four conditions structure the relations between different classes of phenomena and between mind and matter. He emphasises in particular the path condition. He explains that the Right Path, the Noble Eightfold Path factors, leads one to take the Middle Path to the blissful destination of nibbana. The Buddha's teaching is compared to a raft that carries a traveller from one shore to the other. There is no suffering for one who has successfully crossed the ocean of samsara.[4]

The doctrine of dependent origination is the focal part of the Buddha's teaching of cause of suffering. Its twelve-stage formula reveals the causal nexus of how life comes into being and highlights the connection between craving and suffering which occurs at different stages of the life process. The philosophy of relations, on the other hand, complements dependent origination, expressing the interconnected conditions of all phenomena of

---

[3] Ibid. p. 295.
[4] Ibid. p. 312.

existence that elaborate the complex anatomy of the life process, giving an ontological view of life that all phenomena are insubstantial, allowing one to let go of the attachment. According to the doctrine, suffering begins with ignorance.

> With ignorance as a condition, there arise the kamma formations; with kamma formations, consciousness; with consciousness, mentality-materiality; with mentality-materiality, the sixfold sense base; with the sixfold sense base, contact; with contact, feeling; with feeling, craving; with craving, clinging; with clinging, existence; with existence, birth; and with birth, ageing, death, sorrow, lamentation, pain, grief, and despair. Such is the origination of this entire mass of suffering.
>
> The corollary of this formula, which constantly accompanies it, describes the conditioned cessation of suffering. It shows how, when ignorance ceases, all the following conditions successively cease, down to the cessation of the entire mass of suffering.

Given below is the mundane order of arising of suffering and the transcendental order of cessation of suffering:

Mundane Order

Ignorance (*avijja*)
Kamma formations (*sankhara*)
Consciousness (*viññana*)
Mentality-materiality (*nama-rupa*)
Sixfold sense base (*salayatana*)
Contact (*phassa*)
Feeling (*vedana*)
Craving (*tanha*)
Clinging (*upadana*)
Existence (*bhava*)
Birth (*jati*)
Suffering (*dukkha*)

Transcendental Order

Faith (*saddha*)
Joy (*pamojja*)
Rapture (*piti*)
Tranquillity (*passaddhi*)
Happiness (*sukha*)
Concentration (samadhi)
Knowledge and vision of things as they are (*yathabhutañanadassana*)
Disenchantment (*nibbida*)
Dispassion (*viraga*)
Emancipation (*vimutti*)
Knowledge of destruction of the cankers (*asavakkhaye ñana*)[5]

There are four stages of enlightenment: stream entry, once-returning, nonreturning, and Arahantship.

A trainee who undertakes the practice of the Middle Path progresses stage by stage. Each transitional stage involves two types of supramundane consciousness: path consciousness (*magga-citta*) and fruition consciousness (*phala-citta*). These stages are the direct result of the culmination of wisdom—vipassana knowledge that is sharp enough to cut through defilements with surgical precision. Path consciousness is exclusively wholesome and denoted by *kusala-citta*. The fruition is a result denoted by phala-citta instead of *vipaka-citta*, to differentiate it from the mundane vipaka. (*Note:* The law of kamma dictates that action is always followed by a result; according to this principle of kamma, it could be called *kamma vipaka* or just *vipaka*.)

All four supramundane consciousnesses produce four corresponding results. Superior results are dependent on the place in the order of progress, encompassing eight supramundane consciousnesses.[6] This order of progress is governed by the laws of consciousness (*citta niyama*) and Dhamma (*Dhamma Niyama*). At any given time, all operating laws of the universe are applicable equally to everyone. The Four Noble Truths is the

---

[5] Bhikkhu Bodhi, *Transcendental Dependent Arising: A Translation and Exposition of the Upanisa Sutta (Discourse on Supporting Conditions)*, SN 12.23 (1995).
[6] Bhikkhu Bodhi, *A Comprehensive Manual of Abhidhamma*, p. 66.

theory that unfolds the path of liberation and provides a practical guideline to progress and reach nibbana. The truth is presented clearly and in a concise manner without any ambiguity. This is necessary for a worldly or a trainee who is unaware of the laws of the universe. The Buddha revealed what was hidden to us; otherwise, we would have never known these laws. This timeless truth is in conformity with the facts of life for those who have eyes to see.

All supramundane consciousness takes nibbana as its object. The types of consciousness differ according to the types of defilements that are permanently eradicated from the mind in progression. For example, with the development of insight, the faculty of wisdom strengthens to a degree that it dispels greed, hatred, and delusion. This is a gradual development of purification. The corresponding fruition consciousness experiences the bliss of purity gained from the absence of the defilement. Each path consciousness arises only once and lasts for only one mind moment. It will never repeat again. Fruition consciousness arises immediately after the path moment. Higher attainers such as nonreturners and Arahants can develop supramundane absorption known as *phala samapatti*. It is a great skill of meditation practice to attain these sublime states. Some of the higher saints have mastery over these skills and attain complete cessation, known as Nirodha samapatti. This is a still state of mind attained through jhana and comes with insight into the three characteristics of its factors (impermanence, suffering, and nonself). After the attainment of those higher states, the mind lapses into bhavanga (the life continuum).

"The attainment of cessation is a meditative attainment where the stream of consciousness and mental factors are temporarily suspended for a fixed length of time by the wish of the attainer, to a maximum of seven days. It can be obtained only in the sensuous plane or fine material plane of existence and cannot be obtained in the immaterial plane. For example, the human realm of existence is said to be the most favourable because the human mind is receptive to the full range of consciousness."[7]

---

[7] Bhikkhu Bodhi, (1999), *A Comprehensive Manual of Abhidhamma*, Guide 43, BPS *A Comprehensive Manual of Abhidhamma*, p. 364.

CHULAN SAMPATHGE

The First Stage of Enlightenment

The Path Consciousness of Stream Entry (Sotapatti-Maggacitta)

The stream (*sota*) is the Noble Eightfold Path, that is to say a state of mind inclined and firmly established in the eight factors of the path. Entry upon this irreversible path leads to full liberation of Arahantship. The Pali commentators give the following simile:

"As the current of the river Ganges flows uninterrupted from the Himalayas to the ocean, so the supramundane Noble Eightfold Path flows uninterrupted from the arising of right view to the attainment of nibbana." This may not be the case for a worldly being of mundane wholesome consciousness which is not fixed on the destiny. The mind of a worldly being is like a butterfly's wings, changing all the time. The Buddha gave a simile in the Anguttara Nikaya, saying that a log floating in a river may not definitely reach the ocean for one of several reasons: it may be washed to the shore by floods, or taken by villagers, or lifted by gods, or it may rot and sink to the bottom. These reasons were then interpreted in the following manner:

To be washed out by the floods of life, of greed, hatred, and delusion; enter into the household's life and responsibilities; follow the eternality view and be reborn in heavenly realms; get corrupted and go to woeful realms.

The stream enterer has a strong conviction and firm affirmation of the Buddha's teachings. The path consciousness of stream entry has the function of cutting off the delusions that formed as taints of wrong view that bind a prisoner to his or her cell. At the first stage of enlightenment the personality view is eradicated, along with doubt about the path, unwholesome mental factors, and the clinging to rites and rituals. One turns away from these with the correct understanding that they serve no purpose towards enlightenment. The strong conviction uproots greed, which coexists with wrong view, delusion, and doubt. The stream entry consciousness maintains the five precepts that prevent rebirth in woeful planes of existence. One is assured of reaching nibbana within seven lives:

i. those who reach the first stage, who will be reborn seven times in the heavenly and earthly realms.

ii. those who seek birth in noble families, who are reborn two or three times before they attain Arahant magga and pala.

iii. those who are born only once more before attaining nibbana.

The Second Stage of Enlightenment—Once-Returner

Sakadagami-Maggacitta

As the name implies, a once-returner will return to the sense-sphere world only one more time. This may be to complete any important duty undone and to acquire kusala kamma. For example, to look after elderly parents, to perform a duty required by the order of sangha (order of monks), to repair or build a monastery, or to give directions to sangha, one may delay one's own enlightenment. These stages of enlightenment refer to a state of mind gained through insight knowledges and are according to circumstances, not necessarily a new life. Depending on the purity of the mind, one may be reborn with virtues straight into the path of progress. The consciousness of the once-returner is associated with the Noble Eightfold Path. It does not eradicate any of the remaining fetters (*samyojana*), but instead cuts off grosser forms of sensual desire and ill will. This manifests as loving-kindness and compassion that has an attenuated desire for sensory pleasures. After attaining Sakadagami in this life, the once-returner may be born into a heavenly realm. There are five kinds of once-returners; each classification is given according to the realm of birth and the realm of attaining nibbana.

Combinations of defilements operate together in a manner that binds a being to the mundane world.[8] The following list shows the combinations of unwholesome mental factors grouped together which operate and produce unwholesome results. In accordance with the conditions and circumstances, they produce new kamma that is powerful enough to bind a being, like a strong chain binding a prisoner to his cell.

---

[8] Ibid. pp. 125–6.

- Fetters
- Knots
- Bonds
- Defilements
- Hindrances
- Floods
- Yokes
- Clinging

Fetters (Samjoyanar)—Latent Dispositions

Latent dispositions (*anusaya*) are defilements which "lie along with" (*anusenti*) the mental process to which they belong. They rise to the surface as obsessions whenever they meet with suitable conditions. The term *latent dispositions* highlight the fact that the defilements are liable at any future time to arise, so long as they have not been eradicated by the supramundane paths.

The Third Stage of Enlightenment

Nonreturner (Anagami-Maggacitta)

The nonreturner path consciousness cuts off the remaining fetters that bind a being to sensual desire, together with ill will and permanent uprooting of hate. Nonreturners will never return to the sense-sphere world (*kamaloka*, human world, and heaven). Such beings who have not attained nibbana are born in pure abodes (Sabbhavasa). They abide in higher Brahma realms and attain nibbana there. There are five classes of Anāgāmis classified according to the realms they are born into according to the purity and where subsequently they attain nibbana. In Buddhism, an anāgāmin is a partially enlightened person who has cut off the first five fetters that bind the ordinary mind to worldly pleasures. Anāgāmis are the third of the four aspirants. The lifespan of those realms is very long, and these meditators may attain nibbana with the necessary exertion.

The Fourth Stage of Enlightenment

Arahant (Arahant Magga-Citta)

The Arahant path consciousness is the culmination of seven stages of purifications by kusala kamma that sharp wisdom is capable of cutting off all remaining subtle fetters, including the desire for fine material and immaterial existences. It is embedded with insight wisdom, uprooting all other remaining defilements, and is free from ignorance. It has no roots producing the force of rebirth. The Pali word *Arahant* is derived from the root words *ara* and *hata*. *Ara* denotes the spokes of a wheel that enable the wheel to rotate. Its meaning is "desire" and "craving", driving the wheel of life in samsara. *Hata* means "one who has broken the spokes", so that the wheel will not turn again. Such an individual has now won the path and reached the final destination.

Acariya Anuruddha comments that Arahantship should not be understood as mere annihilation but as an achievement. It is to tread the path that destroys the defilements that would otherwise bring suffering, to reach the purity of mind that brings supreme happiness and full liberation. The wisdom associated with Arahant-magga-citta sees things as they are; all phenomena are impermanent, subject to suffering and nonself. The Arahant can experience the bliss of nibbana in this very life, for example attainment of Nirodha samapatti, jhana samapatti, and phala samapatti. These states of mind are unshaken by any worldly conditions.

The Pali commentators prefer to express the term *nibbana* as the negation of, or departure from, craving, that which binds to samsara, the cycle of birth and death. It should be understood as deliverance or victory from the misery of life.

"Nibbana is a single, undifferentiated and ultimate reality. It is exclusively supramundane, and has one intrinsic nature (*sabhava*), which is that of being the unconditioned deathless element, totally transcendent to the conditioned world."[9]

An Arahant who reaches nibbana still lives with the five aggregates, including the physical body. The remaining five aggregates are termed as *residue*. Residue remains for the length of one's lifespan. There are

---

[9] Guide 31, ibid. p. 259.

three aspects of nibbana: it is not a realm of existence and has no rebirth consciousness or death consciousness; it is the pure gold of consciousness.

Nibbana is a consciousness that has understood the three characteristics of all phenomena:

1. The contemplation of nonself becomes the door to emancipation, which discards the clinging to a self. This is one aspect of nibbana, termed "the contemplation of the void".
2. The contemplation of impermanence discards the roots of greed and the signs of permanence. It becomes the door to emancipation and is the second aspect, termed "contemplation of the signless".
3. The contemplation of suffering, which discards desire through craving, becomes the door to emancipation and is the third aspect, termed "contemplation of desirelessness".

These aspects correspond to the faculty of the mind that the meditator has developed. This becomes the dominant faculty which is focused upon by the element of nibbana. These aspects are the work of other faculties working in harmony towards the same goal, namely the faculty of wisdom, mindfulness, concentration faith, and confidence respectively.[10]

The Arahant is also called a "destroyer of the taints" (*khinasava*). The supramundane path consciousness of the Arahant eradicates all remaining unwholesome mental factors (*cetasikas*) that were not destroyed by the earlier path consciousness. The Arahant is known as the Worthy One, having reached a holy state of mind by cultivating sterling virtues.

One who aspires to understand and benefit from the practice of the Buddha's dispensation should develop this twofold meditation of samatha–vipassana (Tranquillity leading to Insight). So excellent is the way that it is explained by ancient teachers and commentators.

Conclusions

Practice of jhana through samatha meditation enables one to develop mental faculties consistently and, at the same time, to burn up grosser

---

[10] Ibid. pp. 357–8.

defilements temporally and form a firm foundation to develop insight in the supramundane path. The word *development* is used in the context of a progressive path. It means intensive training in accordance with the Noble Eightfold Path.

This piece of work concludes the analytical study of the cognitive process as taught by the Buddha leading to full liberation.

SOURCES

Ajhan Chah Quotes: Amaravati Publications. UK.

Allison G. Harvey and Nicole B. Gumport, *"Evidence-Based Psychological Treatments for Mental Disorders: Modifiable Barriers to Access and Possible Solutions",* doi: 10.1016/j.brat.2015.02.004; Golden Bear Sleep and Mood Research Center, https://www.ocf.berkeley.edu/~ahsleep/gbsmrc_mock/.

Ambalatthika-Rahulovada Sutta: Instructions to Rahula at Mango Stone (MN 61), tr. Thanissaro Bhikkhu, Access to Insight (30 Nov. 2013), http://www.accesstoinsight.org/tipitaka/mn/mn.061.than.html.

American Association of Psychology Dictionary.

Angulimala Sutta: About Angulimala (MN 86), tr. Thanissaro Bhikkhu, Access to Insight (2013), http://www.accesstoinsight.org/tipitaka/mn/mn.086.than.html.

Best Medical Degrees, https://www.bestmedicaldegrees.com/experimental-cancer-treatments/.

Biden, Joe, "20 Years of Change: Joe Biden on the Violence Against Women Act," *Time* (10 Sept. 2014), https://time.com/3319325/joe-biden-violence-against-women/.

Blue Cross Blue Shield, https://www.bcbs.com/the-health-of-america/reports/millennial-health-trends-behavioral-health.

Bly, Benjamin, ed., *Cognitive Science* (1999).

Bodhi, Bhikkhu, ed., *The Abhidhammattha Sangaha of Acariya Anuruddha* (Kandy, Sri Lanka, Buddhist Publication Society, 1999).

Bodhi, Bhikkhu, *A Comprehensive Manual of Abhidhamma* (Kandy, Sri Lanka, Buddhist Publication Society,1999).

Bodhi Bhikkhu, *In the Buddha's Words: An Anthology of Discourses from the Pali Canon.* (Simon and Schuster 2005).

Bodhi, Bhikkhu, and Nayanaponika Thera, *Numerical Discourses of the Buddha: An Anthology of Suttas from the Anguttara Nikaya* (Buddhist Publication Society, 1999).

Bodhi, Bhikkhu, *Transcendental Dependent Arising: A Translation and Exposition of the Upanisa Sutta (Discourse on Supporting Conditions) SN 12.23* (1995).

Buddhaghosa, *The Path of Purification: Visuddhimagga*, tr. Bhikkhu Nanamoli (Buddhist Publication Society, 1999).

Cancer.net, https://www.cancer.net/navigating-cancer-care/cancer-basics/cancer-care-team/types-oncologists.

Cancer Research UK, https://www.cancerresearchuk.org/about-cancer/what-is-cancer/genes-dna-and-cancer.

Choudry, Moshin, Aishah Latif, and Katharine G. Warburton, "An Overview of the Spiritual Importances of End-of-Life Care among the Five Major Faiths of the United Kingdom", *Clinical Medicine*, 18/1 (Feb. 2018), 23–31, doi: 10.7861/clinmedicine.18-1-23.

Clear, James, "40 Years of Stanford Research Found that People with This One Quality Are More Likely to Succeed", https://jamesclear.com/delayed-gratification.

Collins English Dictionary. Collins Learning, (HarperCollins Publishers 2022).

Csikszentmihalyi, Mihaly, *Finding Flow: The Psychology of Engagement with Everyday Life* (New York, Basic Books, 1997).

Csikszentmihalyi, Mihaly, *The Psychology of Optimal Experience* (Cambridge, Cambridge University Press, 1990).

Csikszentmihalyi, Mihaly, *Beyond Boredom and Anxiety* (1975).

Damien Keown, *Buddhism a Very Short Introduction,* (Oxford University Press 1996).

Dietrich, Arne, https://www.sciencedirect.com/science/article/abs/pii/S1053810004000583.

Dilullo, John J., "Rethinking the Criminal Justice System towards a New Paradigm", US Department of Justice, Office of Justice Programs, Bureau of Justice Statistics (Dec. 1992), http://cfcj-fcjc.org/sites/default/files/docs/2006/conly-en.pdf.

Dhammapada Verses 338 to 343, Sukarapotika Vatthu, https://www.tipitaka.net/tipitaka/dhp/verseload.php?verse=338

Eisenbruch, Maurice, "Medical Distress Explanatory Model", *The Journal of Nervous and Mental Diseases*.

Encyclopaedia of Human Behaviour (2nd edn), ed. C. M. Weaver and R. G. Meyer (2012).

Encyclopedia of Philosophy, Stanford University (SEP) digital reference.

Garfield, Jay L., and Jan Westerhoff, *Madhyamaka and Yogacara: Allies or Rivals?* (Oxford University Press).

Gunaratna, V. F., Rebirth Explained (Buddhist Publication Society, 2008).

Gold, Joshua, and Joseph Ciorciari, https://www.ncbi.nlm.nih.gov/pmc/articles/PMC7551835/.

Golden Bear Sleep and Mood Research Center, https://www.ocf.berkeley.edu/~ahsleep/gbsmrc_mock/.

Harvey, Allison G., and Nicole B. Gumport, "Evidence-Based Psychological Treatments for Mental Disorders: Modifiable Barriers to Access and Possible Solutions", National Library of Medicine. 2015 Elsevier Ltd., doi: 10.1016/j.brat.2015.02.004.

Henepola Gunaratana, The Jhanas In Theravada Buddhist Meditation, (Buddhist Publication Society,1988), https://www.accesstoinsight.org/lib/authors/gunaratana/wheel351.html

International Atomic Energy Agency, Fukushima Daiichi Accident, (Updated May 2022), https://world-nuclear.org/information-library/safety-and-security/safety-of- plants/fukushima-daiichi-accident.aspx, https://www-pub.iaea.org/mtcd/publications/pdf/pub1710-reportbythedg-web.pdf, accessed 8 July 2022.

IMF, Income Inequality, https://www.imf.org/en/Topics/Inequality.

James Clear, *Atomic Habits*, https://jamesclear.com/.

John Sommers-Flanagan, Rita Sommers-Flanagan, *Clinical Interviewing*, (John Wiley & Sons, Inc. 2016).

Kleinman, A. (1988) *Rethinking Psychiatry: from Cultural Category to Personal Experience,* New York: Free Press, (Published online by Cambridge University Press: 02 January 2018).

Kotler, Steven, *The Art of Impossible* (HarperCollins, 2013, 2020).

Lau, Joe, and Max Deutsch, "Externalism about Mental Content", *Stanford Encyclopaedia of Philosophy* (Fall 2019), ed. Edward N. Zalta, https://plato.stanford.edu/archives/fall2019/entries/content-externalism/.

Lemon, Nancy, "Violence Against Women Act" (1990), Berkeley Law (Boalt Hall School of Law), UC Berkeley, Ca. 94720, In the early 1990's, the US Congress passed the Violence Against Women Act. (VAWA). The act was intended to change attitudes toward domestic violence, foster awareness ...

Loneliness 'increases risk of premature death', Association for Psychological Science, https://www.nicswell.co.uk/health-news/. Content supply by the NHS website, nhs.uk. accessed 8 July 2022.

Lynn et al., *Trauma and Dissociation, Reality versus fantasy*: The Assessment of Alexithymia in Medical Settings: Implications for Understanding and Treating Health Problems. National Library of Medicine. 2015 (Elsevier Ltd. 2014), Contributors; Mark A. Lumley, Lynn C. Neely, Amanda J. Burger. Published in final edited form as: J Pers Assess. 2007, The National Center for Biotechnology Information. https://www.ncbi.nlm.nih.gov/pmc/articles/PMC2931418/.

Maurice Eisenbruch, MD, MPhil (Cambridge), *Classification of Natural and Supernatural Causes of Mental Distress,* The Journal of Nervous and Mental Disease, (Williams & Wilkins 1990).

Mark Williams et al., *The Mindful Way through Depression*, The Guilford Press, (2007).

MedicinePlus, https://ghr.nlm.nih.gov/primer/basics/gene.

Merz, Beverly, "Six Common Depression Types", Harvard Health Publishing, Harvard Medical School (13 Oct. 2020), https://www.health.harvard.edu/mind-and-mood/six-common-depression-types, accessed 8 July 2022.

Metaphysics Research Lab, *Stanford Encyclopedia of Philosophy* (Stanford University, California), https://plato.stanford.edu/10.

Metropolitan Planning Council. Chicago. (2022), https://www.metroplanning.org/index.html.

Monkeypox 2022, WHO, https://www.who.int/multi-media/details/monkeypox--what-you-need-to-know? accessed 8 July 2022.

Namto, Achan Sobin S. (Sopako Bodhi Bhikkhu), *Insight Meditation: Practical Steps to Ultimate Truth*, ed. Venerable Phra Dhamma Theerarach Mahamuni, tr. Helen and Vorasak Jandamit (Bangkok, 1988).

National Health Service, https://www.nhs.uk/conditions/autism/what-is-autism/.

Nikaya, Khuddaka, *Dhammapada*, tr. Jack Austin.

Nikaya, Khuddaka, *Udana Buddhavagga.*

Nina van Gorkom (1999), Chapter 31 - *Six Pairs of Beautiful Cetasikas.*

Nyanatiloka Thera, Buddhist Dictionary: A Manual of Buddhist Terms and Doctrines. (Buddhist Publication Society - BPS 2004).

Oxford Bibliographies, (Oxford University Press 2022).

Philip G. Alston (ed.), Nikki R. Reisch (ed.). Tax, Inequality, and Human Rights. (2019). Oxford University Press.

Ratana Sutta: The Jewel Discourse, http://www.accesstoinsight.org/tipitaka/kn/snp/snp.2.01.piya.html.

Rieu, André, "13 Year Old Girl Playing Il Silenzio (The Silence)" [video], YouTube (uploaded 21 May 2020), https://www.youtube.com/watch?v=DRrTujHaHis&list=RDn4lo4Szzx7g&index=18, accessed 8 July 2022.

Robinson, Austin, "Paradigm Shift Rough Draft: Crime and Punishment" (2 Nov. 2018), https://sites.psu.edu/awr5537/2018/11/07/paradigm-shift-rough-draft-crime-and-punishment/.

Robert E. Buswell Jr. et al, *The Princeton Dictionary of Buddhism,* (Princeton University Press 2013).

Ronen Palan et al, *Tax Havens: How Globalization Really Works,* (Cornell Studies in Money). Ithaca, N.Y. (Cornell University Press 2010).

Stanford Encyclopaedia, The Theory of Two Truths in India, https://plato.stanford.edu/entries/twotruths-india/. Metaphysics, https://plato.stanford.edu/entries/metaphysics/.

Sayadaw Mahathera Ledi, Aggamahapandita, D.Litt., *The Catusacca-Dipani,* The Manual of the Four Noble Truths, translated into English by Sayadaw U Nyana, (Patamagyaw of Masoeyein Monastery Mandalay, Edited by The English Editorial Board 1903).

Sayadaw, Mahathera Ledi, *The Manual of Insight: The Vipassana Dipani* (1965).

Sayadaw, Mahathera Ledi, *The Buddhist Philosophy of Relations: The Patthanuddesa Dipani* (1965).

Sayadaw, Mahathera Ledi, *The Manual of the Four Noble Truths: The Catusacca-Dipani*, tr. Sayadaw U Nyana (Mandalay, Myanmar, Patamagyaw of Masoeyein Monastery).

Sayadaw, Mahathera Ledi, *The Buddhist Philosophy of Relations*, tr. Sayadaw U Nyana (Mandalay, Myanmar, Patamagyaw of Masoeyein Monastery).

Sayaduw U Silananda, *The Four Foundation of Mindfulness* (Wisdom Publications, 2002).

Sayadaw Ven. Taungpulu Tawya, *Vipassana-Insight Meditation - Maha Satipatthana*. Department of Religious Affairs, Rangoon Burma, (1979).

Suárez, L. M., Bennett, S. M., Goldstein, C. R., & Barlow, D. H. *Understanding anxiety disorders from a "triple vulnerability framework*. In M. M. Antony & M. B. Stein (Eds.), Oxford handbook of anxiety and related disorders (pp. 153–172). (Oxford University Press 2009). American Psychological Association (APA) 2022. APA PsycInfo accessed 8 July 2022.

Susan A. Jackson, *Journal of Applied Sport Psychology,* Volume 4 – Issue 2 Athletes in flow: A qualitative investigation of flow states in elite figure skaters (1992).

Thanissaro Bhikkhu, *Upatissa-Pasine: Upatissa's (Sariputta's) Question* (1966). "Upatissa-pasine: Upatissa's (Sariputta's) Question (Mv 1.23.1-10), by Thanissaro Bhikkhu. Access to Insight (BCBS Edition), 30 November 2013, http://www.accesstoinsight.org/tipitaka/vin/mv/mv.01.23.01-10.than.html.

Thera, Nyanaponika, *Anguttara Nikaya: The Heart of Buddhist Meditation* (Buddhist Publication Society, 2005).

Thera, Nyanaponika, *Abhidhamma Studies: Researches in Buddhist Psychology* (Kandy, Sri Lanka, Buddhist Publication Society, 1965).

Thera, Nyanatiloka, *Buddhist Dictionary* (Buddhist Publication Society).

Thera, Nyanatiloka, *Guide through the Abhidhamma Pitaka* (Kandy, Sri Lanka, Buddhist Publication Society).

The Rosicrucian Order, AMORC, https://www.rosicrucian.org/, accessed 8 July 2022.

The Udana and the Itivuttaka, tr. John D. Ireland (Kandy, Sri Lanka, Buddhist Publication Society, 1998).

Top 10 Most Promising Experimental Cancer Treatments. (2022). https://www.bestmedicaldegrees.com/experimental-cancer-treatments/.

Trickle-down economics, International Monetary Fund, Quartz - https://qz.com/, https://www.metroplanning.org/index.html, https://www.metroplanning.org/news/7415/Why-trickle-down-economics-is-still-not- the-answer.

Uposatha Sutta: *Uposatha (discourse delivered by the Buddha to a community of monks on observance day), Ud 5.5.* by Thanissaro Bhikkhu. Access to Insight (BCBS Edition), 3 September2012. http://www.accesstoinsight.org/tipitaka/kn/ud/ud.5.05.than.html.

US Department of Justice, Office of Justice Programs, National Institute of Justice, "Research in Brief", (May 2020).

Van Ginneken, Esther, "The Pain and Purpose of Punishment: A Subjective Perspective", Howard League for Penal Reform (2016), https://howardleague.org/wp-content/uploads/2016/04/HLWP-22-2016.pdf, accessed 8 July 2022.

Weisz, John R., *Comprehensive Clinical Psychology* (1998).

Printed and bound by CPI Group (UK) Ltd, Croydon, CR0 4YY

NOTE ON REFERENCES

References to the discourse in the second collection of the Pali Canon, the Digha Nikaya (DN) and the Majjhima Nikaya (MN), refer to the number of the sutta. References to the Samyutta Nikaya (SN) refer to the number of the chapter followed by the number of the sutta within that chapter. References to the Anguttara Nikaya (AN) refer to the nipata (numerical division) followed by the number of the sutta within that nipata.

Majority of references are from the Abhidhamma, the third collection of the Buddha's teachings. Main source of Abhidhamma is from Nayanaponika Thera, one of the foremost interpreters of Theravada Buddhism in modern times and Bhikkhu Bodhi's edition of A Comprehensive Manual of Abhidhamma published by the Buddhist Publication Society (1999), Kandy Sri Lanka. Bhikkhu Bodhi is a Buddhist monk of American nationality. Since 1984 he has been the editor for the Buddhist Publication Society, and since 1988 its President. He is credited for his scholarly work of editing Buddhist scriptures to modern English.

Pali translations of the textbooks used are compatible with the Pali Text Society of the UK. Throughout the book, author has inserted pali words and their English meanings. The Pali word 'wisdom' is often translated as Panna, however, this word is not available on the online dictionary, therefore its Sanskrit equivalent 'prajna' is used in many occasions for consistency. In many online resources use both mixture of Pali and Sanskrit words when describing Buddha dhamma.

A complete Pali-English Glossary could be found in the A Comprehensive Manual of Abhidhamma, ISBN 955-24-0198-4 edited by Bhikkhu Bodhi, published by the Buddhist Publication Society, Kandy Sri Lanka (1999).